Targeted Nanotechnology

This book examines the role of nanotechnology in advancing cancer treatment, focusing on phytochemicals derived from natural sources and their therapeutic delivery through nanotechnology. The initial chapters provide an overview of cellular and molecular mechanisms of cancer and present both passive and active targeting strategies to enhance the effectiveness of phytochemical-loaded nanosystems. The chapters also explore the nano-delivery systems tailored for potent anticancer agents such as camptothecin, berberine, paclitaxel, and many more. These molecules improve the bioavailability, stability, and efficacy of these bioactive compounds, addressing the latest innovations for agents such as hesperidin, thymoquinone, epigallocatechin gallate, quercetin, and curcumin. This book also investigates the co-delivery of natural and conventional anticancer drugs, exploring potential synergies in cancer therapy. In addition, the chapters focus on nano-delivery systems for essential oils, carotenoids, and phytomolecules targeting drug-resistant cancer cells and the role of natural compounds in cancer prevention.

Key Features:

- Provides foundational insights into the cellular and molecular mechanisms of cancer
- Explores innovative nano-delivery systems for natural anti-cancer agents from phytoconstituents
- Discusses passive and active targeting strategies for phytochemical-loaded nanosystems
- Highlights the therapeutic potential of nano-formulations for various natural compounds
- Examines co-delivery of natural and synthetic anticancer drugs for improved efficacy

This book is useful for researchers and students in oncology, nanotechnology, and pharmaceutical sciences.

Targeted Nanotechnology

Enhancing the Efficacy of Natural Anticancer Drug Therapies

Edited by
Santwana Padhi and Anindita Behera

CRC Press
Taylor & Francis Group
Boca Raton London New York

CRC Press is an imprint of the
Taylor & Francis Group, an **informa** business

Designed cover image: Shutterstock

First edition published 2026
by CRC Press
2385 NW Executive Center Drive, Suite 320, Boca Raton FL 33431

and by CRC Press
4 Park Square, Milton Park, Abingdon, Oxon, OX14 4RN

CRC Press is an imprint of Taylor & Francis Group, LLC

ISBN: 9781041007586 (hbk)
ISBN: 9781041007593 (pbk)
ISBN: 9781003611516 (ebk)

DOI: 10.1201/9781003611516

Typeset in Times
by codeMantra

Contents

About the Editors viii

SECTION I General Insights into the Use of Phytomolecules in Chemotherapy

Chapter 1 Cancer: Cellular and Molecular Basis 3

Pratap Kumar Sahu

Chapter 2 Emerging Phytoconstituents from the Natural Origin in the Battle Against Cancer: Overview of Molecular Mechanisms 11

Deepa S. Mandlik, Satish K. Mandlik, and Dileep Kumar

Chapter 3 Passive and Active Targeting Strategies for the Delivery of Phytochemical-Entrapped Nanosystems for Cancer 26

Ananya Kuanar, Sandeep Kumar Kabi, Bibhudutta Pattnaik, and Dattatreya Kar

Chapter 4 Use of Phytochemical-Entrapped Nanosystems for Cancer 42

Vivek Chavda, Disha Chawnani, Mital Patel, Radhika Pandya, and Ketan M. Ranch

SECTION II Alkaloid Phytomolecules in Cancer Chemotherapy

Chapter 5 Nanoenabled Systems for the Delivery of Camptothecin as an Anticancer Agent 63

Apurba Gohain, Ekta, Ravjit Kaur, Anita Devi, Ajay Sharma, and Garima Bhardwaj

Chapter 6 Nanoenabled Systems for the Delivery of Berberine 72

Padmaja Santosh Kore, Ujwala Shivaji Desai, Anuradha Ghanshyam More, and Minal Tejram Harde

Chapter 7 Nanoenabled Systems for the Delivery of Paclitaxel as an Anticancer Agent 87

Jessica Pandohee, Swati Sharma, Pradeep Kumar, Emmanuel Kyereh, and Saurabh Kulshrestha

Chapter 8 Anticancer Nanodrug Delivery Cargos of Artemisinin 99

Shubhra, Isha Kailash Joshi, Atul Mourya, Onkar Kulkarni, Murali Monohar Pandey, Dharmendra Kumar Khatri, Shashi Bala Singh, Ashish Baldi, and Jitender Madan

Chapter 9 Nanoenabled Systems for the Delivery of Vinca Alkaloids as Anticancer Agents 117

Bhargavi Kulkarni and Dileep Kumar

SECTION III Flavonoid Phytomolecules in Cancer Chemotherapy

Chapter 10 Nanoenabled Systems for the Delivery of Hesperidin as an Anticancer Agent.. 129

Chirag Patel, Sandip Patel, Jayesh Beladiya, Devang Sheth, Ketan Ranch, and Veena Patel

Chapter 11 Nanoenabled Systems for the Delivery of Thymoquinone as an Anticancer Agent.................................. 140

Puja Das, Sayantan Ghosh, Subhashree Mohapatra, and Bismita Nayak

Chapter 12 Nanoenabled Delivery Systems for Epigallocatechin Gallate as an Anticancer Agent.............................. 149

Sweta Priyadarshini Pradhan, Santwana Padhi, Anindita Behera, and Pratap Kumar Sahu

Chapter 13 Nanoenabled Systems for the Delivery of Quercetin as an Anticancer Agent 161

Rahul Lotan Shirole, Vijayeta Pralhad Nage, Nitin Lotan Shirole, and Sushil Vilas Chawre

SECTION IV Phenolic Phytomolecules in Cancer Chemotherapy

Chapter 14 Nanoenabled Systems for the Delivery of Curcumin as an Anticancer Agent.. 181

Saswati Mishra, Sweta Priyadarshini Pradhan, and Anindita Behera

Chapter 15 Nanoenabled Systems for the Delivery of Resveratrol as an Anticancer Agent.. 194

Santwana Palai

Chapter 16 Nanoenabled Systems for the Delivery of Cannabinoids as Anticancer Agents 204

Plachikkattu Parambil Akhila, Aaliya Basheer, and Kappat Valiyapeediyekkal Sunooj

SECTION V Other Natural Drugs in Cancer Chemotherapy

Chapter 17 Nanoenabled Systems for the Delivery of Anthracyclines as Anticancer Agents 219

Bharti Mittu and Abida Bhat

Chapter 18 Nanoenabled Systems for the Delivery of Different Essential Oils as Anticancer Agents......................... 231

Muhammed Navaf, Plachikkattu Parambil Akhila, and Kappat Valiyapeediyekkal Sunooj

Chapter 19 Nanoenabled Systems for the Delivery of Carotenoids as Anticancer Agents 250

Sumit Sharma, Sonali Batra, Lalit Kumar, Valamla Bhavana, and Neelesh Kumar Mehra

Chapter 20 Nanoenabled Systems for the Delivery of Docetaxel as an Anticancer Agent...262

Dipak N. Raut, Sadikali F. Sayyad, and Anita S. Wagh

SECTION VI Co-Delivery of Phytomolecules in Cancer Chemotherapy

Chapter 21 Codelivery of Natural Drugs with Other Anticancer Drugs Entrapping Nanodelivery Systems for Cancer 275

Pooja Mahale and Hrushikesh Velis

Chapter 22 Drug Delivery Systems Entrapping Natural Drugs and Derivatives: Recent Advancements and Future Challenges 293

Vidhi Shah, Radhika Pandya, and Ketan M. Ranch

Chapter 23 Phytomolecules Targeting Resistant Cancer Cells through Nano-Delivery Systems 301

Sopan N. Nangare, Sameer Nadaf, Preeti Karade, Pranali Takale, Prajwala Khapale, and Namdeo R. Jadhav

Chapter 24 Prevention Is Better than Cure: The Role of Phytomolecules in the Prevention of Cancer Development 324

Sandesh Lodha, Gajanan Kalyankar, Shrikant Joshi, Rutvi Vaidya, and Nikita Panchal

Index 347

About the Editors

Anindita Behera is a Professor in the Department of Pharmaceutical Chemistry at Amity Institute of Pharmacy, Amity University, Kolkata. She earned her master's degree from Pune University, India, and completed her PhD at Siksha 'O' Anusandhan (Deemed to be University). She has developed innovative analytical methods for drug formulations used in AIDS treatment and is currently engaged in green synthesis of nanomaterials targeting various life-threatening diseases. With over 19 years of experience in research and academia, she has published numerous research and review articles in peer-reviewed journals. She has also contributed chapters to books published by major international publishers and has authored two books. She has guided two PhD scholars, 16 postgraduate students, and 45 undergraduate students in their research pursuits.

Santwana Padhi serves as the Scientific Communication Manager at Publication Practice Counsel, USA. She obtained her PhD in Pharmaceutics from Jamia Hamdard University, specializing in Cancer Nanotherapeutics. Her research focuses on preclinical and clinical investigations of drug-loaded targeted nanosystems designed to enhance solid tumour treatment. Before her doctoral studies, she gained hands-on experience as a Research Assistant at the National Institute of Immunology (NII), honing her skills in formulation science. Following her PhD, she worked as a Medical Science Liaison (MSL) at Biocon, contributing to advancements in oncology and dermatology. Later, she joined the KIIT-Technology Business Incubator as Scientific Officer, where she played a key role in supporting entrepreneurial initiatives among young researchers and in securing funding for technology development and commercialization. As a Senior Medical Writer at Reseapro Scientific Services, she has contributed to more than 50 publications, medical communication materials, and regulatory documents. Additionally, she has co-edited one book published by Springer Nature. She has received numerous accolades, including the Senior Research Fellowship (SRF) from ICMR, Junior Research Fellowship (JRF) from DBT, and the Award of Excellence for Scientific Contribution from Jamia Hamdard University for two consecutive years.

Section I

General Insights into the Use of Phytomolecules in Chemotherapy

1 Cancer
Cellular and Molecular Basis

Pratap Kumar Sahu
School of Pharmaceutical Sciences, Siksha 'O' Anusandhan
(Deemed to be University), Bhubaneswar, Odisha, India

1.1 CELL CYCLE

The cell cycle is the stage through which a cell passes from one division to the next. The cell cycle is divided into two phases: interphase and mitosis. Interphase is a period of preparation for cell division, while Mitosis, or M phase, is the actual period of cell division. Interphase is further subdivided into G_1 (pre-DNA synthesis), S (DNA synthesis), and G_2 (preparation for mitosis). In the G_1 phase, the cell grows and carries out normal metabolism. Most of the organelle duplication also occurs during this phase. The time required may be non-existent in rapidly dividing cells to 8–10 hours, days, weeks, or years. The S-phase is the phase of DNA replication and chromosome duplication. It takes 6–8 hours. G_2 phase is the period of cytoplasmic growth. It takes 4–6 hours. G_0 (Quiescence) is a phase where the cells are not dividing but can re-enter the cycle (Turajlic et al., 2019) (Figure 1.1).

Mitosis is divided into four stages: prophase, metaphase, anaphase, and telophase (Table 1.1). Chromosome condensation occurs during prophase. Simultaneously, the centrioles move to the opposite poles. The nuclear envelope and the nucleolus disappear, and the spindle fibres appear. Such condensed chromatin can no longer be used as a template for RNA synthesis. So, transcription ceases during mitosis. There is alignment of the chromosomes at the equatorial plate in metaphase. The metaphase chromosome consists of two identical sister chromatids. These sister chromatids are held together at the centromere. As mitosis proceeds, microtubules of the mitotic spindle attach to the centromere, and the two sister chromatids separate and move to opposite poles of the spindle. During anaphase, the centromere divides, and the chromatids move towards the opposite poles. Once the chromatids reach the two poles, the chromosomal elongation (decondensation) starts, and the nucleolus and the nuclear membrane reappear. This stage is called the telophase. Nuclear division is then followed by cytoplasmic division and is called cytokinesis. Mitosis is thus the equational division in which the chromosome number of the parent is conserved in the daughter cell (Mitchison and Salmon, 2001).

TABLE 1.1
Phases of Mitosis and Their Characteristic Features

Phases of Mitosis	Characteristic Features
Prophase	Chromosome condensation, movement of centrioles to opposite poles, disappearance of nuclear envelope and nucleolus, starting of appearance of spindle fibres
Metaphase	Alignment of chromosomes at the equatorial plate
Anaphase	Division of centromeres, movement of chromatids towards opposite poles
Telophase	Chromosome elongation, reappearance of nuclear membrane, and nucleolus

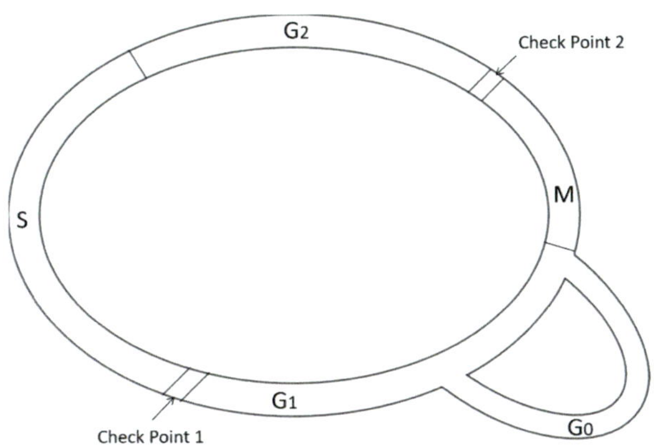

FIGURE 1.1 Phases of cell cycle [Cell cycle consists of G_1 (Pre-DNA synthesis), S (DNA synthesis), G_2 (preparation for mitosis), and M (mitosis). G_0 (Quiescence) is a phase where the cells are not dividing but can re-enter the cycle. Checkpoint 1 at the start of the S-phase, the DNA damage checkpoint, arrests the cell cycle in response to damaged or un-replicated DNA. Checkpoint 2 at the beginning of the M-phase, also called the spindle assembly checkpoint, controls mitosis if the chromosomes are not properly aligned on the mitotic spindle.

1.2 REGULATORS OF THE CELL CYCLE

The positive regulators of the cell cycle are cyclin and cyclin-dependent kinases. The negative regulators of the cell cycle are the p53 gene and the Rb (retinoblastoma) gene (Kamb, 1995). Cyclin is a family of proteins that control the progression of cells through the cell cycle by activating cyclin-dependent kinases. Cyclins were originally named because they undergo a cycle of synthesis and degradation in each cell cycle. All eukaryotic cells require three classes

DOI: 10.1201/9781003611516-2

TABLE 1.2
Cyclins, Their Corresponding Cyclin-Dependent Kinases Controlling the Progression of Cells Through the Cell Cycle; and Cyclin-Dependent Kinase Inhibitor Proteins Inhibiting Cyclin-Cyclin-Dependent Kinase Complex in Cell Cycle

Cyclin	Cyclin-Dependent Kinase (Cdk)	Phases of the Cell Cycle-Regulated	Cdk Inhibitor Proteins
Cyclin A	Cdk2	Active in S-phase	p27
Cyclin A	Cdk1	Active in G_2-phase	p27
Cyclin B	Cdk1	Regulates progression from G_2 to M phase	p21
Cyclin D	Cdk4	Regulates transition from G_1 to S-phase	p16
Cyclin D	Cdk6	Regulates transition from G_1 to S-phase	p16
Cyclin E	Cdk2	Regulates transition from G_1 to S-phase	p21

of cyclins, e.g., G_1/S, S, and M. Each class is defined by the cell cycle stage at which they bind cyclin-dependent kinases. G_1/S cyclins activate cyclin-dependent kinases in late G_1 and start the cell cycle. Their levels fall in the S-phase. S cyclins bind to cyclin-dependent kinases soon after the progression of the cell cycle through start and help stimulate chromosome replication. Their levels remain elevated until mitosis. M cyclins activate cyclin-dependent kinases that promote entry into mitosis at the G_2/M transition. Their levels fall in mid-mitosis. G_1/S cyclins include cyclins A, D, and E. G_2/M cyclins include cyclin B (Alberts et al., 2002).

Cyclin-dependent kinases are a family of protein kinases that control the cell cycle. Cyclins regulate the activity of cyclin-dependent kinases (Table 1.2). Each cyclin-dependent kinase is inactive and becomes active when it binds to the respective cyclin. After binding, they regulate their action. Cyclin-dependent kinase 1 binds to cyclin B or cyclin A. Cyclin-dependent kinase 2 can bind to cyclin A or cyclin E. Cyclin-dependent kinase 4 and cyclin-dependent kinase 6 can bind to cyclin D. Cyclin D/cyclin-dependent kinase 4, Cyclin D/cyclin-dependent kinase 6, and Cyclin E/cyclin-dependent kinase 2 complexes regulate the transition from G_1 to S-phase. Cyclin A/cyclin-dependent kinase 2 complex is active in the S-phase. Cyclin A/cyclin-dependent kinase 1 complex is active in the G_2 phase. Cyclin B/cyclin-dependent kinase 1 regulates G_2 to M phase progression. Cyclin-dependent kinase activity is suppressed by inhibitors of kinases (Ink family) and cyclin-dependent kinase inhibitor proteins (CIP/CKI family). The cyclin-dependent kinase inhibitor proteins (p27, p21, p16, etc.) bind to the cyclin and cyclin-dependent kinase. p27 inhibits cyclin A/cdk1 and cyclin A/cdk2 complex. p21 inhibits cyclin B/cdk1 and cyclin E/cdk2 complex. p16 inhibits cyclin D/cdk4 and cyclin D/cdk6 complex (Viallard et al., 2001).

1.3 CELL CYCLE CHECKPOINTS

Several checkpoints function to ensure that complete genomes are transmitted to daughter cells. An entry of the cell into the S-phase or M-phase is controlled at two checkpoints, e.g., Checkpoint 1 at the start of the S-phase and Checkpoint 2 at the beginning of the M-phase. Checkpoint 1 is the DNA damage checkpoint, which arrests the cell cycle in response to damaged or un-replicated DNA. Checkpoint 2 is called the spindle assembly checkpoint, which controls mitosis if the chromosomes are not properly aligned on the mitotic spindle (Barnum and O'Connell, 2014).

When DNA is damaged, various protein kinases are recruited to the site of damage and initiate a signalling pathway that causes cell cycle arrest. The first kinase at the damage site is either ataxia telangiectasia mutated or ataxia telangiectasia and Rad3-related protein, depending on the type of damage. Ataxia telangiectasia mutated, or ataxia telangiectasia and Rad3-related proteins, are members of the phosphatidyl-inositol 3-kinase-like family of protein kinases. Additional protein kinases called chk1 and chk2 are then recruited and activated, resulting in the phosphorylation of the transcription regulatory protein p53 (Abraham, 2001).

The murine double minute 2 (mdm2) normally binds to p53 (Figure 1.2), promoting its ubiquitylation and destruction in proteasomes. Phosphorylation of p53 blocks its binding to the murine double minute 2; as a result, p53 accumulates to high levels and stimulates transcription of numerous genes, including the gene that encodes the inhibitors of cyclin-dependent kinases, protein p21. The p21 binds and inactivates G_1/S-cyclin-cyclin-dependent kinase and S-cyclin-cyclin-dependent kinase complexes, arresting the cell in the G_1 phase. In some cases, DNA damage also induces either the phosphorylation of murine double minute 2 or a decrease in murine double minute 2 production, which causes a further increase in p53 (Kubbutat et al., 1997).

Excessive stimulation of the mitogenic pathway induces cell cycle arrest or apoptosis. Abnormally high levels of transcription factors (Myc) cause the activation of an alternative reading frame, which binds and inhibits murine double minute 2, thereby increasing p53 levels. Depending on the cell type and extracellular conditions, p53 causes cell cycle arrest or apoptosis. So, p53 stops the cell cycle at checkpoint 1 and allows the DNA to repair if there is DNA damage. If repair fails, apoptosis (programmed cell death) is triggered by p53 by initiating a cascade of proteases called caspases (Woods and Vousden, 2001).

Retinoblastoma protein controls the cell cycle transition from G_1 to the S-phase. Retinoblastoma protein is so named

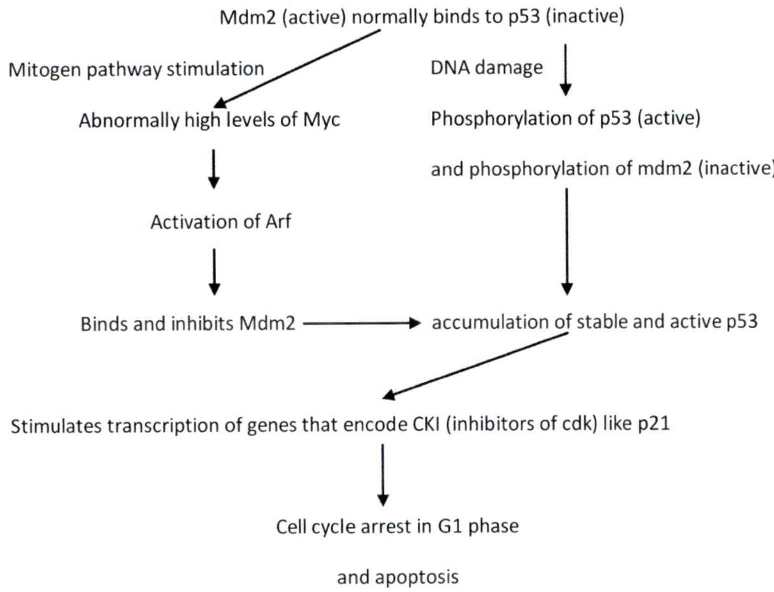

FIGURE 1.2 p53 as a negative regulator of the cell cycle. [Murine double minute 2 (mdm2) normally binds to p53, making p53 inactive. p53 is active in its free form. p53 is activated by excessive stimulation of the mitogenic pathway and by DNA damage. DNA damage causes phosphorylation of p53, thereby dissociating p53 from mdm2. Excessive stimulation of the mitogenic pathway activates an alternative reading frame (Arf), inhibiting mdm2, thereby increasing p53 levels. p53 stimulates transcription of numerous genes, including p21. p21 binds, inactivates cyclin E-cyclin-dependent kinase2 complex, and arrests the cell cycle at checkpoint 1.]

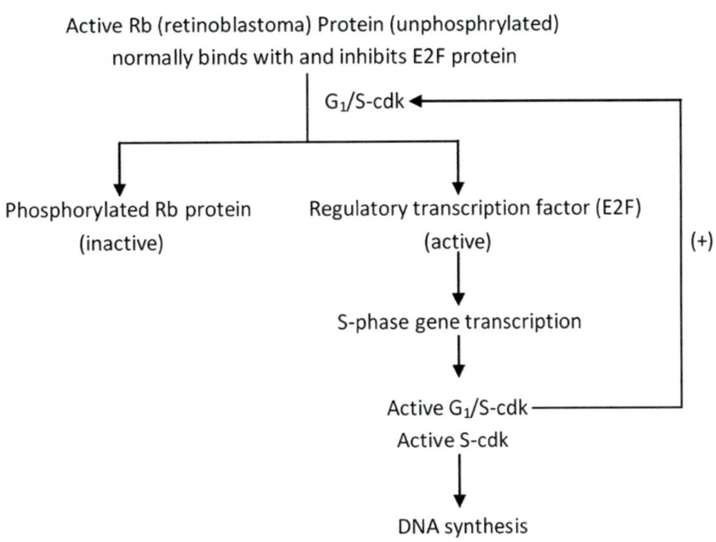

FIGURE 1.3 Retinoblastoma (Rb) gene as a negative regulator of the cell cycle [The gene regulatory protein E2F activates the transcription of G_1/S genes, including the genes for G_1/S cyclin (cyclin E) and S cyclin (cyclin A) and helps in the progression of the cell cycle from G_1 to S-phase. E2F is inactive when bound to retinoblastoma protein and active when free. Retinoblastoma protein (Rb) inhibits E2F and blocks DNA transcription or replication, causing the cell cycle arrest at checkpoint 1. G_1/S cyclin-cyclin-dependent kinase complex phosphorylates retinoblastoma protein to make it inactive and release E2F.]

because mutations of retinoblastoma genes are associated with retinoblastoma tumours. Retinoblastoma protein, by inhibiting the regulatory transcription factor, E2F, acts as a negative regulator of the cell cycle (Figure 1.3). Disruption or deletion of retinoblastoma gene leads to uncontrolled proliferation. Retinoblastoma protein primarily monitors cell size. Retinoblastoma protein binds to and inhibits E2F in the active, dephosphorylated state. E2F turns on the genes encoding proteins necessary for the G_1/S transition. E2F is required for the synthesis of DNA replication enzymes. As the cell increases in size, retinoblastoma protein is slowly phosphorylated until it becomes inactivated and releases E2F (Giacinti and Giordano, 2006).

In the G_0 phase, retinoblastoma protein is not phosphorylated. This unphosphorylated retinoblastoma protein serves as a brake and keeps the cells in G_1 by inhibiting the genes essential for passing into S-phase. In mid-G_1, the growth factors activate cyclin-dependent kinases 4/6 (cdk 4/6). Cyclin-cyclin-dependent kinases complex (Cyclin D/cdk4 or 6 complexes) phosphorylates the retinoblastoma protein to release the break. The phosphorylation of retinoblastoma protein frees the gene regulatory protein E2F to activate the transcription of G1/S genes, including the genes for G1/S cyclin (cyclin E) and S cyclin (cyclin A). The resulting G_1/S-cyclin-dependent and S-cyclin-dependent kinases further enhance retinoblastoma protein phosphorylation. E2F proteins also stimulate the transcription of their genes (Kent and Leone, 2019).

Once checkpoint 1 is cleared, the process cannot be reversed. The cell is committed to entering the S-phase. Any abnormal DNA synthesis or damage in the S-phase or G_2 phase can further stop the cell cycle at checkpoint 2 (spindle assembly checkpoint) by inhibiting the cyclin B/cyclin-dependent kinases 1 complex. Abnormalities in chromosome segregation resulting from failure of the spindle assembly checkpoint are common in cancer cells. They are thought to play an important role in the development of many tumours (Barnum and O'Connell, 2014).

1.4 TUMOUR VS CANCER

Neoplastic cells or tumours are characterized by an abnormal tissue mass formed due to uncontrolled proliferation. Tumours can be benign or malignant (Table 1.3). Benign tumours remain localized and are usually not harmful. They are slow-growing and resemble normal cells. Malignant tumours, in addition to uncontrolled proliferation, also manifest de-differentiation, invasiveness, and metastasis. They are harmful if untreated. These malignant tumours are known as cancers. Cancer is the second leading cause of mortality worldwide. Lung cancer is the most common cancer prevalent throughout the world (Ramzy et al., 2017).

TABLE 1.3
Different Types of Benign and Malignant Tumours and the Affected Tissues

Name of Tumour	Benign/Malignant	Affected Tissue
Papilloma	Benign	Surface epithelial tissue
Carcinoma	Malignant	Surface epithelial tissue
Adenoma	Benign	Glandular epithelial tissue
Adenocarcinoma	Malignant	Glandular epithelial tissue
Melanoma	Benign	Pigment cell
Myoma	Benign	Muscle
Fibroma	Benign	Fibrous tissue
Neurofibroma	Benign	Nerve sheath
Sarcoma	Malignant	Connective tissue (bone, muscle, nerve sheath)
Myeloma	Malignant	Haematopoietic tissue of bone marrow
Lymphoma	Malignant	Lymphatic tissue
Leukaemia	Malignant	Blood

Malignant tumours are either solid tumours or liquid tumours. Solid tumours are localized masses of abnormal tissue that occur in surface epithelial tissue (carcinoma), glandular epithelial tissue (adenocarcinoma), connective tissue (sarcoma), haematopoietic tissue of bone marrow (myeloma), and lymphatic tissue (lymphoma). Solid tumours do not contain cysts or liquid areas. A solid tumour is usually used to distinguish between a localized tissue mass and leukaemia. Leukaemias are examples of liquid tumours or blood cancers resulting from the disorganized proliferation of abnormal leukocytes (DeVita et al.1997).

1.5 OVEREXPRESSED RECEPTORS AND CANCER

Many receptors are overexpressed or specifically expressed in cancers, especially solid tumours. They are epidermal growth factor receptors or human epidermal growth factor receptors (Zhai et al., 2015), fibroblast growth factor receptors (Akhtar et al., 2014), folate receptors (Zwicke et al., 2012), integrin receptors (Wu et al., 2019), bombesin receptor (Sriram et al., 2019), sigma receptor (Collina et al., 2017), transferrin receptors (Daniels et al., 2012), vascular endothelial growth factor receptors (Liang et al., 2015), somatostatin receptor (Sun and Coy, 2011), endothelin receptor (Bagnato and Natali, 2004), follicle-stimulating hormone receptors (Bose, 2008) and biotin receptor (Ren et al., 2015). These receptors and the cancer in which they are overexpressed are listed in Table 1.4.

TABLE 1.4
Overexpressed Receptors in Different Cancers (Solid Tumours)

Overexpressed Receptors	Type of Solid Tumour
Epidermal growth factor receptors	Lungs, breast, ovary, and urinary bladder
Fibroblast growth factor receptors	Breast, prostate, and gastric cancer
Folate receptors	Uterus, ovary, brain, testis, lungs, and pituitary cancers
Integrin receptors	Brain, breast, pancreas, prostate, ovary, and cervix
Bombesin receptors	Lung, prostate, breast, pancreas, head/neck, colon, uterus, ovary, and kidney
Sigma receptors	Breast, pancreas, neuroblastoma, lung, and urinary bladder
Transferrin receptors	Breast, ovary, and brain
Vascular endothelial growth factor receptors	Breast, colon, lung, renal, gastric, and oropharyngeal cancers
Somatostatin receptor	Lung, prostate, breast, colorectal, gastric, and liver cancer
Endothelin receptors	Melanoma
follicle-stimulating hormone receptors	Ovarian cancer
Biotin receptor	Ovarian, breast, colon, lung, renal cancer, and leukaemia

1.6 HALLMARKS OF CANCER

The characteristic features of cancer are:

- Sustaining proliferative signalling
- De-differentiation
- Activating invasion and metastasis
- Reprogrammed energy metabolism
- Immunosuppression.

1.6.1 Sustaining Proliferative Signalling

When the proliferation of cancer cells is not controlled by the processes that normally regulate cell division and tissue growth, it leads to sustained proliferative signalling or uncontrolled proliferation (Nenclares and Harrington, 2020).

Uncontrolled proliferation occurs due to (Hanahan and Weinberg, 2011)

i. Suppression or inactivation of tumour suppressor genes, e.g., p53, retinoblastoma
ii. Activation of proto-oncogenes to oncogenes
iii. Deregulation of normal growth factor signalling
iv. Avoidance of apoptosis
v. Angiogenesis
vi. Telomerase expression.

Deoxyribonucleic acid (DNA) is the genetic material found in each eukaryotic cell's nucleus or the nucleoid of each prokaryotic cell. The nucleotide sequence in DNA is the key to heredity. A gene is a group of nucleotides on a DNA molecule. Gene acts as the master mould for manufacturing specific proteins in a cell. The genes that have been implicated in carcinogenesis are divided into two broad categories, e.g., oncogenes (cell accelerators) and tumour suppressor genes (cell brakes) (Anderson et al., 1992).

Tumour suppressor genes like p53 and retinoblast act as negative cell cycle regulators. If there is DNA damage, the cell cycle stops at checkpoint 1 by the tumour suppressor genes. This allows the cell to repair DNA. If repair fails, apoptosis is triggered. Any abnormal DNA synthesis or damage in the S-phase or G_2 phase can further stop the cell cycle at checkpoint 2 (Figure 1.1). Uncontrolled proliferation occurs when tumour suppression genes are inactivated (Hassanpour and Dehghani, 2017).

Proto-oncogenes are normal genes expressed during embryogenesis, wound healing, tissue regeneration, cell division, etc. They have different functions, like providing signals for cell division and regulating programmed cell death. A list of proto-oncogenes and their functions is given in Table 1.5. When altered by mutation, proto-oncogenes become oncogenes that can contribute to cancer (Anderson et al., 1992).

TABLE 1.5
Different Proto-Oncogenes and Their Functions

Proto-Oncogenes	Functions
sis	Growth factor
erbB, fms, met	Growth factor receptor
src, abl	Tyrosine kinase
raf, mos	Serine/threonine kinase
myc, jun	Nuclear regulatory protein
K-ras, N-ras, Ha-ras	Regulatory protein in signal transduction

Growth factors stimulate a cell in the G_0 phase to enter the G_1 phase. They act through tyrosine kinase-linked receptors to stimulate the production of positive and negative cell cycle regulators. Cancer cells deregulate the normal signalling of growth factors like epidermal growth factor, insulin-like growth factor, platelet-derived growth factor, vascular endothelial growth factor, and transforming growth factor β to promote continuous cell division. Many growth factor receptors are overexpressed or specifically expressed in cancers (Turajlic et al., 2019).

Apoptosis is programmed cell death. Apoptosis prevents cancer by disposing of cells with abnormal DNA that could become malignant. Development of resistance to apoptosis, inactivation of proapoptotic factors, and activation of antiapoptotic factors leads to uncontrolled proliferation (Pfeffer and Singh, 2018). Tumours of 1–2 mm receive nutrients through diffusion, but tumours of more than 2 mm in size require the development of new blood vessels. Angiogenesis is the process of the formation of new blood vessels. Growing tumours promote vascular endothelial growth factor-induced angiogenesis (Turajlic et al., 2019).

Telomeres are the protective caps over the distal extremity of a chromosomal arm. With each cell division round, a telomere portion gets eroded and ultimately becomes nonfunctional. Cancer cells express an enzyme, telomerase, that maintains and provides stability to telomeres so that they proliferate indefinitely (Bukhtoyarov and Samarin, 2015).

1.6.2 De-differentiation

Cellular differentiation is the process by which a cell becomes a specialized cell. The daughter cells mature by differentiation and behave as the parent cell. Cancer cells do not act as normal cells in function or morphology as they are not differentiated (de-differentiation). Poorly differentiated cancers multiply faster than well-differentiated ones (Turajlic et al., 2019).

1.6.3 Activating Invasion and Metastasis

If any cell escapes its tissue of origin to the outside, it undergoes apoptosis and dies. But cancer cells invade nearby tissues due to mutations in their genes. They do so by secreting an enzyme, metalloproteinase, that breaks down the extracellular matrix. Metastases are the secondary tumours released from primary tumours and move to distant places through the blood or lymphatic circulation. Angiogenesis makes the metastasis even faster (Hanahan and Weinberg, 2011).

1.6.4 Reprogrammed Energy Metabolism

Normally, glucose, in aerobic conditions, undergoes glycolysis followed by the Kreb's cycle to produce CO_2 and H_2O. One glucose molecule produces about 38 adenosine triphosphate molecules as an energy source. But cancer cells have a reprogrammed energy metabolism. Even in oxygen, cancer cells promote anaerobic metabolism and convert glucose to lactate, producing only two adenosine triphosphates per glucose molecule. Lactate provides important substrates for cell growth and division (Phan et al., 2014).

1.6.5 Immunosuppression

Chronic immunosuppression is associated with an increased risk of cancers as the immune system fails to recognize any antigen with cancer potential. Using immunosuppressants in different conditions like organ transplantation increases the risk of cancers of viral origin (Engels et al., 2011).

1.7 CAUSES OF CANCER

About 0.3% of the population are carriers of a genetic mutation causing cancer. Three to ten percent of all cancers may be hereditary. However, there can be several other causes besides inherited genetic mutations (Stewart and Wild, 2014; Irigaray et al., 2007).

i. *Infections:* Viral, bacterial, and parasitic infections may cause cancers. Viruses that can cause cancer can be grouped as oncoviruses. The list of viruses, bacteria, and parasites causing different types of cancer is given in Table 1.6 (Irigaray et al., 2007).
ii. *Physical and chemical agents:* They include environmental and occupational hazards like asbestos, tobacco smoke, benzene, etc. Asbestos, like glass wool and rock wool, is also carcinogenic. Eighty percent of lung cancer is associated with tobacco smoking. Electronic cigarettes or e-cigarettes also generate formaldehyde-forming chemicals, thereby increasing the cancer risk by 5–15 times compared to smoking. Metals like cobalt, nickel, arsenic, cadmium, chromium, etc, are also carcinogenic.
iii. *Lifestyle:* A sedentary lifestyle (lack of physical exercise), processed food, red meat, animal fats, and refined carbohydrates are major contributors to cancer risk. Obesity is closely linked with colon cancer, oesophageal cancer, post-menopausal breast cancer, and endometrial cancer. A high salt diet is associated with gastric cancer. About 3.6% of cancer cases and 3.5% of cancer deaths can be related to alcohol consumption worldwide.
iv. *Hormones:* Sex hormones play an important role in the cancer of the breast, ovary, endometrium, testes, prostate, thyroid, and bone. Higher level of oestrogen and progesterone is associated with breast cancer. Similarly, a higher level of testosterone is associated with prostate cancer.
v. *Inflammation:* Chronic inflammation may lead to cancer. *H. pylori*-induced gastritis, when it becomes chronic, may lead to gastric cancer. Inflammatory bowel disease may lead to colorectal cancer.
vi. *Radiation:* Radiation may cause DNA damage, thereby causing cancer. Non-ionizing radiofrequency radiations from mobile phones, electric power transmissions, etc, are carcinogenic. Prolonged exposure to non-ionizing UV (ultraviolet) radiations, X-rays, and γ-rays is carcinogenic. UV radiation is the most common cause of melanoma skin cancers. Radioisotopes treat certain cancers, but exposure to ionizing radiation may cause secondary cancers.
vii. *Other causes:* Organ transplantation rarely results in the transfer of donor-derived tumours to the recipient, thereby causing cancer. Physical trauma, such as chronic heat exposure, may cause skin cancer.

TABLE 1.6
Different Types of Cancers Caused by Microorganisms

Name of Microorganism	Type of Cancer
Virus	
Epstein-Barr virus	Nasopharyngeal carcinoma
Human papillomavirus	Cervical carcinoma
Hepatitis B virus, Hepatitis C virus	Hepatocellular carcinoma
Human T-cell leukaemia virus	T-cell leukaemia
Herpes virus	Primary effusion lymphoma
Bacteria	
Helicobacter pylori	Gastric carcinoma
Mycobacterium tuberculosis	Lung cancer
Parasites	
Schistosoma haematobium	Squamous cell carcinoma of the bladder
Clonorchis sinensis	Cholangio carcinoma

1.8 BIOMARKERS OF CANCER

Biomarkers are the substances present in the blood, body fluids, faecal matter, or body tissues and can be measured. Their levels are correlated with the presence or absence of any disease condition. Cancer biomarkers can be DNA, mRNA, proteins, metabolites, or processes such as apoptosis, angiogenesis, or proliferation. The tumour or other tissues produce the markers in response to cancer or other associated conditions, such as inflammation. They play a key role in diagnosing, prognosing, and managing cancers. In addition, biomarkers are often used throughout the cancer drug discovery process (Idikio, 2011).

TABLE 1.7
Biomarkers in Different Types of Cancers and Their Use

Cancer	Biomarkers	Use	References
Breast cancer	BRCA1, BRCA2	Diagnosis	Ulaner et al. (2016)
	CA 15-3, CA 27–29	Diagnosis and prognosis	
	FGFR2, HER2	Prognosis	
Ovarian cancer	hCG, B2M, M-CSF, VEGF	Diagnosis	Liang et al. (2015) and Coticchia et al. (2008)
	CA 125	Diagnosis and prognosis	
	CD 34, TP53	Prognosis	
Gastric cancer	HSP, CEA, CA19-9	Diagnosis and prognosis	Matsuoka and Yashiro (2018)
Thyroid cancer	TG	Diagnosis and prognosis	Xiao et al. (2020)
Prostate cancer	PSA	Diagnosis and prognosis	Filella et al. (2018)
Endometrial cancer	HSP 27, HSP 70	Diagnosis and prognosis	Idikio (2011)
Cervical cancer	HSP 27, HSP 70	Diagnosis and prognosis	Idikio (2011)
Pancreatic cancer	CA 19-9	Diagnosis and prognosis	Hasan et al. (2019)
Urinary bladder cancer	HSP 27, HSP 70, CA 19-9	Diagnosis and prognosis	Idikio, 2011
Hepatocellular carcinoma	α-foetoprotein (AFP)	Diagnosis and prognosis	Behne and Copur (2012)
Testicular cancer	hCG, α-foetoprotein (AFP), LDH	Diagnosis	Leão et al. (2019)
Colorectal cancer	CEA, EGFR/ HER	Diagnosis and prognosis	Alves Martins et al. (2019)

Biomarkers help diagnose a cancer before it is detectable by conventional methods. Biomarkers also forecast the aggressiveness of the disease progression (prognosis). They also help identify patients who will respond to a particular anticancer drug. Biomarkers used in different types of cancers are given in Table 1.7. They can be classified (Idikio, 2011) as:

a. *Carbohydrates:* carbohydrate antigen
b. *Proteins:* carcinoemryonic antigen, prostate-specific antigen, thyroglobulin, epidermal growth factor receptor or human epidermal growth factor receptor, fibroblast growth factor receptor, heat shock protein, cluster of differentiation, lactate dehydrogenase
c. *Glycoproteins:* α-foetoprotein, human chorionic gonadotropin
d. *Genes:* breast cancer gene

1.9 CONCLUSION

Malignant tumours are called cancers. Sustained proliferation, de-differentiation, invasiveness, metastasis, preferential anaerobic metabolism, and immunosuppression are the characteristic features of cancer. Genetic mutations are a major cause of cancer. Effective use of biomarkers can help in diagnosis, prognosis, and treatment of cancer. Scientists in the drug discovery process also use biomarkers.

REFERENCES

Abraham RT. (2001) Cell cycle checkpoint signaling through the ATM and ATR kinases. *Genes Dev*, 15(17):2177–2196.

Akhtar MJ, Ahamed M, Alhadlaq HA, Alrokayan SA, Kumar S. (2014) Targeted anticancer therapy: Overexpressed receptors and nanotechnology. *Clinica Chimica Acta*, 436:78–92.

Alberts B, Johnson A, Lewis J, Raff M, Roberts K, Walter P. (2002) *Molecular Biology of the Cell*. 4th edition. New York: Garland Science. Intracellular Control of Cell-Cycle Events.

Alves Martins BA, de Bulhões GF, Cavalcanti IN, Martins MM, de Oliveira PG, Martins AMA. (2019) Biomarkers in colorectal cancer: The role of translational proteomics research. *Front Oncol*, 9:1284.

Anderson MW, Reynolds SH, You M, Maronpot RM. (1992) Role of proto-oncogene activation in carcinogenesis. *Environ Health Perspect*, 98:13–24.

Bagnato A, Natali PG. (2004) Endothelin receptors as novel targets in tumour therapy. *J Transl Med*, 2(1):1–9.

Barnum KJ, O'Connell MJ. (2014) Cell cycle regulation by checkpoints. *Methods Mol Biol*, 1170:29–40.

Behne T, Copur MS. (2012) Biomarkers for hepatocellular carcinoma. *Int J Hepatol*, 2012:859076.

Bose CK. (2008) Follicle stimulating hormone receptor in ovarian surface epithelium and epithelial ovarian cancer. *Oncol Res*, 17(5):231–238.

Bukhtoyarov OV, Samarin DM. (2015) Pathogenesis of cancer: Cancer reparative trap. *J Cancer Ther*, 6(05):399.

Collina S, Bignardi E, Rui, M, Rossi D, Gaggeri R, Zamagni A, Cortesi M, Tesei A. (2017). Are sigma modulators an effective opportunity for cancer treatment? A patent overview (1996–2016). *Expert Opin Ther Pat*, 27(5), 565–578.

Coticchia CM, Yang J, Moses MA. (2008). Ovarian cancer biomarkers: Current options and future promise. *J Natl Compr Canc Netw*, 6(8):795–802.

Daniels TR, Bernabeu E, Rodríguez JA, Patel S, Kozman M, Chiappetta DA, Holler E, Ljubimova JY, Helguera G, Penichet ML. (2012) The transferrin receptor and the targeted delivery of therapeutic agents against cancer. *Biochim Biophys Acta (BBA) Gen Subj*, 1820(3):291–317.

DeVita VT, Hellman S, Rosenberg SA. (Eds.). (1997) *Cancer: Principles & Practice of Oncology* (Vol. 2). Lippincott-Raven.

Engels EA, Pfeiffer RM, Fraumeni JF, Kasiske BL, Israni AK, Snyder JJ, Wolfe RA, Goodrich NP, Bayakly AR, Clarke CA, Copeland G. (2011) Spectrum of cancer risk among US solid organ transplant recipients. *JAMA*, 306(17):1891–901.

Filella X, Fernández-Galan E, Fernández Bonifacio R, Foj L. (2018) Emerging biomarkers in the diagnosis of prostate cancer. *Pharmacogenomics Pers Med*, 11:83–94.

Giacinti C, Giordano A. (2006) RB and cell cycle progression. *Oncogene*, 25(38):5220–5227.

Hanahan D, Weinberg RA. (2011) Hallmarks of cancer: The next generation. *Cell*, 144(5):646–74.

Hasan S, Jacob R, Manne U, Paluri R. (2019) Advances in pancreatic cancer biomarkers. *Oncol Rev*, 13(1):410.

Hassanpour SH, Dehghani M. (2017) Review of cancer from perspective of molecular. *J Cancer Res Pract*, 4(4):127–129.

Idikio HA. (2011) Human cancer classification: A systems biology-based model integrating morphology, cancer stem cells, proteomics, and genomics. *J Cancer*, 2:107.

Irigaray P, Newby JA, Clapp R, Hardell L, Howard V, Montagnier L, Epstein S, Belpomme D. (2007) Lifestyle-related factors and environmental agents causing cancer: An overview. *Biomed Pharmacother,* 61(10):640–58.

Kamb A. (1995) Cell-cycle regulators and cancer. *Trends Genet*, 11(4):136–140.

Kent LN, Leone G. (2019) The broken cycle: E2F dysfunction in cancer. *Nat Rev Cancer*, 19(6):326–338.

Kubbutat MH, Jones SN, Vousden, KH. (1997) Regulation of p53 stability by Mdm2. *Nature*, 387(6630):299–303.

Leão R, Ahmad AE, Hamilton RJ. (2019) Testicular cancer biomarkers: A role for precision medicine in testicular cancer. *Clin Genitourin Cancer*, 17(1):e176–e183.

Liang B, He Q, Zhong L, Wang S, Pan Z, Wang T, Zhao Y. (2015) Circulating VEGF as a biomarker for diagnosis of ovarian cancer: A systematic review and a meta-analysis. *Onco Targets Ther*, 8:1075.

Matsuoka T, Yashiro M. (2018) Biomarkers of gastric cancer: Current topics and future perspective. *World J Gastroenterol*, 24(26):2818–2832.

Mitchison TJ, Salmon ED. (2001) Mitosis: A history of division. *Nat Cell Biol*, 3(1):E17–E21.

Nenclares P, Harrington KJ. (2020) The biology of cancer. *Medicine*, 48(2):67–72.

Pfeffer CM, Singh A. (2018). Apoptosis: A target for anticancer therapy. *Int J Mol Sci*, 19(2):448.

Phan LM, Yeung SC, Lee MH. (2014) Cancer metabolic reprogramming: Importance, main features, and potentials for precise targeted anticancer therapies. *Cancer Biol Med*, 11(1):1–19.

Ramzy L, Nasr M, Metwally AA, Awad GA. (2017) Cancer nanotheranostics: A review of the role of conjugated ligands for overexpressed receptors. *Euro J Pharma Sci*, 104:273–292.

Ren WX, Han J, Uhm S, Jang YJ, Kang C, Kim JH, Kim JS. (2015) Recent development of biotin conjugation in biological imaging, sensing, and target delivery. *Chem Commun*, 51(52):10403–10418.

Sriram K, Moyung K, Corriden R, Carter H, Insel PA. (2019) GPCRs show widespread differential mRNA expression and frequent mutation and copy number variation in solid tumours. *PLoS Biol*, 17(11):e3000434.

Stewart BW, Wild CP. (Eds.). (2014) Cancer etiology. In *World Cancer Report 2014*. World Health Organization. pp. 16–54.

Sun LC, Coy DH. (2011) Somatostatin receptor-targeted anticancer therapy. *Curr Drug Del*, 8(1):2–10.

Turajlic S, Sottoriva A, Graham T, Swanton C. (2019) Resolving genetic heterogeneity in cancer. *Nat Rev Genet*, 20(7):404–416.

Ulaner GA, Riedl CC, Dickler MN, Jhaveri K, Pandit-Taskar N, Weber W. (2016) Molecular imaging of biomarkers in breast cancer. *J Nucl Med*, 57(Suppl 1):53S–59S.

Viallard JF, Lacombe F, Belloc F, Pellegrin JL, Reiffers J. (2001) Molecular mechanisms controlling the cell cycle: Fundamental aspects and implications for oncology. *Cancer Radiother*, 5(2):109–129.

Woods DB, Vousden KH. (2001) Regulation of p53 function. *Exp Cell Res*, 264(1):56–66.

Wu PH, Opadele AE, Onodera Y, Nam JM. (2019) Targeting integrins in cancer nanomedicine: Applications in cancer diagnosis and therapy. *Cancers*, 11(11):1783.

Xiao Q, Jia Q, Tan J, Meng Z. (2020) Serum biomarkers for thyroid cancer. *Biomark Med*, 14(9):807–815.

Zhai J, Scoble JA, Li N, Lovrecz G, Waddington LJ, Tran N, Mulet X. (2015) Epidermal growth factor receptor-targeted lipid nanoparticles retain self-assembled nanostructures and provide high specificity. *Nanoscale*, 7(7):2905–2913.

Zwicke GL, Ali Mansoori G, Jeffery CJ. (2012) Utilizing the folate receptor for active targeting of cancer nanotherapeutics. *Nano Rev*, 3(1):18496.

2 Emerging Phytoconstituents from the Natural Origin in the Battle Against Cancer
Overview of Molecular Mechanisms

Deepa S. Mandlik[1], *Satish K. Mandlik*[1], *and Dileep Kumar*[2]
[1]Bharati Vidyapeeth (Deemed to be University), Poona College of Pharmacy, Erandawane, Pune, Maharashtra, India
[2]Department of Pharm Chemistry, Manipal College of Pharmaceutical Sciences, Manipal Academy of Higher Education, Manipal, Karnataka, India

2.1 INTRODUCTION

In the modern age, cancer remains the most dangerous illness. It is an irregular cell proliferation that overcomes other tissues in an uncontrollable way (Antoni et al., 2017; Padhi et al., 2016). There are more than 200 diverse cancer types, the most prevalent of which are breast, colon, lung, and prostate cancers (Gezici and Şekeroğlu, 2019; Behera and Padhi, 2022). Lung cancer is one of the most commonly identified forms of cancer in humans, being the most frequently mentioned in males, while breast cancer is the most common in females (Zhou et al., 2017). Cancer can be caused by internal factors such as immune systems, genetic mutations, hormonal disorders, and external factors like smoking, radiation, cigarettes, certain metals, toxins in food, drinking water, air, chemicals, and infectious agents (Krishnamurthi, 2007). Given this unpredicted growth and higher life expectancies, new cancer cases are expected to rise sharply from 14.1 to 19.3 million from 2012 to 2025 (Torre et al., 2015). Identifying successful novel cancer therapeutic alternatives is still a big challenge for drug discovery scientists and pharmaceutical firms. Present cancer treatment modalities, however, include surgery, different anticancer medications, radiotherapy, chemotherapy, hormone therapy, transplants of stem cells and bone marrow (Padhi et al., 2018).

This debilitating global cancer scenario poses enormous questions about access to treatment due to the high expense of modern cancer care in general and inadequate access in developing countries, in particular, to overpriced medications. Also, improved survival is often followed by enhanced healthcare costs to monitor certain recurrent cancers. The need for economic and new initiatives that could have a wider global impact is underlined by this disturbing evidence. To counter these troubling numbers, the opportunity to use naturally occurring products to treat cancer is an enticing option. Comprehensive research has been carried out to identify new safe medications for cancer treatment that would have fewer adverse effects on the body (Kundu et al., 2022).

The innovation of natural product-based drugs is linked to certain problems, such as the availability and detection of bioactive compounds, wild specimen collection problems, and the unsuitability of natural products with high-throughput selection (Atanasov et al., 2015). A challenging task is further determining the precise mode of action of natural products at the molecular level. As a result of these difficulties, the pharmaceutical industry has shifted its focus to synthetic chemical medicine research. However, the results of the introduced synthetic anticancer drugs in the 1990s did not fulfil expectations (Scannell et al., 2012).

The US Food and Drug Administration has poor approval rates for these products. For these reasons, awareness of the discovery of natural product-based drugs has been revived. Various plant-based phytoconstituents are attracting attention for drug discovery and research for treating different diseases, including cancer (Dutta et al., 2018; Saha et al., 2018). Such phytoconstituents are shown to be significantly potent against various cancer types. They illustrate this efficacy by changing cancer initiation, growth and development and interfering with many processes, such as angiogenesis, differentiation, cell proliferation, apoptosis, and metastasis (Amin et al., 2009; Verma et al., 2017).

Natural products and their derivatives have shown huge structural multiplicity and molecular and pharmacological characteristics that support such growth (Pan et al., 2012). Paclitaxel and docetaxel, both accepted for the treatment of breast cancer in 1990, are a few examples of natural products and their derivatives. Vincristine and vinblastine were approved in the 1960s and are still used successfully for the treatment of breast, testicular, and bladder cancer. Many others were synthesized or reformulated from these

DOI: 10.1201/9781003611516-3

compounds, and new drugs, such as Abraxane®, an injectable paclitaxel nanoformulation, were approved in 2012 for treating small-cell lung carcinoma (Dias et al., 2012).

Marine species are a fantastic source of natural marine products that can be used to treat life-threatening disorders, such as cancer and acquired immunodeficiency syndrome. However, the use of these creatures has not been thoroughly investigated since 79% of the Earth's surface is covered by water. When the Food and Drug Administration approved the ziconotide isolated from a cone snail in 2004, it was obvious that natural products obtained from marine sources may be a potential way of acquiring new entities of tremendous therapeutic value. Marine organisms produce bioactive compounds to shield themselves from the dangerous effects of light and high oxygen concentrations. In the marine world, several kinds of bacteria, microalgae, actinomycetes, fungi, invertebrates, and vertebrates can produce a significant amount of bioactive compounds with high pharmacological activity. A broad range of secondary bioactive metabolites, sterols, peptides, sulfated polysaccharides, chitosan, chito oligosaccharides, carotenoids, and derivatives of chitin are found in marine natural products (Ruiz-Torres et al., 2017).

Phlorotannins are the secondary metabolites primarily found in marine brown algae with many biological properties, including antidiabetic, anti-inflammatory, anti-hypertensive, radioprotective, and antimicrobial activities. Bioactive peptides have powerful antimicrobial, anti-coagulant, and antioxidant properties, and they play a critical role in lowering the risk of life-threatening cardiovascular illnesses (Erdmann et al., 2008). A notable activity against acquired immune deficiency syndrome and cancer is exhibited by sulfated polysaccharides derived from various forms of marine algae. Coloured pigments like carotenoids derived from plants, algae, and fungi have beneficial effects in preventing cardiovascular, cancer, and other acute/chronic diseases (Agarwal and Rao, 2000). All-natural marine products have tremendous pharmacologically significant activities against cancer and viruses. Several molecules obtained from marine sources are useful in the fight against this disease, either by preventing the proliferation of cancer cells or by enhancing apoptosis in human cancer cell lines. As a large area of marine resources is yet to be explored, the potential role of marine natural products in the human health care system is enormous.

Because dietary patterns are consistently shown to be one of the most important determinants of chronic diseases such as diabetes, cardiovascular diseases, neurodegenerative diseases, gallstones, cataracts, and a variety of cancers in a vast field of marine epidemiological studies, the potential role of marine natural products in the human health care system is massive (Nivya et al., 2012). Nutraceuticals, functional foods, and supplemental micronutrients can decrease the growth of cancer cells, inhibit cell proliferation, and induce apoptosis of cancer cells (Tripathi et al., 2005). Various natural dietary items have shown an impending part in cancer prevention and treatment, and an important approach has been considered (Zheng et al., 2016). In several preclinical and clinical experiments, a variety of natural foods, such as broccoli, soybeans, tomatoes, milk thistle, black raspberries, grapes, green tea, pomegranate and isolated compounds from them, have been studied for their anticancer potential (Umar et al., 2012; Maresso et al., 2015). Varied nutraceuticals found in vegetables, fruits, and spices work by suppressing the mechanisms of cancer growth and activating the mechanisms associated with disease prevention. These compounds stimulate antioxidant, anti-tumour, anti-inflammatory, and antiproliferative mechanisms, providing beneficial options as a new cancer treatment regimen. Specific compounds display selectivity in causing cancer cells to become cytotoxic, leaving non-cancer cells unaffected (Banerjee et al., 2017). Epidemiological studies have shown that decreased overall cancer risk can be linked to the daily consumption of a high-fibre, low-fat diet complemented by a large intake of fruits and vegetables (Turati et al., 2015). A variety of anticancer effects of dietary natural products were also found by researchers, like inhibition of tumour growth and metastasis, defence against carcinogens, immunomodulation, and improvement of chemotherapeutic drug effects (Ren et al., 2014).

In this chapter, we will address the causes and recent treatment modalities of cancer, as well as prevention and treatment methods for cancer, such as natural products of plant origin, marine origin, nutraceuticals, and food products with their mechanism of action.

2.2 NEED FOR NATURAL DRUGS FOR CANCER TREATMENT

For cancer treatment, a significant number of attempts have been completed to reduce the adverse effects of medications, such as preventing adverse effects on surrounding cells, improving drug build-up and wound effectiveness, and creating new drug delivery and targeting schemes (Vinogradov and Wei, 2017; Padhi et al., 2022; Padhi and Behera, 2021). Various other cancer treatment approaches include chemotherapy, radiotherapy, tumour surgery, immunotherapy, photodynamic therapy, cancer vaccinations or a combination of stem cell transformation, sometimes followed by serious adverse effects like inadequate bioavailability, non-specificity, toxicity, rapid clearance and metastasis restriction (Patra et al., 2014; Mukherjee and Patra, 2016). Methods of treatment depend on the type, phase, and place of the cancer. The cytotoxic drugs that have exhibited favourable effects combined with other cancer treatments include chemotherapeutic drugs. Such chemotherapy agents consist of microtubule-acting agents like docetaxel, paclitaxel, vincristine and vinblastine, topoisomerase inhibitors like irinotecan (neuropathy, neutropenia and diarrhoea), alkylating agents like cisplatin, melphalan, oxaliplatin, carboplatin (gastrointestinal, cardiovascular, pulmonary, kidney and blood toxicity), doxorubicin (cardiotoxicity), cyclophosphamide (Caruso et al., 2000, 2014; Weaver, 2014). The drugs listed above are extremely effective against a wide variety of cancers. Still, these drugs have certain drawbacks, which include being expensive, toxic, associated with adverse effects, intricate, and non-ecological. In our body, a few cells quickly grow under usual physiological circumstances, such as bone marrow, hair follicles, alimentary tract cells, etc. Such new anti-cancerous medications often point to these normal cells that divide easily, which is a major task, so destructive adverse effects occur. Immunosuppression, hair loss, reduced blood production, gastrointestinal tract inflammation, heart disease, and brain disorders may occur due to these side effects. Another drawback is that these cancerous cells resist these medications when they undergo mutations. For example, drug-resistant genes (ABCA4 and ABCA12) were overexpressed when docetaxel was applied to human breast cancer cells (MCF-7).

Furthermore, drug resistance gene regulation was seen when phytoconstituents like curcumin were combined with docetaxel. Therefore, treating cancer cells using a mono-target chemical agent is not an appropriate strategy. Hence, based on comprehensive research results, phytochemicals and their derivatives are the most favourable alternatives for improved and less toxic cancer treatment (Singh et al., 2016).

2.3 CELL SIGNALLING PATHWAYS IN CANCER

Cell signalling paths are a multifaceted communication linkage composed of several proteins in small three-dimensional signalling cascades. These proteins have distinct actions depending on the site of cells, the types of cells, and intra-molecular interactions. The homeostasis of these proteins is disrupted, resulting in various diseases. Altering cell signalling cascades in cancer patients may increase immune response and amplify cancer cell propagation and existence (Hanahan and Weinberg, 2011).

2.3.1 Cell Death Pathways and Cancer

Apoptosis and autophagy are examples of programmed cell death. Some intracellular programmes can initiate programmed cell death in any pathophysiological situation. Cell death pathways may work together to control cancer cell fate. Autophagy can be pro-survival or pro-death, whereas apoptosis contributes to cell death (Ouyang et al., 2012).

2.3.1.1 Apoptosis

The apoptosis pathway is essential in development because it preserves tissue homeostasis by killing destructive cells. Apoptosis inhibition induces irregular cell propagation and raises the risk of hereditary flaws, which can contribute to tumour development (Sun et al., 2015). Extrinsic signals, such as ligand binding to cell surface receptors, or intrinsic signals, such as genotoxic stress, may cause apoptosis.

Recent research has discovered that cancerous cells defend themselves from apoptosis by unusually acting apoptosis regulatory proteins such as inhibitors of apoptosis proteins, B-cell lymphoma 2 family proteins, cellular FLICE-inhibitory proteins, and others (Siddiqui et al., 2015). By regulating the stimulation of Bcl-2-associated X-protein and Bcl-2 antagonist killer 1, the B-cell lymphoma 2 proteins function as an apoptotic regulator, triggering mitochondrial outer membrane permeabilization (Brahmbhatt et al., 2015). Mitochondrial outer membrane permeabilization induces protein diffusion into the cytosol, usually confined to the mitochondrial intermembrane space. In response to different stimuli, activated mitochondrial outer membrane permeabilization changes the binding properties of pro-apoptotic and anti-apoptotic B-cell lymphoma 2 family proteins, which eventually decide the cancer cell fate, whether it will survive or die (Brahmbhatt et al., 2015). Members of the apoptosis protein family inhibitors are usually overexpressed in cancer, resulting in tumour chemoresistance, cell survival, and disease progression. They prevent apoptosis in various tumour cells by preventing caspase stimulation, or various proteasomes degrade them to enable apoptosis (Obexer and Ausserlechner, 2014). Cellular FLICE-inhibitory protein inhibits caspase-8 activation inside the death-causing signalling complex, which prevents apoptosis. Current research raises the c-FLIP expression in human cancers (like breast, pancreatic, lung, etc.). In human breast and lung cancer cells, c-FLIP also significantly controls drug-prompted apoptosis (Nazim et al., 2016).

2.3.1.2 Autophagy

The intracellular mechanism of lysosomal breakdown, reprocessing of proteins, and cellular organelles is autophagy (self-eating process). It is impaired by a variety of cancerous risk features, including long-term inflammation, obesity, and ageing. Because of its assorted cellular activities, autophagy's function in cancer is context-dependent (Maiuri et al., 2007). Altering autophagic signalling in epithelial cells raises the likelihood of gene variability, oxidative pressure and activation of nuclear factor erythroid 2-related factor 2 that stimulates tumour formation and encourages the expression of many drug-metabolizing enzymes and antioxidant genes (Moscat et al., 2016). Autophagy is impaired, preventing tumour-suppressing gene-mediated cell senescence and leading to cancer progenitor cell proliferation (Dou et al., 2015). Autophagy offers nutrients to fast-growing cancer cells, helping them to live longer (Perera et al., 2015; Ladoire et al., 2016). Enhanced autophagy contributes to the upregulation of antitumor immunity, whereas autophagy downregulation increases cancer by weakening immune surveillance. Oncogene stimulation can prevent autophagy through apoptosis (Amaravadi et al., 2007).

2.3.2 Embryonic Developmental Pathway and Cancer

Cancer stem cells have gained clinical interest in recent decades due to their sluggish growth rates, disputed nature, and resistance to chemotherapeutic or radio-therapeutic agents. Cancer stem cells share many characteristics similar to embryonic stem cells, such as activating conserved developmental signal transduction paths such as the Wnt, Notch, and Hedgehog cascades (Phi et al., 2018).

2.3.2.1 Hedgehog Pathway

It plays a key part in embryonic growth as well as the repair of tissue. Indian Hedgehog, Sonic Hedgehog, and Desert Hedgehog are the three forms of hedgehog ligands (Skoda et al., 2018). Such ligands inhibit smoothened by suppressing the inhibitory effect of the patched cell membrane receptor. The nuclear localization of zinc finger protein glioma-associated oncogene transcription factors is activated due to the activated smoothened glioma-associated oncogene. Consequently, various hedgehog target genes are expressed and connected to survival, angiogenesis, and proliferation (Rimkus et al., 2016). Since an alteration in the hedgehog signalling cascade contributes to tumour growth and carcinogenesis, it can be viewed as a noticeable therapeutic target for cancer treatment. Smoothened inhibitors of hedgehog signalling have been utilized effectively in cancer management in recent years. However, resistance to smoothened inhibitors has previously been identified in patients with basal cell carcinoma (Yang and Dinehart, 2017). More downstream hedgehog signalling inhibitors should be identified and used therapeutically to regulate cancer cell propagation. Glioma-associated oncogene homolog inhibitors are the most powerful strategy for controlling the downstream hedgehog signalling pathway (Lauth and Toftgård, 2007; Mastrangelo and Milani, 2018).

2.3.2.2 Notch Signalling Pathway

This pathway is conserved throughout evolution and is essential for embryonic and postnatal growth. The cascade is vulnerable to the stimulation of dissimilar signalling molecules, which can result in the onset of several human diseases, such as developmental disorders, cancer, and many more (Angulo et al., 2017). Mutations in the notch signalling pathway proteins cause several genetic disorders (Gridley, 2003). Numerous notch receptors and ligands are expressed in different tumours. D'Souza et al. (2010) found that post-translational alterations to its receptors can affect ligand-receptor affinity and intracellular dynamics. Various notch inhibitors (such as anti-dll4 antibodies and γ-secretase inhibitors) have been established for clinical use in cancer treatment (Ran et al., 2017). Achieving an epithelial-mesenchymal transition phenotype subsidizes tumour recurrence and drug resistance, according to new confirmation and notch signalling, which controls this phase. Further research into the functions of notch in numerous cancers and the identification of dissimilar notch is needed to alleviate the notch pathway as a beneficial method for treating cancer.

2.3.2.3 Wnt Signalling Pathway

It is a well-maintained cell signalling cascade involved in a variety of developmental processes (such as cell migration, organogenesis, cell fate determination, stem cell renewal, and so on) as well as the development of diseases such as Robinow syndrome, osteoporosis-pseudoglioma syndrome, and various forms of cancers. Two types of Wnt signalling pathways exist β-catenin dependant and independent signalling (Many and Brown, 2014). Mutations in the Wnt signalling cascade cause several carcinomas. Adenomatous polyposis syndrome (hereditary colon cancer) is often associated with mutations in the adenomatous polyposis coli gene. Because of a mutation in adenomatous polyposis coli, β-catenin is not properly stabilized, creating a complex between β-catenin and transcription factor 7-like 2/T-cell factor 4 (Hankey et al., 2018). Colorectal tumours, colon cancer, hepatocellular carcinomas, and melanoma all have genetic Axin2 mutations (Mazzoni and Fearon, 2014). Deactivating mutations in zinc and ring finger 3 (E3 ligase genes) have recently been discovered in adrenocortical carcinoma and pancreatic cancer (MacDonald and He, 2012; Hao et al., 2016).

2.4 PHYTOCHEMICALS FROM HERBAL SOURCE

Medicinal plants are a gift from the environment to humans, aiding their search for improved well-being. Herbal plants and phytoconstituents isolated from them have been used in traditional medicine since the dawn of time. Phytochemicals present in many medicinal plant species prevent the growth and development of cancer (Aung et al., 2017; Behera and Padhi, 2020). According to a study, the plant kingdom comprises approximately 2,50,000 plant species, of which only about 10% have been explored to manage numerous disorders. Phytoconstituents and their derivatives can be found in several plant parts and execute various pharmacological tasks. Several plant materials, including glycosides, flavonoids, vitamins, alkaloids, saponins, lignans, oils, terpenes, taxanes, minerals, gums, biomolecules and several primary and secondary metabolites, play an important part in either preventing cancer cell-stimulating enzymes, proteins and signalling cascades or stimulating DNA repair mechanisms, such as Bcl-2-associated X-protein, BH3 interacting domain death agonist and B-cell lymphoma 2 antagonist proteins, activating the production of the defensive protein (Thakore et al., 2012; Tariq et al., 2017). Table 2.1 describes these medicinal plants and the list of phytoconstituents used against specific types of cancer (Iqbal et al., 2017).

TABLE 2.1
Important Medicinal Plants in the Treatment of Cancer

Sr. No.	Plant Name	Phytoconstituents	Specific Cancer Suppressed	References
1	*Butea monosperma*	Butrin	Liver cancer	Choedon et al. (2010)
2	*Actaea racemosa*	Actein	Liver and breast cancer	Einbond et al. (2009)
3	*Allium sativum*	Allin	Breast, cervix cancer, lymphoma	Karmakar et al. (2011)
4	*Annona coriacea*	Sinapic and ferulic acid	Lung, renal and ovary cancer, glioma, lymphoid melanoma	Formagio et al. (2015)
5	*Betula utilis*	Betulinic acid	Melanomas and leukaemia	Król et al. (2015)
6	*Broussonetia papyrifera*	2S-abyssinone II verubulin	Glioblastoma and brain cancer	Pang et al. (2014)
7	*Artemisia annua*	Artemisinin	Liver, breast and pancreatic cancer	Efferth (2017)
8	*Andrographis paniculata*	Andrographolide	Colon cancer	Osman et al. (2015)
9	*Curcuma longa*	Curcumin	Lung, breast, colon, oesophagus, liver, prostate and skin cancer	Vallianou et al. (2015)
10	*Broussonetia papyrifera*	2S-abyssinone II verubulin	Glioblastoma and brain cancer	Pang et al. (2014)
11	*Annona squamosa*	Bullatacin	Liver cancer	Vanitha et al. (2017)
12	*Aegle marmelos*	Lupeol, Skimmianine	Lymphoma, leukaemia, breast and liver cancer	Wal et al. (2015)
13	*Aloe vera*	Aloesin, emodin	Anti-angiogenic activity	Shalabi et al. (2015)
14	*Boerrhavia diffusa*	Punarnavine	Malignant melanoma	Mishra et al. (2014)
15	*Panax pseudoginseng*	Protopanaxadiol	Colon cancer	Wang et al. (2015)
16	*Crocus sativus*	Crocin, picrocrocin, crocetin, and safranal	Sarcoma, oral cancer	Hoshyar & Mollaei (2017)
17	*Panax ginseng*	Panaxadiol, Panaxatriol	Breast, ovary, lung, prostate, colon cancer	Du et al. (2013)
18	*Boswellia serrata*	Boswellic acid	Prostate cancer	Garg and Deep 2015
19	*Cannabis sativa*	Cannabinoid	Pancreas, breast, lung, prostate, colorectal cancer	Rakashanda et al. (2013)
20	*Bryophyllum pinnatum*	Bryophyllin A	Cervical cancer	Mahata et al. (2012)
21	*Berberis vulgaris*	Berberine, Cannabisin	Liver, breast, colon cancers	Pierpaoli et al. (2015)
22	*Camelia sinesis*	Epigallocatechin, Theabrownin	Brain, cervical, lung, bladder, skin, prostate, breast cancer	Zhou et al. (2022)
23	*Taxus brevifolia*	Paclitaxel	Ovarian, breast cancer	Caruso et al. (2000)
24	*Bauhinia variegata*	Kaempferol	Lung, breast, liver cancer	Tu et al. (2016)
25	*Gossypium hirsutum*	Gossypol	Breast, stomach, liver, prostate, bladder cancer	Lan et al. (2015)
26	*Citrus limon*	5-hydroxy-6,7,8,30,40-pentamethoxyflavone	Colon cancer	Hirata et al. (2009)
27	*Capsicum annuum*	Luteolin	Colorectal cancer	Osman et al. (2015)
28	*Momordica charantia*	Charantin	Colon, breast cancer	Weng et al. (2013)
29	*Ocimum sanctum*	Eugenol, Vicenin	Breast, liver cancer	Dixit and Ali (2010)
30	*Coccinia grandis*	Protease inhibitors	Colon cancer	Alamgir and Rahman (2014)
31	*Nigella sativa*	Thymoquinone	Colon, prostate, breast, pancreas cancer	Tu et al. (2016)
32	*Ginkgo biloba*	Ginkgetin, Ginkgolide A & B	Hepatocarcinoma, ovary, prostate, colon, liver cancer	Xiong et al. (2016)
33	*C. roseus*	Vincristine, Vinblastine	Leukemias, breast, ovary, cervix, lung, testis cancer	Keglevich et al. (2012)
34	*Moringa oleifera*	Protease inhibitor, Niazinine A	Lung, neuroblastima, blood, colon, abdominal cancer	Shaban et al. (2012)
35	*Saffron crocus*	Saffron	Liver, lung and pancreatic cancer	Ververidis et al. (2007)
36	*Withania somnifera*	Withaferin A, D	Breast, cervix, prostate, colon cancer	Yadav et al. (2010)
37	*Viscum album*	Digallic acid, Viscumin	Breast, cervix, ovary, stomach, colon, kidney, lung cancer	Bhouri et al. (2012)
38	*Pisum sativum*	Protease inhibitors	Colorectal cancer	Rungruangmaitree and Jiraungkoorskul (2017)

(Continued)

TABLE 2.1 (Continued)
Important Medicinal Plants in the Treatment of Cancer

Sr. No.	Plant Name	Phytoconstituents	Specific Cancer Suppressed	References
39	*Nelumbo nucifera*	Neferine	Liver cancer	Yoon et al. (2013)
40	*Xanthium strumarium*	Xanthatin	Leukaemia and liver cancer	Thangapazham et al. (2016)
41	*Vicia faba*	Field bean protease Inhibitors	Skin cancer	Amin and Barkatullah (2016)
42	*Oldenlandia diffusa*	Ursolic acid	Lungs, ovary, uterus, stomach, liver, colon, brain cancer	Woźniak et al. (2015)
43	*Emblica officinalis*	Polyphenols, Tannins	Lymphoma, melanoma	Merina et al. (2012)
44	*Zingiber officinale*	Gingerol, 6-Shogaol	Breast, ovary, cervix, colon, liver cancer	Ghasemzadeh et al. (2015)
45	*Syzygium cumini*	β-sitosterol	Leukaemia	Afify et al. (2011)
46	*Glycyrrhiza glabra*	Licochalcone-A, Licoagrochalcone	Prostate, breast, lung, stomach, kidney cancer	Zhang et al. (2016)
47	*Plumbago zeylanica*	Plumbagin	Liver, leukaemia, breast, blood, skin cancer	Yan et al. (2015)
48	*Sylibum marianum*	Silymarin, Silibinin	Colorectal, lung, liver, skin, colon, prostate cancer	Ramasamy and Agarwal (2008)
49	*Vitex negundo*	Chrysoplenetin	Pancreatic cancer	Awale et al. (2011)
50	*Smilax chinensis*	Flavonoid, Tannin, Saponins	Sarcoma	Madhuri and Pandey (2009)
51	*Solanum nigrum*	Solamargine, solasonine	Breast, liver, lung, skin cancer	Al Sinani et al. (2016)
52	*Vitis vinifera*	Cyanidin, Procyanidins	Colon cancer	Cheah et al. (2014)
53	*Ziziphus jujuba*	Linoleic acids	Breast cancer	Hoshyar and Mollaei (2017)
54	*Tylophora indica*	Tylophorine	Breast cancer	Pratama et al. (2018)
55	*Psoralea corylifolia*	Bavachanin, Corylfolinin, Psoralidin	Stomach, prostate, lung, liver cancer, osteosarcoma	Wang et al. (2011)

2.5 NUTRACEUTICALS AND FOOD PRODUCTS AS AN ANTICANCER AGENT

Tomatoes, broccoli, green tea, grapes, milk thistle, soybeans, pomegranate and black raspberries, as well as isolated compounds from them, have all been tested for their anticancer ability in various preclinical and clinical studies (Wang et al., 2014; Li et al., 2013; Maresso et al., 2015). Patients in emerging countries are eager to investigate natural products as a treatment option for cancer. In a study to evaluate the awareness of taking part in a natural products chemoprevention trial for lung cancer inhibition among heavy smokers, 88% of those surveyed said they wanted to participate (Kumar et al., 2014). Furthermore, one of the latest controversies in natural product science is whether the extract or natural product or its active phytoconstituent is further valuable and whether they are useful in inhibiting or treating cancer. To accomplish this, hypothesis-driven research is required to see whether whole extracts targeting multiple crosstalking and redundant signalling pathways will equal the benefits of isolated phytoconstituents (Melisi et al., 2013). To help with this discussion, a summary of accessible published research that has looked at the efficacy of natural products, isolated constituents, and known molecular targets.

2.5.1 Curcumin

It is derived from *Curcuma longa* persuaded apoptosis by increasing the expression of caspase-3, caspase-9, and p-53 (Ghosh et al., 2015). Curcumin therapy for malignant pleural mesothelioma (asbestos-related cancer) has antiproliferative and antitumor properties with reduced toxicity. It inhibited cellular viability and activated apoptosis in the cell line through caspasein-dependent, mitochondrial and apoptosis-inducing factor-dependent cascades (Zhang et al., 2018). Curcumin induced apoptosis in several cell lines by modifying Bcl-2-associated X-protein, B-cell lymphoma 2-p21 and p53 expression levels (Mu et al., 2019). Curcumin and gallic acid downregulated B-cell lymphoma 2 and upregulated Bcl-2-associated X-protein, cleaved-caspase-3, and poly-ADP ribose polymerase levels in breast cancer cells (MDA-MB-231), inducing apoptosis (Moghtaderi et al., 2018). Curcumin inhibited cell proliferation in various cell lines by blocking the cell cycle and decreasing the expressions of CDC25 and CDC2. It triggered apoptosis by preventing protein kinase B/mammalian target of rapamycin phosphorylation and changing the ratio of Bcl-2-associated X-protein/B-cell lymphoma 2 (Hu et al., 2018). It causes autophagy in several cell lines. The treatment of curcumin (20 M for 24 hours) decreased cell viability because of the stimulation of autophagy. The expression of LC3-II was higher, while that of B-cell lymphoma-extra-large was lower (Naponelli et al., 2015; Rana et al., 2015).

2.5.2 Ferulic Acid

It is a curcumin derivative with extended beneficial significance because of its lower price, great bioavailability, and low adverse effects. It prevents cancerous cells from proliferating, multiplying, and invading. Ferulic acid therapy greatly decreases cervical cancer cell lines' existence in humans. It triggers cell cycle arrest by enhancing p21 and p53 expression while decreasing Cyclin D1 levels. The autophagy markers, like LC3-phosphatidyl-ethanolamine conjugate, Beclin1, and autophagy-related 12 - autophagy-related 5, were also reduced (Gao et al., 2018). Ferulic acid therapy upregulates human epidermal growth factor receptor 2 and estrogen receptor α expression in breast cancer cells in humans in a dose-dependent manner (Zhou et al., 2018). Treatment of prostate cancer cell lines (PC-3 and LNCaP) with ferulic acid prevents the proliferation of cells by inhibiting prostate cancer cell invasion and colony formation (Eroglu et al., 2015).

2.5.3 Resveratrol

It is a natural polyphenol (stilbenoid) that comes in two isomeric forms, cis-resveratrol and trans-resveratrol, from which trans-resveratrol is the most active form (Akinwumi et al., 2018). It can be found in several fruits, including apples, blueberries, red wine, and peanuts (Keylor et al., 2015). According to a published report, it has been shown to prevent the development of human colon cancer cells by increasing pro-apoptotic Bcl-2-associated X-protein and decreasing anti-apoptotic B-cell lymphoma 2 (Peng et al., 2015). Sirt1-mediated inhibition of nuclear factor kappa-light-chain-enhancer of activated B cells also inhibits the development of human colon cancer cells (HCT116) (Mukhtar et al., 2012). Different research showed that providing 5 g of resveratrol daily to nine patients with colon cancer increased the expression of caspase-3 (Singh et al., 2015).

The skin is also a successful target since it can be applied topically. Skin cancer is the most prevalent cancer in the United States, and several research works have shown that resveratrol can help prevent it (Aziz and Aziz, 2018). According to a recent review, resveratrol decreases tumour volume in 7, 12-dimethyl benzanthracene-induced skin cancer in rats by inducing apoptosis and blocking the cell cycle (Hu et al., 2016). It has anticancer properties by inducing apoptosis and autophagy and inhibiting angiogenesis and metastasis (Saunier et al., 2017). It inhibits the development of the nucleotide-binding domain, leucine-rich–containing family, pyrin domain–containing-3 inflammasome by persuading autophagy through the protein kinase B/mammalian target of rapamycin path (Wang et al., 2017b). It also suppresses inflammation by modulating the mitogen-activated protein kinase and nuclear factor erythroid 2 [NF-E2]-related factor 2 pathways (Andrews et al., 2016; Truong et al., 2018). Resveratrol has been shown in studies to prevent cyclooxygenase-1 and 2 enzymes and thus minimize swelling by decreasing the expression of nuclear factor kappa-light-chain-enhancer of activated B (Cianciulli et al., 2012). Moreover, resveratrol exhibited anticancer activity due to the inhibition of cytochrome P_{450} enzyme metabolic activity (Guthrie et al., 2017).

2.5.4 Genistein

It is present abundantly in soy plants and has anticancer properties by inhibiting cancer cell proliferation, angiogenesis, and invasion by altering estrogen receptor-mediated molecular cascades and decreasing expression of cyclin B1. It also prevents protein kinases and nuclear factor kappa-light-chain-enhancer of activated B from being activated (Bhamre et al., 2010; Nakamura et al., 2012). Furthermore, as a phytoestrogen, it prevents mammary gland cancer by binding to the cancerous cells' estrogen receptors with varying estrogen receptor α/estrogen receptor β ratios. Genistein supplementation inhibits uncontrolled colon cancer growth by decreasing the enhanced epidermal growth factor receptor through forkhead box O3 (Qi et al. 2011). The genetic alterations of Sfrp2, Sfrp5, and Wnt5a genes and the promoter-specific modulation of the Wnt/β-catenin signalling cascade by genistein inhibit colon cancer growth (Zhang et al., 2013b). It causes chromatin condensation and DNA fragmentation in colon cancer cells, resulting in morphological alterations (Fan et al., 2010). It slows the development of breast cancer by binding to estrogen receptors due to its structural resemblance to estrogen (McCarty, 2009). It also destroys cancer stem cells by inhibiting the hedgehog-glioma-associated oncogene 1 pathways (Fan et al., 2010). *In-vitro* models of ovarian cancer, genistein induces death by changing the expressions of insulin receptor-1, protein kinase B, estrogen receptor, and insulin-like growth factor 1 receptor (Hwang et al., 2013). On genistein supplementation, decreased expression of kinesin family member 20A causes blockade of the cell cycle and viability loss in gastric cancer cells (Yan et al., 2012). Anti-cancerous medications like tamoxifen (Saha et al., 2014), cisplatin, quercetin (Hu et al., 2014), docetaxel, and selenium (Phillip et al., 2012) may be sensitized by genistein.

2.5.5 Quercetin

It is a polyphenolic flavonoid found in vegetables, fruits and grains, which inhibits E-cadherin expression while suppressing vimentin and N-cadherin, reversing the epithelial-mesenchymal transition. It also downregulates glycogen synthase kinase, matrix metalloproteinases-2, Matrix metalloproteinases-9 and phosphoinositide 3 kinase/protein kinase B expression in breast cancer cells (Balakrishnan et al., 2016), as well as upregulate Bcl-2-associated X-protein, caspases and janus kinase/signal transducer and activator of transcription cascades, potentially enhancing apoptosis of cancer cell (Seo et al., 2016). Quercetin decreases inducible nitric oxide synthase, Tumour necrosis factor-α and interleukin-1 expression in renal cancers, in addition to defending against doxorubicin-induced nephrotoxicity (Heeba and Mahmoud, 2016). It persuades transforming growth factor-β-induced expressions of downstream genes and inhibits vascular endothelial growth factor and Ki67 expressions in prostate cancer (Baruah et al., 2016; Sharma et al., 2016). It also aims to prevent lung cancer by inhibiting cancer cell (A549) migration and encouraging apoptosis mediated by B-cell lymphoma 2/B-cell lymphoma 2-X-protein (Klimaszewska-Wiśniewska et al., 2017). It stimulates caspases and prevents the phosphorylation of c-Jun NH_2-terminal and extracellular signal-related kinases in pancreatic cancer (Zou et al., 2015). This type of cancer also reduces the growth of the tumour and prevents drug resistance. Furthermore, liver cancer decreases reactive oxygen species in cancer cells, increasing B-cell lymphoma 2-X-protein and p53 expression while inhibiting the expressions of protein kinase C, phosphatidylinositol-3 kinase and cyclooxygenase-2 (Guan et al., 2016). Several studies have also shown that nano-mediated targeted delivery of quercetin improved its anticancer efficacy (Bernela et al., 2018). As a result, additional work could soon re-establish this molecule as a hopeful anticancer drug.

2.5.6 Lycopene

It is a red β-carotenoid pigment found in human plasma and tomatoes. Watermelons, pink guavas, pink grapefruits, papaya and apricots also contain this beta-carotene isomer. It is present in microorganisms as well, but it is not found in animals. Two modes, oxidative and non-oxidative, predominantly mediate lycopene's anticancer effects. It can quench single oxygen and preserve cellular redox balance. It inhibits multistage cancer by preventing the degradation of proteins, lipids and DNA. Lycopene also arrests the cell cycle and suppresses carcinogen-induced phosphorylation of retinoblastoma and p53 genes (Gajowik and Dobrzynska, 2014). Lycopene has also been shown in many studies to have radioprotective properties (Andic et al., 2009). It was also found that lycopene, a biologically active carotenoid, decreased inflammatory signals and controlled the Wnt/β-catenin expressions, IGF/Akt and AR signalling. It also inhibited insulin-like growth factor 1 stimulated prostate cancer development by preventing-catenin nuclear localization and modifying the activities of Insulin-like growth factor 1 on glycogen synthase kinase-3 and protein kinase B phosphorylation (Allen et al. 2008).

2.5.7 Piperine

It is the most common alkaloid in the roots and fruits of the *Piper longum L.* (long pepper) and *Piper nigrum* L. plant species of the *Piperaceae* family. It is accountable for the pepper's distinctive pungent odour and biting flavour. Antioxidant, anti-mutagenic, anti-inflammatory, immunomodulatory, anti-convulsant, anti-asthmatic, anti-amoebic, anti-mycobacterial and anticancer properties have been exhibited by piperine. Multiple studies have shown chemopreventive properties against cancerous cells of different backgrounds by stimulating apoptotic signals and inhibiting cell cycle progression (Tanaka, 2013). It is also noted for its effects on estrogen receptor redox homeostasis, stress regulation, self-renewal of cancer stem cells, cancer cell invasion, autophagy, metastasis and angiogenesis. It functions by inhibiting P-gp and thereby reversing multi-drug resistance in cancerous cells. It serves as a bioavailability enhancer for several chemotherapeutic drugs by inducing the drug-metabolizing enzyme mechanism. In a rat model, it was found to reduce the incidence of lung cancer (Rather and Bhagat, 2018). In current studies, piperine has been discovered to disrupt breast cancer stem cells successfully by decreasing the Wnt/-catenin signalling pathway (Tarapore et al., 2012).

2.5.8 Epigallocatechin-3-gallate

Epigallocatechin-3-gallate (EGCG) is majorly found in green tea. At doses of 2–25 μM, the preventive effects of epigallocatechin-3-gallate on Wnt and hedgehog pathways have been documented in several studies. Epigallocatechin-3-gallate inhibited β-catenin/T-cell factor 4 reporter activity and β-catenin expression dose-dependently, suggesting that it inhibits Wnt signalling (Zhu et al., 2019). The treatment of epigallocatechin-3-gallate prevented T-cell factor/lymphoid neoplasia reporter activity and reduced cytoplasmic β-catenin expression levels in lung and colon cancer cell lines (Oh et al., 2014). Treating epigallocatechin-3-gallate and fish oil for nine weeks substantially disrupted the β-catenin-Wnt signalling pathway, decreasing the volume of the tumours (Bose et al., 2007). Using the microarray gene expression technique, researchers discovered two signalling cascades, hedgehog and Wnt signalling, that are associated with cell proliferation and are inhibited by epigallocatechin-3-gallate therapy, suggesting that epigallocatechin-3-gallate harms the proliferation of cells (Laurendeau et al., 2010). Different pathways might be linked in the blockade of Wnt signalling by epigallocatechin-3-gallate. It inhibited the transcription factor HBP1 in breast cancer cells by increasing HMG-box transcription factor 1-messenger RNA stability, inhibiting Wnt signalling.

Moreover, epigallocatechin-3-gallate prevents breast cancer cells' proliferation and invasiveness (Kim et al., 2006). Epigallocatechin-3-gallate inhibited glioma-associated oncogene homolog 1 mRNA expression and downregulation of glioma-associated oncogene homolog 1 reported to be potential in prostate cancer cells, significantly reducing cancer cell development. In transgenic adenocarcinoma mouse prostate mice, epigallocatechin-3-gallate inhibited the growth of prostate cancer cells (Ślusarz et al., 2010). Epigallocatechin-3-gallate was discovered to prevent cell propagation and apoptosis induction in human chondrosarcoma cells, as published in research findings (Tang et al., 2010).

2.5.9 Silymarin

It is a milk thistle extract obtained from *Silybum marianum* that inhibits *N*-Nitrosodiethylamine prompted liver cancer in rats by minimizing lipid peroxidation, reducing expressions of cyclooxygenase-2, matrix metalloproteinases-2 and matrix metalloproteinases-9 with balancing expressions of glutathione and increasing the activity of several antioxidant enzymes (Ramakrishnan et al., 2009). Silibinin, a major constituent of silymarin, inhibits the expression of insulin growth factor receptors, encourages apoptosis in breast cancerous cells, and suppresses breast cancer cell (MDA-MB-231) metastasis (Kim et al., 2009). It also inhibits mitomycin-C-induced cell death in human melanoma A cells by inhibiting apoptosis mediated by B-cell lymphoma-2 and p53. Silibinin is found to be protective against gastrointestinal tract cancers. It prevents colon cancer cell lines from growing by induction of apoptosis via cleavage of poly-ADP ribose polymerase and caspase-3 stimulation. By decreasing the expressions of surviving protein and increasing p53 expression, silymarin inhibits the proliferation of human bladder cells in mice (Boamah et al., 2007).

2.5.10 Gingerol

The most pharmacologically active ingredient of ginger is [6]-gingerol. [6]-Gingerol inhibits cancer growth by activating key cell signalling mediators and various pathways (de Lima et al., 2018). It also can modify several signalling molecules such as cyclin D1, nuclear factor kappa-light-chain-enhancer of activated B cells, B-cell lymphoma-2, tumour necrosis factor-α, signal transducer and activator of transcription3, phosphoinositide-3-kinases, mitogen-activated protein kinase, cellular inhibitor of apoptosis protein-1, extracellular signal-regulated protein kinases 1 and 2, cyclooxygenase-2, protein kinase B, cyclin-dependent kinases, surviving and matrix metalloproteinases-9. Several reports have exhibited that [6]-gingerol has potent anticancer properties, particularly against pancreatic, liver, gastric, cholangiocarcinoma and colorectal cancer (Prasad and Tyagi, 2015).

2.5.11 Emodin

It is derived from the rhubarb *Rheum palmatum* plant. It is found in fungi like *Pyrenochaeta*, *Aspergillus*, and *Pestalotiopsis*. It is an inhibitor of tyrosine kinase that inhibits the progress and alteration of cancerous cells. It is used along with paclitaxel for treating breast and ovarian cancer (Vinod et al., 2013). It is a potent antioxidant agent shown to modulate reactive oxygen species pathways (Yi et al., 2004). Extracellular signal-regulated protein kinases 1 and 2, activating protein-1, *P*-glycoprotein, nuclear factor kappa-light-chain-enhancer of activated B cells, tumour necrosis factor-related apoptosis-inducing ligand, B-cell lymphoma-2, were altered by emodin to enhance chemosensitization and chemoprevention (Vinod et al., 2013).

2.5.12 Sulforaphane

Talalay and Zhang were the first to isolate and decode the structure of sulforaphane, which was present in cruciferous vegetables like Brussels sprouts and broccoli. They were also the first to demonstrate its anticancer ability, as it inhibited breast cancer in rats after adding 9, 10-dimethyl-1, and 2-benzanthracene carcinogens (Zhang et al., 1994). It's worth noting, however, that plants produce very little sulforaphane under normal, unstressed conditions. After mechanical injury to the plant, the precursor of sulforaphane, glucoraphanin, is exposed to the enzyme myrosinase and transformed into the reactive isothiocyanate sulforaphane. Myrosinase is not present in mammalian cells, although it is found in the gastrointestinal tract's bacterial microflora (Yagishita et al., 2019). External factors that alter gut bacteria can thus affect this conversion. Sulforaphane is the most well-known and effective natural product inducer of nuclear factor erythroid 2-related factor 2 (Thimmulappa et al., 2002). In nuclear factor erythroid 2-related factor 2 knock-out mice, the stimulation of these genes has been related to a rise in cancer occurrence. Following exposure to carcinogens, nuclear factor erythroid 2-related factor 2 knock-out mice are prone to several cancers (Becks et al., 2010). Sulforaphane has also been shown to reduce the occurrence of skin and oral cancer in mice (Lan et al., 2016). Sulforaphane also inhibits the nuclear factor kappa-light-chain-enhancer of activated B cells, which decreases inflammation and induces apoptosis by activating the pro-apoptotic cascade while inhibiting anti-apoptotic pathways (Jiang et al., 2018). Sulforaphane has also been shown to suppress epithelial-mesenchymal transition in liver and lung cancer cells (Wang et al., 2017a).

2.6 CONCLUSION AND FUTURE PROSPECTIVE

The current study has shown that phytoconstituents are a hopeful and successful research field with a promising future. The rising incidence of cancer and its extraordinary cost, as well as numerous shortcomings in traditional therapy, such as the high cost and toxicity of current anticancer drugs, have posed a serious challenge to all scientists to develop an environmentally safe, alternate and economical drug in a greener way. Phytoconstituents are expected to transform cancer care in the next years under this scenario.

The effectiveness of these phytoconstituents in cancer treatment has improved due to their high biocompatibility and biodegradability. A wide range of tumour cell lines, including prostate, renal, bladder, melanoma, lung, osteosarcoma, lymphoid, and mammary lines, exhibit anticancer activity *in vitro.* According to recent reports, natural products can be used as anti-cancerous medications with few adverse effects, and they can also control the development and persistence of cancer stem cells. However, the clinical applications of natural compounds are limited due to their rapid catabolism, low water solubility, poor intestinal absorption, and low bioavailability. Various efforts have recently been made to overwhelm the limits of natural products and their derivatives by combining micelles, lipids, nanoparticles, metal complexes and liposomes. More research into natural products and their weaknesses could contribute to producing new anticancer therapeutics with targeted delivery. Even though each of the nutraceuticals mentioned above has extraordinary chemopreventive and chemosensitizing efficiencies, their low bioavailability and retention period within the body are the major disadvantages of the clinical use of these phytoconstituents.

Poor tissue absorption and distribution, low serum levels, small half-life and fast metabolism and elimination are all factors that significantly lead to their clinical failure. Extensive research is being conducted to enhance the bioavailability of such products by introducing chemical alterations such as polymerization, nano-encapsulation, glucuronidation, and so on to boost therapeutic efficacy. The successful usage of non-toxic nutraceuticals will increase the effectiveness of chemotherapeutic medications by lowering the drug's optimum dosage reducing the price and adverse effects of chemotherapy. As a result, it was proposed that supplementing with edible natural agents may assist in the safe use of cancer technology in medicine and many other facets of modern life. This comprehensive book chapter covers medicinal plants and their phytoconstituents that can cure various cancers. We summarized the pharmaceutical predictions and chemical space properties, which offered important insights and useful information on the largely unexplored anticancer leads obtained from natural products. As a result, it is clear that extensive isolation of cancer-causing molecules and the development of their derivatives are crucial and will open up new avenues for developing new pharmacological and therapeutic agents.

REFERENCES

Afify AEMM, Fayed SA, Shalaby EA, El-Shemy HA (2011) *Syzygium cumini* (pomposia) active principles exhibit potent anticancer and antioxidant activities. *Afr J Pharm Pharmacol* 5(7): 948–956.

Agarwal S, Rao AV (2000) Carotenoids and chronic diseases. *Drug MetabDrug Interact* 17: 189–210.

Akinwumi BC, Bordun KA, Anderson HD (2018) Biological activities of stilbenoids. *Int J Mol Sci* 19(3): 792.

Al Sinani SS, Eltayeb EA, Coomber BL, Adham SA (2016) Solamargine triggers cellular necrosis selectively in different types of human melanoma cancer cells through extrinsic lysosomal mitochondrial death pathway. *Cancer Cell Int* 16(1): 1–12.

Alamgir ANM, Rahman M, Rahman A (2014) Phytochemical characteristics, antimitotic, cytotoxic and antiinflamatory activities of *Coccinia grandis* (L.) J. Voigt. *J Pharmacogn Phytochem* 3: 218–225.

Allen JM, Schwartz B, Erdman JW Jr. (2008) Lycopene inhibits androgen receptor expression and signaling in prostate cancer cells. *J Nutr* 138(1): 49–53. doi:10.1093/jn/138.1.49.

Amaravadi RK, Yu D, Lum JJ, Bui T, Christophorou MA, Evan GI, Thomas-Tikhonenko A, Thompson CB (2007) Autophagy inhibition enhances therapy-induced apoptosis in a Myc-induced model of lymphoma. *J Clin Investig* 117(2): 326–336.

Amin AR, Kucuk O, Khuri FR, Shin DM (2009) Perspectives for cancer prevention with natural compounds. *J Clin Oncol* 27: 2712–2725.

Amin S, Barkatullah HK (2016) Pharmacology of Xanthium species: a review. *J Phytopharm* 5: 126–127.

Andic F, Garipagaoglu M, Yurdakonar E, Tuncel N, Kucuk O (2009) Lycopene in the prevention of gastrointestinal toxicity of radiotherapy. *Nutr Cancer* 61: 784–788.

Andrews CS, Matsuyama S, Lee BC, Li JD (2016) Resveratrol suppresses NTHi-induced inflammation via up-regulation of the negative regulator MyD88 short. *Sci Rep* 6: 34445.

Angulo P, Kaushik G, Subramaniam D, Dandawate P, Neville K, Chastain K, Anant S (2017) Natural compounds targeting major cell signaling pathways: a novel paradigm for osteosarcoma therapy. *J Hematol Oncol* 10: 10.

Antoni S, Ferlay J, Soerjomataram I, Znaor A, Jemal A, Bray F (2017) Bladder cancer incidence and mortality: a global overview and recent trends. *Eur Urol* 71: 96–108.

Atanasov AG, Waltenberger B, Pferschy-Wenzig EM, Linder T, Wawrosch C, Uhrin P, Temml V, Wang L, Schwaiger S, Heiss EH, Rollinger JM, Schuster D, Breuss JM, Bochkov V, Mihovilovic, MD, Kopp B, Bauer R, Dirsch VM, Stuppner H (2015) Discovery and resupply of pharmacologically active plant-derived natural products: a review. *Biotechnol Adv* 33: 1582–1614.

Aung TN, Qu Z, Kortschak RD, Adelson DL (2017) Understanding the effectiveness of natural compound mixtures in cancer through their molecular mode of action. *Int J Mol Sci* 18(3): 656.

Awale S., Linn TZ, Li F, Tezuka Y, Myint A, Tomida A, Yamori T, Esumi H, Kadota S (2011) Identification of chrysoplenetin from Vitex negundo as a potential cytotoxic agent against PANC-1 and a panel of 39 human cancer cell lines (JFCR-39). *Phytother Res* 25(12): 1770–1775.

Aziz SW, Aziz MH (2018) Protective molecular mechanisms of resveratrol in UVR-induced Skin carcinogenesis. *Photodermatol Photoimmunol Photomed* 34: 35–41.

Balakrishnan S, Bhat FA, Raja Singh P, Mukherjee S, Elumalai P, Das S, Patra CR, Arunakaran J (2016) Gold nanoparticle–conjugated quercetin inhibits epithelial–mesenchymal transition, angiogenesis and invasiveness via EGFR/VEGFR-2-mediated pathway in breast cancer. *Cell Prolif* 49: 678–697.

Banerjee S, Singh SK, Chowdhury I, Lillard Jr JW, Singh R (2017) Combinatorial effect of curcumin with docetaxel modulates apoptotic and cell survival molecules in prostate cancer. *Front Biosci* 9: 235–245.

Baruah MM, Khandwekar AP, Sharma N (2016) Quercetin modulates Wnt signaling components in prostate cancer cell line by inhibiting cell viability, migration, and metastases. *Tumor Boil* 37: 14025–14034.

Becks L, Prince M, Burson H, Christophe C, Broadway M, Itoh K, Yamamoto M, Mathis M, Orchard E, Shi R, McLarty J (2010) Aggressive mammary carcinoma progression in Nrf2 knock-out mice treated with 7, 12-dimethylbenz[a]anthracene. *BMC Cancer* 10: 540.

Behera A, Padhi S (2020) Passive and active targeting strategies for the delivery of the camptothecin anticancer drug: a review. *Environ Chem Lett* 18(5): 1557–1567.

Behera A, Padhi S (2022) pH-Sensitive polymeric nanoparticles for cancer treatment. In: Padhi S, Behera A, Lichtfouse E (Eds) *Polymeric Nanoparticles for the Treatment of Solid Tumours*. Environmental Chemistry for a Sustainable World, vol 71. Springer, Cham, pp. 401–425.

Bernela M, Kaur P, Ahuja M, Thakur R (2018) Nano-based delivery system for nutraceuticals: the potential future. In: Gahlawat S, Duhan J, Salar R, Siwach P, Kumar S, Kaur P (Eds) *Advances in Animal Biotechnology and Its Applications*. Springer, Singapore, pp. 103–117.

Bhamre S, Sahoo D, Tibshirani R, Dill DL, Brooks JD (2010) Gene expression changes induced by genistein in the prostate cancer cell line LNCaP. *Open Prostate Cancer J* 3: 86–98.

Bhouri W, Boubaker J, Skandrani I, Ghedira K, Chekir Ghedira L (2012) Investigation of the apoptotic way induced by digallic acid in human lymphoblastoid TK6 cells. *Cancer Cell Int* 12: 1–8.

Boamah EK, White DE, Talbott KE, Arva NC, Berman D, Tomasz M, Bargonetti J (2007) Mitomycin–DNA adducts induce p53-dependent and p53-independent cell death pathways. *ACS Chem Biol* 2: 399–407.

Bose M, Hao X, Ju J, Park S, Husain A, Yang CS (2007) Inhibition of intestinal tumorigenesis in ApcMin/+ mice by a combination of (−)-epigallocatechin-3-gallate (EGCG) with fish oil. *J Agric Food Chem* 55: 7695–7700.

Brahmbhatt H, Oppermann S, Osterlund EJ, Leber B, Andrews DW (2015) Molecular pathways: leveraging the BCL-2 interactome to kill cancer cells-mitochondrial outer membrane permeabilization and beyond. *Clin Canc Res* 21: 2671–2676.

Caruso M, Colombo AL, Fedeli L, Pavesi A, Quaroni S, Saracchi M, Ventrella G (2000) isolation of endophytic fungi and actinomycetes taxane producers. *Ann Microbiol* 50(1): 3–14.

Cheah KY, Howarth GS, Bindon KA, Kennedy JA, Bastian SE (2014) Low molecular weight procyanidins from grape seeds enhance the impact of 5-fluorouracil chemotherapy on Caco-2 human colon cancer cells. *PLoS One* 9(6): e98921.

Choedon T, Shukla SK, Kumar V (2010) Chemopreventive and anticancer properties of the aqueous extract of flowers of Butea monosperma. *J Ethnopharmacol* 129(2): 208–213.

Cianciulli A, Calvello R, Cavallo P, Dragone T, Carofiglio V, Panaro MA (2012) Modulation of NF-κB activation by resveratrol in LPS treated human intestinal cells results in downregulation of PGE2 production and COX-2 expression. *Toxicol Vitro* 26: 1122–1128.

de Lima RM, Dos Reis AC, de Menezes AA, Santos JV, Filho JW, Ferreira JR, de Alencar MV, da Mata AM, Khan IN, Islam A, Uddin SJ (2018) Protective and therapeutic potential of ginger (*Zingiber officinale*) extract and [6]-gingerol in cancer: a comprehensive review. *Phytother Res* 32: 1885–1907.

Dias DA, Urban S, Roessner U (2012) A historical overview of natural products in drug discovery. *Metabolites* 2: 303–336.

Dixit S, Ali H (2010) Anticancer activity of medicinal plant extract: a review. *J Chem Cheml Sci* 1(1): 79–85.

Dou Z, Xu C, Donahue G, Shimi T, Pan JA, Zhu J, Ivanov A, Capell BC, Drake AM, Shah PP, Catanzaro JM (2015) Autophagy mediates degradation of nuclear lamina. *Nature* 527: 105–109.

D'Souza B, Meloty-Kapella L, Weinmaster G (2010) Canonical and non-canonical Notch ligands. *Curr Top Dev Biol* 92: 73–129.

Du GJ, Wang CZ, Qi LW, Zhang ZY, Calway T, He TC, Du W, Yuan CS (2013) The synergistic apoptotic interaction of panaxadiol and epigallocatechin gallate in human colorectal cancer cells. Phytother Res 27(2): 272–277.

Dutta S, Saha S, Mahalanobish S, Sadhukhan P, Sil PC (2018) Melatonin attenuates arsenic induced nephropathy via the regulation of oxidative stress and inflammatory signaling cascades in mice. *Food Chem Toxicol* 118: 303–316.

Efferth T (2017) From ancient herb to modern drug: Artemisia annua and artemisinin for cancer therapy. *Seminars Cancer Biol* 46: 65–83. Academic Press.

Einbond LS, Soffritti M, Esposti DD, Park T, Cruz E, Su T, Wu H-A, Wang X, Zhang Y-J, Ham J, Goldberg IJ, Kronenberg F, Vladimirova A (2009) Actein activates stress-and statin-associated responses and is bioavailable in Sprague-Dawley rats. *Fundam Clin Pharmacol* 23(3): 311–321.

Erdmann K, Cheung BW, Schröder H (2008) The possible roles of food-derived bioactive peptides in reducing the risk of cardiovascular disease. *J Nutr Biochem* 19: 643–654.

Eroglu C, Secme M, Bagci G, Dodurga Y (2015) Assessment of the anticancer mechanism of ferulic acid via cell cycle. *Tumour Biol* 36: 9437–9446.

Fan YZ, Li GH, Wang YH, Ren QY, Shi HJ (2010) Effects of genistein on colon cancer cells in vitro and in vivo and its mechanism of action. *Zhonghua Zhong Liu Za Zhi* 32: 4–9.

Formagio ASN, Vieira MC, Volobuff CRF, Silva MS, Matos AI, Cardoso CAL, Foglio MA, Carvalho JE (2015) In vitro biological screening of the anticholinesterase and antiproliferative activities of medicinal plants belonging to Annonaceae. *Braz J Med Biol Res* 48: 308–315.

Gajowik A, Dobrzynska MM (2014) Lycopene-antioxidant with radioprotective and anticancer properties: a review. *Rocz Państw Zakł Hig* 65: 263–271.

Gao J, Yu H, Guo W, Kong Y, Li Q, Yang S, Zhang Y, Wang Y (2018) The anticancer effects of ferulic acid is associated with induction of cell cycle arrest and autophagy in cervical cancer cells. *Cancer Cell Int* 18(1): 102.

Garg P, Deep A (2015) Anticancer potential of boswellic acid: a mini. *J Drugs Med* 7: 18–27.

Gezici S, Şekeroğlu N (2019) Current perspectives in the application of medicinal plants against cancer: novel therapeutic agents. *Anticancer Agents Med Chem* 19: 101–11.

Ghasemzadeh A, Jaafar HZ, Rahmat A (2015) Optimization protocol for the extraction of 6-gingerol and 6-shogaol from Zingiber officinale var. rubrum Theilade and improving antioxidant and anticancer activity using response surface methodology. *BMC Complement Altern Med* 15(1): 1–10.

Ghosh S, Bhattacharyya S, Rashid K, Sil PC (2015) Curcumin protects rat liver from streptozotocin-induced diabetic pathophysiology by counteracting reactive oxygen species and inhibiting the activation of p53 and MAPKs mediated stress response pathways. *Toxicol Rep* 2: 365–376.

Gridley T (2003) Notch signaling and inherited disease syndromes. *Hum Mol Genet* 12: R9–R13.

Guan X, Gao M, Xu H, Zhang C, Liu H, Lv L, Deng S, Gao D, Tian Y (2016) Quercetin-loaded poly (lactic-co-glycolic acid)-d-α-tocopheryl polyethylene glycol 1000 succinate nanoparticles for the targeted treatment of liver cancer. *Drug Deliv* 23: 3307–3318.

Guthrie AR, Chow HH, Martinez JA (2017) Effects of resveratrol on drug-and carcinogen-metabolizing enzymes, implications for cancer prevention. *Pharmacol Res Perspect* 5: e00294.

Hanahan D, Weinberg RA (2011) Hallmarks of cancer: the next generation. *Cell* 144, 646–674.

Hankey W, Frankel WL, Groden J (2018) Functions of the APC tumor suppressor protein dependent and independent of canonical WNT signaling: implications for therapeutic targeting. *Cancer Metastasis Rev* 37: 159–172.

Hao HX, Jiang X, Cong F (2016) Control of Wnt receptor turnover by R-spondin-ZNRF3/RNF43 signaling module and its dysregulation in cancer. *Cancers* 8: 54.

Heeba GH, Mahmoud ME (2016) Dual effects of quercetin in doxorubicin-induced nephrotoxicity in rats and its modulation of the cytotoxic activity of doxorubicin on human carcinoma cells. *Environ Toxicol* 31: 624–636.

Hirata T, Fujii M, Akita K, Yanaka N, Ogawa K, Kuroyanagi M, Hongo D (2009) Identification and physiological evaluation of the components from citrus fruits as potential drugs for anti-corpulence and anticancer. *Bioorg Med Chem* 17(1): 25–28.

Hoshyar R, Mollaei H (2017) A comprehensive review on anticancer mechanisms of the main carotenoid of saffron, crocin. *J Pharm Pharmacol* 69(11): 1419–1427.

Hu S, Xu Y, Meng L, Huang L, Sun H (2018) Curcumin inhibits proliferation and promotes apoptosis of breast cancer cells. *Exp Therapeut Med* 16: 1266–1272.

Hu XJ, Xie MY, Kluxen FM, Diel P (2014) Genistein modulates the antitumor activity of cisplatin in MCF-7 breast and HT-29 colon cancer cells. *Arch Toxicol* 88: 625–635.

Hu YQ, Wang J, Wu JH (2016) Administration of resveratrol enhances cell-cycle arrest followed by apoptosis in DMBA-induced skin carcinogenesis in male Wistar rats. *Eur Rev Med Pharmacol Sci* 20: 2935–2946.

Hwang KA, Park MA, Kang NH, Yi BR, Hyun SH, Jeung EB, Choi KC (2013) Anticancer effect of genistein on BG-1 ovarian cancer growth induced by 17 β-estradiol or bisphenol A via the suppression of the crosstalk between estrogen receptor alpha and insulin-like growth factor-1 receptor signaling pathways. *Toxicol Appl Pharmacol* 272: 637–646.

Iqbal J, Abbasi BA, Mahmood T, Kanwal S, Ali B, Shah SA, Khalil AT (2017) Plant-derived anticancer agents: a green anticancer approach. *Asian Pac J Trop Biomed* 7: 1129–1150.

Jiang X, Liu Y, Ma L, Ji R, Qu Y, Xin Y, Lv G (2018) Chemopreventive activity of sulforaphane. Drug design, development and therapy. *Dove Press* 12: 2905–2913.

Karmakar S, Roy Choudhury S, Banik LN, Ray KS (2011) Molecular mechanisms of anticancer action of garlic compounds in neuroblastoma. *Anticancer Agents Med Chem* 11(4): 398–407.

Keglevich P, Hazai L, Kalaus G, Szántay C (2012) Modifications on the basic skeletons of vinblastine and vincristine. *Molecules* 17(5): 5893–5914.

Keylor MH, Matsuura BS, Stephenson CR (2015) Chemistry and biology of resveratrol-derived natural products. *Chem Rev* 115: 8976–9027.

Kim J, Zhang X, Rieger-Christ KM, Summerhayes IC, Wazer DE, Paulson KE, Yee AS (2006) Suppression of Wnt signaling by the green tea compound (–)-epigallocatechin 3-gallate (EGCG) in invasive breast cancer cells requirement of the transcriptional repressor HBP1. *J Biol Chem* 281: 10865–10875.

Kim S, Kim SH, Hur SM, Lee SK, Kim WW, Kim JS, Kim JH, Choe JH, Nam SJ, Lee JE, Yang JH (2009) Silibinin prevents TPA-induced MMP-9 expression by down-regulation of COX-2 in human breast cancer cells. *J Ethnopharmacol* 126: 252–257.

Klimaszewska-Wiśniewska A, Hałas-Wiśniewska M, Izdebska M, Gagat M, Grzanka A, Grzanka D (2017) Antiproliferative and anti-metastatic action of quercetin on A549 non-small cell lung cancer cells through its effect on the cytoskeleton. *Acta Histochem* 119: 99–112.

Krishnamurthi K (2007) 17-Screening of natural products for anticancer and antidiabetic properties. *Cancer* 3: 69–75.

Król SK, Kiełbus M, Rivero-Müller A, Stepulak A (2015) Comprehensive review on betulin as a potent anticancer agent. *BioMed Res Int* 2015: 584189.

Kumar NB, Quinn GP, Alexandrow MG, Gray J, Schell M, Sutton S, Haura EB (2014) Chemoprevention trial feasibility using botanicals in exceptionally high risk populations for lung cancer. *J Clin Trials* 4: 1–6.

Kundu A, Padhi S, Behera A, Hasnain MS, Nayak AK (2022) Tumour targeting strategies by chitosan-based nanocarriers. In: Hasnain MS, Beg S, Nayak AK (Eds) *Chitosan in Biomedical Applications*. Academic Press, London, pp. 163–188.

Ladoire S, Enot D, Senovilla L, Chaix M, Zitvogel L, Kroemer G (2016) Positive impact of autophagy in human breast cancer cells on local immunosurveillance. *Oncoimmunol* 5: e1174801.

Lan A, Li W, Liu Y, Xiong Z, Zhang X, Zhou S, Palko O, Chen H, Kapita M, Prigge JR, Schmidt EE (2016) Chemoprevention of oxidative stress-associated oral carcinogenesis by sulforaphane depends on NRF2 and the isothiocyanate moiety. *Oncotarget* 7: 53502–53514.

Lan L, Appelman C, Smith AR, Yu J, Larsen S, Marquez RT, Liu H, Wu X, Gao P, Roy A, Anbanandam A, Gowthaman R, Karanicolas J, De Guzman RN, Rogers S, Aubé J, Ji M, Cohen RS, Neufeld KL, Xu L (2015) Natural product (−)-gossypol inhibits colon cancer cell growth by targeting RNA-binding protein Musashi-1. *Mol Oncol* 9(7): 1406–1420.

Laurendeau I, Ferrer M, Garrido D, D'Haene N, Ciavarelli P, Basso A, Vidaud M, Bieche I, Salmon I, Szijan I (2010) Gene expression profiling of the hedgehog signaling pathway in human meningiomas. *Mol Med* 16: 262–270.

Lauth M, Toftgård R (2007) Non-canonical activation of GLI transcription factors: implications for targeted anticancer therapy. *Cell Cycle* 6: 2458–2463.

Li G, Rivas P, Bedolla R, Thapa D, Reddick RL, Ghosh R, Kumar AP (2013) Dietary resveratrol prevents development of high-grade prostatic intraepithelial neoplastic lesions: involvement of SIRT1/S6K axis. *Cancer Prev Res* 6: 27–39.

MacDonald BT, He X (2012) A finger on the pulse of Wnt receptor signaling. *Cell Res* 22: 1410–1412.

Madhuri S, Pandey G (2009) Some anticancer medicinal plants of foreign origin. *Curr Sci* 96: 779–783.

Mahata S, Maru S, Shukla S, Pandey A, Mugesh G, Das BC, Bharti AC (2012) Anticancer property of *Bryophyllum pinnata* (Lam.) Oken. leaf on human cervical cancer cells. *BMC Complement Altern Med* 12(1), 1–11.

Maiuri MC, Zalckvar E, Kimchi A, Kroemer G (2007) Self-eating and self-killing: crosstalk between autophagy and apoptosis. *Nat Rev Mol Cell Biol* 8: 741.

Many AM, Brown AMC (2014) Both canonical and non-canonical Wnt signalling independently promote stem cell growth in mammospheres. *PLoS One* 9: e101800.

Maresso KC, Tsai KY, Brown PH, Szabo E, Lippman S, Hawk ET (2015) Molecular cancer prevention: current status and future directions. Cancer J Clin 65: 345–383.

Mastrangelo E, Milani M (2018) Role and inhibition of GLI1 protein in cancer. *Lung Cancer Targets Ther* 9: 35–43.

Mazzoni SM, Fearon ER (2014) AXIN1 and AXIN2 variants in gastrointestinal cancers. *Cancer Lett* 355: 1–8.

McCarty MF (2009) Isoflavones made simple–agonist activity for the beta-type estrogen receptor may mediate their health benefits. In Watson RR (Eds) *Complementary and Alternative Therapies and the Aging Population: An Evidence-Based Approach.* Academic Press, San Diego, CA, pp. 475–522.

Melisi D, Piro G, Tamburrino A, Carbone C, Tortora G (2013) Rationale and clinical use of multitargeting anticancer agents. *Curr Opin Pharmacol* 13: 536–542.

Merina N, Chandra KJ, Jibon K (2012) Medicinal plants with potential anticancer activities: a review. *Int Res J Pharm* 3(6): 26–30.

Mishra S, Aeri V, Gaur PK, Jachak SM (2014) Phytochemical, therapeutic, and ethnopharmacological overview for a traditionally important herb: Boerhavia diffusa Linn. *BioMed Res Int* 2014: 808302.

Moghtaderi H, Sepehri H, Delphi L, Attari F (2018) Gallic acid and curcumin induce cytotoxicity and apoptosis in human breast cancer cell MDA-MB-231. *BioImpacts* 8(3): 185.

Moscat J, Karin M, Diaz-Meco MT (2016) p62 in cancer: signaling adaptor beyond autophagy. *Cell* 167: 606–609.

Mu J, Wang X, Dong L, Sun P (2019) Curcumin derivative L6H4 inhibits proliferation and invasion of gastric cancer cell line BGC-823. *J Cell Biochem* 120: 1011–1017.

Mukherjee S, Patra CR (2016) Therapeutic application of anti-angiogenic nanomaterials in cancers. *Nanoscale* 8: 12444–12470.

Mukhtar E, Mustafa Adhami V, Khan N, Mukhtar H (2012) Apoptosis and autophagy induction as mechanism of cancer prevention by naturally occurring dietary agents. *Curr Drug Targets* 13: 1831–1841.

Nakamura A, Aizawa J, Sakayama K, Kidani T, Takata T, Norimatsu Y, Miura H, Masuno H (2012) Genistein inhibits cell invasion and motility by inducing cell differentiation in murine osteosarcoma cell line LM8. *BMC Cell Biol* 13: 24.

Naponelli V, Modernelli A, Bettuzzi S, Rizzi F (2015) Roles of autophagy induced by natural compounds in prostate cancer. *Bio Med Res Int* 2015: 121826.

Nazim UMD, Moon JH, Lee JH, Lee YJ, Seol JW, Eo SK, Lee JH, Park SY (2016) Activation of autophagy flux by metformin downregulates cellular FLICE–like inhibitory protein and enhances TRAIL- induced apoptosis. *Oncotarget* 7: 23468–23481.

Nivya MA, Raja K, Kumaravel M, Sasidharan S, Seethapathy GS (2012) Role of nutraceuticals in cancer. *Int J Pharm Pharm Sci* 4: 14.

Obexer P, Ausserlechner MJ (2014) X-linked inhibitor of apoptosis protein: a critical death resistance regulator and therapeutic target for personalized cancer therapy. *Front Oncol* 4: 197.

Oh S, Gwak J, Park S, Yang CS (2014) Green tea polyphenol EGCG suppresses Wnt/β-catenin signaling by promoting GSK-3β-and PP2A-independent β-catenin phosphorylation/degradation. *Biofactors* 40: 586–595.

Osman NH, Said UZ, El-Waseef AM, Ahmed ES (2015) Luteolin supplementation adjacent to aspirin treatment reduced dimethylhydrazine-induced experimental colon carcinogenesis in rats. *Tumor Biol* 36: 1179–1190.

Ouyang L, Shi Z, Zhao S, Wang FT, Zhou TT, Liu B, Bao JK (2012) Programmed cell death pathways in cancer: a review of apoptosis, autophagy and programmed necrosis. *Cell Prolif* 45: 487–498.

Padhi S, Behera A (2021) Advanced drug delivery systems in the treatment of ovarian cancer. In: Dua K, Mehta M, de Jesus Andreoli Pinto T, Pont LG, Williams KA, Rathbone MJ (Eds) *Advanced Drug Delivery Systems in the Management of Cancer*. Academic Press, London, pp. 127–139.

Padhi S, Behera A, Hasnain MS, Nayak AK (2022) Chitosan-based drug delivery systems in cancer therapeutics. In: Hasnain MS, Beg S, Nayak AK (Eds) *Chitosan in Drug Delivery*. Academic Press, London, pp. 159–193.

Padhi S, Kapoor R, Verma D, Panda AK, Iqbal Z (2018) Formulation and optimization of topotecan nanoparticles: in vitro characterization, cytotoxicity, cellular uptake and pharmacokinetic outcomes. *J Photochem Photobiol B Biol* 183, 222–232.

Padhi S, Mirza MA, Verma D, Khuroo T, Panda AK, Talegaonkar S, Khar RK, Iqbal Z (2016) Revisiting the nanoformulation design approach for effective delivery of Topotecan in its stable form: an appraisal of its in vitro behaviour and tumour amelioration potential. *Drug Delivery* 23(8): 2827–2837.

Pan L, Chai HB, Kinghorn AD (2012) Discovery of new anticancer agents from higher plants. *Front Biosci* 4: 142–156.

Pang SQ, Wang GQ, Lin JS, Diao Y, Xu RA (2014) Cytotoxic activity of the alkaloids from *Broussonetia papyrifera* fruits. *Pharm Biol* 52(10): 1315–1319.

Patra CR, Mukherjee S, Kotcherlakota R (2014) Biosynthesized silver nanoparticles: a step forward for cancer theranostics? *Nanomed* 9: 1445–1448.

Peng L, Zhu H, Wang J, Sui H, Zhang H, Jin C, Li L, Xu T, Miao R (2015) MiR-492 is functionally involved in Oxaliplatin resistance in colon cancer cells LS174T via its regulating the expression of CD147. *Mol Cell Biochem* 405: 73–79.

Perera RM, Stoykova S, Nicolay BN, Ross KN, Fitamant J, Boukhali M, Lengrand J, Deshpande V, Selig MK, Ferrone CR, Settleman J (2015) Transcriptional control of autophagy–lysosome function drives pancreatic cancer metabolism. *Nature* 524: 361–365.

Phi LT, Sari IN, Yang YG, Lee SH, Jun N, Kim KS, Lee YK, Kwon HY (2018) Cancer stem cells (CSCs) in drug resistance and their therapeutic implications in cancer treatment. *Stem Cell Int* 2018: 16.

Phillip CJ, Giardina CK, Bilir B, Cutler DJ, Lai YH, Kucuk O, Moreno CS (2012) Genistein cooperates with the histone deacetylase inhibitor vorinostat to induce cell death in prostate cancer cells. *BMC Cancer* 12: 145.

Pierpaoli E, Damiani E, Orlando F, Lucarini G, Bartozzi B, Lombardi P, Salvatore C, Geroni C, Donati A, Provinciali M (2015) Antiangiogenic and antitumor activities of berberine derivative NAX014 compound in a transgenic murine model of HER2/neu-positive mammary carcinoma. *Carcinogenesis* 36(10): 1169–1179.

Prasad S, Tyagi AK (2015) Ginger and its constituents: role in prevention and treatment of gastrointestinal cancer. *Gastroenterol Res Pract* 2015: 142979.

Pratama NP, Wulandari S, Nugroho AE, Fakhrudin N, Astuti P (2018) Tylophorine abrogates G2/M arrest induced by doxorubicine and promotes increased apoptosis in T47D breast cancer cells. *Asian Pacific J Cancer Prevent* 19(11): 3065.

Qi W, Weber CR, Wasland K, Roy H, Wali R, Joshi S, Savkovic SD (2011) Tumor suppressor FOXO3 mediates signals from the EGF receptor to regulate proliferation of colonic cells. *Am J Physiol Gastrointest Liver Physiol* 300: G264–G272.

Rakashanda S, Mubashir S, Qurishi Y, Hamid A, Masood A, Amin S (2013) Trypsin inhibitors from *Lavatera cashmeriana* Camb. seeds: isolation, characterization and in-vitro cytoxicity activity. *Int J Pharm Sci Invention* 2(5): 55–65.

Ramakrishnan G, Lo Muzio LO, Elinos-Báez CM, Jagan S, Augustine TA, Kamaraj S, Anandakumar P, Devaki T (2009) Silymarin inhibited proliferation and induced apoptosis in hepatic cancer cells. *Cell Prolif* 42: 229–240.

Ramasamy K, Agarwal R (2008) Multitargeted therapy of cancer by silymarin. *Cancer Lett* 269(2): 352–362.

Ran Y, Hossain F, Pannuti A, Lessard CB, Ladd GZ, Jung JI, Minter LM, Osborne BA, Miele L, Golde TE (2017) γ-Secretase inhibitors in cancer clinical trials are pharmacologically and functionally distinct. *EMBO Mol Med* 9: 950–966.

Rana C, Piplani H, Vaish V, Nehru B, Sanyal SN (2015) Downregulation of telomerase activity by diclofenac and curcumin is associated with cell cycle arrest and induction of apoptosis in colon cancer. *Tumor Biol* 36: 5999–6010.

Rather RA, Bhagat M (2018) Cancer chemoprevention and piperine: molecular mechanisms and therapeutic opportunities. *Front Cell Dev Biol* 2018: 6–10.

Ren M, Ye L, Hao X, Ren Z, Ren S, Xu K, Li J (2014) Polysaccharides from *Tricholoma matsutake* and *Lentinus edodes* enhance 5-fluorouracil-mediated H22 cell growth inhibition. *J Tradit Chin Med* 34: 309–316.

Ruiz-Torres V, Encinar JA, Herranz-López M, Pérez-Sánchez A, Galiano V, Barrajón-Catalán E, Micol V (2017) An updated review on marine anticancer compounds: the use of virtual screening for the discovery of small-molecule cancer drugs. *Molecules* 22: 1037.

Rungruangmaitree R, Jiraungkoorskul W (2017) Pea, Pisum sativum, and its anticancer activity. *Pharmacognosy Rev* 11(21): 39.

Saha S, Sadhukhan P, Mahalanobish S, Dutta S, Sil PC (2018) Ameliorative role of genistein against age-dependent chronic arsenic toxicity in murine brains via the regulation of oxidative stress and inflammatory signalling cascades. *J Nutr Biochem* 55: 26–40.

Saha S, Sadhukhan P, Sil CP (2014) Genistein: a phytoestrogen with multifaceted therapeutic properties. *Mini Rev Med Chem* 14: 920–940.

Saunier E, Antonio S, Regazzetti A, Auzeil N, Laprévote O, Shay JW, Coumoul X, Barouki R, Benelli C, Huc L, Bortoli S (2017) Resveratrol reverses the Warburg effect by targeting the pyruvate dehydrogenase complex in colon cancer cells. *Sci Rep* 7(1): 1–6.

Scannell JW, Blanckley A, Boldon H, Warrington B (2012) Diagnosing the decline in pharmaceutical R&D efficiency. *Nat Rev Drug Discov* 11: 191–200.

Seo HS, Ku JM, Choi HS, Choi YK, Woo JK, Kim M, Kim I, Na CH, Hur H, Jang BH, Shin YC (2016) Quercetin induces caspase-dependent extrinsic apoptosis through inhibition of signal transducer and activator of transcription 3 signalling in HER2-overexpressing BT-474 breast cancer cells. *Oncol Rep* 36: 31–42.

Shaban A, Mishra GM, Nautiyal R, Srivastava S, Tripathi K, Chaudhary P, Verma SK (2012) In vitro cytotoxicity of Moringa oleifera against different human cancer cell lines. *Asian J Pharm Clin Res* 5(4): 271–272.

Shalabi M, Khilo K, Zakaria MM, Elsebaei MG, Abdo W, Awadin W (2015) Anticancer activity of Aloe vera and Calligonum comosum extracts separately on hepatocellular carcinoma cells. *Asian Pacific J Tropic Biomed* 5(5): 375–381.

Sharma DR, Wani WY, Sunkaria A, Kandimalla RJ, Sharma RK, Verma D, Bal A, Gill KD (2016) Quercetin attenuates neuronal death against aluminum-induced neurodegeneration in the rat hippocampus. *Neurosci* 324: 163–176.

Siddiqui WA, Ahad A, Ahsan H (2015) The mystery of BCL2 family: Bcl-2 proteins and apoptosis: an update. *Arch Toxicol* 89: 289–317.

Singh CK, Ndiaye MA, Ahmad N (2015) Resveratrol and cancer: challenges for clinical translation. *Biochim Biophys Acta* 1852: 1178–1185.

Singh S, Bhupender S, Kanwar SS, Kumar A (2016) Lead phytochemicals for anticancer drug development. *Front Plant Sci* 7: 1667.

Skoda AM, Simovic D, Karin V, Kardum V, Vranic S, Serman L (2018) The role of the Hedgehog signaling pathway in cancer: a comprehensive review. *Bosn J Basic Med Sci* 18: 8.

Ślusarz A, Shenouda NS, Sakla MS, Drenkhahn SK, Narula AS, MacDonald RS, Besch-Williford CL, Lubahn DB (2010) Common botanical compounds inhibit the hedgehog signaling pathway in prostate cancer. *Cancer Res* 70: 3382–3390.

Sun C, Liu Z, Li S, Yang C, Xue R, Xi Y, Wang L, Wang S, He Q, Huang J, Xie S, Jiang W, Li D (2015) Down-regulation of c-Met and Bcl2 by microRNA-206, activates apoptosis, and inhibits tumor cell proliferation, migration and colony formation. *Oncotarget* 6, 25533–25574.

Syed DN, Chamcheu JC, Adhami VM, Mukhtar H (2013) Pomegranate extracts and cancer prevention: molecular and cellular activities. *Anticancer Agents Med Chem* 13: 1149–1161.

Tanaka T (2013) Role of apoptosis in the chemoprevention of cancer. *Int J Clin Exp Med* 5: 89–91.

Tang GQ, Yan TQ, Guo W, Ren TT, Peng CL, Zhao H, Lu XC, Zhao FL, Han X (2010) (−)-Epigallocatechin-3-gallate induces apoptosis and suppresses proliferation by inhibiting the human Indian Hedgehog pathway in human chondrosarcoma cells. *J Cancer Res Clin Oncol* 136: 1179–1185.

Tarapore RS, Siddiqui IA, Mukhtar H (2012) Modulation of Wnt/β-catenin signaling pathway by bioactive food components. *Carcinogenesis* 33: 483–491.

Tariq A, Sadia S, Pan K, Ullah I, Mussarat S, Sun F, Abiodun OO, Batbaatar A, Li Z, Song D, Xiong Q (2017) A systematic review on ethnomedicines of anticancer plants. *Phytother Res* 31: 202–264.

Thakore P, Mani RK, Kavitha SJ (2012) A brief review of plants having anticancer property. *Int J Pharm Res Dev* 3: 129–136.

Thangapazham RL, Sharad S, Maheshwari RK (2016) Phytochemicals in wound healing. *Adv Wound Care* 5(5): 230–241.

Thimmulappa RK, Mai KH, Srisuma S, Kensler TW, Yamamoto M, Biswal S (2002) Identification of Nrf2-regulated genes induced by the chemopreventive agent sulforaphane by oligonucleotide microarray. *Cancer Res* 62: 5196–5203.

Torre LA, Bray F, Siegel RL, Ferlay J, Lortet-Tieulent J, Jemal A (2015) Global cancer statistics, 2012. *CA Cancer J Clin* 65: 87–108.

Tripathi YB, Tripathi P, Arjmandi BH (2005) Nutraceuticals and cancer management. *Front Biosci* 10: 1607–1618.

Truong VL, Jun M, Jeong WS (2018) Role of resveratrol in regulation of cellular defense systems against oxidative stress. *Biofactors* 44: 36–49.

Tu LY, Pi J, Jin H, Cai JY, Deng SP (2016) Synthesis, characterization and anticancer activity of kaempferol-zinc (II) complex. *Bioorganic Med Chem Lett* 26(11): 2730–2734.

Turati F, Rossi M, Pelucchi C, Levi F, La Vecchia C (2015) Fruit and vegetables and cancer risk: a review of southern European studies. *Br J Nutr* 113: S102–S110.

Umar A, Dunn BK, Greenwald P (2012) Future directions in cancer prevention. *Nat Rev Cancer* 12: 835–848.

Vallianou NG, Evangelopoulos A, Schizas N, Kazazis C (2015) Potential anticancer properties and mechanisms of action of curcumin. *AnticancerRes* 35(2): 645–651.

Vanitha V, Hemalatha S, Pushpabharathi N, Amudha P, Jayalakshmi M (2017) Fabrication of nanoparticles using Annona squamosa leaf and assessment of its effect on liver (Hep G2) cancer cell line. *IOP Conf Series Mater Sci Eng* 191(1): 012010. IOP Publishing.

Verma D, Thakur PS, Padhi S, Khuroo T, Talegaonkar S, Iqbal Z (2017) Design expert assisted nanoformulation design for co-delivery of topotecan and thymoquinone: optimization, in vitro characterization and stability assessment. *J Mol Liquids* 242: 382–394.

Ververidis F, Trantas E, Douglas C, Vollmer G, Kretzschmar G, Panopoulos N (2007) Biotechnology of flavonoids and other phenylpropanoid-derived natural products. Part I: Chemical diversity, impacts on plant biology and human health. *Biotechnol J Healthcare Nutr Technol* 2(10): 1214–1234.

Vinod BS, Maliekal TT, Anto RJ (2013) Phytochemicals as chemosensitizers: from molecular mechanism to clinical significance. *Antioxid Redox Signal* 18: 1307–1348.

Vinogradov S, Wei X (2017) Cancer stem cells and drug resistance: the potential of nanomedicine. *Nanomed* 7: 597–615.

Wal A, Srivastava RS, Wal P, Rai A, Sharma S (2015) Lupeol as a magical drug. *Pharm Biol Eval* 2(5): 142–151.

Wang CZ, Zhang Z, Wan JY, Zhang CF, Anderson S, He X, Yu C, He T-C, Qi L-W, Yuan CS (2015) Protopanaxadiol, an active ginseng metabolite, significantly enhances the effects of fluorouracil on colon cancer. *Nutrients* 7(2): 799–814.

Wang DX, Zou YJ, Zhuang XB, Chen SX, Lin Y, Li WL, Lin JJ, Lin ZQ (2017a) Sulforaphane suppresses EMT and metastasis in human lung cancer through miR-616-5p-mediated GSK3β/β-catenin signaling pathways. *Acta Pharmacol Sinica* 38: 241–251.

Wang LS, Burke CA, Hasson H, Kuo CT, Molmenti CL, Seguin C, Liu P, Huang TH, Frankel WL, Stoner GD (2014) A phase Ib study of the effects of black raspberries on rectal polyps in patients with familial adenomatous polyposis. *Cancer Prev Res* 7: 666–674.

Wang WR, Li TT, Jing T, Li YX, Yang XF, He YH, Zhang W, Zhang JY, Lin R (2017b) SIRT1 regulates the inflammatory response of vascular adventitial fibroblasts through autophagy and related signaling pathway. *Cell Physiol Biochem* 41: 569–582.

Wang Y, Hong C, Zhou C, Xu D, Qu HB (2011) Screening antitumor compounds psoralen and isopsoralen from Psoralea corylifolia L. seeds. *Evid Based Complement Alternat Med* 2011: 363052.

Weaver BA (2014) How taxol/paclitaxel kills cancer cells. *Mol Biol Cell* 25: 2677–2681.

Weng JR, Bai LY, Chiu CF, Hu JL, Chiu SJ, Wu CY (2013) Cucurbitane triterpenoid from Momordica charantia induces apoptosis and autophagy in breast cancer cells, in part, through peroxisome proliferator-activated receptor γ activation. *Evid Based Complement Alternat Med* 2013: 935675.

Woźniak Ł, Skąpska S, Marszałek K (2015) Ursolic acid: a pentacyclic triterpenoid with a wide spectrum of pharmacological activities. *Molecules* 20(11): 20614–20641.

Xiong M, Wang L, Yu HL, Han H, Mao D, Chen J, Zeng Y, He N, Liu Z-G, Wang Z-Y, Xu S-J, Guo L-Y, Wang YA (2016) Ginkgetin exerts growth inhibitory and apoptotic effects on osteosarcoma cells through inhibition of STAT3 and activation of caspase-3/9. *Oncol Rep* 35(2): 1034–1040.

Yadav B, Bajaj A, Saxena M, Saxena AK (2010) In vitro anticancer activity of the root, stem and leaves of Withania somnifera against various human cancer cell lines. *Indian J Pharm Sci* 72(5): 659.

Yagishita Y, Fahey JW, Dinkova-Kostova AT, Kensler TW (2019) Broccoli or sulforaphane: is it the source or dose that matters? *Molecules* 24: 3593.

Yan CH, Li F, Ma YC (2015) Plumbagin shows anticancer activity in human osteosarcoma (MG-63) cells via the inhibition of S-phase checkpoints and down-regulation of c-myc. *Int J Clin Exp Med* 8(8): 14432.

Yan GR, Zou FY, Dang BL, Zhang Y, Yu G, Liu X, He QY (2012) Genistein-induced mitotic arrest of gastric cancer cells by downregulating KIF 20 A, a proteomics study. *Proteomics* 12: 2391–2399.

Yang X, Dinehart MS (2017) Triple hedgehog pathway inhibition for basal cell carcinoma. *J Clin Aesth Dermatol* 10: 47–49.

Yi J, Yang J, He R, Gao F, Sang H, Tang X, Richard DY (2004) Emodin enhances arsenic trioxide-induced apoptosis via generation of reactive oxygen species and inhibition of survival signaling. *Cancer Res* 64: 108–116.

Yoon JS, Kim HM, Yadunandam AK, Kim NH, Jung HA, Choi JS, Kim C-Y, Kim GD (2013) Neferine isolated from Nelumbo nucifera enhances anticancer activities in Hep3B cells: molecular mechanisms of cell cycle arrest, ER stress induced apoptosis and anti-angiogenic response. *Phytomedicine* 20(11): 1013–1022.

Zhang C, Hao Y, Wu L, Dong X, Jiang N, Cong B, Liu J, Zhang W, Tang D, De Perrot M, Zhao X (2018) Curcumin induces apoptosis and inhibits angiogenesis in murine malignant mesothelioma. *Int J Oncol* 53: 2531–2541.

Zhang Y, Kensler TW, Cho CG, Posner GH, Talalay P (1994) Anticarcinogenic activities of sulforaphane and structurally related synthetic norbornyl isothiocyanates. *Proc Natl Acad Sci USA* 91: 3147–3150.

Zhang Y, Li Q, Zhou D, Chen H (2013) Genistein, a soya isoflavone, prevents azoxymethane-induced up-regulation of WNT/β-catenin signalling and reduces colon pre-neoplasia in rats. *Br J Nutr* 109: 33–42.

Zhang YY, Huang CT, Liu SM, Wang B, Guo J, Bai JQ, Fan X-J, Jia YS (2016) Licochalcone A exerts antitumor activity in bladder cancer cell lines and mice models. *Tropic J Pharm Res* 15(6): 1151–1157.

Zheng J, Zhou Y, Li Y, Xu DP, Li S, Li HB (2016) Spices for prevention and treatment of cancers. *Nutrients* 8(8): 495.

Zhou J, Li R, Jia Y, Wang Y, Liu J, Panichayupakaranant P, Chen H (2022) Recent progress in natural anticancer agents discovery from tea (Camellia sinensis): a review. *Recent Patents Anticancer Drug Disc* 17(4): 343–357.

Zhou Q, Gong X, Kuang GE, Jiang R, Xie T, Tie H, Chen X, Li K, Wan J, Wang B (2018) Ferulic acid protected from kidney ischemia reperfusion injury in mice: possible mechanism through increasing adenosine generation via HIF-1α. *Inflammation* 41: 2068–2078.

Zhou Z, Tang M, Liu Y, Zhang Z, Lu R, Lu J (2017) Apigenin inhibits cell proliferation, migration, and invasion by targeting Akt in the A549 human lung cancer cell line. *Anticancer Drugs* 28: 446–456.

Zhu J, Jiang Y, Yang X, Wang S, Xie C, Li X, Li Y, Chen Y, Wang X, Meng Y, Zhu M, Wu R, Huang C, Ma X, Geng S, Wu J, Zhong C (2019) Wnt/betacatenin pathway mediates (-)-epigallocatechin-3-gallate (EGCG) inhibition of lung S. *Food Chem Toxicol* 128: 240–255.

Zou Y, Yan F, Zheng T, Shi D, Sun F, Yang N, Chen L (2015) Highly luminescent organosilane-functionalized carbon dots as a nanosensor for sensitive and selective detection of quercetin in aqueous solution. *Talanta* 135: 145–148.

3 Passive and Active Targeting Strategies for the Delivery of Phytochemical-Entrapped Nanosystems for Cancer

Ananya Kuanar[1], *Sandeep Kumar Kabi*[2], *Bibhudutta Pattnaik*[1], *and Dattatreya Kar*[3]

[1]Centre for Biotechnology, Siksha O Anusandhan (Deemed to be University), Kalinga Nagar, Ghatikia, Bhubaneswar, India

[2]Environment & Sustainability Department, CSIR-Institute of Minerals Materials Technology, Bhubaneswar, India

[3]Department of Medical Research, Health Science, IMS & SUM Hospital, Siksha O Anusandhan University, Bhubaneswar, India

3.1 INTRODUCTION

The deadliest disease and the leading cause of morbidity is cancer, which is on the rise all over the world (Sharma et al., 2014; Padhi et al., 2016). In advanced countries, cancer is the second deadliest disease, next to cardiovascular diseases (Siegel et al., 2016; Kundu et al., 2022). The five common cancers in India are lung, breast, mouth, uteri, and cervical (Mathur et al., 2020). The unchecked multiplication and de-differentiation of a cell converts a healthy cell into a cancerous cell. A cancerous cell turns down the signals for differentiation and proliferation, leading to proliferation, eventually overcoming the apoptosis process and causing invasion and angiogenesis.

The alteration in the genetic sequence of a cell causes genetic instabilities. It converts a normal cell into a malignant cell (Padhi et al., 2018). The alteration in the genes/genetic sequence causes gene mutations (Oncogenes, tumour suppressors), DNA repair genes, and genes involved in cell growth and development. Besides these alterations, external and internal factors also lead to cancer. The external factors are radiation, pollution, and smoking, whereas the internal factors are gene mutations, immune systems, and hormones (Ferlay et al., 2010). Colorectal, lung, prostate, liver, and stomach cancer are commonly reported in men, while cervical, lung, colorectal, breast, and thyroid cancer are noted in women (Horn et al., 2012; Behera & Padhi, 2022). Normal cell growth, division, and development depend upon the coordination of several genes. The alteration of these genes turns normal cells into cancerous cells (Biswas et al., 2015). For the development of anticancer therapeutics, these genes can be targeted. Alterations of epigenetic pathways involved in cell growth and division regulation may lead to cancerous cell development.

A new study age has begun in the treatment of many cancers. Chemotherapy, radiation therapy, and surgery are among the many traditional and contemporary techniques for the treatment of cancer. The drugs used in the treatment have adverse effects that lead to more toxic products. The failure of conventional chemotherapeutic methods endures, paving the path for introducing novel cancer-control strategies (Padhi & Behera, 2022). Phytochemical compounds isolated from medicinal plants are safe, non-toxic, and easily available resources for healing cancer. Anticancer drug discovery can alter these epigenetic variations, which is also a favourable choice in therapeutic interventions (Schnekenburger et al., 2014). In this regard, vorinostat, Azacitidine, romidepsin, and decitabine are some perfect epigenetic anticancer drugs. Among other anticancer medicinal plants, *Phaleria macrocarpa* and *Fagonia indica* have been traditionally used as anticancer agent for their bioactive compounds (Shehab et al., 2011; Faried et al., 2007).

Plant metabolites or bioactive compounds are used to activate apoptosis in cancerous cells (Table 3.1). Gallic acid, a bioactive compound extracted and refined from the fruit extract of *P. macrocarpa,* has shown apoptotic induction in cell lines of colon adenocarcinoma, lung cancer, and leukaemia (Sohi et al., 1994). Gallic acid is a polyhydroxy phenolic compound and a normal antioxidant derived from various natural products such as grapes, bananas, strawberries, green tea, and green vegetables (Sun et al., 2002). This phytochemical compound plays a crucial role in inhibiting malignancy development and the transformation of cancer cells (Taraphdar et al., 2001). Similarly, other phytochemical compounds like vinca alkaloids, podophyllotoxin, and camptothecin have been explored for cancer treatment.

DOI: 10.1201/9781003611516-4

TABLE 3.1
Major Phytochemical Constituents of Plants Showing Anticancer Activity

Phytochemical Constituents	Plant Name	Anticancer Activity	References
Columbin, Cordioside	*Tinospora cardifolia*	Cytotoxic against cervical cancer cells (HeLa cells)	Jagetia et al. (2006)
Betulinic acid, Betulin	*Ziziphus nummularia*	1. Selectively cytotoxic towards cancerous cells 2. Reactive oxygen species generation, topoisomerase-I inhibition, and mitogen-activated protein kinase activation are all used to induce apoptosis. 3. Angiogenesis is inhibited, transcriptional activators are modulated, and p53 and CD95 mediate cell death. 4. Causes the depletion of mitochondrial membrane potential and the release of cytochrome C.	Gauthier et al. (2006), Eiznhamer and Xu (2004) and Fulda and Debatin (2005)
Andrographolide	*Andrographis paniculata*	1. A panel of cancer cell lines is cytotoxic by alcoholic extract. 2. Superoxide dismutase, catalase, and glutathione-*S*-transferase antioxidant enzymes are increased, while lactate dehydrogenase and malonaldehyde decrease.	Kumar et al. (2004) and Jada et al. (2007)
Asiaticoside, hydrocotyline, stigmasterol, vallerine, flavonoids and thankunosides	*Centella asiatica*	1. Leaf extract inhibits the proliferation of transformed cell lines (dose-dependent) 2. Inhibits DNA synthesis	Babu et al. (1995)
Curcumin	*Curcuma longa*	1. Cell proliferation inhibition in a broad range of cell lines 2. Early growth response protein-1, Nuclear factor kappa-light-chain-enhancer of activated B cells, Activator protein 1, nitric oxide synthase, cyclooxygenase-2, Lipoxygenase, matrix metalloproteinase-9, epidermal growth factor receptor, tumour necrosis factor, chemokines, and human epidermal growth factor receptor 2 are all downregulated. 3. Blocks the c-Jun N-terminal kinase and serine/threonine kinase pathways. 4. Reduces matrix metalloproteinase-2, which inhibits metastasis. 5. Decreases mitochondrial membrane perspective, releases Cytochrome C, activates caspase 3 and 9 pathways and downregulates antiapoptotic proteins B-cell lymphoma-extra large and integrin-linked protein in vitro, inducing apoptosis.	Aggarwal et al. (2003), Shao et al. (2002), and Mitra et al. (2006)
Niranthrin, Nirtetralin, phyltetralin phyllanthin	*Phyllanthus amarus*	1. Suppress the action of cell division cycle 25 A tyrosine phosphatase, which suppresses cell multiplication in various cancer cells. 2. In *Saccharomyces cerviacae* mutant cell cultures inhibit topoisomerase I and II activity. 3. It inhibits DNA repair and induces cell cycle arrest. 4. Purified lignins function as a multi-drug resistant reversal agent. 5. In mice with Dalton's lymphoma ascites (DLA) and Erlich ascites carcinoma, oral administration of extract prolongs life and reduces tumour size of Ehrlich ascites carcinoma. 6. In mice with Lewis lung carcinoma, plant extract reduces *n*-nitroso diethylamine-induced tumour incidence and has anti-angiogenic effects.	Rajeshkumar and Kuttan (2000) and Leite et al. (2006)
Bullatacin	*Annona atemoya*	1. It is cytotoxic to tumour cell lines. 2. Tumour cell condensation and chromatin margination caused cell death.	Chiu et al. (2003), Wu et al. (1995) and Chang et al. (2003)
Camptothecin	*Mappia foetida*	1. Nucleic acid synthesis inhibitors that are selective in cervical (HeLa) cells. 2. Stops topoisomerase-I (topo I) from working. 3. Prevents the development of colon cancer and rhabdomyo sarcoma cells in humans. 4. A leukaemia-fighting chemotherapy drug 5. The xenograft model method causes partial or total remission of breast carcinoma. 6. The lymphoma, leukaemia, and solid epithelial tumour phase II trials	Wall and Wani (1996) and Potmesil (1994)

(*Continued*)

TABLE 3.1 (*Continued*)
Major Phytochemical Constituents of Plants Showing Anticancer Activity

Phytochemical Constituents	Plant Name	Anticancer Activity	References
Withaferin A	*Withania somnifera*	1. The rapid production of reactive oxygen species activates apoptosis in a mixture of cancer cells. 2. In human leukaemia (HL - 60) cells, upregulation of apical death receptors and modified ratios of members of the B-cell lymphoma 2 protein family contribute to the downstream initiation of caspase-3 and PARP cleavage, resulting in nucleosomal DNA cleavage. 3. Withaferin root and leaf extracts show cytotoxicity against a panel of cancer cell lines and activate apoptosis via reactive oxygen species generation. 4. Improves T cell and cluster of differentiation 40 expression in the mouse tumour model by increasing Th1 cytokine expression, the cluster of differentiation 3.	Malik et al. (2007) and Singh (2007)
Triterpenic acid	*Boswellia serrata*	1. Apoptosis is induced in a variety of cancer cell lines 2. Stops human leukaemia (HL-60) cells from making DNA and growing. 3. Topoisomerase I and II are inhibited. 4. Effective in the treatment of brain tumours.	Syrovets et al. (2005) and Hostanska et al. (2002)

With new techniques and research interests, using common ingredients in pharmaceutical businesses is becoming more common (Koehn & Carter, 2005; Saklani & Kutty, 2008). According to the World Health Organization, 80 % of the world's population uses traditional therapies for a variety of maladies (Wang et al., 2012). Modern biomolecular science has made it easier to comprehend the effects of herbs on diverse targets by identifying several key features such as anticancer, antiviral, and anti-inflammatory capabilities, as well as numerous tumours. Hepatocellular carcinoma has become the world's fifth most common cancer, and many studies have been conducted on the prevention and treatment of hepatocellular carcinoma using herbal medicine, showing that various constituents of herbs can harm all stages of this cancer, from initiation to promotion and development (Ruan et al., 2006).

Natural products or their derivatives account for around half of all internationally authorised anticancer medications. The knowledge gained from tiny molecules to macromolecules found in nature was used to construct the drugs (Song et al., 2014). Anticancer medicines in clinical use have major drawbacks. Low water solubility, inability to absorb orally, limited half-life inside the body, low specificity, and various side effects are among the disadvantages (Cho et al., 2008). By overexpressing the transporter proteins of the ATP-binding cassette, the therapeutic process induces severe systemic toxicity and leads to the formation of multidrug-resistant cells. P-glycoprotein, a multidrug-resistant protein family member, decreases the amount of chemotherapeutic agents required in cancer cells through energy-dependent drug efflux. Several studies have focused on ABC blockers to combat MDR in cancer over the last few decades. Verapamil, cyclosporine A, and quinidine are among the most thoroughly researched first-generation blockers. These proteins have been cleared for use in new clinical trials.

Clinical trials on breast cancer patients show that combining vincristine (vinca alkaloids) and verapamil (Pgp blocker) boosts antitumour efficacy. The development of effective, safe, and new drug transport systems against breast cancer is highlighted in these publications. Similar research has been undertaken to increase the therapeutic efficacy of numerous anticancer phytochemicals in conjunction with Pgp blockers against multidrug-resistant cancer cell lines (Granja et al., 2016). The tumour microenvironment is a key element in the development of MDR and impacts a patient's immunogenic response to treatment. Nanodelivery of carrier molecules provides promising anticancer therapeutic alternatives for various malignancies. Nanotechnology's promise has changed cancer diagnosis and treatment. Because tumour angiogenesis is poorly regulated, tumours are associated with a faulty vascular architecture.

Because the lymphatic system fails to drain the tumour interstitial fluid, submicron-sized particulate matter is retained within the tumour, causing the "enhanced permeability and retention" effect (Padhi & Behera, 2020). This enhanced permeability and retention effect is critical for passive targeting because it permits storing a well-designed nanocarrier with cytotoxic factors in tumour tissues. The conjugation of nanocarriers with chemotherapeutic compounds binds to the targeted cells' more-expressed antigens/receptors in the active targeting phase (Verma et al., 2017). The biological distribution of different nanocarriers might be properly tuned by modification of surface physico-chemical properties with the help of advanced surface-engineering technology to target a particular tissue. These nanocarriers enhance drug delivery to targeted tumour sites, thereby reducing systemic toxicity (Padhi et al., 2022).

With the progress of nanotechnology, drug and drug delivery system modification is becoming more common in cancer treatment, with positive clinical outcomes (Behera & Padhi, 2020). By enhancing cancer patients' clinical care, this technology enables better detection, effective drug delivery to malignant cells, and molecularly targeted cancer therapy (Aliosmanoglu & Basaran, 2012). Chemotherapy platforms now include various sized nanodelivery vehicles containing phytochemical anticancer drugs (Patnaik et al., 2021). The application of this technology has improved the formulation of drugs and its delivery of selected phytochemical anticancer agents like vincristine, etoposide, capsaicin, genistein, resveratrol, quercetin, paclitaxel, and curcumin (Taratula et al., 2009) (Figure 3.1).

3.2 NANOCARRIERS IN DRUG DELIVERY SYSTEMS

The key hurdles in cancer detection and treatment are building drug/gene delivery systems that successfully target unhealthy cells without affecting typical healthy tissues. This could be accomplished by effectively delivering anticancer medicines to the tumour microenvironment and thus to tumour cells (Wang & Thanou, 2010, Mussi & Torchilin, 2013). The formed nanoparticles, on the other hand, must overcome numerous physiological and biological barriers. To avoid unspecific interactions and to add specific binds to their objectives, their employment as delivery systems uses requirements for their size, biocompatibility, and surface chemistry. These nanocarriers must be effective.

1. To be stable until they reach tumour microenvironment in the blood;
2. To avoid clearance from the reticuloendothelial system and not to be caught by the mononuclear phagocyte system. To improve their hydrophilic effects and confer stealth features, to evade identification by the immune system, and increase the possibility of targeting the tissues/cells desired, both points were accomplished by PEGylating the nanoparticles' surface.
3. To accumulate through abnormal tumour vasculature in the tumour microenvironment.
4. To penetrate the high-pressure tumour microenvironment interstitial fluid into the tumour (Barar & Omidi, 2013; Omidi & Barar, 2014).
5. Access the active sites exclusively associated with the selected cells.

The optimal approach for promoting the accumulation of nanoparticles in the place of interest is active/passive targeting. Its surface functionalisation, physicochemical effects and the pathophysiological properties of the tumour microenvironment are the key parts that regulate nanoparticles drug targeting. Nanocarriers must prevent cargo molecules from degrading or releasing until they achieve their objectives, resulting in long-term toxicity issues. Nanoparticles are also attractive options for improving API loading performance through covalent binding or encapsulation (Padhi & Behera, 2020). Nanoparticles can be made from various materials, including lipids, composite metals or semiconductors, polymers, and proteins (Behera et al., 2023). Tubes, solid spheres, rods, and other precise-form nanoparticles have lately been created. Nanomaterials have a metallic or polymeric centre filled with single or many lipid layers and are often tailored to increase the system's biocompatibility (Huang et al., 2010) (Figure 3.2).

Tumour-targeted drug delivery systems have many benefits, such as

1. To protect normal cells from cytotoxic compounds
2. To reduce dose-reduce unfavourable effects and
3. To combat drug-immune cancer cells.

The nucleus is essentially the end objective for many treatments that treat diseases such as cancer, neurological problems, and heart failure. Due to their unique cell intake and trafficking mechanisms, Nanoparticles enable the delivery of specific therapies to their targeted lesions in a vital form and sufficient concentration, reducing the amount that accumulates in undesirable tissues. Though it is becoming increasingly obvious that the cytosolic internalisation of a therapeutic molecule does not require its connection with its subcellular target, careful design and optimisation of nanoparticles are necessary to allow nuclear targeting.

3.3 PHARMACOLOGICAL TARGETING

It is desirable to have pharmacological agents acting mainly on the diseased cells. Initially, chemotherapeutics is targeted to kill cancer cells that proliferate quickly. It is possible to design these agents to influence various details of the mitosis process. DNA covalently binds to alkylating agents, such as mustine and cisplatin, and blocks DNA replication. For example, gemcitabine and 5-fluorouracil are antimetabolites that mimic nucleobases and might be introduced into the cell's DNA, prohibiting enzymes engaged in DNA synthesis. Topoisomerase inhibitors control DNA's unwinding and result in DNA cleavage. Antibiotics intercalate within the DNA, as do anthracyclines. Unique receptor pathways can also be blocked by drug molecules such as folate inhibitors, like methotrexate, which was primarily designed to bind acute lymphoblastic leukaemia (ALL) cells to folate receptors (Miller, 2006). Tamoxifen competes with naturally occurring oestrogen, known as anti-hormonal therapy, to bind to the oestrogen receptor to prevent the development of estrogen-mediated breast cancer (Table 3.1).

3.3.1 Active Targeting

3.3.1.1 Magic Bullet Theory

In the early 1900s, Paul Ehrlich proposed the "magic bullet hypothesis," which suggested that a molecule might selectively convey a toxin to a disease-causing organism.

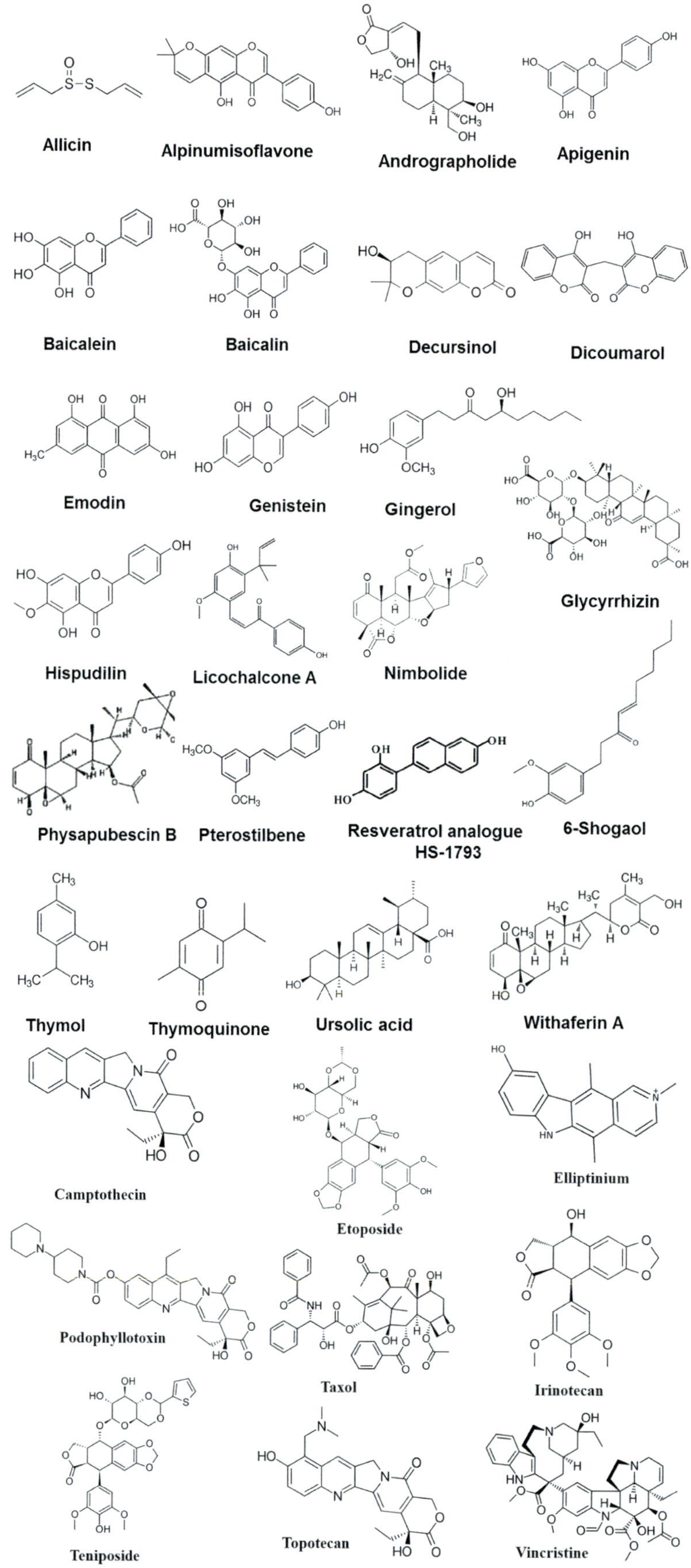

FIGURE 3.1 Different phytomolecules having anticancer activity.

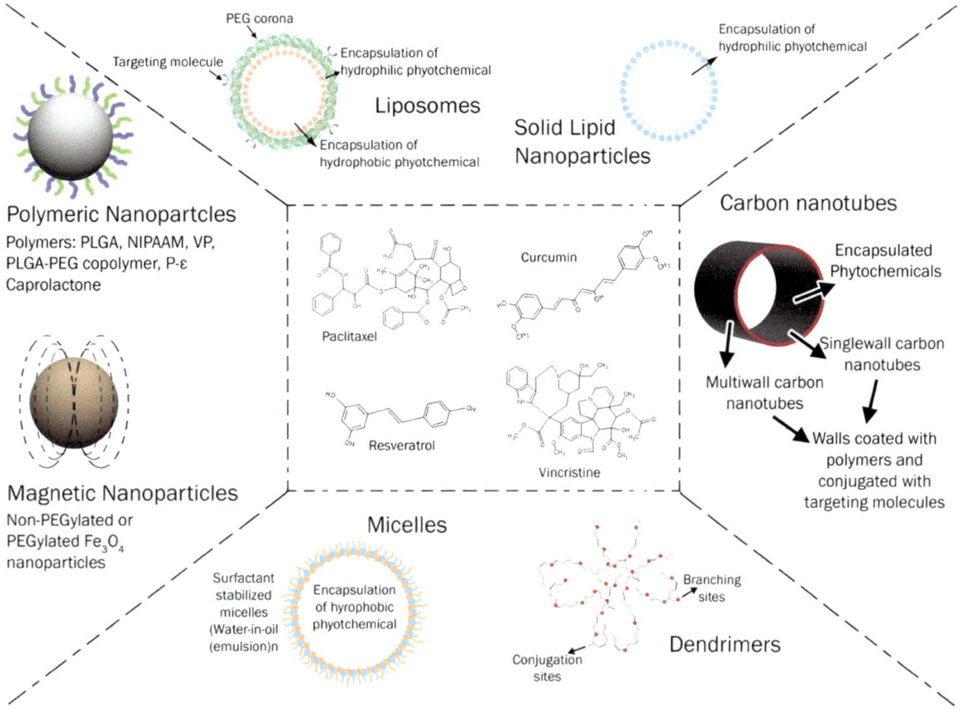

FIGURE 3.2 Multifunctional drug delivery systems for anticancer phytochemicals [A few natural phytomolecules like curcumin, paclitaxel, resveratrol, and vincristine are formulated as different nanoformulations like polymeric nanoparticles, magnetic nanoparticles, micelles, dendrimers, carbon nanotubes, liposomes, and solid lipid nanoparticles for targeted delivery to tumour cells].

Linus Pauling later confirmed his proposal in 1940 that set out the lock and key antibody process via developments in hybridoma technology; monoclonal antibodies have revolutionised medicine, allowing protein detection. Kohler and Milstein developed the primary monoclonal antibody for lymphoma through the mixture of mouse myeloma and mouse spleen cells from an immunised donor. Monoclonal lymphoma antibodies were initially used in a laboratory in 1982. In 1997, rituximab was accepted for cancer therapy (Strebhardt & Ullrich, 2008).

3.3.1.2 Key Cell Receptors

Several novel antibodies targeting membrane proteins in lymphomas, such as CD52 (alemtuzumab), cluster of differentiation 40, and cluster of differentiation 80, as well as solid tumours, such as carcinoembryonic antigen, epidermal growth factor sensory receptor (cetuximab), epithelial cell adherence molecule, and tumour necrosis factor, have sped up clinical progress with rituximab. The overexpression of integrins such as integrin alpha v beta 3 or 1. Mucin-1, folate receptor, prostate cancer lipid antigen, prostate-specific membrane antigen, and transferrin receptor have also been demonstrated to target tumour accumulation in vivo (Weinstein & van Osdol,1992; Schmidt & Wittrup, 2009).

3.3.1.3 Peptides and Aptamers

Aptamers and peptides, which can be utilised to identify a binding molecule, are short amino acid or oligonucleotide sequences. The use of peptide and aptamer targeting agents provides several advantages. They exhibit minor immunogenicity when compared to antibodies. All can be made synthetically and in large quantities for a fraction of the cost of antibodies. They may have enhanced stability due to their small size and lack of a complex, three-dimensional conformational structure. Peptides and aptamers, in contrast to antibodies, can have low joining affinities for their target, which can exacerbate off-target effects (Cerchia & de Franciscis, 2010).

3.3.1.4 Dual Targeting

The circumstance when a plateau in cell joining is passed over a particular surface ligand density is the only major issue with the chosen drug delivery. Viruses have instructed researchers to overcome this constraint, which occurs when two or more receptors are engaged in an associated connection. The human immunodeficiency virus impacts mammalian cells by the gp120 and gp40 anchoring the C-X-C motif chemokine receptor 4 and C-C chemokine receptor type 5 receptors, respectively. In vitro, a dual-targeting approach to prevent HIV using an adeno-connected viral antisense vector showed potential. This selective targeting strategy has been highlighted as a successful technique for reducing unwanted off-target effects of drugs. Multivalent nanocarriers can achieve higher joining avidity than single receptor focusing. First, the homo or heterodimerisation of cell surface receptors may play a major role in carcinogenic signalling (Hynes & Lane, 2005). Selective breast cancer therapy can be utilised to prevent oestrogen growth factor receptor dimerisation. The dimeric ligand, vascular endothelial growth factor receptor 1–2, can bind to the heterodimer. Dual targeting provides the most up-to-date anticancer targeting strategy and a theoretical foundation for cellular adherence. Constructed hetero bivalent ligands of cholecystokinin and melanocortin can crosslink complex cell surface receptors with 12 times greater dual targeting selectivity than any ex vivo single receptor, as validated in vivo. Double-targeting doxorubicin encapsulating liposomes gly-asn-gly-arg-gly and ala-pro-arg-pro-gly (APRPG) have been found to decrease tumour formation in 26NL-17 colon carcinoma-bearing mice (Yu et al., 2012).

3.3.1.5 Backpacking

Unsusceptible cells have the intrinsic expertise to detect foreign matter and inflammation regions. Associations of nanoparticles with dendritic cells might be used to transport anticancer agents straight to tumours. The patient's cells might be separated and functionalised by drug-encapsulating liposomes to avoid internalisation (Wall et al., 2010). Intravenous and home to tumours, the cells are reintroduced and added to the drug in the procedure. Even though the procedure of how cells point the tumour stays a secret, backpacking shows the potential to dramatically raise aggregation within solid tumours as a current targeting strategy (Parente-Pereira et al., 2014). Bacteria can also house tumours and carry nanoparticles intracellularly for nucleic acid transporting to cancer cells in a similar capacity. It is possible to use T cells that can identify tumour antigens as cell therapy and for drug transport (Huang et al., 2015).

3.3.1.6 Smart Drug Delivery

Smart drug transport systems might be tuned to release drugs within a targeted area, independently or through external manipulation. Modification in the tumour microenvironment, like tumour pH, enzyme activity, or redox, is used by autonomous systems. For example, a polyvalent melanoma vaccine, presently in clinical trials, is used to deliver pH-sensitive liposomes (NCI-G98-1488). In addition, light, ultrasound, and magnetic fields can influence the location of nanoparticles and subsequent drug transport. Liposomal vector heating results in the transportation of doxorubicin locally. Magnetofection has contributed to a modern method for resolving the essential limitations of in vivo gene therapy. The logical design of intelligent nanocarriers may provide selected drug delivery through independent or external stimuli (Torchilin, 2014).

3.3.1.7 Multifunctional Drug Delivery Vehicles

Developments in targeting, tracking, diagnostic, and therapeutic functions have helped lay the framework for integration into a single multifunctional drug delivery system. The first instances of cell-determined targeting and imaging appeared in 1980. Nano theranostics, which provide targeted therapy and intranuclear radio sensitisation based on images, are currently popular. The background of different therapeutically translational functions for a contrast diagnostic magnetic resonance imaging factor is superparamagnetic magnetic nanoparticles of iron oxide (Vander Heiden, 2011).

Using various arrangements of targeting ligands, materials, and anticancer drugs, it is now possible to target, monitor, and supply a therapeutic agent. Multifunctional, pH-sensitive nanoparticles, for example, have been developed for bimodal imaging and cure of immune heterogeneous tumours. Multifunctional nanocarriers can also benefit from the tumour microenvironment's inherent differences, whereas other particles might explode using outside energy sources, including magnetism, ultrasound, and light (Feng et al., 2015).

3.3.2 Passive Targeting

3.3.2.1 Enhanced Permeability and Retention Effect

Owing to the unregulated multiplication of a single cell, solid tumours develop. Due to nutrient transport limitations, solid tumours can exhibit a necrotic core Vascular absorbency factors like nitric oxide, vascular endothelial growth factor, matrix metalloproteinases, bradykinin, and peroxynitrite are elevated in response to tumours. Blood flow variations in tumours compared to healthy tissues were recorded in the 1960s. The term enhanced permeation and retention effect of macromolecules was then coined in solid tumours, which is also used to study the passive transport of anticancer drugs to tumours (Matsumura & Maeda, 1986; Allen & Cullis, 2004). Angiogenesis, or the growth of new blood vessels, is characterised by abnormally formed arteries with large vascular fenestrae (as large as 600 nm) and disruption to lymphatic drainage. As a result, nanoparticles smaller than 200 nm are more likely to accumulate in tumours (Noble et al., 2014). The liver, kidney, and spleen all have large fenestrae, which allow chemotherapeutic nanoparticle aggregation and toxicity (Gaumet et al., 2008). Furthermore, phagocytosis of particles by monocytes in the liver (e.g., Kupffer cells) and spleen contributes to particle aggregation in the reticuloendothelial system.

3.3.2.2 Composition

Various materials are used to construct nanocarriers because of localising chemotherapeutics between tumours through the enhanced permeability and retention effect. These materials comprise nanogold, semiconductors, carbon (nanotubes), porous silica, iron oxide (Jain et al., 2012), lipids (liposomes, exosomes), polymers, dendrimers, proteins (albumin, antibodies), cyclodextrins, carbohydrates (Maruyama, 2011; Hubbell & Hubbell, 2013; Ohno et al., 2013), and their mixture or conjugation (Figure 3.3). Polymeric nanoparticles, for example, are rigid amorphous matrices; liposomes are bilayer spheres that encapsulate a volume of aqueous or gas, and certain inorganic structures have crystalline lattices that may emit light, whereas nanoparticles of silicon have directional dispersion (Behera et al., 2020b). Drug loading and stabilisation are also influenced by how each particle is synthesised.

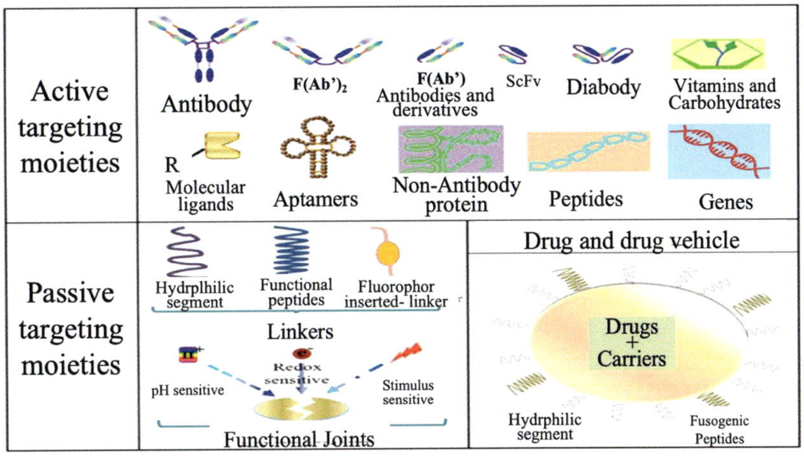

FIGURE 3.3 Different active and passive targeting moieties in the drug delivery system.

3.3.2.3 Size

Size is perhaps the most well-studied parameter in nanoparticle transport. Many essential in vivo functions of particles, including circulation duration, protein absorption, immunogenicity, biodistribution, extravasation, internalisation, intracellular trafficking, payload transfer, and destruction, are influenced by particle size (Bertrand et al., 2014; Behera et al., 2020a). As previously stated, carriers can extravasate through gaps in the peritumoural tissue in a size-dependent manner. Studies using liposomes of varied mean sizes show that the vesicle size threshold for extravasation into tumours is 400 nm. However, because defective lymphatic drainage cannot efflux fluid or carriers correctly, interstitial fluid pressure rises, lowering the convective interstitial transport driving force. When the kinetics of intra-tumour accumulation were examined over 30 minutes in mice xenograft models, tiny 40 macromolecules (40–70-kDa dextrans, 11.2–14.6 nm in diameter) passed through 15 nm from the vessel wall, whereas 2 MDa dextran (50 nm) was established to be 5 nm from the vessel wall (Dreher et al., 2006). This buildup was only temporary since small molecules easily diffused back into the circulatory compartment.

3.3.2.4 Shape

Shape is a particle's critical property in mitigating cellular responses. In addition, particle shape is influenced by the particle transportation in the body, greatly affecting their performance as drug carriers. As 150 nm rod-like particles display a higher rate of internalisation in HeLa cells than nanospheres, elongated shapes may also provide advantages for internalisation. Higher tumour accumulation was observed in gold nanorods and nanospheres relative to the non-spherical form (Chauhan et al., 2011).

Different preclinical, clinical trials conducted on the anticancer activities of phytomolecules are listed in Tables 3.2 and 3.3, respectively. Similarly, the patents filed for the anticancer properties of phytomolecules are listed in Table 3.4.

3.4 FUTURE PROSPECTIVE

The chemical structure of the drug, as well as the features of the tumour cells, is used to target cancerous cells. Plain encapsulation in 'stealth' nanosystems (which hit the target passively) is enough for drugs without problems with cell penetration. Encapsulating these drugs in stealth nanocarriers with a long blood half-life would give the best results, raising their chances of accumulating directly inside tumourous tissues and lowering drug toxicity in extremely perfused organs like the liver, heart, and kidneys. Active targeting, however, should be the preferred strategy for therapeutic molecules that have difficulty crossing cell membranes and can cause serious harm to normal cells. This includes coating nanocarriers with ligands that target overexpressed receptors on the surface of cancer cells. Even though the outcome is higher, the scientific knowledge for surface decoration with ligands might be complicated chemistry. Therefore, in a few cases, the global strategy can adapt drug selection by prioritising only the easily enhanced permeation and retention effect without active targeting. Although non-targeted cationic nanocarriers can cause interactions with cell membranes, they are non-selective and toxic.

On the other hand, nanocarriers adorned with ligands can induce receptor-mediated endocytosis, resulting in drug delivery to the site of action, the cytoplasm, or even the nucleus. Off-target side effects and general toxicity would be reduced as a result. An 'intelligent' theranostic system will be created by combining an active or passive drug transporter with an imaging or diagnostic agent. This system will be able to track disease development and determine the drug's therapeutic effectiveness in real-time.

TABLE 3.2
Preclinical Trials of Phytochemicals in Cancer Treatment

Phytochemicals with Botanical Names	Molecular Mechanisms or Targets	References
6-Shogaol *Zingiber officinale*	Signal transducers and activators of transcription and Protein kinase B signalling pathway	Kim et al. (2014)
Allicin (organosulfurs) *Allium sativum*	Signal transducers and activators of transcription 3 signalling pathway	Huang et al. (2017)
Epigallocatechin (flavonoids) *Camellia sinensis*	Inhibit cell proliferation and apoptosis	Thangapazham et al. (2007)
Dicumarol *Melilotus officinalis*	Intrinsic apoptosis pathway	Zhang et al. (2017)
Curcumin (phytopolyphenol) *Curcuma longa*	Alterations of gene expression and cell signalling regulatory pathways	Kunnumakkara et al. (2017)
Baicalein *Scutellaria baicalensis*	Signalling pathways like extracellular signal-regulated kinase, p38, and Mitogen-activated protein kinase	Dou et al. (2018)
Apigenin (flavonoid) *Petroselinum crispum*	Intrinsic apoptosis pathway	Chang et al. (2018)
Alpinumisoflavone (pyranoisoflavone) *Derris eriocarpa*	Protein kinase B and micro RNA-10 signalling pathways	Wang et al. (2017)
Emodin *Rheum palmatum L.*	Signalling pathways like Mitogen-activated protein kinase and phosphoinositide 3-kinase	Su et al. (2017)
Genistein *Glycine max*	Canonical Wnt pathway signalling pathway	Hsiao et al. (2019)
Gingerol (polyphenol) *Zingiber officinale*	Intrinsic apoptosis pathway	Martin et al. (2017)
Licochalcone A *Glycyrrhiza glabra*	Cyclins and cyclin-dependent kinases	Lu et al. (2018)
Nimbolide *Azadirachta indica*	Signalling of mammalian target of rapamycin and extracellular signal-regulated kinase	Subramani et al. (2016)
Resveratrol *Polygonum cuspidatum*	Regulates cell division and apoptosis	Banerjee et al., 2002
Sulforaphane *Brassica oleracea*	Arrests apoptosis and cell cycle. It targets caspase 8 and p21	Qazi et al. (2010)
Withaferin-A *Withania somnifera*	p38 pathway and forkhead transcription factors 3a-Par-4 cell death pathway	Kuppusamy et al. (2017)

TABLE 3.3
Currently Used Phytochemicals in Clinical Trials on Various Cancers

Phytochemicals	Source	Type of Cancer	Assessment	References
Berberine, (alkaloid)	*Berberis sp.*	Colorectal cancer	Prevention of recurrence of	NCT03281096 Xiaohua (2017)
Curcumin, (polylphenol)	*Curcuma longa*	Breast cancer	Time to illness, quality of life, combined protection, progression-free survival period to treatment failure and improvement	NCT03072992 National Center of Oncology, Armenia (2017)
Epigallocatechin (flavonoids)	*Camellia sinensis*	Colorectal cancer	In contrast to the baseline, there has been a change in methylation trends.	NCT02891538 The University of Texas Health Science Center at San Antonio (2016)
Lycopene (carotenoids)	*Solanum lycopersicum*	Colorectal cancer (metastatic)	When used alone or in combination with panitumumab, it is effective in reducing skin toxicity. Pharmacokinetics.	NCT03167268 DR. Mauro Moroni, Ospedale San Carlo Borromeo (2017)

(Continued)

TABLE 3.3 (*Continued*)
Currently Used Phytochemicals in Clinical Trials on Various Cancers

Phytochemicals	Source	Type of Cancer	Assessment	References
Quercetin, (carotenoids)	—	Prostate cancer	In prostate tissue and urine, epigallocatechin gallate and quercetin, as well as their methylated metabolites, are used. Catechol-*O*-methyltransferase, DNA (cytosine-5)-methyltransferase 1, and multi-drug resistance proteins enzyme activity expression. Catechol-*O*-methyltransferase genotype differences between individuals	NCT01912820 Jonsson Comprehensive Cancer Center & National Cancer Institute (NCI) (2013)
Resveratrol (stilbenoid)	*Polygonum cuspidatum*	Low-grade gastrointestinal and neuroendocrine tumours	Notch1 activation, toxicity	NCT01476592 University of Wisconsin, Madison (2011)
Sulforaphane (isothiocyanate)	*Brassica oleracea*	High-risk Lung cancer	Mitotic index (Ki-67) cell proliferation marker, caspase-3 and terminal deoxynucleotidyl transferase dUTP nick-end labelling apoptosis markers and Bronchial dysplasia index	NCT03232138 Jian-Min Yuan, MD, University of Pittsburgh & National Cancer Institute (NCI) (2017)

Reference found from www.clinicaltrials.gov database with corresponding identifier code (NCT)

The architecture of such systems necessitates a thorough understanding of tumour biology. Despite many published reports demonstrating their clinical potential in preclinical models, only 15 passively targeted nanocarriers have been licensed for therapeutic use, and none of the successful targeted nanocarriers have led to therapeutical trials. Physiological impediments (hypoxia, endosomal escape, tumour heterogeneity, and penetration) caused issues and slowed growth. Aside from regulatory hurdles, scaling up the production process of actively targeted nanocarriers is challenging. As a result, much effort and new techniques are needed to establish nanocarriers with predictable biological identities to speed up nanocarriers' clinical translation.

3.5 CONCLUSION

Both active and passive targeting have their flaws. Passive targeting faces major challenges, resulting in low drug concentrations within the tumour and reduced efficacy. Passive targeting, like chemotherapeutic drugs, can fail to distinguish between healthy and diseased tissue, raising possible toxicity issues. In the case of active targeting, increasing the payload inside tumour cells is no guarantee of transport of the actual clinical agent to the target site, as the release of the therapeutic agent may be hampered by cell components. Endosomal escape and subsequent payload delivery are often a problem for targeted receptor-mediated cellular delivery. Since PEGylation may not be feasible to a sufficient degree, tweaking the surface chemistry of these conjugated nanoparticles may compromise the stealth capability of these polymers.

It is a major challenge for these targeted nanocarrier systems to come across tumour cells that are overexpressing receptors or proteins, or other regions of interest, without any barriers. Suppose these carriers' stealth ability is impaired. In that case, they will ultimately be cleared by the liver, spleen, and other reticuloendothelial system organs, resulting in very little accumulation of drug-loaded targeted nanocarriers in tumour cells. Extensive research is needed to improve these targeted nanocarriers to favourably alter biodistribution and increase overall efficacy in the region of interest. Improving the carrier's specificity, optimising the packing, and optimally tailoring the release of these nanocarriers are critical for dramatically improving the efficiency of cancer therapy. Extensive research is required to develop these targeted nanotherapeutics with a favourable biodistribution study and increased power at the targeted site. To bridge the space between bench and bedside, powerful and more predictive pre-therapeutical animal models, an understanding of the tumour microenvironment, and the execution of good laboratory practice and standardisation policies in academia are needed. A major advancement in nanotechnology could improve the effectiveness of cure modalities for various diseases, as well as cancer, and add to an increasing arsenal of passive and active targeted nanomedicines to combat debilitating diseases.

TABLE 3.4
Patents Filed for Phytochemical Chemical Constituents Showing Cytotoxicity

Patent Number and Date	Title	Bioactive Constituents	Patent Description	Cancer Type and Its Mechanism	Type of Targeting	References
US20100197619 08/05/2010	Cyclin-reliant protein kinases barrier of *Scutellaria* flavonoid organic amine derivatives, incorporate and use	*Scutellaria* flavonoid organic amine derivatives	The patent includes several Cdks inhibitors and the synthesis and application of Scutellaria flavonoid organic amine derivatives. Anticancer drugs with more efficacy and less toxicity are expected to be produced from this product.	Various cancer Apoptosis and Cdks inhibitors	Passive	Zhang et al. (2010)
US20100197584 08/05/2010	Application of curcumin to inhibit brain tumour development in mice	Curcumin and its derivatives	The current development relates to compositions and methods for activating the pro-apoptotic enzymes caspase-3/7 in cancer cells using curcumin or its derivatives or analogues.	Brain tumour Prevent metastasis and tumour formation	Active	Banerjee and Raja (2010)
US7662976 02/16/2010	Triptolide derivatives for attenuation of apoptosis and Immune-suppression	Triptolide derivatives	Triptolide compounds with differently changed carbonate and carbamate derivatives have strong aqueous solubility and transform into naturally active compounds in vivo. Prodrugs can be used as immunorepressing, cancer-fighting agents and anti-inflammatory.	Colon cancer, breast cancer and lung cancer Apoptosis and immune-suppression Modulation	Active	Dai et al. (2010)
US7592367 09/22/2009	Compounds from *Garcinia hanburyi*, their application in healing cancer and technique of isolating epimers	C-2 Epimeric xanthones	Garcinia hanburyi was used to separate three pairs of C-2 epimeric xanthones and accurately isolate the xanthone compounds. The compounds are valuable because of their anticancer properties.	Various cancer Cytotoxicity	Passive	Han et al. (2009)
US7514412 04/07/2009	Anticancer biangeloyl saponins	Saponins isolated from *Xanthoceras sorbifolia*	Novel compounds are disclosed, including those designated as Y or Y3, Y8, Y9, Y1, Y2, and Y10. Those compounds have anticancer effects. The compounds of the current invention can be extracted from plants or chemically synthesised.	Various cancer Anticancer	Passive	Chan et al. (2009)
US20090298938 12/03/2009	Apply of semi-synthetic cognate of boswellic acids for anticancer activity	Analogues of boswellic acids	The current discovery describes the use of compounds of general formula 1 for anticancer activity, in which the said compound is extracted semi-synthetically from biological triterpenoic acids known as boswellic acids by inducing anticancer activity, apoptosis, and cytotoxicity exhibited by semi-synthetic boswellic acids.	Various cancer Induces apoptosis	Passive	Qazi et al. (2009)
US20070298132 12/27/2007	Berberine as a choosy lung cancer factor	Berberine	Berberine or its salts or derivatives have been analysed as the active compound for inhibiting lung cancer selectively and potentially without toxicity.	Lung cancer Cytotoxicity	Active	Maung (2007)
US20070099993 05/03/2007	Anticancer and antimicrobial quality of methyl-beta orcinol carboxylate from lichen (*Everniastrum cirrhatum*)	Methyl-beta orcinolcarboxylate	The present discovery relates to a novel application of an established biomolecule, methyl-orcinol carboxylate of formula I, derived from a lichen (Everniastrum cirrhatum), for healing human pathogenic fungal infections and cancers.	Various cancer Cytotoxicity	Passive	Khanuja et al. (2007)

(*Continued*)

TABLE 3.4 (*Continued*)
Patents Filed for Phytochemical Chemical Constituents Showing Cytotoxicity

Patent Number and Date	Title	Bioactive Constituents	Patent Description	Cancer Type and Its Mechanism	Type of Targeting	References
US7183289 02/27/2007	39-Desmethoxy rapamycin, composition and methods of use thereof	39-Desmethoxy rapamycin	The present invention is concerned with the medical application of 39-desmethoxy rapamycin, especially in healing of cancer and/or B-cell malignancies, as well as the induction and maintenance of immune suppression.	Various cancer Cytotoxicity	Passive	Zhang and Sheridan (2007)
US6933315 08/23/2005	Derivatives of isoindigo, indigo and indirubin and techniques of healing cancer	Isoindigo indigo and indirubin	NATURA, a derivative of Isoindigo, helps improve a patient's overall health and, in particular, prevents or treats cancer.	Various cancer Inhibits cell growth pathway	Passive	Wang et al. (2005)
US6787161 09/07/2004	Anticancer compounds	An angeloyl alternate ingenane and its derivatives	This invention is concerned with a compound or group of compounds contained in an active theory obtained from Euphorbia peplus, Euphorbia hirta and Euphorbia drummondii plants. These plants' extracts have been shown to have selective cytotoxicity against several cancer cell lines.	Skin cancer Cytotoxicity	Active	Aylward (2004)
US20040024055 02/05/2004	Pharmaceutical constitution carrying the extract of *Saururus chinensis* baill helpful as an anticancer agent and a procedure for the preparation	Eepimanassantin A and manassantin A	HNP-98701C, HNP-98701A, and HNP-98701B are carcinostatis substances that selectively affect cancer cell line cells and induce apoptosis-type cell death.	Various cancer Apoptosis	Passive	Cragg and Newman (2005)
US6589567 07/08/2003	Anticancer treatment agent of arsenic hexoxide (As_4O_6) of a biological chemical substance and its pharmaceutical constitution.	Arsenic hexoxide	The HD-2, a biological chemical substance isolated and distilled from a biological Arsenic hexoxide (As_4O_6), a biological chemical substance derived from Sinsuk after removing the toxic property, has a good anticancer effect through direct cyto-toxicity on tumour cells and inhibits the development of new blood vessels in tumour masses, according to this invention.	Various cancer Cytotoxic to tumour cells and inhibit the development of new blood vessels of tumour masses	Passive	Weiss and McMichael (2004)
US6498195 12/24/2002	Application of 1-propanone-1-(2,4-dihydroxy phenyl)-3-hydroxy-3-(4-dihydroxyphenyl) as an anti-carcinogenic factor.	1-Propanone-1-(2,4-di-hydroxy phenyl)-3-hydroxy-3-(4-hydroxyphenyl)	Cancer prevention and treatment compositions and procedures are given. Pure hydroxylated chalcone compounds of liquorice root (Glycyrrhizaglabra),aswellas1-propanone-1-(2,4-dihydroxyphenyl)-3-hydroxy-3-(4-hydroxyphenyl), are used in the formulations.	Many cancer Suppress tumour cell growth and induce apoptosis.	Passive	Rosen et al. (2002)
US6395745 05/28/2002	Alkaloid halide salts of swainsonine and techniques of use	Swainsonine	The current discovery is associated with synthetic halide salts of swainsonine that are stable and significantly purified. Swainsonine and other Golgi mannosidase II inhibitors have anticancer properties in several tumour forms, comprising direct anti-metastatic and anti-invasion effects on tumour cells, as well as other anticancer activities including immune stimulatory effects and myeloproliferative and hemorestorative activities, as mentioned herein.	Various cancer Inhibits Golgi mannosidase II	Passive	Dennis et al. (2002)

REFERENCES

Aggarwal BB, Kumar A, Bharti AC (2003) Anticancer potential of curcumin: preclinical and clinical studies. *Anticancer Res.* 23(1/A), 363–398.

Aliosmanoglu A, Basaran I (2012) Nanotechnology in cancer treatment. *J. Nanomed. Biotherapeut. Discov.* 2, 1–3.

Allen TM, Cullis PR (2004) Drug delivery systems: entering the mainstream. *Science* 303, 1818–1822.

Aylwar, JH (2004) Anticancer compounds. US6787161.

Babu TD, Kuttan G, Padikkala J (1995) Cytotoxic and anti-tumour properties of certain taxa of Umbelliferae with special reference to *Centella asiatica* (L.) Urban. *J. Ethnopharmacol.* 48(1), 53–57.

Banerjee P, Raja KS (2010) Use of curcumin to block brain tumour formation in mice. US20100197584.

Banerjee S, Bueso-Ramos C, Aggarwal BB (2002) Suppression of 7, 12-dimethylbenz (a) anthracene-induced mammary carcinogenesis in rats by resveratrol: role of nuclear factor-κB, cyclooxygenase 2, and matrix metalloprotease 9. *Cancer Res.* 62(17), 4945–4954.

Barar J, Omidi Y (2013) Dysregulated pH in tumor microenvironment checkmates cancer therapy. *BioImpacts: BI* 3(4), 149.

Behera A, Mittu B, Padhi S, Patra N, Singh J (2020a). Bimetallic nanoparticles: green synthesis, applications, and future perspectives. In: Abd-Elsalam KA (Ed.) *Multifunctional Hybrid Nanomaterials for Sustainable Agri-food and Ecosystems*. Elsevier, Amsterdam, pp. 639–682.

Behera A, Mittu B, Padhi S, Singh A (2020b). Antimicrobial efficacy of essential oil nanoemulsions. In: Dhull, SB, Chawla P, Kaushik R (Eds.) *Nanotechnological Approaches in Food Microbiology*. CRC Press, Boca Raton, FL, pp. 293–309.

Behera A, Padhi S (2020) Passive and active targeting strategies for the delivery of the camptothecin anticancer drug: a review. *Environ. Chem. Lett.* 18(5), 1557–1567.

Behera A, Padhi S (2022) pH-Sensitive polymeric nanoparticles for cancer treatment. In: Padhi S, Behera A, Lichtfouse E (Eds.) *Polymeric Nanoparticles for the Treatment of Solid Tumours*. Environmental Chemistry for a Sustainable World, vol 71. Springer, Cham, pp. 401–425.

Behera A, Padhi S, Nayak AK (2023) Engineered liposomes as drug delivery and imaging agents. In: Ray, S, Nayak AK (Eds.) *Design and Applications of Theranostic Nanomedicines*. Woodhead Publishing, Cambridge, MA, pp. 75–108.

Bertrand N, Wu J, Xu X, Kamaly N, Farokhzad OC (2014) Cancer nanotechnology: the impact of passive and active targeting in the era of modern cancer biology. *Adv. Drug Deliv. Rev.* 66, 2–25.

Biswas J, Roy M, Mukherjee A (2015) Anticancer drug development based on phytochemicals. *J. Drug Disc. Dev. Delivery* 2, 1012.

Cerchia L, de Franciscis V (2010) Targeting cancer cells with nucleic acid aptamers. *Trends Biotechnol.* 28, 517–525.

Chan PK, Mak MS, Wang Y (2009) Anticancer biangeloyl saponins. US7514412.

Chang FR, Liaw CC, Lin CY, Chou CJ, Chiu HF, Wu YC (2003) New adjacent bis-tetrahydrofuran annonaceous acetogenins from Annona muricata. *Planta Medica* 69(03), 241–246.

Chang JH, Cheng CW, Yang YC, Chen WS, Hung WY, Chow JM, Chen P-S, Hsiao M, Lee W-J, Chien M-S (2018) Downregulating CD26/DPPIV by apigenin modulates the interplay between Akt and Snail/Slug signaling to restrain metastasis of lung cancer with multiple EGFR statuses. *J. Exp. Clin. Cancer Res.* 37(1), 1–16.

Chauhan VP, Popovic Z, Chen O, Cui J, Fukumura D, Bawendi MG, Jain RK (2011) Fluorescent nanorods and nanospheres for real-time in vivo probing of nanoparticle shape-dependent tumour penetration. *Angew. Chem. Int. Ed. Engl.* 50, 11417–11420.

Chiu HF, Chih TT, Hsian YM, Tseng CH, Wu MJ, Wu YC (2003) Bullatacin, a potent antitumor Annonaceous acetogenin, induces apoptosis through a reduction of intracellular cAMP and cGMP levels in human hepatoma 2.2. 15 cells. *Biochem. Pharmacol.* 65(3), 319–327.

Cho KJ, Wang X, Nie SM, Chen Z, Shin DM (2008) Therapeutic nanoparticles for drug delivery in cancer. *Clin. Cancer Res.* 14, 1310–1316.

Cragg GM, Newman DJ (2005) Plants as a source of anti-cancer agents. *J. Ethnopharmacol.* 100(1–2), 72–79.

Dai D, Musser JH, Lennox ES (2010) Triptolide derivatives for modulation of apoptosis and immunosuppression. US7662976.

Dennis JW, Shah RN, Ziser L (2002) Alkaloid halide salts of swainsonine and methods of use. US6395745.

Dou J, Wang Z, Ma L, Peng B, Mao K, Li C (2018) Baicalein and baicalin inhibit colon cancer using two distinct fashions of apoptosis and senescence. *Oncotarget* 9(28), 20089–20102. doi:10.18632/oncotarget.24015.

Dreher MR, Liu W, Michelich CR, Dewhirst MW, Yuan F, Chilkoti A (2006) Tumour vascular permeability, accumulation, and penetration of macromolecular drug carriers. *J. Natl. Cancer Inst.* 98, 335–344.

Eiznhamer DA, Xu ZQ (2004) Betulinic acid: a promising anticancer candidate. *IDrugs* 7(4), 359–373.

Faried A, Kurnia D, Faried L, Usman N, Miyazaki T, Kato H, Kuwano H (2007) Anticancer effects of gallic acid isolated from Indonesian herbal medicine, *Phaleria macrocarpa* Boerl, on human cancer cell lines. *Int. J. Oncol.* 30, 605–613.

Feng X, Wang P, Liu Q, Zhang T, Mai B, Wang X (2015) Glycolytic inhibitors 2-deoxyglucose and 3-bromopyruvate synergize with photodynamic therapy respectively to inhibit cell migration. *J. Bioenerg. Biomembr.* 47, 189–197.

Ferlay J, Shin HR, Bray F, Forman D, Mathers C, Parkin DM (2010) Estimates of worldwide burden of cancer in 2008: GLOBOCAN 2008. *Int. J. Cancer* 127, 2893–2917. doi:10.1002/ijc.25516.

Fulda S, Debatin KM (2005) Sensitization for anticancer drug-induced apoptosis by betulinic acid. *Neoplasia* 7(2), 162–170.

Gaumet M, Vargas A, Gurny R, Delie F (2008) Nanoparticles for drug delivery: the need for precision in reporting particle size parameters. *J. Pharmacokinet. Biopharm.* 69, 1–9.

Gauthier C, Legault J, Lebrun M, Dufour P, Pichette A (2006) Glycosidation of lupane-type triterpenoids as potent in vitro cytotoxic agents. *Bioorg. Med. Chem.* 14(19), 6713–6725.

Granja A, Pinheiro M, Reis S (2016) Epigallocatechin gallate nanodelivery systems for cancer therapy. *Nutrients* 8, E307.

Han QB, Son JZ, Qiao CF, Yang L, Xu HX (2009) Compounds from *Garcinia hanburyi*, their use in treating cancer and method of separating epimers thereof. US7592367.

Horn L, Pao W, Johnson DH (2012) Neoplasms of the lung. In: Longo DL, Fauci AS, Kasper DL, Hauser SL, Jameson JL, Loscalzo J (Eds.) *Harrisons Principles of Internal Medicine*, 18th Edn. McGraw-Hill, New York, NY, pp. 825–836.

Hostanska K, Daum G, Saller R (2002) Cytostatic and apoptosis-inducing activity of boswellic acids toward malignant cell lines in vitro. *Anticancer Res.* 22(5), 2853–2862. [PubMed: 12530009].

Hsiao YC, Peng SF, Lai KC, Liao CL, Huang YP, Lin CC (2019) Genistein induces apoptosis in vitro and has antitumour activity against human leukemia HL-60 cancer cell xenograft growth in vivo. *Environ. Toxicol.* 34(4), 443–456.

Huang X, Teng X, Chen D, Tang F, He J (2010) The effect of the shape of mesoporous silica nanoparticles on cellular uptake and cell function. *Biomaterials* 31(3):438–448. doi:10.1016/j.biomaterials.2009.09.060.

Huang L, Song Y, Lian J, Wang Z (2017) Allicin inhibits the invasion of lung adenocarcinoma cells by altering tissue inhibitor of metalloproteinase/matrix metalloproteinase balance via reducing the activity of phosphoinositide 3-kinase/AKT signaling. *Oncol. Lett.* 14(1), 468–474.

Huang WD, Abraham Y, Zheng SC, Bustamante L, Luo SS, Irvine DJ (2015) Active targeting of chemotherapy to disseminated tumours using nanoparticle carrying T cells. *Sci. Transl. Med.* 7, 291–294.

Hubbell JA, Langer R (2013) Translating materials design to the clinic. *Nat. Mater.* 12, 963–966.

Hynes NE, Lane HA (2005) ERBB receptors and cancer: the complexity of targeted inhibitors. *Nat. Rev. Cancer* 5, 341–354.

Jada SR, Subur GS, Matthews C, Hamzah AS, Lajis NH, Saad MS, Stevens MFG, Stanslas J (2007) Semisynthesis and in vitro anticancer activities of andrographolide analogues. *Phytochemistry* 68(6), 904–912.

Jagetia GC, Rao SK (2006) Evaluation of the antineoplastic activity of guduchi (Tinospora cordifolia) in Ehrlich ascites carcinoma bearing mice. *Biol. Pharm. Bull.* 29(3), 460–466.

Jain S, Hirst DG, O'Sullivan JM (2012) Gold nanoparticles as novel agents for cancer therapy. *Br. J. Radiol.* 85, 101–113.

Khanuja SPS, Tiruppadiripuliyur RSK, Gupta, VK, Chand P, Garg A, Srivastava SK, Verma SC, Saikia D, Darokar MP, Shasany AK, Pal A (2007) Antimicrobial and anticancer properties of methyl-beta-orcinolcarboxylate from lichen. US20070099993.

Kim MO, Lee MH, Oi N, Kim SH, Bae KB, Huang Z (2014) [6]-Shogaol inhibits growth and induces apoptosis of non-small cell lung cancer cells by directly regulating Akt1/2. *Carcinogenesis* 35(3), 683–691.

Koehn FE, Carter GT (2005) The evolving role of natural products in drug discovery. *Nat. Rev. Drug Discov.* 4, 206–220.

Kumar RA, Sridevi K, Kumar VN, Nanduri S, Rajagopal S (2004) Anticancer and immunostimulatory compounds from *Andrographis paniculata*, *J. Ethnopharmacol.* 92 (2–3), 291–295.

Kundu A, Padhi S, Behera A, Hasnain MS, Nayak AK (2022) Tumour targeting strategies by chitosan-based nanocarriers. In: Hasnain MS, Sarwar Beg S, Nayak AK (Eds.) *Chitosan in Biomedical Applications*, Academic Press, London, pp. 163–188.

Kunnumakkara AB, Bordoloi D, Harsha C, Banik K, Gupta SC, Aggarwal BB (2017) Curcumin mediates anticancer effects by modulating multiple cell signaling pathways. *Clin. Sci. (Lond)* 131(15), 1781–1799.

Kuppusamy P, Nagalingam A, Muniraj N, Saxena NK, Sharma D (2017) Concomitant activation of ETS-like transcription factor-1 and death receptor-5 via extracellular signal-regulated kinase in withaferin a-mediated inhibition of hepatocarcinogenesis in mice. *Sci. Rep.* 7(1), 17943.

Leite DF, Kassuya CA, Mazzuco TL, Silvestre A, de Melo LV, Rehder VL, Rumjanek VM, Calixto JB (2006) The cytotoxic effect and the multidrug resistance reversing action of lignans from *Phyllanthus amarus. Planta Med.* 72(15), 1353–1358.

Lu L, Zhao Z, Liu L, Gong W, Dong J (2018) Combination of baicalein and docetaxel additively inhibits the growth of non-small cell lung cancer in vivo. *Tradit. Med. Modern Med.* 01(03), 213–218.

Malik F, Singh J, Khajuria A, Suri KA, Satti NK, Singh S, Kaul MK, Kumar A, Bhatia A, Qazi GN (2007) A standardized root extract of Withania somnifera and its major constituent withanolide-A elicit humoral and cell-mediated immune responses by upregulation of Th1-dominant polarization in BALB/c mice. *Life Sci.* 80(16), 1525–1538.

Martin A, Fuzer AM, Becceneri AB, da Silva JA, Tomasin R, Denoyer D (2017) [10]-Gingerol induces apoptosis and inhibits metastatic dissemination of triple negative breast cancer in vivo. *Oncotarget* 8(42), 72260–72271.

Maruyama K (2011) Intracellular targeting delivery of liposomal drugs to solid tumours based on EPR effects. *Adv. Drug Deliv. Rev.* 63, 161–169.

Mathur P, Krishnan SK, Chaturvedi M, Das P, Sudarshan KL, Santhappan S, Vinodh N, Anish J, Sandeep N (2020) Report from National Cancer Registry Programme, India. *JCO Global Oncol.* 6, 1063–1075.

Matsumura Y, Maeda H (1986) A new concept for macromolecular therapeutics in cancer chemotherapy: mechanism of tumouritropic accumulation of proteins and the antitumour agent smancs. *Cancer Res.* 46, 6387–6392.

Maung, TW (2007) Berberine as a selective lung cancer agent. US20070298132.

Miller DR (2006) A tribute to Sidney Farber: the father of modern chemotherapy. *Br. J. Haematol.* 134, 20–26.

Mitra A, Chakrabarti J, Banerji A, Chatterjee A, Das BR (2006) Curcumin, a potential inhibitor of MMP-2 in human laryngeal squamous carcinoma cells HEp2. *J. Environ. Pathol. Toxicol. Oncol.* 25(4), 679–690.

Mussi SV, Torchilin VP (2013) Recent trends in the use of lipidic nanoparticles as pharmaceutical carriers for cancer therapy and diagnostics. *J. Mater. Chem.* B1, 5201–5209.

Noble GT, Stefanick JF, Ashley JD, Kiziltepe T, Bilgicer B (2014) Ligand-targeted liposome design: challenges and fundamental considerations. *Trends Biotechnol.* 32, 32–45.

Ohno S, Takanashi M, Sudo K, Ueda S, Ishikawa A, Matsuyama N, Fujita K, Mizutani T, Ohgi T, Ochiya T, Gotoh N, Kuroda M (2013) Systemically injected exosomes targeted to EGFR deliver antitumour microRNA to breast cancer cells. *Mol. Ther.* 21, 185–191.

Omidi Y, Barar J (2014) Targeting tumour microenvironment: crossing tumour interstitial fluid by multifunctional nanomedicines. *Bioimpacts* 4, 55–67.

Padhi S, Behera A (2020) Nanotechnology based targeting strategies for the delivery of camptothecin. In: Saneja A, Panda A, Lichtfouse E (Eds.) *Sustainable Agriculture Reviews,* vol 44. Springer, Cham, pp. 243–272.

Padhi S, Behera A (2022) Cellular internalization and toxicity of polymeric nanoparticles. In: Padhi S, Behera A, Lichtfouse E (Eds.) *Polymeric Nanoparticles for Treating Solid Tumours*. Environmental Chemistry for a Sustainable World, vol 71. Springer, Cham, pp. 473–488.

Padhi S, Behera A, Hasnain MS, Nayak AK (2022) Chitosan-based drug delivery systems in cancer therapeutics. In: Hasnain MS, Beg S, Nayak AK (Eds.) *Chitosan in Drug Delivery*. Academic Press, London, pp. 159–193.

Padhi S, Kapoor R, Verma D, Panda AK, Iqbal Z (2018) Formulation and optimization of topotecan nanoparticles: in vitro characterization, cytotoxicity, cellular uptake and pharmacokinetic outcomes. *J. Photochem. Photobiol. B Biol.* 183, 222–232.

Padhi S, Mirza MA, Verma D, Khuroo T, Panda AK, Talegaonkar S, Khar RK, Iqbal Z (2016) Revisiting the nanoformulation design approach for effective delivery of topotecan in its stable form: an appraisal of its in vitro behavior and tumour amelioration potential. *Drug Deliv.* 23(8), 2827–2837.

Parente-Pereira AC, Shmeeda H, Whilding LM, Zambirinis CP, Foster J, van der Stegen SJ, Beatson R, Zabinski T, Brewig N, Sosabowski JK, Mather S, Ghaem-Maghami S, Gabizon A, Maher J (2014) Adoptive immunotherapy of epithelial ovarian cancer with Vgamma9Vdelta2 T cells, potentiated by liposomal alendronic acid. *J. Immunol.* 193, 5557–5566.

Patnaik S, Gorain B, Padhi S, Choudhury H, Gabr GA, Md S, Kumar Mishra D, Kesharwani P (2021) Recent update of toxicity aspects of nanoparticulate systems for drug delivery. *Eur. J. Pharm. Biopharm.* 161, 100–119.

Potmesil M (1994) Camptothecins: from bench research to hospital wards. *Cancer Res.* 54(6), 1431–1439.

Qazi A, Pal J, Maitah M, Fulciniti M, Pelluru D, Nanjappa P (2010) Anticancer activity of a broccoli derivative, sulforaphane, in barrett a denocarcinoma: potential use in chemoprevention and as adjuvant in chemotherapy. *Trans. Oncol.* 3(6), 389–399.

Qazi GN, Taneja SC, Singh J, Saxena AK, Sethi VK, Shah BA, Kapahi BK, Andotra SS, Kumar A, Bhushan S, Malik F, Mondhe DM, Muthiah S, Singh S, Verma M, Singh SK (2009) Use of semi synthetic analogues of boswellic acids for anticancer activity. US20090298938.

Rajeshkumar NV, Kuttan R (2000) *Phyllanthus amarus* extract administration increases the life span of rats with hepatocellular carcinoma. *J. Ethnopharmacol.* 73(1–2), 215–219.

Rosen RT, Ho CT, Dipaola RS, Rafi MM, Vastano BC, Ghai G (2002) Use of 1-propanone-1-(2,4-dihydroxyphenyl)-3-hydroxy-3-(4-hydroxyphenyl) as an anti-carcinogenic agent. US6498195.

Ruan WJ, Lai MD, Zhou JG (2006) Anticancer effects of Chinese herbal medicine, science or myth? *J. Zhejiang Univ. Sci. B* 7, 1006–1014.

Saklani A, Kutty SK (2008) Plant-derived compounds in clinical trials. *Drug Discov. Today* 13, 161–171.

Schmidt MM, Wittrup KD (2009) A modeling analysis of the effects of molecular size and binding affinity on tumour targeting. *Mol. Cancer Ther.* 8, 2861–2871.

Schnekenburger M, Dicato M, Diederich M (2014) Plant derived epigenetic modulators for cancer treatumour microenvironmentnt and prevention. *Biotechnol. Adv.* 32, 1123–1132.

Shao ZM, Shen ZZ, Liu CH, Sartippour MR, Go VL, Heber D, Nguyen M (2002) Curcumin exerts multiple suppressive effects on human breast carcinoma cells. *Int. J. Cancer* 98(2), 234–240.

Sharma B, Singh S, Kanwar SS (2014) L-Methionase: a therapeutic enzyme to treat malignancies. *Biomed. Res. Int.* 2014, 1–13.

Shehab NG, Mahdy A, Khan SA, Noureddin SM (2011) Chemical constituents and biological activities of *Fagonia indica* Burm F. *Res. J. Med. Plant.* 5, 531–546.

Siegel RL, Miller KD, Jemal A (2016) Cancer statistics 2016. *CA Cancer J. Clin.* 66, 7–30.

Singh, J (2007) Indo-US symposium on botanicals organized by CSIR, IIIM, Jammu & NCNPR, University of Mississippi. New Delhi: IGH, NASC complex; 2007. A novel standardized herbal formulation of *Withania somnifera* useful for anticancer land Th-1 immune upregulation. Indian Patent: 0202NF2006; Del 01321 dated 19.06.2007.

Sohi KK, Mittal N, Hundal MK, Khanduja KL (1994) Gallic acid, an antioxidant, exhibits antiapoptotic potentInoue gallic acid, induces apoptosis in HL-60RG cells. *Biochem. Biophys. Res. Commun.* 204, 898–904.

Song YH, Sun H, Zhang A, Yan G, Han Y, Wang X (2014) Plant derived natural products as leads to anticancer drugs. *J. Med. Plant Herb. Ther. Res.* 2, 6–15.

Strebhardt K, Ullrich A (2008) Paul Ehrlich's magic bullet concept: 100 years of progress. *Nat. Rev. Cancer* 8, 473–480.

Su J, Yan Y, Qu J, Xue X, Liu Z, Cai H (2017) Emodin induces apoptosis of lung cancer cells through ER stress and the TRIB3/NF-kappaB pathway. *Oncol. Rep.* 37(3), 1565–1572.

Subramani R, Gonzalez E, Arumugam A, Nandy S, Gonzalez V, Medel J (2016) Nimbolide inhibits pancreatic cancer growth and metastasis through ROS-mediated apoptosis and inhibition of epithelial-to-mesenchymal transition. *Sci. Rep.* 6, 19819.

Sun J, Chu YF, Wu X, Liu RH (2002) Antioxidant and antiproliferative activities of common fruits. *J. Agric. Food Chem.* 50, 7449–7454.

Syrovets T, Gschwend JE, Büchele B, Laumonnier Y, Zugmaier W, Genze F, Simmet T (2005) Inhibition of IκB kinase activity by acetyl-boswellic acids promotes apoptosis in androgen-independent PC-3 prostate cancer cells in vitro and in vivo. *J. Biol. Chem.* 280(7), 6170–6180.

Taraphdar AK, Roy M, Bhattacharya R (2001) Natural products as inducers of apoptosis: implication for cancer therapy and prevention. *Curr. Sci.* 80, 1387–1396.

Taratula O, Garbuzenko OB, Kirkpatrick P, Pandya I, Savla R, Pozharov VP (2009) Surface-engineered targeted PPI dendrimer for efficient intracellular and intratumoural siRNA delivery. *J. Control. Release.* 140, 284–293.

Thangapazham RL, Singh AK, Sharma A, Warren J, Gaddipati JP, Maheshwari RK (2007) Green tea polyphenols and its constituent epigallocatechin gallate inhibits proliferation of human breast cancer cells in vitro and in vivo. *Cancer Lett.* 245, 232–241.

The University of Texas Health Science Center at San Antonio. (2016) A study of the effect of metformin on biomarkers in endometrial cancer. NCT02891538.

Torchilin VP (2014) Multifunctional stimuli-sensitive nanoparticulate systems for drug delivery. *Nat. Rev. Drug Discov.* 13, 813–827.

University of Wisconsin, Madison. (2011) A biological study of resveratrol's effects on notch-1 signaling in subjects with low grade gastrointestinal tumors. NCT01476592.

Vander Heiden MG (2011) Targeting cancer metabolism: a therapeutic window opens. *Nat. Rev. Drug Discov.* 10, 671–684.

Verma D, Thakur PS, Padhi S, Khuroo T, Talegaonkar S, Iqbal Z (2017) Design expert assisted nanoformulation design for co-delivery of topotecan and thymoquinone: optimization, in vitro characterization, and stability assessment. *J. Mol. Liquids* 242, 382–394.

Wall DM, Srikanth CV, McCormick BA (2010) Targeting tumours with *Salmonella typhimurium*: potential for therapy. *Oncotarget* 1, 721–728.

Wall ME, Wani MC (1996) Discovery to Clinic A. *Ann. N.Y. Acad. Sci.* 803(1), 1–12.

Wang CZ, Calway T, Yuan CS (2012) Herbal medicines as adjuvant for cancer therapeutics. *Am. J. Chin. Med.* 40, 657–669.

Wang L, Liu X, Chen R (2005) Derivatives of isoindigo, indigo and indirubin and methods of treating cancer. US6933315.

Wang M, Thanou M (2010) Targeting nanoparticles to cancer. *Pharmacol. Res.* 62, 90–99.

Wang T, Jiang Y, Chu L, Wu T., You J (2017) Alpinumisoflavon suppresses tumour growth and metastasis of clear-cell renal cell carcinoma. *Am. J. Cancer Res.* 7(4), 999–1015.

Weinstein JN, van Osdol W (1992) Early intervention in cancer using monoclonal antibodies and other biological ligands: micropharmacology and the "binding site barrier". *Cancer Res.* 52, 2747s–2751s.

Weiss RA, McMichael AJ (2004) Social and environmental risk factors in the emergence of infectious diseases. *Nat. Med.* 10, S70–6.

Wu FE, Gu ZM, Zeng LU, Zhao GX, Zhang Y, McLaughlin JL, Sastrodihardjo S (1995) Two new cytotoxic monotetrahydrofuran Annonaceous acetogenins, annomuricins A and B, from the leaves of Annona muricata. *J. Nat. Products* 58(6), 830–836.

Xiaohua li, Xijing Hospital of Digestive Diseases. (2017) A randomized trial of berberine hydrochloride to prevent colorectal adenomas in patients with previous colorectal cancer. NCT03281096.

Yu MK, Park J, Jon S (2012) Targeting strategies formultifunctional nanoparticles in cancer imaging and therapy. *Theranostics* 2, 3–44.

Zhang MQ, Sheridan RM (2007) 39-Desmethoxyrapamycin, compositions and methods of use thereof. US7183289.

Zhang S, Bao Y, Sun Y, Li K, Zou L, Ma J, Sun, X, Shang H, Li J (2010) Cyclin-dependent protein kinases inhibitor of *Scutellaria* flavonoid organic amine derivatives, synthesis and use thereof. US20100197619.

Zhang W, Su J, Xu H, Yu S, Liu Y, Zhang Y (2017) Dicumarol inhibits PDK1 and targets multiple malignant behaviors of ovarian cancer cells. *PLoS One* 12(6), e0179672.

4 Use of Phytochemical-Entrapped Nanosystems for Cancer

Vivek Chavda, Disha Chawnani, Mital Patel, Radhika Pandya, and Ketan M. Ranch
Department of Pharmaceutics, L. M. College of Pharmacy, Ahmedabad, India

4.1 INTRODUCTION

Cancer is defined as abnormal cell growth that can cause life-threatening malignancies with high financial costs for patients as well as the health care system. Cancer is a leading cause of death worldwide. It is an uncontrolled and excessive growth of cells that can metastasize to several organs. Cancer therapy has been challenging for years, although therapeutic approaches, such as surgery, chemotherapy, and radiotherapy, have been developed. Till now, the leading clinical treatment has been chemotherapy in addition to surgery. However, most chemotherapy medications have high toxicity and low specificity and are associated with painful adverse side effects. In addition, multidrug resistance could be induced by chemotherapeutics, causing treatment failure upon disease recurrence (Padhi and Behera, 2020, 2021).

Medicinal plants have been used for therapeutic purposes since the beginning of human civilization. The therapeutic efficacy of medicinal plants is due to the wide range of phytochemical constituents or secondary metabolites. The medicinal plants are traditionally used for several types of illnesses. Many secondary metabolites have been proven to exhibit chemopreventive properties when tested on cell lines or in vivo. Phytoconstituents have emerged as a prospective approach for cancer treatment and are also widely used, especially in many Asian countries (Behera, 2024). These phytochemical molecules are of great attention due to their wide range of biological activity, minimal side effects, and low cost. This chapter discusses the pharmacological significance of phytochemicals entrapped as nanosystems for cancer. Nanotechnology is the study of particles within the range of 100 nm. A high surface area-to-volume ratio of atoms or molecules is one of the important benefits of nanoparticles. This property enhances the surface activity and changes the physical and biological properties of the nanomaterials (Behera et al., 2020a,b). The influence of nanosystems for phytoconstituents on tumour cells has gained broader attention due to improvements in targeted therapy and bioavailability and stability (Padhi et al., 2016). Today, several nanoformulations are designed to deliver phytoconstituents, including nanosuspensions, solid lipid nanoparticles, liposomes, gold nanoparticles, and polymeric nanoparticles, resulting in better anticancer activity (Behera & Padhi, 2020; Pradhan et al., 2022; Behera et al., 2023a,b; Sa et al., 2023).

Various plant-derived anticancer metabolites from different groups presented in this chapter, including polyphenols, terpenes, alkaloids, and sulphur-containing compounds, underline the multidirectional nature of natural products. Phytoconstituents demonstrate notable anticancer potential in pharmacotherapy; however, the difficulties in terms of their bioavailability and toxicity to normal cells, as well as targeted drug delivery to malignant cells, can be overcome using nano-system technologies, which optimize the bioefficacy of these phytoconstituents.

4.2 PHYTOCHEMICALS WITH ANTICANCER ACTIVITY

Phytochemicals have strong anticancer potential. According to a scientific research study conducted from 1940 to 2014, it was observed that over half of all licensed anticancer medicines were generated directly or indirectly from natural sources (Newman & Cragg, 2020). This chapter will elaborate on some of the most notable anticancer phytochemicals. The most promising phytochemical categories for cancer prevention and therapy include alkaloids, terpenoids, polyphenols, and organosulfur compounds. Various phytochemicals such as capsaicin, cucurbitacin B, isoflavones, catechins, lycopene, benzyl isothiocyanate, phenethyl isothiocyanate, and piperlongumine have shown inhibitory effects on cancer cells, indicating that they might be used as anticancer medicines (Ranjan et al., 2019).

In the study by Lee et al. (2013), various phytoagents counteract carcinogen-induced oxidative stress by scavenging free radicals. Furthermore, metal ion chelation by phytoagents aids in the prevention of oxidative DNA damage produced by transition metal ions.

4.2.1 Capsaicin

Capsaicin is a strong alkaloid and active component of chilli peppers from the *Capsicum* plant genus (Ranjan et al., 2019). Capsaicin therapy inhibits the activation of tumour-promoting signalling pathways such as activator

DOI: 10.1201/9781003611516-5

protein 1, nuclear factor kappa B, and signal transducer and activator of transcription 3. Capsaicin has also been demonstrated to produce reactive oxygen species, depolarize mitochondria, and cause cell cycle arrest, leading to death (Oyagbemi et al., 2010). Capsaicin inhibits bladder cancer cell migration by directly binding to sirtuin 1 and inhibiting sirtuin 1 deacetylase (Ranjan et al., 2019).

The research found that capsaicin-induced apoptosis in pancreatic cancer cells was linked to -catenin signalling suppression. In mice, oral treatment of 5 mg/kg capsaicin effectively inhibited the development of implanted pancreatic tumours. Within an hour of oral administration, the maximum concentration of capsaicin in the blood is obtained, as is the maximal distribution in many organs, such as the kidneys, lungs, and gut (Suresh & Srinivasan, 2010).

4.2.2 Vinca Alkaloids

Vinca alkaloids are a class of medications derived from the periwinkle plant of Madagascar. They're made from the pink periwinkle plant *Catharanthus roseus*. They've been used to treat cancer, diabetes, high blood pressure, and disinfectants, among other things. The cancer-fighting properties of vinca alkaloids are also essential. Vinblastine, vinorelbine, vincristine, and vindesine are the four main vinca alkaloids used in clinical practice. Vincristine, vinblastine, and vinorelbine have been approved for usage in the United States. Vinca alkaloids are the second most often used family of cancer medications, and they will continue to be among the first cancer treatments (Moudi et al., 2013). The two naturally separated alkaloids, vinblastine and vincristine, have been utilized in clinical oncology for almost 50 years. Leukaemia, Hodgkin and non-Hodgkin lymphomas, advanced testicular carcinoma, breast and lung malignancies, and Kaposi's sarcoma have all been treated with these drugs in combination with chemotherapy (Madunić et al., 2018).

The major mechanisms of vinca alkaloid cytotoxicity include interactions with tubulin and disruption of microtubule function, especially microtubules that make up the mitotic spindle machinery, resulting in metaphase arrest. Because microtubules are involved in numerous nonmitotic processes, vinca alkaloids and other antimicrotubule drugs affect nonmalignant and malignant cells in the nonmitotic cell cycle. Vinca alkaloids bind to tubulin binding sites distinct from those seen in taxanes, colchicine, podophyllotoxin, and guanosine 5′-triphosphate. Binding happens quickly and may also be reversed. Vinca alkaloids attach to microtubules and impede cell growth, resulting in a mitotic block and death, binding to tubulin and inhibiting polymerization (Ronnett & Moon, 2002). Vinca alkaloids have long been used in medical treatments as part of combination chemotherapy regimens. They have no cross-resistance with medicines that alkylate deoxyribonucleic acid (DNA) and act differently. Vinblastine has been utilized in the treatment of testicular cancer, as well as hodgkin and non-hodgkin lymphomas, breast cancer and germ cell cancers are also treated with it. Vinca alkaloids attach to microtubules and impede cell growth, resulting in a mitotic block and death. By binding to tubulin and inhibiting polymerization, vincristine and similar chemicals cause microtubule instability (Wang et al., 1999).

4.2.3 Aloe Vera

Aloe barbadensis (aloe vera) belongs to the Liliaceae family. Various pharmacologic characteristics of aloe vera whole gel extract have been documented, including hypoglycaemic, hypolipidemic, antifungal, anticancer, antioxidant, and immunoprotective effects. Aloe vera has several active components extracted and researched for their biological activity. The most common active components are glycoproteins, anthraquinones, polysaccharides, and low-molecular-weight species. Aloe-emodin, barbaloin, and aloesin are pharmacologically active anthraquinones in aloe vera. Aloe vera leaves modulate the growth of certain tumours to produce anticancer efficacies (El-Shemy et al., 2010). Aloe vera leaves have anticancer properties by slowing the progression of certain malignancies. Aloe vera therapy, including aloin and aloe-emodin, can greatly suppress the growth of Ehrlich ascites carcinoma cells. According to Lin et al. (Kuo et al., 2002). Aloe-cellular emodin's growth suppression effectiveness on a variety of tumour cell lines, including lung carcinoma (Lee et al., 2001), hepatoma (Kuo et al., 2002), and leukaemia cell lines (Chen et al., 2004), has been extensively researched. Aloe-emodin has anticancer properties and can limit the creation and expansion of tumour blood vessels by inhibiting signal transducers and transcription activators, making it a viable cancer therapy option. The anticancer effects of numerous aloe vera extracts have led to speculation that these compounds may alter multiple pharmacological processes involved in cancer progression (Gao et al., 2019).

4.2.4 Podophyllotoxins

Podophyllotoxin is a naturally occurring toxin in *Podophyllum peltatum* and *Podophyllum emodi* (Berberidaceae). Podophyllotoxin reversibly binds to tubulin, whereas its main derivatives, like etoposide and teniposide, block topoisomerase II, causing DNA breakage by topoisomerase II. It has been suggested to treat cancers and venereal warts (Ardalani et al., 2017). Moreover, podophyllotoxin showed potential anti-multidrug resistance action against various drug-resistant tumour cells. For example, CIP-36, a podophyllotoxin derivative, has overcome the multidrug resistance of adriamycin-resistant human leukaemic cell line (K562/ADR) by modulating the activity of topoisomerase-IIa (Cao et al., 2015). According to Liu et al. (2015), podophyllotoxin-derived pharmaceuticals are being studied, including the approval for clinical use of etoposide and teniposide, as well as numerous additional drug candidates with C-4 alterations, such as NK611, TOP53, and GL-331. However, the discovery of safe, cost-effective, and site-specific anticancer medicines (Liu et al., 2015).

4.2.5 Silymarin

Silymarin is a flavonolignan from the *Sylibum marianum* that exhibits anticancer effects against various cancers. Silydianin, silychrystin, silibinin (silybin A and B), and isosilybin (A and B) were found in this flavonolignan combination. Silymarin protects against human hepatic and pancreatic disorders by acting as an anti-inflammatory, antioxidant, and proapoptotic agent. It has also been proven to have anticancer properties in malignancies of the liver, breast, cervix, skin, lung, and prostate (Campos et al., 1989). Silymarin, which acts on cyclin-dependent kinases to cause cell cycle arrest and death, has been used to treat colorectal cancer in combination with paclitaxel and doxorubicin. According to research by Won et al. (2018), silymarin decreased the growth of oral cancer cells (HSC-4) and enhanced caspase-dependent death. It also elevated cleaved caspase-8 and shortened BH3-only protein, resulting in cytochrome c buildup. In two oral cancer cell lines, silymarin activated death receptor 5/caspase-8, causing apoptotic cell death. Silymarin also inhibited tumour development and volume without causing damage to the liver or kidneys (Won et al., 2018).

4.2.6 Lactones of Sesquiterpenes

Sesquiterpene lactones are a wide and varied collection of bioactive chemicals identified from various plant families (e.g. Asteraceae). According to recent investigations, they are active molecules with different pharmacological and anticancer effects. Deoxyelephantopin has anticancer activity in nasopharyngeal, cervical, breast, lung, colon, and liver cancer, whereas its isomer isodeoxyelephantopin has potentiated leukaemia, breast cancer, and lung cancer by targeting numerous pathways. Deoxyelephantopin has been shown to have lower cytotoxic action against normal colon cells (CCD 841 CoN) and lymphocytes at doses lethal to lung cancer cells (A549). Isodeoxyelephantopin has also demonstrated a specific cytotoxic effect against breast cancer cells (T47D) and lung cancer cells (A549) but not normal lymphocytes (Mehmood et al., 2017).

Elephantopus carolinianus and *Elephantopus scaber* contain sesquiterpene lactones, such as deoxyelephantopin and isodeoxyelephantopin, which have been shown to induce apoptosis through a variety of mechanisms, including the production of reactive oxygen species, mitochondrial dysfunction, and cell death (Shoaib et al., 2017).

4.2.7 Thymoquinone

Thymoquinone is obtained from *Nigella sativa* or black cumin. It belongs to the Ranunculaceae family of plants. Its therapeutic benefits are linked to thymoquinone in plant extracts, which has been shown to have a variety of medicinal qualities, including anticancer, anti-inflammatory, antioxidant, and antimodulatory. Thymoquinone has been shown to slow the growth of diseases such as leukaemia, breast adenocarcinoma, colorectal, pancreatic, prostate, and hepatic cancer (Khan et al., 2017). Lei et al. (2012) reported thymoquinone's anticancer efficacy against gastric cancer cells for the first time. Thymoquinone may trigger apoptosis in gastric cancer cell lines (BGC-823, SGC-7901, MGC-803, and HGC-27) in a dose- and time-dependent manner. Thymoquinone has also been demonstrated to make gastric cancer cells more sensitive to chemotherapy drugs like 5-fluorouracil and cisplatin. Studies have shown that combining thymoquinone with 5-FU or cisplatin improves chemotherapeutic-induced antitumour effects on gastric cancer by increasing immune system activation (Fatfat et al., 2021).

4.2.8 *Allium sativum*

Allium species are medicinal plants high in bioactive compounds with anticancer activities. This work aimed to examine the molecular makeup and in vitro anticancer properties of *Allium willeanum* Holmboe, an indigenous *Allium* species from Cyprus belonging to the Amaryllidaceae family. *Allium willeanum* H. belongs to the same section as *Allium sativum*, which already has well-established antitumoural activities. Bioactive compounds such as octadecanoic acid 2-hydroxy-1-(hydroxymethyl)ethyl ester (21.99%), hexadecanoic acid (20.42%), pentadecanoic acid (9.19%), and 1,2-benzene dicarboxylic acid, diethyl ester (8.79%) were found in ethanolic extracts of *Allium willeanum* H. bulb (Isbilen & Volkan, 2021). The anticancer properties of *Allium sativum* have also been widely demonstrated, with studies showing that increased consumption of garlic lowers the risk of several cancer types, including colon, pancreas, and breast cancer, and increased consumption of *Allium* class vegetables reduces the risk of stomach cancer (Mikaili et al., 2013).

Furthermore, *Allium*-related molecular activities are involved in the control of several cancers, including the formation of DNA adducts, activation of metabolizing enzymes responsible for carcinogen detoxification, and induction of apoptotic mechanisms, all of which are involved in the control of human breast cancer (BCa). According to Isbilen and Volkan (2021), there is high antitumoural activity against human breast cancer cells that are very metastatic. The cytotoxic effects of *A. willeanum* H. bulb point to a more powerful activity of *A. willeanum* H. bulb in reducing cell viability via mitochondrial activity in highly metastatic human breast cancer cells (MDA-MB-231) in a concentration-dependent manner. Allicin has been shown to suppress the proliferation of gastric cancer cell lines by causing apoptosis and cell cycle arrest during the G2/M phase (Nakonieczna et al., 2020).

4.3 LIMITATIONS FOR DRUG DELIVERY OF PHYTOCHEMICALS AGAINST CANCER

More than a thousand plants with notable anticancer potential have been identified. Separation of Vinca alkaloids, Madagascar periwinkle vinblastine, and *Catharanthus roseus* G. are some leading examples of anticancer drugs (Qu et al., 2019; Martino et al., 2018) (Figure 4.1). This is used to administer a cancer chemotherapy drug, along with vincristine and other cancer chemotherapeutic medications (Wong & Chiu, 2010). *Taxus brevifolia* Nutt, the discovery of paclitaxel (taxol) from pacific yew bark (Family: *Taxaceae*), is another major source of success in finding medicines for natural products (Cragg & Newman, 2005). The usage by many native American tribes of different parts of *Taxus brevifolia* from which paclitaxel and other Taxus species have been discovered ignites the concept of historical medicinal plants (Ashraf, 2020). Conversely, one important cytotoxic phytochemical, Homoharringtonine, was isolated from the Chinese tree *Cephalotaxus harringtonia* var. *Drupaceae* (Sieb and Zucc.) (Family: *Cephalotaxaceae*) and has long been successfully applied for the treatment of acute myelogenous leukaemia in China in a racemic blend with harringtonin (Zhu et al., 2020; Bohlander, 2020).

This is just an overview of the entire picture of the anticancer phytochemical, which will surely grow in the future (Abotaleb et al., 2020). To achieve better therapeutic effectiveness, there should be good quality of the constituents with sufficient quantity, which varies with latitude, altitude, climate, and season (Samaniego et al., 2020). Various sections of the medicinal plant can have different phytoconstituents with different pharmacological activity and levels (Ege et al., 2020). The pharmacological activity of particular phytoconstituents may have additive or synergistic effects along with other constituents than the pure extract (Ansari et al., 2020). These bioactive phytoconstituents may be produced in antitumour therapeutic entities but require intense effort. Multiple methods are used for getting purified products. These include isolation and assay, fractionation driven by combinatorial chemistry and bioassay (Aznar & Rai, 2020). In the case of bioassay, various analytical techniques are used to separate active molecules from a mixture of compounds (Ahmad et al., 2020).

FIGURE 4.1 Limitations for drug delivery of phytochemicals against cancer.

One such technique for isolating active constituents is fractionation, after which the separated moieties are tested for bioactivity, and different analytical techniques are used to analyse active fractions. High-performance thin layer chromatography, reverse phase high-performance liquid chromatography, Fourier transform infrared spectroscopy, and mass spectroscopy are techniques used for the same purpose. However, reaching the desired quality product is a very tedious process with a lot of effort, time, and money, as these compounds pose many limitations (Figure 4.2). The purification of antimicrobial, antilipolytic, and antioxidant compounds can also be used in this approach (Ingle et al., 2017).

The molecule(s) must be tested *in vivo* for the effects of the anticancer after purification. The drug molecule must be investigated for other parameters, i.e. safety and adverse effects, dosage concentration, pharmacokinetics, drug interactions, etc. Using natural bioactive functional leads as initials or turning them into products is possible. Outrageous costs involved in the drug discovery process, with a few out of millions reaching the market due to the complexity of the process, are serious obstacles to the discovery of new drugs (Prasad et al., 2017). The main task of searching for new phytoconstituents from the plant is to select the correct candidate species for pharmacological activity.

A further obstacle for plant-based drugs is their low bioavailability. The concerns associated with bioavailability problems are mainly because of higher degradation by drug metabolic enzymes (Rousseau et al., 2020). A Plant constituent has a high concentration, a large number of chiral centres, and a large number of hydrogen bond donors and receptors, varying in odour and high cell size & durability; it poses a challenge to their growth. The US Food and Drug Administration and the Dietary Supplement Health and Education Act of 1994 control therapeutic phytomolecules and dietary supplements of natural origin for their pharmacological effects and related safety concerns (Sayed et al., 2019). Although there is a desire to work with the ADMET profile (absorption, distribution, metabolism, excretion and toxicity), most naturally occurring chemical compounds do not meet Lipinski's "fifth rule", i.e., the drug candidates must have a molecular weight less than 500 Da, five hydrogen bond providers, who receive less than ten hydrogen bonds and a coefficient less than 5 (log*P*) acting as a drug molecule, depending on the drug-like nature. These rules aim to achieve maximum bioavailability (Campos et al., 1989). Therefore, an urgent need exists for an acceptable chemical/biochemical entity. To improve phytomedicine availability, various coating materials, micelles, liposomes, phospholipid structures, and nanomaterials can be used (Mirhadi et al., 2020).

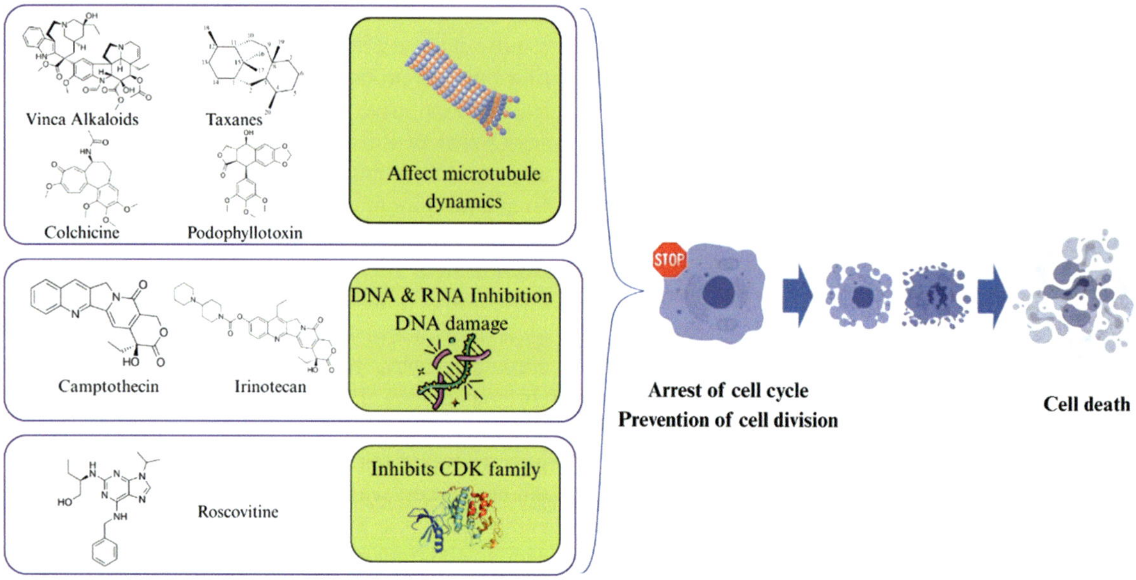

FIGURE 4.2 Common pathway by which phytochemicals exert anticancer action. (Obtained from an open-access source from Garcia-Oliveira et al. (2021).)

4.4 CHALLENGES IN THE DELIVERY OF ANTICANCER PHYTOCHEMICALS

Chemicals found in plants have been a major source of beneficial medicinal properties since ancient times (Padhi et al., 2018). To determine their chemical activity and chemical treatment, the performance of several plants has been tested in various systems (Abd El-Kaream, 2019). Since the advent of advanced chromatography systems to separate significant amounts of compounds from complex mixtures and the progress of chemical synthesis, there have been *in vitro* and *in vivo* samples of many isolated phytochemicals and synthetic chemicals/metabolites (Ingle et al., 2017). Several chemicals have been identified as promising agents successfully converted into commercial drugs. Many researchers have turned to plants as a source of new and powerful agents, such as cancer cells (Singh et al., 2019).

There are also major drawbacks to most of the clinically used anticancer drugs (formulated as traditional delivery systems) (Cho et al., 2008). These include their low water solubility, inadequacy for oral administration, limited body half-life, poor specificity associated with serious side effects, etc. (Colone et al., 2020).

The therapeutic regimen is associated with severe systemic toxicity and multidrug resistance growth, primarily mediated by overexpression of the transporter superfamily, i.e., ATP-binding cassette proteins. These multidrug resistance proteins, particularly *P*-glycoprotein, are responsible for energy-dependent drug efflux, decreasing chemotherapy in cancer cells. Many *P*-glycoprotein inhibitors have been researched to resolve multidrug resistance in cancer over the last few decades. As these are already licenced for other applications, the first generation of ATP-binding cassette blockers [verapamil, cyclosporine A, and quinidine] are the most commonly studied and can be repositioned and clinically tested for new therapeutic uses (Callaghan et al., 2014). Breast cancer clinical trials have shown that combining vincristine and *P*-glycoprotein blocker verapamil can increase an antitumour function (Alfarouk et al., 2015). In conjunction with *P*-glycoprotein blockers, similar studies can be performed on several anticancer phytochemicals to enhance their clinical efficacy in multidrug-resistant cancer cell lines (Khan & Gurav, 2018).

However, to satisfy the medical community, beneficial drug action alone against the disease is insufficient; however, it is equally necessary to prevent adverse drug actions on normal tissues, as well as to reduce the side effects of therapy (Singh et al., 2019). Clinically, an anticancer drug's therapeutic effectiveness depends on its inherent anticancer function and bioavailability (Lee et al., 2011). The ability to devise an appropriate pharmaceutical formulation for delivery is paramount in developing novel therapeutics. Therefore, delivering chemopreventive drugs is essential for successful cancer prevention and care. New technologies have created considerable interest in developing new drug delivery systems to advance parenterally administered drugs' pharmacological and therapeutic properties (Pucek et al., 2020). Applying drug-loaded nanosized drug carriers, such as polymeric nanoparticles, liposomes, and dendrimers, is a promising approach to overcome low bioavailability and systemic toxicity (Naqvi et al., 2020). Compared to systemic chemotherapy, using such carriers has many benefits, may modulate the pharmacokinetics of existing drugs, and can help enhance the delivery of anticancer agents to target sites (Zeinali et al., 2020).

4.5 NANOCARRIERS EXPLORED FOR THE DELIVERY OF ANTICANCER PHYTOCHEMICALS

There is an opportunity for nanotechnology to transform cancer detection and treatment. Typically, there is a defective, leaky vascular system connected with a tumour due to the poorly regulated existence of tumour angiogenesis. A poorly developed lymphatic system drains the interstitial fluid inside a tumour. As a result, micrometre-range particulate matter can be easily extravasated into the cancer and retained, resulting in an "enhanced permeability and retention" effect. This enhanced permeability and retention effect is useful for the passive targeting of a precisely engineered nanoparticulate system that allows tumour tissues to focus on nanocarriers packed with cytotoxic agents. Successful strategies for targeting work by combining nanocarriers containing chemotherapy with molecules that bind more to the target cells' antigens or receptors. Due to recent advances in surface science, the biodistribution profile of nanocarriers could be close to perfect by changing their surface physicochemical properties to target the tissue of interest. This leads to a greater quantity of drugs with minimal systemic drug toxicity accessing the intended tumour sites. In cancer care, nano-technologically altered drugs and drug delivery systems are growing; some are used medically with a favourable performance each day. Nanotechnology may enhance clinical management in cancer patients for improved cancer detection, more effective drug delivery to tumour cells, and chemotherapeutic agents for cancer treatment (Sutradhar & Amin, 2014; Chidambaram et al., 2011).

Most current chemotherapy nanotechnology technologies have included repurposing conventional anticancer agents into different types of delivery vehicles of nanometer scale. Using new ideas and carriers that existing technologies cannot attain, nanotechnology has created a forum to transform drug delivery. This analysis discusses the use of nanotechnology to formulate drug delivery mechanisms for certain identified plant-based anticancer agents [vincristine, paclitaxel, etoposide, curcumin, resveratrol, genistein, quercetin, and capsaicin, to name a few], resulting in an improved anticancer profile. Various nanocarriers explored for the successful targeted delivery of anticancer phytochemicals are summarized in Table 4.1.

TABLE 4.1
Nanocarriers Utilized for Anticancer Phytochemicals Delivery

Formulations	Drug	Composition	Concluding Remarks	References
Nanoparticles	Vincristine and verapamil	Vincristine-verapamil poly(lactic-co-glycolic acid) nanoparticles	*In vitro* and *in vivo* studies have also shown strong antitumour potency due to synergistic co-encapsulation of vincristine and verapamil as nanoparticles.	Chen et al. (2014)
	Apigenin	Apigenin poly(lactic-co-glycolic acid) nanoparticles	It is used for the therapeutic management of skin cancer	Das et al. (2013)
	Polymeric nanoparticles with paclitaxel and etoposide	Poly(lactic-co-glycolic acid) and polyethylene glycol	These studies were performed by paclitaxel and etoposide co-loaded polymeric nanoparticles, and their antitumour ability on the osteosarcoma cell lines was analysed (MG63 and Saos-2). The paclitaxel and etoposide poly(lactic-co-glycolic acid) nanoparticles exhibited good anticancer activity in osteosarcoma with synergism	Wang et al. (2015)
	Epigallocatechin-3-gallate	Polyethylene glycol-poly lactic acid derivative	These nanoparticles have been tested on human melanoma cells and are eight times more effective than the free drug	Siddiqui et al. (2015)
	Etoposide	Poly(lactic-co-glycolic acid) nanoparticles/P188-blend encapsulating etoposide	It was tested on murine glioblastoma cell lines (C6 and F98), demonstrating enhanced etoposide antitumour effectiveness	Callewaert et al. (2013)
	Paclitaxel	Poly(lactic-co-glycolic acid) nanoparticles	Using the 3-[4,5-dimethylthiazol-2-yl]-2,5 diphenyl tetrazolium bromide assay, paclitaxel-loaded polymeric nanoparticles exhibited substantial cytotoxicity on colon cancer cells (Caco-2), compared to paclitaxel alone	Abotaleb et al. (2020)
	Curcumin	*N*-isopropylacrylamide, *N*-vinyl-2-pyrrolidone and poly(ethyleneglycol) monoacrylate	The outcome of drug distribution indicated that nanoformulation had increased absorption by the pancreatic cell lines. It has been observed that this nanoformulation increased curcumin solubility	Bisht et al. (2007)

(*Continued*)

TABLE 4.1 (*Continued*)
Nanocarriers Utilized for Anticancer Phytochemicals Delivery

Formulations	Drug	Composition	Concluding Remarks	References
	Noscapine	Human serum albumin nanoparticles	These are designed for targeted drug delivery in human breast cancer cells (SK-BR-33)	Sebak et al. (2010)
	Paclitaxel	Paclitaxel-loaded nanoparticles containing cholic acid-polylactic acid-tocopheryl polyethylene glycol succinate	As per the study results, the star-shaped nanoparticles of the paclitaxel proved to be the better option for breast cancer management	Tang et al. (2013)
	Luteolin	Luteolin is encapsulated into polymeric nanoparticles consisting of polylactic acid-polyethylene glycol.	The analysis indicated that nano luteolin had greater antitumour activity for lung cancer cells. The half-maximal inhibitory concentration is also lower than the free drug.	Majumdar et al. (2014)
	Resveratrol	Poly(epsilon-caprolactone) and poly(D,L-lactic-co-glycolic acid), poly(ethylene glycol) conjugate	The prostate cancer cell lines (PCa) effectively took up the polymeric nanoparticles. Compared to free resveratrol, the nanoparticles showed improved cytotoxicity in prostate cancer cells (DU-145, LNCaP, and PC-3)	Sanna et al. (2013)
	Ferulic acid	Poly(lactic-co-glycolic acid) nanoparticles	The polymeric nanoparticles have better cytotoxicity against colon and lung cancer than the free drug due to DNA damage and other synergistic actions	Merlin et al. (2012)
	Curcumin	Nanoparticles of silk fibre and chitosan	The prepared nanoparticles have shown better anticancer activity in breast cancer by reducing cell viability compared to the free drug	Gupta et al. (2009)
Magnetic nanoparticles	Camptothecin-loaded iron oxide magnetic nanoparticles	Iron oxide nanoparticles coated with polyethylene glycol	It had low toxicity in lung cancer cell lines and a longer circulatory half-life due to PEGylation	Castillo et al. (2014)
	Quercetin	Magnetic nanoparticle with a surface coating of poly(lactic-co-glycolic acid)	The study results proved the efficacy of the quercetin-loaded magnetic nanoparticles by nasal delivery as it reduced the number of viable lung cancer cells (A549). The nebulization of the prepared nanoparticles has shown better-targeted drug delivery	Verma et al. (2013)
Liposomes	Folic acid and vincristine	PEGylated liposomes of folic acid and vincristine	These folic acid and vincristine conjugated liposomes have demonstrated very good anticancer activity in multidrug-resistant variants of epidermoid nasopharyngeal carcinoma cell lines. The PEGylated liposomes have improved the efficacy of Vincristine action	Lu and Park (2013)
	Ursolic acid	pH-sensitive ursolic acid liposomes	The study's findings showed the liposomal version of the drug had better antitumour activity than free ursolic acid, which meant that nano-liposomes had increased anticancer activity	Ashraf (2020)
	Vincristine	Liposomes are made up of cholesterol, egg yolk phosphatidylcholine, and distearoylphosphatidylcholine	Vincristine liposomes exhibited higher durability with good circulatory half-life and antitumour activity. A very good targeted vincristine delivery can be achieved with this approach	Bisht et al. (2007)
	Thymoquinone (2-isopropyl-5-methyl-1, 4-benzoquinone)	Thymoquinone liposomes modified with Triton X-100	The liposomes have shown promising results in the breast cancer cell lines (MCF-7). It has demonstrated very low toxicity to the periodontal fibroblast cells	Odeh et al. (2012)
	Berberine	Liposomes using various compositions of lipids	The findings showed better cytotoxicity relative to free berberine	Khan and Gurav (2018) and Hou et al. (2011)
Carbon nanotubes	Paclitaxel using folic acid	Multi-walled carbon nanotubes	These functionalized multi-walled carbon nanotubes showed good aqueous solubility, biocompatibility, and high targeting capacity in the immortal cell line. The complex has demonstrated good anticancer activity	Ghasemvand et al. (2016)

(*Continued*)

TABLE 4.1 (*Continued*)
Nanocarriers Utilized for Anticancer Phytochemicals Delivery

Formulations	Drug	Composition	Concluding Remarks	References
Dendrimers	Chondrocyte affinity peptide	Dendrimers	The anticancer activity of the prepared dendrimer was evaluated in kidney (VERO), liver (HepG2), and breast (MCF-7) cell lines. It showed high cytotoxicity at a lower dose	Tunki et al. (2020)
	Gallic acid and polyamidoamine	Dendrimers	The results demonstrated the benefit of dendrimers as an efficient platform for nanotechnology to improve cytotoxicity in breast cancer cells (MCF-7)	Sharma et al. (2011)
Micelles	Luteolin	Monomethoxy poly (ethylene glycol)-poly (3-caprolactone)	It was evaluated in the colon cancer cell line. A good anticancer activity has been shown in *in vitro* cytotoxicity studies, indicating that luteolin encapsulation into monomethoxypoly (ethylene glycol)-poly (3-caprolactone) micelles potentially improved the drug availability at a target site and antitumour activity	Qiu et al. (2013)
	β-lapachone	Poly(ethylene glycol)-poly(lactic-co-glycolic acid) polymeric β-lapachone micelles	It mainly acts by DNA damage and poly (ADP-ribose) polymerase-1 hyperactivation. The prepared micelles have demonstrated good circulatory half-life for β-lapachone with better anticancer activity in mice models	Khan and Gurav (2018)
	Curcumin	Methoxy-poly(ethylene glycol)-poly(D/L-lactide) block copolymer micelles	In methoxy-poly(ethylene glycol)-poly(D/L-lactide) micelles, the cytotoxicity of curcumin was greater, suggesting that methoxy-poly(ethylene glycol)-poly(D/L-lactide) polymer micelles are an important curcumin nanocarrier	Kumari et al. (2016)
	Honokiol	Monomethoxy poly(ethylene glycol) and poly(3-caprolactone) self-assembled biodegradable star-shaped polymeric honokiol micelles	They were tested on murine colon carcinoma cells from CT26. The findings showed that the polymer micelles substantially improved the dose-related cytotoxicity of the CT26 cells	Gong et al. (2011)
Solid lipid nanoparticles	Paclitaxel	Paclitaxel-loaded solid lipid nanoparticles	The group of animals treated with paclitaxel-loaded solid lipid nanoparticles was found to have substantially smaller tumour sizes and lower inhibition percentages	Serpe et al. (2004)
	Silibinin	Phosphatidylcholines	Targeted delivery of the drug with double the amount of silibinin cellular uptake	Xu et al. (2013)
	Hyaluronic acid	Coated hyaluronic acid solid lipid nanoparticles of paclitaxel	Targeted delivery of the paclitaxel to the tumour-containing tissues required a low dose as compared to the conventional delivery	Elhissi et al. (2012)

4.6 NANO-DRUG DELIVERY OF PHYTOCHEMICALS AGAINST CANCER

Cancer is the most common cause of death and disease in humans, with 7.6 million deaths per year and a predicted 13.1 million by 2030. Many phytochemicals used in traditional medicine have been used to treat cancer. Although conventional chemotherapy is widely recognized as the most effective treatment for metastatic cancer, these malignant cells may develop resistance to a variety of anticancer medications over time, resulting in treatment failure. This chapter attempted to depict the progress in anticancer and chemopreventive effects of numerous phytochemicals and some of their members encapsulated in the drug's nano-based delivery system. It discusses the concerns associated with the enhanced use of each phytochemical compound in human treatment for cancer. The rewards of encapsulation into nanocarriers are analysed in terms of drug loading efficiency, nanocarrier size, the dissolution rate of the drug, and in vitro and/or *in vivo* therapeutic strategies (Behera et al., 2020a). Although the enhanced therapy options are accessible to cancer patients, no one approach can entirely cure cancer. This encourages scientists to continue working on anticancer medicines and effective administration methods. In addition, many phytochemicals have been shown to have anticancer properties against various cancers (Chidambaram et al., 2011).

Scientists use nanotechnology to improve their small backlogs (Bulbake et al., 2017). Different nanocarriers are utilized to deliver drugs in fatal illnesses such as cancer. The function of nanocarriers in cancer is unavoidable since they play a key role in both imaging and treatment. These nanocarriers may be easily produced to target cancer cells while ignoring normal cells (Ferrari, 2005). Micelles, liposomes, dendrimers, carbon nanotubes, nanoshells, and nanocages are some of the most common nanocarriers employed in cancer research (Behera et al., 2023b; Padhi & Behera, 2022). Because of their site-specificity and tiny size, nanocarriers carefully carry medications to specific sites. Passive, active, pH sensitivity, and temperature sensitivity are nanocarriers' target properties (Behera & Padhi, 2022). The nanocarrier can target delivery depending on pH and temperature variations in cancer cells, and it can also be cloaked by coating materials like PEG to correlate to circulation time (Subramanian et al., 2016).

4.6.1 Liposomes

Liposomes have been studied extensively for decades and are the first nanoscale drug delivery method to be effectively converted into clinical use. Liposomes are spherical vesicles with an internal aqueous core comprising one or more phospholipid bilayer membranes. Traditional liposomes are generated in aqueous conditions by self-assembling amphiphilic lipids, which form bilayers around an aqueous core for drug encapsulation. Liposomes' biphasic nature has permitted the trapping of lipophilic and hydrophilic pharmaceuticals, with hydrophilic medicines delivered in the inner aqueous core and lipophilic medications in the phospholipid bilayers. The phospholipid bilayers can be made from natural or synthesized phospholipids, changing the liposome's physicochemical characteristics, surface charge, composition, and size (Barea et al., 2010).

Liposomal formulations have been widely used as nano-drug carriers in various applications, including the delivery of antifungal, anti-inflammatory, and anticancer medicines. Doxil®, the first FDA-approved liposome-based anticancer nano-drug, was authorized in 1995 (Barenholz, 2012). Subsequently, a few other products for treating different tumours were accessible. SkyPharma Inc.'s Depocyt®, Elan Pharmaceuticals' Myocet®, Takeda Pharmaceuticals' Mepact®, and Talon Therapeutics' Marqibo® are among these drugs (Gajera & Patel, 2022). A fluorouracil and leucovorin combination therapy-based product was recently licenced for metastatic pancreatic cancer. Merrimack Pharmaceuticals, Inc. markets this medicine as Onivyde™ (Zhang, 2016).

Although cancer has been the most extensively researched subject regarding clinically approved liposome products (Bulbake et al., 2017), the capacity of liposomes to carry a variety of medicines, huge surface-to-volume ratio, and customizable surface for targeting and liposomes may be the best alternatives for combinational chemotherapy (Zhao et al., 2018). Vyxeos™ (daunorubicin and cytarabine) is a liposomal injectable formulation licenced by the FDA for combination chemotherapy for acute myeloid leukaemia (Alfayez et al., 2020). The goal of the study by Mahira et al. (2019) was to create cationic liposomes of cabazitaxel and silibinin to kill cancer cells and cancer stem cells in prostate cancer. Because of their capacity to carry a variety of medications, high surface-to-volume ratio, and customizable surface for targeting, liposomes may be an ideal alternative for combinational chemotherapy. Hyaluronic acid-coated liposomal nanocarriers improved intracellular drug distribution to CD44-expressing cells, inhibiting CD44+ cells via dual anticancer medicines' distinct cellular actions. Overall, it can be stated that an effective drug delivery system with promising prospects for combination treatment was established, with liposomes having a possible method for eradicating prostate cancer by simultaneously targeting cancer cells and cancer stem cells. Nanotechnology, targeted medicine delivery, and cancer biology have all advanced significantly (Crain, 2018).

4.6.2 Phytosomes

Oral bioavailability, physico-chemical characteristics, interactions, solubility, stability, intrinsic and extrinsic interactions, absorption permeability, physiological pH and temperature, and protein binding are all factors that influence phytomolecule oral bioavailability (Lahlou, 2013). Phyto-phospholipid complexes, also known as herbosomes or phytosomes, increase the bioavailability of insoluble plant extracts and phytoconstituents. Phytosomes are recommended over basic herbal extracts to increase membrane absorption and permeability. Phytosomes administer various medications and phytoconstituents to cure diverse disorders via active and passive targeting (Karpuz et al., 2020). Because of their important role in promoting several cancer therapy techniques, nanoparticulate and nanomedicine drug delivery technologies are advancing. Their use is fraught with difficulties, including toxicity and high costs. Manufacturing costs and complications. There is a fresh and feasible way.

Green nanomedicines are a viable and new technique rapidly expanding as a tumour medication delivery system and substantially impacting cancer therapy. Encapsulation of cancer cells using phytosomes can encourage and successfully develop cancer treatment (Babazadeh et al., 2018). The therapeutic benefits of encapsulating 5-fluorouracil and phytosomal curcumin in a mouse model of colitis-related collagenous colitis were investigated. The study revealed that combining 5-fluorouracil with phytosomes might decrease colorectal cancer cell proliferation and invasive behaviour. In colitis-related collagenous colitis, the combination might reduce the size and quantity of tumours in the intermediate and distal sections of the colon, lowering disease activity. In addition, curcumin has been shown to reduce colonic inflammation, thiol levels, and catalase activity (Marjaneh et al., 2018; Wu et al., 2014). The present chemotherapy suffers as a result of multidrug resistance, and therapeutic efficiency suffers. Bioengineering or nanocarrier polychemotherapy has been created to target the cancer cell. Phytochemicals are employed to reduce multidrug resistance. Fisetin was laced with geldanamycin and radicicol to boost its anticancer properties and reduce collagenous colitis cell lines. Multiple routes initiate this combination treatment, which targets protein expression and acts through DNA degradation.

Due to its potency to inhibit selectively the nuclear factor erythroid 2-related factor 2 signalling pathway and sensitize non-small lung cancer cell lines (A549) to anticancer agents, luteolin, a natural component, has been found to have anticancer activities by acting on multiple molecular targets to kill cancer cells (Zhao et al., 2011). Nuclear factor erythroid 2-related factor 2 expression is high in cancer cells, and its function in drug resistance in cancer treatment has been shown. Sabzichi et al. used luteolin-loaded nanophytosome complexes to sensitize human breast adenocarcinoma cells (MDA-MB 231) to doxorubicin by blocking nuclear factor erythroid 2-related factor 2-mediated signalling. On breast cancer cells (MDA-MB 231), cytotoxicity testing revealed that nano-phytosomes of luteolin in conjunction with doxorubicin outperformed nano-dox alone. To summarise, phytosomes might enhance the therapeutic impact of dox by increasing the efficacy of luteolin on nuclear factor erythroid 2-related factor 2 inhibition by raising the bioavailability of luteolin (Sabzichi et al., 2014).

4.6.3 Micelles

Micelles are lipid molecules that form spherical structures in aqueous liquids. Polymeric micelles are often quite thin and have a size range of 10–100 nm. Polymeric micelles' critical association concentration is much lower than the usual essential concentration values for surfactant micelles, making them more stable in biological fluids. Because the medication is effectively shielded from probable inactivation by its micellar surrounds, it can boost drug bioavailability and retention. Several parameters influence drug release from micelles, including micelle stability, drug diffusion rate, partition coefficient, and copolymer biodegradation rate. The drug concentration within the micelles and the drug's molecular weight, physicochemical properties, and placement within the micelles can influence drug release (Teng et al., 1998).

Micelles are lipid molecules that form spherical structures in aqueous liquids. Matsumura and coworkers found that paclitaxel micellar formulation, including polyethylene glycol and modified polyaspartate as a hydrophobic block, had equal cytotoxicity in 12 human tumour cell lines (lung, stomach, oesophagal, colon, breast, and ovarian) compared to paclitaxel alone (Shenoy et al., 2005). Another study found that paclitaxel delivered by pH-responsive nanoparticles was more effective in vivo against breast tumours (SKOV-3) than free paclitaxel (Devalapally et al., 2007).

Polymeric biomaterials, recently created to deliver drugs to cancer cells, are used in many diagnostic and medication delivery systems (Le & Kim, 2019). Oerlemans et al. (2010) have demonstrated that bio-applicable polymers may easily form micelles with sizes ranging from a few tens to a few hundred nanometers. Drugs are encapsulated and delivered to particular target locations to enhance efficacy and decrease cancer therapy's adverse effects. The Wnt/-catenin signalling pathway was used to explore the release of curcumin from biodegradable polyaspartamide and biocompatible micelles in collagenous colitis therapy. Curcumin was coupled with a cleavable acid bond, folic acid was used to target collagenous colitis cells, and hydrazone was used to release the pH-sensitive medication. According to the findings, folic acid curcumin micelles have higher anticancer potential than colon anticancer drugs because they block the Wnt/-catenin pathway (Dragojevic et al., 2015).

4.6.4 Nanoparticles

Nanoparticles are extremely small materials ranging from 1 to 100 nm. They can be classified based on their forms, sizes, or qualities. Magnetic nanoparticles, carbon nanotubes, and polymeric nanoparticles are three types of nanoparticles. Nanoparticles have unique chemical and physical characteristics because of their nanoscale size and surface area (Khan et al., 2019). Nanoparticles have special physicochemical and pharmacokinetic features, making cancer therapy more effective and less harmful than standard medicines (Jiang et al., 2017). Nanoparticles deliver medications to malignant cells more accurately than chemotherapy agents. Nanoparticles have numerous advantages, including increased drug effectiveness with fewer side effects, protection of medicinal compounds from the harsh microenvironment before entry into targeted tumour cells, sustained drug release for a specific period with required administrative doses, and low exposure of unaffected cells.

Sun et al. (2021) found that nanoparticles are beneficial for the co-delivery of various medications and diagnostic agents for successful therapy. Sufi et al. (2020) expected that nano-encapsulation of curcumin and indole-incorporated curcumin analogue biodegradable polymeric nanoparticles would improve bioavailability while extending medication retention time. The nanoparticles' anticancer potential and physicochemical characterizations were studied in the colorectal cancer cell lines (SW480). The diameter of the nanoparticles synthesized ranged from 50 to 150 nm. Nanoformulation protected the medication from degradation in a wide variety of pH conditions. Nuclear fragmentation, apoptosis inhibition, cell cycle blockage, and metastatic biomarkers were induced using nanoparticles against the colorectal cancer cell lines (SW480). Because it can endure severe pH conditions, nanoformulations containing drug-loaded nanoparticles can potentially treat colorectal cancer. As a result, it can be employed to administer oral drugs (Sufi et al., 2020). To investigate the safety and stability of zinc oxide nanoparticles, researchers synthesized them from the leaves of *Artocarpus heterophyllus*. The reduction of zinc ions into zinc oxide nanoparticles in *Artocarpus heterophyllus* was remarkable. Zinc oxide nanoparticles with sizes ranging from 12 to 26 nm were discovered to have strong anticancer efficacy against colorectal cancer cells (HCT-116) by transmission electron microscopy. Apoptotic activity has been demonstrated to cause morphological alterations and cell membrane blebbing (Majeed et al., 2019).

Drug and gene delivery, tissue engineering, and biosensors have all benefited from stimuli-responsive polymer-based nanoparticles. In response to changes in ambient factors such as pH, temperature, light, magnetic field, or glucose, nanoparticles undergo sudden physical or chemical changes. Colby et al. (2013) have published a study of physiologically sensitive polymeric nanoparticles for drug delivery that release their drug payload in response to pH or oxidative stress. These would be of great therapeutic importance since they allow medicine administration to be linked to a specific area or illness state. Paclitaxel administration by pH-responsive nanoparticles is one example. Compared to paclitaxel administration on non-responsive polycaprolactone nanoparticles, this approach shows higher cytotoxicity against human breast cancer cells (MDA-MB-231) in vitro (Shenoy et al., 2005).

4.6.5 Dendrimers

Curcumin, carotenoids, numerous phenolic compounds, and other bioactive phytochemical substances are hydrophobic; therefore, they can't be formed freely at a desirable dose. Furthermore, roughly 15% of phytochemical bioactive components function poorly due to their low solubility. Low bioavailability of phytochemicals is caused by a lack of solubility and permeability (Azzi et al., 2018). Surface modification and shrinking the size of these chemicals are standard ways to increase their bioavailability, but the use of organic solvents and stability issues restrict their applicability. Recent studies have used delivery strategies to increase the bioavailability of phytochemical bioactive components (Jafari & McClements, 2017). Dendrimers provide a unique polyvalency and nano-encapsulation feature for the delivery of active biomolecules, giving a new technique for solubilizing these components.

Curcumin has received a lot of interest due to its ability to induce apoptosis in various cancer cell lines. However, inactivity in clinical trials, quick breakdown in alkaline and neutral environments, and low oral and vascular absorption due to its hydrophobic nature have limited its pure form. These two bioactive molecules were effectively conjugated with fourth-generation polyamidoamine, which has acetyl terminal groups, and fifth-generation polyamidoamine, which has amine terminal groups. The results showed that resveratrol and curcumin had a 40- and 200-fold increase in the solubility of dendrimer formulation over their pure forms, respectively (Wang et al., 2013).

Dendritic macromolecules' growing role in anticancer therapies and diagnostic imaging has heightened interest in these well-defined materials as the newest class of macromolecular nanoscale delivery techniques (Wolinsky & Grinstaff, 2008). Dendrimers contribute excellent delivery mechanisms for chemotherapeutic drugs in combination. Shukla et al. (2019) define dendrimers as a collection of macromolecules with various chemical characteristics and structures.

Due to its cardiotoxicity, epirubicin has a restricted therapeutic use. As a result, a modified dendrimer with three types of aptamers was created for epirubicin administration. Epirubicin was released pH-sensitively in this delivery method (Shukla et al., 2019). The aptamers-dendrimer-epirubicin combination has been demonstrated to be less cytotoxic and can effectively prevent tumour development in vivo. To test the curcumin-loaded dendrimer-gold hybrid structure's theranostic capacity, gold chloride ions were complexed with PEGylated amine-terminated generation 5 poly (amidoamine) dendrimer. Curcumin was added to the system to produce in vitro findings for colorectal cancer cells (HT29 and C26). In HT29 and C26 cells, the targeted theranostic agent had more cellular cytotoxicity than the non-targeted system in vitro. The colorectal cancer cells (HT29 and C26) accumulated the designated theranostic substance *in vitro*. It had stronger cellular cytotoxicity and a superior therapeutic index against colorectal cancer adenocarcinoma than the non-targeted system (Fridman et al., 2017; Alibolandi et al., 2018).

4.6.6 Microbubbles

As a possible therapeutic strategy for fighting solid tumours, immunotherapy tries to stimulate endogenous immune responses to target cancer cells precisely while maintaining the integrity of normal or healthy cells (Fridman et al., 2017). Cancer immunotherapy has the promise to eliminate solid tumours, but its therapeutic efficacy is strongly contingent on sufficient CD8+ T cell infiltration. Ultrasound-produced microbubble cavitation has been found to promote programmed cell death ligand-1 inhibition of mice colon adenocarcinoma cell lines (MC38) by facilitating tumour perfusion, increasing CD8+ T cell penetration and anti-programmed cell death ligand-1 antibody distribution. It was coupled in this trial to optimize ultrasound-produced microbubble cavitation as a therapy option with an anti-programmed cell death ligand-1 antibody. The combination therapy increased tumour perfusion and helped tumour vascularity return to normal. Flow cytometry further showed that the combination treatment increased the proportion and total number of tumour-infiltrating CD8+ T cells and the expression of Ki67 antibody, interferons, and granzyme B. Microbubbles. The combination therapy increased tumour perfusion and helped tumour vascularity return to normal. Flow cytometry also revealed that the combination therapy increased the percentage and total number of tumour-infiltrating CD8+ T cells, as well as the expression of Ki67 antibody, interferons, and granzyme B secretions, implying that the drug combination inhibited tumour growth and prolonged survival when compared to monotherapies. These data suggest that when paired with immune checkpoint inhibition, ultrasound-produced microbubble cavitation might be a viable therapy option for solid cancers (Li et al., 2021).

He et al. (2011) found that multidrug resistance mutation 1 small interfering RNA delivery through ultrasonic microbubbles was safe and effective in yolk sac carcinoma cells (L2-RYC). Multidrug resistance mutation 1 silencing reduced *P*-glycoprotein activity and drug resistance in L2-RYC cells, suggesting that it might be used as part of a new combined gene and chemotherapeutic strategy for yolk sac cancer. The *P*-glycoprotein-encoding multidrug resistance mutation 1 gene is an ATP-dependent drug efflux transporter linked to treatment resistance in yolk sac cancer. A unique and successful gene delivery mechanism has been developed using ultrasound microbubbles.

4.7 EXPLORING MARKET AND PATENT SCENARIO OF ANTICANCER PHYTOMOLECULES WORLD OVERVIEW

In 2021, there was an increase in the number of new cancer patients, which was more than 18.3 million, and more than 9 million deaths were registered. According to the report published by The International Agency for Research on Cancer, the number of cases will be more than 24 million by 2030. Cancer treatment using nanoparticle-based approaches is being thoroughly explored to address the disadvantages associated with traditional therapy. Progress in nanotechnology for cancer therapy has resulted in the creation of smart nanoformulations that enhance the intracellular delivery of drugs because of their improved multi-functionality and targeting ability. Smart nanoformulations are nano-range particles that release the drug with the pre-existing biological stimulus at the disease site. This section summarizes many of the latest patents for cancer therapy related to smart nanopreparations. The therapeutic impact of current conventional treatment modalities, such as chemotherapy and radiotherapy, is improved by such a smart method. Several polymeric nanoparticles that kill tumours by thermal energy deposition and thus improve tumour therapy have been patented. Patented solid lipid thermoresponsive nanoformulations serving as smart drug delivery systems demonstrated temperature sensitivity (39°C–45°C) and payload release at cancer cell target sites.

The pH-sensitive polymeric drug delivery system comprising decorative acidic or simple groups to accept or contribute protons in response to environmental pH showed accelerated drug release at the tumour site. A novel thermosensitive peptide-linked liposome showing temperature-dependent peptide shrinkage and drug release was invented. Similarly, as a stimuli-sensitive drug delivery device, they are generally gaining interest because of the unique properties of metallic nanoparticles in the imaging and diagnostic sector. Gold nanoparticles can strongly absorb light and rapidly and efficiently transform photon energy into heat, raising the temperature of a tumour cell (41°C–47°C) and thus specifically killing cancer cells. The combination of pH-sensitive targeting of metallic nanoparticle properties resulted in an improved antitumour effect, as found in recently granted patents.

Some patents (US20170157005A1, US20160228362A1, and US20150050357A1) have reservations about phytopharmaceuticals being trapped in nanoparticles. Curcumin, quercetin, silymarin, paclitaxel, etc., are prevalent among different phytochemicals, effectively manufactured as nanoparticles. The justification for the availability of their marvellous compositions lies in the addition of low stability, weak solubility in water and, therefore, poor bioavailability (Table 4.2).

Diseases such as cancer, Alzheimer's, diabetes, liver disease, etc., may be specifically treated if bioavailability limitations are resolved. While doing so, culture, which is overwhelmingly affected, will find relaxation. The current report attempts to expand and summarize how nanoparticles, along with in vivo targeting, have improved the efficacy of phytochemicals through solubilization, dissolution, and resultant improvements in bioavailability. The analysis examines disorder and phytochemicals, focusing on their nanoparticles to raise awareness.

4.8 CONCLUSION

Phytomolecules in anticancer therapy have been booming in the current scenario due to their advantages and increased efficacy and safety. Incorporation of nanoformulations in designing phytomolecules for chemotherapy has additional benefits of delivering the drug molecules through a targeted strategy. Passive and active targeting of anticancer phytomolecules increases the efficacy of the phytomolecules by overcoming the limitations of low solubility, low permeability, and lesser bioavailability. The enhanced retention and permeation effect in passive targeting increases the blood circulation time of the phytomolecules' higher accumulation, with some adverse effects of off-target cytotoxicity. Meanwhile, in active targeting, more targeted delivery can be achieved through overexpressed receptor-mediated endocytosis, and the extent of off-target toxicity is lessened with enhanced anticancer activity. The phytomolecules with other anticancer drugs also exhibit efficient anticancer activity. Multidrug resistance can also be reversed using phytomolecules as anticancer drugs. The availability of phytomolecules opens a wide field of research for cancer chemotherapy.

TABLE 4.2
A Patent Overview of the Anticancer Phytochemicals

Sr No.	Patent Number	Phytomolecule	Description	References
1.	WO2010074553A1	Piper betel, tropical seaweeds, elaeisguineensis	Alcohol-soluble extracts from plant parts of *piper* species, *palm* species, tropical marine algae, and/or *solanum* species comprise an edible composition with anticancer properties.	Aigner et al. (2011)
2.	WO2006134609A2	*Sphaeranthus indicus* (sesquiterpenes)	An herbal anticancer agent comprising the extract of the Sphaeranthusindicus plant or a group of compounds obtained from the *Sphaeranthusindicus* plant is disclosed herein. A pharmaceutical composition containing the said agent, methods for preparing the composition, and methods for treating all forms of cancer in mammals, including humans, are also disclosed in the current invention. Methods for extracting the active constituents are also discussed in the plant extract methods.	Mitra et al. (2009)
3.	WO2015193741A1	*Boswellia* spp extract, *Curcuma longa* extract, *Vitis vinifera* extract	The new disclosure relates to a cancer prevention and care formulation that prevents the adverse side effects found when traditional treatments are used. In compliance with the present disclosure, the formula contains active constituents consisting of *Boswellia* spp, *Curcuma longa*, and *Vitis vinifera* extracts.	Thakkar and Lárusson (2020)
4.	WO1999008994A1	*Euphorbia peplus*	The whole invention concerns a compound or group of active substances extracted from plants of the group *Euphorbia peplus*, *Euphorbia hirta,* and *Euphorbia drummondii,* including pharmaceutical formulations of such compounds. It was discovered that extracts from these plants exhibit selective cytotoxicity against various lines of cancer cells. Compounds are effective for efficiently treating tumours, especially malignant melanomas and carcinomas of squamous cells. A substance composed of jatrophanes, pepluanes, paralyanes, and ingenanes, and pharmaceutically appropriate salts or esters, and, more specifically, jatrophanes of conformation II, is chosen as the favoured aspect of the invention.	Aylward (1999)
5.	US 20160137663 A1	Sesquiterpenoid	The present invention offers a method for purifying pharmaceutical formulations composed primarily of a combination of compound-based signal transducer and activator of transcription-3 inhibitors, pharmaceutical formulations comprising signal transducer and activator of transcription-3 inhibitors used to inhibit signal transducer and activator of transcription-3 in cancer cells and some pharmaceutically suitable salts, and procedures of use.	Chang et al. (2014)

REFERENCES

Abd El-Kaream, S. A. (2019). Biochemical and biophysical study of chemopreventive and chemotherapeutic anti-tumor potential of some Egyptian plant extracts. *Biochemistry and Biophysics Reports*, 18, 100637.

Abotaleb, M., Liskova, A., Kubatka, P., & Büsselberg, D. (2020). Therapeutic potential of plant phenolic acids in the treatment of cancer. *Biomolecules*, 10(2), 221.

Ahmad Dar, A., Sangwan, P. L., & Kumar, A. (2020). Chromatography: an important tool for drug discovery. *Journal of Separation Science*, 43(1), 105–119.

Aigner, A., Gross, R., Schmidt, R., Braun, M., & Mauer, S. (2011). Patent Application Publication (10) Pub. No.: US 2011/0207128A1. Patent, 1(19), 10–13.

Alfarouk, K. O., Stock, C. M., Taylor, S., Walsh, M., Muddathir, A. K., Verduzco, D., … & Rauch, C. (2015). Resistance to cancer chemotherapy: failure in drug response from ADME to P-gp. *Cancer Cell International*, 15(1), 1–13.

Alfayez, M., Kantarjian, H., Kadia, T., Ravandi-Kashani, F., & Daver, N. (2020). CPX-351 (vyxeos) in AML. *Leukemia & Lymphoma*, 61(2), 288–297.

Alibolandi, M., Hoseini, F., Mohammadi, M., Ramezani, P., Einafshar, E., Taghdisi, S. M., … & Abnous, K. (2018). Curcumin-entrapped MUC-1 aptamer targeted dendrimer-gold hybrid nanostructure as a theranostic system for colon adenocarcinoma. *International Journal of Pharmaceutics*, 549(1–2), 67–75.

Ansari, P., Flatt, P. R., Harriott, P., & Abdel-Wahab, Y. H. (2020). Evaluation of the antidiabetic and insulin-releasing effects of *A. squamosa*, including isolation and characterization of active phytochemicals. *Plants*, 9(10), 1348.

Ardalani, H., Avan, A., & Ghayour-Mobarhan, M. (2017). Podophyllotoxin: a novel potential natural anticancer agent. *Avicenna Journal of Phytomedicine*, 7(4), 285.

Ashraf, M. A. (2020). Phytochemicals as potential anticancer drugs: time to ponder nature's bounty. *BioMed Research International*, 2020, 8602879.

Aylward J. H. (1999). Anti-cancer compounds. WO1999008994A1

Aznar, R., & Rai, D. K. (2020). Purification and isolation techniques for enrichment of bioactive phytochemicals from herbs and spices. In: Goyal, M. R., Birwal, P., Chauhan, D. N. (Eds.) *Herbs, Spices and Medicinal Plants: Processing, Health Benefits and Safety*. Apple Academic Press, Palm Bay, FL, pp. 177–206.

Azzi, J., Jraij, A., Auezova, L., Fourmentin, S., & Greige-Gerges, H. (2018). Novel findings for quercetin encapsulation and preservation with cyclodextrins, liposomes, and drug-in-cyclodextrin-in-liposomes. *Food Hydrocolloids*, 81, 328–340.

Babazadeh, A., Zeinali, M., & Hamishehkar, H. (2018). Nano-phytosome: a developing platform for herbal anti-cancer agents in cancer therapy. *Current Drug Targets*, 19(2), 170–180.

Barea, M. J., Jenkins, M. J., Gaber, M. H., & Bridson, R. H. (2010). Evaluation of liposomes coated with a pH responsive polymer. *International Journal of Pharmaceutics*, 402(1–2), 89–94.

Barenholz, Y. C. (2012). Doxil®—The first FDA-approved nano-drug: lessons learned. *Journal of Controlled Release*, 160(2), 117–134.

Behera, A. (2024). Fungal metabolites as anticancer agent and their agricultural applications. In: Abd-Elsalam, K. A. (Ed.) *Fungal Secondary Metabolites*. Elsevier, Amsterdam, pp. 259–282.

Behera, A., & Padhi, S. (2020). Passive and active targeting strategies for the delivery of the Camptothecin anticancer drug: a review. *Environmental Chemistry Letters*, 18(5), 1557–1567.

Behera, A., & Padhi, S. (2022). pH-Sensitive polymeric nanoparticles for cancer treatment. In: Padhi, S., Behera, A., Lichtfouse, E. (Eds.) *Polymeric Nanoparticles for the Treatment of Solid Tumours*. Environmental Chemistry for a Sustainable World, vol 71. Springer, Cham, pp. 401–425.

Behera, A., Mittu, B., Padhi, S., & Singh, A. (2020a). Antimicrobial efficacy of essential oil nanoemulsions. In: Dhull S. B., Chawla, P., Kaushik, R. (Eds.) *Nanotechnological Approaches in Food Microbiology*. CRC Press, Boca Raton, FL, pp. 293–309.

Behera, A., Mittu, B., Padhi, S., Patra, N., & Singh, J. (2020b). Bimetallic nanoparticles: green synthesis, applications, and future perspectives. In: Abd-Elsalam, K. A. (Ed.) *Multifunctional Hybrid Nanomaterials for Sustainable Agri-food and Ecosystems*. Elsevier, Amsterdam, pp. 639–682.

Behera, A., Pradhan, S. P., Tejaswani, P., Sa, N., Pattnaik, S., & Sahu, P. K. (2023a). Ameliorative and neuroprotective effect of core-shell type Se@ Au conjugated hesperidin nanoparticles in diabetes-induced cognitive impairment. *Molecular Neurobiology*, 60(12), 7329–7345.

Behera, A., Sa, N., Pradhan, S. P., Swain, S., & Sahu, P. K. (2023b). Metal nanoparticles in Alzheimer's disease. *Journal of Alzheimer's Disease Reports*, (Preprint), 1–20.

Bisht, S., Feldmann, G., Soni, S., Ravi, R., Karikar, C., Maitra, A., & Maitra, A. (2007). Polymeric nanoparticle-encapsulated curcumin. *Journal of Nanobiotechnology*, 5(1), 1–18.

Bohlander, S. K. (2020). A new kid on the block for acute myeloid leukemia treatment? Homoharringtonine interferes with key pathways in acute myeloid leukemia cells. *Haematologica*, 105(1), 7.

Bulbake, U., Doppalapudi, S., Kommineni, N., & Khan, W. (2017). Liposomal formulations in clinical use: an updated review. *Pharmaceutics*, 9(2), 12.

Callaghan, R., Luk, F., & Bebawy, M. (2014). Inhibition of the multidrug resistance P-glycoprotein: time for a change of strategy? *Drug Metabolism and Disposition*, 42(4), 623–631.

Callewaert, M., Dukic, S., Van Gulick, L., Vittier, M., Gafa, V., Andry, M. C., … & Roullin, V. G. (2013). Etoposide encapsulation in surface-modified poly (lactide-co-glycolide) nanoparticles strongly enhances glioma antitumor efficiency. *Journal of Biomedical Materials Research Part A*, 101(5), 1319–1327.

Campos, R., Garrido, A., Guerra, R., & Valenzuela, A. (1989). Silybin dihemisuccinate protects against glutathione depletion and lipid peroxidation induced by acetaminophen on rat liver. *Planta Medica*, 55(05), 417–419.

Cao, B., Chen, H., Gao, Y., Niu, C., Zhang, Y., & Li, L. (2015). CIP-36, a novel topoisomerase II-targeting agent, induces the apoptosis of multidrug-resistant cancer cells in vitro. *International Journal of Molecular Medicine*, 35(3), 771–776.

Castillo, P. M., de la Mata, M., Casula, M. F., Sánchez-Alcázar, J. A., & Zaderenko, A. P. (2014). PEGylated versus non-PEGylated magnetic nanoparticles as camptothecin delivery system. *Beilstein Journal of Nanotechnology*, 5(1), 1312–1319.

Chang, L. C., Turkson, J., Wongwiwatthananukit, S., Youn, U. J., & Sun, D. (2014). Novel sesquiterpenoid stat3 inhibitors, US 20160137663 A1.

Chen, H. C., Hsieh, W. T., Chang, W. C., & Chung, J. G. (2004). Aloe-emodin induced in vitro G2/M arrest of cell cycle in human promyelocytic leukemia HL-60 cells. *Food and Chemical Toxicology*, 42(8), 1251–1257.

Chen, Y., Zheng, X. L., Fang, D. L., Yang, Y., Zhang, J. K., Li, H. L., … & Song, X. R. (2014). Dual agent loaded PLGA nanoparticles enhanced antitumor activity in a multidrug-resistant breast tumor xenograft model. *International Journal of Molecular Sciences*, 15(2), 2761–2772.

Chidambaram, M., Manavalan, R., & Kathiresan, K. (2011). Nanotherapeutics to overcome conventional cancer chemotherapy limitations. *Journal of Pharmacy & Pharmaceutical Sciences*, 14(1), 67–77.

Cho, K., Wang, X. U., Nie, S., Chen, Z., & Shin, D. M. (2008). Therapeutic nanoparticles for drug delivery in cancer. *Clinical Cancer Research*, 14(5), 1310–1316.

Colby, A. H., Colson, Y. L., & Grinstaff, M. W. (2013). Microscopy and tunable resistive pulse sensing characterization of the swelling of pH-responsive, polymeric expansile nanoparticles. *Nanoscale*, 5(8), 3496–3504.

Colone, M., Calcabrini, A., & Stringaro, A. (2020). Drug delivery systems of natural products in oncology. *Molecules*, 25(19), 4560.

Cragg, G. M., & Newman, D. J. (2005). Plants as a source of anti-cancer agents. *Journal of Ethnopharmacology*, 100(1–2), 72–79.

Crain, M. L. (2018). Daunorubicin & cytarabine liposome (VyxeosTM). *Oncology Times* 40, 30.

Das, S., Das, J., Samadder, A., Paul, A., & Khuda-Bukhsh, A. R. (2013). Efficacy of PLGA-loaded apigenin nanoparticles in Benzo[a]pyrene and ultraviolet-B induced skin cancer of mice: mitochondria mediated apoptotic signalling cascades. *Food and Chemical Toxicology*, 62, 670–680.

Devalapally, H., Shenoy, D., Little, S., Langer, R., & Amiji, M. (2007). Poly (ethylene oxide)-modified poly (beta-amino ester) nanoparticles as a pH-sensitive system for tumor-targeted delivery of hydrophobic drugs: part 3. Therapeutic efficacy and safety studies in ovarian cancer xenograft model. *Cancer Chemotherapy and Pharmacology*, 59, 477–484

Dragojevic, S., Ryu, J. S., & Raucher, D. (2015). Polymer-based prodrugs: improving tumor targeting and the solubility of small molecule drugs in cancer therapy. *Molecules*, 20(12), 21750–21769.

Ege, B., Yumrutas, O., Ege, M., Pehlivan, M., & Bozgeyik, I. (2020). Pharmacological properties and therapeutic potential of saffron (Crocus sativus L.) in osteosarcoma. *Journal of Pharmacy and Pharmacology*, 72(1), 56–67.

Elhissi, A., Ahmed, W., Hassan, I. U., Dhanak, V., & D'Emanuele, A. (2012). Carbon nanotubes in cancer therapy and drug delivery. *Journal of Drug Delivery*, 2012, 837327.

El-Shemy, H. A., Aboul-Soud, M. A. M., Nassr-Allah, A. A., Aboul-Enein, K. M., Kabash, A., & Yagi, A. (2010). Antitumor properties and modulation of antioxidant enzymes' activity by Aloe vera leaf active principles isolated via supercritical carbon dioxide extraction. *Current Medicinal Chemistry*, 17(2), 129–138.

Fatfat, Z., Fatfat, M., & Gali-Muhtasib, H. (2021). Therapeutic potential of thymoquinone in combination therapy against cancer and cancer stem cells. *World Journal of Clinical Oncology*, 12(7), 522.

Ferrari, M. (2005). Cancer nanotechnology: opportunities and challenges. *Nature Reviews Cancer*, 5(3), 161–171.

Fridman, W. H., Zitvogel, L., Sautès–Fridman, C., & Kroemer, G. (2017). The immune contexture in cancer prognosis and treatment. *Nature Reviews Clinical Oncology*, 14(12), 717–734.

Gajera, K., & Patel, A. (2022). An overview of FDA approved liposome formulations for cancer therapy. *Journal of Advances in Medical and Pharmaceutical Sciences*, 24(3), 1–7.

Gao, Y., Kuok, K. I., Jin, Y., & Wang, R. (2019). Biomedical applications of Aloe vera. *Critical Reviews in Food Science and Nutrition*, 59(sup1), S244–S256.

Garcia-Oliveira, P., Otero, P., Pereira, A. G., Chamorro, F., Carpena, M., Echave, J., … & Prieto, M. A. (2021). Status and challenges of plant-anticancer compounds in cancer treatment. *Pharmaceuticals*, 14(2), 157.

Ghasemvand, F., Biazar, E., Tavakolifard, S., Khaledian, M., Rahmanzadeh, S., Momenzadeh, D., … & Keshel, S. H. (2016). Synthesis and evaluation of multi-wall carbon nanotube–paclitaxel complex as an anti-cancer agent. *Gastroenterology and Hepatology from Bed to Bench*, 9(3), 197.

Gong, C., Wang, Y., Wang, X., Wei, X., Wu, Q., Wang, B., … & Qian, Z. (2011). Biodegradable self-assembled PEG-PCL-PEG micelles for hydrophobic drug delivery, part 2: in vitro and in vivo toxicity evaluation. *Journal of Nanoparticle Research*, 13, 721–731.

Gupta, V., Aseh, A., Ríos, C. N., Aggarwal, B. B., & Mathur, A. B. (2009). Fabrication and characterization of silk fibroin-derived curcumin nanoparticles for cancer therapy. *International Journal of Nanomedicine*, 4, 115–122.

He, Y., Bi, Y., Hua, Y., Liu, D., Wen, S., Wang, Q., … & Wei, G. (2011). Ultrasound microbubble-mediated delivery of the siRNAs targeting MDR1 reduces drug resistance of yolk sac carcinoma L2 cells. *Journal of Experimental & Clinical Cancer Research*, 30(1), 1–11.

Hou, Q., Tang, X., Liu, H., Tang, J., Yang, Y., Jing, X., … & Wang, Z. (2011). Berberine induces cell death in human hepatoma cells in vitro by downregulating CD147. *Cancer Science*, 102(7), 1287–1292.

Ingle, K. P., Deshmukh, A. G., Padole, D. A., Dudhare, M. S., Moharil, M. P., & Khelurkar, V. C. (2017). Phytochemicals: extraction methods, identification and detection of bioactive compounds from plant extracts. *Journal of Pharmacognosy and Phytochemistry*, 6(1), 32–36.

Isbilen, O., & Volkan, E. (2021). Allium willeanum Holmboe exerts anticancer activities on metastatic breast cancer cells MCF-7 and MDA-MB-231. *Heliyon*, 7(8), e07730.

Jafari, S. M., & McClements, D. J. (2017). Nanotechnology approaches for increasing nutrient bioavailability. *Advances in Food and Nutrition Research*, 81, 1–30.

Jiang, T., Jin, K., Liu, X., & Pang, Z. (2017). Nanoparticles for tumor targeting. In: Jana, S., Maiti, S., Jana, S. (Eds.) *Biopolymer-Based Composites*. Woodhead Publishing, Duxford, pp. 221–267.

Karpuz, M., Gunay, M. S., & Ozer, A. Y. (2020). Liposomes and phytosomes for phytoconstituents. In: Singh, M. R., Singh, D., Kanwar, J., Chauhan, N. S. (Eds.) *Advances and Avenues in the Development of Novel Carriers for Bioactives and Biological Agents*. Academic Press, London, pp. 525–553.

Khan, I., Saeed, K., & Khan, I. (2019). Nanoparticles: properties, applications and toxicities. *Arabian Journal of Chemistry*, 12(7), 908–931.

Khan, M. A., Tania, M., Fu, S., & Fu, J. (2017). Thymoquinone, as an anticancer molecule: from basic research to clinical investigation. *Oncotarget*, 8(31), 51907.

Khan, T., & Gurav, P. (2018). PhytoNanotechnology: enhancing delivery of plant based anti-cancer drugs. *Frontiers in Pharmacology*, 8, 1002.

Kumari, P., Swami, M. O., Nadipalli, S. K., Myneni, S., Ghosh, B., & Biswas, S. (2016). Curcumin delivery by poly (Lactide)-based co-polymeric micelles: an in vitro anticancer study. *Pharmaceutical Research*, 33, 826–841.

Kuo, P. L., Lin, T. C., & Lin, C. C. (2002). The antiproliferative activity of aloe-emodin is through p53-dependent and p21-dependent apoptotic pathway in human hepatoma cell lines. *Life Sciences*, 71(16), 1879–1892.

Lahlou, M. (2013). The success of natural products in drug discovery. *Pharmacology & Pharmacy*, 4(03), 17–31.

Le, T. T., & Kim, D. (2019). Folate-PEG/Hyd-curcumin/C18-g-PSI micelles for site specific delivery of curcumin to colon cancer cells via Wnt/β-catenin signaling pathway. *Materials Science and Engineering: C*, 101, 464–471.

Lee, H. Z., Hsu, S. L., Liu, M. C., & Wu, C. H. (2001). Effects and mechanisms of aloe-emodin on cell death in human lung squamous cell carcinoma. *European Journal of Pharmacology*, 431(3), 287–295.

Lee, K. W., Bode, A. M., & Dong, Z. (2011). Molecular targets of phytochemicals for cancer prevention. *Nature Reviews Cancer*, 11(3), 211–218.

Lee, W. L., Huang, J. Y., & Shyur, L. F. (2013). Phytoagents for cancer management: regulation of nucleic acid oxidation, ROS, and related mechanisms. *Oxidative Medicine and Cellular Longevity*, 2013, 925804.

Lei, X., Lv, X., Liu, M., Yang, Z., Ji, M., Guo, X., & Dong, W. (2012). Thymoquinone inhibits growth and augments 5-fluorouracil-induced apoptosis in gastric cancer cells both in vitro and in vivo. *Biochemical and Biophysical Research Communications*, 417(2), 864–868.

Li, N., Tang, J., Yang, J., Zhu, B., Wang, X., Luo, Y., … & Wang, Z. (2021). Tumor perfusion enhancement by ultrasound stimulated microbubbles potentiates PD-L1 blockade of MC38 colon cancer in mice. *Cancer Letters*, 498, 121–129.

Liu, Y. Q., Tian, J., Qian, K., Zhao, X. B., Morris-Natschke, S. L., Yang, L., … & Lee, K. H. (2015). Recent progress on C-4-modified podophyllotoxin analogs as potent antitumor agents. *Medicinal Research Reviews*, 35(1), 1–62.

Lu, Y., & Park, K. (2013). Polymeric micelles and alternative nanonized delivery vehicles for poorly soluble drugs. *International Journal of Pharmaceutics*, 453(1), 198–214.

Madunić, J., Madunić, I. V., Gajski, G., Popić, J., & Garaj-Vrhovac, V. (2018). Apigenin: a dietary flavonoid with diverse anticancer properties. *Cancer Letters*, 413, 11–22.

Mahira, S., Kommineni, N., Husain, G. M., & Khan, W. (2019). Cabazitaxel and silibinin co-encapsulated cationic liposomes for CD44 targeted delivery: a new insight into nanomedicine based combinational chemotherapy for prostate cancer. *Biomedicine & Pharmacotherapy*, 110, 803–817.

Majeed, S., Danish, M., Ismail, M. H. B., Ansari, M. T., & Ibrahim, M. N. M. (2019). Anticancer and apoptotic activity of biologically synthesized zinc oxide nanoparticles against human colon cancer HCT-116 cell line-in vitro study. *Sustainable Chemistry and Pharmacy*, 14, 100179.

Majumdar, D., Jung, K. H., Zhang, H., Nannapaneni, S., Wang, X., Amin, A. R., ... & Shin, D. M. (2014). Luteolin nanoparticle in chemoprevention: in vitro and in vivo anticancer activity. *Cancer Prevention Research*, 7(1), 65–73.

Marjaneh, R. M., Rahmani, F., Hassanian, S. M., Rezaei, N., Hashemzehi, M., Bahrami, A., ... & Khazaei, M. (2018). Phytosomal curcumin inhibits tumor growth in colitis-associated colorectal cancer. *Journal of Cellular Physiology*, 233(10), 6785–6798.

Martino, E., Casamassima, G., Castiglione, S., Cellupica, E., Pantalone, S., Papagni, F., ... & Collina, S. (2018). Vinca alkaloids and analogues as anti-cancer agents: looking back, peering ahead. *Bioorganic & Medicinal Chemistry Letters*, 28(17), 2816–2826.

Mehmood, T., Maryam, A., Ghramh, H. A., Khan, M., & Ma, T. (2017). Deoxyelephantopin and isodeoxyelephantopin as potential anticancer agents with effects on multiple signaling pathways. *Molecules*, 22(6), 1013.

Merlin, J. J., Prasad, N. R., & Shibli, S. M. A. (2012). Ferulic acid loaded poly-d,l-lactide-co-glycolide nanoparticles: systematic study of particle size, drug encapsulation efficiency and anticancer effect in non-small cell lung carcinoma cell line in vitro. *Biomedicine & Preventive Nutrition*, 2(1), 69–76.

Mikaili, P., Maadirad, S., Moloudizargari, M., Aghajanshakeri, S., & Sarahroodi, S. (2013). Therapeutic uses and pharmacological properties of garlic, shallot, and their biologically active compounds. *Iranian Journal of Basic Medical Sciences*, 16(10), 1031.

Mirhadi, E., Nassirli, H., & Malaekeh-Nikouei, B. (2020). An updated review on therapeutic effects of nanoparticle-based formulations of saffron components (safranal, crocin, and crocetin). *Journal of Pharmaceutical Investigation*, 50, 47–58.

Mitra, S., Saxena, E., Dixit, M., Uddagiri, V., Marikunte, V., Mathad, S., & Shanbhag, S. (2009). U.S. Patent Application No. 11/917,269.

Moudi, M., Go, R., Yien, C. Y. S., & Nazre, M. (2013). Vinca alkaloids. *International Journal of Preventive Medicine*, 4(11), 1231–1235.

Nakonieczna, S., Grabarska, A., & Kukula-Koch, W. (2020). The potential anticancer activity of phytoconstituents against gastric cancer: a review on in vitro, in vivo, and clinical studies. *International Journal of Molecular Sciences*, 21(21), 8307.

Naqvi, S., Panghal, A., & Flora, S. J. S. (2020). Nanotechnology: a promising approach for delivery of neuroprotective drugs. *Frontiers in Neuroscience*, 14, 494.

Newman, D. J., & Cragg, G. M. (2020). Natural products as sources of new drugs over the nearly four decades from 01/1981 to 09/2019. *Journal of Natural Products*, 83(3), 770–803.

Odeh, F., Ismail, S. I., Abu-Dahab, R., Mahmoud, I. S., & Al Bawab, A. (2012). Thymoquinone in liposomes: a study of loading efficiency and biological activity towards breast cancer. *Drug Delivery*, 19(8), 371–377.

Oerlemans, C., Bult, W., Bos, M., Storm, G., Nijsen, J. F. W., & Hennink, W. E. (2010). Polymeric micelles in anticancer therapy: targeting, imaging and triggered release. *Pharmaceutical Research*, 27, 2569–2589.

Oyagbemi, A. A., Saba, A. B., & Azeez, O. I. (2010). Capsaicin: a novel chemopreventive molecule and its underlying molecular mechanisms of action. *Indian Journal of Cancer*, 47(1), 53–58.

Padhi, S., & Behera, A. (2020). Nanotechnology based targeting strategies for the delivery of camptothecin. In: Saneja, A., Panda, A., Lichtfouse, E. (Eds.) *Sustainable Agriculture Reviews*, vol 44. Springer, Cham, pp. 243–272.

Padhi, S., & Behera, A. (2021). Advanced drug delivery systems in the treatment of ovarian cancer. In: Dua, K., Mehta, M., de Jesus Andreoli Pinto, T., Pont, L. G., Williams, K. A., Rathbone, M. J. (Eds.) *Advanced Drug Delivery Systems in the Management of Cancer*. Academic Press, London, pp. 127–139.

Padhi, S., & Behera, A. (2022). Cellular internalization and toxicity of polymeric nanoparticles. In: Padhi, S., Behera, A., Lichtfouse, E. (Eds.) *Polymeric Nanoparticles for Treating Solid Tumours*. Environmental Chemistry for a Sustainable World, vol 71. Springer, Cham, pp. 473–488.

Padhi, S., Kapoor, R., Verma, D., Panda, A. K., & Iqbal, Z. (2018). Formulation and optimization of topotecan nanoparticles: in vitro characterization, cytotoxicity, cellular uptake and pharmacokinetic outcomes. *Journal of Photochemistry and Photobiology B: Biology*, 183, 222–232.

Padhi, S., Mirza, M. A., Verma, D., Khuroo, T., Panda, A. K., Talegaonkar, S., Khar, R. K., & Iqbal, Z. (2016). Revisiting the nanoformulation design approach for effective delivery of topotecan in its stable form: an appraisal of its in vitro behavior and tumour amelioration potential. *Drug Delivery*, 23(8), 2827–2837.

Pradhan, S. P., Sahoo, S., Behera, A., Sahoo, R., & Sahu, P. K. (2022). Memory amelioration by hesperidin conjugated gold nanoparticles in diabetes induced cognitive impaired rats. *Journal of Drug Delivery Science and Technology*, 69, 103145.

Prasad, V., De Jesús, K., & Mailankody, S. (2017). The high price of anticancer drugs: origins, implications, barriers, solutions. *Nature Reviews Clinical Oncology*, 14(6), 381–390.

Pucek, A., Tokarek, B., Waglewska, E., & Bazylińska, U. (2020). Recent advances in the structural design of photosensitive agent formulations using "soft" colloidal nanocarriers. *Pharmaceutics*, 12(6), 587.

Qiu, J. F., Gao, X., Wang, B. L., Wei, X. W., Gou, M. L., Men, K., ... & Huang, M. J. (2013). Preparation and characterization of monomethoxy poly (ethylene glycol)-poly (ε-caprolactone) micelles for the solubilization and in vivo delivery of luteolin. *International Journal of Nanomedicine*, 8, 3061–3069.

Qu, Y., Safonova, O., & De Luca, V. (2019). Completion of the canonical pathway for assembly of anticancer drugs vincristine/vinblastine in Catharanthus roseus. *The Plant Journal*, 97(2), 257–266.

Ranjan, A., Ramachandran, S., Gupta, N., Kaushik, I., Wright, S., Srivastava, S., ... & Srivastava, S. K. (2019). Role of phytochemicals in cancer prevention. *International Journal of Molecular Sciences*, 20(20), 4981.

Ronnett, G. V., & Moon, C. (2002). G proteins and olfactory signal transduction. *Annual Review of Physiology*, 64(1), 189–222.

Rousseau, S., Kyomugasho, C., Celus, M., Hendrickx, M. E., & Grauwet, T. (2020). Barriers impairing mineral bioaccessibility and bioavailability in plant-based foods and the perspectives for food processing. *Critical Reviews in Food Science and Nutrition*, 60(5), 826–843.

Sa, N., Tejaswani, P., Pradhan, S. P., Alkhayer, K. A., Behera, A., & Sahu, P. K. (2023). Antidiabetic and antioxidant effect of magnetic and noble metal nanoparticles of Clitoria ternatea. *Journal of Drug Delivery Science and Technology*, 84, 104521.

Sabzichi, M., Hamishehkar, H., Ramezani, F., Sharifi, S., Tabasinezhad, M., Pirouzpanah, M., … & Samadi, N. (2014). Luteolin-loaded phytosomes sensitize human breast carcinoma MDA-MB 231 cells to doxorubicin by suppressing Nrf2-mediated signalling. *Asian Pacific Journal of Cancer Prevention*, 15(13), 5311–5316.

Samaniego, I., Brito, B., Viera, W., Cabrera, A., Llerena, W., Kannangara, T., … & Carrillo, W. (2020). Influence of the maturity stage on the phytochemical composition and the antioxidant activity of four Andean blackberry cultivars (Rubus glaucus Benth) from Ecuador. *Plants*, 9(8), 1027.

Sanna, V., Siddiqui, I. A., Sechi, M., & Mukhtar, H. (2013). Resveratrol-loaded nanoparticles based on poly(epsilon-caprolactone) and poly(d,l-lactic-co-glycolic acid)–poly (ethylene glycol) blend for prostate cancer treatment. *Molecular Pharmaceutics*, 10(10), 3871–3881.

Sayed, N., Khurana, A., & Godugu, C. (2019). Pharmaceutical perspective on the translational hurdles of phytoconstituents and strategies to overcome. *Journal of Drug Delivery Science and Technology*, 53, 101201.

Sebak, S., Mirzaei, M., Malhotra, M., Kulamarva, A., & Prakash, S. (2010). Human serum albumin nanoparticles as an efficient noscapine drug delivery system for potential use in breast cancer: preparation and in vitro analysis. *International Journal of Nanomedicine*, 5, 525–532.

Serpe, L., Catalano, M. G., Cavalli, R., Ugazio, E., Bosco, O., Canaparo, R., … & Zara, G. P. (2004). Cytotoxicity of anticancer drugs incorporated in solid lipid nanoparticles on HT-29 colorectal cancer cell line. *European Journal of Pharmaceutics and Biopharmaceutics*, 58(3), 673–680.

Sharma, A., Gautam, S. P., & Gupta, A. K. (2011). Surface modified dendrimers: synthesis and characterization for cancer targeted drug delivery. *Bioorganic & Medicinal Chemistry*, 19(11), 3341–3346.

Shenoy, D., Little, S., Langer, R., & Amiji, M. (2005). Poly(ethylene oxide)-modified poly(β-amino ester) nanoparticles as a pH-sensitive system for tumor-targeted delivery of hydrophobic drugs: part 2. In vivo distribution and tumor localization studies. *Pharmaceutical Research*, 22, 2107–2114.

Shoaib, M., Shah, I., Ali, N., Adhikari, A., Tahir, M. N., Shah, S. W. A., … & Umer, M. N. (2017). Sesquiterpene lactone! a promising antioxidant, anticancer and moderate antinociceptive agent from Artemisia macrocephala jacquem. *BMC Complementary and Alternative Medicine*, 17(1), 1–11.

Shukla, R., Singh, A., Pardhi, V., Kashyap, K., Dubey, S. K., Dandela, R., & Kesharwani, P. (2019). Dendrimer-based nanoparticulate delivery system for cancer therapy. In: Kesharwani, P., Paknikar, K. M., Gajbhiye, V. (Eds.) *Polymeric Nanoparticles as a Promising Tool for Anti-cancer Therapeutics*. Academic Press, London, pp. 233–255.

Siddiqui, I. A., Sanna, V., Ahmad, N., Sechi, M., & Mukhtar, H. (2015). Resveratrol nanoformulation for cancer prevention and therapy. *Annals of the New york Academy of Sciences*, 1348(1), 20–31.

Singh, V. K., Arora, D., Ansari, M. I., & Sharma, P. K. (2019). Phytochemicals based chemopreventive and chemotherapeutic strategies and modern technologies to overcome limitations for better clinical applications. *Phytotherapy Research*, 33(12), 3064–3089.

Subramanian, A. P., Jaganathan, S. K., Manikandan, A., Pandiaraj, K. N., Gomathi, N., & Supriyanto, E. (2016). Recent trends in nano-based drug delivery systems for efficient delivery of phytochemicals in chemotherapy. *RSC Advances*, 6(54), 48294–48314.

Sufi, S. A., Hoda, M., Pajaniradje, S., Mukherjee, V., Coumar, S. M., & Rajagopalan, R. (2020). Enhanced drug retention, sustained release, and anti-cancer potential of curcumin and indole-curcumin analog-loaded polysorbate 80-stabilizied PLGA nanoparticles in colon cancer cell line SW480. *International Journal of Pharmaceutics*, 588, 119738.

Sun, T., Zhang, Y. S., Pang, B., Hyun, D. C., Yang, M., & Xia, Y. (2021). Engineered nanoparticles for drug delivery in cancer therapy. In: Raliya, R., & Pathak, Y. (Eds.) *Nanomaterials and Neoplasms*. Springer Nature, Cham, pp. 31–142.

Suresh, D., & Srinivasan, K. (2010). Tissue distribution & elimination of capsaicin, piperine & curcumin following oral intake in rats. *Indian Journal of Medical Research*, 131(5), 682–691.

Sutradhar, K. B., & Amin, M. L. (2014). Nanotechnology in cancer drug delivery and selective targeting. *International Scholarly Research Notices*, 2014, 1–12.

Tang, X., Cai, S., Zhang, R., Liu, P., Chen, H., Zheng, Y., & Sun, L. (2013). Paclitaxel-loaded nanoparticles of star-shaped cholic acid-core PLA-TPGS copolymer for breast cancer treatment. *Nanoscale Research Letters*, 8, 1–12.

Teng, Y., Morrison, M. E., Munk, P., Webber, S. E., & Procházka, K. (1998). Release kinetics studies of aromatic molecules into water from block polymer micelles. *Macromolecules*, 31(11), 3578–3587.

Thakkar, J., & Lárusson, G. Ö. (2020). U.S. Patent No. 10,537,606. Washington, DC: U.S. Patent and Trademark Office.

Tunki, L., Kulhari, H., Sistla, R., & Pooja, D. (2020). Dendrimer-based targeted drug delivery. In: *Pharmaceutical Applications of Dendrimers*. Applied Biology Division, CSIR-Indian Institute of Chemical Technology, Hyderabad, India. Elsevier, pp. 107–129.

Verma, N. K., Crosbie-Staunton, K., Satti, A., Gallagher, S., Ryan, K. B., Doody, T., … & Gun'ko, Y. K. (2013). Magnetic core-shell nanoparticles for drug delivery by nebulization. *Journal of Nanobiotechnology*, 11(1), 1–12.

Wang, B., Yu, X. C., Xu, S. F., & Xu, M. (2015). Paclitaxel and etoposide co-loaded polymeric nanoparticles for the effective combination therapy against human osteosarcoma. *Journal of Nanobiotechnology*, 13(1), 1–11.

Wang, L. G., Liu, X. M., Kreis, W., & Budman, D. R. (1999). The effect of antimicrotubule agents on signal transduction pathways of apoptosis: a review. *Cancer Chemotherapy and Pharmacology*, 44, 355–361.

Wang, L., Xu, X., Zhang, Y., Zhang, Y., Zhu, Y., Shi, J., … & Huang, Q. (2013). Encapsulation of curcumin within poly (amidoamine) dendrimers for delivery to cancer cells. *Journal of Materials Science: Materials in Medicine*, 24, 2137–2144.

Wolinsky, J. B., & Grinstaff, M. W. (2008). Therapeutic and diagnostic applications of dendrimers for cancer treatment. *Advanced Drug Delivery Reviews*, 60(9), 1037–1055.

Won, D. H., Kim, L. H., Jang, B., Yang, I. H., Kwon, H. J., Jin, B., … & Cho, S. D. (2018). In vitro and in vivo anti-cancer activity of silymarin on oral cancer. *Tumor Biology*, 40(5), 1010428318776170.

Wong, M. Y., & Chiu, G. N. (2010). Simultaneous liposomal delivery of quercetin and vincristine for enhanced estrogen-receptor-negative breast cancer treatment. *Anti-cancer Drugs*, 21(4), 401–410.

Wu, M. S., Lien, G. S., Shen, S. C., Yang, L. Y., & Chen, Y. C. (2014). N-acetyl-L-cysteine enhances fisetin-induced cytotoxicity via induction of ROS-independent apoptosis in human colonic cancer cells. *Molecular Carcinogenesis*, 53(S1), E119–E129.

Xu, P., Yin, Q., Shen, J., Chen, L., Yu, H., Zhang, Z., & Li, Y. (2013). Synergistic inhibition of breast cancer metastasis by silibinin-loaded lipid nanoparticles containing TPGS. *International Journal of Pharmaceutics*, 454(1), 21–30.

Zeinali, M., Abbaspour-Ravasjani, S., Ghorbani, M., Babazadeh, A., Soltanfam, T., Santos, A. C., … & Hamblin, M. R. (2020). Nanovehicles for co-delivery of anticancer agents. *Drug Discovery Today*, 25(8), 1416–1430.

Zhang, H. (2016). Onivyde for the therapy of multiple solid tumors. *OncoTargets and Therapy*, 9, 3001–3007.

Zhao, C. Y., Cheng, R., Yang, Z., & Tian, Z. M. (2018). Nanotechnology for cancer therapy based on chemotherapy. *Molecules*, 23(4), 826.

Zhao, Y., Yang, G., Ren, D., Zhang, X., Yin, Q., & Sun, X. (2011). Luteolin suppresses growth and migration of human lung cancer cells. *Molecular Biology Reports*, 38, 1115–1119.

Zhu, M., Gong, Z., Wu, Q., Su, Q., Yang, T., Yu, R., … & Zhang, Y. (2020). Homoharringtonine suppresses tumor proliferation and migration by regulating EphB4-mediated β-catenin loss in hepatocellular carcinoma. *Cell Death & Disease*, 11(8), 632.

Section II

Alkaloid Phytomolecules in Cancer Chemotherapy

5 Nanoenabled Systems for the Delivery of Camptothecin as an Anticancer Agent

Apurba Gohain[1], *Ekta*[2], *Ravjit Kaur*[3], *Anita Devi*[4], *Ajay Sharma*[5], *and Garima Bhardwaj*[6]

[1]Department of Chemistry, Assam University Silchar, Dorgakona, Silchar, India

[2]Department of Chemistry, DAV University, Jalandhar-Pathankot, National Highway, Jalandhar, India

[3]SBMDSOE School Longowal, Sangrur, India

[4]Dr Shanti Swarup Bhatnagar University Institute of Chemical Engineering and Technology, Panjab University, Sector 14, Chandigarh, India

[5]Medicinal and Natural Product Laboratory, Department of Chemistry, Chandigarh University, Gharuan, Mohali, India

[6]Department of Chemistry, Sant Longowal Institute of Engineering and Technology, Longowal, Sangrur, India

5.1 INTRODUCTION

Medicinal and aromatic plants are a rich reservoir of bioactive phytoconstituents and have been used by more than 80% of the world's inhabitants for their primary health care for decades. In recent years, once again, a significant rise in the utilization of plant-based natural medicines has been seen, firstly because, in many instances, synthetic or modern drugs have proven to be dangerous due to their toxicity and, secondly, the rising societal consciousness about the side effects of different synthetic medicines. The demand for the world's herbal medicinal products is expanding at an annual pace of 10%–15% (Kulkova et al. 2023). The global market for herbal medicinal products is hitting a value of US$9.21 billion and is projected to rise to US$17.7 bn by 2026. The tremendous value of natural molecules has redeployed enormous phytomolecules (approx. one lakh), and about 4,000 new bioactive compounds from various plant species are discovered yearly. These bioactives from numerous medicinal and aromatic plants have become even more meaningful and a basis for creating new therapies for different illnesses. Moreover, because of the social perception of danger associated with synthetic components, these bioactive phytomolecules also have a promising area in the perfumery, nutraceuticals, food, and flavours industries. This is a sign that demand for plant-based medicines will rise in the years to come (Kussmann et al. 2023). Due to low amounts and few side effects, the use of the plant species and its active molecules for human well-being cannot be prevented. Phytomolecules have improved the quality of life and lowered human health care expenses by utilizing phytoms as a nutritional substitute (Ayoob et al. 2017).

Generally, bioactive compounds or phytomolecules are secondary plant metabolites that influence man and animals, either pharmacologically or toxicologically, and they play no major role in plant primary metabolism. Subsequent metabolites may, therefore, be viewed as biochemical "sidetrack" products in plant cells and not essential for the everyday functioning of plants. However, they still have main roles in plant safety against different forms of stress. Bioactive compounds such as flavonoids (protection from free radicals), terpenoids (attraction and inhibition of competing plants), and alkaloids are the essential secondary metabolites (defence from predators) (Seker and Erdogan 2023). Few metabolites of secondary plants often play a significant role in the production of cell signals and the metabolism process in biological bodies, and one such notable phytomolecule is camptothecin, which shows great anticancer properties (Ashfaq et al. 2017). A detailed scheme of the extraction and screening of phytomolecules for anticancer activity is illustrated in Figure 5.1.

Camptothecin is one of the human success stories in the history of drug discovery from natural resources. The compound camptothecin, extracted from a medicinal plant named *Camptotheca acuminata*, is mainly obtained from the bark and stem of the plant, which is native to China (Adak et al. 2024). Camptothecin showed promising anticancer activity against a wide range of cancers, including breast, lung, ovarian, pancreatic, and colon cancers in animal models, leading to clinical evaluation (Bondarev et al. 2024). Thousands of camptothecin derivatives have been synthesized; among them, topotecan and irinotecan are commercially approved antitumour agents still in

DOI: 10.1201/9781003611516-7

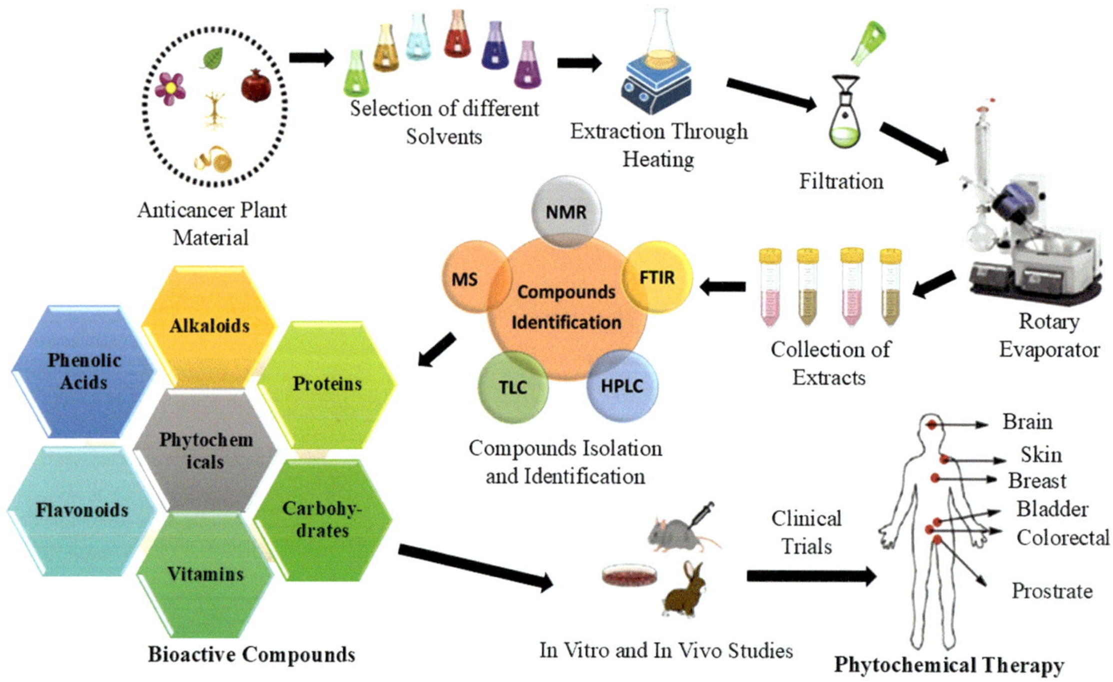

FIGURE 5.1 Optimization of phytochemicals, synthesis, characterization, and screening for anticancer activity. [The phytomolecules are optimized, extracted, characterized by mass spectroscopy, nuclear magnetic resonance, Fourier transform infrared spectroscopy, separated by chromatographic methods, synthesized into nanoformulation, and screened by in-vitro and in-vivo models followed by clinical evaluation for their efficacy and safety as anticancer agent.]

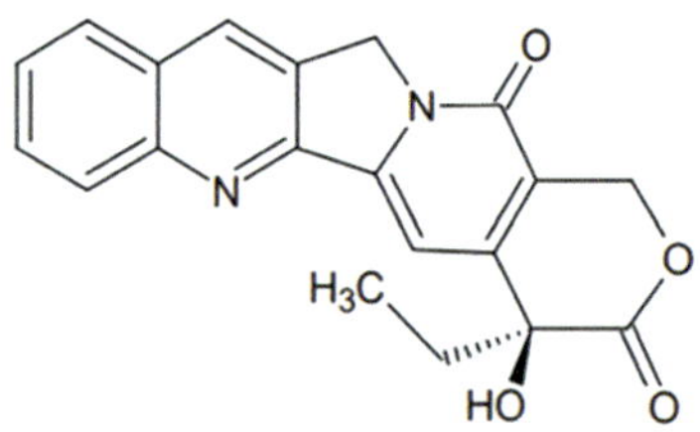

FIGURE 5.2 Chemical structure of camptothecin.

clinical trials (Padhi et al. 2018; Padhi and Behera 2020). Therapeutic use of the unchanged parent camptothecin molecule is hindered by its poor solubility in aqueous and oil phases and shows high toxicity. Camptothecin is a pale-yellow crystalline powder with a planar pentacyclic ring structure, empirical formula $C_{20}H_{16}N_2O_4$ (Figure 5.2). It is an optically active compound; under ultraviolet light, it gives intense blue fluorescence. Its IUPAC name is (*S*)-4-ethyl-4-hydroxy-1*H*-pyrano-indolizino quinoline-3,14 (4H, 12H)-dione. Camptothecin should be stored in air-tight containers at 2°C–8°C.

Camptothecin can exist either as lactone (at acidic pH) or carboxylate (at alkaline pH) based on the pH of the surrounding medium (Andreani et al. 2024). It is susceptible to spontaneous, reversible hydrolysis, so the biologically active lactone form predominates at acidic pH, and the inactive open-ring carboxylate species is favoured at neutral or alkaline pH. Hence, camptothecin should be delivered so that the lactone ring predominates at the site of action for effective cancer treatment (Wang et al. 2023).

Camptothecin is a potent inhibitor of topoisomerase-I. Camptothecin has been shown to capture the cleavable complex between topoisomerase-I and DNA (Hertzberg et al. 1989). Camptothecin is a novel inhibitor of topoisomerase-I, which, while quite powerful, is not yet well known. Furthermore, camptothecin prevents HIV replication and other viruses (Krishnan et al. 2023). Camptothecin can exist in two forms, the closed and open lactone forms, depending upon the pH of the aqueous solution. The closed lactone form (stable in acid) exhibits potent anticancer effects. The closed form may be hydrolysed to the biologically inactive free lactone form at a higher pH (>5.5) (Ziomkowska et al. 2006). Camptothecin binds to mammalian DNA topoisomerase-I and inhibits its function (Hsiang et al. 1985). Topoisomerase-I is an enzyme essential for the replication of DNA. Topoisomerase-I unwinds and divides the two strands of DNA during replication to create two single strands. In topoisomerase-I and cytosine from the+DNA strand, camptothecin attaches to Asp533 and Arg364, making a topoisomerase cleavage complex (Pommier et al. 2003). This complex prevents DNA replication and activates apoptosis in the S step of cell development (Hsiang et al. 1989).

Weak water solubility and aqueous instability of the lactone ring moiety have reduced the maximum therapeutic efficacy of camptothecin (Shaw et al. 2024). Opening the ring would result in the depletion of this compound's biological function. Preferential attachment with human serum albumin of the biologically inactive carboxylate type further hinders the drug's efficacy (Burke and Mi 1994).

The closed-ring lactone form of camptothecin is water-insoluble. In early clinical trials, the administration of sodium salt camptothecin proved unsuccessful and hindered the potency of the anticancer drug.

Liposome-encapsulated camptothecin served as an efficient drug delivery system by solubilizing camptothecin. In experiments with camptothecin and liposomal bilayers, it was observed that the open lactone form of camptothecin was stable when associated with a phosphate-buffered saline solution inside a liposomal bilayer for 48 hours (Burke et al. 1992). Researchers at Glaxo have published the *in-vivo* behaviour of two new analogues of water-soluble camptothecin (Emerson et al. 1995).

Camptothecin is a pentacyclic ring bearing a pyrrolo[3,4-β]-quinoline moiety. This compound has essential anticancer activities and has acted as a mother molecule for many variants and analogues for enhanced anticancer applications (Padhi et al. 2016). Its planar arrangement is the primary explanation for the potent topoisomerase inhibitory effect. Multiple camptothecin resistance phenomena have also been reported in many tumours. The basic processes have not yet been clarified, although numerous research studies indicated that the pharmacological causes and mechanisms linked to the tumour could be involved. In particular, the proposed resistance mechanisms involve the reduction of the drug's cellular aggregation and topoisomerase-I structural changes. At the same time, novel pathways have been suggested, such as cellular reaction variations following the development of the camptothecin-DNA complex and potential micro-RNA shifts during the early stages of cell degeneration (Beretta et al., 2013).

Camptothecin derivatives are a broad class of clinically effective human antitumour drugs. Camptothecin primarily aims at topoisomerase-I, a catalytic enzyme that catalyses DNA replication. Camptothecin can stabilize DNA-linked topoisomerase-I and potentially cause damage to DNA synthesis. However, the most important issues associated with cancer treatment are the systematic toxicity of traditional anticancer medications and their poor accumulation at the tumour site. It is observed that camptothecin has a variety of consequences on malignant tumours. Subsequently, in 1985, Y. H. Hsiang found camptothecin to inhibit the synthesis of the cell division-inclusive topoisomerase-I. Blocking the development of topoisomerase-I by camptothecin can inhibit the growth of cancer cells, granting it special anticancer properties. The study on camptothecin started a new stage from this observation. Many camptothecin analogues have been produced and listed, including irinotecan, topotecan, and 10-hydroxy camptothecin. In addition, separate active drugs are in the therapeutic stage now. However, camptothecin is restricted to its use, and intrinsic deficits include weak water solubility, low biocompatibility, toxic side effects on healthy tissue, and a broad spectrum of risks comparable to other chemotherapeutic agents.

Nanotechnology can rapidly and sensibly diagnose tumours, allowing scientists to detect molecular shifts in a limited percentage of cells. Nanotechnology is a union between different fields of research. It involves not only astronomy, chemistry, and molecular biology. It also includes medicine. Nanotechnology may solve typical drug delivery pitfalls by tuning pharmacokinetics, reducing toxic effects, and improving performance (Sharaf et al. 2014). The surface area is a significant consideration in considering the properties of nanoparticles concerning drug delivery. The volume of anticancer agents that can be attached rises with an expanded surface region: this and many other features to be discussed render possible effective candidates as vectors for drug distribution.

The advancement of nanotechnology improves the capacity of drug delivery and imaging applications by developing complex nanosystems composed of two or more nanoparticles. Nanoparticle sizes range from 10 to 400 nm, and depending on the surface characteristics, different pharmacologically active medicinal products may either encapsulate or bind to their surface. Nanotechnological cancer therapy experiments extend to the creation of novel therapies by utilizing nanomaterial properties outside the delivery of medications. While the nanoparticles are tiny compared with cells, they are big enough to encapsulate many small molecule compounds of different kinds. At the same time, ligands such as small molecules, DNA or RNA, peptides, and aptamers may be used in the comparatively wide region of the nanoparticles. These ligands may be used for therapy or for guiding the destiny of nanoparticles *in-vivo*. These characteristics allow the combination of medications, multi-modal treatment, and clinical and diagnostic combinations, known as theranostics, to function. Nanoparticles' physical characteristics, such as energy absorption and radiation, may also be used for diseased tissue, such as laser ablation or hyperthermia (Weng et al. 2017).

The benefits of employing nanoparticles for therapy include:

- Increased circulation times by reducing renal excretion.
- Improved drug delivery by overcoming chemical and anatomical barriers within the tumour micro-environment. This is because nanoparticles can be designed with strategies for crossing these barriers to transport the drug to the tumour mass.
- Providing a theranostic approach where therapy and diagnostics are combined into one agent.
- Reducing the drug volume allows for decreasing accumulation in healthy tissues.
- Providing site-specific drug delivery.

In 2015, the FDA approved Onivyde, a nano liposomal irinotecan that is currently being evaluated clinically for other malignancies. Camptothecin in nano-drug supply structures enabled by the carrier has also been thoroughly tested. Furthermore, nano-drug supply networks of drug couples are best equipped to hold and monitor the camptothecin-based narcotics efficiently. Pure nano-drug delivery systems, self-assembled carrier-free, offer more effective medicinal dosage for greater medication efficacy. All three nano-drug delivery systems focused on camptothecins are discussed in this study (Kumari et al. 2017).

5.1.1 Different Nanocarriers for Delivering Camptothecin

5.1.1.1 Polymer-Based Nanocarriers for Delivering Camptothecin

Camptothecin in polymer-based drug delivery can be designed with natural polymers (such as proteins and polysaccharides) and synthetic polymers (such as poly (lactic-co-glycolic acid), polyethylene glycol, polycaprolactone, etc.). In glycol-modified chitosan nano-drug delivery, camptothecin demonstrated a long circulation period in human breast cancer therapy (Torre et al. 2015). Xiao et al. (2015) designed a chitosan-based delivery system of camptothecin along with curcumin in different weight ratios. The weight ratio of camptothecin/curcumin at 4:1 exhibited the highest anticancer potency in colon cancer cells. Sepehri et al. (2014) conjugated a potent metabolite of irinotecan, SN38 (7-ethyl-10-hydroxy-camptothecin), with human serum albumin and assessed the anticancer activity in human colon cancer cells (HT-29). The protein-bound nanoconjugate exhibited excellent distribution in different organs without toxicity, with a prompt release within 4 hours.

5.1.1.2 Lipid-Based Nanocarriers for Delivering Camptothecin

A liposome is a small artificial vesicle, spherical in shape, having at least one lipid bilayer. This arrangement enables the water-insoluble drugs to be loaded in the hydrophobic portion of the liposome (the lipid layer partition) or the water-soluble ones in the hydrophilic part (the aqueous phase). This function makes liposome nanocarriers promising in drug delivery applications (Gatto and Johnson 2024).

In 1999, irinotecan encapsulation (camptothecin-11), a novel method for stabilizing anticancer drugs in liposomes, was first reported (Accardo et al. 2012). The most significant discovery that followed was the FDA's approval of irinotecan nanoliposomes (Onivyde) in the year 2015 for the treatment of metastatic pancreatic cancer.

The appearance of polyethylene glycol on the nanocarrier's surface may also inhibit liposome absorption by the endothelial vascular system. Researchers also observed that the blood concentrations of PEGylated nanoliposomes carrying SN-38 were about 50% higher than non-PEGylated liposomes, and the levels were smaller than those of non-PEGylated nanomicelles in the liver, spleen, lung, kidneys, and other organs.

5.1.1.3 Inorganic Nanoparticles for Delivering Camptothecin

Inorganic nanoparticles are commonly utilized in imaging cancers, radiation, and the distribution of therapeutic drugs. Low immunocompetence single-walled carbon nanotubes demonstrated successful cell endocytosis and were used to deliver SN-38 for targeted distribution and controlled release of drugs. Ji et al. (2021) fabricated the mesoporous silica structure with ferric oxide and labelled it with folate to load camptothecin, which showed low cytotoxicity in lung cancer cell lines (A549 cells) but high cytotoxicity in folate-positive epithelial cancer cell lines (KB cells). The loading capacity, targeting ability, and biocompatibility of the nanoformulation demonstrated a potential delivery system for camptothecin.

5.2 CAMPTOTHECIN AS AN ANTICANCER AGENT: MECHANISM

Despite many interventions available, several patients die every year from cancer.

The development of effective anticancer treatments with minimal adverse effects is driving research in healthcare pharmacies. Phytochemicals have shown great promise in this regard.

Phytochemicals are the most effective candidates as anticancer drugs (Lee et al. 2013). Efforts are being made to establish leading phytochemical candidates to inhibit or delay cancer progression without side effects.

Cancer treatment requires tumour surgery, nuclear therapy, and the use of chemotherapeutics. The method of treatment depends on the tumour stage and location. Cytotoxic and cytostatic medications are used in chemotherapy and are proven very effective when used following other procedures. Alkylating compounds, topoisomerase inhibitors, acting agents, and antimetabolites are the most widely used chemotherapeutics (Manzoor et al. 2012).

Camptothecin exhibits anticancer activity by targeting the human deoxyribonucleic acid topoisomerase-I. Inhibition of the enzyme causes the deposition of the reaction intermediate during the cleavage reaction of topoisomerase-I. The deposition of the intermediate causes cell death in the S phase of the cell cycle, which involves DNA replication. Camptothecin brings about the collisions between the intermediate topoisomerase-I cleavable complex and advancing replication forks of DNA (Ryan et al. 1994). The collision results in three phenomena, i.e., the breakdown of the double-strand of DNA, the irreversible arrest of the replication fork, and the collapse of the topoisomerase-I-linked DNA at the collision site. The formation of prolonged covalent topoisomerase-I – deoxyribonucleic acid complexes is the main mechanism of cytotoxicity of camptothecin (Avemann et al. 1988). When the concentration of camptothecin is high, the non-S-phase cells die due to cytotoxicity (Liu et al. 2000).

Camptothecin also interferes with the synthesis of RNA at transcription elongation. Camptothecin leads to the accumulation of RNA-polymerase elongation complexes. The inhibition of transcription by the collision of the replication fork depends on the alignment of the cleavable ternary complex relative to the transcribing RNA-polymerase.

The collision between ternary complexes and ribonucleic acid polymerase on the transcription template strand inhibits RNA transcription. Subsequently, the reversible ternary cleavable complexes are transformed into irreversible topoisomerase-I-linked single-strand breaks (Wu & Liu 1997).

5.3 NANOENABLED TARGETING STRATEGIES OF CAMPTOTHECIN FOR CANCER TREATMENT

Conventional chemotherapy is appended with bioavailability challenges (Tyner et al. 2004; Singh and Lillard 2009; Wen et al. 2017; Behera and Padhi 2020).

Anticancer drugs administered through traditional methods are absorbed by cell membranes, undergo first-pass metabolism, bind firmly to proteins, and are rapidly eliminated by the kidneys, all of which impair the therapeutic efficacy (Bae and Park 2011). To enhance drug retention in tumour tissues with localized activity, a novel targeted drug delivery strategy is required.

According to Vasir and Labhasetwar (2005), targeted drug delivery focuses on methods that deliver a small dose of the drug to the body's malignant cells over an extended period. This preserves the drug levels in the tissue and plasma without affecting healthy or non-cancerous tissues. Thus, a focused distribution method minimizes complications and boosts efficiency. As a result, targeting systems ensure that enough nanoparticles reach the tumour cells. Utilizing a nano-based approach, drugs can be delivered to specific regions either passively or actively.

5.3.1 Passive Targeting Strategy

Since apoptosis in cancerous cells is halted, the blood vessels creating large and leaky blood vessels around the angiogenesis-related cells begin to suck nutrients abnormally. Due to underpinning membrane defects and reduced pericyte numbers of endothelial cells, leaky blood vessels are generated. As a result, molecules passing through the vascular wall and the interstitial tumour cells have increased permeability. Leaky endothelial cells include pores that range in size from 100 to 780 nm. Moreover, nanoparticles smaller than that will swiftly pass through these openings.

As a consequence, nanoparticles may be distributed across neoplastic cells. Nanoparticles can be aimed at the particular capillary endothelium region, localized in one organ, and perforated through passive propagation or convection in tumour cells (Ooko et al. 2017). Lymphatic drainage insufficiency promotes diffusion. A collagen network and gel-like fluid are developed in the tumour interstitium. Drugs penetrating the interstitial region can also have prolonged tumour interstitial retention periods. This function is called the enhanced permeability and retention effect and promotes the accumulation of drugs. By improving permeability and retention, nanoparticles will quickly aggregate and deliver the drugs.

Instead, it depends on the disease; thus, the drug-loaded nano delivery system should ideally gather at the disease site to prevent non-specific dissemination. When polymer-drug conjugates are administered intravenously, it was found that the well-known effect of enhanced permeability and retention could be combined with the ability to achieve 10–100 times higher tumour concentrations in solid tumours when compared to free drug administration. As a result, the permeability and retention of the infected tissues will increase the tumour concentration by 70 times, which will help even macromolecules aggregate. This leaky vasculature found in pathologically affected cancerous tissues is the biggest benefit in the management of cancer with advanced non-solution-dependent therapies.

Due to the rapid vascularisation, this leaky and deficient architecture is a key element in enhancing the ever-growing malignant tumours and weak lymphatic drainage, which enables enhanced permeability and retention effect. These particles' form, scale, and surface area depend on many important factors, such as circulation period, targeting, and ability to bypass barriers. The upper limit for the easy diffusion of nanoparticles is usually around 400 nm. A selection of optimum sizes of 40–200 nm guarantees longer circulating time, improved tumour mass aggregation, and decreased renal clearing. Similar to particle size, particle structure largely determines the direction by which nanoparticles may be ingested in the tumour (Ferlay et al. 2015). The spherical particles are easily internalized only because of their symmetry. It has conventionally been found that a nanoparticle shall have a diameter of at least 10 nm to prevent clearance by renal filtration by the first-pass metabolism. Multiple considerations decide the acceptable size of the nano part.

Passive targeting is of utmost significance because of diffusion-mediated transport through the tumour.

Dreher and associates have shown that particles with a diameter of hundreds of nanometers have a greater capacity to passively accumulate in tumour tissues.

With dextran as a macromolecule model, the permeability and level of penetration of medication movement are severely decreased due to a rise in molecular weight from 3.3 kDa to 2 MDa (Dreher et al. 2006). Larger molecules can aggregate superficially to the tumour's vascular surface. However, smaller molecules may considerably more extensively infiltrate the tumour interstitium and distribute more evenly.

Non-spherical particle design can significantly increase the particles' in-vivo circulation time to cover their biodiversity profile (Qin et al. 2017). The cellular absorption mechanism has a strong effect on particulate size: either phagocytosis, macropinocytosis, caveolar-mediated endocytosis or clathrin-mediated endocytosis are mediated via the cell's internalization phase, which is one of the key determinants for the internalization-driven nanoparticles.

Surface characteristics often play an important role in internalizing these nanoparticles in cells. The surface may be changed relative to the polymer structure such that certain particles may have additional hydrophobicity or hydrophilicity. Passive targeting, however, may be controlled by changing the surface dimensions of nanoparticles, either in shape or size or by surface modifications. One of the key disadvantages of passive targeting is that the normal or healthy tissues cannot be separated from the diseased ones, like in a chemotherapy regimen (Rizvi and Saleh 2018). From the brief discussion above about the passive targeting strategy, we can summarise the whole concept into two categories-

5.3.1.1 Enhanced Permeability and Retention Effect

Passive targeting of nanoparticles is due to the uniqueness of solid tumours (i.e., leaky vasculature and defective lymphatic drainage) that allow nanoparticles to accumulate in the neoplastic tissue. This phenomenon is called the enhanced permeability and retention effect (Koo et al. 2005; Torchilin 2011; Kudgus et al. 2015).

To achieve an unequal distribution of nanoparticles by concentrating more on tumours, the improved permeability and retention effect is used for "passive targeting" (Maeda et al. 2009). The size and physical properties of the nanoparticles must be controlled to avoid reticuloendothelial system absorption as a prerequisite for such a passive targeting mechanism to function.

PEGylated nanoparticles enhance cellular uptake, leading to high passive absorption due to the enhanced permeability and retention effects (Upponi and Torchilin 2014).

5.3.1.2 Direct Local Delivery

The passive targeted strategy is accompanied by the direct local delivery of anticancer drugs to tumour cells. Direct administration can be highly invasive since it involves parenteral or other surgical techniques. This technique is almost difficult to use for a few tumours that are not easy to reach, like lung cancers (Wei and Zhao 2014).

5.3.2 Active Targeting Strategy

The initial task in active targeting is to identify an ideal target agent for delivering nanoparticle systems selectively and effectively to cancerous tissues, thus preventing any toxicity in the process. Such techniques often depend on the ability of the target agents or ligands to bind with a strong affinity to cause receptor endocytosis at the tumour cell surface (Sharaf et al. 2014). Active targeting combines anticancer drugs with receptor-specific ligands (Bazak et al. 2015). Numerous targeting ligands have been used to selectively bind nanoparticles, including antibodies, antibody fragments, aptamers, peptides, entire proteins (e.g. transferrin), as well as various receptor ligands (e.g., folic acid) (Byrne et al. 2008; Li et al. 2016; Salahpour Anarjan 2019). The technique often focuses on complex interactions such as aptamer-receptor, ligand-receptor, and antibody-antigen (Emami and Ansarypour 2019).

Nanoparticles carrying the chemotherapeutic agents have been formulated to associate specifically with over-expressed receptors on the surface of cancer cells for successful targeting. The nanoparticles are conjugated with suitable ligands into modified nanostructures to target cancer cells. Crafted nanoparticles target cancer cells by associating ligands and receptors or detecting antibody antigens (Siegel et al. 2017).

Cancer cells have certain specific characteristics that distinguish them from healthy cells. On the surface, the receptors are over-expressed and render a distinctive function. Adjusting complementary ligands to the nanoparticle's surface only allows them to target cancerous cells. When they are bound to the receptors, receptor-mediated endocytosis or phagocytosis by the cell happens easily, and the entrapped chemotherapeutic agent gets internalized. A plethora of studies have been conducted to analyse these associations of the ligand-receptor and to use them for therapeutic purposes (Siegel et al. 2017).

To increase the precision of the distribution process, active targeting makes use of interactions between ligands with receptors or other molecular moieties. Eventually, it reduces the drug's non-specific interactions with distant tissues. The overexpression of certain receptors, such as folate, on the surface of the tumour cell, promotes an active targeting process.

Nanosystems, including polymeric micelles, can be modified to provide more precision with surface chemistry. Conventionally, targeted nanocarriers are more efficient at their distribution site and often reduce possible unwanted toxicities over their non-targeted equivalent. Several cancer forms, including ovarian carcinomas, osteosarcomas, and non-Hodgkin lymphomas, over-express the folate receptor.

Nanoparticles conjugated to folate receptors are also more prone to internalisation through folate receptors. A glutathione-responsive prodrug delivery system was developed by conjugating folic acid with camptothecin via disulfide bonds. The glutathione triggered the release of the drug up to 82% in 60 hours, from 12% in the physiological environment. The presence of high levels of glutathione inside tumour cells triggered the specificity and enhanced cytotoxicity in folic acid receptor-positive human epithelial carcinoma cells (KB tumour cells) than in folic acid-negative lung cancer cells (A549) (Xu et al. 2017). Liu et al. (2014) designed a folic acid-polymer-drug conjugate of camptothecin with amphiphilic poly(ethylene glycol)-b-poly(ε-caprolactone) copolymers. The nanomicelle was assessed for receptor-mediated uptake, cytotoxicity, drug release, and cell apoptosis in breast cancer cells with over-expressed folic acid receptors. The folic acid decorated polymer-drug conjugate didn't show premature release during circulation and showcased enhanced cellular internalization with intracellular drug release.

Aptamers can also be used along with nanoparticles for targeting. Aptamers are short ligand-binding oligonucleotides of DNA or RNA. This ligand can be screened and tested toward potential goals through numerous high-capacity processes, such as the systematic evolution of ligands via exponential amplification. Alibolandi et al. (2017) designed a PEGylated polyamidoamine dendrimer loaded with camptothecin, then functionalized with AS1411 anti-nucleolin aptamers targeting nucleolin-positive receptors in colorectal cancer cells. The drug exhibited a sustained release for four days. The *in-vitro* cytotoxicity study revealed higher antiproliferation activity in nucleolin-positive colorectal cancer cells (C26 and HT29) than in nucleolin-negative cell lines (CHO). In contrast, the *in-vivo* study in C26 tumour-bearing mice (BALB/C) revealed that the aptamer-functionalized PEGylated camptothecin dendrimers showed enhanced survival rate and antitumour activity.

Similarly, Jabbari et al. (2022) designed a star-like polymeric micelle using a diblock polymer of β-cyclodextrin-{poly(ε-caprolactone)-poly(2-aminoethylmethacrylate)}$_{21}$, decorated with nucleolin-responsive aptamer (AS1411) followed by loading of camptothecin. The drug release study showed 50% drug release in 72 hours in an acidic tumour microenvironment. The *in-vitro* cytotoxicity study by (3-[4,5-dimethylthiazol-2-yl]-2,5-diphenyl tetrazolium bromide) assay showed a significant cell death of breast cancer cells (4T1 and MCF-7). The flow cytometry analysis showcased an enhanced internalization of the camptothecin complex into the targeted cells with improved antitumour activity.

In the case of the delivery of numerous hydrophobic chemotherapeutic compounds, peptides have often demonstrated considerable promise. These peptides are very economical and simple to synthesize. The arginyl-glycyl-aspartic acid sequence is strongly affixed to the αvβ3 integrins, which are highly expressed in tumour cells. Koh et al. (2019) conjugated cyclic arginyl-glycyl-aspartic acid peptide with carboxylic acid-functionalized carbon nanotubes and encapsulated camptothecin. The peptide-conjugated camptothecin was delivered to αvβ3-expressing epithelial cancer cells (A375). Apoptotic gene expression study revealed the increased expressions of cleaved caspase-3, Bax, and nuclear factor kappa-light-chain-enhancer of activated B cells in cell lines. Singh et al. (2023) designed a nanoformulation conjugating camptothecin with cyclic arginyl-glycyl-aspartic acid and targeted to colon cancer cells. The peptide-conjugated camptothecin suppressed the progression of tumour growth, with high distribution and a potential increase in anticancer activity compared to free camptothecin.

Antibodies are target-specific agents used in nuclear medicine for diagnostic and therapeutic purposes. Antibodies were one of the first nano-vehicle agents to target particular cell lines based on the antigens on their surface (Alley et al. 2010). Lyski et al. (2021) reported the conjugation of 7-aminomethyl-10,11-methylenedioxy camptothecin attached to a hydrophilic protease-cleavable valine–lysine–glycine tripeptide linker. Polyethylene glycol was used as a stretcher, which improved the properties of the drug with the linker, promoting an increased drug loading to the antibody-drug conjugate and reduced aggregation. The camptothecin-antibody conjugate with eight drug-linkers/monoclonal antibodies exhibited enhanced serum stability. The antibody-camptothecin conjugate showcased a strong anticancer activity in in-vitro and mouse xenograft models, causing tumour growth suppression at a very low dose (≤3 mg/kg, single administration).

5.4 CONCLUSION

Since cancer is usually fatal and is spreading aggressively around the world, there is an urgent need for novel treatments. Medicinal plants have long been the focus of efforts to discover contemporary cures for human ailments. Another promising option for creating novel anticancer drugs may be medicinal plants. Only a small portion of the hundreds of plants that are screened for cancer undergo in vitro trials, laboratory tests, and clinical evaluation.

A vast number of researchers are working to develop more economical, biocompatible, effective, less toxic, environmentally friendlier, and greener alternative cancer treatment methods due to the rising incidence of cancer and the high expense and increased toxicity associated with conventional treatments. In such a pursuit, phytochemicals offer promising anticancer potential.

Camptothecin is the first class of antineoplastic drugs acting as topoisomerase-I inhibitor. It is widely used for the treatment of several tumours (i.e. ovarian cancer, small-cell lung cancer, and cervical cancer). Poor solubility, erratic side effects, and drug resistance limit the wide use of camptothecin. Phytomolecule-based nano delivery systems, however, can revolutionize cancer treatment in the coming years. The utility of these nano-based phytomolecules in cancer treatment has been improved by strong permeability, biocompatibility, bioavailability, and biodegradability. Further, nanoparticles-based delivery of anticancer drugs increases the selectivity of delivered drugs and retention time at targeted site.

REFERENCES

Accardo A, Salsano G, Morisco A, et al. (2012) Peptide-modified liposomes for selective targeting of bombesin receptors overexpressed by cancer cells: A potential theranostic agent. *Int J Nanomedicine* 7, 2007–2017.

Adak D, Ray P, Setua S (2024) Unlocking therapeutic precision: "*Camptotheca acuminata*, a traditional Chinese herb tailored for phytonano-cancer theranostics". *Pharmacol Res-Mod Chin Med* 11, 100447.

Alibolandi M, Taghdisi SM, Ramezani P, et al. (2017) Smart AS1411-aptamer conjugated pegylated PAMAM dendrimer for the superior delivery of camptothecin to colon adenocarcinoma in vitro and in vivo. *Int J Pharm* 519(1–2), 352–364.

Alley SC, Okeley NM, Senter PD (2010) Antibody–drug conjugates: Targeted drug delivery for cancer. *Curr Opin Chem Biol* 14, 529–537.

Andreani T, Cheng R, Elbadri K, et al. (2024) Natural compounds-based nanomedicines for cancer treatment: Future directions and challenges. *Drug Deliv Transl Res* 14(10), 2845.

Ashfaq UA, Riaz M, Yasmeen E, Yousaf MZ (2017) Recent advances in nanoparticle-based targeted drug-delivery systems against cancer and role of tumor microenvironment. *Crit Rev Ther Drug Carrier Syst* 34, 317–353.

Avemann K, Knippers R, Koller T, Sogo J (1988) Camptothecin, a specific inhibitor of type I DNA topoisomerase, induces DNA breakage at replication forks. *Mol Cell Biol* 8, 3026–3034.

Ayoob I, Hazari YM, Lone SH, et al. (2017) Phytochemical and cytotoxic evaluation of peganum harmala: Structure activity relationship studies of harmine. *Chem Select* 2, 2965–2968.

Bae YH, Park K (2011) Targeted drug delivery to tumors: Myths, reality and possibility. *J Control Release* 153, 198–205

Bazak R, Houri M, El Achy S, et al. (2015) Cancer active targeting by nanoparticles: A comprehensive review of literature. *J Cancer Res Clin Oncol* 141, 769–784.

Behera A, Padhi S (2020) Passive and active targeting strategies for the delivery of the camptothecin anticancer drug: A review. *Environ Chem Lett* 18, 1557–1567.

Beretta LG, Gatti L, Perego P, Zaffaroni N (2013) Camptothecin resistance in cancer: Insights into the molecular mechanisms of a DNA-damaging drug. *Curr Med Chem* 20, 1541–1565.

Bondarev AD, Jonsson J, Chubarev VN, Tarasov VV, Lagunas-Rangel FA, Schiöth HB (2024) Recent developments of topoisomerase inhibitors: Clinical trials, emerging indications, novel molecules and global sales. *Pharmacol Res* 209, 107431.

Burke TG, Mi Z (1994) The structural basis of camptothecin interactions with human serum albumin: Impact on drug stability. *J Med Chem* 37, 40–46.

Burke TG, Staubus AE, Mishra AK, Malak H (1992) Liposomal stabilization of camptothecin's lactone ring. *J Am Chem Soc* 114, 8318–8319

Byrne JD, Betancourt T, Brannon-Peppas L (2008) Active targeting schemes for nanoparticle systems in cancer therapeutics. *Adv Drug Deliv Rev* 60, 1615–1626.

Dreher MR, Liu W, Michelich CR, Dewhirst MW, Yuan F, Chilkoti A (2006) Tumor vascular permeability, accumulation, and penetration of macromolecular drug carriers. *J Nat Cancer Inst* 98(5), 335–344.

Emami J, Ansarypour Z (2019) Receptor targeting drug delivery strategies and prospects in the treatment of rheumatoid arthritis. *Res Pharm Sci* 14, 471.

Emerson DL, Besterman JM, Brown HH, et al. (1995) In vivo antitumor activity of two new seven-substituted water-soluble camptothecin analogues. *Cancer Res* 55, 603–609.

Ferlay J, Soerjomataram I, Dikshit R, et al. (2015) Cancer incidence and mortality worldwide: Sources, methods and major patterns in GLOBOCAN 2012. *Int J Cancer* 136, E359–E386.

Gatto MS, Johnson MP (2024) Targeted liposomal drug delivery: Overview of the current applications and challenges. *Life* 14(6), 672.

Hertzberg RP, Caranfa MJ, Hecht SM (1989) On the mechanism of topoisomerase I inhibition by camptothecin: Evidence for binding to an enzyme-DNA complex. *Biochemistry* 28, 4629–4638.

Hsiang YH, Hertzberg R, Hecht S, Liu LF (1985) Camptothecin induces protein-linked DNA breaks via mammalian DNA topoisomerase I. *J Biol Chem* 260, 14873–14878.

Hsiang YH, Lihou MG, Liu LF (1989) Arrest of replication forks by drug-stabilized topoisomerase I-DNA cleavable complexes as a mechanism of cell killing by camptothecin. *Cancer Res* 49, 5077–5082.

Jabbari A, Yaghoobi E, Azizollahi H, et al. (2022) Design and synthesis of a star-like polymeric micelle modified with AS1411 aptamer for targeted delivery of camptothecin for cancer therapy. *Int J Pharm* 611, 121346.

Ji Y, Song S, Li X, et al. (2021) Facile fabrication of nanocarriers with yolk-shell mesoporous silica nanoparticles for effective drug delivery. *J Drug Deliv Sci Technol* 63, 102531.

Koh B, Park SB, Yoon E, et al. (2019) $\alpha V\beta 3$-targeted delivery of camptothecin-encapsulated carbon nanotube-cyclic RGD in 2D and 3D cancer cell culture. *J Pharm Sci* 108(11), 3704–3712.

Koo OM, Rubinstein I, Onyuksel H (2005) Role of nanotechnology in targeted drug delivery and imaging: A concise review. *Nanomed Nanotechnol Biol Med* 1, 193–212.

Krishnan N, Iyer R, Priyadarshini S, et al. (2023) Effect of abiotic elicitor on the production of camptothecin from *Nothapodytes nimmoniana* (J. Graham) Mabb. *Biocatal Agric Biotechnol* 53, 102846.

Kudgus RA, Walden CA, McGovern RM, et al. (2015) Tuning pharmacokinetics and biodistribution of a targeted drug delivery system through incorporation of a passive targeting component. *Sci Rep* 4, 5669.

Kulkova J, Kulkoy I, Rohrbeck R, et al. (2023) Medicine of the future: How and who is going to treat us? *Futures* 146, 103097

Kumari A, Singh D, Kumar S (2017) Biotechnological interventions for harnessing podophyllotoxin from plant and fungal species: Current status, challenges, and opportunities for its commercialization. *Crit Rev Biotechnol* 37, 739–753.

Kussmann M, Abe Cunha DH, Berciano S (2023) Bioactive compounds for human and planetary health. *Front Nutr* 10, 1193848.

Lee PC, Chiou YC, Wong JM, et al. (2013) Targeting colorectal cancer cells with single-walled carbon nanotubes conjugated to anticancer agent SN-38 and EGFR antibody. *Biomaterials* 34, 8756–8765.

Li M, Zhang W, Wang B, et al. (2016) Ligand-based targeted therapy: A novel strategy for hepatocellular carcinoma. *Int J Nanomedicine* 11, 5645–5669

Liu C, Yuan J, Luo X, et al. (2014) Folate-decorated and reduction-sensitive micelles assembled from amphiphilic polymer–camptothecin conjugates for intracellular drug delivery. *Mol Pharm* 11(11), 4258–4269.

Liu L, Desai S, Li T, Mao Y, Sun M, Sim S (2000) Mechanism of action of camptothecin. *Ann NY Acad Sci* 922, 1–10.

Lyski RD, Bou LB, Lau UY, et al. (2021) Development of novel antibody–camptothecin conjugates. *Mol Cancer Ther* 20(2), 329–339.

Maeda H, Bharate GY, Daruwalla J (2009) Polymeric drugs for efficient tumor-targeted drug delivery based on EPR-effect. *Eur J Pharm Biopharm* 71, 409–419.

Manzoor AA, Lindner LH, Landon CD, et al. (2012) Overcoming limitations in nanoparticle drug delivery: Triggered, intravascular release to improve drug penetration into tumors. *Cancer Res* 72, 5566–5575.

Ooko E, Kadioglu O, Greten HJ, Efferth T (2017) Pharmacogenomic characterization and isobologram analysis of the combination of ascorbic acid and curcumin-two main metabolites of Curcuma longa-in cancer cells. *Front Pharmacol* 8, 38.

Padhi S, Behera A (2020) Nanotechnology based targeting strategies for the delivery of camptothecin. In: Saneja A, Panda A, Lichtfouse E (Eds.) *Sustainable Agriculture Reviews*, vol 44. Springer, Cham, pp. 243–272.

Padhi S, Kapoor R, Verma D, Panda AK, Iqbal, Z (2018) Formulation and optimization of topotecan nanoparticles: In vitro characterization, cytotoxicity, cellular uptake and pharmacokinetic outcomes. J *Photochem Photobiol B Biol* 183, 222–232.

Padhi S, Mirza MA, Verma D, et al. (2016) Revisiting the nanoformulation design approach for effective delivery of topotecan in its stable form: An appraisal of its in vitro behavior and tumour amelioration potential. *Drug Deliv* 23(8), 2827–2837.

Pommier Y, Redon C, Rao VA, et al. (2003) Repair of and checkpoint response to topoisomerase I-mediated DNA damage. *Mutat Res Fundam Mol Mech Mutagen* 532, 173–203.

Qin W, Huang G, Chen Z, Zhang Y (2017) Nanomaterials in targeting cancer stem cells for cancer therapy. *Front Pharmacol* 8, 1.

Rizvi SAA, Saleh AM (2018) Applications of nanoparticle systems in drug delivery technology. *Saudi Pharm J* 26, 64–70.

Ryan A, Squires S, Strutt H, Evans A, Johnson R (1994) Different fates of camptothecin-induced replication fork-associated double-strand DNA breaks in mammalian cells. *Carcinogenesis* 15, 823–828.

Salahpour Anarjan F (2019) Active targeting drug delivery nanocarriers: Ligands. *Nanostruct Nano-Objects* 19, 100370.

Seker ME, Erdogan A (2023) Phenolic and carotenoid composition of rhododendron luteum sweet and ferula communis l. subsp. communis flowers. *Front Life Sci RT* 4(1), 37–42.

Sepehri N, Rouhani H, Ghanbarpour AR, et al. (2014) Human serum albumin conjugates of 7-ethyl-10-hydroxycamptothecin (SN38) for cancer treatment. *BioMed Res Int* 2014, 963507.

Sharaf MG, Cetinel S, Heckler L, et al. (2014) Nanotechnology-based approaches for ophthalmology applications. *Asia-Pacific J Ophthalmol* 3, 172–180.

Shaw PA, Klausen M, Bradley M (2024) A dual action coumarin-camptothecin polymer for light responsive drug release and photodynamic therapy. *Polym Chem* 15, 54–58.

Siegel RL, Miller KD, Fedewa SA, et al. (2017) Colorectal cancer statistics, 2017. *CA Cancer J Clin* 67, 177–193.

Singh R, Lillard JW (2009) Nanoparticle-based targeted drug delivery. *Exp Mol Pathol* 86, 215–223.

Singh T, Kim TW, Murthy AS, et al. (2023) Tumor–homing peptide iRGD-conjugate enhances tumor accumulation of camptothecin for colon cancer therapy. *Eur J Med Chem* 265, 116050.

Torchilin V (2011) Tumor delivery of macromolecular drugs based on the EPR effect. *Adv Drug Deliv Rev* 63, 131–135.

Torre LA, Bray F, Siegel RL, et al. (2015) Global cancer statistics, 2012. *CA Cancer J Clin* 65, 87–108.

Tyner KM, Schiffman SR, Giannelis EP (2004) Nanobiohybrids as delivery vehicles for camptothecin. *J Control Release* 95, 501–514.

Upponi JR, Torchilin VP (2014) *Passive vs. Active Targeting: An Update of the EPR Role in Drug Delivery to Tumors*. Springer, Cham, pp. 3–45.

Vasir JK, Labhasetwar V (2005) Targeted drug delivery in cancer therapy. *Technol Cancer Res Treat* 4, 363–374.

Wang X, Zhuang Y, Wang Y, Jiang M, Yao L (2023) The recent developments of camptothecin and its derivatives as potential antitumor agents. *Eur J Med Chem* 260, 115710.

Wei Y, Zhao L (2014) Passive lung-targeted drug delivery systems via intravenous administration. *Pharm Dev Technol* 19, 129–136.

Wen Y, Wang Y, Liu X, et al. (2017) Camptothecin-based nanodrug delivery systems. *Cancer Biol Med* 14, 363–370.

Weng Y, Liu J, Jin S, et al. (2017) Nanotechnology-based strategies for treatment of ocular disease. Acta Pharm. Sin. B 7:281–291

Wu J, Liu L (1997) Processing of topoisomerase I cleavable complexes into DNA damage by transcription. *Nucleic Acids Reticulo Endothelial Syst* 25, 4181–4186.

Xiao B, Si X, Han MK, Viennois E, Zhang M, Merlin D (2015) Co-delivery of camptothecin and curcumin by cationic polymeric nanoparticles for synergistic colon cancer combination chemotherapy. *J Mater Chem B* 3(39), 7724–7733.

Xu Z, Hou M, Shi X, et al. (2017) Rapidly cell-penetrating and reductive milieu-responsive nanoaggregates assembled from an amphiphilic folate-camptothecin prodrug for enhanced drug delivery and controlled release. *Biomater Sci* 5(3), 444–454.

Ziomkowska B, Kruszewski S, Siuda R, Cyrankiewicz M (2006) Deactivation rate of camptothecin determined by factor analysis of steady-state fluorescence and absorption spectra. *Opt Appl* 36(1), 137–146.

6 Nanoenabled Systems for the Delivery of Berberine

Padmaja Santosh Kore, Ujwala Shivaji Desai, Anuradha Ghanshyam More, and Minal Tejram Harde
Progressive Education Society's Modern College of Pharmacy, Yamunanagar, Nigdi, India

6.1 INTRODUCTION

Phytochemicals have recently attracted much interest as a viable anticancer therapy strategy. Cancer is thought to be the world's second-greatest cause of death. The latest global burden of cancer database, 'Globocan 2018,' has been released by the International Agency for Cancer Research. According to this data, the number of new cancer cases increased to 18.1 million in 2018, and the rate of mortality increased to 9.6 million, culminating in a global cancer prevalence of 23.6 million by 2030 (Bray et al., 2020). Colorectal cancer, lung cancer, liver cancer, stomach cancer, and breast cancer are the most common forms and the leading causes of cancer-related fatalities globally. Phytomolecules and their derivatives are employed alone or in combination with other anticancer medications or as nanoformulations for better chemotherapeutic efficacy. Plant products, particularly secondary metabolites such as alkaloids, phenols, flavonols, flavonoids, terpenes, terpenoids, steroids, and other secondary metabolites, have recently emerged as one of the most promising sources of anticancer chemicals in clinical trials and therapeutic applications (Seca and Pinto, 2018).

Berberine is a quaternary benzylisoquinoline alkaloid (5,6-dihydro-9,10-dimethoxybenzo[*g*]-1,3-benzodioxolo [5,6-*a*]) that has the potential to treat a variety of diseases, including cancer (Chen et al., 2005). Berberine is a member of the Berberidaceae family, with 450–500 species in the genus Berberis, which is the primary natural source of berberine (Singh et al., 2010). The oldest evidence for the use of barberry fruit (*Berberis vulgaris*) as a blood purifier is from 650 BC, and it has been used in Ayurveda to treat various illnesses of the eye, ear, and mouth. It has demonstrated a promising wound-healing effect (Zhou et al., 2021). Berberine is also used to treat gastrointestinal issues like indigestion and dysentery, as well as a variety of uterine and vaginal diseases. It also works as an anti-obesity agent and an antidote for scorpion stings and snakebites (Dev, 2006). Berberine has been shown to have hypolipidemic, hypoglycaemic, hepatoprotective, antioxidant, anti-microbial, antitumour, and neuroprotective activities reported by several researchers in recent years (Wang et al., 2017; Tillhon et al., 2012).

Despite several advantages berberine offers, some serious limitations restrict its clinical applications. Poor aqueous solubility and absorption by the gastrointestinal tract are the most critical limitations to be solved for its efficient delivery. Its bioavailability is very low (about 5%), and is classified as a biopharmaceutical classification class-IV drug. The low bioavailability of berberine is due to its high affinity for protein binding. Hence, there is a low unbound fraction of berberine to reach target tissues. Additionally, high first-pass metabolism is another reason for its low bioavailability (Shen et al., 2016; Xue et al., 2015). Considering all the above-mentioned pitfalls, employing nanoparticulate drug delivery systems is essential to overcome the limitations and improve berberine's therapeutic benefits.

6.2 ANTICANCER MECHANISM OF ACTION OF BERBERINE

Berberine exhibits various mechanisms as an anticancer agent (Figure 6.1). The detailed anticancer mechanisms are explained in Figure 6.1.

6.2.1 Inhibition of Migration and Invasion of Tumour Cells

Tumour cells are thought to have two basic characteristics: migration and invasion. E-cadherin and N-cadherin are two well-known proteins that regulate cell migration and invasion. E-cadherin is vital for maintaining the structural integrity of epithelial cells and their shape and regulating cell-cell adhesion (Shi et al., 2017; Qi et al., 2014). Berberine downregulates the expression of N-cadherin and upregulates the expression of E-cadherin protein in human lung cancer cell lines (A549) in a time and concentration-dependent manner (Li et al., 2018). This is crucial for the prevention of invasion and metastasis. The destruction of the extracellular matrix barrier is the initial stage of tumour cell metastasis. Matrix metalloproteinase plays a key role in the metastasis of cancer cells (Hao et al., 2017). Berberine inhibits the expression of matrix metalloproteinase 2 and matrix metalloproteinase 9 in a concentration and time-dependent manner. Berberine regulates the expression

DOI: 10.1201/9781003611516-8

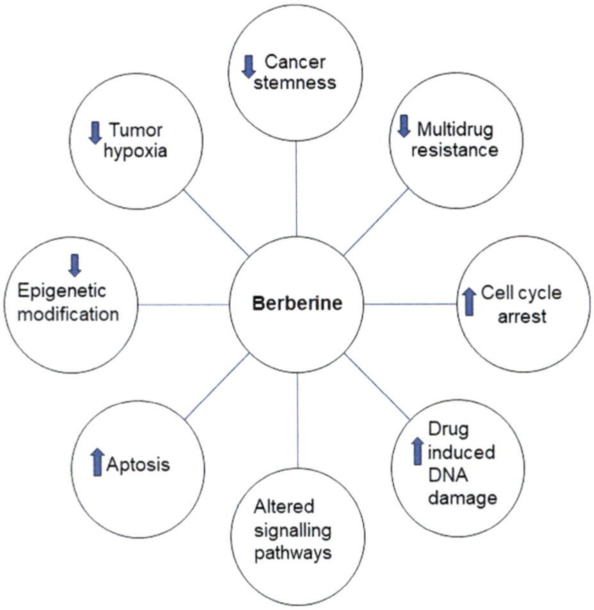

FIGURE 6.1 Different anticancer mechanisms for berberine [The anticancer mechanism of action of berberine involves several mechanisms of anticancer activities, such as decreased tumour hypoxia, multidrug resistance, and epigenetic modifications. The anticancer mechanism also involves increased apoptosis, drug-induced DNA damage, and cell cycle arrest. It also acts through the alteration of signalling pathways.

of matrix metalloproteinase by preventing signal transducer and activator of transcription 3. Berberine was an efficient inhibitor of hepatocellular carcinoma cell invasion and migration. The downregulation of different proteins like cyclooxygenase 2, nuclear factor kappa B, urokinase plasminogen activator, and matrix metalloproteinase 9 in a dose-dependent manner by berberine is the therapeutic mechanism in hepatocellular cancer cells (Sengupta et al., 2017). Berberine significantly controls matrix metalloproteinases and cadherin-mediated pathways, preventing cancer cell migration and invasion.

Jin and colleagues reported that berberine inhibited the migration, adhesion, and invasion of cancer cells by activating the adenosine monophosphate-activated protein kinase signalling pathway, which reduced the activity of extracellular signal-regulated enzymes and the expression of cyclooxygenase 2 enzyme (Jin et al., 2017a,b). Berberine suppressed the nuclear factor kappa B signalling pathway and production of interleukin-8, which inhibited endometrial cancer metastasis and colon cancer cell migration by targeting the adenosine monophosphate-activated protein kinase signalling pathway (Li et al., 2015a,b).

6.2.2 Inhibition of Metastasis

The main causes of cancer mortality are uncontrolled cellular proliferation and tumour cell invasion, and metastasis. Berberine targets the ephrin-B2 protein and inhibits the expression of matrix metalloproteinase 2 and matrix metalloproteinase 9 in breast cancer, which prevents cell proliferation and metastasis (Ma et al., 2017). Berberine also suppresses epithelial-mesenchymal transition in melanoma by upregulating retinoic acid receptor-ß and downregulating retinoic acid receptor-α, which stops the phosphatidylinositol 3-kinase/protein kinase B signalling pathway (Naveen et al., 2016). Berberine also reduces the expression of the transcription factor Snail-1, which prevents tumour metastasis (Chu et al., 2014). Finally, berberine alters the expression of tumour-related signalling pathways and proteins and prevents metastasis.

The angiogenic factor, the vascular endothelial growth factor, is secreted by tumour cells and induces tumour neovascularization by increasing vascular endothelial cells' mitogenic and survival capabilities. Angiogenesis is well established to have a crucial part in tumour growth, as progression and metastasis are two of the most important factors in solid tumour growth. According to some studies, berberine can lower the levels of inflammatory markers (interleukin-1β, interleukin-6, and tumour necrosis factor) and granulocyte-macrophage colony-stimulating factor in the serum of tumour-inoculated rodents, as well as inhibit the elevation of inflammatory cytokine tumour necrosis factor-α and nitric oxide, inflammatory mediators involved in angiogenesis (Ma et al., 2008).

Berberine in the prevention of chemotherapy-induced ovarian cancer metastases has recently been investigated. Berberine can reverse epithelial-mesenchymal transition and migration and downregulate cancer stem cell-like traits by blocking the Gli1/BMI1 signalling pathway (Zhao et al., 2021). During chemotherapy for ovarian cancer, berberine has been used as an adjuvant therapeutic agent in combination with other chemotherapeutics to prevent metastases.

6.2.3 Inhibition of Tumour Cell Proliferation (Apoptosis and Autophagy)

In cancer research, apoptosis, a sort of planned cell death, also believed to be a well-regulated process, has generated much interest. The apoptosis process progresses through inner mitochondrial signalling channels and extracellular death receptor pathways, activating multiple caspase cascades, cell cycle arrest, and cell death (Pistritto et al., 2016). The cell cycle development and proliferation operations are divided into three phases: growth phases (G1/G2), S-phase (DNA synthesis), and M-phase (mitosis) (Kar, 2016). Catalytic cyclin-dependent kinases govern several cell cycle phases by interacting with regulatory cyclin subunits, and cyclin-dependent kinase inhibitors regulate cyclin-dependent kinases. Inhibiting complex cyclin-dependent kinases/cyclin phosphorylation to cause cell cycle arrest during the G1/S, S/G2, and G2/M phase transitions results in apoptotic cell death and delay in the initiation, development, and progression of the malignant phase (Hema et al., 2018).

Berberine causes apoptosis by decreasing BCL-2 expression through caspase-3 activation and increasing the production of caspase-8 and caspase-9 (Okubo et al., 2017). Berberine activates caspases by boosting cytochrome-C levels and increasing reactive oxygen species generation (Yang and Huang, 2013; Li et al., 2015a,b). Mitochondria play a crucial role in apoptosis regulation. External stimulation increases the mitochondrial membrane's permeability, leading to the caspase cascade and apoptosis. (Jeong and Seol, 2008). Berberine increased phosphorylation of p53 and c-Jun N-terminal protein kinase/p38 mitogen-activated protein kinase activation, allowing two apoptotic proteins, 'bim' and 'bax', to enter mitochondria more easily (Park et al., 2016). Furthermore, several studies have found that berberine increases cellular death by increasing the rate of acetylation of forkhead box protein O1 and forkhead box protein O3a and increasing the expression of bim and bax (Shukla et al., 2014, 2016)

In glioblastoma, berberine inhibits the adenosine monophosphate-activated protein kinase/mammalian target of rapamycin/ULK1 pathway, causing autophagy to be activated (Wang et al., 2016a,b) (Figure 6.2). Berberine stimulates autophagy in breast cancer cells by modulating c-Jun N-terminal protein kinase phosphorylation, causing the BCL-2/beclin-1 complex to dissociate (Wang et al., 2016a,b). Berberine activates autophagy in hepatoma cells by promoting beclin-1 release from the BCL-2/beclin-1 complex. (Wang et al., 2010). It promotes autophagy by indirectly inhibiting sterol regulatory element-binding protein 1 expression through adenosine monophosphate-mediated protein kinase activation, which activates the ULK1/mechanistic target of rapamycin complex 1 signalling pathway (Mihaylova and Shaw, 2011). Induction of cytostatic autophagy *in vitro* and *in vivo* was used to investigate the role of berberine in gastric cancer (Zhang et al., 2020). Berberine was potent against gastric cancer cells (BGC-823). The punctuate distribution of mammalian expression vectors containing light chain 3 genes fused with green fluorescent proteins and monodansyl cadaverine staining, as well as the promotion of a lipid-modified form of light chain 3 protein-II, phospho-ULK1 antibody, beclin-1, and protein (p62) degradation, all showed that berberine treatment promoted cellular autophagy (La et al., 2017). The improved efficacy of berberine-exposed gastric carcinoma cells is due to autophagy by chloroquine, 3-methyladenine, beclin-1, small interfering RNA, and bafilomycin A1, demonstrating the anticancer impact of berberine.

Berberine has also been shown to play an important function in inhibiting the proliferation of human prostate cancer (PC3) cells (Huang et al., 2015). Recently, it was found that berberine can successfully inhibit renal cell cancer cell multiplication. A twofold staining approach was used to detect the total apoptotic rate of renal carcinoma cells. It was reported that treatment of renal cancer cell lines (A498 and 786-0) with different dose levels of berberine improved the apoptosis as the drug concentration increased (Wang et al., 2015a; Liu et al., 2017a, b, 2018). The primary mechanism of action of berberine is the upregulation of proapoptotic genes and downregulation of apoptosis-inhibiting genes.

6.2.4 Berberine Arrests Tumour Cell Cycle

Inhibition of cyclin D1 and cyclin E1 expression by berberine increases the G1 phase of the cell cycle in lung cancer cells (A549) (Xiao et al., 2018). In colorectal cancer cells, berberine inhibits the expression of cyclin-dependent kinase 4 and cyclin-dependent kinase 1 modulation when combined with a chaperone protein (Hsp90) inhibitor (Su et al., 2015). Berberine modulates the activation of the mitogen-activated protein kinases (p38) and phosphatidylinositol 3-kinase/protein kinase B signalling pathways, resulting in G2/M phase arrest in chondrosarcoma cells (HBT-94), upregulating the expression of tumour protein (p53) and cyclin-dependent kinase inhibitor protein (p21) (Eo et al., 2014).

Berberine stops the progress of human osteosarcoma (U20S) cells in the G1 phase at low concentration, causing DNA double-strand breaks that activate the p53–p21 pathway (Yang et al., 2015; Li et al., 2017). Berberine significantly inhibits ovarian cancer cells (HEY and SKOV3 cells) in various studies (Liu et al., 2019a,b). Berberine has also

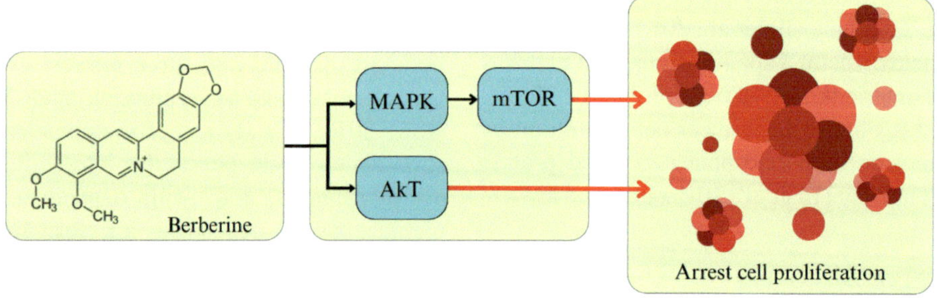

FIGURE 6.2 Role of berberine in inhibition of tumour cell proliferation [Berberine inhibits the mammalian target of rapamycin, protein kinase B, and mitogen-activated protein kinase (extracellular-signal-regulated kinase, c-Jun N-terminal kinase, and p38) pathways, promoting autophagy, according to mechanistic studies. Berberine-induced autophagy and cytotoxicity are increased when these mechanisms are inhibited. Interestingly, the mitogen-activated protein kinase, but not protein kinase B, suppressed the mammalian target of rapamycin.

been shown to have a dose-dependent inhibitory effect on gastric cancer cell lines (MGC803) (Zhang et al., 2015). According to current research, berberine inhibits hepatoma cell proliferation. It induces cell death by arresting the human hepatic (HepG2) cell cycle in the G1 phase via overexpression of B-cell translocation gene 2 and downregulation of cyclin-dependent kinase 1.

6.2.5 Epigenetic Effects of Berberine on Tumours

Berberine exhibits various dynamic effects, such as induction of high apoptosis and cell cycle arrest, but lower production of reactive oxygen species in chimeric antigen receptor overexpressed constitutive androstane receptor in hepatocellular carcinoma cells (HepG2) (Zhang et al., 2016a,b). Also, berberine inhibited the overexpression of chimeric antigen receptors and their associated target genes, cytochrome P450 3A4 and cytochrome P450 2B6 (Figure 6.3). Furthermore, berberine enhances DNA methylation throughout the genome. Still, the rate is lowered in the promoter regions of CpG sites of the target genes cytochrome P450 3A4 and cytochrome P450 2B6 when a chimeric antigen receptor is present (Zhang et al., 2016a,b). These data showed the strong antiproliferative efficacy of berberine. As a result of its different pharmacological features in clinical perspectives, berberine could be a good choice in anticancer adjuvant chemotherapy.

6.2.6 Regulation of Tumour Microenvironment

Tumour cells are well known to secrete cytokines that are key factors in changing the adjacent tumour microenvironment, promoting cancer cell metastasis and proliferation (Cheng et al., 2015; Delloye-Bourgeois et al., 2017; De Palma et al., 2017). In osteosarcoma, berberine changed the microenvironment by downregulating the caspase-1/interleukin-1β signalling pathway, resulting in cell apoptosis (Jin et al., 2017a,b). Berberine regulates the cancer microenvironment by influencing immune molecules and inflammatory responses (Casey et al., 2015).

6.2.7 Berberine in Tumour Immunotherapy

Immunotherapy approaches for managing tumours have gained huge recognition from many researchers. Berberine is believed to exert positive effects on tumour immunotherapy. By acting as an antagonist at dopamine 1 and dopamine 2-like receptors, berberine inhibits the release of proteins (tumour necrosis factor-α, interferon-γ, interleukin-1β, and interleukin-6) from lipopolysaccharide-stimulated lymphocytes (Kawano et al., 2015). Moreover, berberine prevents the phosphorylation of signal transducer and activator of transcription-1 protein, resulting in the downregulation of expression of interferon-γ-induced Indolamine-2,3-dioxygenase 1 enzyme (Wang et al., 2018a,b). These results suggest that berberine is a potent remedial contender for immunotherapy against the tumour.

6.2.8 Effect of Berberine on MicroRNA

MicroRNA belongs to a category of single-stranded RNA, with its main function being the regulation of post-transcriptional events (Bhaskaran and Mohan, 2014). Berberine is thought to inhibit cancer metastasis and invasion by reacting with microRNAs in multiple myeloma cells and endometrial cancer cells (Ayati et al., 2017; Wang et al., 2018a,b; Yin et al., 2018). Berberine is proven to mediate the transcriptional activation of cyclin-dependent kinase 1

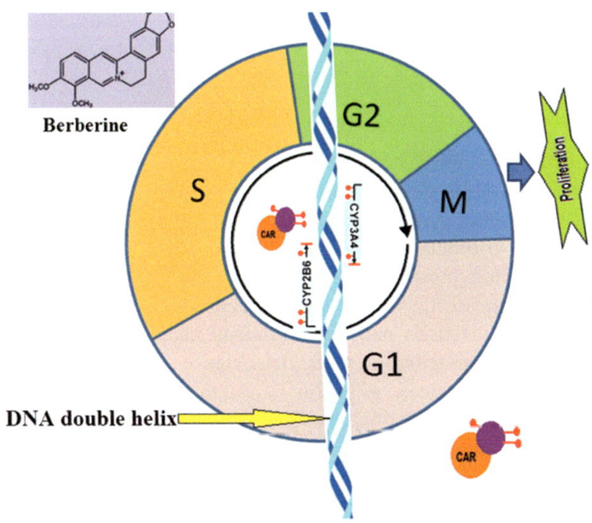

FIGURE 6.3 Epigenetic modifications of tumour: Overexpression of chimeric antigen receptor, as well as its related target genes cytochrome P450 3A4 and cytochrome P450 2B6, were found to be inhibited by berberine. Furthermore, berberine enhances DNA methylation throughout the genome. Still, the rate is lowered in the promoter regions of CpG sites of the genes mentioned above when a chimeric antigen receptor is present. The antiproliferative efficacy of berberine could be related to the chimeric antigen receptor metabolic pathway's unique epigenetic-altering mechanism.

inhibitor (p21) and nuclear cell cycle-regulated protein i.e. growth arrest and DNA-damage-inducible protein 45 alpha by upregulating the expression of the gene (microRNA-23a) in hepatoma cells (Wang et al., 2014a). In colon cancer cells, berberine inhibits the expression of E-cadherin and downregulates the tumour suppressor gene (microRNA-429) to demonstrate its anticancer properties (Liu et al., 2016). Berberine inhibits the invasion of cancer and metastasis by regulating microRNA expression (Figure 6.4).

6.3 NANOENABLED TARGETING STRATEGIES FOR BERBERINE

Nanoenabled targeting is superior to conventional dosage forms since it prevents damage to healthy cells from cytotoxic compounds, reduces the adverse effects due to high doses, and battles against drug-resistant cancerous cells (Padhi et al., 2016; Kundu et al., 2022). Nanoenabled drug targeting allows the delivery of phytochemicals to the

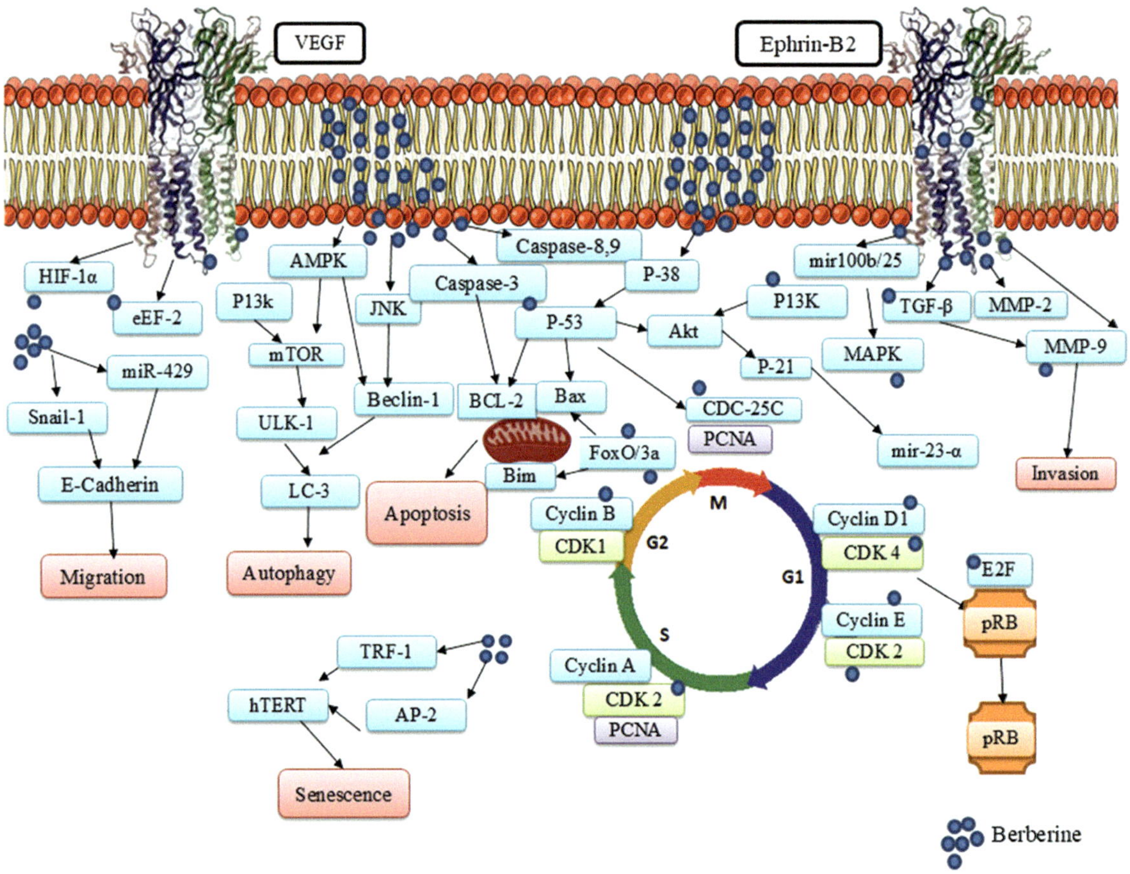

FIGURE 6.4 Intracellular mechanisms of action of berberine [Berberine suppressed vascular endothelial growth factor mRNA expression in tumour cells, promoted eukaryotic elongation factor 2 phosphorylation, down-regulated hypoxia-inducible factor activity, resulting in lower vascular endothelial growth factor expression, and inhibited the phosphoinositide 3-kinases/protein kinase B signalling pathway. Berberine also reduced the expression of the transcription factor snail-1, which hindered tumour metastasis. Berberine decreased E-Cadherin expression by downregulating miR-429. Berberine suppresses cancer invasion and metastasis by regulating or interfering with the expression of microRNAs. Berberine activates autophagy by targeting the activated protein kinase/mammalian target of rapamycin/ULK1 pathway. Berberine caused autophagic death in breast cancer by regulating c-Jun N-terminal Kinase phosphorylation and contributing to the dissociation of the BCL-2/beclin-1 complex. Berberine triggered autophagy in hepatoma cells by encouraging the release of beclin-1 from the BCL-2/beclin-1 complex. Berberine causes apoptosis by various pathways, including increased phosphorylation of p53 via Jun N-terminal kinase/p38 activation, which promotes the entry of the apoptotic proteins bax and bim into mitochondria. Berberine-induced apoptosis via increasing foxo1/3a acetylation and increasing bim and bax expression. Berberine stops the cell cycle by inhibiting the overexpression of cyclins like cyclin-D1, cyclin-B, cyclin-A, and cyclin-E. By regulating the activation of the Phosphoinositide 3-kinases/Protein kinase-B and p38 signalling pathways, berberine increased the expression of p53 and p21, resulting in G2/M phase arrest. By targeting ephrin-B2 and decreasing the production of metalloproteinases-2 and metalloproteinases-9, berberine reduced cell proliferation and metastasis. In addition, berberine reduced metalloproteinases-2 and metalloproteinases-9 production by downregulating transforming growth factor beta 1. Berberine mediated the transcriptional activation of p21 by upregulating the expression of miR-23a, leading to inhibition of cancer invasion and metastasis.]

target site in active form and accurate dose due to the specific cell uptake mechanisms (Padhi et al., 2021, 2022a). However, cytosolic incorporation of drug molecules does not ensure interaction with its target; hence, nanoparticle design and optimization are necessary (Attia et al., 2019). Two approaches are used to deliver anticancer drugs for targeted delivery at the site of tumour cells (Behera and Padhi, 2020). Passive targeting delivers the drug through the leaky vasculature of cancer cells by enhanced permeation and retention effect. In contrast, active targeting delivers anticancer drugs by interacting with drug-conjugated ligands with overexpressed receptors on tumour cells (Padhi et al., 2018). The active targeting of anticancer drugs delivers the drug more efficiently than passive targeting (Padhi and Behera, 2020).

6.3.1 Passive Targeting Strategy for Berberine

In cancerous patients, the permeability of blood vessel endothelium is higher than that of healthy individuals (Torchilin, 2011). The rapidly growing tumours generate new blood vessels or destroy the existing ones. The newly developed blood vessels are leaky, and thus, they allow selective permeation of macromolecules (greater than 40 kDa) and nano drugs to tumour stroma. In addition, the tumour lacks normal lymphatic drainage, which enhances nanoparticle retention. This property is absent in small drug molecules with short circulation time and speedy flush from tumours. Hence, small drug molecules incorporated in nanosized drug carriers elevate their pharmacokinetic parameters (prolonged systemic circulation), provide specificity to the tumour, and reduce adverse effects (Padhi et al., 2022b). A passive targeting strategy is executed by enhanced permeability and retention effect (Hu et al., 2010; Xin et al., 2017; Behera and Padhi, 2022). Passive targeting of phytomolecules depends on carrier molecule characteristics (size, shape, surface charge, circulation time) and tumour morphology (vascularity index, size, type, leakiness, stage) and does not rely on a ligand for selective tissue or organ binding (Maeda, 2001; Fang et al., 2011; Lai et al., 2018; Padhi and Behera, 2022).

6.3.1.1 Mesoporous Silica Nanoparticles

Silica is called mesoporous as it contains small internal pores with uniform size and shape where the drug can be incorporated, thus protecting it from degradation and early release. In addition, mesoporous nanoparticles can be customized for lipophilic or lipophobic drug encapsulation and enhance cellular internalization (Sharma et al., 2018). Mesoporous nanoparticles increase the wetting property and the high dispersibility of encapsulated drugs, leading to increased solubility and bioavailability of poorly soluble drugs (Zhou et al., 2018).

Yue et al. (2018) formulated tailored mesoporous organosilica nanoparticles containing disulfide linkage to deliver berberine into the liver cancer cell line (HepG2). These mesoporous nanoparticles exhibited enhanced drug-loading ability, superior biocompatibility, and monodisperse morphology. When the nanoparticles were exposed to tumour glutathione, the disulfide bond broke, releasing berberine to elicit the antitumour activity. They were found to be easily biodegradable due to the breakage of the disulfide bonds, leading to berberine release. This bio-simulated nanoenabled targeted drug delivery exhibited maximum cytotoxicity compared to free berberine and mesoporous silica nanoparticles alone.

6.3.1.2 Cyclodextrin Nanoparticles

Cyclodextrin comprises internal lipophilic cavities surrounded by a hydrophilic outer region. Cyclodextrin can be used in cancer therapy due to its inherent features, such as easy chemical modification and lower immunogenicity (Mellet et al., 2011). Popiołek and colleagues (2019) formulated a nanoenabled drug delivery system using γ-cyclodextrin to deliver berberine. The berberine-loaded cyclodextrin nanoparticles were found to be mucoadhesive and resistant to enzymatic degradation and hence acted as an efficient vector for oral drug delivery. When the berberine–cyclodextrin–propylene diamine complex was exposed to the breast cancer cell line (4T1), the cytotoxicity was doubled compared to free berberine due to the accumulation of the complex in lysosomes. The toxicity was not detected in normal cells, as the complex accumulated in the cytoplasm of normal cells.

6.3.1.3 Solid Lipid Nanoparticle

Solid lipid nanoparticles are biocompatible and belong to the first generation of lipid nanoparticles (Riaz et al., 2018). Solid lipid nanoparticles have special characteristics like controlled drug release capability, incorporation of water-insoluble drugs, and a large surface-to-mass ratio (Ghaffari and Dolatabadi, 2019). Solid lipid nanoparticles offer advantages in cancer therapy through prolonged and sustained drug release, incorporating a broad range of drugs for sustained release in the systemic circulation (Pattni et al., 2015).

Wang and colleagues prepared berberine hydrochloride-loaded solid lipid nanoparticles and checked its antitumour efficacy against liver, breast, and lung cancer cell lines (HepG2, MCF-7, and A549), respectively. It was observed that cell proliferation was inhibited by berberine-loaded solid lipid nanoparticles (Wang et al., 2014b).

Yu and colleagues conjugated berberine with phospholipid, namely soybean phosphatidylcholine, to form a complex (which offered drug-loading capacity and controlled release characteristics) and loaded this complex in self-assembled polymeric lipid hybrid nanoparticles (polyethylene glycol – lipid – polylactic-*co*-glycolic acid nanoparticles) to increase the oral bioavailability of berberine. Due to the lipid environment, the solubility of berberine increased, leading to superior biological stability (Yu et al., 2017).

6.3.1.4 Nanolipid Carrier

Nanolipid carriers contain both liquid and solid lipids and surfactants. It acts as a reservoir for lipophilic drugs. The second generation of lipid carriers includes nano lipid carriers (Dolatabadi and Omidi, 2016). Nanolipid carriers have better physical stability, high water content, and low drug release than solid lipid nanoparticles (Beloqui et al., 2016).

The anti-hepatocarcinoma effect of berberine was studied by Wang and colleagues, who developed nano lipid carriers loaded with berberine and evaluated it both *in vitro* and *in vivo*. The proliferation of hepatic cancer cell line (H22) cells was curbed, and higher antitumour activity was observed for nano lipid-loaded berberine (Wang et al., 2015b). Meng and colleagues formulated berberine-loaded nano lipid carriers and evaluated them against hepatocarcinoma. Methylthiazol tetrazolium assay results indicated that berberine-loaded nano lipid carriers inhibited the rapid growth of human liver cell lines (HepG2, Huh7) and oesophageal cancer cell lines (EC9706). There was an enhancement in the therapeutic efficiency of berberine due to nano lipid carriers because of high bioavailability (Meng et al., 2016). Yin and colleagues prepared nanostructured lipid carriers coated with selenium to increase the oral bioavailability of berberine. The size of the selenium-coated nanostructured lipid carrier was 160 nm, with high encapsulation efficacy and prolonged release of berberine compared to berberine nano lipid carriers (Yin et al., 2017).

6.3.1.5 Micelles

Micelles comprise lipophilic polymers suitable for drug delivery systems for poor soluble compounds. Shen et al. (2016) prepared micelles by mixing 1,2-distearoyl-sn-glycero-3-phosphoethanolamine–*N*-[methoxy (polyethylene glycol)-2000] with d-α-tocopheryl polyethylene glycol 1000 succinate (polyethylene glycol-succinate ester of vitamin E) in a 3:1 molar ratio and tested them against tumours. The stability and solubility of berberine were remarkably increased by micelles with low toxicity and controlled release of berberine. Apoptosis was induced in cancerous cells, and the cytotoxic effect was observed against human prostate carcinomas due to the ability of tocopheryl polyethylene glycol succinate to combat drug resistance by inhibiting *p*-glycoprotein efflux pumps.

6.3.1.6 Liposomes

Liposomes are bilayer vesicles of phospholipids that can incorporate hydrophilic and lipophilic drugs. The structure of liposomes is similar to that of the cell membrane; hence, it can easily merge with the cell membrane and deliver drugs to cells (Díaz and Vivas-Mejja, 2013; Behera et al., 2023). Nanoliposomes have been used to deliver peptides, proteins, nucleic acids, and small molecules. Nanoliposomes improve drug stability by limiting the interactions of the dosage form with biological components and reducing possible side effects (Khan et al., 2015).

Lin and colleagues encapsulated berberine in liposomal nanovesicles and evaluated the activity of the liposomes against human liver carcinoma cells (HepG2). The results of *in vitro* studies exhibited that apoptosis was induced through the caspase/mitochondria-dependent pathway and restricted the growth of hepatic cell lines (HepG2). The inhibitory potential of berberine was increased up to 2.5 times in liposome formulation. The results indicated a significant decrease in the excretion of berberine in tissues and plasma and a significant reduction in the size and weight of tumours (Lin et al., 2013). Oral delivery of berberine was improved by Nguyen and co-workers, who formulated a nanoliposome of berberine coated with chitosan. *In vitro* and *in vivo* studies indicated the stability and controlled release of chitosan-coated nanoliposomes of berberine as compared to non-coated liposomes (Nguyen et al., 2014).

Ma and colleagues fabricated two types of liposomes, one loaded with berberine and the other with paclitaxel, to determine the effect of liposomes on drug resistance in breast cancer stem cells (MCF-7). The *in vivo* and *in vitro* studies were carried out in xenografts of cancer stem cells in nude mice and human breast cancer stem cells. It was found that berberine liposomes had an increased circulation time in the body by evading the rapid clearance by the liver and reticuloendothelial cells. The small size of the liposomes favoured the permeability through cancer stem cells, which induced acute cytotoxicity injury and apoptosis. The liposomes activated the proapoptotic proteins and suppressed the anti-apoptotic protein, further activating the apoptotic enzymes in the cytosol. The efficiency of apoptosis and maximum *in vivo* anticancer activity was observed when a combination of paclitaxel and berberine liposomes was administered simultaneously; hence, berberine can be a good lead for co-therapy with paclitaxel in the prevention and relapse of breast cancer from stem cells (Ma et al., 2013).

6.3.1.7 Dendrimer

Dendrimers are heavily branched polymeric nanostructures (Choudhary et al., 2017). Gupta and colleagues (2017) designed the two berberine dendrimer formulations investigated for anticancer effects against breast cancer cell lines (MCF-7 and MDA-MB-468). Both the formulations (conjugated and encapsulated) exhibited anticancer activity. It was observed that berberine-conjugated dendrimers had a low particle size and more drug-loading capacity, showing more therapeutic efficacy.

6.3.1.8 Metal Nanoparticles

The metal nanoparticles possess characteristics like surface modification, high surface plasmon resonance, and implementation for biomedical applications (Behera et al., 2020). Pandey and colleagues used the peel of *Trapa bispinosa* to prepare gold nanoparticles (monodispersed) with specific functional groups, which acted as conjugates for attachment of folic acid and berberine hydrochloride against cervical cancer cells (HeLa). The cytotoxic effect of the complex was higher in cervical cancer cells than in normal Vero cells. Expression of folate receptors in Vero cells is known to be low and hence can justify the low toxicity of the complex towards Vero cells. The drug-loading capacity and release of berberine were higher at low pH (Pandey et al., 2013).

Dziedzic and colleagues (2016) evaluated the effect of berberine-conjugated silver nanoparticles in head and neck cancer cell lines (SCC-25). It was discovered that oral administration of silver nanoparticles alone had an antiproliferative impact on cancer treatment, and this effect was amplified for silver nanoparticles containing berberine. Berberine instigated pro-proliferative gene expression and maintained cancer cell lines' vitality (SCC-25). The electrostatic forces between berberine (positive charge) and silver nanoparticles (negative charge) were responsible for this non-synergistic phenomenon.

High drug loading can be obtained by surface modification with electrostatic charges on the surface of nanoparticles. Photothermal excitation is one of the features associated with metal nanoparticles that can track cancer growth (Sharma et al., 2018). Kim and co-workers (2018) formulated zinc oxide nanoparticles of berberine against lung cancer. The anticancer activity was investigated against alveolar cancer cell lines (A549). It was noted that berberine zinc oxide nanoparticles showed greater cytotoxicity to alveolar cancer cell lines (A549) than free berberine and zinc oxide nanoparticles. Intravenous administration of berberine-loaded zinc oxide nanoparticles did not show hepatotoxicity or hemotoxicity.

Iron oxide nanoparticles are used extensively for biomedical applications like diagnosis and targeted drug or gene therapy (Tietze et al., 2015). The excitation of iron oxide nanoparticles leads to the production of heat and reactive oxygen species, leading to the death of tumour cells (Mody et al., 2010).

Sreeja and co-workers formulated iron oxide nanoparticles loaded with berberine and sanazole (hypoxic-stimulating drug). The berberine iron oxide nanoparticles reduced the volume of the tumour. Hypoxia in tumours leads to hypoxia-inducible factor 1α activation, resulting in overexpression of responsive genes responsible for angiogenesis. Thus, berberine iron oxide nanoparticles induced apoptosis by targeting hypoxic tumours and inhibiting hypoxia-inducible factor 1α activation (Sreeja and Nair, 2018).

6.3.1.9 Carbon Dots Nano Drug Delivery System

Carbon dots possess unique characteristics like greater biocompatibility, optical characteristics, and increased optical and chemical stability (Zhu et al., 2015; Hola et al., 2014; Zhu et al., 2016; Zheng et al., 2015). Carbon dots are more optically photostable than organic dyes and show lower toxicity than quantum dots. Zhang and co-workers prepared berberine-based carbon dots and evaluated their bio-imaging and antitumour effects (Zhang et al., 2018). Thakur and colleagues (2016) prepared graphene quantum dots from cow's milk and incorporated berberine in the graphene quantum dots using cysteamine hydrochloride. The cytotoxic effects were observed against cervical and breast cancer cells (HeLa and MDA-MB-231). The complex induced apoptosis and arrested the cell cycle (G1 phase).

6.3.1.10 Polymeric Nanoparticles

Yu and co-workers (2017) conjugated berberine with phospholipid, namely soybean phosphatidylcholine, to form a complex (which offered drug-loading capacity and controlled release characteristics) and loaded the complex in self-assembled polymeric lipid hybrid nanoparticles (polyethylene glycol–lipid–poly lactic-co-glycolic acid nanoparticles) to increase the oral bioavailability of berberine. Due to the lipidic environment, the solubility of berberine increased, leading to supreme biological stability.

6.3.1.11 Others

Pund and colleagues prepared a self-nano-emulsifying drug delivery system of berberine to increase its therapeutic efficacy using excipients (Capmul MCM, polyethylene glycol-400, and acrysol K-150). The nanoformulation showed anti-angiogenic and anti-inflammatory Activity (Pund et al., 2014).

6.3.2 Active Targeting Strategy for Berberine

Active targeting includes a ligand-receptor-facilitated drug delivery system, which results in maximum drug concentration at the tumour site with greater selectivity (Faraji and Wipf, 2009; Senapati et al., 2018). Remarkably, the active targeting strategy is necessary to deliver drugs, genes, and theranostics to the tumour site, evading the healthy cells and tissues, finally upsurging therapeutic activity and restraining the side effects.

Active targeting can deliver more drugs to the target cells than free drugs or nanosystems that have drugs passively. Higher drug concentration at the targeted site due to active targeting subsequently leads to greater therapeutic efficacy. This is accomplished by decorating the surfaces of the nanocarrier with ligands specific to receptors present in the tumour cells (Wolfram and Ferrari, 2019). Nanocarriers delivering the drugs by active targeting strategy get internalized into the tumour cells through overexpressed receptor-induced endocytosis, enhancing the drug efficacy and reducing off-target adverse effects (Behera and Padhi, 2020).

6.3.2.1 Polymeric Nanoparticles

Bhatnagar and his co-workers (2018) formulated poly (lactic-co-glycolic acid) copolymer nanoparticles loaded with berberine, further surface modified with hyaluronic acid. High cytotoxicity was observed against cervical and breast cancer cell lines (HeLa, MCF7). Higher expression of cell surface receptors on breast cancer cell lines (MCF7) led to greater cellular uptake of the targeted complex. An *in-vivo* study showed increased apoptosis, raised reactive oxygen species levels, and decreased mitochondrial membrane potentials in tumour-bearing mice. Wang and co-workers (2018b) prepared folic acid-modified chitosan nanoparticles loaded with berberine and evaluated the efficacy against human nasopharyngeal carcinoma (CNE-1 cells). The nanoparticles inhibited the growth of tumours and migration of cancer cells and induced apoptosis.

Bhanumathi and co-workers designed citrate-capped silver nanoparticles of berberine conjugated with polyethylene glycol and further functionalized them with folic acid and evaluated their anticancer efficacy in breast cancer cell lines (MCF-7 and MDA-MB-231). The nanoformulation induced apoptosis by increasing the reactive oxygen species, and western blotting revealed the expression of different genes favouring the process of apoptosis. The *in vivo* study of the berberine nanoformulation showed a significant decrease in tumour progression (Bhanumathi et al., 2017).

6.3.2.2 pH-Sensitive Mesoporous Nanoparticles

The microenvironment of the tumour is acidic. Hence, using pH-sensitive carriers is an effective approach for drug targeting (Behera and Padhi, 2022). Wang and co-workers (2016c) formulated mesoporous nanoparticles (pH-sensitive), including iron oxide (head group) and silicon dioxide (body group). These nanoparticles possessed superior paramagnetic characteristics and high drug-loading capacity. They were effective for hepatocellular carcinoma cells and allowed the release of berberine to the cytosol by destabilizing the lysosomal membranes.

6.3.2.3 Graphene-Based Nanocarriers

The properties of graphene and graphene oxide, such as a two-dimensional planar structure, stability (both chemical and mechanical), high surface area, high conductivity, and large biocompatibility, have made graphene one of the deserving carriers in drug delivery. Yu and co-workers (2015) formulated graphene oxide-bearing berberine, which was pH and redox-sensitive to the tumour microenvironment. The redox-sensitive graphene oxide carrier showed a more controlled release of berberine than the pH-sensitive graphene oxide due to the "skin effect" of the graphene oxide sheet and polyethylene glycol.

6.4 BERBERINE IN COMBINATION CANCER THERAPY

Nanomaterials can simultaneously deliver drugs, pharmacological agents, genes, proteins, and imaging probes in a single carrier system. The advantage of the co-delivery system is the synergistic effect of drug-drug or drug-gene is observed in an optimal single dose. Due to the co-delivery of drugs, the therapeutic index is improved due to the combating effect on biological barriers, increased drug permeation in tumours, increased targeting, and reduced unwanted toxicity (Li et al., 2019; Liu et al., 2019a,b).

Combination therapy involves administering two or more anticancer drugs, leading to additive or synergistic effects resulting in enhanced therapeutic Activity (Vahed et al., 2017; Kang et al., 2015; Gupta et al., 2017). Combination chemotherapy is beneficial over a single drug due to different pharmacological mechanisms involved in lowering overall toxicity (Sun et al., 2016). Multiple drug therapy targets different signalling pathways inside the cell, combating the barriers offered by genetic mutations and removing the possibility of the formation of multiple drug-resistant tumours (Hu et al., 2010; Yang and Gao, 2017).

Tuo et al. (2016) designed a conjugation of modified berberine (a 16-carbon aliphatic chain was incorporated at the C-9th position of berberine) to doxorubicin-loaded folic acid-conjugated polyethylene glycol liposomes for mitochondria-targeted delivery to breast cancer cells and to bypass the multidrug resistance development. The *in-vitro* study showed significant cytotoxicity and apoptotic activity in drug-resistant breast cancer cells (MCF-7/ADR) compared to free doxorubicin and liposomal doxorubicin. The cellular uptake of the nanoformulation increased by 15-fold as compared to free doxorubicin. The *in-vivo* and *ex-vivo* studies showed increased drug distribution and decreased tumour growth. The nanoformulation exhibited triple targeting ability by passive targeting with enhanced permeation and retention effect, active targeting due to folic acid, and novel mitochondria-targeted delivery of doxorubicin by conjugation with berberine derivatives.

Zhang and co-workers (2019) formulated hyaluronic acid-modified Janus-type mesoporous silica nanoparticles to co-deliver berberine and doxorubicin against hepatocellular carcinoma (HCC) cells. Berberine was loaded with doxorubicin and targeted to overexpressed CD44 receptors on hepatocellular cancer cells. Berberine inhibited the caspase pathway (caspase-3-iPLA2-COX-2), and both drugs showed a controlled release towards the pH-sensitive tumour microenvironment of hepatocellular cancer cells.

Kabary and colleagues (2018) formulated layer-by-layer lipid nanoparticles for the dual delivery of berberine and rapamycin to lung cancer using lactoferrin and hyaluronic acid. Prior to release from the nanoparticles, berberine was hydrophobically coupled with sodium lauryl sulphate to control its release. The nanoparticles exhibited a faster release of berberine, followed by a controlled release of rapamycin, allowing lung tumour cells to become sensitized to rapamycin's anticancer properties. The drug cytotoxicity was enhanced by covering the nanoparticles with lactoferrin. *In-vivo* investigations in mice with lung cancer have indicated that the berberine rapamycin nanoparticles have better therapeutic efficacy than free vascular endothelial growth factor levels.

Using a new co-delivery approach, Zhang and colleagues formulated a co-loaded liposome containing berberine and doxorubicin. Compared to berberine alone, the combination treatment resulted in a significant increase in apoptotic cells in 4T1. The prepared co-loaded liposome was stable and had a high encapsulation efficiency. The nanoliposome carrier regulated the drugs' biological fate. Compared to Doxil, it significantly decreased tumour development in the mouse mammary cancer model (4T1) and reduced the toxicity of the cardiac rupture of Doxil in mice. The co-loaded liposome delivery platform technology offered a new avenue for triple-negative breast cancer clinical treatment and the safe use of doxorubicin (Zhang et al., 2019).

Khan and co-workers formulated berberine in a co-delivery system with doxorubicin (synthetic origin) using poly (lactic-co-glycolic) by conjugation/encapsulation approach through carbodiimide chemistry. Antiproliferative activity against breast cancer cell lines (MDA-MB-231 and T47D) was considerably greater than the bio-actives. It was found that the nanoparticles depicted a small rise in the pattern of reactive oxygen species in the MDA-MB-231 cell line in a dose-dependent way. However, no such effect was seen in T47D cells. The nanoparticles showed a change in the permeability of the mitochondrial membrane (depolarization) and cell cycle progression arrest in the sub-G1 phase. *In-vivo* experiments demonstrated a nearly 14-fold increase in half-life with a large increase in plasma drug concentration (Khan et al., 2015).

Vanti and colleagues (2021) formulated co-delivery of berberine chloride and tariquidar-loaded nanoliposome to improve solubility and desired chemotherapeutic effects. Both human erythroleukemic cell lines (K562 and K562/DOXO) showed receptor-mediated endocytosis. Nanoliposomes were loaded with tariquidar, berberine chloride, or both and retained chemical and physical properties such as size, homogeneity, and encapsulation efficiency. In the K562/DOXO cell line, a *P*-glycoprotein inhibitor (tariquidar) was able to reverse multiple drug resistance. Nanoliposomes containing tariquidar and berberine chloride demonstrated a considerable increase in berberine chloride accumulation in tumour cells. This could be linked to the re-sensitization of resistant cells to the anticancer drug.

6.5 RECENT PATENTS ON BERBERINE AS AN ANTICANCER AGENT

Berberine-conjugated nanoformulations as anticancer agents have been patented by different researchers, making them a promising candidate for future work. Some of the patents are summarized in Table 6.1.

TABLE 6.1
List of Patents of Berberine-Conjugated Nanosystems

Sr. No	Patent No.	Patent Title	Inventors/Applicants	Year	Description
1	CN111298126A	Folic acid-berberine nano drug and preparation method and application thereof	Cheng yu; Ji yuanhui; Wu dongxu	2020	The folic acid-conjugated berberine provided a specific antibody response and showed targeted drug release of the drug at a tumour site. The nanoformulation was low in cost, high in stability, free of immunogenicity, capable of reducing toxicity accumulation and drug resistance, capable of exerting a good treatment effect due to the low drug-loading capacity, and capable of being selectively delivered to tumour tissues.
2	CN109125327A	Application of gefitinib and berberine in preparing medicine for treating lung cancer	Han shouwei; Tang qing; Wu jingjing; Zheng fang	2019	The invention discloses an application of gefitinib and berberine in treating lung cancer. Berberine can improve the susceptibility of lung cancer cells to gefitinib, and berberine and gefitinib can synergistically treat lung cancer.
3	CN110105402A	Novel Berberine antineoplastic platinum (II) complex and synthetic method and application thereof	Huang xiaoling; Qin qipin; Wang zhenfeng	2019	The invention relates to a novel berberine antineoplastic platinum (II) complex and a synthetic method and application of the complex to prepare antineoplastic drugs.
4	CN107550906A	Application of Berberine ion pair compound in the preparation of an anti-breast cancer drug	Jia benzhen; Lin jing; Tang jianwu; Wang xi	2018	The invention discloses an application of a berberine ion pair compound to prepare an anti-breast-cancer drug.
5	CN105853416A; CN105853416B	Application of berberine to the preparation of Hedgehog signalling pathway inhibitor	Bang yuanqiu; Liu yuan; Tan wenfu; Wang juan; Yang jun	2016	The present invention applies berberine as the hedgehog pathway inhibitor, which affects anti-medulloblastoma growth.
6	US2011086872A1	Berberine as a specific lung cancer agent, and other formulations and methods	Tin-wa maung [US]	2011	Berberine, its salts, and its derivatives are recognized as active constituents for treating lung cancer without adverse effects.
7	CN101683322A	Method for preparing nano-berberine hydrochloride liposome by the supercritical carbon dioxide method	Dan qiu; Min zhang; Weiguo shen; Xueqin an	2010	The invention discloses a method for preparing nano-berberberine hydrochloride liposomes by a supercritical carbon dioxide method.
8	KR20090032617A	An anticancer medicine including berberine	Choi myoung suk [KR]; Hong jin tae [KR]	2009	The present invention was related to a composition for anticancer chemotherapy comprising berberine as an active ingredient, particularly inhibiting cancer cell growth through induction of tumour protein p53-dependent cell death in human neuroblastoma and prostate cancer cells.

6.6 CONCLUSION AND FUTURE PERSPECTIVES

For thousands of years, berberine has been used for various disorders. Several clinical and preclinical studies demonstrated that conventional herbs, such as berberine, quercetin, and curcumin, significantly affected the tumour metastatic phase. (Cragg and Newman, 2013; SithrangaBoopathy and Kathiresan, 2010). Despite several limitations, berberine and many other naturally occurring phytocompounds, when incorporated in formulations, may prove their effectiveness against multiple types of carcinomas. Slow dissolution rate, limited cell permeability, and short plasma half-life have restricted its effective use. In such a prevalent situation, applying nanotechnology has proven to improve berberine's antitumour and antimetastatic efficacy.

Nanopharmaceuticals with improved and advanced technologies would be the succeeding trend of pharmaceutics globally. The innovations in artificial intelligence and computational biology will play a crucial role in selectively targeting the malignant cells in our system. According to recently published research, berberine delivery via nano vectors at specific cancer sites increases the response of bax, tumour protein p53, beclin-1, caspases activity, including mitogen-activated protein kinase signalling cascade, caspase 3/9, and decreases the response of the expression of mammalian target of rapamycin, B-cell lymphoma 2, and Wnt/-catenin. Although berberine has been nanoformulated to increase its effectiveness as an anticancer agent, several clinical studies still need to be conducted to overcome cancer progression.

REFERENCES

Attia, M. F., Anton, N., Wallyn, J., Omran, Z., & Vandamme, T. F. (2019). An overview of active and passive targeting strategies to improve the nanocarriers efficiency to tumour sites. *Journal of Pharmacy and Pharmacology*, 71(8), 1185–1198.

Ayati, S. H., Fazeli, B., Momtazi-Borojeni, A. A., Cicero, A. F., Pirro, M., & Sahebkar, A. (2017). Regulatory effects of berberine on microRNome in cancer and other conditions. *Critical Reviews in Oncology/Hematology*, 116, 147–158.

Behera, A., & Padhi, S. (2020). Passive and active targeting strategies for the delivery of the camptothecin anticancer drug: a review. *Environmental Chemistry Letters*, 18(5), 1557–1567.

Behera, A., Mittu, B., Padhi, S., Patra, N., & Singh, J. (2020). Bimetallic nanoparticles: green synthesis, applications, and future perspectives. In: Abd-Elsalam, K. A. (Ed.) *Multifunctional Hybrid Nanomaterials for Sustainable Agri-food and Ecosystems*. Elsevier, Amsterdam, pp. 639–682.

Behera, A., & Padhi, S. (2022). pH-Sensitive polymeric nanoparticles for cancer treatment. In: Padhi, S., Behera, A., & Lichtfouse, E. (Eds.) *Polymeric Nanoparticles for the Treatment of Solid Tumours*. Environmental Chemistry for a Sustainable World, vol 71. Springer, Cham, pp. 401–425.

Behera, A., Padhi, S., & Nayak, A. K. (2023). Engineered liposomes as drug delivery and imaging agents. In: Ray, S., & Nayak, A. K. (Eds.) *Design and Applications of Theranostic Nanomedicines*. Woodhead Publishing, Cambridge, MA, pp. 75–108.

Beloqui, A., Solinís, M. Á., Rodríguez-Gascón, A., Almeida, A. J., & Préat, V. (2016). Nanostructured lipid carriers: promising drug delivery systems for future clinics. *Nanomedicine: Nanotechnology, Biology and Medicine*, 12(1), 143–161.

Bhanumathi, R., Vimala, K., Shanthi, K., Thangaraj, R., & Kannan, S. (2017). Bioformulation of silver nanoparticles as berberine carrier cum anticancer agent against breast cancer. *New Journal of Chemistry*, 41(23), 14466–14477.

Bhaskaran, M., & Mohan, M. (2014). MicroRNAs: history, biogenesis, and their evolving role in animal development and disease. *Veterinary Pathology*, 51(4), 759–774.

Bhatnagar, P., Kumari, M., Pahuja, R., Pant, A. B., Shukla, Y., Kumar, P., & Gupta, K. C. (2018). Hyaluronic acid-grafted PLGA nanoparticles for the sustained delivery of Berberine chloride for an efficient suppression of Ehrlich ascites tumours. *Drug Delivery and Translational Research*, 8(3), 565–579.

Bray, F., Ferlay, J., Soerjomataram, I., Siegel, R. L., Torre, L. A., & Jemal, A. (2020). Global cancer statistics 2018: GLOBOCAN estimates of incidence and mortality worldwide for 36 cancers in 185 countries (vol 68, pg 394, 2018). *Cancer Journal For Clinicians*, 70(4), 313–313.

Casey, S. C., Amedei, A., Aquilano, K., Azmi, A. S., Benencia, F., Bhakta, D., Bilsland, A. E., Boosani, C. S., Chen, S., Ciriolo, M. R., & Crawford, S. (2015). Cancer prevention and therapy through the modulation of the tumour microenvironment. *Seminars in Cancer Biology*, 35, S199–S223. Academic Press.

Chen, W. H., Pang, J. Y., Qin, Y., Peng, Q., Cai, Z., & Jiang, Z. H. (2005). Synthesis of linked Berberine dimers and their remarkably enhanced DNA-binding affinities. *Bioorganic & Medicinal Chemistry Letters*, 15(10), 2689–2692.

Cheng, C. J., Bahal, R., Babar, I. A., Pincus, Z., Barrera, F., Liu, C. & Slack, F. J. (2015). MicroRNA silencing for cancer therapy targeted to the tumour microenvironment. *Nature*, 518(7537), 107–110.

Choudhary, S., Gupta, L., Rani, S., Dave, K., & Gupta, U. (2017). Impact of dendrimers on solubility of hydrophobic drug molecules. *Frontiers in Pharmacology*, 8, 61.

Chu, S. C., Yu, C. C., Hsu, L. S., Chen, K. S., Su, M. Y., & Chen, P. N. (2014). Berberine reverses epithelial-to-mesenchymal transition and inhibits metastasis and tumour-induced angiogenesis in human cervical cancer cells. *Molecular Pharmacology*, *86*(6), 609–623.

Cragg, G. M., & Newman, D. J. (2013). Natural products: a continuing source of novel drug leads. *Biochimica et Biophysica* Acta, 1830, 3670–3695.

Kabarya, D. M., Helmy, M. W., Elkhodairya, K. A., Fang, J.-Y., Elzoghby, A. O. (2018). Hyaluronate/lactoferrin layer-by-layer-coated lipid nanocarriers for targeted co-delivery of rapamycin and berberine to lung carcinoma. *Colloids and Surfaces B: Biointerfaces*, 169, 183–194.

De Palma, M., Biziato, D., & Petrova, T. V. (2017). Microenvironmental regulation of tumour angiogenesis. *Nature Reviews Cancer*, 17(8), 457.

Delloye-Bourgeois, C., Bertin, L., Thoinet, K., Jarrosson, L., Kindbeiter, K., Buffet, T., et al. (2017). Microenvironment-driven shift of cohesion/detachment balance within tumours induces a switch toward metastasis in neuroblastoma. *Cancer Cell*, 32(4), 427–443.

Dev, S. (2006). *A Selection of Prime Ayurvedic Plants Drugsancient-Modern Concordance*. New Delhi: Anamaya Publishers.

Díaz, M. R., &Vivas-Mejia, P. E. (2013). Nanoparticles as drug delivery systems in cancer medicine: emphasis on RNAi-containing nanoliposomes. *Pharmaceuticals*, 6(11), 1361–1380.

Dolatabadi, J. E. N., & Omidi, Y. (2016). Solid lipid-based nanocarriers as efficient targeted drug and gene delivery systems. *TrAC Trends in Analytical Chemistry*, 77, 100–108.

Dziedzic, A., Kubina, R., Bułdak, R. J., Skonieczna, M., & Cholewa, K. (2016). Silver nanoparticles exhibit the dose-dependent antiproliferative effect against human squamous carcinoma cells attenuated in the presence of berberine. *Molecules*, 21(3), 365.

Eo, S. H., Kim, J. H., & Kim, S. J. (2014). Induction of G2/M arrest by berberine via activation of PI3K/Akt and p38 in human chondrosarcoma cell line. *Oncology Research Featuring Preclinical and Clinical Cancer Therapeutics*, 22(3), 147–157.

Fang, J., Nakamura, H., & Maeda, H. (2011). The EPR effect: unique features of tumour blood vessels for drug delivery, factors involved, and limitations and augmentation of the effect. *Advanced Drug Delivery Reviews*, 63(3), 136–151.

Faraji, A. H., & Wipf, P. (2009). Nanoparticles in cellular drug delivery. *Bioorganic & Medicinal Chemistry*, 17(8), 2950–2962. doi:10.1016/j.bmc.2009.02.043.

Ghaffari, M., & Dolatabadi, J. E. N. (2019). Nanotechnology for pharmaceuticals. In: Thomas, S., Grohens, Y., & Pottathara, Y. B. (Eds.) *Industrial Applications of Nanomaterials*. Elsevier, Amsterdam, pp. 475–502.

Gupta, L., Sharma, A. K., Gothwal, A., Khan, M. S., Khinchi, M. P., Qayum, A., & Gupta, U. (2017). Dendrimer encapsulated and conjugated delivery of berberine: a novel approach mitigating toxicity and improving in vivo pharmacokinetics. *International Journal of Pharmaceutics*, 528(1–2), 88–99.

Hao, D. C., He, C. N., Shen, J., & Xiao, P. G. (2017). Anticancer chemodiversity of Ranunculaceae medicinal plants: molecular mechanisms and functions. *Current Genomics*, 18(1), 39–59.

Hema, S., Thambiraj, S., & Shankaran, D. R. (2018). Nanoformulations for targeted drug delivery to prostate cancer: an overview. *Journal of Nanoscience and Nanotechnology*, 18(8), 5171–5191.

Hola, K., Zhang, Y., Wang, Y., Giannelis, E. P., Zboril, R., & Rogach, A. L. (2014). Carbon dots: emerging light emitters for bioimaging, cancer therapy and optoelectronics. *Nano Today*, 9(5), 590–603.

Hu, C. M. J., Aryal, S., & Zhang, L. (2010). Nanoparticle-assisted combination therapies for effective cancer treatment. *Therapeutic Delivery*, 1(2), 323–334. doi:10.4155/tde.10.13.

Huang, Z. H., Zheng, H. F., Wang, W. L., Wang, Y., Zhong, L. F., Wu, J. L., & Li, Q. X. (2015). Berberine targets epidermal growth factor receptor signalling to suppress prostate cancer proliferation in vitro. *Molecular Medicine Reports*, 11(3), 2125–2128.

Jeong, S. Y., & Seol, D. W. (2008). The role of mitochondria in apoptosis. *BMB Reports*, 41(1), 11–22.

Jin, H., Jin, X., Cao, B., & Wang, W. (2017a). Berberine affects osteosarcoma via downregulating the caspase-1/IL-1β signaling axis. *Oncology Reports*, 37(2), 729–736.

Jin, Y., Liu, S., Ma, Q., Xiao, D., & Chen, L. (2017b). Berberine enhances the AMPK activation and autophagy and miti gates high glucose-induced apoptosis of mouse podocytes. *European Journal of Pharmacology*, 794, 106–114.

Kabary, D. M., Helmy, M. W., Elkhodairy, K. A., Fang, J. Y., & Elzoghby, A. O. (2018). Hyaluronate/lactoferrin layer-by-layer-coated lipid nanocarriers for targeted co-delivery of rapamycin and berberine to lung carcinoma. *Colloids and Surfaces B: Biointerfaces*, 169, 183–194. doi:10.1016/j.colsurfb.2018.05.008.

Kang, L., Gao, Z., Huang, W., Jin, M., & Wang, Q. (2015). Nanocarrier-mediated co-delivery of chemotherapeutic drugs and gene agents for cancer treatment. *Acta Pharmaceutica Sinica B*, 5(3), 169–175.

Kar, S. (2016). Unraveling cell-cycle dynamics in cancer. *Cell Systems*, 2(1), 8–10.

Kawano, M., Takagi, R., Kaneko, A., & Matsushita, S. (2015). Berberine is a dopamine D1-and D2-like receptor antagonist and ameliorates experimentally induced colitis by suppressing innate and adaptive immune responses. *Journal of Neuroimmunology*, 289, 43–55.

Khan, D. R., Webb, M. N., Cadotte, T. H., & Gavette, M. N. (2015). Use of targeted liposome-based chemotherapeutics to treat breast cancer. *Breast Cancer: Basic and Clinical Research*, 9, BCBCR–S29421.

Kim, S., Lee, S. Y., & Cho, H. J. (2018). Berberine and zinc oxide-based nanoparticles for the chemo-photothermal therapy of lung adenocarcinoma. *Biochemical and Biophysical Research Communications*, 501(3), 765–770.

Kundu, A., Padhi, S., Behera, A., Hasnain, M. S., & Nayak, A. K. (2022). Tumour targeting strategies by chitosan-based nanocarriers. In: Hasnain, M. S., Beg, S., & Nayak, A. K. (Eds.) *Chitosan in Biomedical Applications*. Academic Press, London, pp. 163–188.

La, X., Zhang, L., Li, Z., Yang, P., & Wang, Y. (2017). Berberine-induced autophagic cell death by elevating GRP78 levels in cancer cells. *Oncotarget*, 8(13), 20909.

Lai, Y. H., Chiang, C. S., Kao, T. H., & Chen, S. Y. (2018). Dual-drug nanomedicine with hydrophilic F127-modified magnetic nanocarriers assembled in amphiphilic gelatin for enhanced penetration and drug delivery in deep tumour tissue. *International Journal of Nanomedicine*, 13, 3011.

Li, L., Wang, X., Sharvan, R., Gao, J., Qu, S., Li, L., et al. (2017). Berberine could inhibit thyroid carcinoma cells by inducing mitochondrial apoptosis, g0/g1 cell cycle arrest and suppressing migration via PI3K-Akt and MAPK signalling pathways. *Biomedicine & Pharmacotherapy*, 95, 1225–1231.

Li, W., Li, Q., Kang, S., Same, M., Zhou, Y., Sun, C., et al. (2018). Cancer detector: ultrasensitive and non-invasive cancer detection at the resolution of individual reads using cell-free DNA methylation sequencing data. *Nucleic Acids Res*, 46, e89.

Li, D. X., Zhang, J., Zhang, Y., Zhao, P. W., & Yang, L. M. (2015a). Inhibitory effect of berberine on human skin squamous cell carcinoma A431 cells. *Genetics and Molecular Research*, 14(3), 10553–10568.

Li, W., Saud, S. M., Young, M. R., Chen, G., & Hua, B. (2015b). Targeting AMPK for cancer prevention and treatment. *Oncotarget*, 6(10), 7365.

Li, Z., Jiang, T., Lu, Q., Xu, K., He, J., Xie, L., & Hu, A. (2019). Berberine attenuated the cytotoxicity induced by t-BHP via inhibiting oxidative stress and mitochondria dysfunction in PC-12 cells. In: *Cellular and Molecular Neurobiology*. Springer, New York, pp. 1–16.

Lin, Y. C., Kuo, J. Y., Hsu, C. C., Tsai, W. C., Li, W. C., Yu, M. C., & Wen, H. W. (2013). Optimizing manufacture of liposomal berberine with evaluation of its antihepatoma effects in a murine xenograft model. *International Journal of Pharmaceutics*, 441(1–2), 381–388.

Liu, H. C., Lämke, J., Lin, S. Y., Hung, M. J., Liu, K. M., Charng, Y. Y., et al. (2018). Distinct heat shock factors and chromatin modifications mediate the organ-autonomous transcriptional memory of heat stress. *The Plant Journal*, 95, 401–413.

Liu, L., Chen, L., Jiang, C., Guo, J., Xie, Y., Kang, L., et al. (2017a). Berberine inhibits the LPS-induced proliferation and inflammatory response of stromal cells of adenomyosis tissues mediated by the lps/tlr4 signaling pathway. *Experimental and Therapeutic Medicine*, 14, 6125.

Liu, L., Luo, N., Guo, J., Xie, Y., Chen, L., & Cheng, Z. (2017b). Berberine inhibits growth and inflammatory invasive phenotypes of ectopic stromal cells: imply the possible treatment of adenomyosis. *Journal of Pharmacological Sciences*, 137, 5–11.

Liu, D., Meng, X., Wu, D., Qiu, Z., and & Luo, H. (2019a). A natural isoquinoline alkaloid with antitumour activity: studies of the biological activities of berberine. *Frontiers in Pharmacology*, 10, 9.

Liu, H., Huang, C., Wu, L., & Wen, B. (2016). Effect of evodiamine and Berberine on miR-429 as an oncogene in human colorectal cancer. *OncoTargets and Therapy*, 9, 4121.

Liu, L., Fan, J., Ai, G., Liu, J., Luo, N., Li, C., & Cheng, Z. (2019b). Berberine in combination with cisplatin induces necroptosis and apoptosis in ovarian cancer cells. Biological *Research*, 52(1), 1–14.

Ma, C. Y., Shen, S. C., Huang, D. W., Chang, H. M., & Wu, J. B. (2008). Growth inhibition and induction of apoptosis in U937 cells by Coptischinensis extract. *Journal of Food Science*, 73(6), H127–H133.

Ma, W., Zhu, M., Zhang, D., Yang, L., Yang, T., Li, X., & Zhang, Y. (2017). Berberine inhibits the proliferation and migration of breast cancer ZR-75-30 cells by targeting Ephrin-B2. *Phytomedicine*, 25, 45–51.

Ma, X., Zhou, J., Zhang, C. X., Li, X. Y., Li, N., Ju, R. J., & Lu, W. L. (2013). Modulation of drug-resistant membrane and apoptosis proteins of breast cancer stem cells by targeting Berberine liposomes. *Biomaterials*, 34(18), 4452–4465.

Maeda, H. (2001). The enhanced permeability and retention (EPR) effect in tumour vasculature: the key role of tumour-selective macromolecular drug targeting. *Advances in Enzyme Regulation*, 41, 189–207.

Mellet, C. O., Fernández, J. M. G., & Benito, J. M. (2011). Cyclodextrin-based gene delivery systems. *Chemical Society Reviews*, 40(3), 1586–1608.

Meng, X. P., Fan, H., Wang, Y. F., Wang, Z. P., & Chen, T. S. (2016, October). Anti-hepatocarcinoma effects of Berberine-nanostructured lipid carriers against human HepG2, Huh7, and EC9706 cancer cell lines. *Proc. SPIE*, 10024, 1002417.

Mihaylova, M. M., & Shaw, R. J. (2011). The AMPK signalling pathway coordinates cell growth, autophagy and metabolism. *Nature Cell Biology*, 13(9), 1016–1023.

Mody, V. V., Siwale, R., Singh, A., & Mody, H. R. (2010). Introduction to metallic nanoparticles. *Journal of Pharmacy and Bioallied Sciences*, 2(4), 282.

Naveen, C. R., Gaikwad, S., & Agrawal-Rajput, R. (2016). Berberine induces neuronal differentiation through inhibition of cancer stemness and epithelial-mesenchymal transition in neuroblastoma cells. *Phytomedicine*, 23(7), 736–744.

Nguyen, T. X., Huang, L., Liu, L., Abdalla, A. M. E., Gauthier, M., & Yang, G. (2014). Chitosan-coated nano-liposomes for the oral delivery of Berberine hydrochloride. *Journal of Materials Chemistry B*, 2(41), 7149–7159.

Okubo, S., Uto, T., Goto, A., Tanaka, H., Nishioku, T., Yamada, K., & Shoyama, Y. (2017). Berberine induces apoptotic cell death via activation of caspase-3 and-8 in HL-60 human leukemia cells: nuclear localization and structure–activity relationships. *The American Journal of Chinese Medicine*, 45(07), 1497–1511.

Padhi, S., & Behera, A. (2020). Nanotechnology based targeting strategies for the delivery of Camptothecin. *Sustainable Agriculture Reviews*, 44, 243–272.

Padhi, S., Behera, A., Hasnain, M. S., & Nayak, A. K. (2022a). Chitosan-based drug delivery systems in cancer therapeutics. In: Hasnain, M. S., Beg, S., & Nayak, A. K. (Eds.) *Chitosan in Drug Delivery*. Academic Press, London, pp. 159–193.

Padhi, S., Behera, A., Hasnain, M. S., & Nayak, A. K. (2022b). Uses of chitosan in drug delivery. In: Hasnain, M. S., Beg, S., & Nayak, A. K. (Eds.) *Chitosan in Drug Delivery*. Academic Press, London, pp. 139–162.

Padhi, S., Dash, M., & Behera, A. (2021). Nanophytochemicals for the treatment of type II diabetes mellitus: a review. *Environmental Chemistry Letters*, 19, 4349–4373.

Padhi, S., Kapoor, R., Verma, D., Panda, A. K., & Iqbal, Z. (2018). Formulation and optimization of topotecan nanoparticles: in vitro characterization, cytotoxicity, cellular uptake and pharmacokinetic outcomes. *Journal of Photochemistry and Photobiology B: Biology*, 183, 222–232.

Padhi, S., Mirza, M. A., Verma, D., Khuroo, T., Panda, A. K., Talegaonkar, S., Khar, R. K., & Iqbal, Z. (2016). Revisiting the nanoformulation design approach for effective delivery of topotecan in its stable form: an appraisal of its in vitro behavior and tumour amelioration potential. *Drug Delivery*, 23(8), 2827–2837.

Padhi, S., & Behera, A. (2022). Cellular internalization and toxicity of polymeric nanoparticles. In: Padhi, S., Behera, A., & Lichtfouse, E. (Eds.) *Polymeric Nanoparticles for Treating Solid Tumours*. Environmental Chemistry for a Sustainable World, vol 71. Springer, Cham, pp. 473–488.

Pandey, S., Mewada, A., Thakur, M., Shah, R., Oza, G., & Sharon, M. (2013). Biogenic gold nanoparticles as fotillas to fire Berberine hydrochloride using folic acid as molecular road map. *Materials Science and Engineering: C*, 33(7), 3716–3722.

Park, G. B., Park, S. H., Kim, D., Kim, Y. S., Yoon, S. H., & Hur, D. Y. (2016). Berberine induces mitochondrial apoptosis of EBV-transformed B cells through p53-mediated regulation of XAF1 and GADD45alpha. *International Journal of Oncology*, 49(1), 411–421.

Pattni, B. S., Chupin, V. V., & Torchilin, V. P. (2015). New developments in liposomal drug delivery. *Chemical Reviews*, 115(19), 10938–10966.

Pistritto, G., Trisciuoglio, D., Ceci, C., Garufi, A., & D'Orazi, G. (2016). Apoptosis as anticancer mechanism: function and dysfunction of its modulators and targeted therapeutic strategies. *Aging (Albany NY)*, 8(4), 603.

Popiołek, I., Niziołek, A., Kamiński, K., Kwolek, U., Nowakowska, M., & Szczubiałka, K. (2019). Cellular delivery and enhanced anticancer activity of berberine complexed with a cationic derivative of γ–cyclodextrin. *Bioorganic & Medicinal Chemistry*, 27(7), 1414–1420.

Pund, S., Borade, G., & Rasve, G. (2014). Improvement of anti-inflammatory and anti-angiogenic activity of berberine by novel rapid dissolving nanoemulsifying technique. *Phytomedicine*, 21(3), 307–314.

Qi, H. W., Xin, L. Y., Xu, X., Ji, X. X., & Fan, L. H. (2014). Epithelial-to-mesenchymal transition markers to predict response of berberine in suppressing lung cancer invasion and metastasis. *Journal of Translational Medicine*, 12(1), 1–10.

Riaz, M. K., Riaz, M. A., Zhang, X., Lin, C., Wong, K. H., Chen, X. & Yang, Z. (2018). Surface functionalization and targeting strategies of liposomes in solid tumour therapy: a review. *International Journal of Molecular Sciences*, 19(1), 195.

Seca, A. M., & Pinto, D. C. (2018). Plant secondary metabolites as anticancer agents: successes in clinical trials and therapeutic application. *International Journal of Molecular Sciences*, 19(1), 263.

Senapati, S., Mahanta, A. K., Kumar, S., & Maiti, P. (2018). Controlled drug delivery vehicles for cancer treatment and their performance. *Signal Transduction and Targeted Therapy*, 3(1), 1–19.

Sengupta, D., Chowdhury, K. D., Chatterjee, S., Sarkar, A., Paul, S., Sur, P. K., & Sadhukhan, G. C. (2017). Modulation of adenylate cyclase signaling in association with MKK3/6 stabilization under combination of SAC and Berberine to reduce HepG2 cell survivability. *Apoptosis*, 22(11), 1362–1379.

Sharma, A., Goyal, A. K., & Rath, G. (2018). Recent advances in metal nanoparticles in cancer therapy. *Journal of Drug Targeting*, 26(8), 617–632.

Shen, R., Kim, J. J., Yao, M., & Elbayoumi, T. A. (2016). Development and evaluation of vitamin E D-α-tocopheryl polyethylene glycol 1000 succinate-mixed polymeric phospholipid micelles of berberine as an anticancer nanopharmaceutical. *International Journal of Nanomedicine*, 11, 1687.

Shi, Y., Zhao, Y., Shao, N., Ye, R., Lin, Y., Zhang, N., & Wang, S. (2017). Overexpression of microRNA-96-5p inhibits autophagy and apoptosis and enhances the proliferation, migration and invasiveness of human breast cancer cells. *Oncology Letters*, 13(6), 4402–4412.

Shukla, S., Rizvi, F., Raisuddin, S., & Kakkar, P. (2014). FoxO proteins' nuclear retention and BH3-only protein Bim induction evoke mitochondrial dysfunction-mediated apoptosis in Berberine-treated HepG2 cells. *Free Radical Biology and Medicine*, 76, 185–199.

Shukla, S., Sharma, A., Pandey, V. K., Raisuddin, S., & Kakkar, P. (2016). Concurrent acetylation of FoxO1/3a and p53 due to sirtuins inhibition elicit Bim/PUMA mediated mitochondrial dysfunction and apoptosis in Berberine-treated HepG2 cells. *Toxicology and Applied Pharmacology*, 291, 70–83.

Singh, A., Duggal, S., Kaur, N., & Singh, J. (2010). Berberine: alkaloid with wide spectrum of pharmacological activities. *Journal of Natural Products*, 3, 64–75.

SithrangaBoopathy, N., & Kathiresan, K. (2010). Anticancer drugs from marine flora: an overview. *Journal of Oncology*, 2010, 214186.

Sreeja, S., & Nair, C. K. (2018). Tumour control by hypoxia-specific chemotargeting of iron-oxide nanoparticle–Berberine complexes in a mouse model. *Life Sciences*, 195, 71–80.

Su, Y. H., Tang, W. C., Cheng, Y. W., Sia, P., Huang, C. C., Lee, Y. C., et al. (2015). Targeting of multiple oncogenic signaling pathways by Hsp90 inhibitor alone or in combination with berberine for treatment of colorectal cancer. *Biochimica et Biophysica Acta (BBA)-Molecular Cell Research*, 1853(10), 2261–2272.

Sun, Y., Kang, C., Zhang, A., Liu, F., Hu, J., Zhong, X., & Xie, J. (2016). Co-delivery of dual-drugs with nanoparticle to overcome multidrug resistance. *European Journal of BioMedical Research*, 2(2), 12–18.

Thakur, M., Mewada, A., Pandey, S., Bhori, M., Singh, K., Sharon, M., & Sharon, M. (2016). Milk-derived multi-fluorescent graphene quantum dot-based cancer theranostic system. *Materials Science and Engineering: C*, 67, 468–477.

Tietze, R., Zaloga, J., Unterweger, H., Lyer, S., Friedrich, R. P., Janko, C., & Alexiou, C. (2015). Magnetic nanoparticle-based drug delivery for cancer therapy. *Biochemical and Biophysical Research Communications*, 468(3), 463–470.

Tillhon, M., Ortiz, L. M. G., Lombardi, P., & Scovassi, A. I. (2012). Berberine: new perspectives for old remedies. *Biochemical Pharmacology*, 84(10), 1260–1267.

Torchilin, V. (2011). Tumour delivery of macromolecular drugs based on the EPR effect. *Advanced Drug Delivery Reviews*, 63(3), 131–135.

Tuo, J., Xie, Y., Song, J., Chen, Y., Guo, Q., Liu, X., & Hu, H. (2016). Development of a novel Berberine-mediated mitochondria-targeting nano-platform for drug-resistant cancer therapy. *Journal of Materials Chemistry B*, 4(42), 6856–6864.

Vahed, S. Z., Salehi, R., Davaran, S., & Sharifi, S. (2017). Liposome-based drug co-delivery systems in cancer cells. *Materials Science and Engineering: C*, 71, 1327–1341.

Vanti, G., Coronnello, M., Bani, D., Mannini, A., Bergonzi, M. C., & Bilia, A. R. (2021), Co-delivery of berberine chloride and tariquidar in nanoliposomes enhanced intracellular berberine chloride in a doxorubicin-resistant K562 cell line due to P-gp overexpression. *Pharmaceutics*, 13(3), 306. doi:10.3390/pharmaceutics13030306.

Wang, X., Li, Y., Chen, G., Zhang, Y., Zhang, X., & Sun, Z., et al. (2016c) Recent developments in mesoporous silica nanoparticles for tumor theranostic applications. *Current Pharmaceutical Design*, 22(23), 3518–3531. doi:10.2174/1381612822666160321114728.

Wang, J., Qi, Q., Feng, Z., Zhang, X., Huang, B., Chen, A., et al. (2016a). Berberine induces autophagy in glioblastoma by targeting the AMPK/mTOR/ULK1-pathway. *Oncotarget*, 7(41), 66944.

Wang, K., Feng, X., Chai, L., Cao, S., & Qiu, F. (2017). The metabolism of berberine and its contribution to the pharmacological effects. *Drug Metabolism Reviews*, 49(2), 139–157.

Wang, K., Zhang, C., Bao, J., Jia, X., Liang, Y., Wang, X., & He, C. (2016b). Synergistic chemopreventive effects of curcumin and berberine on human breast cancer cells through induction of apoptosis and autophagic cell death. *Scientific Reports*, 6(1), 1–14.

Wang, L., Li, H., Wang, S., Liu, R., Wu, Z., Wang, C., & Chen, M. (2014a). Enhancing the antitumour activity of Berberine hydrochloride by solid lipid nanoparticle encapsulation. *AAPS PharmSciTech*, 15(4), 834–844.

Wang, N., Feng, Y., Cheung, F., Wang, X., Zhang, Z., & Feng, Y. (2015a). A Chinese medicine formula GegenQinlian decoction suppresses expansion of human renal carcinoma with inhibition of matrix metalloproteinase-2. *Integrative Cancer Therapies*, 14(1), 75–85.

Wang, N., Feng, Y., Zhu, M., Tsang, C. M., Man, K., Tong, Y., & Tsao, S. W. (2010). Berberine induces autophagic cell death and mitochondrial apoptosis in liver cancer cells: the cellular mechanism. *Journal of Cellular Biochemistry*, 111(6), 1426–1436.

Wang, N., Zhu, M., Wang, X., Tan, H. Y., Tsao, S. W., & Feng, Y. (2014b). Berberine-induced tumour suppressor p53 up-regulation gets involved in the regulatory network of MIR-23a in hepatocellular carcinoma. *Biochimica et Biophysica Acta (BBA)-Gene Regulatory Mechanisms*, 1839(9), 849–857.

Wang, W., Lu, Y. C., Huang, H., Feng, J. J., Chen, J. R., & Wang, A. J. (2014c). Facile synthesis of water-soluble and biocompatible fluorescent nitrogen-doped carbon dots for cell imaging. *Analyst*, 139(7), 1692–1696.

Wang, Y. X., Pang, W. Q., Zeng, Q. X., Deng, Z. S., Fan, T. Y., Jiang, J. D., & Song, D. Q. (2018a). Synthesis and biological evaluation of new Berberine derivatives as cancer immunotherapy agents through targeting IDO1. *European Journal of Medicinal Chemistry*, 143, 1858–1868.

Wang, Y., Wen, B., Yu, H., Ding, D., Zhang, J., Zhang, Y., & Zhang, W. (2018b). Berberine hydrochloride-loaded chitosan nanoparticles effectively targets and suppresses human nasopharyngeal carcinoma. *Journal of Biomedical Nanotechnology*, 14(8), 1486–1495.

Wang, Z. P., Wu, J., Chen, T. S., Zhou, Q., & Wang, Y. F. (2015b, March). In vitro and in vivo antitumour efficacy of berberine-nanostructured lipid carriers against H22 tumour. *In Biophotonics and Immune Responses X* (Vol. 9324, p. 9324). International Society for Optics and Photonics.

Wolfram, J., & Ferrari, M. (2019). Clinical cancer nanomedicine. *Nano Today*, 25, 85–98.

Xiao, Y., Tian, C., Huang, T., Han, B., Wang, M., Ma, H., & Li, X. (2018). 8-Cetyl berberine inhibits growth of lung cancer in vitro and in vivo. *Life Sciences*, 192, 259–269.

Xin, Y., Yin, M., Zhao, L., Meng, F., & Luo, L. (2017). Recent progress on nanoparticle-based drug delivery systems for cancer therapy. *Cancer Biology & Medicine*, 14(3), 228.

Xue, M., Zhang, L., Yang, M. X., Zhang, W., Li, X. M., Ou, Z. M., & Yang, S. Y. (2015). Berberine-loaded solid lipid nanoparticles are concentrated in the liver and ameliorate hepatosteatosis in db/db mice. *International Journal of Nanomedicine*, 10, 5049.

Yang, M., Yang, Y., Cui, H., Guan, Z., Yang Y., Zhang H., et al. (2015). The natural compound gambogic acid radiosensitizes nasopharyngeal carcinoma cells under hypoxic conditions. *Tumouri*, 102, 135.

Yang, S., & Gao, H. (2017). Nanoparticles for modulating tumour microenvironment to improve drug delivery and tumour therapy. *Pharmacological Research*, 126, 97–108.

Yang, X., & Huang, N. (2013). Berberine induces selective apoptosis through the AMPK-mediated mitochondrial/caspase pathway in hepatocellular carcinoma. *Molecular Medicine Reports*, 8(2), 505–510.

Yin, J., Hou, Y., Yin, Y., & Song, X. (2017). Selenium-coated nanostructured lipid carriers used for oral delivery of berberine to accomplish a synergic hypoglycemic effect. *International Journal of Nanomedicine,* 12, 8671.

Yin, Z., Yang, J., Ning, R., Liu, Y., Feng, M., Gu, C., & Li, Y. (2018). Signal pathways, diseases, and functions associated with the miR-19a/92a cluster and the use of berberine to modulate the expression of this cluster in multiple myeloma cells. *Journal of Biochemical and Molecular Toxicology*, 32(6), e22057.

Yu, D., Ruan, P., Meng, Z., & Zhou, J. (2015). The structure-dependent electric release and enhanced oxidation of drug in graphene oxide-based nanocarrier loaded with anticancer herbal drug berberine. *Journal of Pharmaceutical Sciences*, 104(8), 2489–2500.

Yu, F., Ao, M., Zheng, X., Li, N., Xia, J., Li, Y., & Chen, X. D. (2017). PEG–lipid–PLGA hybrid nanoparticles loaded with berberine–phospholipid complex to facilitate the oral delivery efficiency. *Drug Delivery*, 24(1), 825–833.

Yue, J., Wang, Z., Shao, D., Chang, Z., Hu, R., Li, L., et al. (2018). Cancer cell membrane-modified biodegradable mesoporous silica nanocarriers for berberine therapy of liver cancer. *RSC Advances*, 8(70), 40288–40297.

Zhang, J. Y., Lin, M. T., Zhou, M. J., Yi, T., Tang, Y. N., Tang, S. L., et al. (2015). Combinational treatment of curcumin and quercetin against gastric cancer mgc-803 cells in vitro. *Molecules*, 20 11524–11534.

Zhang, F., Jia, Y., Zheng, X., Shao, D., Zhao, Y., Wang, Z., & Chen, L. (2019). Janus nanocarrier-based co-delivery of doxorubicin and berberine weakens chemotherapy-exacerbated hepatocellular carcinoma recurrence. *Acta Biomaterialia*, 100, 352–364.

Zhang, F., Zhang, M., Zheng, X., Tao, S., Zhang, Z., Sun, M. & Chen, L. (2018). Berberine-based carbon dots for selective and safe cancer theranostics. *RSC Advances*, 8(3), 1168–1173.

Zhang, L., Miao, X. J., Wang, X., Pan, H. H., Li, P., Ren, H., et al. (2016a). Antiproliferation of berberine is mediated by epigenetic modification of constitutive androstane receptor (CAR) metabolic pathway in hepatoma cells. Scientific *R*eports, 6(1), 1–10.

Zhang, L., Miao, X.J., Wang, X., Pan, H.H., Li, P., Ren, H., et al. (2016b). Antiproliferation of berberine is mediated by epigenetic modification of constitutive androstane receptor (CAR) metabolic pathway in hepatoma cells. *Scientific Reports*, *6*(1), 1–10.

Zhang, Q., Wang, X., Cao, S., Sun, Y., He, X., Jiang, B., et al. (2020). Berberine represses human gastric cancer cell growth in vitro and in vivo by inducing cytostatic autophagy via inhibition of MAPK/mTOR/p70S6K and Akt signaling pathways. *Biomedicine & Pharmacotherapy*, 128, 110245.

Zhao, Y., Yang, X., Zhao, J., Gao, M., Zhang, M., Shi, T., et al. (2021). Berberine inhibits chemotherapy-exacerbated ovarian cancer stem cell-like characteristics and metastasis through GLI1. *European Journal of Pharmacology*, 895, 173887.

Zheng, X., Shao, D., Li, J., Song, Y., Chen, Y., Pan, Y., & Chen, L. (2015). Single and repeated dose toxicity of citric acid-based carbon dots and a derivative in mice. *RSC Advances*, 5(111), 91398–91406.

Zhou, R., Xiang, C., Cao, G., Xu, H., Zhang, Y., Yang, H., & Zhang, J. (2021). Berberine accelerated wound healing by restoring TrxR1/JNK in diabetes. *Clinical Science*, *135*(4), 613–627.

Zhou, Y., Quan, G., Wu, Q., Zhang, X., Niu, B., Wu, B., & Wu, C. (2018). Mesoporous silica nanoparticles for drug and gene delivery. *Acta Pharmaceuticasinica B*, 8(2), 165–177.

Zhu, S., Song, Y., Zhao, X., Shao, J., Zhang, J., & Yang, B. (2015). The photoluminescence mechanism in carbon dots (graphene quantum dots, carbon nanodots, and polymer dots): current state and future perspective. *Nano Research*, 8(2), 355–381.

Zhu, S., Zhao, X., Song, Y., Lu, S., & Yang, B. (2016). Beyond bottom-up carbon nanodots: citric-acid derived organic molecules. *Nano Today*, 11(2), 128–132.

7 Nanoenabled Systems for the Delivery of Paclitaxel as an Anticancer Agent

Jessica Pandohee[1], *Swati Sharma*[2], *Pradeep Kumar*[2], *Emmanuel Kyereh*[3], *and Saurabh Kulshrestha*[2]
[1]Centre for Crop and Disease Management, School of Molecular and Life Sciences, Curtin University, Bentley, Australia
[2]School of Biotechnology, Faculty of Applied Sciences and Biotechnology, Shoolini University, Solan, India
[3]CSIR-Food Research Institute, Box M20, Accra, Ghana

7.1 INTRODUCTION

In the 1960s, the National Cancer Institute endeavoured to screen thousands of plants to discover new compounds with anticancer activity. Paclitaxel, a natural product isolated from *Taxus brevifolia*, showed promising anti-cancerous properties. Chemically, paclitaxel is a pseudoalkaloid, a diterpenoid containing a tetracyclic taxane ring, with molecular formula $C_{47}H_{51}NO_{14}$, a monoisotopic mass of 853.33095530, which is insoluble in water (Figure 7.1). Among the chemotherapeutic drugs, paclitaxel is one of the most effective and potent against various cancers, such as ovarian, lung, and breast. It was formulated using cremophor EL (an excipient) and dehydrated ethanol (50:50, v/v). Despite the efficacy of paclitaxel in preclinical testing, some of the hurdles in approving its use in clinical trials (Walker, 1993) include its hypersensitivity reaction, haematologic toxicity, neurotoxicity (Scripture et al., 2006), cardiotoxicity (Bird & Swain, 2008), etc.

During the initial trials of paclitaxel, about 25% of the cohort had developed Type 1 hypersensitivity as side effects, including dyspnoea with bronchospasm, urticaria, and hypotension, observed from 2 to 10 minutes post administration after the first or second dose (Kloover et al., 2004). It was later found that the vehicle used (cremophor EL) was responsible for the side effects. Moreover, paclitaxel caused neutropenia, decreasing neutrophil counts to below 500 mm^3 after most infusions at a dose of 200–250 mg/m^2 used in phase II trials (He et al., 2020). Paclitaxel has been tested in breast (Holmes, 1996), lung (Pilkington et al., 2015), and ovarian cancer (Kumar et al., 2010), with the median endurance of response to paclitaxel being 9 months in breast cancer, 1–15 months with median 6 months with ovarian cancer, and 6.5 months (1-year survival rate of 41.7% and median survival rate of 24.1 weeks) in lung cancer. Thus, there is a need to find

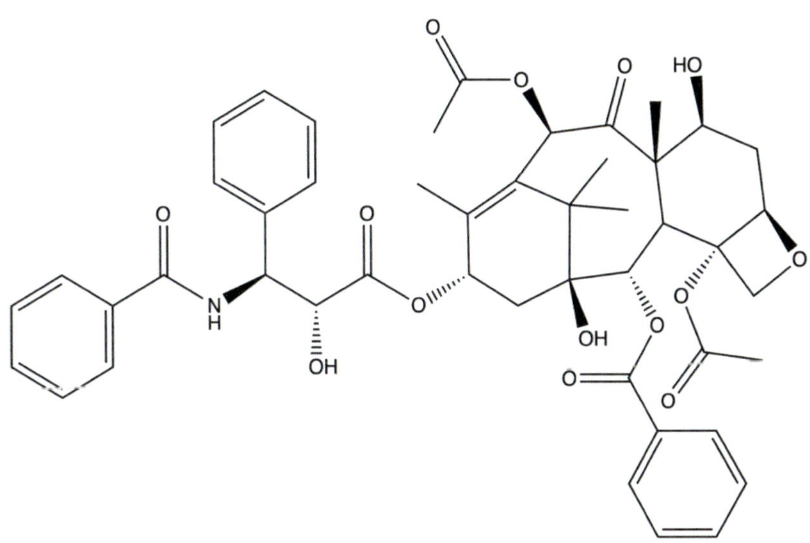

FIGURE 7.1 Chemical structure of paclitaxel [IUPAC name - (2α,4α,5β,7β,10β,13α)-4,10-Bis(acetyloxy)-13-{[(2*R*,3*S*)-3-(benzoylamino)-2-hydroxy-3-phenylpropanoyl]oxy}-1,7-dihydroxy-9-oxo-5,20-epoxytax-11-en-2-yl benzoate].

DOI: 10.1201/9781003611516-9

alternate ways to formulate this drug to reduce toxicity (Ma & Mumper, 2013).

Nanotechnology is a tool that is reforming approaches toward healthcare and is projected to have a positive effect by offering better healthcare amenities (Padhi et al., 2016; Kundu et al., 2022; Padhi et al., 2021; Padhi et al., 2022a). This technology makes drug delivery feasible and improves diagnoses (Ali et al., 2021). Cremophor EL-free solutions and local drug delivery approaches, including intra-tumoural or local drug delivery, are becoming more adapted for managing and treating cancer therapy due to the significant decline in systemic toxicity accompanying these approaches. Researchers are exploring novel delivery strategies like liposomes, nanoparticles, and other nanoformulations to administer paclitaxel through cremophor EL and dehydrated ethanol system (Marupudi et al., 2007). The degradation of paclitaxel during circulation has been reduced by encapsulating the drug in non-toxic and biodegradable nano-delivery systems, which also lessened the drug's toxic side effects in the body. Hence, the nanoformulations of paclitaxel lowered its toxicity, increased circulation half-life, exhibited improved pharmacokinetic profiles, and demonstrated improved patient compliance (Ma & Mumper, 2013).

One of the nano-encapsulated formulations of paclitaxel available in the market is Abraxane®. This formulation gained a lot of attention in 2005 after the successful delivery of paclitaxel with reduced toxicity in the body. The success of Abraxane® inspired new nanoformulations of paclitaxel keeping the generics aside. This led to preclinical and clinical evaluations and patentability leading to a significant increase in the regulatory approval of drugs using nanotechnology (Harshita et al., 2019).

7.2 MECHANISM OF ACTION OF PACLITAXEL AS AN ANTICANCER DRUG

The mechanism of action of paclitaxel was elucidated by Horwitz (1994), who demonstrated that paclitaxel inhibits cell replication and migration at a tubulin/microtubule level. Microtubules are biological polymers, a protofilament chain, of alpha and beta proteins up to 50 μm in length. Microtubules have important roles in cell movement, cell division, cell transport, signal transmission, and other cytoskeletal functions due to their chemical structures, dynamicity, and fluidity (Brouhard & Rice, 2018). After treatment with paclitaxel, the latter binds to the guanosine triphosphate, colchicine, vinblastine, and podofilox binding site of microtubules to form solid bundles that lose their dynamic behaviour and, therefore, are rendered inoperable. At low concentrations (about 10 nM), cell death occurs due to induced mitosis, aneuploidy and asymmetric cell division, whereas at high concentrations (20–50 nM) of paclitaxel, the formation of protein complexes and mitotic block occurs (Komlodi-Pasztor et al., 2011; Torres & Horwitz, 1998).

7.3 NANOENABLED TARGETING STRATEGIES FOR PACLITAXEL AS AN ANTICANCER AGENT

7.3.1 Passive Targeting Strategies

The targeting of tumour cells by nanocarriers increases with their cell uptake and accumulation (Padhi et al., 2018). Passive tumour targeting does not require acquiring a ligand for particular tissue or organ binding but depends on carrier characteristics such as circulation time and tumour biology (Fang et al., 2011; Padhi & Behera, 2022). Compared to typical organs, it offers only moderate tumour specificity, with a 20%–30% improvement in chemotherapy efficacy. The passive targeting effect is highly dependent on intrinsic tumour biology, specifically: (1) intratumor strain, (2) the degree of vasculature around the tumour and stromal response density, and (3) the degree of lymph angiogenesis and angiogenesis (Attia et al., 2019). Passive targeting of anticancer drugs follows the enhanced permeability and retention effect (Behera & Padhi, 2020). These considerations and nanocarriers' physical and chemical properties will determine the efficacy of drug delivery (Padhi et al., 2022a; Behera & Padhi, 2022). It is well known that the blood vessel endothelium has more permeability in conditions like hypoxia or inflammation, which is common in tumour cells (Torchilin, 2011).

In hypoxia, quickly developing tumours engulf established blood vessels or recruit new ones. These newly developed vessels enable macromolecules (generally larger than 40 kDa) and nano-systems selective entry into the tumour stroma. In addition, the nanomaterial's retention in the tumour is aided by a lack of natural lymphatic drainage. This distinguishing feature does not apply to small-molecule medications, which have a limited circulation period and rapid clearance from the tumour (Behera et al., 2023). As a result, encapsulating small molecules in nanocarriers increases their pharmacokinetics, offers tumour selectivity, and reduces side effects (Fang et al., 2011). Size is essential for retention and permeation in the tumour. It is thus restricted by tumour vessel fenestrations (200–800 nm) (Khuroo et al., 2014). However, the diameter of nanoformulations affects their renal excretion (6 nm) or via the reticuloendothelial system (500 nm) (Kobayashi et al., 2013). Charge and surface chemistry also significantly impact circulation time (Verma et al., 2017). As a result, the hydrophilic surface of the nanoparticle is favoured, and water-soluble polymers such as polyethylene glycols are grafted on the surface of nanocarriers to achieve this (Immordino et al., 2006; Padhi et al., 2022b). PEGylation avoids non-specific interactions and nanoparticle aggregation by hydration and changing surface charge (Noble et al., 2014; Padhi & Behera, 2021).

7.3.2 Active Targeting Strategies

Active targeting of anticancer drugs is critical for delivering the active moiety to the desired location without affecting normal tissues, improving efficacy, and decreasing side effects. Compared to conventional drug delivery, active targeting nano-systems can increase the drug distribution to the target cell (Patnaik et al., 2021). The nanocarrier surfaces can be conjugated with ligands that bind to overexpressed receptors on tumour cells. This technique increases the surface binding of the nano-systems to cancer cells and increases the cellular uptake of anticancer drugs (Attia et al., 2019). This was first suggested in 1980 by grafting antibodies on liposomes' surfaces and other ligands like aptamers, nucleic acids, and peptides. Transferrin and nicotinic acetylcholine receptors are other traditional targets (Kamaly et al., 2012; Wang et al., 2014). The nanocarriers targeting transferrin receptors and nicotinic acetylcholine receptors enter brain tumour cells through endothelial cells, also known as vascular targeting. Transferrin-targeting ligands were grafted on superparamagnetic iron oxide nanoparticles, dendrimers, micelles, and solid lipid nanoparticles and applied to glioma, distribution of anticancer drugs, or biomedical imaging (Attia et al., 2019). Furthermore, the literature describes cases in which glioblastoma and the central nervous system have been approached by treating nicotinic acetylcholine with micelles. Many antibody-targeting receptors on cancer cells have been identified, and accordingly, different ligand-conjugated nanoformulations have been synthesized and tested *in-vivo* and *in-vitro* (Zhan et al., 2011, 2012).

Targeting drugs to tumour cells with overexpressed receptors through nanocarriers ligand/receptor binding is the primary mechanism of active targeting technology. The arginyl-glycyl-aspartic acid peptide has been shown to bind to alpha V beta3 integrin. These receptors are abundant in both glioma cells and the vasculature of the tumour microenvironment. The F3 peptide binds to the nucleolin receptor, which is overexpressed on angiogenic endothelial cells. Similarly, aminopeptidase N (CD 13) has been identified as a possible overexpressed receptor ligated by a tripeptide (Asparagine-Glycine-Arginyl peptide). Among the most well-known ligands, folic acid is one of the most widely used ligands that directly bind to the folate receptor overexpressed in different cancer cells (Christian et al., 2003; Pasqualini et al., 2000).

7.4 Advanced Target Drug Delivery Systems for Paclitaxel as an Anticancer Agent

7.4.1 Nanoemulsion

Nanoemulsions are colloidal dispersions of safe-grade excipients that can be used as drug vehicles for molecules with low water solubility (Ganta et al., 2014; McClements, 2012). Nanoemulsions can encapsulate the drug inside its core, enhancing payload delivery and reducing toxicity (Behera et al., 2020). In the field of cancer therapy, there has been an increased focus on the behaviour of nanoemulsions to provide a better understanding of their characteristics. This is necessary to attain an effective therapeutic effect regarding specific targeting, superficial charge, the elevated half-life of circulation, large surface area, and the imaging capacity of the formulation. In an early study, Nornoo et al. (Nornoo et al., 2008; Nornoo & Chow, 2008) developed two micro-emulsion systems; the first system was a mixture of capmul: myvacet: water, and the second system was a mixture of lecithin: butanol: myvacet: water as the carriers for paclitaxel delivery and showed an increase in half-life as well as wide tissue distribution.

Moreover, Pawar et al. (2014) reported higher cytotoxicity in breast cancer cell lines (MCF-7 cells) by paclitaxel-loaded vitamin E nanoemulsion compared to free and marketed Taxol® formulations. The developed nanoemulsion efficiently enhanced cell cycle arrest, apoptosis, and modulated tumour cell immunology. An *in-vitro* study reported by Najlah et al. (2016) found that Clinoleic® and Intralipid® (parenteral nutrition nanoemulsion) when loaded with paclitaxel, showed anticancer activity against grade 4 glioma cell (U87-MG) and normal glial cell (SVG-P12 cell lines). Therefore, the nutritional lipid nanoemulsion has a good potential to serve as an alternate nanocarrier of paclitaxel for cancer treatment.

Kim and Park (2017a, b) developed a hyaluronan-coated nanoemulsion carrier loaded with paclitaxel, demonstrating an increased solubilizing capacity and an excellent hyaluronic acid-based cell-targeting ability. Synthesized paclitaxel-loaded hyaluronan solid nanoemulsions have thus been used to deliver paclitaxel to improve ovarian cancer treatment *via* active tumour targeting. Abu-Fayyad et al. (2018) recently used Tocosol™, a tocopherol-based non-ionic surfactant, to formulate paclitaxel nanoemulsion. The tocopherol (α-T) oil core of TocosolTM was substituted with vitamin E isomer γ-tocotrienol (γ-T3) isomer to develop a novel emulsion. The new γ-T_3 conjugated paclitaxel was loaded into these nanoemulsions and tested against pancreatic tumour cells (Bx-PC-3 and PANC-1) and showed a promising response to be used for cancer chemotherapy as they were found to have anticancer activity.

7.4.2 Liposomes

Liposomes are ideal drug delivery system candidates and are extensively studied as nanocarrier systems due to their safety, biocompatibility, high efficacy, biodegradability, and low toxicity (Yan et al., 2019). Liposomes are round-shaped structures of a hydrophilic core surrounded by a phospholipid bilayer, usually made of amphiphilic lipid materials. Liposomes are also one of the first drug delivery carriers to be studied for improving the pharmacokinetics of new anticancer drugs, as well as other liposomal medicines that have been approved for tumour therapy (Allen & Cullis, 2013; Barenholz, 2012; O'Brien et al., 2010; Zhang et al., 2009). The main advantage of liposomes is that they have the potential to target cancer tissues through active and passive targeting strategies. Additionally, as liposomes are primarily made of phospholipids, they are biocompatible carriers capable of incorporating both lipophilic and hydrophilic drugs (Koudelka & Turánek, 2012; Torchilin, 2005).

Early studies carried out by Yang et al. (2007) demonstrated that PEGylation of paclitaxel-loaded liposomes led to improved tumour and plasma levels of paclitaxel, inhibition of tumour growth, and, more importantly, reduced spleen and liver paclitaxel levels compared to conventional paclitaxel liposomes and Taxol®. To increase the solubility and specific targetability of paclitaxel, Meng et al. (2011) developed neovessel-targeted PEGylated paclitaxel liposomes that showed more efficacy than the Taxol®-treated group in decreasing tumour density in mice bearing lungs tumour (A549 cells). Tian et al. (2018) reported a new paclitaxel liposome system that targeted mitochondria and CD44 cells to improve drug accumulation in cell organelles and activate apoptosis of drug-resistant cancer cells.

A7R-cysteine peptide conjugated paclitaxel liposomes were prepared to evaluate their anticancer efficacy against high and low neuropilin-1 breast cancer cell lines (MDA-MB-231; MCF-7) and human umbilical vein endothelial cells (HUVECs) (Cao et al., 2015). This modified conjugation resulted in enhanced delivery of paclitaxel for treating breast cancer in the future. Still, more *in vivo* studies on the inhibition rate of tumour cells are required (Cao et al., 2015). In 2016, breast cancer cell line T47D and 4T1 tumour-bearing mice were examined by a hyaluronic acid-coated paclitaxel liposome formulation by Ravar et al. (2016). Dose-dependently, the developed formulation could elicit antitumour activity and be used further for cancer chemotherapy. Estrone-conjugated long-lasting paclitaxel liposomes containing polyethylene glycol long-acting and estrone-targeting fragments were formed. *In-vivo* and *in-vitro* studies evaluated that the prolongation of drug action, increase in tumour cell drug uptake, increase in drug accumulation in target sites, and targeting of breast cancer cells (MCF-7) by developed liposomes made it a potential source of breast cancer chemotherapy for a healthy life (Han et al., 2020).

7.4.3 Carbon Nanotubes

Carbon nanotubes enhance the drug delivery system because of their biocompatibility, selectivity, solubility, and decreased retention and toxicity in the tumour. As they are chemically inert and undissolved in solvents, they bind to covalent and noncovalent interactions to render solubility and attach to chemical moieties (Hashemzadeh & Raissi, 2017). Carbon nanotubes are chemically stable nanomaterials. They are molecules of cylindrical shape and are made of rolled-up sheets of single-layer carbon atoms (graphene). Carbon nanotubes can be classified into two categories: in the first instance, single-walled carbon nanotubes having a diameter of less than 1 nm and multi-walled carbon nanotubes consisting of numerous interlinked nanotubes with more than 100 nm diameter. In 2008, Liu and co-workers (2008) conjugated paclitaxel to branched polyethylene glycol on single-walled carbon nanotubes to form water-soluble carbon nanotubes. These paclitaxel-loaded polyethylene glycol-functionalized carbon nanotubes showed great stability, biocompatibility, and little toxicity. In 2010, Lay and co-workers (Lay et al., 2010) used polyethylene glycol-functionalized single or multiple-walled carbon nanotubes to load paclitaxel. *In-vitro* cytotoxicity studies in HeLa and MCF-7 cells exhibited that both paclitaxel-loaded carbon nanotubes extensively improved cell death compared to free paclitaxel.

Furthermore, single-walled carbon nanotubes can protect the drug against enzymatic degradation before entering the tumour cells (Zhou et al., 2012). A novel drug delivery system was constructed based on single-walled carbon nanotubes to realize the controlled release of paclitaxel. Single-walled carbon nanotubes provide several advantages as drug delivery vehicles (Yu et al., 2016). Single-walled carbon nanotubes are pre-functionalized covalently with folic acid (a ligand) and paclitaxel (an anticancer drug) for many tumours. The outcome showed good conjugation of the anticancer drug and targeting molecule on the surface of the carbon nanotubes (Tavakolifard et al., 2016).

Sobhani and co-workers (2011) functionalized multiple-walled carbon nanotubes using hyperbranched polycitric acid and conjugated paclitaxel on the carbon nanotubes. The multiple-walled carbon nanotubes-paclitaxel complex enhanced the effects of paclitaxel on the cell cycle or apoptosis compared to paclitaxel alone. Multi-walled carbon nanotubes/paclitaxel complex encouraged apoptosis in the gastric adenocarcinoma cell line (MKN-45) (Ghasemvand et al., 2016). Rathod et al. (2019) reported the synthesized multi-walled carbon nanotubes from xylene/ferrocene using the catalytic chemical vapour deposition method, showing improved aqueous dispersity and biocompatibility. Carbon nanotubes, including both multi-walled and single-walled carbon nanotubes, show exceptionally good properties in mechanical, electrical, and thermal applications (Hashemzadeh & Raissi, 2017; Pop et al., 2006).

7.4.4 Nanodiamonds

Diamond nanoparticles, identified as nanodiamonds, are single-crystal diamonds comprising carbon as the basic constituent with high chemical and physical properties (Chauhan et al., 2020). Nanodiamonds are novel nano-sized carbon building blocks with important active moiety carriers for biomedical use. The nanodiamonds have a size of 4–5 nm. There are numerous techniques for producing nanodiamonds (Chaudhary et al., 2010; Kharisov et al., 2010). Nanodiamonds are well-known as ultra-disperse diamonds having special characteristics of diamond core, viz., superior thermal conductivity, chemically inert core electrical resistivity, prominent hardness, resistance to harsh environments, and biocompatibility (Georgakilas et al., 2015; Mochalin et al., 2012). Nanodiamonds can track and label cancer cells (Liu et al., 2009). Liu et al. (2010) reported a functional covalently bonded nano-diamond-paclitaxel conjugate developed for an innovative cancer therapy and drug delivery strategy. The nanodiamond/hydrophobic drug complex exhibited the physical adsorption or chemically conjugated form (Cui et al., 2016; Zhao et al., 2014). Compared to the 47.33% release of bare crystalline, hydroxylated nanodiamond/paclitaxel complexes retained the sustained release of paclitaxel by up to 97.32% at 70 hours. The IC_{50} values of nanodiamond/paclitaxel complex and paclitaxel complexes in HeLa cancer cells were 0.037 and 0.137 μg/mL, respectively (Lim et al., 2016).

A nanocomposite was designed using nanodiamonds conjugated with paclitaxel and cetuximab as a feasible approach for triple-negative breast cancer. Breast cancer cells (MDA-MB-231) were analysed for nanodiamond/paclitaxel-cetuximab efficacy. Nanodiamond/paclitaxel-cetuximab was found to bind with epidermal growth factor receptor and elicited anticancer effects as drug uptake levels in epidermal growth factor receptor-expressed MDA-MB-231 cells were enhanced, leading to apoptosis and mitotic catastrophe. So, this study suggested nanodiamond/paclitaxel-cetuximab as a lead for human triple-negative breast cancer therapy (Liao et al., 2019).

7.4.5 Nanoparticles

Nanoparticles are the ultrafine particles or smallest microscopic particles with sizes of 1–100 nm (Saxena et al., 2010). Nanoparticles are one of the most commonly studied materials due to their exclusive chemical and physical characteristics that begin from their nanoscale dimensions (Giljohann et al., 2010). Nanoparticles are gaining popularity in cancer therapy because they can co-encapsulate multiple therapeutic agents in target-specific drug delivery (Hu & Zhang, 2012). In 2005, Jain et al. (2005) synthesized pluronic-stabilized and oleic acid-coated metal nanoparticles. Two anticancer drugs, paclitaxel and doxorubicin, were incorporated into metal nanoparticles, and a combination of the two drugs exhibited synergistic antiproliferative effects.

Gold-chitosan-pluronic composite nanoparticles were successfully synthesized and utilized for paclitaxel delivery. In 2008, Oh and co-workers (2008) reported that gold nanoparticles served as the core of the composite nanoparticles, and paclitaxel could be easily released from the nanocomposite. A photoinduced gold nanoparticle-capped mesoporous silica nanosphere has been designed as a paclitaxel delivery system in which the surface charge of gold nanoparticles was shifted from positive charge to negative charge in the presence of ultraviolet irradiation.

In 2009, Vivero and co-workers (2009) observed that these nanoparticles effectively reduced cell viability in combination with UV irradiation. Partha and colleagues (2008) reported buckysome, a water-soluble round-shaped nanostructure (100–200 nm diameter) with an amphiphilic fullerene AF-1, loaded with paclitaxel. Its therapeutic efficiency was evaluated *in-vitro* in breast cancer cells (MCF-7). Paclitaxel-loaded polymer-based metal nanoparticles were developed as a targeting delivery system. The *in vitro* study revealed that paclitaxel-loaded metal nanoparticles had the same antiproliferative effect on rat aortic smooth muscle cells (A10) as free paclitaxel. The magnetic field amplified the effects of metal nanoparticles (Chorny et al., 2010).

Poly(lactic-co-glycolic) acid, polylactide, poly(ε-caprolactone), chitosan, and hyaluronic acid nanoparticles are some of the paclitaxel-loaded polymeric nanoparticles that are biocompatible, superior, and robust for functioning/construction of paclitaxel (des Rieux et al., 2006; Panyam & Labhasetwar, 2003). Following these reports, a study in 2015 reported that paclitaxel-loaded poly(lactic-co-glycolic) acid-b-polyethylene glycol-b-poly(lactic-co-glycolic) acid nanoparticles had significantly stronger anticancer efficacy than Taxol, and also surface decoration combined with epidermal growth factor receptor was one of the methods for achieving paclitaxel-poly(lactic-co-glycolic) acid-b-polyethylene glycol nanoparticles (Gao et al., 2015; Milane et al., 2011).

In-vitro and *in-vivo* studies have shown that poly(lactide) nanoformulation has higher biomedical efficacy and fewer side effects than Taxol®. Polyethylene glycol, D-alpha-tocopheryl polyethylene glycol 1000 succinate, polyethylene glycol-b-poly(lactide)-b-polyethylene glycol, poly(lactide)-b-polyethylene glycol-b-poly(lactide), poly(lactide)/Tween 80, poly(γ-glutamic acid)-poly(lactide), and poly[(amine-ester)-co-(d,l-lactide)] are various polymers used for effective paclitaxel delivery (Dong & Feng, 2004; He et al., 2007; Liang et al., 2006; Zhang et al., 2012; Zhang & Feng, 2006). For the metastatic breast cancer treatment paclitaxel-loaded poly(ε-caprolactone) nanoparticles were developed by Su et al. (2016, 2020), where poly(ε-caprolactone) acted as a reservoir for paclitaxel and tumour targeting and long circulation, endowed in the cell membrane (Anselmo & Mitragotri, 2014).

Abraxane® is an albumin-based nanoparticle of paclitaxel, which is Food and Drug Administration-approved for non-small cell lung cancer, metastatic pancreatic cancer, and breast cancer. It is superior to Taxol® in adjuvant treatment and has also decreased breast cancer stem cell frequency ninefold in orthotopic breast cancer xenografts in nonobese diabetic/severe combined immunodeficient mice. Abraxane® can be an effective treatment for eliminating breast cancer stem cells (Yuan et al., 2020).

7.4.6 Dendrimers

Dendrimers are spherical polymeric particles with a large surface area to encapsulate therapeutically active moieties (Duncan & Izzo, 2005). The formulation of dendrimers exploits malignant tumour characteristics and morphology like specific cell surface antigen expression, leaky vasculature, and rapid proliferation (Soni et al., 2017). Choi and co-workers (2005) described DNA-accumulated polyamidoamine dendrimers for tumour-cell-specific targeting. With active targeting dendrimers, there are promising preclinical results *in vitro* and *in vivo* (Wolinsky & Grinstaff, 2008). In 2006, Shukla et al. (2006) reported enhanced efficacy in the antibody-dendrimer conjugates compared to the free antibodies. As passive anticancer nanocarriers, these dendrimers have already been used (Dhanikula & Hildgen, 2007; Khandare et al., 2006; Li et al., 2012; Wang et al., 2013). Yao and Ma (2018) also observed that the dendrimer-paclitaxel complex was more effective than paclitaxel alone in ovarian cancer and proved a potential drug carrier for targeting ovarian cancer cells.

Teow and co-workers (2013) reported novel dendrimer-based drug delivery systems consisting of third-generation polyamidoamine dendrimer-based carriers to overcome biological barriers and improve Paclitaxel's permeability. The paclitaxel-conjugated, fifth-generation polyamidoamine dendrimers have the potential to change the structure of microtubules by two mechanisms: (1) by bundling preformed microtubules and (2) by stabilizing polymerizing microtubules (Cline et al., 2013). Paclitaxel-conjugated fifth-generation polyamidoamine dendrimers have demonstrated cytotoxicity against cancerous cells (Majoros et al., 2006). Rompicharla et al. (2018) developed a dendrimer system with great potential as a productive chemotherapeutic agent for cancer and potent for the intracellular delivery of Paclitaxel in tumour cells.

7.4.7 Quantum Dots

Quantum dots are ultra-small-sized semiconductor nanocrystals of 100–10,000 atoms (Bera et al., 2010). Probst and colleagues (2013) reported that the therapeutic and diagnostic fields are quantum dots. Quantum dots have been explored extensively for therapeutic and medical applications (Pawar et al., 2018). In 2010, it was shown that quantum dots demonstrated significant results in cancer detection or tumours (Gao et al., 2010). Kuo and co-workers (2009) concluded that compared to free paclitaxel drug, the poly (D,L lactide-co-glycolide conjugated paclitaxel-loaded in quantum dots showed more cytotoxic. Paclitaxel-loaded poly (D,L-lactide-*co*-glycolide)-quantum dots are a new therapeutic tool for curing cancers.

In 2017, Olerile and co-workers (2017) reported a novel multifunctional co-loaded quantum dots and Paclitaxel in nanostructured lipid carriers. Olerile (2020) concluded that the co-loaded nanostructured lipid carriers could be a splendid parenteral drug delivery system for cancer therapy and the most suitable formulation for treating breast cancer. Kang and co-workers (2018) developed a tumour-targeted theranostic nanocarrier system, anti-endothelial growth factor receptor antibodies or anti-endothelial growth factor receptor aptamers were coupled to micelle nanoparticles co-encapsulating paclitaxel and fluorescence quantum dots (called as immuno-paclitaxel-quantum dots and aptamer-paclitaxel-quantum dots). They concluded that the aptamer-paclitaxel-quantum dots or immuno-paclitaxel-quantum dots could be used as tumour-targeting theranostic nanoparticles. A report by Maxwell and co-workers (2020) concluded that quantum dots exhibit unique optical properties due to changes in band gap energy caused by quantum confinement effects.

7.5 REPORT ON CLINICAL TRIALS USING PACLITAXEL

Paclitaxel clinical trials were delayed for many years because of raw material shortages due to slow growth from the Pacific yew tree, *Taxus brevifolia*. However, a clinical study on ovarian cancer concluded that 30% of patients with advanced ovarian cancer responded to paclitaxel therapy (Weaver, 2014). This has led to numerous clinical trials to understand the effectiveness of paclitaxel as a single dose or combined with other treatments in the fight against cancer. However, finding an ideal dosage for effective relief has been challenging. Although the dose ranges from 60 to 250 mg/m^2, over 1–96 hours, and from 1 to 3 weekly intervals used in practice, there is no ideal dosing strategy for paclitaxel in treatments such as ovarian cancer (Kumar et al., 2010). Intense clinical trials have been conducted to find the least expensive dosing approach, hence the evolution of a broad range of more convenient dosing schedules (Kumar et al., 2010).

Recent reports suggest that stomach cancer is the third most common cause of death from cancer in males and fourth in females (Xu et al., 2013), with approximately 1,089,103 new cancer cases and 768,793 deaths in 2020 (Sung et al., 2021). Chemotherapy, by far, has led to a significant survival difference compared to the best supportive care and could help relieve gastric cancer-related symptoms and improve quality of life (Xu et al., 2013). However, a study by Xu et al. (2013) involving 62 patients with advanced gastric cancer receiving chemotherapy was recruited to receive two treatment regimens of paclitaxel. Thus, patients in group A received paclitaxel liposomes at a dose of 135 mg/m^2 on day 1 of each cycle; patients in group B were given paclitaxel at the same dose with the same timing. In response, all the patients received tegafur at a dose of 500 mg/m^2 on days 1–5 and oxaliplatin at 80–100 mg/m^2 on day 1 for 2 cycles. Their findings suggested that paclitaxel liposome is as effective as conventional paclitaxel when combined with tegafur and oxaliplatin in treating patients with advanced gastric cancer.

Fujita et al. (2017) reported on the safety and efficacy of treatment with nab-paclitaxel and carboplatin for patients with advanced squamous non-small cell lung cancer. They treated eight patients with inoperable-stage lung squamous cell carcinoma and idiopathic interstitial pneumonia with nab-paclitaxel plus carboplatin and observed that the overall response rate was 50%. However, no patient experienced a chemotherapy-related exacerbation of the IIPs in the first-line treatment with nab-paclitaxel plus carboplatin, and their findings suggested that treatment with nab-paclitaxel may be an effective and feasible regimen for patients with lung squamous cell carcinoma and idiopathic interstitial pneumonia.

There has been a different treatment regimen for the use of paclitaxel in cancer treatment, which has gotten a lot of attention. In randomized, multicentre, open-label, and phase III trials on carboplatin plus paclitaxel once a week versus every 3 weeks in patients with advanced ovarian cancer (MITO-7), Pignata et al. (2014) concluded that paclitaxel given once a week compared with the standard regimen administered every 3 weeks did not significantly prolong progression-free survival in women with ovarian cancer. Besides, the once-a-week regimen augmented the quality of life and had a better toxicity profile. However, they noted that this positive effect on the quality of life recorded with a once-a-week dosage might be accounted for simply by the frequency and severity of side effects (Pignata et al., 2014).

7.6 CONCLUSION AND FUTURE PERSPECTIVES

This chapter discusses how nanotechnology is fundamental in developing new biomedical applications. With the rise of cancer worldwide, there has been an urgent and life-dependent demand for new approaches to diagnosis and drug delivery systems in anticancer therapy. Undoubtedly, a plethora of research is being performed to develop and test a wide range of novel nanocarriers, such as inorganic nanoparticles, liposomes, carbon-nanomaterials, and vesicles, to evaluate their role, efficiency, and toxicity as an anticancer agent *in-vitro* and *in-vivo* (Yang et al., 2016). These novel approaches for paclitaxel-based nano-delivery systems as an alternative to the standard and first formulation of paclitaxel (Taxol®) have led to a safer and more efficient approach in cancer therapy. Various paclitaxel-combined formulations have been found to reduce toxicity levels compared to paclitaxel alone. In other words, using highly potent paclitaxel alone will not be as useful as combining therapies such as nanoenabled technologies with paclitaxel for cancer chemotherapeutics.

Abraxane® and Taxol® are the most promising nanoformulations developed for chemotherapy against various cancers. Abraxane® is the most recommended treatment for patients by medical analysts, but its dose is five-fold higher than the paclitaxel formulation (Taxol®). Genexol-PM®, Lipusu®, and Nanoxel® are the alternative paclitaxel nanoformulations that are successfully available in the market; however, these new formulations have not been approved by the Food and Drug Administration for use. The dilemma is that while paclitaxel nanoformulations are effective, their mechanism of action, complex synthetic routes, multi-stepped modification, and separation remain unknown and risky to some extent.

In the future, molecular profiling of paclitaxel-based nanoparticles would bridge the knowledge gap between laboratory testing and clinical trials, thus life-changing for cancer behaviour, outcome, treatment response, and individual therapy predictions. Another area of research that could positively impact is the use of caveolin transcytosis and cell-based strategies, which can potentially increase the absorption of paclitaxel nanoparticles into solid tumours. Metabolism and pharmacological studies using *in-vivo* models should be examined for cancer imaging and therapy. The current evaluation scenario will be a cornerstone for developing an efficient paclitaxel nano-delivery system with a well-known mechanism and a vivid prospect of biomedical applications in oncology.

REFERENCES

Abu-Fayyad, A., Kamal, M. M., Carroll, J. L., Dragoi, A.-M., Cody, R., Cardelli, J., & Nazzal, S. (2018). Development and in-vitro characterization of nanoemulsions loaded with paclitaxel/γ-tocotrienol lipid conjugates. *International Journal of Pharmaceutics*, 536(1), 146–157.

Ali, E. S., Sharker, S. M., Islam, M. T., Khan, I. N., Shaw, S., Rahman, M. A., Uddin, S. J., Shill, M. C., Rehman, S., Das, N., Ahmad, S., Shilpi, J. A., Tripathi, S., Mishra, S. K., & Mubarak, M. S. (2021). Targeting cancer cells with nanotherapeutics and nanodiagnostics: Current status and future perspectives. *Seminars in Cancer Biology*, 69, 52–68.

Allen, T. M., & Cullis, P. R. (2013). Liposomal drug delivery systems: From concept to clinical applications. *Advanced Drug Delivery Reviews*, 65(1), 36–48.

Anselmo, A. C., & Mitragotri, S. (2014). Cell-mediated delivery of nanoparticles: Taking advantage of circulatory cells to target nanoparticles. *Journal of Controlled Release: Official Journal of the Controlled Release Society*, 190, 531–541.

Attia, M. F., Anton, N., Wallyn, J., Omran, Z., & Vandamme, T. F. (2019). An overview of active and passive targeting strategies to improve the nanocarriers efficiency to tumour sites. *Journal of Pharmacy and Pharmacology*, 71(8), 1185–1198.

Barenholz, Y. (2012). Doxil®–the first FDA-approved nano-drug: Lessons learned. *Journal of Controlled Release: Official Journal of the Controlled Release Society*, 160(2), 117—134.

Behera, A., & Padhi, S. (2020). Passive and active targeting strategies for the delivery of the camptothecin anticancer drug: A review. *Environmental Chemistry Letters*, 18(5), 1557–1567.

Behera, A., & Padhi, S. (2022). pH-Sensitive polymeric nanoparticles for cancer treatment. In: Padhi, S., Behera, A., & Lichtfouse, E. (Eds.) Polymeric nanoparticles for the treatment of solid tumours. *Environmental Chemistry for a Sustainable World*, vol 71. Springer, Cham, pp.401–425.

Behera, A., Mittu, B., Padhi, S., & Singh, A. (2020). Antimicrobial efficacy of essential oil nanoemulsions. In: Dhull, S. B., Chawla, P., & Kaushik, R. (Eds.) *Ravinder Nanotechnological Approaches in Food Microbiology*. CRC Press, Boca Raton, FL, pp. 293–309.

Behera, A., Padhi, S., & Nayak, A. K. (2023). Engineered liposomes as drug delivery and imaging agents. In: Ray, S., & Nayak, A. K. (Eds.) *Design and Applications of Theranostic Nanomedicines*. Woodhead Publishing, Elsevier, Cambridge, MA, pp. 75–108.

Bera, D., Qian, L., Tseng, T.-K., & Holloway, P. H. (2010). Quantum dots and their multimodal applications: A review. *Materials*, 3(4), 2260–2345.

Bird, B. R. J. H., & Swain, S. M. (2008). Cardiac toxicity in breast cancer survivors: Review of potential cardiac problems. *Clinical Cancer Research*, 14(1), 14 LP – 24.

Brouhard, G. J., & Rice, L. M. (2018). Microtubule dynamics: An interplay of biochemistry and mechanics. *Nature Reviews. Molecular Cell Biology*, 19(7), 451–463.

Cao, J., Wang, R., Gao, N., Li, M., Tian, X., Yang, W., Ruan, Y., Zhou, C., Wang, G., Liu, X., Tang, S., Yu, Y., Liu, Y., Sun, G., Peng, H., & Wang, Q. (2015). A7RC peptide modified paclitaxel liposomes dually target breast cancer. *Biomaterials Science*, 3(12), 1545–1554.

Chaudhary, A., Welch, J. O., & Jackman, R. B. (2010). Electrical properties of monodispersed detonation nanodiamonds. *Applied Physics Letters*, 96(24), 242903.

Chauhan, S., Jain, N., & Nagaich, U. (2020). Nanodiamonds with powerful ability for drug delivery and biomedical applications: Recent updates on in vivo study and patents. *Journal of Pharmaceutical Analysis*, 10(1), 1–12.

Choi, Y., Thomas, T., Kotlyar, A., Islam, M. T., & Baker, J. R. (2005). Synthesis and functional evaluation of DNA-assembled polyamidoamine dendrimer clusters for cancer cell-specific targeting. *Chemistry & Biology*, 12(1), 35—43.

Chorny, M., Fishbein, I., Yellen, B. B., Alferiev, I. S., Bakay, M., Ganta, S., Adamo, R., Amiji, M., Friedman, G., & Levy, R. J. (2010). Targeting stents with local delivery of paclitaxel-loaded magnetic nanoparticles using uniform fields. *Proceedings of the National Academy of Sciences of the United States of America*, 107(18), 8346—8351.

Christian, S., Pilch, J., Akerman, M. E., Porkka, K., Laakkonen, P., & Ruoslahti, E. (2003). Nucleolin expressed at the cell surface is a marker of endothelial cells in angiogenic blood vessels. *Journal of Cell Biology*, 163(4), 871–878.

Cline, E. N., Li, M.-H., Choi, S. K., Herbstman, J. F., Kaul, N., Meyhöfer, E., Skiniotis, G., Baker, J. R., Larson, R. G., & Walter, N. G. (2013). Paclitaxel-conjugated PAMAM dendrimers adversely affect microtubule structure through two independent modes of action. *Biomacromolecules*, 14(3), 654–664.

Cui, Z., Zhang, Y., Zhang, J., Kong, H., Tang, X., Pan, L., Xia, K., Aldalbahi, A., Li, A., Tai, R., Fan, C., & Zhu, Y. (2016). Sodium alginate-functionalized nanodiamonds as sustained chemotherapeutic drug-release vectors. *Carbon*, 97, 78–86.

des Rieux, A., Fievez, V., Garinot, M., Schneider, Y.-J., & Préat, V. (2006). Nanoparticles as potential oral delivery systems of proteins and vaccines: A mechanistic approach. *Journal of Controlled Release : Official Journal of the Controlled Release Society*, 116(1), 1—27.

Dhanikula, R. S., & Hildgen, P. (2007). Influence of molecular architecture of polyether-co-polyester dendrimers on the encapsulation and release of methotrexate. Biomaterials, 28(20), 3140–3152.

Dong, Y., & Feng, S.-S. (2004). Methoxy poly(ethylene glycol)-poly(lactide) (MPEG-PLA) nanoparticles for controlled delivery of anticancer drugs. *Biomaterials*, 25(14), 2843–2849.

Duncan, R., & Izzo, L. (2005). Dendrimer biocompatibility and toxicity. *Advanced Drug Delivery Reviews*, 57(15), 2215—2237.

Fang, J., Nakamura, H., & Maeda, H. (2011). The EPR effect: Unique features of tumour blood vessels for drug delivery, factors involved, and limitations and augmentation of the effect. *Advanced Drug Delivery Reviews*, 63(3), 136–151.

Fujita, T., Hiroishi, T., Yanagisawa, A., Hayama, N., Amano, H., Nakamura, M., Hirano, S., Tabeta, H., & Nakamura, S. (2017). The safety and efficacy of treatment with nab-paclitaxel and carboplatin for patients with advanced squamous non-small cell lung cancer with interstitial lung disease. *European Respiratory Journal*, 50(61), OA1474.

Ganta, S., Talekar, M., Singh, A., Coleman, T. P., & Amiji, M. M. (2014). Nanoemulsions in translational research-opportunities and challenges in targeted cancer therapy. *AAPS PharmSciTech*, 15(3), 694—708.

Gao, J., Chen, X., & Cheng, Z. (2010). Near-infrared quantum dots as optical probes for tumour imaging. *Current Topics in Medicinal Chemistry*, 10(12), 1147–1157.

Gao, N., Chen, Z., Xiao, X., Ruan, C., Mei, L., Liu, Z., & Zeng, X. (2015). Surface modification of paclitaxel-loaded tri-block copolymer PLGA-b-PEG-b-PLGA nanoparticles with protamine for liver cancer therapy. *Journal of Nanoparticle Research*, 17(8), 347.

Georgakilas, V., Perman, J. A., Tucek, J., & Zboril, R. (2015). Broad family of carbon nanoallotropes: Classification, chemistry, and applications of fullerenes, carbon dots, nanotubes, graphene, nanodiamonds, and combined superstructures. *Chemical Reviews*, 115(11), 4744–4822.

Ghasemvand, F., Biazar, E., Tavakolifard, S., Khaledian, M., Rahmanzadeh, S., Momenzadeh, D., Afroosheh, R., Zarkalami, F., Shabannezhad, M., Hesami Tackallou, S., Massoudi, N., & Heidari Keshel, S. (2016). Synthesis and evaluation of multi-wall carbon nanotube-paclitaxel complex as an anticancer agent. *Gastroenterology and Hepatology from Bed to Bench*, 9(3), 197–204.

Giljohann, D. A., Seferos, D. S., Daniel, W. L., Massich, M. D., Patel, P. C., & Mirkin, C. A. (2010). Gold nanoparticles for biology and medicine. *Angewandte Chemie International Edition*, 49(19), 3280–3294.

Han, B., Yang, Y., Chen, J., Tang, H., Sun, Y., Zhang, Z., Wang, Z., Li, Y., Li, Y., Luan, X., Li, Q., Ren, Z., Zhou, X., Cong, D., Liu, Z., Meng, Q., Sun, F., & Pei, J. (2020). Preparation, characterization, and pharmacokinetic study of a novel long-acting targeted paclitaxel liposome with antitumour activity. *International Journal of Nanomedicine*, 15, 553–571.

Harshita, B., Barkat, M. A., Beg, S., Pottoo, F. H., & Ahmad, F. J. (2019). Nanopaclitaxel therapy: An evidence based review on the battle for next-generation formulation challenges. *Nanomedicine*, 14(10), 1323–1341.

Hashemzadeh, H., & Raissi, H. (2017). The functionalization of carbon nanotubes to enhance the efficacy of the anticancer drug paclitaxel: A molecular dynamics simulation study. *Journal of Molecular Modeling*, 23(8), 222.

He, G., Ma, L. L., Pan, J., & Venkatraman, S. (2007). ABA and BAB type triblock copolymers of PEG and PLA: A comparative study of drug release properties and "stealth" particle characteristics. *International Journal of Pharmaceutics*, 334(1), 48–55.

He, Y., Li, T., Liu, J., Ou, Q., & Zhou, J. (2020). Early onset neutropenia: A useful predictor of chemosensitivity and favorable prognosis in patients with serous ovarian cancer. *BMC Cancer*, 20(1), 116.

Holmes, F. A. (1996). Paclitaxel combination therapy in the treatment of metastatic breast cancer: A review. *Seminars in Oncology*, 23(5 Suppl 11), 46–56.

Horwitz, S. B. (1994). Taxol (paclitaxel): Mechanisms of action. *Annals of Oncology*, 5, S3.

Hu, C.-M. J., & Zhang, L. (2012). Nanoparticle-based combination therapy toward overcoming drug resistance in cancer. *Biochemical Pharmacology*, 83(8), 1104–1111.

Immordino, M. L., Dosio, F., & Cattel, L. (2006). Stealth liposomes: Review of the basic science, rationale, and clinical applications, existing and potential. *International Journal of Nanomedicine*, 1(3), 297–315.

Jain, T. K., Morales, M. A., Sahoo, S. K., Leslie-Pelecky, D. L., & Labhasetwar, V. (2005). Iron oxide nanoparticles for sustained delivery of anticancer agents. *Molecular Pharmaceutics*, 2(3), 194–205.

Kamaly, N., Xiao, Z., Valencia, P. M., Radovic-Moreno, A. F., & Farokhzad, O. C. (2012). Targeted polymeric therapeutic nanoparticles: Design, development and clinical translation. *Chemical Society Reviews*, 41(7), 2971–3010.

Kang, S. J., Jeong, H. Y., Kim, M. W., Jeong, I. H., Choi, M. J., You, Y. M., Im, C. S., Song, I. H., Lee, T. S., & Park, Y. S. (2018). Anti-EGFR lipid micellar nanoparticles co-encapsulating quantum dots and paclitaxel for tumour-targeted theranosis. *Nanoscale*, 10(41), 19338–19350.

Khandare, J. J., Jayant, S., Singh, A., Chandna, P., Wang, Y., Vorsa, N., & Minko, T. (2006). Dendrimer versus linear conjugate: Influence of polymeric architecture on the delivery and anticancer effect of paclitaxel. *Bioconjugate Chemistry*, 17(6), 1464–1472.

Kharisov, B. I., Kharissova, O. V., & Chávez-Guerrero, L. (2010). Synthesis techniques, properties, and applications of nanodiamonds. *Synthesis and Reactivity in Inorganic, Metal-Organic, and Nano-Metal Chemistry*, 40(2), 84–101.

Khuroo, T., Verma, D., Talegaonkar, S., Padhi, S., Panda, A. K., & Iqbal, Z. (2014). Topotecan–tamoxifen duple PLGA polymeric nanoparticles: Investigation of in vitro, in vivo and cellular uptake potential. *International Journal of Pharmaceutics*, 473(1–2), 384–394.

Kim, J. E., & Park, Y. J. (2017b). Paclitaxel-loaded hyaluronan solid nanoemulsions for enhanced treatment efficacy in ovarian cancer. *International Journal of Nanomedicine*, 12, 645–658.

Kim, J.-E., & Park, Y.-J. (2017a). High paclitaxel-loaded and tumour cell-targeting hyaluronan-coated nanoemulsions. *Colloids and Surfaces B: Biointerfaces*, 150, 362–372.

Kloover, J. S., den Bakker, M. A., Gelderblom, H., & van Meerbeeck, J. P. (2004). Fatal outcome of a hypersensitivity reaction to paclitaxel: A critical review of premedication regimens. *British Journal of Cancer*, 90(2), 304–305.

Kobayashi, H., Watanabe, R., & Choyke, P. L. (2013). Improving conventional enhanced permeability and retention (EPR) effects; what is the appropriate target? *Theranostics*, 4(1), 81–89.

Komlodi-Pasztor, E., Sackett, D., Wilkerson, J., & Fojo, T. (2011). Mitosis is not a key target of microtubule agents in patient tumours. *Nature Reviews Clinical Oncology*, 8(4), 244–250.

Koudelka, Š., & Turánek, J. (2012). Liposomal paclitaxel formulations. *Journal of Controlled Release*, 163(3), 322–334.

Kumar, S., Mahdi, H., Bryant, C., Shah, J. P., Garg, G., & Munkarah, A. (2010). Clinical trials and progress with paclitaxel in ovarian cancer. *International Journal of Women's Health*, 2(1), 411–427.

Kundu, A., Padhi, S., Behera, A., Hasnain, M. S., & Nayak, A. K. (2022). Tumour targeting strategies by chitosan-based nanocarriers. In: Hasnain, M. S., Beg, S., & Nayak, A. K. (Eds.) *Chitosan in Biomedical Applications*. Academic Press, London, pp. 163–188.

Kuo, W.-S., Ku, Y.-C., Sei, H.-T., Cheng, F.-Y., & Yeh, C.-S. (2009). Paclitaxel-loaded stabilizer-free poly(D,L-lactide-co-glycolide) nanoparticles conjugated with quantum dots for reversion of anticancer drug resistance and cancer cellular imaging. *Journal of the Chinese Chemical Society*, 56(5), 923–934.

Lay, C. L., Liu, H. Q., Tan, H. R., & Liu, Y. (2010). Delivery of paclitaxel by physically loading onto poly(ethylene glycol) (PEG)-graftcarbon nanotubes for potent cancer therapeutics. *Nanotechnology*, 21(6), 65101.

Li, M.-H., Choi, S. K., Thomas, T. P., Desai, A., Lee, K.-H., Kotlyar, A., Banaszak Holl, M. M., & Baker Jr, J. R. (2012). Dendrimer-based multivalent methotrexates as dual acting nanoconjugates for cancer cell targeting. *European Journal of Medicinal Chemistry*, 47(1), 560–572.

Liang, H.-F., Chen, S.-C., Chen, M.-C., Lee, P.-W., Chen, C.-T., & Sung, H.-W. (2006). Paclitaxel-loaded poly(γ-glutamic acid)-poly(lactide) nanoparticles as a targeted drug delivery system against cultured HepG2 cells. *Bioconjugate Chemistry*, 17(2), 291–299.

Liao, W.-S., Ho, Y., Lin, Y.-W., Naveen Raj, E., Liu, K.-K., Chen, C., Zhou, X.-Z., Lu, K.-P., & Chao, J.-I. (2019). Targeting EGFR of triple-negative breast cancer enhances the therapeutic efficacy of paclitaxel- and cetuximab-conjugated nanodiamond nanocomposite. *Acta Biomaterialia*, 86, 395–405.

Lim, D. G., Jung, J. H., Ko, H. W., Kang, E., & Jeong, S. H. (2016). Paclitaxel–nanodiamond nanocomplexes enhance aqueous dispersibility and drug retention in cells. *ACS Applied Materials & Interfaces*, 8(36), 23558–23567.

Liu, K.-K., Wang, C.-C., Cheng, C.-L., & Chao, J.-I. (2009). Endocytic carboxylated nanodiamond for the labeling and tracking of cell division and differentiation in cancer and stem cells. *Biomaterials*, 30(26), 4249–4259.

Liu, K.-K., Zheng, W.-W., Wang, C.-C., Chiu, Y.-C., Cheng, C.-L., Lo, Y.-S., Chen, C., & Chao, J.-I. (2010). Covalent linkage of nanodiamond-paclitaxel for drug delivery and cancer therapy. *Nanotechnology*, 21(31), 315106.

Liu, Z., Chen, K., Davis, C., Sherlock, S., Cao, Q., Chen, X., & Dai, H. (2008). Drug delivery with carbon nanotubes for in vivo cancer treatment. *Cancer Research*, 68(16), 6652–6660.

Ma, P., & Mumper, R. J. (2013). Paclitaxel nano-delivery systems: A comprehensive review. *Journal of Nanomedicine & Nanotechnology*, 4(2), 1000164.

Majoros, I. J., Myc, A., Thomas, T., Mehta, C. B., & Baker, J. R. (2006). PAMAM Dendrimer-Based Multifunctional Conjugate for Cancer Therapy: Synthesis, Characterization, and Functionality. *Biomacromolecules*, 7(2), 572–579.

Marupudi, N. I., Han, J. E., Li, K. W., Renard, V. M., Tyler, B. M., & Brem, H. (2007). Paclitaxel: A review of adverse toxicities and novel delivery strategies. *Expert Opinion on Drug Safety*, 6(5), 609–621.

Maxwell, T., Nogueira Campos, M. G., Smith, S., Doomra, M., Thwin, Z., & Santra, S. (2020). Quantum dots. In Chung, E. J., Leon, L., C. B. T.-N. for Rinaldi, B. A. (Eds.), *Micro and Nano Technologies*. Elsevier, Amsterdam, pp. 243–265.

McClements, D. J. (2012). Nanoemulsions versus microemulsions: Terminology, differences, and similarities. *Soft Matter*, 8(6), 1719–1729.

Meng, S., Su, B., Li, W., Ding, Y., Tang, L., Zhou, W., Song, Y., & Caicun, Z. (2011). Integrin-targeted paclitaxel nanoliposomes for tumour therapy. *Medical Oncology*, 28(4), 1180–1187.

Milane, L., Duan, Z., & Amiji, M. (2011). Pharmacokinetics and biodistribution of lonidamine/paclitaxel loaded, EGFR-targeted nanoparticles in an orthotopic animal model of multi-drug resistant breast cancer. *Nanomedicine: Nanotechnology, Biology, and Medicine*, 7(4), 435–444.

Mochalin, V. N., Shenderova, O., Ho, D., & Gogotsi, Y. (2012). The properties and applications of nanodiamonds. *Nature Nanotechnology*, 7(1), 11–23.

Najlah, M., Kadam, A., Wan, K.-W., Ahmed, W., Taylor, K. M. G., & Elhissi, A. M. A. (2016). Novel paclitaxel formulations solubilized by parenteral nutrition nanoemulsions for application against glioma cell lines. *International Journal of Pharmaceutics*, 506(1), 102–109.

Noble, G. T., Stefanick, J. F., Ashley, J. D., Kiziltepe, T., & Bilgicer, B. (2014). Ligand-targeted liposome design: Challenges and fundamental considerations. *Trends in Biotechnology*, 32(1), 32–45.

Nornoo, A. O., & Chow, D. S.-L. (2008). Cremophor-free intravenous microemulsions for paclitaxel: II. Stability, in vitro release and pharmacokinetics. *International Journal of Pharmaceutics*, 349(1), 117–123.

Nornoo, A. O., Osborne, D. W., & Chow, D. S.-L. (2008). Cremophor-free intravenous microemulsions for paclitaxel: I: Formulation, cytotoxicity and hemolysis. *International Journal of Pharmaceutics*, 349(1), 108–116.

O'Brien, S. M., Aulitzky, W., Ben Yehuda, D., Lister, J., Schiller, G. J., Seiter, K., Smith, S. E., Stock, W., Silverman, J. A., & Kantarjian, H. (2010). Phase II study of marqibo in adult patients with refractory or relapsed philadelphia chromosome negative (Ph-) acute lymphoblastic leukemia (ALL). *Journal of Clinical Oncology*, 28(15_suppl), 6507.

Oh, K. S., Kim, R. S., Lee, J., Kim, D., Cho, S. H., & Yuk, S. H. (2008). Gold/chitosan/pluronic composite nanoparticles for drug delivery. *Journal of Applied Polymer Science*, 108(5), 3239–3244.

Olerile, L. D. (2020). Further development of near-infrared mediated quantum dots and paclitaxel co-loaded nanostructured lipid carrier system for cancer theragnostic. *Technology in Cancer Research & Treatment*, 19, 1533033820914308.

Olerile, L. D., Liu, Y., Zhang, B., Wang, T., Mu, S., Zhang, J., Selotlegeng, L., & Zhang, N. (2017). Near-infrared mediated quantum dots and paclitaxel co-loaded nanostructured lipid carriers for cancer theragnostic. *Colloids and Surfaces B: Biointerfaces*, 150, 121–130.

Padhi, S., & Behera, A. (2021). Advanced drug delivery systems in the treatment of ovarian cancer. In: Dua, K., Mehta, M., de Jesus Andreoli Pinto, T., Pont, L. G., Williams, K. A., and Rathbone, M. J. (Eds.) *Advanced Drug Delivery Systems in the Management of Cancer*. Academic Press, London, pp. 127–139.

Padhi, S., & Behera, A. (2022). Cellular internalization and toxicity of polymeric nanoparticles. In: Padhi, S., Behera, A., & Lichtfouse, E. (Eds.) *Polymeric Nanoparticles for Treating Solid Tumours*. Environmental Chemistry for a Sustainable World, vol 71. Springer, Cham, pp. 473–488.

Padhi, S., Behera, A., Hasnain, M. S., & Nayak, A. K. (2022a). Uses of chitosan in drug delivery. In: Hasnain, M. S., Beg, S., & Nayak, A. K. (Eds.) *Chitosan in Drug Delivery*. Academic Press, London, pp. 139–162.

Padhi, S., Behera, A., Hasnain, M. S., & Nayak, A. K. (2022b). Chitosan-based drug delivery systems in cancer therapeutics. In: Hasnain, M. S., Beg, S., & Nayak, A. K. (Eds.) *Chitosan in Drug Delivery*. Academic Press, London, pp. 159–193.

Padhi, S., Dash, M., & Behera, A. (2021). Nanophytochemicals for the treatment of type II diabetes mellitus: A review. *Environmental Chemistry Letters*, 19, 4349–4373.

Padhi, S., Kapoor, R., Verma, D., Panda, A. K., & Iqbal, Z. (2018). Formulation and optimization of topotecan nanoparticles: In vitro characterization, cytotoxicity, cellular uptake and pharmacokinetic outcomes. *Journal of Photochemistry and Photobiology B: Biology*, 183, 222–232.

Padhi, S., Mirza, M. A., Verma, D., Khuroo, T., Panda, A. K., Talegaonkar, S., Khar, R. K., & Iqbal, Z. (2016). Revisiting the nanoformulation design approach for effective delivery of Topotecan in its stable form: An appraisal of its in vitro behavior and tumour amelioration potential. *Drug Delivery*, 23(8), 2827–2837.

Panyam, J., & Labhasetwar, V. (2003). Biodegradable nanoparticles for drug and gene delivery to cells and tissue. *Advanced Drug Delivery Reviews*, 55(3), 329—347.

Partha, R., Mitchell, L. R., Lyon, J. L., Joshi, P. P., & Conyers, J. L. (2008). Buckysomes: Fullerene-based nanocarriers for hydrophobic molecule delivery. *ACS Nano*, 2(9), 1950–1958.

Pasqualini, R., Koivunen, E., Kain, R., Lahdenranta, J., Sakamoto, M., Stryhn, A., Ashmun, R. A., Shapiro, L. H., Arap, W., & Ruoslahti, E. (2000). Aminopeptidase N is a receptor for tumour-homing peptides and a target for inhibiting angiogenesis. *Cancer Research*, 60(3), 722–727.

Patnaik, S., Gorain, B., Padhi, S., Choudhury, H., Gabr, G. A., Md, S., Kumar Mishra, D., & Kesharwani, P. (2021). Recent update of toxicity aspects of nanoparticulate systems for drug delivery. *European Journal of Pharmaceutics and Biopharmaceutics*, 161, 100–119.

Pawar, R. S., Upadhaya, P. G., & Patravale, V. B. (2018). Quantum dots: Novel realm in biomedical and pharmaceutical industry. In: C. B. T.-H. of N. for Mustansar Hussain, I. A. (Ed.), *Micro and Nano Technologies*. Elsevier, London, pp. 621–637.

Pawar, V. K., Panchal, S. B., Singh, Y., Meher, J. G., Sharma, K., Singh, P., Bora, H. K., Singh, A., Datta, D., & Chourasia, M. K. (2014). Immunotherapeutic vitamin E nanoemulsion synergies the antiproliferative activity of paclitaxel in breast cancer cells via modulating Th1 and Th2 immune response. *Journal of Controlled Release*, 196, 295–306.

Pignata, S., Scambia, G., Katsaros, D., Gallo, C., Pujade-Lauraine, E., De Placido, S., Bologna, A., Weber, B., Raspagliesi, F., Panici, P. B., Cormio, G., Sorio, R., Cavazzini, M. G., Ferrandina, G., Breda, E., Murgia, V., Sacco, C., Cinieri, S., Salutari, V., … Perrone, F. (2014). Carboplatin plus paclitaxel once a week versus every 3 weeks in patients with advanced ovarian cancer (MITO-7): A randomized, multicentre, open-label, phase 3 trial. *The Lancet Oncology*, 15(4), 396–405.

Pilkington, G., Boland, A., Brown, T., Oyee, J., Bagust, A., & Dickson, R. (2015). A systematic review of the clinical effectiveness of first-line chemotherapy for adult patients with locally advanced or metastatic non-small cell lung cancer. *Thorax*, 70(4), 359 LP–367.

Pop, E., Mann, D., Wang, Q., Goodson, K., & Dai, H. (2006). Thermal conductance of an individual single-wall carbon nanotube above room temperature. *Nano Letters*, 6(1), 96–100. https://doi.org/10.1021/nl052145f

Probst, C. E., Zrazhevskiy, P., Bagalkot, V., & Gao, X. (2013). Quantum dots as a platform for nanoparticle drug delivery vehicle design. *Advanced Drug Delivery Reviews*, 65(5), 703–718.

Rathod, V., Tripathi, R., Joshi, P., Jha, P. K., Bahadur, P., & Tiwari, S. (2019). Paclitaxel encapsulation into dual-functionalized multiwalled carbon nanotubes. *AAPS PharmSciTech*, 20(2), 51.

Ravar, F., Saadat, E., Gholami, M., Dehghankelishadi, P., Mahdavi, M., Azami, S., & Dorkoosh, F. A. (2016). Hyaluronic acid-coated liposomes for targeted delivery of paclitaxel, in-vitro characterization and in-vivo evaluation. *Journal of Controlled Release: Official Journal of the Controlled Release Society*, 229, 10–22.

Rompicharla, S. V. K., Kumari, P., Ghosh, B., & Biswas, S. (2018). Octa-arginine modified poly(amidoamine) dendrimers for improved delivery and cytotoxic effect of paclitaxel in cancer. *Artificial Cells, Nanomedicine, and Biotechnology*, 46(sup2), 847–859.

Saxena, A., Tripathi, R. M., & Singh, R. P. (2010). Biological synthesis of silver nanoparticles by using onion (*Allium cepa*) extract and their antibacterial activity. *Digest Journal of Nanomaterials and Biostructures*, 5(2), 427–432.

Scripture, C. D., Figg, W. D., & Sparreboom, A. (2006). Peripheral neuropathy induced by paclitaxel: Recent insights and future perspectives. *Current Neuropharmacology*, 4(2), 165–172.

Shukla, R., Thomas, T. P., Peters, J. L., Desai, A. M., Kukowska-Latallo, J., Patri, A. K., Kotlyar, A., & Baker, J. R. (2006). HER2 specific tumour targeting with dendrimer conjugated anti-HER2 mAb. *Bioconjugate Chemistry*, 17(5), 1109–1115.

Sobhani, Z., Dinarvand, R., Atyabi, F., Ghahremani, M., & Adeli, M. (2011). Increased paclitaxel cytotoxicity against cancer cell lines using a novel functionalized carbon nanotube. *International Journal of Nanomedicine*, 6, 705–719.

Soni, N., Tekade, M., Kesharwani, P., Bhattacharya, P., Maheshwari, R., Dua, K., Hansbro, P. M., & Tekade, R. K. (2017). Recent advances in oncological submissions of dendrimer. *Current Pharmaceutical Design*, 23(21), 3084–3098.

Su, J., Sun, H., Meng, Q., Yin, Q., Tang, S., Zhang, P., Chen, Y., Zhang, Z., Yu, H., & Li, Y. (2020). Long circulation red-blood-cell-mimetic nanoparticles with peptide-enhanced tumour penetration for simultaneously inhibiting growth and lung metastasis of breast cancer. *Advanced Functional Materials*, 30(15), 1910229.

Su, J., Sun, H., Meng, Q., Yin, Q., Zhang, P., Zhang, Z., Yu, H., & Li, Y. (2016). Bioinspired nanoparticles with NIR-controlled drug release for synergetic chemophotothermal therapy of metastatic breast cancer. *Advanced Functional Materials*, 26(41), 7495–7506.

Sung, H., Ferlay, J., Siegel, R. L., Laversanne, M., Soerjomataram, I., Jemal, A., & Bray, F. (2021). Global cancer statistics 2020: GLOBOCAN estimates of incidence and mortality worldwide for 36 cancers in 185 countries. *CA: A Cancer Journal for Clinicians*, 71(3), 209–249.

Tavakolifard, S., Biazar, E., Pourshamsian, K., & Moslemin, M. H. (2016). Synthesis and evaluation of single-wall carbon nanotube-paclitaxel-folic acid conjugate as an anticancer targeting agent. *Artificial Cells, Nanomedicine, and Biotechnology*, 44(5), 1247–1253.

Teow, H. M., Zhou, Z., Najlah, M., Yusof, S. R., Abbott, N. J., & D'Emanuele, A. (2013). Delivery of paclitaxel across cellular barriers using a dendrimer-based nanocarrier. *International Journal of Pharmaceutics*, 441(1–2), 701–711.

Tian, Y., Zhang, H., Qin, Y., Li, D., Liu, Y., Wang, H., & Gan, L. (2018). Overcoming drug-resistant lung cancer by paclitaxel-loaded hyaluronic acid-coated liposomes targeted to mitochondria. *Drug Development and Industrial Pharmacy*, 44(12), 2071–2082.

Torchilin, V. (2011). Tumour delivery of macromolecular drugs based on the EPR effect. *Advanced Drug Delivery Reviews*, 63(3), 131–135.

Torchilin, V. P. (2005). Recent advances with liposomes as pharmaceutical carriers. *Nature Reviews Drug Discovery*, 4(2), 145–160.

Torres, K., & Horwitz, S. B. (1998). Mechanisms of taxol-induced cell death are concentration dependent. *Cancer Research*, 58(16), 3620 LP–3626.

Verma, D., Thakur, P. S., Padhi, S., Khuroo, T., Talegaonkar, S., & Iqbal, Z. (2017). Design expert assisted nanoformulation design for co-delivery of topotecan and thymoquinone: Optimization, in vitro characterization and stability assessment. *Journal of Molecular Liquids*, 242, 382–394.

Vivero-Escoto, J. L., Slowing, I. I., Wu, C.-W., & Lin, V. S.-Y. (2009). Photoinduced intracellular controlled release drug delivery in human cells by gold-capped mesoporous silica nanosphere. *Journal of the American Chemical Society*, 131(10), 3462–3463.

Walker, F. E. (1993). Paclitaxel (TAXOL®): Side effects and patient education issues. *Seminars in Oncology Nursing*, 9(4, Supplement 2), 6–10.

Wang, L., Xu, X., Zhang, Y., Zhang, Y., Zhu, Y., Shi, J., Sun, Y., & Huang, Q. (2013). Encapsulation of curcumin within poly(amidoamine) dendrimers for delivery to cancer cells. *Journal of Materials Science: Materials in Medicine*, 24(9), 2137–2144.

Wang, X., Li, S., Shi, Y., Chuan, X., Li, J., Zhong, T., Zhang, H., Dai, W., He, B., & Zhang, Q. (2014). The development of site-specific drug delivery nanocarriers based on receptor mediation. *Journal of Controlled Release*, 193, 139–153.

Weaver, B. A. (2014). How Taxol/paclitaxel kills cancer cells. *Molecular Biology of the Cell*, 25(18), 2677–2681.

Wolinsky, J. B., & Grinstaff, M. W. (2008). Therapeutic and diagnostic applications of dendrimers for cancer treatment. *Advanced Drug Delivery Reviews*, 60(9), 1037–1055.

Xu, X., Wang, L., Xu, H.-Q., Huang, X.-E., Qian, Y.-D., & Xiang, J. (2013). Clinical comparison between paclitaxel liposome (Lipusu®) and paclitaxel for treatment of patients with metastatic gastric cancer. *Asian Pacific Journal of Cancer Prevention: APJCP*, 14(4), 2591–2594.

Yan, W., Leung, S. S. Y., & To, K. K. W. (2019). Updates on the use of liposomes for active tumour targeting in cancer therapy. *Nanomedicine*, 15(3), 303–318.

Yang, T., Cui, F.-D., Choi, M.-K., Cho, J.-W., Chung, S.-J., Shim, C.-K., & Kim, D.-D. (2007). Enhanced solubility and stability of PEGylated liposomal paclitaxel: In vitro and in vivo evaluation. *International Journal of Pharmaceutics*, 338 (1–2), 317–326.

Yang, Y., Zhang, Y.-M., Chen, Y., Chen, J.-T., & Liu, Y. (2016). Polysaccharide-based noncovalent assembly for targeted delivery of taxol. *Scientific Reports*, 6(1), 19212.

Yao, H., & Ma, J. (2018). Dendrimer-paclitaxel complexes for efficient treatment in ovarian cancer: Study on OVCAR-3 and HEK293T cells. *Acta Biochimica Polonica*, 65(2), 219–225.

Yu, B., Tan, L., Zheng, R., Tan, H., & Zheng, L. (2016). Targeted delivery and controlled release of paclitaxel for the treatment of lung cancer using single-walled carbon nanotubes. *Materials Science and Engineering: C*, 68, 579–584.

Yuan, H., Guo, H., Luan, X., He, M., Li, F., Burnett, J., Truchan, N., & Sun, D. (2020). Albumin nanoparticle of paclitaxel (abraxane) decreases while taxol increases breast cancer stem cells in treatment of triple negative breast cancer. *Molecular Pharmaceutics*, 17(7), 2275–2286.

Zhan, C., Li, B., Hu, L., Wei, X., Feng, L., Fu, W., & Lu, W. (2011). Micelle-based brain-targeted drug delivery enabled by a nicotine acetylcholine receptor ligand. *Angewandte Chemie International Edition*, 50(24), 5482–5485.

Zhan, C., Wei, X., Qian, J., Feng, L., Zhu, J., & Lu, W. (2012). Co-delivery of TRAIL gene enhances the anti-glioblastoma effect of paclitaxel in vitro and in vivo. *Journal of Controlled Release: Official Journal of the Controlled Release Society*, 160(3), 630—636.

Zhang, Q., Huang, X.-E., & Gao, L.-L. (2009). A clinical study on the premedication of paclitaxel liposome in the treatment of solid tumours. *Biomedicine & Pharmacotherapy = Biomedecine & Pharmacotherapie*, 63(8), 603–607.

Zhang, Z., & Feng, S.-S. (2006). In vitro investigation on poly(lactide)–Tween 80 copolymer nanoparticles fabricated by dialysis method for chemotherapy. *Biomacromolecules*, 7(4), 1139–1146.

Zhang, Z., Tan, S., & Feng, S.-S. (2012). Vitamin E TPGS as a molecular biomaterial for drug delivery. *Biomaterials*, 33(19), 4889–4906.

Zhao, L., Xu, Y.-H., Qin, H., Abe, S., Akasaka, T., Chano, T., Watari, F., Kimura, T., Komatsu, N., & Chen, X. (2014). Platinum on nanodiamond: A promising prodrug conjugated with stealth polyglycerol, targeting peptide and acid-responsive antitumour drug. *Advanced Functional Materials*, 24(34), 5348–5357.

Zhou, F., Wu, S., Song, S., Chen, W. R., Resasco, D. E., & Xing, D. (2012). Antitumour immunologically modified carbon nanotubes for photothermal therapy. *Biomaterials*, 33(11), 3235–3242.

8 Anticancer Nanodrug Delivery Cargos of Artemisinin

Shubhra[1], *Isha Kailash Joshi*[2], *Atul Mourya*[3], *Onkar Kulkarni*[1], *Murali Monohar Pandey*[2], *Dharmendra Kumar Khatri*[4], *Shashi Bala Singh*[4], *Ashish Baldi*[5], *and Jitender Madan*[3]

[1]Department of Pharmacy, Birla Institute of Technology and Science, Hyderabad, Telangana, India

[2]Department of Pharmacy, Birla Institute of Technology and Science, Pilani, Rajasthan, India

[3]Department of Pharmaceutics, National Institute of Pharmaceutical Education and Research, Hyderabad, Telangana, India

[4]Department of Biological Sciences, National Institute of Pharmaceutical Education and Research, Hyderabad, Telangana, India

[5]Department of Pharmaceutical Sciences and Technology, Maharaja Ranjit Singh Punjab technical University, Bathinda, Punjab, India

8.1 INTRODUCTION

Cancer is one of the key determinants of global mortality, causing one in six deaths worldwide. A 2020 report by the WHO projected 1,806,590 new cancer cases and 606,520 cancer deaths to occur in the US (Siegel et al., 2020). Similarly, the Indian Council of Medical Research predicted the number of cancer cases to rise to 13.9 lacs in India in 2020 (NCDIR, 2020). The cancers with the highest mortality rates are lung, breast, colorectal, stomach, and liver cancers. The number of anticipated deaths due to cancer worldwide is predicted to reach over 13 million in 2030 (WHO, 2018).

Cancer is a group of related diseases all characterized by uncontrolled proliferation of cells, which form a mass of tissue referred to as a tumour. There are three main tumour types: benign, premalignant, and malignant. Benign tumours do not spread throughout the body and are not cancerous. Surgery usually removes them (e.g., adenomas, lipomas, etc.). In the premalignant stage, the cells are poised to become malignant but are not so at the given time (e.g. cervical dysplasia, leukoplakia, etc.). On the other hand, malignant tumours, being cancerous, spread and invade other body tissues by a process called metastasis (e.g. carcinoma, sarcoma, etc.) (Cooper, 2000).

Early diagnosis and treatment of cancer are essential to improve the patients' chance of survival. Three main strategies are employed in cancer treatment: chemotherapy, radiation therapy, and immunotherapy. Chemotherapy is one of the extensively used approaches to treat cancer. However, drug resistance presents itself as a major hurdle to be overcome for the successful application of chemotherapy. Hence, one of the alternative approaches that can be useful in cancer treatment is phytoconstituents (Padhi et al., 2016). Naturally derived products, like many single compounds, herbal extracts, and formulas, have an untapped potential in the domain of cancer treatment, which is currently being investigated. There are multiple examples of anticancer natural drugs, including curcumin, vincristine, vinblastine, epipodophyllotoxin, paclitaxel, and artemisinin. (Greenwell & Rahman, 2015). Among these, artemisinin is of particular interest as, along with having antitumour activity, it is shown to be especially effective in cells with chemotherapeutic drug resistance, and hence, there have been endeavours to repurpose it for cancer treatment (Zhang et al., 2018).

Since the discovery of artemisinin's anticancer potential, many targets and mechanisms have been identified for the same. However, certain pharmacokinetic properties obstruct its ability as a clinical candidate, resulting in subpar drug delivery (Das, 2015). Recent advances in nanotechnology have paved the way for it to be viewed as an effective medium for unravelling the complications associated with artemisinin and augmenting its delivery (Padhi et al., 2018). Consequently, the rising number of clinical trials of artemisinin and its derivatives in different forms in recent years is a testament to its commercial viability as an anticancer drug (Liu et al., 2020). Owing to their promising results, oncologists should explore the strengths of artemisinin combined with nanotechnology and consider it for cancer treatment.

8.2 ARTEMISININ: A WONDER MOLECULE

Artemisinin has a long history of being used in Chinese medicine to treat chills and fevers. In 340 BC, Ge Hong, a medical expert and alchemist from the East Jin Dynasty, first described the antimalarial properties of *Artemisia*

DOI: 10.1201/9781003611516-10

annua plants (Hsu, 2014). Project 523, a military operation of the Chinese Government headed by Tu Youyou in 1970, discovered the potential of artemisinin as a frontline in antimalarial treatment, for which she was awarded the Nobel Prize in 2015 (Park, 2019).

Further, in 1993, scientists discovered the cytotoxic action of artemisinin on Ehrlich ascites tumour (EAT) cells, which gave an impetus to research on the properties of artemisinins beyond antimalarial effects, which include anti-inflammatory, antifungal and antiviral, and anti-asthmatic effects (Woerdenbag et al., 1993). To date, various studies have demonstrated the pronounced and highly selective cytotoxic effects of artemisinin and its derivatives against a gamut of cancer types, both *in vitro* and *in vivo*. This attribute, paired with proven safety, highlights the capability of artemisinin to be an excellent candidate for drug repurposing.

8.2.1 Structural Insights on Artemisinin

The structure of artemisinin was determined early in 1979 by using X-ray analysis and subsequently by poly arithmetic analysis and mass spectroscopy. Artemisinin is a sesquiterpene lactone with the empirical formula $C_{15}H_{22}O_5$ having a 1,2,4-trioxane ring system (Meunier & Robert, 2010). It does not contain nitrogen in the heterocyclic ring but has an endoperoxide linkage in its structure (Klayman, 1983). The iron-mediated cleavage of this linkage is one of the major factors responsible for the anticancer activity of the compound (Zhang et al., 2018). Scientific advancements have synthesized three generations of semi-synthetic and synthetic artemisinin-like endoperoxide compounds. The first generation of artemisinins comprises artemisinin and its semi-synthetic derivatives, including dihydroartemisinin, artesunate, artemether, arteether, and artelinic acid (Figure 8.1). The derivatives, artemether and arteether, are soluble in oil, whereas artesunate is water soluble. The second generation is composed of the synthetic derivative with enhanced stability, potency, and lower toxicity such as artemisone. These derivatives are superior to artemisinin in terms of pharmacokinetic properties as well as in effectiveness at μM and nM concentrations (Crespo-ortiz & Wei, 2012). Recurrent studies have cited the cytotoxicity of artemisinin as a function of its endoperoxide linkage. However, the absence of this linkage does not completely repeal its antineoplastic activity, although it does markedly diminish the cytotoxic activity of artemisinin to 1/50th in comparison with other compounds bearing a trioxane ring. The residual anticancer activity may exist due to a peroxide-independent mechanism (Crespo-ortiz & Wei, 2012).

Preceding studies have conveyed the importance of oxidative stress regulation for artemisinin activity. These bring the function of peroxide as a source of reactive oxidative species to attention. Reductive scission with the help of low valency transition metals (Fe^{2+} as ferrous heme) is posited as a general mechanism for the peroxides to generate O-centred (oxygen-centred radical), which might produce C-centred radicals via abstraction of intra-molecular hydrogen. It is understood that C-centred radicals interact with biomolecules. The alternative hypothesis states that protonation or metal ion (e.g. Mn^{2+}) complex formation mediated C–O bond cleavage leads to metal or hydroperoxide formation. These hydroperoxides bring about the auto-oxidation and hydroxylation of various biomolecules, for instance, membrane bilayers and lipids (Olliaro et al., 2001). Caspase-dependent apoptosis is also known to be triggered by C-centred radicals leading to cell death (Mercer et al., 2007).

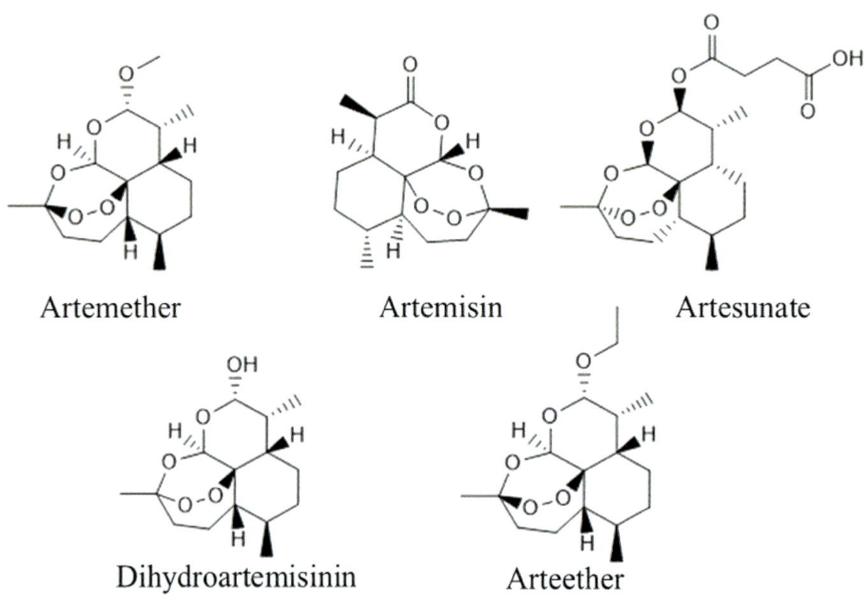

FIGURE 8.1 Schematic representations of artemisinins.

8.2.2 Artemisinin Hits Multiple Targets in Cancer Cells to Induce Apoptosis: Preclinical Analysis

Over the past three decades, artemisinins have exhibited extensive anticancer properties in various *in vitro* and *in vivo* studies (Augustin et al., 2020). Multiple comprehensive reviews have also evidenced these, paving the way for developing artemisinins as an effective anticancer therapy. Figure 8.2 is a schematic representation of the mechanism of action of artemisinin.

8.2.2.1 Cell Signalling Pathways

Artemisinins trigger various key pathways involved in signal transduction in cancer. For instance, the Wnt/β-catenin pathway, identified as a critical cell signalling pathway in colorectal carcinogenesis, has been reported to be inhibited by artesunate in a dose-dependent manner. Xu et al. recognized the role of artesunate in inducing apoptosis in myelodysplastic syndromes cells by inhibiting the Wnt/β-catenin pathway (Xu et al., 2015). Besides their role in the Wnt/β-catenin pathway, artemisinins have been observed to down-regulate the expression of cellular-myelocytomatosis,

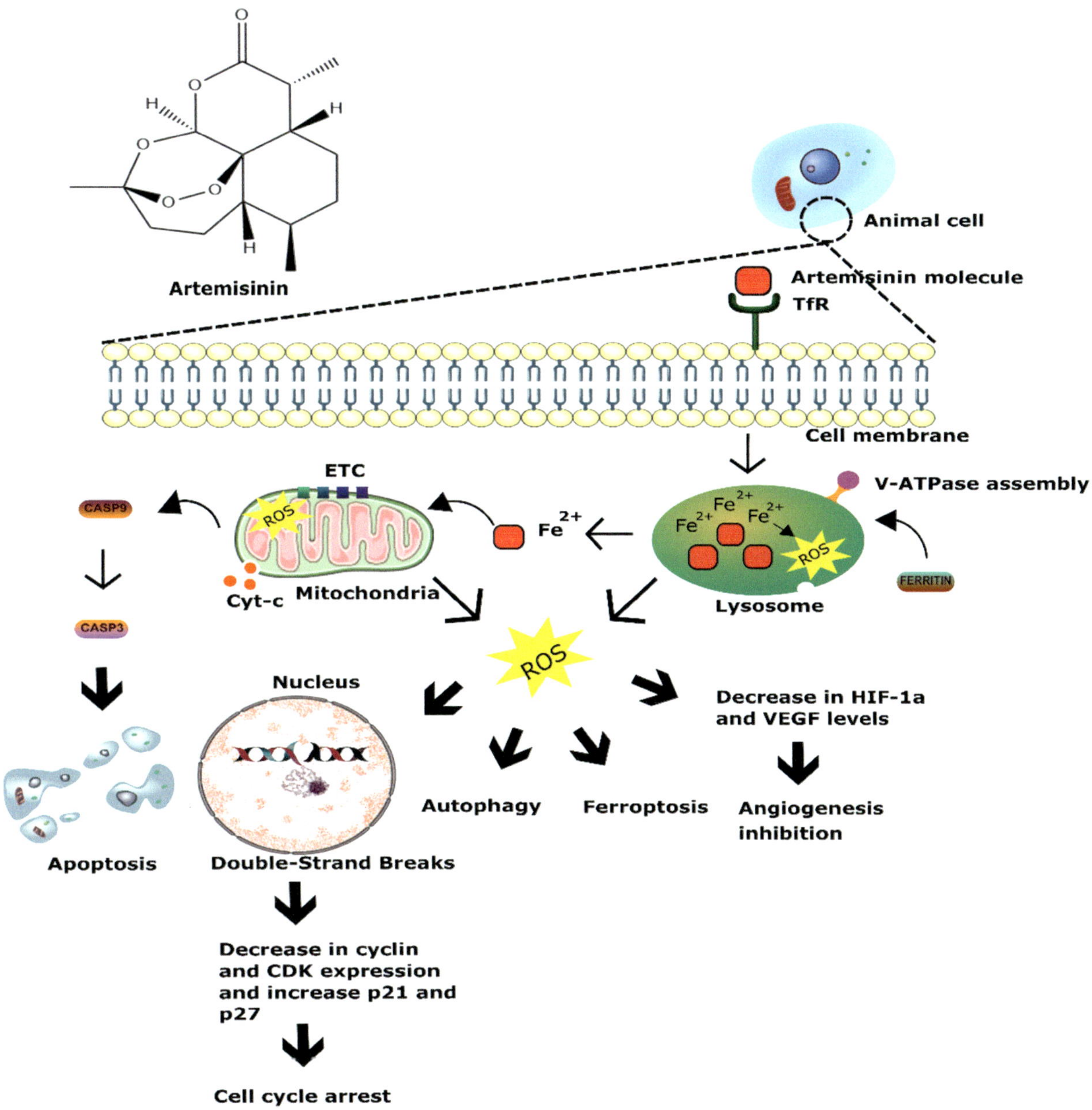

FIGURE 8.2 Schematic representation of the molecular mechanism of action of artemisinin in cancer cells. (This image is a derivative created by the use of illustrations available through creative common rights (CC BY 4.0) on Reactome [Designer-CristofferSevilla, curated by Steve Jupe ('Cell generic', 'ferritin', 'caspase9', 'Lysosome', 'Ionotropic receptor', 'VLDLR', 'DNA damaged'), Veronica Shamovsky ('CASPASE3', 'Apoptotic cell') and BijayJassal ('Electron Transport Chain')] and Smart Servier Medical Art (CC BY 3.0)-('Mitochondria', 'Nucleus', and 'Cell membrane').)

survivin, cyclin D1, mammalian target of rapamycin, mesenchymal-epithelial transition factor, steroid receptor coactivator, α-tubulin and focal adhesion kinase, nuclear factor kappa-light-chain-enhancer of activated B cells, activator protein 1, cyclic-AMP response element binding and inhibit epidermal growth factor receptor (Efferth, 2017; Ho et al., 2014; Lee et al., 2012; Lu et al., 2011).

8.2.2.2 Cancer Metastasis

Cancer metastasis is characterized by the invasion and spread of malignant cells through the extracellular matrix, linked with elevated morbidity and mortality. Artemisinin enhances the activation of E-cadherin and cell division control protein 42 homolog, thereby increasing cell-to-cell adhesion (Hwang et al., 2010). It influences vascular endothelial growth factor C inhibition to resist lymphangiogenesis and lymph node metastasis in lung cancer, as discovered in various animal investigations (Steeg, 2006). Similarly, in non-small cell lung cancer and fibrosarcoma, decreased levels of matrix metalloproteinases-2, 7 and 9, driven by the inactivation or modulation of nuclear factor kappa B and activator protein-1, were induced by dihydroartemisinin treatment (Wang et al., 2008). Artesunate, generally known as a more effective alternative in less differentiated cell lines, attenuates metastasis by upregulating the expression of neural cell adhesion molecule and integrin β1 in embryonal rhabdomyosarcoma cells (Beccafico et al., 2015).

8.2.2.3 Immunosuppression

Numerous studies have reported the ability of artemisinin to induce immunosuppression in tumour cells. Artemisinin constrained murine mammary carcinoma growth by aiding the activation of T-cells and curbing immunosuppression from myeloid-derived suppressor cells and regulatory T cells by decreasing their frequencies (Cao et al., 2019), and it increased those of cytotoxic T lymphocytes and D4+ interferon γ+T cells. Also, treatment with artemisinin increased the messenger RNA levels of interferon γ, T-bet and tumour necrosis factor-α and decreased the transforming growth factor-beta. In an outcome beneficial for colorectal cancer patients with high levels of interleukin-10 and transforming growth factor-beta, artesunate was observed to diminish the concentration of these cytokines in colorectal cancer (Cui et al., 2015). Additionally, micro-RNA expression was influenced by treatment with artesunate, consequently regulating immunosuppression. For instance, micro-RNA-142 expression was upregulated in ovarian cancer, suppressing the sirtuin level, enhancing differentiation of type 1 T helper cells, and promoting cell apoptosis (Chen et al., 2019).

8.2.2.4 DNA Damage

Artemisinin induces DNA damage by the formation of oxidative DNA lesions, resulting in the emergence of formamidopyrimidine DNA glycosylase-sensitive sites and the production of 8-oxoguanine and 1, *N*-6-ethenoadenine. It acts as a robust inducer of oxidative DNA damage in a dose-dependent manner. This leads to the provocation of cellular response, including the phosphorylation of multiple signal transducers such as ataxia telangiectasia mutated, Rad3-related protein, and checkpoint kinase 1 and 2 (Berdelle et al., 2011). Furthermore, Rad3-related protein has been known to induce dose-dependent double-strand breaks, identified by running the comet assay. This was verified after the dose-dependent increase of H2A histone family member X, a DNA-related histone induced when double-strand breaks occur in DNA (Li et al., 2008). Artesunate was found to down-regulate homologous recombination repair and RAD51 recombinase (a protein-coding gene) foci, thus impairing double-strand breaks repair along with inducing double-strand breaks (Wang et al., 2015).

8.2.2.5 Cancer Cell Metabolism

Artemisinins act as good modulators of cancer-specific glucose metabolism. In non-small cell lung cancer, artemisinin obstructs glycolytic metabolism via inhibiting glucose uptake by way of reducing glucose transporter 1, decreasing the number of ATPs, and interfering with lactate production (Martinez-outschoorn et al., 2016; Mi et al., 2015). Inhibition of glycolysis by dihydroartemisinin was also demonstrated by the downregulation of pyruvate kinase and hypoxia-inducible factor-1α expression and the inhibition of phosphoinositide-3-kinase–protein kinase B pathway (Gao et al., 2020; Li et al., 2019; Zhu et al., 2019). In colon cancer cells (HCT116), it was discovered to suppress fatty acid biosynthetic pathways as well, primarily by down-regulating three proteins named hydroxyacyl-coenzyme A dehydrogenase, acyl-coenzyme A synthetase 5, and fatty acid synthase (Chen et al., 2017).

8.2.2.6 Inflammatory Pathways

Chronic inflammation is one of the fundamental attributes of carcinogenesis. Recently, artemisinin has been identified to inhibit signal transducer and activator of transcription 3 directly. A recent investigation on the effect of artesunate on multi-organ failure in a mouse model led to the finding that it diminished the haemorrhagic shock-induced stimulation of nuclear factor kappa B along with that of the expression of nitric oxide, interleukin-6 and tumour necrosis factor-α lending protection against multi-organ failure (Sordi et al., 2016). Similar studies have been carried out in animal models of myocardial acute lung injury, ischaemia, and nephritis (Khan et al., 2017; Wan & Li, 2017). These findings covering the anti-inflammatory properties of artesunate in a range of diseases should be utilized as a base for designing specific studies exploring the anti-inflammatory effects of artemisinin in cancer.

8.2.2.7 Oxidative Stress

Tumour cells express lower levels of antioxidant enzymes, including catalase, superoxide dismutase, and glutathione peroxidase, than normal cells, endangering them to the attack of reactive oxygen species (Bostwick et al., 2000; May & Cuatrecasas, 1985). It is hypothesized that reactive oxygen species and highly alkylating carbon-centred radicals act on tumour cells by DNA damage, growth arrest, enhanced apoptosis, and reduced angiogenesis (Mercer et al., 2011, 2007). A study on the human leukaemia cell line (HL-60) demonstrated that artemisinin-mediated cellular damage and apoptosis might be a function of early and fast reactive oxygen species generation (Zhou et al., 2012). Reactive oxygen species have been incriminated as triggers of cytotoxicity owing to their antecedence to it. The generation of reactive oxygen species has also been found to be influenced by the electron transport chain in mitochondria (Hansen et al., 2008). However, there are mechanisms for generating reactive oxygen species independent of the electron transport chain, as proved in cervical cells (HeLa) (Lu et al., 2011). Artemisin-induced necrosis may generally operate through multiple reactive oxygen species-independent pathways.

8.2.2.8 Anti-angiogenesis

For a cancer cell to grow and develop, it must undergo the process of angiogenesis, which is mediated by a balance between pro-angiogenic factors like matrix metalloproteinases, vascular endothelial growth factor and fibroblast growth factor, its receptors and cytokines and anti-angiogenic factors such as thrombospondin, angiostatin, tissue inhibitors of metalloproteinases, plasminogen activator inhibitor-1, endostatin, and others. This class of drugs exhibits its effects via the downregulation of growth factors (vascular endothelial growth factor, fibroblast growth factor), some metalloproteinases (bone morphogenetic protein 1, matrix metalloproteinases 9 and 11), new vessel mediator angiogenin (angiotensinogen), collagens, the cysteine-rich angiogenic inducer (CYR61), and hypoxia-inducible factor-1α. Akin to this, artemisinins upregulated the expression of anti-angiogenic factors (Crespo-ortiz & Wei, 2012). Dihydroartemisinin has a multimodal action in nuclear factor kappa B regulation, reduced levels of which are linked with inhibition of metastasis and proliferation (Li et al., 2009; Wang et al., 2010). Furthermore, dihydroartemisinin impedes angiogenesis by depleting vascular endothelial growth factor receptor-1 and kinase insert domain receptors in number. Artesunate has also been shown to suppress tumour growth and angiogenesis in Kaposi's sarcoma (Anfosso et al., 2004) and glioma in mouse and rat models, respectively (Wu et al., 2009).

8.2.2.9 Cell Cycle Arrest

Artemisinins exert cytostatic and cytotoxic effects on tumour cells in all cell cycle phases (Gravett et al., 2011). They disrupt the cell cycle kinetics, and they are also known to mediate proliferation interacting pathways, thereby inducing cell cycle arrest in cancer cell lines. Retinoblastoma protein, a regulator of cell cycle progression, is known to be deactivated by artemisinins, along with disruption in cyclin-dependent kinase 4 promoter transcription in prostate cancer cell lines, inducing G1 cell cycle arrest. Artemisinin has also been observed to directly inhibit the expression of the cyclin-dependent kinase 4 gene (Willoughby et al., 2009).

Artesunate was shown to induce G1 cell cycle arrest in nasopharyngeal and human breast cancer cell lines. In contrast, it brought about cell cycle arrest at the G2/M phase in leukaemia, colon cancer, glioma, small cell lung cancer, and cervical cancer cell lines. It has been noted that artesunate-induced G2/M cell cycle arrest occurs by obstruction of the regulators of the mitotic spindle, such as mitotic arrest deficiency 2, 3, and budding uninhibited by benzimidazoles 3 (Steinbrück, Pereira, & Efferth, 2010). Interestingly, artesunate treatment has also been observed to reduce expression of cyclin D1, cyclin B1, and transcription factor E2F1 (Augustin et al., 2020). Dihydroartemisinin-like compounds are more potent as an anticancer agent than other artemisinin-like compounds. Dihydroartemisinin was noted to arrest the cell cycle at the G2/M phase. In pancreatic cell lines, it was also reported to cease the cell cycle at G0/G1 to S phase transition brought by decreasing cyclin-dependent kinase 2, 4, 6, nuclear factor kappa B, and cyclin E (Crespo-ortiz & Wei, 2012).

8.2.2.10 Cell Death Pathways

- *Apoptosis:* One of the key mechanisms underlined in working artemisinins as an anticancer drug is the induction of apoptosis. For instance, artemisinins can induce apoptosis in prostate cancer and lung adenocarcinoma by activating Bax (Lu et al., 2009; Nakase et al., 2009). In leukaemic T cells and breast cancer cell lines, artesunate activates apoptotic cell death through iron-dependent reactive oxygen species generation, leading to the release of cyt-c and eventual procaspase-2,3,8 cleavage (Efferth et al., 2007; Hamacher-brady et al., 2011). It suppresses the fatty acid biosynthesis pathway nuclear factor kappa-light-chain-enhancer of the nuclear factor kappa B pathway (Chen et al., 2017). The apoptotic effects of artemisinin are mainly accredited to intrinsic pathway activation. Therefore, damage to the mitochondrial membrane has been regarded as a decisive player in the events responsible for artemisinin-induced cell death. Increased levels of c-myc have been associated with dihydroartemisinin-induced apoptotic response in lung cancer cell lines. Reportedly, colon tumour cells (HCT116) and human leukaemia cells (HL-60) are markedly sensitive to the pro-apoptotic action of dihydroartemisinin, as they express high levels of c-myc (Lu et al., 2010a,b). In addition, dihydroartemisinin's apoptotic effect has also been correlated to increased oxidative stress, upregulation of NOXA, and activation of caspase 3 in Jurkat T lymphoma cells and metastatic melanoma (A375, G361) cell lines (Cabello et al., 2013; Handrick et al., 2010). However, some evidence is present on an extrinsic apoptosis pathway as dihydroartemisinin is observed to upregulate the transcription of the cell death receptor (DR-5) promoter, inducing it in various prostate cancer cell lines (Lu et al., 2010a,b). Between apoptosis and necrosis, apoptosis is more promoted by artemisinin in most cases, although it induces both apoptosis and necrosis in some systems. Necrosis triggered by artemisinin is linked to diminished ATP levels and flawed apoptotic pathways selected in cell lines (Mercer et al., 2011).

- *Oncosis:* Artemisinins also induce other forms of non-programmed cell death in cancer cells. i.e. anoikis (anchorage-dependent cell death) and oncosis (ischaemic cell death) (Zhang et al., 2016a,b; Zhou et al., 2013). Upon treatment with artesunate, calpain 1 and calpain 2 expression were upregulated, while α-tubulin was down-regulated in certain cell lines (Augustin et al., 2020). Artemisinin-induced oncosis-based cell death in renal cancer cells (cell death due to reactive oxygen species generation and depletion of ATP) (Jeong et al., 2015) and pancreatic cancer cells (cell death due to a change in morphological characteristics of the cell) (Zhang & Ji, 2010). Notably, in gastric cancer, cell-oncosis was induced by artemisinin via the reduction in vascular endothelial growth factor expression and upregulation in calcium levels and calpain-2 expression (Zhou et al., 2013).
- *Ferroptosis:* Ferroptosis is an unconventional, caspase-independent, non-apoptotic type of cell death, characterized by a deadly build-up of lipid-based reactive oxygen species (Dixon et al., 2012). It was first reported that the expression of various iron-associated genes, including the ones encoding ceruloplasmin, transferrin receptors 1 and 2, transferrin, and lactoferrin, was linked to the artemisinin s' $\log_{10}$ (IC_{50}) values, proving their ferroptosis-inducing role (Ooko et al., 2015). It was discovered in the head and neck squamous cell carcinomas model that a mechanism of artesunate-induced ferroptosis was a decrease in cellular glutathione levels and an increase in lipid reactive oxygen species levels (Chen et al., 2020; Roh et al., 2017). Recent findings suggest that artemisinin cytotoxicity is triggered by preloading of tumour cells with iron/iron-saturated holotransferrin (Lu et al., 2011, 2010a; Mercer et al., 2011), which may result in a 100 times increase in the activity of artemisinin in a few cancer cell lines (Lai et al., 2009).
- *Autophagy:* Autophagy is a process that degrades damaged cellular organelles and cytoplasmic content to be used for recycling and maintaining the cell's homeostasis via various signalling pathways. Artemisinins have been proven effective in autophagy-regulated cell death (Chen et al., 2015a, 2015b, 2020; Galluzzi et al., 2015). For example, artesunate increases the expression of beclin-1, which initiates autophagy, promoting cell death in the G2/M phase (Chen et al., 2014). Dihydroartemisinin can upregulate autophagy in multiple cancer cell lines by inhibiting nuclear factor kappa B. Inhibition of nuclear factor kappa B also functions to accumulate reactive oxygen species, which further stimulates autophagy. However, the mechanisms of autophagy-induced cell death and enhancement of cell survival after artemisinin treatment remain contradictory. Dihydroartemisinin-induced autophagy functions to promote cancer cell death (Qu et al., 2017), while autophagy that dihydroartemisinin and artemisinin serve to protect cancer cells (Jia et al., 2014; Jiang et al., 2018).

8.3 CLINICAL TRIALS OF ARTEMISININ IN CANCER

With the recent discovery of the anticancer properties of artemisinins, the clinical trials related to the same are gradually increasing. Most studies are concentrated on solid tumours: colorectal carcinoma, lung cancer, hepatocellular carcinoma, and breast cancer. Most of the clinical trials on artemisinins are focused on the compound artesunate. In the phase 2 clinical trial conducted on 120 non-small cell lung cancer patients (60 in the trial and control group each), 120 mg of intravenous artesunate was administered once daily for 8 days. While the trial group exhibited a disease control rate of 88.2% and a progression-free survival period of 24 weeks, the same was true for the control group, which was 72.7% and 20 weeks, respectively. No dose-limiting side effects were observed, and artesunate was well tolerated in the patients (Zhang et al., 2008).

Similarly, in a phase one dosage escalation study on 10 biopsy-confirmed cervical intraepithelial neoplasia patients, vaginal inserts of artesunate were administered as per the following dosage regimen: One, two, or three 5-day treatment rounds at zero, 2, and 4 weeks, respectively, before a scheduled standard-of-care resection at 15th week. Histologic regression in lesions was detected in 67.9% of patients, and human papillomavirus genotype clearance was observed at baseline, occurring in 47.4% of subjects. No serious side effects were observed. Another Phase 1 (accelerated titration) dose-escalation study gave 19 patients with Stage IV advanced solid tumour malignancies 8, 12, 18, 25, 34, and 45 mg/kg of intravenous artesunate on the 1st and 8th days of 21 days. The maximal tolerated dose was determined to be 18 mg/kg, while the disease control rate was observed at 27%. Four patients had stable disease conditions, and three suffered from extended stable disease for eight, ten, and 11 cycles. The dose-limiting toxicities among liver dysfunction and myelosuppression, including a hypersensitivity reaction and unfettered vomiting/nausea, were noted. Treatment-related side effects (≥ grade 1) were fatigue (60%), anaemia (66%), anorexia (34%), nausea/vomiting (61%), and dizziness

(28%). Electrolyte disturbances were also observed in many patients (Deeken et al., 2018).

Hagens et al. (2017) also conducted a clinical trial on artesunate (oral administration). It was a phase 1 compassionate open-label use study in which 13 breast cancer patients (out of the 23 participating) were administered oral artesunate 100, 150, or 200 mg daily as add-on therapy to individual medications for 4±1 week. The treatment period lasted for a cumulative total of 3,825 days. Individually, patients were given up to 167.3 total doses for a cumulative period of 1,115 days. Twenty-five treatment-related adverse effects graded ≥2 were reported. However, no serious concerns related to safety were documented. In an additional Phase-I dose-escalation study using the same drug and route of administration, 14 daily doses (200 mg in the trial group and equivalent placebo in the control group) were administered to 234 patients suffering from stage 1–3 colorectal cancer. More than 7% cell apoptosis was observed in 12 patients (67%) in the artesunate group and 11 (55%) in the placebo group. Immunohistochemical outcomes were found to influence the antitumour activities of artesunate. As determined by Bayesian analysis, the probability of an increase in CD31 was reported to be 0.79, and that of a decrease in Ki67 was 0.89. One patient from the artesunate group and six from the placebo group developed recurrent colorectal cancer, and two adverse events possibly linked to artesunate were reported (Krishna et al., 2015).

As a part of compassionate therapy, a patient with Stage IV Uveal melanoma was administered 50 mg of oral artesunate two times a day in combination with dacarbazine. This patient survived for over 47 months, surpassing the median survival (2–5 months) (Berger et al., 2005). Another compassionate therapy included the utilization of artesunate tablets and injections to treat a 72-year-old male suffering from laryngeal squamous cell carcinoma. Sixty mg of artesunate injection was administered/day from day 1–15, and one 50 mg tablet/day was administered for 9 months from day 16 onwards. This was in addition to 150 mg of ferrous sulphate and 0.5 mg of folic acid given daily. No side effects were observed, along with a 70% reduction in tumour and enhanced quality of life for around 1 year and 8 months after 2 months of treatment. Hence, it was concluded that artesunate requires long-term treatment (Singh & Verma, 2002). One more compassionate therapy involved the treatment of prostate carcinoma patients by artesunate capsules and artesunate injections in which 5×50 mg/day capsules were administered daily after previous treatment with bicalutamide, and 2×150 mg artesunate injections were given twice weekly. At the end of the 5th week, prostate-specific antigen levels had fallen to 0.98 µg/land. After 6 months, a continuous increase in prostate-specific antigen level was observed, implying skeletal metastases and the recurrence of the tumour. Mediator release test results indicated the development of resistance in tumours, though no side effects were observed (Michaelsen et al., 2015).

Lastly, Jansen et al. (2011) conducted a phase 1 pilot study for the treatment of stage III-IV advanced cervical cancer by oral dihydroartemisinin/artenimol-R. Ten patients were administered 2×50 mg/day of dihydroartemisinin for one week, and later, the dose was increased to 200 mg/day for three weeks after no adverse event was observed in week 1. The patients reported tolerance for treatment and a reduction in vaginal pain and discharge. A decrease in p53, epidermal growth factor, and antigen kiel 67 levels was noted after immunohistochemical staining. No treatment-related grade III-IV toxicities were recorded.

There are only a handful of clinical trials on applying artemisinin in cancer. While these studies are promising, they have been performed on small groups of patients, imploring the conduction of large-scale clinical trials for the corroboration of these optimistic results.

8.4 LIMITATIONS OF ARTEMISININ

Despite its potential, the candidature of artemisinin as a commercial anticancer drug is hampered by certain limitations concerning its pharmacokinetic and toxicological parameters. Artemisinin is a biopharmaceutical classification system class II drug that exhibits low water solubility (133±4 µg/mL) and bioavailability, transient effect, rapid metabolism and decomposition *in vivo*. Due to its short half-life (2–5 hours), the complete clearance is slow (Medhi et al., 2009). Absolute absorption of artemisinin takes place in oral administration; however, low bioavailability (30%) is achieved as a consequence of high first-pass metabolism. It is observed that the distribution of artemisinin is rapid (occurring in tissues, salivary gland, blood-brain barrier, and placenta), and so is the excretion, which takes place by biotransformation, mainly happening in the liver (Titulaer et al., 1991). Both artesunate and artemether are metabolized to a great extent and transformed into the metabolite, dihydroartemisinin, having a 1–2 hour plasma half-life (Medhi et al., 2009). More critically, as it is a radical-induced remedy, achieving local treatment with the help of artemisinins is challenging (Ding et al., 2018a). From the perspective of toxicological parameters, concerns regarding neurotoxicity, fetotoxicity, and auditory toxicity have been associated with artemisinin; however, not much evidence is available on the same. Accordingly, a multitude of issues need to be addressed for artemisinin to be used as an anticancer drug. This is where targeted drug delivery using nanoparticles enters the landscape of artemisinin-mediated cancer therapy, aiding the compound to achieve its full potential as a powerful anticancer drug (Song et al., 2015).

8.5 NANO-ENABLED TARGETING STRATEGIES FOR ARTEMISININ IN CANCER

Nanotechnology can be understood as an umbrella term encompassing all the fields of science, technology, engineering, etc., that operate at the nanoscopic scale (1–1,000 nm) to investigate and control the properties of materials. These properties propel the development of new applications, including nanodrug development systems (Charlie-Silva et al., 2018).

Nanoparticles, an already established crucial tool in cancer therapy, can be harnessed for targeted drug delivery, especially in the case of artemisinin, as it is challenging to achieve a local effect with it due to it being a radical-induced therapeutic drug, as well as to improve on many of its limitations (Behera & Padhi, 2022; Pradhan et al., 2024; Sa et al., 2023). The generally recognized benefits of nanoparticles a function of their small size and large surface area, comprise enhanced aqueous solubility and cellular uptake, ameliorated drug biodegradability and biocompatibility, sustained and controlled release, increased suitability for codelivery and improved drug activity (Bragta et al., 2018; Padhi & Behera, 2020). Moreover, it is possible to manipulate the surface of nanoparticles by integrating target ligands to prevent rapid clearance and aggregation to increase their stability in vivo, enabling targeted delivery and controlled release of the drug (Sykes et al., 2016; Garg et al., 2012; Padhi et al., 2022; Padhi & Behera, 2021).

Recent years have seen an increase in the research endeavour to develop artemisinin-derived nanoplatforms to improve the current forms of cancer treatment (Liu et al., 2017; Natesan et al., 2017). Exploring artemisinin-based nanodrug development systems is imperative in introducing artemisinin as a blossoming clinical candidate. Targeted delivery using nanodrug delivery systems is a complex phenomenon that can be accomplished via passive and active targeting (Figure 8.3).

8.5.1 Passive Targeting

Passive targeting is a function of the physicochemical characteristics of the nanoparticles. It involves the conveyance of nanocarriers employing passive diffusion across leaky tumour capillary fenestrations (pore diameter: 40-few hundreds of nm) and convection to the cell and interstitial tumours (Maeda et al., 2001; Behera & Padhi, 2020). In recent years, anticancer drugs have been developed, showing the enhanced permeability and retention effect, predominantly exploited by all nanocarriers as primary guidance (Unezaki et al., 1996). The enhanced permeability and retention effects capitalize on two factors-first and foremost, the fact that there is an irregular presence of endothelium surrounding the tumour vasculature promoting the movement of large particles (diameter less than 200 nm) and the fact that the tumour microenvironment is characterized by a dearth of lymphatic drainage allowing these particles to remain in the diseased tissue for a longer time (Gref et al., 1994; Sharma et al., 2013; M. Zhu et al., 2013). Table 8.1 enlists all the types of novel drug delivery systems devised to enhance the antitumour effects of artemisinins.

8.5.2 Active Targeting

Active targeting is a specific recognition process that relies on overexpressed receptors on the surface of cancer cells

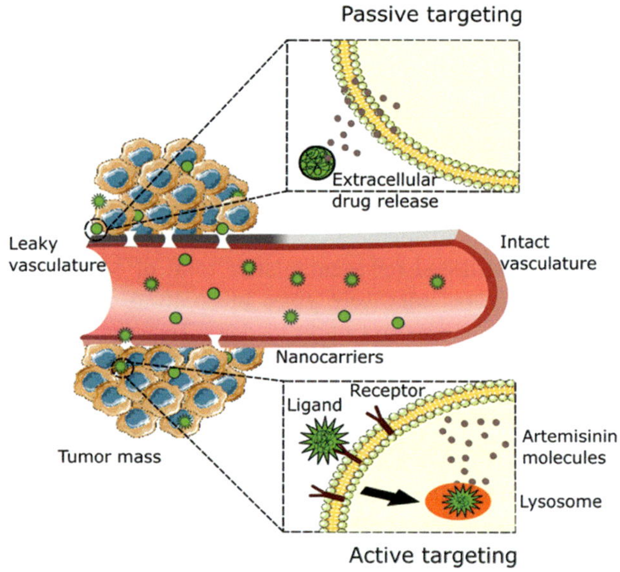

FIGURE 8.3 Illustration showing passive and active targeting of nanoparticles in tumour cells. (This image is a derivative created by the use of various illustrations available through creative common rights (CC BY 4.0) on Reactome [CristofferSevilla, Steve Jupe ('Blood vessel section' and 'Receptor generic') and Smart Servier Medical Art (CC BY 3.0)-('Cell membrane' and 'Cancerous cell').)

to reach and selectively bind and destroy them. The attachment of the ligand to a nanoparticle permits it to enter the cancer cell, over-expressing the corresponding receptor on its surface, thereby enhancing the cellular uptake of the drug and, as a result, its therapeutic efficacy (Patnaik et al., 2021). The widely used technique to attain an active target is the linking of an agent, usually proteins, peptides, monoclonal antibodies, and oligonucleotides, on the surface of the nanoparticle (Gref et al., 1994; Sharma et al., 2013; Zhu et al., 2013). This phenomenon is used to carry drugs into the tumour cells by endocytosis. This targeted drug delivery facilitates the release of higher drug concentrations into the specific tumour cells, evading side effects and lowering systemic toxicity (Nagaraju et al., 2021). Table 8.2 gives a brief description of the active targeting strategies for the delivery of artemisinin to the tumour cells.

8.6 FUTURE PROSPECTS

Altogether, artemisinins are potent anticancer agents; however, despite many attempts at inspecting their antitumour effects, we still have a limited understanding of them. Here, we have mentioned a few domains that require research in the years to come for the successful clinical application of artemisinin as an antitumour drug. The eradication of cancer cells is possible due to the propensity of artemisinins to target a myriad of molecular events. The exact mechanism of action and molecular targets are, nevertheless, contentious. Artemisinins have been documented to target survivin, phorbol-12-myristate-13-acetate-induced protein 1 (NOXA), B-lymphoma Mo-MLV insertion region 1 homolog

TABLE 8.1
Artemisinin-Loaded Novel Drug Delivery Systems Employed for Passive Targeting

S. No.	Compound Drug(s)	Composition/Carriers	Type of Novel Drug Delivery Systems	Targeted Cancer Type	Mechanism of Action	References
1	Dihydroartemisinin and doxorubicin	Poly(lactic-co-glycolic acid), dipalmitoylphosphatidylcholine	Core–shell-type lipid/polymer hybrid nanoparticles	Cervical cancer cells (Hela) and hepatocellular carcinoma cells (HepG 2)	Endocytosis by "lipid raft/caveolae-mediated pathway", bypassing lysosomes	Ma et al. (2015)
2	Artemisinin	Chitosan, magnetite	Polymer/inorganic-based nanoparticles	Breast cancer (4T1)	Application of external magnetic field to aid targeting of nanoparticles and to escape reticuloendothelial system.	Natesan et al. (2017)
3	Artesunate	Bovine serum albumin	Polymer-based nanoparticles	Hepatocellular carcinoma (SMMC-7721 and PLC cells)	Accumulation of bovine serum albumin nanoparticles in the mitochondria due to change in the membrane potential due to the acidic environment of cancer cells.	Liu et al. (2017)
4	Dimeric artemisinin and piperazine	Dipalmitoylphosphatidylcholine, Methoxypoly (ethylene glycol) 2000-1,2-bis(diphenylphosphino) ethane	Lipid-based nanoparticles	Breast carcinoma cells (MDA-MB-23)	pH-Responsive drug release from the nanoparticles	Zhang et al. (2016)
5	Artemether	Soybean lecithin, cholesterol, and vitamin E	Lipid-based nanoparticles	Hepatocellular carcinoma cells (HepG2)	Utilization of the sonodynamic potential of artemether for cytotoxic effect and to enhance reactive oxygen species generation	Chen et al. (2015a)
6	Artemether	Soybean, lecithin, and cholesterol	Liposomes, lipid-based nanoparticles	Hepatocellular carcinoma cell (H22)	Drug accumulation at the tumour site by enhanced permeability and retention effect.	Tian et al. (2017)
7	Dihydroartemisinin	Carboxyl-functionalized superparamagnetic iron oxide Nanoparticles, Fe_3O_4@C@MIL-100(Fe)	Inorganic-based nanoparticles	Human cervical carcinoma (HeLa) and human lung carcinoma cells (A549)	Targeting nanoparticles to the target site via an external magnetic field and pH-responsive drug release.	Wang et al. (2016)a
8	Artemisinin	FeOOH@poly(norepinephrine)	Inorganic-based nanoparticles	Breast cancer cells (4T1)	Near-infrared absorption for production of hydroxyl radicals due to Fenton reaction between iron ions and tumour H_2O_2.	He et al. (2019)

(Continued)

TABLE 8.1 (*Continued*)
Artemisinin-Loaded Novel Drug Delivery Systems Employed for Passive Targeting

S. No.	Compound Drug(s)	Composition/Carriers	Type of Novel Drug Delivery Systems	Targeted Cancer Type	Mechanism of Action	References
9	Artemisinin	Mesoporous nickel oxide - terbium complexes	Inorganic-based nanoparticles	Cervical cancer cells (HeLa)	Mesoporous nickel oxide is a photothermal agent and pH-responsive agent. Also, Ni^{2+} cleaves the endoperoxide bridge of artemisinin by the Fenton reaction.	Liu et al. (2018)
10	Artemisinin	Prussian blue @MIL-100(Fe) dual metal-organic-framework	Inorganic-based nanoparticles	Cervical cancer cells (HeLa)	Photothermal therapy and pH-responsive degradation of nanoparticles.	Wang et al. (2016b)
11	Arteether	Polytetramethylene ether glycol, toluene diisocyanates	Nanomicelles: polymer-based nanoparticles	Breast cancer cell (4T1)	pH-Responsive degradation at endolysosomal pH at 5.4	Jabbarzadegan et al. (2016)
12	Artesunate	Lecithin, labrafil, and tween 80, chitosan	Lipid-based nanoparticles/ polymer-based nanoparticles	Breast cancer cells (MCF-7 and MDA-MB-231)	pH-Responsive chitosan degradation, its protonation in the extracellular matrix aids nanoparticles to attach to tumour cells by electrostatic interaction. It also imparts stealth properties to the nanoparticles.	Tran et al. (2015)
13	Artemisinin	Magnetic supraparticles@ artemisinin nanoparticles	Inorganic-based nanoparticles	Ovarian cancer cells (SKOV-3)	Utilization of near-infrared light to enhance Fenton reaction catalysed by ferrous ions to increase reactive oxygen species generation.	Ding et al. (2018b)
14	Dihydroartemisinin	Dihydroartemisinin@ hydrophobic polydopamine-iron	Inorganic-based nanoparticles	Cervical cancer cells (HeLa)	pH-Responsive degradation activation of dihydroartemisinin by ferrous ions	Dong et al. (2019)
15	Artesunate	Iron oxide@manganese silicate-folic acid	Inorganic-based nanoparticles	Human lung adenocarcinoma cells (A549)	Iron oxide cores comprise magnetic nanocrystals that release ferrous ions in an acidic environment.	Chen et al. (2015b)
16	Artemisinin	Polycaprolactone with polyethylene glycol triblock copolymers	Nanomicelles: polymer-based nanoparticles	Breast cancer cell lines (MCF7 and 4T1)	The copolymer degrades in an acidic environment characteristic of tumour/ lysosomes by hydrolysis to release artemisinin.	Manjili et al. (2017)

TABLE 8.2
Artemisinin-Loaded Novel Drug Delivery Systems Employed for Active Targeting

S. No.	Compound	Composition/Carrier	Type of Novel Drug Delivery Systems	Targeted Cancer Type	Mechanism of Action	References
1	Dihydroartemisinin	Graphene oxide and transferrin	Carbon-based nanoparticles	Mammary carcinoma (EMT 6)	Endosome formation containing Tf-receptor complex by clathrin-mediated endocytosis	Liu et al. (2015)
2	Artemisinin	Hyaluronic acid derivatized multi-walled carbon nanotubes, coupled with transferrin	Carbon-based nanoparticles	Breast adenocarcinoma cells (MCF-7)	Receptor-mediated endocytosis.	Zhang et al. (2015)
3	Artesunate	Transferrin, mesoporous copper sulphide nanoparticles	Inorganic-based nanoparticles	Breast adenocarcinoma cells (MCF-7)	Transferrin receptor-mediated endocytosis. Near-infrared absorption for photothermal therapy coupled with reactive oxygen species generation for photodynamic therapy.	Hou et al. (2017)
4	Artemisinin	Hyaluronic acid grafted on mesoporous iron oxide core	Inorganic-based nanoparticles	Breast cancer cell (MCF7)	Hyaluronic acid-CD44 receptor-mediated endocytosis and further accumulation in the lysosome	Zhang et al. (2016b)
5	Artemisinin with cisplatin	Arginyl-glycyl-aspartic acid tripeptide, iron oxide core core	Inorganic-based nanoparticles	Resistant lung cancer cells (A549R)	Arginyl-glycyl-aspartic acid peptide-integrin receptor-mediated endocytosis, accumulation in the lysosome and reactive oxygen species generation by Fenton reaction	Gao et al. (2018)
6	Artemisinin	Hyaluronic acid, iron oxide core, titanium dioxide	Photodynamic therapy: Inorganic-based nanoparticles	Breast cancer cell (MCF7)	HA-Receptor-mediated endocytosis. Tumour Photodynamic therapy by stimulation of reactive oxygen species generation due to visible light absorption by titanium dioxide - iron oxide nanoparticles	Zhang et al. (2017)
7	Dihydroartemisinin	polyD,L-lactic-co-glycolic acid-polyethylene glycol$_{2000}$-COOH, in situ apolipoprotein E-recruitment as an active targeting ligand	Polymer-based nanoparticles	Breast cancer cell (4T1)	*In-situ* recruiting of endogenous apolipoprotein E on the surface of nanoparticles and facilitation of cancer cell uptake by apolipoprotein E-low density lipoprotein receptor interaction	Zheng et al. (2019)
8	Artemisinin	Polyethylene glycol-polycaprolactone, LyP-1 peptide	Polymer-based nanoparticles	Breast cancer cells (MDA-MB-435S)	Selective tumour targeting by Lyp-1 peptide and its receptor P32/gClqr, the peptide also binds directly to tumour lymphatic vessel facilitating simultaneous destruction if tumour lymph system	Wang et al. (2012)

(Continued)

TABLE 8.2
Artemisinin-Loaded Novel Drug Delivery Systems Employed for Active Targeting

S. No.	Compound	Composition/Carrier	Type of Novel Drug Delivery Systems	Targeted Cancer Type	Mechanism of Action	References
9	Dihydroartemisinin	Transferrin-eight-arm-polyethylene glycol	Polymer-drug conjugate; polymer-based nanoparticles	Lewis lung carcinoma cells (LLC)	Receptor-mediated endocytosis further lysosome accumulation and acid-facilitated release.	Liu et al. (2016)
10	Dihydroartemisinin	Soybean lecithin, cholesterol and distearoylphosphatidylethanolamine-polyethylene glycol$_{2000}$, alkyl glucosides (*n*-octyl-β-D19 glucoside, *n*-β-D-decanyl glucoside and dodecyl glucoside)	Liposomes, lipid-based nanoparticles	Human liver carcinoma cells (HepG2)	Targeting of overexpressed glucose transporter 1 (due to Warburg effect) by alkyl glycoside	Shen et al. (2020)
11	Artemisinin	Transferrin, egg phosphatidylcholine plus cholesterol and 1,2-distearoyl-sn-glycero-3-phosphoethanolamine-*N*-[carboxy(polyethylene glycol)-2000] ammonium salt	Liposomes: lipid-based nanoparticles	Colon cancer cells (HCT-8)	Endosome formation containing transferrin-receptor complex by clathrin-mediated endocytosis	Leto et al. (2016)
12	Dihydroartemisinin	Buthionine sulfoximine and CellROX, decorated with transferrin	Liposomes: lipid-based nanoparticles	Hepato carcinoma cell (HepG2)	Transferrin-receptor mediated endocytosis and inhibition of glutathione (adaptive antioxidant) by L-buthionine-sulfoximine.	Yu et al. (2020)
13	Dihydroartemisinin and doxorubicin	Soybean phosphatidylcholine, cholesterol, and distearoyl phosphatidyl ethanolamine-polyethylene glycol$_{2000}$-amine, mannose	Liposomes; lipid-based nanoparticles	HCT8/ADR human colon cancer cell	Mannose receptors overexpressed cells lead to uptake of NPs inside the cancer cell and further preferential accumulation inside nucleus	Kang et al. (2017)
14	Dihydroartemisinin and Sorafenib	Cholesteryl oleate, glyceryl trioleate (triolein), cholesteryl	Solid-lipid nanoparticles; lipid-based nanoparticles	Liver cancer cells (HepG2)	Surface modification of nanoparticles by attaching low-density lipoprotein (ApoB100) and targeting cancer cells via low-density lipoprotein receptor-mediated endocytosis	Wang et al. (2019)
15	Artemether and paclitaxel	Cholesterol, tocopheryl polyethylene glycol succinate$_{1000}$-mannose, dequalinium-polyethylene glycol$_{2000}$-distearoyl phosphatidyl ethanolamine	Lipid-based nanoparticles	Invasive brain glioma (C6 cells)	Receptor-mediated endocytosis (tocopheryl polyethylene glycol succinate$_{1000}$-mannose) and glucose transporters for transport across the blood-brain barrier	Li et al. (2014)

(BMI-1), nuclear factor kappa-light-chain-enhancer of activated B cells, and hypoxia-inducible factor-1α (Crespo-ortiz & Wei, 2012).

However, numerous additional targets that may have a role in drug interactions, drug response, and mechanisms of resistance are yet to be discovered. The findings of recent research disclosing the ability of artemisinin to bind with 860 proteins could be a step in the right direction to identify more of its molecular targets (Zhang et al., 2016a,b,c). The identification of quantifiable biomarkers, including vascular endothelial growth factor, p53, transferrin receptor 1, *P*-glycoprotein, and the cluster of differentiation 31 upon treatment with artemisinins, might pave the way for their clinical application.

The knotty problem of elucidating the mechanism of action of artemisinins can be solved by employing genomics, proteomics, metabolomics, and immunology studies (Michaelsen et al., 2015). Unfortunately, the pharmacokinetic studies that should ideally follow the development of nanodrugs to enhance the physicochemical properties of artemisinins are rarely reported. Hence, it is unknown whether or not these newly developed artemisinin-based novel drug delivery systems aid in upgrading the pharmacokinetic parameters of the compounds. Although the anticancer property of artemisinins has been verified through *in vitro* and *in vivo* analyses, the dose (usually in μM) used is high as compared to that for the identification of antimalarial activity (usually in nM) (Xu et al., 2020). It is commonly known that the dose levels and side effects vary conjointly. This raises concerns about dose-dependent neurotoxicity and is a major hurdle in developing artemisinin as a clinical drug (Genovese et al., 2000; Q. Li et al., 2005). Hitherto, most anticancer artemisinin-related research has been focused on simple monomers such as artemisinin, dihydroartemisinin, and artesunate, i.e. first-generation derivatives. Monomers are not effective enough as antitumour agents by themselves. Hence, further studies on the next-generation derivatives and polymers of artemisinin (dimers, trimers, etc.) would mark a quantum leap in the research on the anticancer activity of artemisinins (Li et al., 2016). The structure-activity relationship between artemisinins and their antitumour activity must also be investigated. The antineoplastic activity can be intensified by the combination of artemisinin and its derivatives with conventionally available chemotherapy drugs, which might be useful to minimize the resistance of cancer cells to the chemotherapeutic drugs (owing to the property of artemisinins as chemosensitizing agents) with an added advantage of no obvious side effects. However, the determination of appropriate doses and drug combinations is an obstacle to be overcome. At the same time, close surveillance is required to assess artemisinin's long-term therapy effect. More extensive study is needed to understand the immunomodulatory mechanisms and response of artemisinins and possibly couple them with the chemo response. With the intention of utilising artemisinins as adjuvant agents in the treatment of malignant tumours, studies must also prominently concentrate on developing oral formulations of artemisinins, which calls for more research on the safety aspects of these drugs. Artemisinins, in addition to the capability of selectively causing the death of cancer cells, are also known to modulate the tumour environment. Regardless, the exact mechanism for the same needs elaboration is needed. This brings into focus a novel area of research, i.e. exploration of the mechanism of adjustment of tumour microenvironment by artemisinins. Shortly anticipated inclusion of artemisinins in the repository of anticancer remedies has necessitated propitious research in this domain.

8.7 CONCLUSION

Artemisinins, already established as the cardinal drugs in antimalarial therapy, have recently been making headway into the realm of anticancer treatment by possessing a host of qualities such as good therapeutic index, high specificity, broad-spectrum activity, and preferential toxicity against tumour cells, giving them an edge over the pre-existing treatment options. However, their more inapt characteristics for clinical use, such as low bioavailability, short half-life, etc., could be their undoing, warranting the need for novel systems to compensate for their shortcomings. In this chapter, we have summarised the state-of-the-art technology, namely nanoformulations, for improving the drawbacks of artemisinins for their translation from bench to bedside. Nanodrug delivery systems may aid the chemotherapeutic outcome by resolving artemisinin's bioavailability issues, resulting in increased cellular uptake and reduced dosage, enhancing the toxicity profile and patient compliance.

This chapter also contains an exhaustive summary of the clinical trials and case studies conducted thus far to analyse the formulations of artemisinin for cancer treatment. While these trials are fewer in number, they have served to rule out the toxicities of artemisinins and have underlined the requirement for their long-term administration to prevent the recurrence of cancer. They have also shed some light on the administration dosage that will depend on the status, stage, and progression of the tumour, as a low dosage of artemisinins favours cell cycle arrest. At the same time, a high dose could lead to apoptosis. With a surge in clinical trials coupled with encouraging results, it may not be long before we witness the emergence of artemisinins as a substitute for adjuvant and neoadjuvant treatment of solid tumours, especially lung, breast, colorectal, pancreas, and other types of cancers. The extensive research on artesunate and dihydroartemisinin has identified them as chemosensitizing agents, marking their potential use as an add-on therapy combined with current chemotherapeutics, which has a scope of being examined in future clinical trials. In conclusion, the discovery of the antitumour activity of artemisinins has unveiled new avenues in treating highly metastatic and aggressive cancers with the development of drug resistance. The amalgamation of artemisinins and nanotechnology, representing the best of both worlds, the healing power of nature and technology, presents itself as a benefit to the well-being of humanity.

REFERENCES

Artesunate attenuates breast cancer cell migration via CD44 down-regulation and inhibition of hyaluronic acid-induced signaling. *Cancer Cell International* 2016;16, 89. doi:10.1186/s12935-016-0357-7.

Anfosso, L., Dell, R., Pfeffer, U., Vene, R., Forlani, A., Albini, A., & Efferth, T. (2004). Inhibition of angiogenesis in vivo and growth of Kaposi's sarcoma xenograft tumours by the antimalarial artesunate. *Biochemical Pharmacology*, 68(12), 2359–2366.

Augustin, Y., Staines, H. M., & Krishna, S. (2020). Artemisinins as a novel anti-cancer therapy: Targeting a global cancer pandemic through drug repurposing. *Pharmacology and Therapeutics*, 216, 107706.

Beccafico, S., Morozzi, G., Marchetti, M. C., Riccardi, C., Sidoni, A., Donato, R., & Sorci, G. (2015). Artesunate induces ROS- and p38 MAPK-mediated apoptosis and counteracts tumour growth in vivo in embryonal rhabdomyosarcoma cells. *Carcinogenesis*, 36(9), 1071–1083.

Behera, A., & Padhi, S. (2020). Passive and active targeting strategies for the delivery of the camptothecin anticancer drug: A review. *Environmental Chemistry Letters*, 18(5), 1557–1567.

Behera, A., & Padhi, S. (2022). pH-Sensitive polymeric nanoparticles for cancer treatment. In: Padhi, S., Behera, A., & Lichtfouse, E. (Eds.) *Polymeric Nanoparticles for the Treatment of Solid Tumours*. Environmental Chemistry for a Sustainable World, vol 71. Springer, Cham, pp. 401–425.

Berdelle, N., Nikolova, T., Quiros, S., Efferth, T., & Kaina, B. (2011). Artesunate induces oxidative DNA damage, sustained DNA double-strand breaks, and the ATM/ATR damage response in cancer cells. *Molecular Cancer Therapeutics*, 10(12), 2224–2233.

Berger, T. G., Dieckmann, D., Efferth, T., Schultz, E. S., Funk, J.-O., Baur, A., & Schuler, G. (2005). Artesunate in the treatment of metastatic uveal melanoma: First experiences. *Oncology Reports*, 14(6), 1599–1603.

Bostwick, D. G., Alexander, E. E., Singh, R., Shan, A., Qian, J., Santella, R. M., ... & Ph, D. (2000). Antioxidant enzyme expression and reactive oxygen species damage in prostatic intraepithelial neoplasia and cancer. *Cancer*, 89(1), 123–134.

Bragta, S., Sharma, A., & Bhardwaj, R. (2018). Nanoparticles: A promising approach for drug delivery. *International Journal of Pharmaceutical Sciences and Research*, 9(12), 4845–4855. https://doi.org/10.22159/ajpcr.2018.v11i1.22035.

Cabello, C. M., Lamore, S. D., Iii, W. B. B., Qiao, S., Azimian, S., Lesson, J. L., & Wondrak, G. T. (2013). The redox antimalarial dihydroartemisinin targets human metastatic melanoma cells but not primary melanocytes with induction of NOXA-dependent apoptosis. *Investigational New Drugs*, 30(4), 1289–1301.

Cao, Y., Feng, Y. H., Gao, L. W., Li, X. Y., Jin, Q. X., Wang, Y. Y., ... & Wei, M. J. (2019). Artemisinin enhances the antitumour immune response in 4T1 breast cancer cells in vitro and in vivo. *International Immunopharmacology*, 70, 110–116.

Charlie-Silva, I., Fraceto, L. F., & de Melo, N. F. S. (2018). Progress in nano-drug delivery of artemisinin and its derivatives: Towards to use in immunomodulatory approaches. *Artificial Cells, Nanomedicine and Biotechnology*, 46(3), S611–S620.

Chen, G., Benthani, F. A., Wu, J., Liang, D., Bian, Z., & Jiang, X. (2020). Artemisinin compounds sensitize cancer cells to ferroptosis by regulating iron homeostasis. *Cell Death & Differentiation*, 27(1), 242–254.

Chen, H., Huang, X., Zhou, X., Zheng, B., & Huang, J. (2015a). Potential sonodynamic anticancer activities of artmether and liposome- encapsulated artmether. *Chemical Communications*, 51(22), 4681–4684.

Chen, J., Zhang, W., Zhang, M., Guo, J., Wang, H., He, M., ... Chen, Q. (2015b). Mn (II) mediated degradation of artemisinin based on Fe3O4@MnSiO3- FA nanospheres for cancer therapy in vivo. *Nanoscale*, 7(29), 12542–12551.

Chen, K., Shou, L., Lin, F., Duan, W., Wu, M., Xie, X., ... & Tao, M. (2014). Artesunate induces G2/M cell cycle arrest through autophagy induction in breast cancer cells. *Anticancer Drugs*, 25(6), 652–662.

Chen, X., He, L. Y., Lai, S., & He, Y. (2020). Dihydroartemisinin inhibits the migration of esophageal cancer cells by inducing autophagy. *Oncology Letters*, 20(4), 1–9.

Chen, X., Wong, Y. K., Lim, T. K., Lim, W. H., Lin, Q., Wang, J., & Hua, Z. (2017). Artesunate activates the intrinsic apoptosis of HCT116 cells through the suppression of fatty acid synthesis and the NF-κB pathway. *Molecules*, 22(8), 1272.

Chen, X., Zhang, X. L., Zhang, G. H., & Gao, Y. F. (2019). Artesunate promotes Th1 differentiation from CD4+ T cells to enhance cell apoptosis in ovarian cancer via miR-142. *Brazilian Journal of Medical and Biological Research*, 52(5), 1–8.

Cooper, G. M. (2000). The development and causes of cancer. In: Cooper, G. M., & Adams, K. W. (Eds.) *The Cell: A Molecular Approach*, 2nd Edition. Oxford University Press, New York, pp. 713–722.

Crespo-ortiz, M. P., & Wei, M. Q. (2012). Antitumour activity of artemisinin and its derivatives: From a well-known antimalarial agent to a potential anticancer drug. *Journal of Biomedicine & Biotechnology*, 2012, 247597.

Cui, C., Feng, H., Shi, X., Wang, Y., Feng, Z., Liu, J., ... & Tong, H. (2015). Artesunate down-regulates immunosuppression from colorectal cancer Colon26 and RKO cells in vitro by decreasing transforming growth factor β1 and interleukin-10. *International Immunopharmacology*, 27(1), 110–121.

Das, A. (2015). Anticancer effect of antimalarial artemisinin compounds. *The Annals of Medical and Health Sciences Research*, 5(2), 93–102.

Deeken, J. F., Wang, H., Hartley, M., Cheema, A. K., Smaglo, B., Hwang, J. J., ... & Pishvaian, M. J. (2018). A phase I study of intravenous artesunate in patients with advanced solid tumour malignancies. *Cancer Chemotherapy and Pharmacology*, 81(3), 587–596.

Ding, Q., Xu, X., Yue, Y., Mei, C., Huang, C., Jiang, S., ... & Han, J. (2018a). Nanocellulose-mediated electroconductive self-healing hydrogels with high strength, plasticity, viscoelasticity, stretchability, and biocompatibility toward multifunctional applications. *ACS Applied Materials and Interfaces*, 10(33), 27987–28002.

Ding, Y., Wan, J., Zhang, Z., Wang, F., Guo, J., & Wang, C. (2018b). Localized Fe(II)-induced cytotoxic reactive oxygen species generating nano-system for enhanced anticancer therapy. *Applied Materials & Interfaces*, 10(5), 4439–4449.

Dixon, S. J., Lemberg, K. M., Lamprecht, M. R., Skouta, R., Eleina, M., Gleason, C. E., ... & Stockwell, B. R. (2012). Ferroptosis: An iron-dependent form of non-apoptotic cell death. *Cell*, 149(5), 1060–1072.

Dong, L., Wang, C., Zhen, W., Jia, X., An, S., Xu, Z., ... & Jiang, X. (2019). Biodegradable iron-coordinated hollow polydopamine nanospheres for dihydroartemisinin delivery and selectively enhanced therapy in tumour cells. *Journal of Materials Chemistry B*, 7(40), 6172–6180.

Efferth, T. (2017). From ancient herb to modern drug: Artemisia annua and artemisinin for cancer therapy. *Seminars in Cancer Biology*, 46, 65–83.

Efferth, T., Giaisi, M., Merling, A., Krammer, P. H., & Li-weber, M. (2007). Artesunate induces ROS-mediated apoptosis in doxorubicin-resistant T leukemia cells. *PLoS ONE*, 2(8), 693.

Galluzzi, L., Pietrocola, F., Pedro, J. M. B., Ravi, K., Maiuri, M. C., Martin, S. J., ... & Piacentini, M. (2015). Autophagy in malignant transformation and cancer progression. *The EMBO Journal*, 34, 1–26.

Gao, P., Shen, S., Li, X., Liu, D., Meng, Y., Liu, Y., ... & Gu, L. (2020). Dihydroartemisinin inhibits the proliferation of leukemia cells K562 by suppressing PKM2 and GLUT1 mediated aerobic glycolysis. *Drug Design, Development and Therapy*, 14, 2091–2100.

Gao, Z., Li, Y., You, C., Sun, K., An, P., Sun, C., ... & Sun, B. (2018). Iron oxide nanocarrier-mediated combination therapy of cisplatin and artemisinin for combating drug resistance through highly increased toxic reactive oxygen species generation iron oxide nanocarrier-mediated combination therapy of cisplatin and artemisin. *Applied Biomaterials*, 1(2), 270–280.

Genovese, R. F., Newman, D. B., & Brewer, T. G. (2000). Behavioral and neural toxicity of the artemisinin antimalarial, arteether, but not artesunate and artelinate, in rats. *Pharmacology, Biochemistry, and Behavior*, 67, 37–44.

Gravett, A. M., Liu, W. M., Krishna, S., Chan, W. C., Haynes, R. K., Wilson, N. L., & Dalgleish, A. G. (2011). In vitro study of the anti-cancer effects of artemisone alone or in combination with other chemotherapeutic agents. *Cancer Chemotherapy and Pharmacology*, 67(3), 569–577.

Greenwell, M., & Rahman, P. K. S. M. (2015). Medicinal plants : Their use in anticancer treatment. *International Journal of Pharmaceutical Sciences and Research*, 6(10), 4103–4112.

Gref, R., Minamitake, Y., Peracchia, M. T., Trubetskoy, V., Torchilin, V., & Langerll, R. (1994). Biodegradable long-circulating polymeric nanospheres. *Science*, 263(January), 1600–1603.

Hagens, C. V., Maren, I. W., & Munzinger, J. (2017). Prospective open uncontrolled phase I study to define a well-tolerated dose of oral artesunate as add-on therapy in patients with metastatic breast cancer (ARTIC M33/2). *Breast Cancer Research and Treatment*, 164(2), 359–369.

Hamacher-brady, A., Stein, H. A., Turschner, S., Toegel, I., Mora, R., Jennewein, N., ... & Brady, N. R. (2011). Artesunate activates mitochondrial apoptosis in breast cancer cells via iron-catalyzed lysosomal reactive oxygen species production. *Journal of Biological Chemistry*, 286(8), 6587–6601.

Handrick, R., Ontikatze, T., Bauer, K., Freier, F., Rübel, A., Dürig, J., ... & Jendrossek, V. (2010). Dihydroartemisinin induces apoptosis by a Bak-dependent intrinsic pathway. *Molecular Cancer Therapeutics*, 9(9), 2497–2510.

Hansen, J. B., Fisker, N., Westergaard, M., Kjærulff, L. S., Hansen, H. F., Thrue, C. A., ... & Ørum, H. (2008). SPC3042: A pro-apoptotic survivin inhibitor. *Molecular Cancer Therapeutics*, 7(9), 2736–2745.

He, Z., Su, H., Shen, Y., Shi, W., Liu, X., Liu, Y., ... & Ge, D. (2019). Poly(norepinephrine)-coated FeOOH nanoparticles as carriers of artemisinin for cancer photothermalchemical combination therapy. *RSC Advances*, 9(18), 9968–9982.

Ho, W. E., Peh, H. Y., Chan, T. K., & Wong, W. S. F. (2014). Artemisinins: Pharmacological actions beyond antimalarial. *Pharmacology and Therapeutics*, 142(1), 126–139.

Hou, L., Shan, X., Hao, L., Feng, Q., & Zhang, Z. (2017). Copper sulfide nanoparticle-based localized drug delivery system as an effective cancer synergistic treatment and theranostic platform. *Acta Biomaterialia*, 54, 307–320.

Hsu, E. (2014). How techniques of herbal drug preparation affect the therapeutic outcome: Reflections on Qinghao 青蒿 (Herba Artemisiae annuae) in the history of the Chinese materia medica. In: Aftab, T., Ferreira, J., Khan, M., & Naeem, M. (Eds.) *Artemisia annua: Pharmacology and Biotechnology*. Springer, Berlin, Heidelberg.

Hwang, Y. P., Yun, H. J., Kim, H. G., Han, E. H., Lee, G. W., & Jeong, H. G. (2010). Suppression of PMA-induced tumour cell invasion by dihydroartemisinin via inhibition of PKCα/Raf/MAPKs and NF-κB/AP-1-dependent mechanisms. *Biochemical Pharmacology*, 79(12), 1714–1726.

Jabbarzadegan, M., Rajayi, H., Ali, M., Jahromi, M., Yousefi, M., Hassan, Z. M., & Majidi, J. (2016). Application of arteether-loaded polyurethane nanomicelles to induce immune response in breast cancer model. *Artificial Cells, Nanomedicine and Biotechnology*, 45(4), 808–816.

Jansen, F. H., Adoubi, I., C, K. C. J., Cnodder, T. D. E., Jansen, N., Tschulakow, A., & Efferth, T. (2011). First study of oral artenimol-R in advanced cervical cancer: Clinical benefit. *Tolerability and Tumour Markers*, 4422, 4417–4422.

Jeong, D. E., Song, H. J., Lim, S., Lee, S. J., & Lim, J. E. (2015). Repurposing the antimalarial drug artesunate as a novel therapeutic agent for metastatic renal cell carcinoma due to its attenuation of tumour growth, metastasis, and angiogenesis. *Oncotarget*, 6(32), 33046–33064.

Jia, G., Kong, R., Ma, Z. B., Han, B., Wang, Y. W., Pan, S. H., ... & Sun, B. (2014). The activation of c-Jun NH2-terminal kinase is required for dihydroartemisinin-induced autophagy in pancreatic cancer cells. *Journal of Experimental and Clinical Cancer Research*, 33(1), 1–10.

Jiang, F., Zhou, J. Y., Zhang, D., Liu, M. H., & Chen, Y. G. (2018). Artesunate induces apoptosis and autophagy in HCT116 colon cancer cells, and autophagy inhibition enhances the artesunate-induced apoptosis. *International Journal of Molecular Medicine*, 42(3), 1295–1304.

Kang, X., Wang, H., Peng, H., Chen, B., Zhang, W., Wu, A., & Xu, Q. (2017). Codelivery of dihydroartemisinin and doxorubicin in mannosylated liposomes for drug-resistant colon cancer therapy. *Nature Publishing Group*, 38(6), 885–896.

Khan, A. I., Kapoor, A., Che, J., Martin, L., Rogazzo, M., Mercier, T., ... & Thiemermann, C. (2017). The antimalarial drug artesunate attenuates cardiac injury in a rodent model of myocardial infarction. *Shock*, 44(6), 675–681.

Klayman, D. L. (1983). Qinghaosu (artemisinin): An antimalarial drug from China. *Science*, 228, 1049–1055.

Krishna, S., Ganapathi, S., Chis, I., Saeed, M. E. M., Cowan, M., Finlayson, C., ... & Kumar, D. (2015). A randomised, double blind, placebo-controlled pilot study of oral artesunate therapy for colorectal cancer. *Ebiomedicine*, 2(1), 82–90.

Lai, H., Nakase, I., Lacoste, E., Singh, N. P., & Sasaki, T. (2009). Artemisinin-transferrin conjugate retards growth of breast tumours in the rat. *Anticancer Research*, 29(10), 3807–3810.

Lee, J., Zhang, G., Wu, X., Xu, F., Zhou, J., & Zhang, X. (2012). Growth inhibitory effect of dihydroartemisinin on Bcr/Abl+chronic myeloid leukemia K562 cells involve AKT, ERK and NF-κB modulation. *Journal of Cancer Research and Clinical Oncology*, 138(12), 2095–2102.

Leto, I., Coronnello, M., Righeschi, C., Bergonzi, C., Mini, E., & Rita, A. (2016). Enhanced efficacy of artemisinin loaded in transferrin-conjugated liposomes versus stealth liposomes against HCT-8 colon cancer cells. *Chemmedchem*, 11(16), 1745–1751.

Li, P. C. H., Lam, E., Roos, W. P., Zdzienicka, M. Z., Kaina, B., & Efferth, T. (2008). Artesunate derived from traditional Chinese medicine induces DNA damage and repair. *Cancer Research*, 68(11), 4347–4351.

Li, Q., Xie, L. H., Si, Y., Wong, E., Upadhyay, R., Yanez, D., & Weina, P. J. (2005). Toxicokinetics and hydrolysis of artelinate and artesunate in malaria-infected rats. *International Journal of Toxicology*, 24(4), 241–250.

Li, S., Huang, P., Gan, J., Ling, X., Du, X., Liao, Y., … & Bai, Y. (2019). Dihydroartemisinin represses esophageal cancer glycolysis by down-regulating pyruvate kinase M2. *European Journal of Pharmacology*, 854, 232–239.

Li, S., Xue, F., Cheng, Z., Yang, X., Wang, S., Geng, F., & Pan, L. (2009). Effect of artesunate on inhibiting proliferation and inducing apoptosis of SP2/0 myeloma cells through affecting NF-KB p65. *International Journal of Heamtology*, 90(4), 513–521.

Li, X., Zhao, Y., Sun, M., Shi, J., Ju, R., Zhang, C., … & Lu, W.-L. (2014). Multifunctional liposomes loaded with paclitaxel and artemether for treatment of invasive brain glioma. *Biomaterials*, 35(21), 5591–5604.

Li, Z., Li, Q., Wu, J., Wang, M., & Yu, J. (2016). Artemisinin and its derivatives as a repurposing anticancer agent: What else do we need to do? *Molecules*, 21, 1–14.

Liu, K., Dai, L., Li, C., Liu, J., Wang, L., & Lei, J. (2016). Self-assembled targeted nanoparticles based on transferrin-modified eight-arm-polyethylene glycol-dihydroartemisinin conjugate. *Scientific Reports*, 6, 29461.

Liu, K., Zuo, H., Li, G., Yu, H., & Hu, Y. (2020). Global research on artemisinin and its derivatives: Perspectives from patents. *Pharmacological Research*, 169, 105048.

Liu, L., Liu, Y., Mao, F., Jiang, A., Liu, D., Wang, L., … & Zhou, J. (2018). Artemisinin-loaded mesoporous nanoplatform for pH-responsive radical generation synergistic tumour theranostics. *ACS Applied Materials & Interfaces*, 10(7), 6155–6167.

Liu, L., Wei, Y., Zhai, S., Chen, Q., & Xing, D. (2015). Dihydroartemisinin and transferrin dual-dressed nano-graphene oxide for a pH-triggered chemotherapy. *Biomaterials*, 62, 35–46.

Liu, R., Yu, X., Su, C., Shi, Y., & Zhao, L. (2017). Nanoparticle delivery of artesunate enhances the antitumour efficiency by activating mitochondria-mediated cell apoptosis. *Nanoscale Research Letters*, 12(1), 403.

Lu, J. J., Chen, S. M., Zhang, X. W., Ding, J., & Meng, L. H. (2011). The anti-cancer activity of dihydroartemisinin is associated with induction of iron-dependent endoplasmic reticulum stress in colorectal carcinoma HCT116 cells. *Investigational New Drugs*, 29(6), 1276–1283.

Lu, J., Meng, Q., Ji, T., Yang, H., Ma, W., Li, Y., & Xu, X. (2010a). Iron-loaded transferrin (holotransferrin) enhances the cytotoxicity of artemisinin toward cancer cells. *Life Sciences*, 86(25–26), 879–885. https://doi.org/10.1016/j.lfs.2010.03.018.

Lu, J., Meng, L., Shankavaram, U. T., Zhu, C., Tong, L., Chen, G., … & Ding, J. (2010b). Dihydroartemisinin accelerates c-MYC oncoprotein degradation and induces apoptosis in c-MYC-overexpressing tumour cells. *Biochemical Pharmacology*, 80(1), 22–30.

Lu, Y. Y., Chen, T. S., Qu, J. L., Pan, W. L., Sun, L., & Wei, X. B. (2009). Dihydroartemisinin (DHA) induces caspase-3-dependent apoptosis in human lung adenocarcinoma ASTC-a-1 cells. *Journal of Biomedical Science*, 16(1), 1–15.

Lu, Y., Chen, T., & Wang, X. (2010). Single-cell analysis of dihydroartemisinin-induced apoptosis through reactive oxygen species-mediated caspase-8 activation and mitochondrial pathway in ASTC-a-1 cells using fluorescence imaging techniques. *Journal of Biomedical Optics*, 15(4), 046028.

Ma, W., Xu, A., Ying, J., Li, B., & Jin, Y. (2015). Biodegradable core-shell copolymer-phospholipid an in vitro study. *Journal of Biomedical Nanotechnology*, 11(7), 1193–1200.

Maeda, H., Sawa, T., & Konno, T. (2001). Mechanism of tumour-targeted delivery of macromolecular drugs, including the EPR effect in solid tumour and clinical overview of the prototype polymeric drug SMANCS. *Journal of Control Release*, 74(123), 47–61.

Manjili, H. K., Malvandi, H., Mousavi, M. S., Attari, E., & Danafar, H. (2017). In vitro and in vivo delivery of artemisinin loaded PCL-PEG-PCL micelles and its pharmacokinetic study. *Artificial Cells, Nanomedicine, and Biotechnology*, 46(5), 926–936.

Martinez-outschoorn, U. E., Pestell, R. G., Sotgia, F., & Lisanti, M. P. (2016). Cancer metabolism: A therapeutic. *Clinical Oncology*, 14, 11–31.

May, W. S., & Cuatrecasas, P. (1985). Transferrin receptor: Its biological significance. *The Journal of Membrane Biology*, 215, 205–215.

Medhi, B., Patyar, S., Rao, R. S., Byrav Ds, P., & Prakash, A. (2009). Pharmacokinetic and toxicological profile of artemisinin compounds: An update. *Pharmacology*, 84(6), 323–332.

Mercer, A. E., Copple, I. M., Maggs, J. L., Neill, P. M. O., & Park, B. K. (2011). The role of heme and the mitochondrion in the chemical and molecular mechanisms of mammalian cell death induced by the artemisinin antimalarials. *Journal of Biological Chemistry*, 286(2), 987–996.

Mercer, A. E., Maggs, J. L., Sun, X., Cohen, G. M., Chadwick, J., Neill, P. M. O., & Park, B. K. (2007). Evidence for the involvement of carbon-centered radicals in the induction of apoptotic cell death by artemisinin compounds. *Journal of Biological Chemistry*, 282(13), 9372–9382.

Meunier, B., & Robert, A. (2010). Heme as trigger and target for trioxane-containing antimalarial drugs. *Accounts of Chemical Research*, 43(11), 1444–1451.

Mi, Y., Geng, G., Zou, Z., Gao, J., Luo, X., & Liu, Y. (2015). Dihydroartemisinin inhibits glucose uptake and cooperates with glycolysis inhibitor to induce apoptosis in non-small cell lung carcinoma cells. *PLoS ONE*, 10, 1–21.

Michaelsen, F. W., Saeed, M. E. M., Schwarzkopf, J., & Efferth, T. (2015). Activity of Artemisia annua and artemisinin derivatives, in prostate carcinoma. *Phytomedicine*, 22(14), 1223–1231.

Nagaraju, G. P., Srivani, G., Dariya, B., Chalikonda, G., Farran, B., Behera, S. K., … & Kamal, M. A. (2021, February). Nanoparticles guided drug delivery and imaging in gastric cancer. In: Theresa Vincent, C., Barbacid, M., & Gottesman, M. M. (Eds.) *Seminars in Cancer Biology*, vol 69. Academic Press, London, pp. 69–76.

Nakase, I., Gallis, B., Takatani-nakase, T., Oh, S., Lacoste, E., Singh, N. P., … & Sasaki, T. (2009). Transferrin receptor-dependent cytotoxicity of artemisinin-transferrin conjugates on prostate cancer cells and induction of apoptosis. *Cancer Letters*, 274(2), 290–298.

Natesan, S., Ponnusamy, C., Sugumaran, A., Chelladurai, S., Palaniappan, S. S., & Palanichamy, R. (2017). Artemisinin loaded chitosan magnetic nanoparticles for theefficient targeting to the breast cancer. *International Journal of Biological Macromolecules*, 104, 1853–1859.

NCDIR. (2020). Executive Summary. Retrieved from https://www.ncdirindia.org/All_Reports/Report_2020/resources/B1ExecutiveSummary.pdf

Olliaro, P. L., Haynes, R. K., Meunier, B., & Yuthavong, Y. (2001). Possible modes of action of the artemisinin-type compounds. *Trends in Parasitology*, 17(3), 122–126.

Ooko, E., Saeed, M. E. M., Kadioglu, O., Sarvi, S., Colak, M., Elmasaoudi, K., Janah, R., … & Efferth, T. (2015). Artemisinin derivatives induce iron-dependent cell death (ferroptosis) in tumour cells. *Phytomedicine*, 22(11), 1045–1054.

Padhi, S., & Behera, A. (2020). Nanotechnology based targeting strategies for the delivery of camptothecin. In: Ankit, S., Panda Amulya, K., & Eric, L. (Eds.) *Pharmaceutical Technology for Natural Products Delivery, Impact of Nanotechnology*. Springer Nature, Switzerland, pp. 243–272.

Padhi, S., & Behera, A. (2021). Advanced drug delivery systems in the treatment of ovarian cancer. In: Dua, K., Mehta, M., de Jesus Andreoli Pinto, T., Pont, LG, Williams, K. A., & Rathbone, M. J. (Eds.) *Advanced Drug Delivery Systems in the Management of Cancer*. Academic Press, London, pp. 127–139.

Padhi, S., Behera, A., Hasnain, M. S., & Nayak, A. K. (2022). Chitosan-based drug delivery systems in cancer therapeutics. In: Hasnain, M. S., Beg, S., & Nayak, A. K. (Eds.) *Chitosan in Drug Delivery*. Academic Press, London, pp. 159–193.

Padhi, S., Kapoor, R., Verma, D., Panda, A. K., & Iqbal, Z. (2018). Formulation and optimization of topotecan nanoparticles: In vitro characterization, cytotoxicity, cellular uptake and pharmacokinetic outcomes. *Journal of Photochemistry and Photobiology B: Biology*, 183, 222–232.

Padhi, S., Mirza, M. A., Verma, D., Khuroo, T., Panda, A. K., Talegaonkar, S., Khar, R. K., & Iqbal, Z. (2016). Revisiting the nanoformulation design approach for effective delivery of topotecan in its stable form: An appraisal of its in vitro behavior and tumour amelioration potential. *Drug Delivery*, 23(8), 2827–2837.

Park, J. Y. (2019). Artemisinin and the nobel prize in physiology or medicine 2015. *Korean Journal of Pain*, 32(3), 145–146.

Patnaik, S., Gorain, B., Padhi, S., Choudhury, H., Gabr, G. A., Md, S., Kumar Mishra, D., & Kesharwani, P. (2021). Recent update of toxicity aspects of nanoparticulate systems for drug delivery. *European Journal of Pharmaceutics and Biopharmaceutics*, 161, 100–119.

Pradhan, S. P., Tejaswani, P., Behera, A., & Sahu, P. K. (2024). Phytomolecules from conventional to nano form: Next-generation approach for Parkinson's disease. *Ageing Research Reviews*, 93, 102136.

Qu, C., Ma, J., Liu, X., Xue, Y., Zheng, J., Liu, L., … & Liu, Y. (2017). Dihydroartemisinin exerts antitumour activity by inducing mitochondrion and endoplasmic reticulum apoptosis and autophagic cell death in human glioblastoma cells. *Frontiers in Cellular Neuroscience*, 11(September), 1–14.

Roh, J., Kim, E. H., Jang, H., & Shin, D. (2017). Redox biology Nrf2 inhibition reverses the resistance of cisplatin-resistant head and neck cancer cells to artesunate-induced ferroptosis. *Redox Biology*, 11(December 2016), 254–262.

Sa, N., Tejaswani, P., Pradhan, S. P., Alkhayer, K. A., Behera, A., & Sahu, P. K. (2023). Antidiabetic and antioxidant effect of magnetic and noble metal nanoparticles of Clitoria ternatea. *Journal of Drug Delivery Science and Technology*, 84, 104521.

Sharma, A., Jain, N., & Sareen, R. (2013). Nanocarriers for diagnosis and targeting of breast cancer. *Biomedical Research International*, 2013, 10.

Shen, S., Du, M., Liu, Q., Gao, P., Wang, J., Liu, S., & Gu, L. (2020). Development of GLUT1-targeting alkyl glucoside-modified dihydroartemisinin liposomes for cancer therapy. *Nanoscale Research Letters*, 12, 21901–21912.

Siegel, R. L., Miller, K. D., Goding Sauer, A., Fedewa, S. A., Butterly, L. F., Anderson, J. C., … & Jemal, A. (2020). Colorectal cancer statistics, 2020. *CA: A Cancer Journal for Clinicians*, 70(3), 145–164.

Singh, N. P., & Verma, K. B. (2002). Case report of a laryngeal squamous cell carcinoma treated with artesunate. *Archive on Oncology*, 10(4), 279–280.

Song, F., Li, X., Wang, Q., Liao, L., & Zhang, C. (2015). Nanocomposite hydrogels and their applications in drug delivery and tissue engineering. *Journal of Biomedical Nanotechnology*, 11(1), 40–52.

Sordi, R., Nandra, Ã. K. K., Chiazza, Ã. F., Johnson, F. L., Cabrera, C. P., Torrance, H. D., … & Thiemermann, C. (2016). Artesunate protects against the organ injury and dysfunction induced by severe hemorrhage and resuscitation. *Annals of Surgery*, 265(2), 408–417.

Steeg, P. S. (2006). Tumour metastasis: Mechanistic insights and clinical challenges. *Nature Medicine*, 12(8), 895–904.

Steinbrück, L., Pereira, G., & Efferth, T. (2010). Effects of artesunate on cytokinesis and G2/M cell cycle progression of tumour cells and budding yeast. *Cancer Genomics and Proteomics*, 7(6), 337–346.

Sykes, E. A., Dai, Q., Sarsons, C. D., Chen, J., Rocheleau, J. V., Hwang, D. M., … & Chan, W. C. W. (2016). Tailoring nanoparticle designs to target cancer based on tumour pathophysiology. *Proceedings of the National Academy of Sciences of the United States of America*, 113(9), E1142–E1151.

Tian, L., Liu, J., Jia, Q., Ying, Y., Yang, Z., & Huang, G. (2017). Preparation and evaluation of artemether liposomes for enhanced antitumour therapy. *AAPS PharmScitech*, 19(2), 512–521.

Titulaer, H. A. C., Zuidema, J., & Lugt, C. B. (1991). Formulation and pharmacokinetics of artemisinin and its derivatives. *International Journal of Pharmaceutics*, 69(2), 83–92.

Tran, T. H., Nguyen, T. D., Poudel, B. K., Nguyen, H. T., & Kim, J. O. (2015). Development and evaluation of artesunate-loaded chitosan-coated lipid nanocapsule as a potential drug delivery system against breast cancer. *AAPS PharmSciTech*, 16(6), 1307–1316.

Unezaki, S., Maruyama, K., Hosoda, J., Nagae, I., Koyanagi, Y., Nakata, M., … & Tsuchiya, S. (1996). Direct measurement of the extravasation of polyethyleneglycol-coated liposomes into solid tumour tissue by in vivo fluorescence microscopy. *International Journal of Pharmaceutics*, 144(1), 11–17.

Wan, R. J., & Li, Y. H. (2017). Effects of artesunate prevent nephritis via the toll-like receptor 4/nuclear factor-κB signaling pathway in rats. *Molecular Medicine Reports*, 16(5), 6389–6395.

Wang, B., Hou, D., Liu, Q., Wu, T., Guo, H., Zhang, X., … & Shao, C. (2015). Artesunate sensitizes ovarian cancer cells to cisplatin by downregulating RAD51. *Cancer Biology and Therapy*, 16(10), 1548–1556.

Wang, D., Zhou, J., Chen, R., Shi, R., Xia, G., Zhou, S., … & Chen, Q. (2016a). Magnetically guided delivery of DHA and Fe ions for enhanced cancer therapy based on pH-responsive degradation of DHA-loaded Fe_3O_4@C@MIL-100(Fe) nanoparticles. *Biomaterials*, 107, 88–101.

Wang, D., Zhou, J., Chen, R., Shi, R., Zhao, G., Xia, G., … & Chen, Q. (2016b). Controllable synthesis of dual-MOFs nanostructures for pH-responsive artemisinin delivery, magnetic resonance and optical dual-model imaging-guided chemo/photothermal combinational cancer therapy. *Biomaterials*, 100, 27–40.

Wang, J., Zhang, B., Guo, Y., Li, G., Xie, Q., Zhu, B., … & Chen, Z. (2008). Artemisinin inhibits tumour lymphangiogenesis by suppression of vascular endothelial growth factor C. *Pharmacology*, 82(2), 148–155.

Wang, S., Gao, Y., Chen, H., Kong, R., Jiang, H., Pan, S., … & Sun, B. (2010). Dihydroartemisinin inactivates NF-jB and potentiates the antitumour effect of gemcitabine on pancreatic cancer both in vitro and in vivo. *Cancer Letters*, 293(1), 99–108.

Wang, X., Yu, Y., Ma, J., Zhang, H., Zhang, H., Wang, X., … & Zhang, Q. (2012). LyP-1 modification to enhance delivery of artemisinin or fluorescent probe loaded polymeric micelles to highly metastatic tumour and its lymphatics. *Molecular Pharmaceutics*, 9(9), 2646–2657.

Wang, Z., Duan, X., Lv, Y., & Zhao, Y. (2019). Low density lipoprotein receptor (LDLR)-targeted lipid nanoparticles for the delivery of Sorafenib and Dihydroartemisinin in liver cancers. *Life Sciences*, 117013.

WHO. (2018). Cancer. Retrieved from https://www.who.int/health-topics/cancer#tab=tab_1

Willoughby, J. A., Sundar, S. N., Cheung, M., Tin, A. S., Modiano, J., & Firestone, G. L. (2009). Artemisinin blocks prostate cancer growth and cell cycle progression by disrupting Sp1 interactions with the cyclin-dependent kinase-4 (CDK4) promoter and inhibiting. *The Journal of Biological Chemsitry*, 284(4), 2203–2213.

Woerdenbag, H. J., Moskal, T. A., Pras, N., Malingré, T. M., El-Feraly, F. S., Kampinga, H. H., & Konings, A. W. T. (1993). Cytotoxicity of artemisinin-related endoperoxides to Ehrlich ascites tumour cells. *Journal of Natural Products*, 56(6), 849–856.

Wu, Z., Gao, C., Wu, Y., Zhu, Q., Chen, Y., Liu, X., & Liu, C. (2009). Inhibitive effect of artemether on tumour growth and angiogenesis in the rat C6 orthotopic brain gliomas model. *Integrative Cancer Therapies*, 8, 88–92.

.

Xu, C., Zhang, H., Mu, L., Yang, X., & Celia, C. (2020). Artemisinins as anticancer drugs: Novel therapeutic approaches, molecular mechanisms, and clinical trials. *Frontiers in Pharmacology*, 11(October), 1–15.

Xu, N., Zhou, X., Wang, S., Xu, L. L., Zhou, H. S., & Liu, X. L. (2015). Artesunate induces SKM-1 cells apoptosis by inhibiting hyperactive β-catenin signaling pathway. *International Journal of Medical Sciences*, 12(6), 524–529.

Yu, X., Lu, M., Luo, Y., Hu, Y., Zhang, Y., Xu, Z., … & Tian, J. (2020). A cancer-specific activatable theranostic nanodrug for enhanced therapeutic efficacy via amplification of oxidative stress. *Theranostics*, 10(1), 371–383.

Zhang, C., Wang, J., Zhang, J., Lee, Y. M., Feng, G., Lim, T. K., … & Liu, B. (2016a). Mechanism-guided design and synthesis of a mitochondria-targeting artemisinin analogue with enhanced anticancer activity communications angewandte. *Angewandte Chemie*, 55(44), 13770–13774.

Zhang, H., Chen, Q., Zhang, X., Zhu, X., Chen, J., Zhang, H., … & Zhang, Z. (2016b). An intelligent and tumor-responsive Fe^{2+} donor and Fe^{2+}-dependent drugs cotransport system. *ACS Applied Materials & Interfaces*, 8(49), 33484–33498.

Zhang, H., Ji, Y., Chen, Q., Jiao, X., Hou, L., Zhu, X., & Zhang, Z. (2015). Enhancement of cytotoxicity of artemisinin toward cancer cells by transferrin-mediated carbon nanotubes nanoparticles. *Journal of Drug Targeting*, 23(6), 552–567.

Zhang, H., Zhang, H., Zhu, X., Zhang, X., Chen, Q., Chen, J., … & Zhang, Z. (2017). Visible-light-sensitive titanium dioxide nanoplatform for tumour- responsive Fe^{2+} liberating and artemisinin delivery. *Oncotarget*, 8(35), 58738–58753.

Zhang, J. D. H., & Ji, Z. M. K. (2010). Artesunate induces oncosis-like cell death in vitro and has antitumour activity against pancreatic cancer xenografts in vivo. *Cancer Chemotherapy and Pharmacology*, 65(5), 895–902.

Zhang, S., Shi, L., Ma, H., Li, H., Li, Y., Lu, Y., & Li, W. (2016c). Dihydroartemisinin induces apoptosis in human gastric cancer cell line BGC-823 through activation of JNK1/2 and p38 MAPK signaling pathways. *Journal of Receptor and Signal Transduction Research*, 37(2), 174–180.

Zhang, Y. J., Zhan, X. I., Wang, L., Ho, R. J. Y., & Sasaki, T. (2016). pH-Responsive artemisinin dimer in lipid nanoparticles are effective against human breast cancer in a xenograft model. *Journal of Pharmaceutical Science*, 104(5), 1815–1824.

Zhang, Y., Xu, G., Zhang, S., Wang, D., Saravana Prabha, P., & Zuo, Z. (2018). Antitumour research on artemisinin and its bioactive derivatives. *Natural Products and Bioprospecting*, 8(4), 303–319.

Zhang, Z.-Y., Yu, S.-Q., Miao, L.-Y., Huang, X.-Y., Zhang, X.-P., Zhu, Y.-P., … & Li, D.-Q. (2008). Artesunate combined with vinorelbine plus cisplatin in treatment of advanced non-small cell lung cancer: A randomized controlled trial. *Journal of Chinese Integrative Medicine*, 6(2), 134–138.

Zheng, S., Luo, C., Zhang, H., Zhong, L., & Li, W. (2019). In situ apolipoprotein E-enriched corona guides dihydroartemisinin-decorating nanoparticles towards LDLr-mediated tumour-homing chemotherapy. *Asian Journal of Pharmaceutical Sciences*, 15, 482–491.

Zhou, C., Pan, W., & Wang, X. P. (2012). Artesunate induces apoptosis via a Bak-mediated caspase-independent intrinsic pathway in human lung adenocarcinoma cells. *Journal of Cellular Physiology*, 227(12), 3778–3786.

Zhou, X., Sun, W., Wang, W., Chen, K., Zheng, J., Lu, M., … & Zheng, Z. (2013). Artesunate inhibits the growth of gastric cancer cells through the mechanism of promoting oncosis both in vitro and in vivo. *Anticancer Drugs*, 24(9), 920–927.

Zhu, M., Nie, G., Meng, H., Xia, T., Nel, A., & Zhao, Y. (2013). Physicochemical properties determine nanomaterial cellular uptake, transport, and fate. *Accounts of Chemical Research*, 46(3), 622–631.

Zhu, W., Li, Y., Zhao, D., Li, H., Zhang, W., Xu, J., … & Wang, H. (2019). Dihydroartemisinin suppresses glycolysis of LNCaP cells by inhibiting PI3K/AKT pathway and downregulating HIF-1α expression. *Life Sciences*, 233, 116730.

9 Nanoenabled Systems for the Delivery of Vinca Alkaloids as Anticancer Agents

Bhargavi Kulkarni[1] *and Dileep Kumar*[2]
[1]Poona College of Pharmacy, Bharati Vidyapeeth (Deemed to be University) Erandawane, Pune, Maharashtra, India
[2]Department of Pharmaceutical Chemistry, Manipal College of Pharmaceutical Sciences, Manipal Academy of Higher Education, Manipal, Karnataka, India

9.1 INTRODUCTION

Robert Noble and Charles Thomas Beer extracted vinca alkaloids from the Madagascar periwinkle plant *Catharanthus roseus* (basonym Vinca rosea), from the family of Apocynaceae in the mid-twentieth century. This plant is widely available in India and Madagascar. This evergreen subshrub has long been used as an herbal medicine in the Ayurvedic and Chinese systems of medicine, where it was primarily thought to have anti-diabetic and anti-malarial properties (Martino et al., 2018).

Initially, vinca alkaloids were found to be hypoglycaemic agents (Johnson, 1968). This theory was quickly disproved when there was no decrease in blood sugar levels when tested in rabbits. As a result of the leukopenia, the animals died of septicaemia. This unexpected result encouraged researchers to dig deeper into this topic and look for a possible link between vinca alkaloids and cancer (Johnson et al., 1963). As a result of phytochemical analysis, vinblastine, a type of vinca alkaloid, was isolated and identified. This compound was noted to initiate myelosuppression in leukaemia xenograft mice (Johnson et al., 1963; Johnson, 1968). This revolutionary breakthrough paved the way for a novel cancer treatment strategy.

Hence, the Food and Drug Administration has approved these effective drugs to fight against various tumour types, including leukaemia, breast cancer, Hodgkin's lymphoma, and lung cancer (Lucas et al., 2010). Five of the vinca alkaloids are currently in clinical use: both the natural alkaloids like vinblastine and vincristine (approved by the FDA in 1961 and 1963) and the semisynthetic derivatives like vindesine and vinorelbine (approved by the FDA in 1994) (Fahy, 2001), and vinflunine, a bis-fluorinated vinca alkaloid derivative (approved by the European medicines agency in 2012) for second-line treatment of metastasised urothelial cancer (Bennouna et al., 2008) (Table 9.1). The structure of vinca alkaloids contains an indole nucleus known as catharanthine and a dihydroindole nucleus known as vindoline (Figure 9.1) (Martino et al., 2018).

Throughout the years, vinca alkaloids have been effective against cancers like haematological and lymphatic cancers and solid tumours like breast and non-small cell lung cancer (Rowinsky and Donehower, 1991). Vinca alkaloids trigger the antiproliferative activity at the molecular level; however, it is yet to be understood well, although some experimental results have demonstrated the role of p53, B-cell leukaemia/lymphoma 2 protein, and B-cell lymphoma-extra-large, and other genes which can be directly associated with the regulation of the equilibrium between apoptosis (cell death) and proliferation (Howard et al., 1980). This can be well understood as they interfere with the depolymerisation of microtubules and the destruction of mitotic spindles (Martino et al., 2018).

Adverse effects of vinca alkaloids are fatigue, nausea, headaches, peripheral neuropathy, dizziness, ataxia, constipation, diarrhoea, dysphagia, urinary retention, and suppression of the bone marrow. Also, vinca alkaloids can lead to multi-drug resistance. The adverse effects and multi-drug resistance have hindered the clinical applications of vinca alkaloids (Lee et al., 2015).

Side effects and treatment limitations are related to all chemotherapeutic agents; hence, they are a substantial limitation to the use of vinca alkaloids. These limitations

TABLE 9.1
Recent Vinca Alkaloids as an Anticancer Agent

Vinca Alkaloid Name	Substitution at R1	Substitution at R2	Substitution at R3	Substitution at R4	Substitution at R5
Vincristine	$-CHO$	$-COOCH_3$	$-OCOCH_3$	$-OH$	$-CH_2CH_3$
Vinblastine	$-CH_3$	$-COOCH_3$	$-OCOCH_3$	$-OH$	$-CH_2CH_3$
Vindesine	$-CH_3$	$-CONH_2$	$-OH$	$-OH$	$-CH_2CH_3$
Vinflunine	$-CH_3$	$-COOCH_3$	$-OCOCH_3$	—	$-CF_2CH_3$
Vinorelbine	$-CH_3$	$-COOCH_3$	$-OCOCH_3$	$-H$	$-CH_2CH_3$

DOI: 10.1201/9781003611516-11

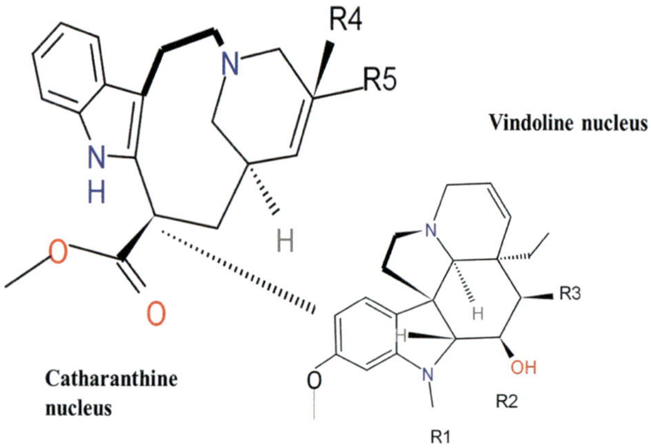

FIGURE 9.1 Basic structure of vinca alkaloids consisting of nucleus catharanthine and vindoline.

were well managed by using vinflunine as an alternative to the other vinca alkaloids, proving to be an improved and novel antineoplastic agent as a vinca alkaloid derivative (Martino et al., 2018). Furthermore, when studies on vinflunine were carried out, it was found that this compound was not sensitive enough to permeability glycoprotein. On the contrary, this type of multi-drug resistance causes the efflux of other vinca alkaloids, leading to low activity against cancer cells. If we consider this, we can conclude that these cytotoxic molecules could be susceptible to other drug resistance mechanisms that would decrease the accumulation and activity of these compounds intracellularly. These mechanisms include lung-resistance protein, multidrug-resistant protein, and human ATP-binding cassette transporter (MXR1/BCRP/ABCP) (Poruchynsky et al., 2001).

A pertinent problem that limits and eases the accomplishment of vinca alkaloids in cancer treatment is identified as the mechanism inside the cells that prohibit vinca alkaloids, hence protecting the cell from their cytotoxic action (Ueda et al., 1987). The action of permeability glycoprotein makes vinca alkaloids fundamentally sensitive, which reduces their antiproliferative effects. In addition, a thorough investigation by Zhang et al. (2017a) revealed various mechanisms of multi-drug resistance, such as changes to the β-tubulin isotype, multidrug-resistant proteins, and the disturbed response of apoptosis, that can contribute to lessening the effect of vinca alkaloids' antineoplastic activity. Neutropenia is one of the major limitations restricting the application of vinca alkaloids. The other toxic manifestations are elevated doses of vinca alkaloids, especially vincristine, gastrointestinal toxicity, and mucositis, which is often identified with vinblastine rather than vinorelbine or vindesine and is more uncommon with vincristine. Other optional limitations with vinca alkaloids include nausea, cerebral pain, vomiting, dizziness, weariness, dysphagia, dryness, urinary retention, blockage, and diarrhoea (Martino et al., 2018). Vincristine, vinblastine, and the semi-synthesised vinca alkaloid, vinorelbine, are known to be potential anticancer agents. When administered, these agents are highly cytotoxic and may enhance extravasation (Bertelli et al., 1994).

Most vinca alkaloids utilised clinically as conventional anticancer agents have numerous critical constraints. These involve low solubility in water, unsatisfactory oral administration, shorter half-life, and extreme side effects (Cho et al., 2008). The treatment regimen of these agents is related to serious systemic toxicity and multi-drug resistance, which is associated with the increase in the expression of ATP-binding cassettes and the efflux of permeability glycoproteins. A few permeability glycoprotein inhibitors have been studied in a couple of years to inhibit the development of multi-drug resistance. The pre-existing multi-drug resistance blockers, such as verapamil, cyclosporine A, and quinidine, are typically the subject of experiments since they have already been approved for various uses and may be repurposed for additional therapeutic settings. Clinical examinations on breast cancer have demonstrated that a blend of vincristine and permeability glycoprotein blocker, verapamil, can improve its activity in combating cancer by reducing multi-drug resistance (Khan and Gurav, 2018; Taylor et al., 1997). A combination of phytochemicals with anticancer activity and permeability glycoprotein blockers may be developed through comparative research to augment the activity decreased by multi-drug resistance in neoplastic cells (Granja et al., 2016; Khan and Gurav, 2018).

Drug delivery via the blood-brain barrier and bypassing the multi-drug resistance mechanism, including permeability glycoprotein efflux, works hand in hand. The blood-brain barrier and efflux pumps protect the brain from hazardous compounds, which explains the connection (Figure 9.2). Therefore, this relationship may be inferred except for vinca alkaloids (Krause, 2019).

There are several ways in which nanoparticles can help overcome the limitations possessed by vinca alkaloids, such as multi-drug resistance and many other serious adverse effects, including extra vascularisation upon intravenous administration.

Liposomes are possibly the most promising nanocarriers for treating malignant growth in preclinical turn of events and clinical applications of vinca alkaloids. They demonstrate various benefits compared to the conventional delivery of vinca alkaloids, including great compatibility with biological systems, diminished toxicity, and increased stability in the biological milieu. Formulation into liposomes may give a reasonable strategy to overcome the limitations, decrease the involvement of the reticuloendothelial system, and lessen adverse effects (Liu, 2015). Liu et al. (2015) reported a UV-responsive delivery system entrapping vincristine. Moreover, exposure to UV light improved the intracellular release of medication, cytotoxicity, and cell death in cervical cancer cells (HeLa cells). Studies in tumour-bearing mice affirmed the reduction in viable cells on light exposure, with a higher rate of decrease in tumour survival. Accumulation of drugs at the cancer site by passively targeting was an impact of liposomes, along with light-responsive controlled delivery. The liposomal

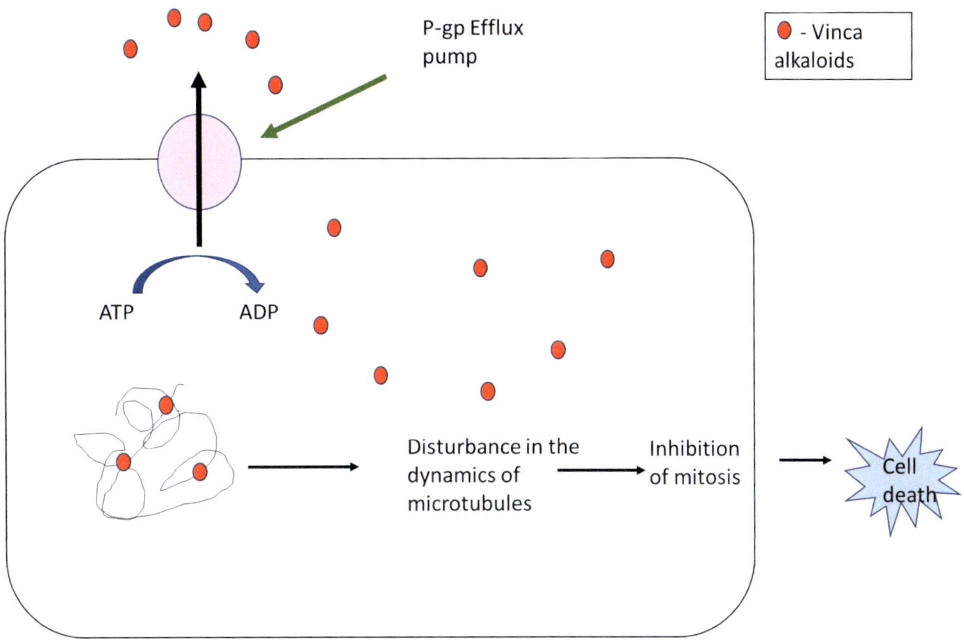

FIGURE 9.2 Mechanism of multi-drug resistance of vinca alkaloid in cancerous cells by permeability glycoprotein efflux pump.

formulation created a more prominent and faster delivery of drugs to tumours and improved its therapeutic efficiency.

These active and passive targeting systems of vinca alkaloid would help overcome most drawbacks. Some of the advantages of vinca nanoparticles are:

- Being compatible with the biological system
- Reduction of toxicity of these agents in the body and long-term stability of these drugs
- Escaping the permeability glycoprotein efflux pump, which enables dealing with the tumour's multi-drug resistance, is the key benefit of the nano-drug delivery of vinca alkaloids. This boosts the malignant cells' cytotoxic, apoptotic, and anti-proliferative activity.

9.2 MECHANISM OF ACTION OF VINCA ALKALOIDS AS AN ANTICANCER AGENT

According to Nogales (1999), the cells are made up of α- and β-tubulin heterodimers that can change into one another during polymerisation. Microtubules are the most important component of various cellular processes; they are responsible for maintaining cell shape and motility, distinct chromosome segregation during mitosis, and movement of macromolecules and cell organelles during the interphase of the cell cycle (Field et al., 2014; Goodson and Jonasson, 2018; Yanagisawa et al., 2018). They are involved in interphase and mitotic cellular functions, making them a promising target for anticancer drugs. Hence, microtubule-targeting agents act by inhibiting the proliferation of cancer cells by disturbing cell signalling events in the interphase and/or inhibiting the spindle microtubules accurately, both of which lead to the apoptosis mechanism in the cells (Field et al., 2014). It is significant to consider that there also exist additional mechanisms that contribute to the action of microtubule-targeting agents on a tumour, like the interaction of microtubule-targeting agents along with secondary tumour targets, which includes microtubule-related proteins and various signal transduction mechanisms (Barbier et al., 2014). However, they are broadly differentiated into microtubule stabilising and destabilising agents. Microtubule-stabilising anticancer drugs increase the polymerisation of tubulin, and microtubules are stabilised from depolymerisation. The microtubule destabilising agents that carry out the depolymerisation of cell microtubules inhibit heterodimers of tubulin from polymers (Kanakkanthara et al., 2013).

The vinca alkaloids have been extensively and commonly used for antitumour therapy because of their biological effects, mediated by interactions with tubulin and microtubules. Tubulin heterodimers (alpha and beta tubulin subunits) form the primary skeleton structure of microtubules (Kavallaris, 2010). Microtubules are formed, which grow and develop into the mitotic spindle during mitosis, where they attach to chromosomes. Microtubules function by pushing or pulling chromosomes toward the cell poles, depending on whether they switch between deliberate growth and fast shrinkage (Zhang et al., 2017a). These agents get attached to one of the subunits of the tubulin dimer at the site of binding in the outer part of microtubules (Kavallaris, 2010). The vinca binding domain on tubulin is close to the binding site where guanosine triphosphate exchange occurs. On microtubules, vinca alkaloids have been shown to have two sites where binding occurs, namely, a binding site with high affinity and a low-binding affinity. The high-affinity site of tubulin is present on the microtubule with plus ends, and the low-binding affinity site of tubulin is present on the lateral surface of the microtubule (Perez, 2009). Vinca alkaloids are differentiated as destabilising agents because they cause depolymerisation, resulting in crosslinking of two microtubules along the sides, which disrupts microtubule polymerisation and eventually stops mitotic progression (Bai et al., 1990). These agents also possess anti-angiogenesis properties (Vacca et al., 1999).

However, the mechanism of action is yet to be fully understood. Vinca alkaloids are thought to hinder the migration of cancer cells and angiogenesis, specifically in endothelial cells, by interfering with microtubule dynamics (Holwell et al., 2001; Pasquier et al., 2006). According to Nogales (1999), the cells are made up of α- and β-tubulin heterodimers that can change into one another during polymerisation. Microtubules are the most important component of various cellular processes; they are responsible for maintaining cell shape and motility, distinct chromosome segregation during mitosis, and movement of macromolecules and cell organelles during the interphase of the cell cycle. In tumour-bearing mice, metronomic administration of vinblastine or vinorelbine causes a persistent decrease in circulating endothelial progenitor mobilisation and viability (Pasquier et al., 2006).

The activity of vinca alkaloids to inhibit angiogenesis may be due to the impairment of circulating endothelial progenitors. Furthermore, it has been proposed that vinca alkaloids have antiangiogenic effects by impairing the vascular endothelial growth factor/vascular endothelial growth factor receptor pathway for signalling (Meissner et al., 2008). Markus and his colleagues hypothesised that vinblastine inhibited endothelial vascular endothelial growth factor receptor-2 protein expression significantly (Meissner et al., 2008). Vincristine suppresses vascular endothelial growth factor expression in human leukaemia T-cell lines (Avramis et al., 2001). Desacetylvinblastine monohydrazide, a vinblastine analogue, also prevents angiogenesis upon inhibition of both the Gas6/Axl and the vascular endothelial growth factor/vascular endothelial growth factor receptor-2 pathways (Zhang et al., 2017a)

For a long time, the exact position of the binding site of vinca alkaloids on microtubules was not understood before the X-ray of vinblastine-tubulin complexed with a neuron-specific protein, i.e., RB3 protein stathmin similar domain, published by Gigant et al. (2005). In a rearrangement from head to toe, the binding site of vinblastine was found to be at the interface along the two tubulin heterodimers, like residues in the lumen present inside and those involved in protofilament longitudinal connections (Gigant et al., 2005). Following structural research, it was discovered that all vinca alkaloids are associated with tubulin dimers and that 16–17 high-affinity binding sites are present at the apex of every microtubule (Cormier et al., 2010). Interestingly, many phytoconstituents, like rhizoxin, cryptophycins, or dolastatins, have a similar binding site and act upon the previously considered vinca domain of tubulin. Vinca alkaloids have a concentration-dependent effect on tubulin polymerisation dynamics. At low concentrations (sub-micromolar), vinca alkaloids inhibit microtubule elongation in two ways: directly through heterodimers at the end, thus preventing the absorption of guanosine triphosphate (Pellegrini and Budman, 2005), and indirectly through protein crosslinking, which promotes the stabilisation of microtubules. Large concentrations of vinca alkaloids, on the other hand, facilitate microtubular depolymerisation. Microtubules have additional alkaloid binding sites, increasing the number of interactions between vinca alkaloids and tubules. Protofilaments, spiral aggregates, and highly organised crystals are formed due to the subsequent changes in microtubular conformation (Figure 9.3) (Gigant et al., 2005; Martino et al., 2018).

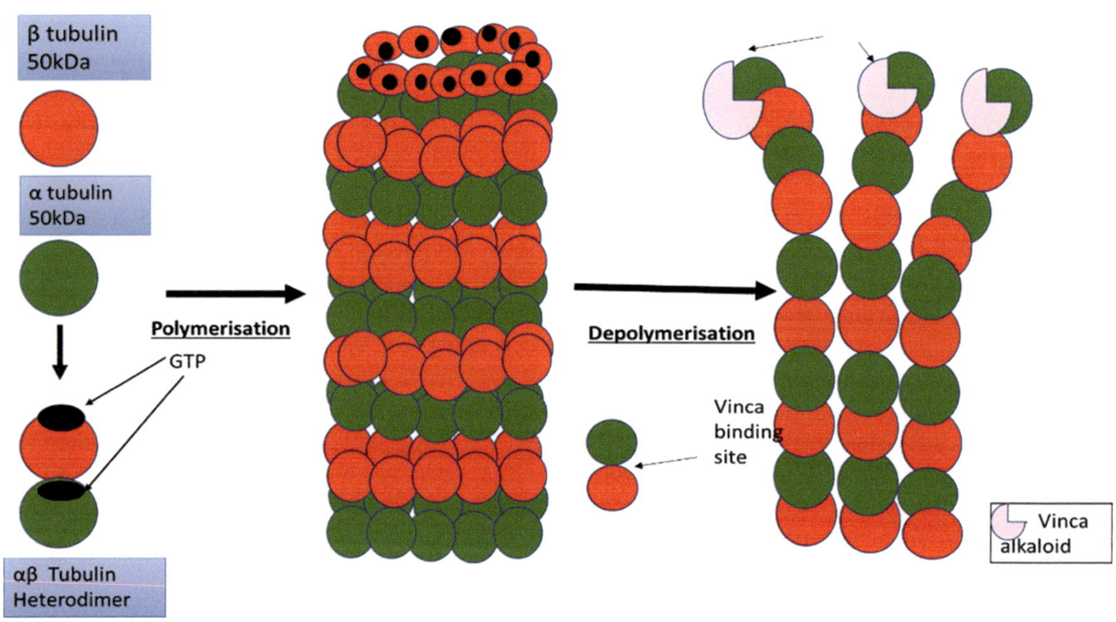

FIGURE 9.3 Mechanism of action of vinca alkaloids for cytotoxic activity.

The finishing event is mitotic spindle disintegration, which causes chromosomes to accumulate in abnormal shapes (balls and stars), which induces cell apoptosis through the p53-dependent and/or - independent cell apoptosis pathways (Zhang and Kanakkanthara, 2020). Vinca alkaloids have non-mitotic toxic effects, such as inhibiting axon transport, microtubules, processes of secretion, and structural disorders and inhibiting the functions of the platelet functions and their biochemical activities (Martino et al., 2018).

9.3 NANO-ENABLED TARGETING STRATEGIES FOR VINCA ALKALOIDS

Several nanocarriers have been developed over the years to deliver anticancer agents, like liposomes, micelles, and polymeric nanoparticles, which are proven to be very efficient (Behera et al., 2020; Hassan et al., 2021; Behera et al., 2023). Researchers are still working on developing tumour-specific nano-drug delivery systems to combat the drawbacks of traditional vinca alkaloid dosing, making the drugs more selective and avoiding their aggregation at the tumour site (Liu et al., 2016; Padhi et al., 2016; Kundu et al., 2022; Behera and Padhi, 2022). Nanoformulations of anticancer drugs such as vinca alkaloids can address significant drawbacks in vinca alkaloid therapy.

However, due to their low ability to target, short half-lives in which they remain inside the body, and their clearance by the reticuloendothelial system, their use in antitumour therapy is very limited (Liu et al., 2016; Hainfeld et al., 2004; Kim et al., 2007). Liposomes are found to be one of the promising nanocarriers as the most efficacious drug carriers for cancer treatment. Liposomes have strong biological compatibility, decreased toxicity, and increased drug stability (Liu et al., 2016).

Chemotherapy is one of the effective approaches to cancer treatment (Padhi et al., 2018; Padhi and Behera, 2022). However, the development of multi-drug resistance is the main obstacle that limits therapeutic efficacy (Gao et al., 2011). The main mechanism of multi-drug resistance has been identified as the permeability glycoprotein encoded by the multi-drug resistance-1 gene in cancer cells (Takahashi et al., 2006; Wang et al., 2014). The permeability glycoprotein can pump a broad variety of chemotherapeutic drugs out of the cells, which results in a lower accumulation of the drug in the cell that is way below the normal effective dose value (Takahashi et al., 2006; Wang et al., 2014; Patnaik et al., 2021). Chemotherapeutic agents can invade normal and tumour tissues without discrimination, causing unwanted adverse effects and even mortality in patients (Fan et al., 2010; Padhi and Behera, 2020; Kundu et al., 2022).

In addition to inhibiting permeability glycoprotein expression in the tumour cell membrane or bypassing permeability glycoprotein efflux pumps by the consequence of endocytosis mediated by the nanoformulations are important techniques for overcoming drug resistance (Han et al., 2012; Liu et al., 2012; Ren et al., 2011). Multifunctional delivery systems for drugs are employed to transfer the drug to the tumour specifically by passive and active targeting, reducing side effects and improving therapeutic effectiveness (Fan et al., 2010; Behera and Padhi, 2022; Padhi et al., 2022; Khuroo et al., 2014).

The use of ligand-modified nanoformulations as an active targeting mechanism of drug delivery has received much attention in recent years. The folate receptor is extensively expressed in a wide range of carcinomas and cancers like endometrial, ovarian, and breast cancer (Wang et al., 2014). To obtain specific tumour targeting and elevation of cellular uptake through receptor-mediated effect, polymeric drug delivery systems and cell-penetrating peptides conjugated nano delivery systems are commonly employed to increase cellular uptake (Chen et al., 2012). Some of the nanoformulations are prepared in such a way that they become sensitive to the tumour microenvironment, like pH, temperature, and redox potential in the cancer cells, and deliver the drug at the site specifically (Wang et al., 2014). pH-sensitive drug delivery systems are configured to minimise leakage of the drug with a sustained or controlled release (Guo et al., 2013).

9.3.1 Passive Targeting of Vinca Alkaloids

Ling et al. (2010) prepared hybrid polymeric nanoparticles of vincristine with dextran sulphate and poly lactic-co-glycolic acid. On oral administration, the nanoformulation demonstrated a 3.3-fold improvement in bioavailability by avoiding the permeability of glycoprotein efflux. The study of the hybrid polymeric nanoparticle in breast cancer cell lines (MCF-7/ADR cells) showcased an enhanced cellular uptake by fourfold.

Ma et al. (2017) studied size-dependent drug distribution, accumulation, and antitumour efficacy of liposomal vincristine in hepatocellular cancer cells (HepG2 cells). The smaller liposomes of 30 nm size were found to be distributed in the cells faster and stronger than 100 nm dimension liposomes entrapping vincristine. The smaller liposomes showed higher drug accumulation as compared to bigger ones.

Ghosh et al. (2021) designed a nanoformulation for the codelivery of vincristine in PEGylated liposomal doxorubicin against non-small cell lung cancer and triple-negative breast cancer cells. The codelivery of vincristine and doxorubicin significantly increased the cell cycle arrest at the G_2/M phase, leading to apoptosis and decreased cell viability in both cell lines. The formulation showed an increased regression of tumour growth compared to liposomal doxorubicin alone. The codelivery showcased a synergistic anticancer effect in both cancer cells.

Ni et al. (2017) designed a doxorubicin-gemcitabine prodrug and vincristine using a nanostructured lipid carrier system. They studied the antitumour activity in human Burkitt's lymphoma cell line and a mice-bearing cancer model. Nanoformulation exhibited higher antitumour efficacy due to stable drug loading than individual drug systems with reduced toxicity.

Maia et al. (2018) prepared hydroxyapatite-based ceramic nanoparticles of vincristine for bone cancer therapy. The synthesised nanoparticles exhibited a strong anti-angiogenic effect *in vitro* by the chorioallantoic membrane assay model.

Amiri et al. (2018) reported the niosome formulation of vinblastine against murine lung cancer cells (TC-1); an *in-vitro* study exhibited the sustained release of the formulation. The *in vivo* study revealed a significant antitumour activity with increased circulation times compared to free vinblastine. The niosomes entrapping vinblastine possessed higher stability, high entrapment efficacy, controlled drug release rate, and stronger cytotoxic activity against lung cancer cells (TC-1) than free vinblastine.

Lakshmi et al. (2022) reported the synergistic activity of vinorelbine and resveratrol-loaded polymeric nanocapsules against breast cancer cells. Nanoformulation reduced not only the tumour volume but also systemic toxicity.

9.3.2 Active Targeting of Vinca Alkaloid

Wang et al. (2014) developed pH-sensitive folate-ligated polymeric nanoparticles (poly lactic-co-glycolic acid – polyethylene glycol) and cell-penetrating peptide (R7) conjugated polymeric nanoparticles of vincristine. The antitumour efficacy of the nanoparticles was evaluated in human breast cancer cells (MCF-7) and multidrug-resistant breast cancer cells (MCF-7/ADR cells). The multifunctional folate–R7 conjugated vincristine nanoparticles showed 65% drug release at pH 5, and the nanoparticles showed enhanced drug internalisation by folate receptor-mediated endocytosis and drug accumulation in resistant cells by escaping permeability glycoprotein efflux.

Wang et al. (2013) developed folic acid-conjugated liposomal vincristine against multi-drug resistance. The folic acid-conjugated PEGylated liposomes of vincristine exhibited a significant inhibiting effect for tumour growth both *in vitro* and *in vivo* in classical multidrug-resistant cell lines (KBv200 cells) overexpressed with permeability glycoprotein. The nanoformulation induced superior cell apoptosis.

Song et al. (2017) developed transferrin-modified liposomes entrapping vincristine and tetrandrine, which exhibited enhanced permeability through the blood-brain barrier with higher cellular uptake. The liposomes inhibited multi-drug resistance and restricted the invasion of cancer cells. The liposomes prolonged the drug's circulation time, leading to a greater accumulation in the brain glioma.

Zhang et al. (2017b) designed a liposome formulation to co-deliver vincristine and doxorubicin conjugated with two-cell-penetrating peptides like T7 and ^{D}A7R against brain glioma. T7 (seven-peptide) is a transferrin receptor-specific ligand, and ^{D}A7R is a D-peptide ligand of vascular endothelial growth factor receptor 2 overexpressed in angiogenesis. The dual peptide-modified liposomes exhibited an anti-glioma effect *in-vivo* with enhanced cellular uptake. The synergistic effect of both drugs suppressed the glioma and inhibited toxicity.

Noble et al. (2014) designed a liposomal formulation of vincristine and vinblastine using intraliposomal triethylammonium sucroseoctasulfate and studied the efficacy against breast cancer cells. The vinblastine liposomes showed increased clearance as compared to vincristine liposomes. The liposomal vincristine exhibited 60- and 253-fold increased efficacy towards breast cancer cells (BT474 and SKBR3, respectively).

Dandamudi and Campbell (2007) reported the effect of the incorporation of magnetite in cationic liposomes of vinblastine sulphate in higher and lower concentrations. However, no significant drug loading efficacy was observed. The cytotoxicity of the liposomes was evaluated against endothelial cells of the microvasculature (HMEC-1 and B16–F10 cells). The magnetite could not influence the cytotoxicity and targeting of tumour vasculature in the liposomes containing chemotherapeutic agents.

9.4 REPORTS ON CLINICAL TRIALS OF VINCA ALKALOIDS

Various vinca alkaloids have been developed, out of which vincristine and vinblastine are the ones that are extensively studied. However, the novel drugs synthesised from vinca, and the combination therapies of vinca and other anticancer drugs are under investigation in preclinical and clinical trials.

Vinorelbine (Navelbine) is a semisynthetic vinca alkaloid with higher lipid solubility and concentration in tissue than its common alkaloidal analogues, vincristine and vinblastine. When vinorelbine is used as a single-drug therapy to act on metastasised breast carcinoma as a first-line drug, it has an average response rate of 35%–60%, and in the second line, it is 15%–36%, with excellent tolerance clinically (Duflos et al., 2002; Fumoleau et al., 1993; Weber et al., 1995). Due to its potency and favourable drug safety profile, it is used with other chemotherapeutic agents targeting the DNA, such as alkylating agents, anthracyclines, and anti-metabolite compounds (Chan, 2007). Vinca alkaloids' susceptibility to multi-drug resistance pumps, like permeability glycoprotein efflux and the growth of drug resistance, are two of the most serious drawbacks of metastatic breast cancer (Fojo and Menefee, 2007). As a result, new vinca alkaloidal drugs with increased efficiency, response, and restricted and controllable toxicity are required (Fumoleau and Guiu, 2013)

Vinflunine, another alkaloidal drug from vinca, tends to have benefits over vinorelbine in terms of potency, tolerability, and duration of action. The opportunity to integrate this agent with other chemotherapeutic drugs or direct therapy to increase efficacy in breast carcinoma would be a key approach for future growth. The first clinical results for previously treated women with metastatic breast cancer are good, and further phase III trials are currently underway. In addition, the possible use of vinflunine in breast cancer therapy would have to be compared against eribulin.

This microtubule-targeting agent is currently licensed and used to treat metastasised breast cancer after the failure of anthracyclines and taxanes (Fumoleau and Guiu, 2013).

Malignant pleural mesothelioma is a neoplasm with a poor prognosis that is becoming more common worldwide (Ceresoli and Zucali, 2015). Just a small percentage of patients are acquiescent to multimodality treatment, which includes surgery (Zucali et al., 2009). Steele et al. (2000) performed a single-centre, open-label, phase II clinical study involving vinorelbine at a dosage of 30 mg/m^2 given weekly once for 6 weeks, with the key objective of exposure of drug therapy in solid tumours and its effect towards the standard of life. The relatively low toxicity of vinorelbine suggests that the co-administration of other chemotherapeutic drugs may be effective for malignant pleural mesothelioma. Vinca alkaloids tend to be active in malignant pleural mesothelioma. Most clinical evidence concerns vinorelbine; vinflunine has only been tested in a first-line trial, and no other trials assessing its efficacy in malignant pleural mesothelioma have been published. Across all the studies, the toxicity profile was found to be moderate (Ceresoli and Zucali, 2015).

9.5 FUTURE PROSPECTIVE

Naturally derived vinca alkaloids have a multitude of applications in treating cancer. However, some limitations hinder their widespread use. Nano-enabled drug strategies that actively and passively target tumour cells can reduce the major side effects of vinca alkaloids given as a free drug. These adverse effects, like neutropenia, phlebitis, and irritation of the veins, can be reduced by employing nano-enabled drug delivery systems and combination therapy for vinca alkaloids.

Although much work in nano-enabled drug delivery systems is dedicated to natural vinca alkaloids, minor modifications in the vindoline and catharanthine nucleus derive certain emerging semisynthetic compounds. Moreover, studies on active targeting using antibody-drug conjugates are still in progress. One of the drugs that is extensively explored is vinblastine conjugated with trastuzumab (Martino et al., 2018). This conjugation has demonstrated great pharmacokinetics and efficacy depending on the stoichiometry of the antibody and drug conjugation ratio against breast cancer cell lines (BT-474, SK-BR-3, and MCF7) and the gastric cancer cell line (NCI-N87) (Yurkovetskiy et al., 2015).

9.6 CONCLUSION

Vinca alkaloids are anticancer drugs used as chemotherapeutic agents for different tumours; however, their side effects limit their use. These side effects may be reduced with drug delivery devices and combination therapy. This chapter discusses drug delivery mechanisms, including liposome-encapsulated drugs, chemically or peptide-modified drugs, polymeric nanoformulations, and combination therapy. Vinca alkaloids encapsulated in liposomes have antioxidant properties, are nontoxic to tissues at the injection site, and hence could increase the half-life of drugs. Combination therapy and active targeting can help to mitigate side effects.

The vinca alkaloids like vindesine, vinflunine, and vinorelbine are not yet used extensively for cancer treatment, and their nanoformulations have yet to be developed and studied thoroughly. However, these agents, along with some conventional vinca alkaloids like vincristine and vinblastine, whether in combination therapy or a nano-enabled drug delivery system, can promise better management of adverse effects observed in a patient treated with conventional formulations. As mentioned above, the nano-enabled drug delivery of vinca alkaloids would also decrease resistance in cancer treatment. All of these clinical applications for vinca alkaloid nanoformulations are excellent for improving the quality of life of cancer patients.

REFERENCES

Amiri, B., Ahmadvand, H., Farhadi, A., Najmafshar, A., Chiani, M., & Norouzian, D. (2018). Delivery of vinblastine-containing niosomes results in potent in vitro/in vivo cytotoxicity on tumour cells. *Drug Development and Industrial Pharmacy*, 44(8), 1371–1376.

Avramis, I. A., Kwock, R., & Avramis, V. I. (2001). Taxotere and vincristine inhibit the secretion of the angiogenesis-inducing vascular endothelial growth factor (VEGF) by wild-type and drug-resistant human leukaemia T-cell lines. *Anticancer Research*, 21(4A), 2281–2286.

Bai, R. L., Pettit, G. R., & Hamel, E. (1990). Binding of dolastatin 10 to tubulin at a distinct site for peptide antimitotic agents near the exchangeable nucleotide and vinca alkaloid sites. *Journal of Biological Chemistry*, 265(28), 17141–17149.

Barbier, P., Tsvetkov, P. O., Breuzard, G., & Devred, F. (2014). Deciphering the molecular mechanisms of anti-tubulin plant derived drugs. *Phytochemistry Reviews*, 13, 157–169.

Behera, A., Mittu, B., Padhi, S., Patra, N., & Singh, J. (2020). Bimetallic nanoparticles: Green synthesis, applications, and future perspectives. In: Abd-Elsalam, K. A. (Ed.) *Multifunctional Hybrid Nanomaterials for Sustainable Agri-Food and Ecosystems*. Elsevier, London, pp. 639–682.

Behera, A., & Padhi, S. (2022). pH-Sensitive polymeric nanoparticles for cancer treatment. In: Padhi, S., Behera, A., & Lichtfouse, E. (Eds.) *Polymeric Nanoparticles for the Treatment of Solid Tumours*. Environmental Chemistry for a Sustainable World, vol 71. Springer, Cham, pp. 401–425.

Behera, A., Padhi, S., & Nayak, A. K. (2023). Engineered liposomes as drug delivery and imaging agents. In: Ray, S., & Nayak, A. K. (Eds.) *Design and Applications of Theranostic Nanomedicines*, Woodhead Publishing, Cambridge, MA, pp. 75–108.

Bennouna, J., Delord, J. P., Campone, M., & Nguyen, L. (2008). Vinflunine: A new microtubule inhibitor agent. *Clinical Cancer Research*, 14(6), 1625–1632.

Bertelli, G., Dini, D., Forno, G. B., Gozza, A., Silvestro, S., Venturini, M., … & Pronzato, P. (1994). Hyaluronidase as an antidote to extravasation of Vinca alkaloids: Clinical results. *Journal of Cancer Research and Clinical Oncology*, 120, 505–506.

Ceresoli, G. L., & Zucali, P. A. (2015). Vinca alkaloids in the therapeutic management of malignant pleural mesothelioma. *Cancer Treatment Reviews*, 41(10), 853–858.

Chan, A. (2007). A review of the use of trastuzumab (Herceptin®) plus vinorelbine in metastatic breast cancer. *Annals of Oncology*, 18(7), 1152–1158.

Chen, J., Li, S., & Shen, Q. (2012). Folic acid and cell-penetrating peptide conjugated PLGA–PEG bifunctional nanoparticles for vincristine sulfate delivery. *European Journal of Pharmaceutical Sciences*, 47(2), 430–443.

Cho, K., Wang, X. U., Nie, S., Chen, Z., & Shin, D. M. (2008). Therapeutic nanoparticles for drug delivery in cancer. *Clinical Cancer Research*, 14(5), 1310–1316.

Cormier, A., Knossow, M., Wang, C., & Gigant, B. (2010). The binding of vinca domain agents to tubulin: Structural and biochemical studies. *Methods in Cell Biology*, 95, 373–390.

Dandamudi, S., & Campbell, R. B. (2007). The drug loading, cytotoxicty and tumour vascular targeting characteristics of magnetite in magnetic drug targeting. *Biomaterials*, 28(31), 4673–4683.

Duflos, A., Kruczynski, A., & Barret, J. M. (2002). Novel aspects of natural and modified vinca alkaloids. *Current Medicinal Chemistry-Anti-Cancer Agents*, 2(1), 55–70.

Fahy, J. (2001). Modifications in the upper or velbenamine part of the vinca alkaloids have major implications for tubulin interacting activities. *Current Pharmaceutical Design*, 7(13), 1181–1197.

Fan, L., Li, F., Zhang, H., Wang, Y., Cheng, C., Li, X., … & Zhang, S. (2010). Codelivery of PDTC and doxorubicin by multifunctional micellar nanoparticles to achieve active targeted drug delivery and overcome multi-drug resistance. *Biomaterials*, 31(21), 5634–5642.

Field, J. J., Kanakkanthara, A., & Miller, J. H. (2014). Microtubule-targeting agents are clinically successful due to both mitotic and interphase impairment of microtubule function. *Bioorganic & Medicinal Chemistry*, 22(18), 5050–5059.

Fojo, T., & Menefee, M. (2007). Mechanisms of multi-drug resistance: The potential role of microtubule-stabilising agents. *Annals of Oncology*, 18, v3–v8.

Fumoleau, P., & Guiu, S. (2013). New vinca alkaloids in clinical development. *Current Breast Cancer Reports*, 5, 69–72.

Fumoleau, P., Delgado, F. M., Delozier, T., Monnier, A., Gil Delgado, M. A., Kerbrat, P., … & Closon, M. T. (1993). Phase II trial of weekly intravenous vinorelbine in first-line advanced breast cancer chemotherapy. *Journal of Clinical Oncology*, 11(7), 1245–1252.

Gao, Y., Chen, L., Zhang, Z., Chen, Y., & Li, Y. (2011). Reversal of multi-drug resistance by reduction-sensitive linear cationic click polymer/iMDR1-pDNA complex nanoparticles. *Biomaterials*, 32(6), 1738–1747.

Ghosh, S., Lalani, R., Maiti, K., Banerjee, S., Bhatt, H., Bobde, Y. S., … & Misra, A. (2021). Synergistic co-loading of vincristine improved chemotherapeutic potential of pegylated liposomal doxorubicin against triple negative breast cancer and non-small cell lung cancer. *Nanomedicine: Nanotechnology, Biology and Medicine*, 31, 102320.

Gigant, B., Wang, C., Ravelli, R. B., Roussi, F., Steinmetz, M. O., Curmi, P. A., … & Knossow, M. (2005). Structural basis for the regulation of tubulin by vinblastine. *Nature*, 435(7041), 519–522.

Goodson, H. V., & Jonasson, E. M. (2018). Microtubules and microtubule-associated proteins. *Cold Spring Harbor Perspectives in Biology*, 10(6), a022608.

Granja, A., Pinheiro, M., & Reis, S. (2016). Epigallocatechin gallate nanodelivery systems for cancer therapy. *Nutrients*, 8(5), 307.

Guo, H., Lai, Q., Wang, W., Wu, Y., Zhang, C., Liu, Y., & Yuan, Z. (2013). Functional alginate nanoparticles for efficient intracellular release of doxorubicin and hepatoma carcinoma cell targeting therapy. *International Journal of Pharmaceutics*, 451(1–2), 1–11.

Hainfeld, J. F., Slatkin, D. N., & Smilowitz, H. M. (2004). The use of gold nanoparticles to enhance radiotherapy in mice. *Physics in Medicine & Biology*, 49(18), N309.

Han, M., Lv, Q., Tang, X. J., Hu, Y. L., Xu, D. H., Li, F. Z., … & Gao, J. Q. (2012). Overcoming drug resistance of MCF-7/ADR cells by altering intracellular distribution of doxorubicin via MVP knockdown with a novel siRNA polyamidoamine-hyaluronic acid complex. *Journal of Controlled Release*, 163(2), 136–144.

Hassan, N., Firdaus, S., Padhi, S., Ali, A., & Iqbal, Z. (2021). Investigating natural antibiofilm components: A new therapeutic perspective against candidal vulvovaginitis. *Medical Hypotheses*, 148, 110515.

Holwell, S. E., Hill, B. T., & Bibby, M. C. (2001). Anti-vascular effects of vinflunine in the MAC 15A transplantable adenocarcinoma model. *British Journal of Cancer*, 84(2), 290–295.

Howard, S. M., Theologides, A., & Sheppard, J. R. (1980). Comparative effects of vindesine, vinblastine, and vincristine on mitotic arrest and hormonal response of L1210 leukemia cells. *Cancer Research*, 40(8_Part_1), 2695–2700.

Johnson, I. S. (1968). Historical background of Vinca alkaloid research and areas of future interest. *Cancer Chemotherapy Reports*, 52(4), 455–461.

Johnson, I. S., Armstrong, J. G., Gorman, M., & Burnett Jr, J. P. (1963). The vinca alkaloids: A new class of oncolytic agents. *Cancer Research*, 23(8_Part_1), 1390–1427.

Kanakkanthara, A., Teesdale-Spittle, H. P., & Miller, H. J. (2013). Cytoskeletal alterations that confer resistance to anti-tubulin chemotherapeutics. *Anticancer Agents in Medicinal Chemistry (Formerly Current Medicinal Chemistry-Anti-Cancer Agents)*, 13(1), 147–158.

Kavallaris, M. (2010). Microtubules and resistance to tubulin-binding agents. *Nature Reviews Cancer*, 10(3), 194–204.

Khan, T., & Gurav, P. (2018). PhytoNanotechnology: Enhancing delivery of plant based anticancer drugs. *Frontiers in Pharmacology*, 8, 1002.

Khuroo, T., Verma, D., Talegaonkar, S., Padhi, S., Panda, A. K., & Iqbal, Z. (2014). Topotecan–tamoxifen duple PLGA polymeric nanoparticles: Investigation of in vitro, in vivo and cellular uptake potential. *International Journal of Pharmaceutics*, 473(1–2), 384–394.

Kim, D., Park, S., Lee, J. H., Jeong, Y. Y., & Jon, S. (2007). Antibiofouling polymer-coated gold nanoparticles as a contrast agent for in vivo X-ray computed tomography imaging. *Journal of the American Chemical Society*, 129(24), 7661–7665.

Krause, W. (2019). Resistance to anti-tubulin agents: From vinca alkaloids to epothilones. *Cancer Drug Resistance*, 2(1), 82.

Kundu, A., Padhi, S., Behera, A., Hasnain, M. S., & Nayak, A. K. (2022). Tumour targeting strategies by chitosan-based nanocarriers. In: Hasnain, M. S., Beg, S., & Naya, A. K. (Eds.) *Chitosan in Biomedical Applications*. Academic Press, London, pp. 163–188.

Lakshmi, Singh, S., Shah, K., & Dewangan, H. K. (2022). Dual vinorelbine bitartrate and resveratrol loaded polymeric aqueous core nanocapsules for synergistic efficacy in breast cancer. *Journal of Microencapsulation*, 39(4), 299–313.

Lee, C. T., Huang, Y. W., Yang, C. H., & Huang, K. S. (2015). Drug delivery systems and combination therapy by using vinca alkaloids. *Current Topics in Medicinal Chemistry*, 15(15), 1491–1500.

Ling, G., Zhang, P., Zhang, W., Sun, J., Meng, X., Qin, Y., ... & He, Z. (2010). Development of novel self-assembled DS-PLGA hybrid nanoparticles for improving oral bioavailability of vincristine sulfate by P-gp inhibition. *Journal of Controlled Release*, 148(2), 241–248.

Liu, Y. (2015). Liposome-based delivery systems for vinca alkaloids in cancer therapy: Advances and challenges. *Journal of Drug Targeting*, 23(7–8), 579–590. doi:10.3109/1061186X.2015.1052084.

Liu, J., Zhao, Y., Guo, Q., Wang, Z., Wang, H., Yang, Y., & Huang, Y. (2012). TAT-modified nanosilver for combating multidrug-resistant cancer. *Biomaterials*, 33(26), 6155–6161.

Liu, T., Liu, X., & Li, W. (2016). Tetrandrine, a Chinese plant-derived alkaloid, is a potential candidate for cancer chemotherapy. *Oncotarget*, 7(26), 40800.

Liu, Y., He, M., Niu, M., Zhao, Y., Zhu, Y., Li, Z., & Feng, N. (2015). Delivery of vincristine sulfate-conjugated gold nanoparticles using liposomes: A light-responsive nanocarrier with enhanced antitumour efficiency. *International Journal of Nanomedicine*, 10, 3081.

Lucas, M. D., Still, C. P., Bueno Perez, L., Grever, R. M., & Douglas Kinghorn, A. (2010). Potential of plant-derived natural products in the treatment of leukemia and lymphoma. *Current Drug Targets*, 11(7), 812–822.

Ma, S., Li, M., Liu, N., Li, Y., Li, Z., Yang, Y., ... & Mei, X. (2017). Vincristine liposomes with smaller particle size have stronger diffusion ability in tumour and improve tumour accumulation of vincristine significantly. *Oncotarget*, 8(50), 87276

Maia, A. L. C., Ferreira, C. D. A., Barros, A. L. B. D., e Silva, A. T. M., Ramaldes, G. A., Silva Cunha Júnior, A. D., ... & Ferreira Soares, D. C. (2018). Vincristine-loaded hydroxyapatite nanoparticles as a potential delivery system for bone cancer therapy. *Journal of Drug Targeting*, 26(7), 592–603.

Martino, E., Casamassima, G., Castiglione, S., Cellupica, E., Pantalone, S., Papagni, F., Rui, M., Marika, A., & Collina, S. (2018). Vinca alkaloids and analogues as anticancer agents: Looking back, peering ahead. *Bioorganic & Medicinal Chemistry Letters*, 28(17), 2816–2826.

Meissner, M., Pinter, A., Michailidou, D., Hrgovic, I., Kaprolat, N., Stein, M., ... & Gille, J. (2008). Microtubule-targeted drugs inhibit VEGF receptor-2 expression by both transcriptional and post-transcriptional mechanisms. *Journal of Investigative Dermatology*, 128(8), 2084–2091.

Ni, S., Qiu, L., Zhang, G., Zhou, H., & Han, Y. (2017). Lymph cancer chemotherapy: Delivery of doxorubicin–gemcitabine prodrug and vincristine by nanostructured lipid carriers. *International Journal of Nanomedicine*, 12, 1565.

Noble, G. T., Stefanick, J. F., Ashley, J. D., Kiziltepe, T., & Bilgicer, B. (2014). Ligand-targeted liposome design: Challenges and fundamental considerations. *Trends in Biotechnology*, 32(1), 32–45.

Nogales, E. (1999). A structural view of microtubule dynamics. *Cellular and Molecular Life Sciences CMLS*, 56, 133–142.

Padhi, S., & Behera, A. (2020). Nanotechnology based targeting strategies for the delivery of camptothecin. In: Saneja, A., Panda, A., & Lichtfouse, E. (Eds.) *Sustainable Agriculture Reviews*, vol 44. Springer, Cham, pp. 243–272.

Padhi, S., & Behera, A. (2022). Cellular internalization and toxicity of polymeric nanoparticles. In: Padhi, S., Behera, A., & Lichtfouse, E. (Eds.) *Polymeric Nanoparticles for Treating Solid Tumours*. Environmental Chemistry for a Sustainable World, vol 71. Springer, Cham, pp. 473–488.

Padhi, S., Behera, A., Hasnain, M. S., & Nayak, A. K. (2022). Chitosan in colon-targeted drug delivery. In: Hasnain, M. S., Beg, S., & Nayak, A. K. (Eds.) *Chitosan in Drug Delivery*. Academic Press, London, pp. 107–132.

Padhi, S., Kapoor, R., Verma, D., Panda, A. K., & Iqbal, Z. (2018). Formulation and optimisation of topotecan nanoparticles: In vitro characterisation, cytotoxicity, cellular uptake and pharmacokinetic outcomes. *Journal of Photochemistry and Photobiology B: Biology*, 183, 222–232.

Padhi, S., Mirza, M. A., Verma, D., Khuroo, T., Panda, A. K., Talegaonkar, S., Khar, R. K., & Iqbal, Z. (2016). Revisiting the nanoformulation design approach for effective delivery of Topotecan in its stable form: An appraisal of its in vitro behavior and tumour amelioration potential. *Drug Delivery*, 23(8), 2827–2837.

Pasquier, E., Honoré, S., & Braguer, D. (2006). Microtubule-targeting agents in angiogenesis: Where do we stand? *Drug Resistance Updates*, 9(1–2), 74–86.

Patnaik, S., Gorain, B., Padhi, S., Choudhury, H., Gabr, G. A., Md, S., Kumar Mishra, D., & Kesharwani, P. (2021). Recent update of toxicity aspects of nanoparticulate systems for drug delivery. *European Journal of Pharmaceutics and Biopharmaceutics*, 161, 100–119.

Pellegrini, F., & Budman, D. R. (2005). Tubulin function, action of antitubulin drugs, and new drug development. *Cancer Investigation*, 23(3), 264–273.

Perez, E. A. (2009). Microtubule inhibitors: Differentiating tubulin-inhibiting agents based on mechanisms of action, clinical activity, and resistance. *Molecular Cancer Therapeutics*, 8(8), 2086–2095.

Poruchynsky, M. S., Giannakakou, P., Ward, Y., Bulinski, J. C., Telford, W. G., Robey, R. W., & Fojo, T. (2001). Accompanying protein alterations in malignant cells with a microtubule-polymerising drug-resistance phenotype and a primary resistance mechanism. *Biochemical Pharmacology*, 62(11), 1469–1480.

Ren, F., Chen, R., Wang, Y., Sun, Y., Jiang, Y., & Li, G. (2011). Paclitaxel-loaded poly (n-butylcyanoacrylate) nanoparticle delivery system to overcome multi-drug resistance in ovarian cancer. *Pharmaceutical Research*, 28, 897–906.

Rowinsky, E. K., & Donehower, R. C. (1991). The clinical pharmacology and use of antimicrotubule agents in cancer chemotherapeutics. *Pharmacology & Therapeutics*, 52(1), 35–84.

Song, X. L., Liu, S., Jiang, Y., Gu, L. Y., Xiao, Y., Wang, X., ... & Li, X. T. (2017). Targeting vincristine plus tetrandrine liposomes modified with DSPE-PEG2000-transferrin in treatment of brain glioma. *European Journal of Pharmaceutical Sciences*, 96, 129–140.

Steele, J. P., Shamash, J., Evans, M. T., Gower, N. H., Tischkowitz, M. D., & Rudd, R. M. (2000). Phase II study of vinorelbine in patients with malignant pleural mesothelioma. *Journal of Clinical Oncology*, 18(23), 3912–3917.

Takahashi, S., Ito, Y., Hatake, K., & Sugimoto, Y. (2006). Gene therapy for breast cancer: Review of clinical gene therapy trials for breast cancer and MDR1 gene therapy trial in Cancer Institute Hospital. *Breast Cancer*, 13(1), 8–15.

Taylor, C. W., Dalton, W. S., Mosley, K., Dorr, R. T., & Salmon, S. E. (1997). Combination chemotherapy with cyclophosphamide, vincristine, adriamycin, and dexamethasone (CVAD) plus oral quinine and verapamil in patients with advanced breast cancer. *Breast Cancer Research and Treatment*, 42, 7–14.

Ueda, K., Cardarelli, C., Gottesman, M. M., & Pastan, I. R. A. (1987). Expression of a full-length cDNA for the human "MDR1" gene confers resistance to colchicine, doxorubicin, and vinblastine. *Proceedings of the National Academy of Sciences*, 84(9), 3004–3008.

Vacca, A., Iurlaro, M., Ribatti, D., Minischetti, M., Nico, B., Ria, R., ... & Dammacco, F. (1999). Antiangiogenesis is produced by nontoxic doses of vinblastine. *Blood, The Journal of the American Society of Hematology*, 94(12), 4143–4155.

Wang, C., Feng, L., Yang, X., Wang, F., & Lu, W. (2013). Folic acid-conjugated liposomal vincristine for multi-drug resistant cancer therapy. *Asian Journal of Pharmaceutical Sciences*, 8(2), 118–127.

Wang, Y., Dou, L., He, H., Zhang, Y., & Shen, Q. (2014). Multifunctional nanoparticles as nanocarrier for vincristine sulfate delivery to overcome tumour multi-drug resistance. *Molecular Pharmaceutics*, 11(3), 885–894.

Weber, B. L., Vogel, C., Jones, S., Harvey, H., Hutchins, L., Bigley, J., & Hohneker, J. (1995). Intravenous vinorelbine as first-line and second-line therapy in advanced breast cancer. *Journal of Clinical Oncology*, 13(11), 2722–2730.

Yanagisawa, M., Alonso, J. M., & Szymanski, D. B. (2018). Microtubule-dependent confinement of a cell signaling and actin polymerisation control module regulates polarised cell growth. *Current Biology*, 28(15), 2459–2466.

Yurkovetskiy, A. V., Yin, M., Bodyak, N., Stevenson, C. A., Thomas, J. D., Hammond, C. E., … & Lowinger, T. B. (2015). A polymer-based antibody–vinca drug conjugate platform: Characterization and preclinical efficacy preparation and efficacy of a polymer-based vinca ADC. *Cancer Research*, 75(16), 3365–3372.

Zhang, D., & Kanakkanthara, A. (2020). Beyond the paclitaxel and vinca alkaloids: Next generation of plant-derived microtubule-targeting agents with potential anticancer activity. *Cancers*, 12(7), 1721.

Zhang, Y., Yang, S. H., & Guo, X. L. (2017a). New insights into Vinca alkaloids resistance mechanism and circumvention in lung cancer. *Biomedicine & Pharmacotherapy*, 96, 659–666.

Zhang, Y., Zhai, M., Chen, Z., Han, X., Yu, F., Li, Z., … & Mei, X. (2017b). Dual-modified liposome codelivery of doxorubicin and vincristine improve targeting and therapeutic efficacy of glioma. *Drug Delivery*, 24(1), 1045–1055.

Zucali, P. A., De Vincenzo, F., Simonelli, M., & Santoro, A. (2009). Future developments in the management of malignant pleural mesothelioma. *Expert Review of Anticancer Therapy*, 9(4), 453–467.

Section III

Flavonoid Phytomolecules in Cancer Chemotherapy

10 Nanoenabled Systems for the Delivery of Hesperidin as an Anticancer Agent

Chirag Patel[1], *Sandip Patel*[1], *Jayesh Beladiya*[1], *Devang Sheth*[1], *Ketan Ranch*[2], *and Veena Patel*[3]
[1]Department of Pharmacology, L. M. College of Pharmacy, Ahmedabad, India
[2]Department of Pharmaceutics, L. M. College of Pharmacy, Ahmedabad, India
[3]Department of Pharmacognosy, Indukaka Ipcowala College of Pharmacy, CVM University, New Vallabh Vidyanagar, Anand, India

10.1 INTRODUCTION

Cancer is one of the most challenging diseases in the present world (Padhi et al., 2016). It accounts for one of the major determinants of mortality worldwide. Chemotherapy is the mainstay of the treatment. However, it is accompanied by undesired toxic effects on healthy cells (Behera & Padhi, 2020; Behera et al., 2020a). Nanotechnology advances the incorporation of chemotherapeutic drugs in biodegradable polymeric matrices, lowering the associated toxic effects and enhancing therapeutic effectiveness (Subramani et al., 2017; Patnaik et al., 2021).

A broad category of low molecular weight polyphenolic chemicals called flavonoids are secondary metabolites that exist naturally. They are one of the most prevalent and fascinating bioactive substances, and they exist in free and glycoside forms. They are derived from the flavone (2-phenyl chromone or 2-phenyl benzopyrone), which has a C_6-C_3-C_6 carbon skeleton with aromatic rings in the C_6 position. Flavonoids can be found in almost any plant part (Garg & Garg, 2001; Barreca et al., 2017). A therapeutically active subclass of flavonoids, flavanones consist mainly of three aglycones: hesperetin (4′-methoxy-3′,5,7-trihydroxy flavanone), naringenin (5,7,4′-trihydroxyflavanone), and eriodictyol (5,7,3′,4′-tetrahydroxyflavanone).

Hesperidin ($C_{28}H_{34}O_{15}$), a flavanone glycoside (Figure 10.1), is abundant in the rind of citrus fruits. *Agathosma serratifolia* (Curtis), *Citrus aurantium* (L.), *Citrus sinensis* (L.), and *Citrus limon* (L.) are all good

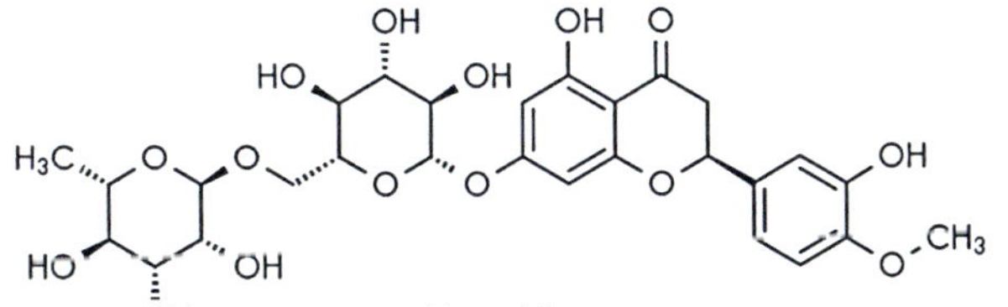

FIGURE 10.1 Structure of hesperidin [Hesperidin ($C_{28}H_{34}O_{15}$), containing aglycone part hesperetin bonded to rutinose [6-*O*-(α-l-Rhamnopyranosyl)-D-glucopyranose] and/or [6-*O*-(α-l-Rhamnosyl)-D-glucose].

sources of hesperidin (L.) (Garg & Garg, 2001). Hesperidin is abundantly found in sweet oranges and lemons as a byproduct of citrus cultivation. It may account for up to 14% of the weight of young, immature oranges. Hesperidin is degraded in the presence of heavy oxidizing agents. It is generally stable below 75°C in an oxygen-free, neutral, or acidic environment. In the presence of metal ions (Cu^{2+}, Fe^{3+}) and intense light, the degradation of hesperidin is accelerated (Ganeshpurkar & Saluja, 2019).

Hesperidin cannot be hydrolyzed in the small intestine after oral administration but is hydrolyzed to an aglycone-hesperetin by colonic microflora before absorption. The absorbed hesperetin is metabolized in the liver at the 3′- and 7-positions by UDP-glucuronosyl transferases and sulfotransferases (Guo et al., 2020) (Figure 10.2).

Hesperidin has been extensively studied and has demonstrated therapeutic effects in numerous studies. It has proven efficacy in treating epilepsy (Kumar et al., 2013, 2014; Haddadi et al., 2017; Jung et al., 2018), parkinsonism (Antunes et al., 2020b, a; Kesh et al., 2020), Alzheimer's disease (Chakraborty et al., 2016; Thenmozhi et al., 2016; Ikram et al., 2019), gout (Ota-kontani et al., 2019), rheumatoid arthritis (Umar et al., 2012; Qi et al., 2019), hypertension (Ikemura et al., 2012; Wunpathe et al., 2018), hepatotoxicity (Tabeshpour et al., 2020), diabetes and its associated complications (Agrawal et al., 2014; Samie et al., 2018; Pradhan et al., 2022), complications of radiotherapy (Karimi et al., 2017; Musa et al., 2019), hyperlipidemia (Sun et al., 2017; Kumar et al., 2020), hyperthyroidism (Panda & Kar, 2014), peptic ulcer (Mota et al., 2019), immunological reactions (Haghmorad et al., 2016; Kilic et al., 2018), microbial infections (Mehrdad et al., 2015; Zhao et al., 2020), obesity (Xiong, 2019), psoriasis (Li et al., 2019a), non-alcoholic fatty liver disease (Haghmorad et al., 2016), varicocoele (Shokoohi et al., 2020), acute renal failure (Hanedan et al., 2018; Park et al., 2019), pancreatitis (Köksoy et al., 2013; Aja et al., 2020), endometriosis (Melekoglu et al., 2018), cytotoxicity (Sahu et al., 2013; Hemanth Kumar et al., 2017; Geng et al., 2020), multiple sclerosis (Ciftci et al., 2015; Haghmorad et al., 2017; Baradaran et al., 2018) and solid tumours (Miguel et al., 2019; Aggarwal et al., 2020). The safety of hesperidin

DOI: 10.1201/9781003611516-13

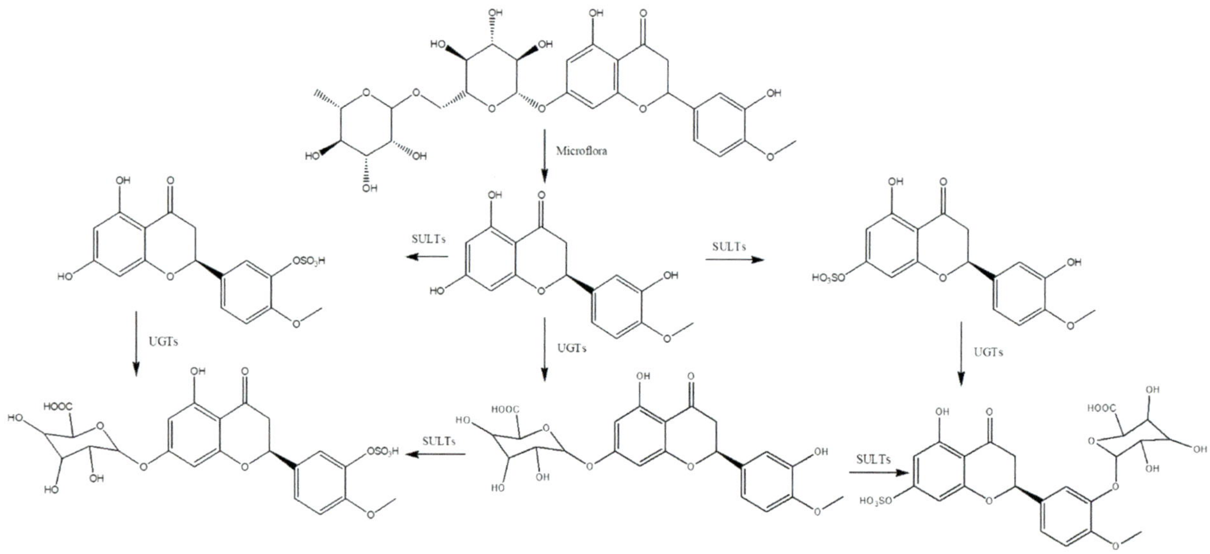

FIGURE 10.2 The metabolic pathways of hesperidin. [1 - Hesperidin; 2 - Hesperitin; 3 - hesperetin-3-*O*-sulfate; 4- hesperetin-7-*O*-sulfate; 5- hesperetin-3-*O*-glucuronide; 6 - hesperetin-7-*O*-glucuronide-3-*O*-sulfate; 7- hesperetin-3-*O* glucuronide-3-*O*-sulfate; UGTs, UDP glucuronosyl transferases; SUTs, Sulfotransferase]. (Figure used from an open source by Boonpawa et al. 2017.)

in the herbal formulation is revealed by an LD_{50} value of more than 2,000 mg/kg in acute toxicity studies and no observed adverse effect level at more than 2,000 mg/kg in subacute and subchronic toxicity studies.

Hesperidin has been largely employed in the treatment and prevention of human ailments. However, its usage is limited by unfavourable features such as low water solubility, precipitation in an acidic environment, poor permeability, intestinal efflux, intestinal and hepatic metabolism, and poor/irregular bioavailability. Various scientific groups attempted different approaches to improve the water solubility of hesperidin (Yamada et al., 2014). The Glucosyl derivative of hesperidin (G-Hesperidin) is one of the water-soluble forms of hesperidin synthesized by regioselective transglycosylation with cyclodextrin glucanotransferase from *Bacillus stearothermophilus* (Ohara et al., 2016).

Nanotechnology is a promising technique to improve many drugs' bioavailability and aqueous solubility (Padhi et al., 2021, 2020). Nanocarriers are small particles (10–100 nm) that increase drug solubility while delivering the drug to targets. These nanocarriers grow and maintain the steady-state levels of the loaded drug moiety and can accumulate at the target site (Behera & Padhi, 2022; Padhi et al., 2022). Metal nanoparticles, polymeric nanoparticles, liposomes, protein conjugates, hydrogels, nanoemulsions, magnetic, and cyclodextrin nanoparticles have all been identified as nanodelivery systems for enhancing drug bioavailability (Behera et al., 2020b; Kundu et al., 2022; Behera et al., 2023a). Hence, hesperidin bioavailability can be improved significantly using novel nanotechnology-mediated delivery strategies (Hijiya & Miyake, 1997). This chapter has attempted to cover current developments in hesperidin-loaded nanocarrier systems, including their potential application in treating various malignancies.

10.2 MECHANISM OF HESPERIDIN AS AN ANTICANCER AGENT

As shown in Figure 10.3, earlier studies revealed the anticancer potential of hesperidin. Also, it suggested mechanisms of anticancer properties, including promoting apoptosis, arresting the cell cycle, attenuating angiogenesis, combating metastasis, and having antioxidant and anti-inflammatory properties (Aggarwal et al., 2020).

10.2.1 Induction of Apoptosis

Apoptosis is the most important mechanism for anticancer activity. The induction of apoptosis is one of the most significant mechanisms for the anticancer activity of hesperidin. Hesperidin produces pro-apoptotic activity by inhibiting the various kinases, increasing the reactive oxygen species (ROS) production, and upregulating caspase activity (Shahbazi et al., 2018). It has been shown that hesperidin produces cytotoxic and pro-apoptotic activity by attenuating nuclear factor kappa B activation by suppressing the phosphoinositide 3-kinase phosphoinositide 3-kinases/protein kinase B/inhibitor of nuclear factor-κB kinase signalling pathway in human B cell precursor leukaemia (NALM-6 cells) (Shahbazi et al., 2018).

A previous study demonstrated that hesperidin inhibits calcium/calmodulin-dependent protein kinase IV, upregulates pro-apoptotic protein, BCL2-associated X protein (apoptosis regulator), and activates the caspase-3 pathway (Naz et al., 2019). It has also been reported that hesperidin induces caspase-dependent apoptosis in non-small cell lung cancer cells (NCI-H358 and A549 cell lines) (Birsu Cincin et al., 2015). Further, studies demonstrated that the pro-apoptotic activity of hesperidin is mediated through increasing the ROS generation and stimulating the apoptosis signal-controlling kinase 1/Jun N-terminal kinase

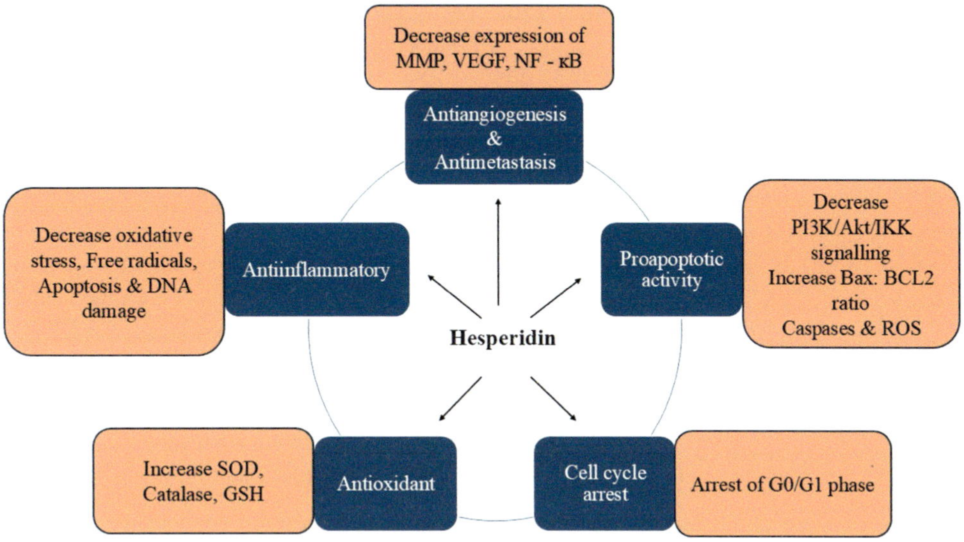

FIGURE 10.3 Mechanism of hesperidin as an anticancer agent [The anticancer activity of hesperidin can be summarized by the antioxidant, anti-inflammatory, anti-angiogenesis, antimetastatic, pro-apoptotic activity and inhibition of cell cycle mechanisms. MMP, matrix metalloproteinase; VEGF, vascular endothelial growth factor; NF-κB, nuclear factor kappa B; PI3K, Phosphoinositide 3-kinases; Akt, protein kinase B; IKK, inhibitor of nuclear factor-κB kinase; BCL2, B cell lymphoma 2 protein; BAX, BCL 2 associated X protein; ROS, reactive oxygen species; SOD, Superoxide dismutases; GSH, glutathion].

(ASK1/JNK) (Palit et al., 2015). This is a key mechanism for the pro-apoptotic activity of hesperidin.

It has been reported that hesperidin increased ROS generation with adenosine triphosphate (ATP) and calcium, which induce apoptosis through the activation of the mitochondrial pathway in hepatocellular carcinoma cells (Zhang et al., 2015a). In line with this finding, it has been reported elsewhere that hesperidin induces apoptosis through stimulation of the mitochondrial pathway. It enhanced the ROS production in gastric and oesophageal cancer cells (Zhang et al., 2015b; Wu et al., 2016). It has been found that hesperidin modulated the anti-apoptotic circumstances by enhancing the Bax/Bcl-2 ratio, along with increasing the liberation of cytochrome-C and activation of caspase-3/9 in an azoxymethane-induced experimental model of colon cancer in mice (Aggarwal et al., 2020).

Another report suggested that hesperidin induced apoptosis by attenuating the activated aurora-A-mediated phosphoinositide 3-kinases/protein kinase B/glycogen synthase kinase 3beta and mammalian target of rapamycin pathway associated with the stimulation of autophagy in colon cancer model (Saiprasad et al., 2014). Various reports suggested that the anticancer or pro-apoptotic activity of hesperidin might be due to increasing the generation of ROS, Bax: BCL2 ratio, and caspases activity and suppression of phosphoinositide 3-kinases/protein kinase B/inhibitor of nuclear factor-κB kinase signalling pathway.

10.2.2 Cell Cycle Arrest

Hesperidin induced the cell cycle arrest at the G0/G1 phase by upregulating the p21 and p53 expression and reducing the expression of cyclinD1 in lung cancer cell lines (A549 cells, a non-small cell lung cancer) (Birsu Cincin et al., 2015; Xia et al., 2018a). Furthermore, it has also been observed that hesperidin induced the cell cycle arrest at phase G0/G1 by reducing the expression of cyclin E1, cyclin D1, and cyclin-dependent kinase 2 (Cdk2) at the protein level (Wang et al., 2015). It has also been reported that hesperidin-induced cell cycle arrest occurs during the G2/M phase (Pandey et al., 2019).

10.2.3 Antiangiogenic Activity

The generation of fresh blood vessels from pre-existing blood vessels characterizes angiogenesis. In instances of cancer, these blood arteries provide a route for tumour cells to enter the systemic circulation (Pandey et al., 2019). Therefore, angiogenesis is an important indicator of the metastasis stage in cancer. Flavonoids, such as hesperidin, showed anti-metastasis by inhibiting tumour angiogenesis (Siveen et al., 2014; Shanmugam et al., 2017). This anti-angiogenesis activity might be produced by modulating matrix metalloproteinases, vascular endothelial growth factor, endothelial cell proliferation, basic fibroblast growth factor, and migration. Hesperidin showed antimetastatic activity by reducing expression of cytoplasmic matrix metalloproteinases-9 through attenuation of activator protein-1 (AP-1) and nuclear factor kappa B signalling in human hepatocellular carcinoma cells (HepG2 cells) (Roohbakhsh et al., 2015). In line with this finding, hesperidin attenuated progression to metastasis stage and tumour cell invasion mediated through nuclear factor kappa B and activator protein-1 by modulating the expression of matrix metalloproteinases-2 and 9, and cyclooxygenase-2 in hepatocellular carcinoma as well as lung cancer cells (Yeh et al., 2009; Lee et al., 2010).

Further, reports showed that hesperidin-mediated anti-metastatic activity is regulated through signal transduction pathways such as mitogen-activated protein kinase 3/6 and p38 in pancreatic cancer cells (Lee et al., 2018). Apart from this, another study demonstrated that the effect of hesperidin on angiogenesis is mediated by protein kinase B and mammalian target of the rapamycin signalling pathway in human umbilical vascular endothelial cells and endothelial resembling cells developed from mouse embryonic stem cells. Hesperidin induced inhibition of the mammalian target of rapamycin and protein kinase B transduction pathway resulting in attenuation of cell migration, reduction of micro-vessel development, and tube generation in human umbilical vascular endothelial cells (Lee et al., 2018).

Hesperidin-induced inhibition of cell migration and tube generation is mediated through the downregulation of nuclear factor of activated T-cell expression in human breast cancer cells and human umbilical vascular endothelial cells (Zhao et al., 2018). Hesperidin inhibited tumour growth and vascular endothelial growth factor, reduced vascular density, and downregulated nuclear factor of activated T cells, vascular endothelial growth factor, and vascular endothelial growth factor receptor2 expression via the nuclear factor of activated T cells signalling in female bagg albino (BALB/c) nude mice (Zhao et al., 2018). Hesperidin attenuated the tumour migration in non-small cell lung cancer cells (A549). This attenuation of tumour migration of hesperidin was mediated by the stromal cell-derived factor 1α, resulting in downregulation of p-Akt, phosphorylated-I kappa B (p-IƙB), C-X-C chemokine receptor type (4CXCR-4), and p-p65 expression (SDF-1/CXCR-4 signalling cascade) (Xia et al., 2018b). Hesperidin attenuated invasion and movement of a cell in male bagg albino mice, a xenograft experimental model, and in MG-63 cells of human osteosarcoma (Du et al., 2018).

10.2.4 Anti-inflammatory

Inflammation is a protective response against noxious stimuli such as allergens, microorganisms, chemical irritants, and particles. Inflammation involves the overexpression of various inflammatory mediators like chemokines, cytokines, and proinflammatory mediators. The overexpression of different inflammatory mediators activates the various signalling pathways such as mitogen-activated protein kinases, nuclear factor kappa B, intercellular adhesion molecule-1, and vascular cell adhesion molecule-1 (Mantovani, 2010; Goldberg & Schwertfeger, 2010; Shanmugam & Sethi, 2013; Chai et al., 2015; Barber, 2015; Rajendran et al., 2018). A close interrelationship exists between the ROS generation and inflammation (Lawrence et al., 2001; Zarubin & Han, 2005; Minami & Aird, 2005; Nizamutdinova et al., 2008; Zhu et al., 2012; Blaser et al., 2016). Furthermore, previously reported studies also suggested that an increase in oxidative stress and inflammation cascade promotes carcinogenesis by suppressing tumour suppressor genes and stimulating oncogenes (Colotta et al., 2009; Grivennikov et al., 2010).

Numerous reports confirmed the beneficial effect of hesperidin on inflammation, oxidative stress, free radicals, apoptosis, proliferation, DNA damage, and carcinogenesis (Li et al., 2019b; Lorzadeh et al., 2019; Pradhan et al., 2022; Behera et al., 2023b). Various previous studies documented the possible mechanism of hesperidin for anti-inflammatory activity, including the suppression of multiple mediators levels like vascular cell adhesion molecule-1, cyclooxygenase-2, matrix metalloproteinase-2 and 9, prostaglandin E2, interleukin-4 and 6, inducible nitric oxide synthase, and nitrogen dioxide (Parhiz et al., 2015; Lorzadeh et al., 2019).

10.2.5 Antioxidant

Flavonoid derivatives are well-reported for their antioxidant activity. Hesperidin possesses potent antioxidants, free radical scavenging, and chelating activity. The antioxidative, chelating, and free radical scavenging of hesperidin might be due to hydrogen donor property (Ahmadi & Shadboorestan, 2016). Previous reports showed that hesperidin inhibits damage to cellular components such as DNA and protein, which may be the reason for its antioxidant and free radical scavenging activity (Parhiz et al., 2015). It has been documented that the protective effect of hesperidin might be due to neutralizing the detrimental actions of hydrogen peroxide, peroxynitrite, carbon tetrachloride, cadmium, acetaminophen, nicotine, cyclophosphamide, acrylonitrile, technetium, and gamma radiation (Kamaraj et al., 2011).

Furthermore, hesperidin also increased the levels and activity of antioxidant enzymes (reduced glutathione, glutathione *S*-transferase, superoxide dismutase, and catalase) (Kamaraj et al., 2011; Pradeep et al., 2012). The antioxidant activity of hesperidin might be mediated through the nuclear factor erythroid 2-related factor 2 (Keapl-Nrf2) pathway, which is most important in governing oxidative and electrophilic stress (Li et al., 2019b). Thus, hesperidin increases the levels of the antioxidant enzymes by increasing the expression and translocation of nuclear factor erythroid 2-related factor 2 by dissociating the Keapl-Nrf2 complex. The consequently published evidence indicated that the anticancer activity of hesperidin might be due to its pro-apoptotic, antiangiogenic, antimetastatic, anti-inflammatory, and antioxidant potential.

10.3 NANOENABLED TARGETING STRATEGIES FOR THE HESPERIDIN

Successful cancer therapy aims to obtain the optimal concentration of therapeutic agents at the tumour site to destroy individual cancer cells (Jahan et al., 2017). Despite the proven anticancer effects of hesperidin, therapeutic success is still reliant on improving the compound's pharmacokinetic profile and efficiently targeting cancerous cells without damaging normal cells. Because they address several of the problems that native hesperidin has, like stability, solubility, and toxicity, hesperidin nanoformulations show greater anticancer activity than natural hesperidin (Wei et al., 2019). Additionally, developing customized nanoparticles helps deliver hesperidin to tumour cells while not affecting healthy cells (Wei et al., 2019).

There are two nanoenabled tumour targeting strategies: active and passive targeting. Passive targeting takes advantage of the tumour's leaky vasculature and is accomplished by the "enhanced permeability and retention" impact (Yin & Liao, 2014; Blanco et al., 2015; Padhi & Behera, 2020). Active targeting, on the other hand, relies on ligand-receptor binding, in which the outer surface of these nanocarriers can be decorated with various bioorganic fractions that directly attach to tumour cell membrane receptors and increase the internalization of nanoparticles in malignant cells. This technique is called "cell or secondary targeting" (Din et al., 2017; Jahan et al., 2017; Villaverde & Baeza, 2019; Padhi & Behera, 2021). Figure 10.4 shows schematic representations of nanoenabled active and passive strategies for hesperidin.

10.3.1 Passive Targeting Strategy for the Delivery of Hesperidin

As previously stated, passive targeting is a key feature of nanoparticles, promoting tumour accumulation through the enhanced permeability and retention effect. Particle size, zeta potential, and solubility of nanoparticles are important factors that could influence passive targeting (De Jong & Borm, 2008; Padhi & Behera, 2022).

In passive targeting, surface coating of ligands on the surface of nanoparticles significantly impacts their solubility, interactions with other biological entities, and pharmacokinetic properties of nanoparticles (Shahbazi et al., 2012; Gera et al., 2017). For nanoparticle encapsulation, polymeric materials such as poly-(lactic-co-glycolic)-acid, polyethylene glycol, and surfactants are used to preserve the higher stability of the nanoparticle framework in the biological setting (De Jong & Borm, 2008; Vijayalakshmi et al., 2019; Ali et al., 2019; Essa et al., 2020). Hesperidin-loaded poly-(lactic-co-glycolic)-acid nanoparticles have been fabricated by Balakrishnan et al. using a single emulsion solvent evaporation technique (Figure 10.5a) that shows the apoptosis of cancer cells. Furthermore, in human epidermoid carcinoma-2 (HEp-2) cells, the biodegradable hesperidin-loaded poly-(lactic-co-glycolic)-acid nanoparticles effectively increased pro-apoptotic protein (Bax, Bad) expression while decreasing anti-apoptotic protein (Bcl-2 and Bcl-xl) expression (Balakrishnan et al., 2020). Hesperidin encapsulated within poly-(lactic-co-glycolic)-acid nanoparticles significantly reduced tumour volume, weight, haematological, and tumour multidrug resistance protein-1 gene regulation in solid Ehrlich carcinoma-bearing mice compared to the control group (Siveen et al., 2014).

Polymeric hesperidin nanoparticles (hesperidin – poly-(lactic-co-glycolic)-acid - poloxamer 407) were successfully fabricated using the nanoprecipitation method (Figure 10.5a) to improve the delivery of hesperidin. The results showed that polymeric hesperidin nanoparticles inhibited cancer cell growth, enabled DNA fragmentation, and induced apoptotic cell death in breast cancer cells (MCF-7 cells) through caspase-3 and p53-dependent pathways (El-Sisi et al., 2020).

The nanoemulsion of hesperidin with a particle size of less than 400 nm could enter the tumour microenvironment through the enhanced permeability and retention effect, thereby improving targeting and enhancing efficacy (El-Sisi et al., 2020). Hesperidin-loaded nanoemulsion was prepared using an emulsification technique

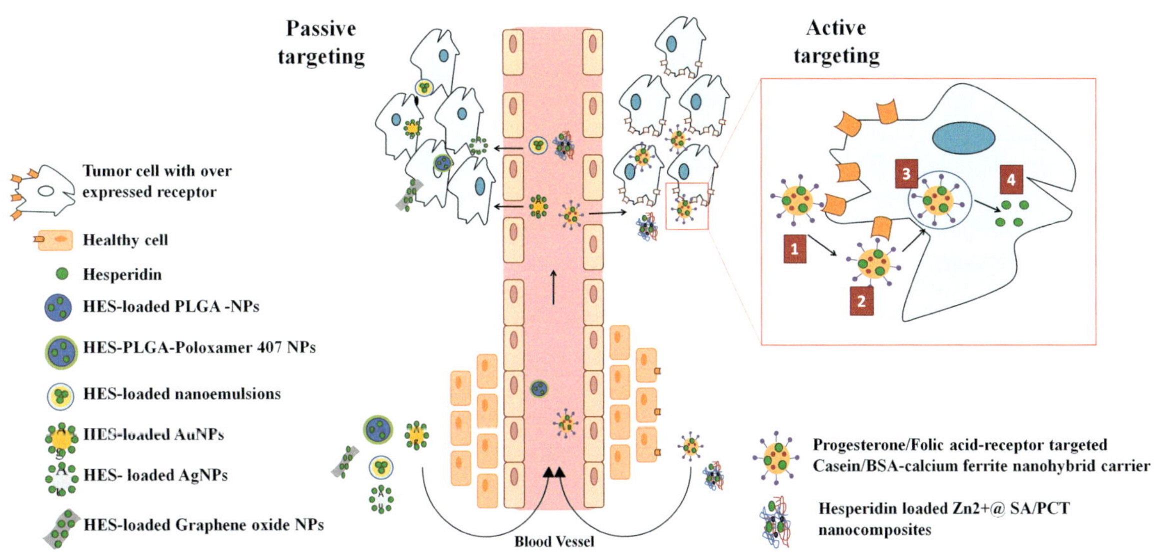

FIGURE 10.4 Schematic representations of nanoenabled active and passive strategies for hesperidin. [1- cellular recognition; 2- receptor-mediated endocytosis; 3- endosome formation; 4- release hesperidin into cytoplasm; HES, Hesperidin; Nanoparticles, nanoparticles; PLGA, poly(lactic-co-glycolic) acid; SA, sodium alginate; PCT, pectin].

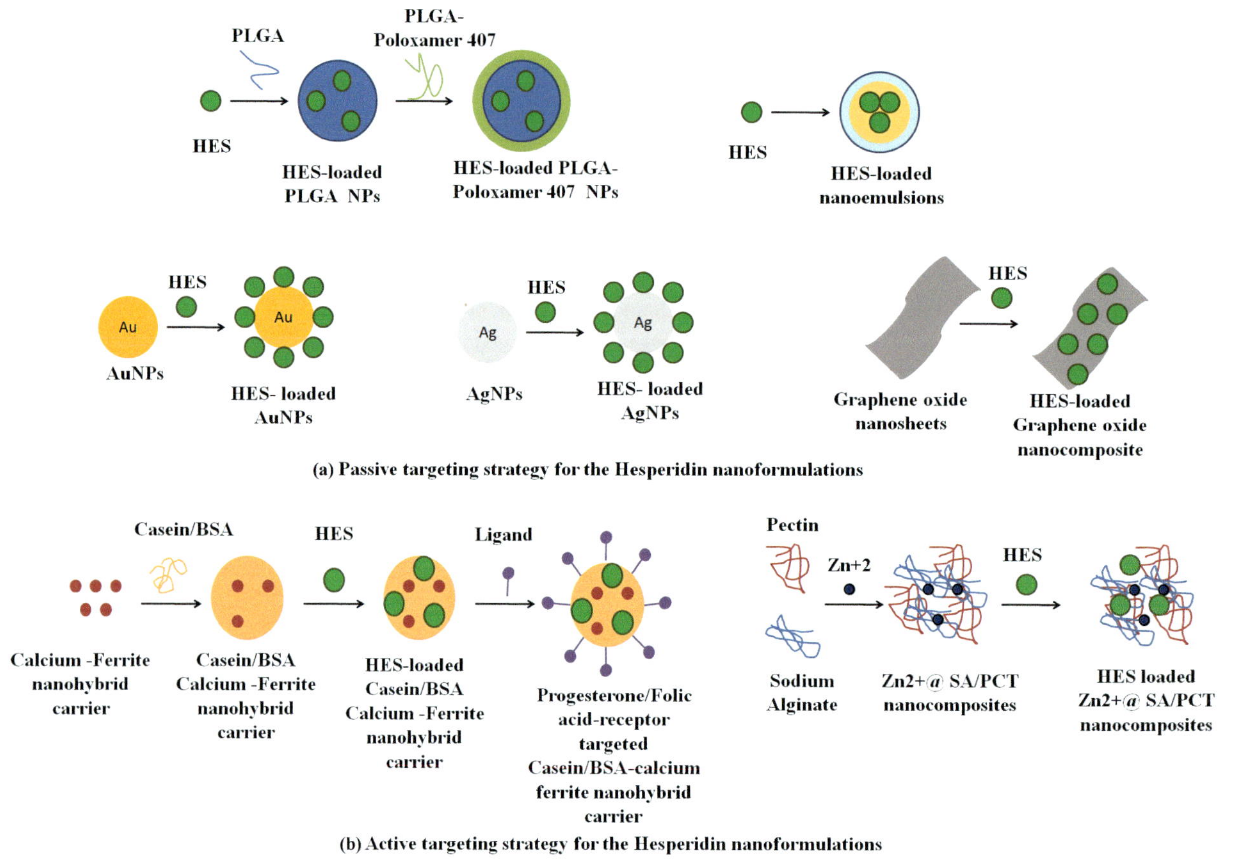

FIGURE 10.5 Different types of hesperidin nanoparticles for passive (a) and active (b) targeting with their preparation methods.

(Figure 10.5a), with nanoemulsion sizes ranging from 245 to 285 nm. Hesperidin-loaded nanoemulsion induced cell death via apoptosis, cell cycle arrest in the G2/M phase, and downregulated microRNA (miR-21 and miR-155) expression in breast cancer cells (MCF-7 cells) (Magura, 2020).

Sulaiman et al. chemically synthesized hesperidin gold nanoparticles (Figure 10.5a) and observed a substantial decrease in the proliferation of human breast cancer (MDA-MB-231) cells treated with hesperidin gold nanoparticles relative to normal human breast epithelial cells (HBL-100). Multiple surface functionalities of gold nanoparticles were noted to improve the drug delivery efficacy as gold nanoparticles are biocompatible and have a surface that is sufficiently large to load higher concentrations of drug or even multiple drugs in a single nanoparticulate system (Mahato et al., 2019; Sulaiman et al., 2020).

In another study, hesperidin silver nanoparticles with a size of 5–50 nm were synthesized using a chemical synthesis method (Figure 10.5a) that showed substantial cytotoxicity against HL-60 cells (Pritha Bosea, Amiya Priyamb, 2020)

Furthermore, hesperidin nanocomposite-loaded graphite oxide (Figure 10.5a) inhibited the inducible nitric oxide synthase and cycloxygenase-2 pathway in 1,2-dimethylhydrazine-induced colon tumour in rats. The broad surface area, electrostatic interactions, and easy functioning made graphene and its derivatives an effective drug delivery candidate (Tonga et al., 2014; Liu & Ma, 2020).

10.3.2 Active Targeting Strategy for the Hesperidin

"Active targeting" describes a particular ligand-receptor interaction at the target site (Bae & Park, 2011). In addition, active targeting depends on the abundance of receptors on target cells but not on normal cells (Blanco et al., 2015). Nanoparticles are designed with ligands specifically targeted to overexpressed receptors in cancerous cells, thereby taking drugs within the cancer cell via receptor-mediated endocytosis (Sutradhar & Amin, 2014).

The progesterone receptor is a membrane receptor that is overexpressed in most of the tumour cells. However, its expression in healthy tissue and organs is limited (Sutradhar & Amin, 2014). Casein-calcium ferrite nanohybrid carrier was developed by desolvation method followed by an ionic gelation process (Figure 10.5b) to provide the progesterone receptor targeted nanohybrid carrier dependent hesperidin delivery. Progesterone nanohybrid carriers showcased cytotoxic activity against ovarian (SKOV-3) and breast cancer cells (MDA-MB-231) (Purushothaman et al., 2020).

Folic acid is required for cell proliferation and differentiation. Folic acid receptors are overexpressed in various tumour cells, such as the breast and ovary (Garcia-Bennett et al., 2011).

For targeting overexpressed folate receptors in breast cancer cells, hesperidin encapsulated-folic acid conjugated-bovine serum albumin-calcium ferrite nanohybrid was formulated using the desolvation method. The hesperidin-encapsulated nanohybrid carrier exhibited enhanced cytotoxicity with a 30-fold reduction in IC_{50} values in breast cancer cell lines (MCF-7) (Uma Maheswari et al., 2021).

In the active targeted drug delivery method, the natural blends of biopolymers consisting of polysaccharides are also widely used. The bioavailability of drugs is improved by encapsulating drugs with polysaccharide blends, such as sodium alginate and pectin (Mary Lazer et al., 2018). Yang et al. have developed a nanocomposite delivery system by encapsulating hesperidin in a zinc cross-linked nanopolymer mixture (polysaccharides pectin and sodium alginate) (Figure 10.5b) that efficiently inhibited cell growth and induced apoptosis in human colorectal carcinoma cell lines (HCT116) via excessive generation of reactive oxygen species. Hesperidin nanocomposites led to an increase in mitochondrial membrane permeability in human colorectal carcinoma cell lines. The antitumor activity was mainly due to pro-apoptotic caspase activation and Bax protein and BCL2 protein inhibition, which induced apoptosis (Yang et al., 2020).

10.4 CONCLUSION AND FUTURE PERSPECTIVES

Hesperidin, flavanone glycoside, is a powerful antioxidant with excellent anticancer activity. However, its inherent low solubility, increased metabolic activity, and poor pharmacokinetics impede its potential to develop as a potent anticancer medicine. Thus, applying nanoparticles as delivery systems can increase the therapeutic capability of hesperidin. In addition, passive and active targeting strategies deliver large amounts of hesperidin without affecting healthy cells. *In-vitro* studies emphasized the anticancer effect of hesperidin-laden nanoparticles in different types of cancer cells. However, most of these formulations were conceptually proven, and very limited *in vivo* studies were performed to determine the safety and effectiveness of hesperidin to allow translation of these preclinical findings to a clinical perspective. Hesperidin nanoformulations are, therefore, expected to undergo thorough analyses in preclinical and clinical research to become pharmacological candidates for cancer treatment either alone or in conjunction with other therapeutic molecules.

REFERENCES

Aggarwal V, Tuli HS, Thakral F, Singhal P, Aggarwal D, Srivastava S, Pandey A, Sak K, Varol M, Khan MA, Sethi G (2020) Molecular mechanisms of action of hesperidin in cancer: Recent trends and advancements. *Exp Biol Med* 245:486–497.

Agrawal YO, Sharma PK, Shrivastava B, Ojha S, Upadhya HM, Arya DS, Goyal SN (2014) Hesperidin produces cardioprotective activity via PPAR-c pathway in ischemic heart disease model in diabetic rats. *PLoS One* 9:1–13.

Ahmadi A, Shadboorestan A (2016) Oxidative stress and cancer; the role of hesperidin, a citrus natural bioflavonoid, as a cancer chemoprotective agent. *Nutr Cancer* 68:29–39.

Aja PM, Izekwe FI, Famurewa AC, Ekpono EU, Nwite FE, Igwenyi IO, Awoke JN, Ani OG, Aloke C, Obasi NA, Udeh KU, Ale BA (2020) Hesperidin protects against cadmium-induced pancreatitis by modulating insulin secretion, redox imbalance and iNOS/NF-κB signaling in rats. *Life Sci* 259:118268.

Ali SH, Sulaiman GM, Al-Halbosiy MMF, Jabir MS, Hameed AH (2019) Fabrication of hesperidin nanoparticles loaded by poly lactic co-glycolic acid for improved therapeutic efficiency and cytotoxicity. *Artif Cells Nanomed Biotechnol* 47:378–394.

Antunes MS, Souza LC, Vagner F, Ladd L, Antunes A, Lobo B, Moreira AL, Bortolotto VC, Rósula M, Silva P, Araújo SM, Prigol M, Nogueira CW, Boeira SP (2020a) Hesperidin ameliorates anxiety-depressive-like behavior in 6-OHDA model of Parkinson's disease by regulating striatal cytokine and neurotrophic factors levels and dopaminergic innervation loss in the striatum of mice. *Mol Neurobiol* 57:3027–3041.

Antunes MS, Vagner F, Ladd L, Antunes A, Lobo B, Moreira AL, Boeira SP, Souza LC (2020b) Hesperidin protects against behavioral alterations and loss of dopaminergic neurons in 6-OHDA-lesioned mice: The role of mitochondrial dysfunction and apoptosis. *Metab Brain Dis* 36:153–167.

Bae YH, Park K (2011) Targeted drug delivery to tumors: Myths, reality and possibility. *J Control Release* 153:198–205.

Balakrishnan K, Casimeer SC, Ghidan A, Ghethan FY, Venkatachalam K, Singaravelu A (2020) Bioformulated hesperidin-loaded PLGA nanoparticles counteract the mitochondrial-mediated intrinsic apoptotic pathway in cancer cells. *J Inorg Organomet Polym Mater* 31:331–343.

Baradaran S, Hajizadeh Moghaddam A, Ghasemi-Kasman M (2018) Hesperetin reduces myelin damage and ameliorates glial activation in lysolecithin-induced focal demyelination model of rat optic chiasm. *Life Sci* 207:471–479.

Barber GN (2015) STING: Infection, inflammation and cancer. *Nat Rev Immunol* 15:760–770.

Barreca D, Calderaro A, Trombetta D, Smeriglio A, Daglia M, Meneghini S (2017) Review article flavanones: Citrus phytochemical with health-promoting properties. *BioFactors* 43:1–12.

Behera A, Mittu B, Padhi S, & Singh A (2020b). Antimicrobial efficacy of essential oil nanoemulsions. In: Dhull SB, Chawla P, Kaushik R (Eds.) *Nanotechnological Approaches in Food Microbiology*. CRC Press, Boca Raton, FL, pp. 293–309.

Behera A, Mittu B, Padhi S, Patra N, Singh J (2020a) Bimetallic nanoparticles: Green synthesis, applications, and future perspectives. In: Abd-Elsalam KA (Ed.) *Multifunctional Hybrid Nanomaterials for Sustainable Agri-food and Ecosystems*. Elsevier, Amsterdam, pp. 639–682.

Behera A, Padhi S (2020) Passive and active targeting strategies for the delivery of the camptothecin anticancer drug: A review. *Environ Chem Lett* 18(5):1557–1567.

Behera A, Padhi S (2022). pH-Sensitive polymeric nanoparticles for cancer treatment. In: Padhi S, Behera A, Lichtfouse E (Eds.) *Polymeric Nanoparticles for the Treatment of Solid Tumours*. Environmental Chemistry for a Sustainable World, vol 71. Springer, Cham, pp. 401–425.

Behera A, Padhi S, Nayak AK (2023a). Engineered liposomes as drug delivery and imaging agents. In: Ray S, Nayak AK (Eds.) *Design and Applications of Theranostic Nanomedicines*. Woodhead Publishing, Cambridge, MA, pp. 75–108.

Behera A, Pradhan SP, Tejaswani P, Sa N, Pattnaik S, Sahu PK (2023b). Ameliorative and neuroprotective effect of core-shell type Se@Au conjugated hesperidin nanoparticles in diabetes-induced cognitive impairment. *Mol Neurobiol* 60:7329–7345.

Birsu Cincin Z, Unlu M, Kiran B, Sinem Bireller E, Baran Y, Cakmakoglu B (2015) Anti-proliferative, apoptotic and signal transduction effects of hesperidin in non-small cell lung cancer cells. *Cell Oncol (Dordr)* 38:195–204.

Blanco E, Shen H, Ferrari M (2015) Principles of nanoparticle design for overcoming biological barriers to drug delivery. *Nat Biotechnol* 33:941–951.

Blaser H, Dostert C, Mak TW, Brenner D (2016) TNF and ROS crosstalk in inflammation. *Trends Cell Biol* 26:249–261.

Boonpawa R, Spenkelink A, Punt A, Rietjens IMCM (2017) Physiologically based kinetic modelling of hesperidin metabolism and its use to predict in vivo effective doses in humans. *Mol Nutr Food Res* 61(8):1600894.

Chai EZP, Siveen KS, Shanmugam MK, Arfuso F, Sethi G (2015) Analysis of the intricate relationship between chronic inflammation and cancer. *Biochem J* 468:1–15.

Chakraborty S, Bandyopadhyay J, Chakraborty S (2016) Multi-target screening mines hesperidin as a multi-potent inhibitor: Implication in Alzheimer's disease therapeutics. *Eur J Med Chem* 121:810–822.

Ciftci O, Ozcan C, Kamisli O, Cetin A, Basak N, Aytac B (2015) Hesperidin, a citrus flavonoid, has the ameliorative effects against experimental autoimmune encephalomyelitis (EAE) in a C57BL/J6 mouse model. *Neurochem Res* 40:1111–1120.

Colotta F, Allavena P, Sica A, Garlanda C, Mantovani A (2009) Cancer-related inflammation, the seventh hallmark of cancer: Links to genetic instability. *Carcinogenesis* 30:1073–1081.

De Jong WH, Borm PJA (2008) Drug delivery and nanoparticles: Applications and hazards. *Int J Nanomed* 3:133–149.

Din FU, Aman W, Ullah I, Qureshi OS, Mustapha O, Shafique S, Zeb A (2017) Effective use of nanocarriers as drug delivery systems for the treatment of selected tumours. *Int J Nanomed* 12:7291–7309.

Du G-Y, He S-W, Zhang L, Sun C-X, Mi L-D, Sun Z-G (2018) Hesperidin exhibits in vitro and in vivo antitumor effects in human osteosarcoma MG-63 cells and xenograft mice models via inhibition of cell migration and invasion, cell cycle arrest and induction of mitochondrial-mediated apoptosis. *Oncol Lett* 16:6299–6306.

El-Sisi AE, Sokkar SS, Ibrahim HA, Hamed MF, Abu-Risha SE (2020) Targeting MDR-1 gene expression, BAX/BCL2, caspase-3, and Ki-67 by nanoencapsulated imatinib and hesperidin to enhance anticancer activity and ameliorate cardiotoxicity. *Fundam Clin Pharmacol* 34:458–475.

Essa D, Kondiah PPD, Choonara YE, Pillay V (2020) The design of poly(lactide-co-glycolide) nanocarriers for medical applications. *Front Bioeng Biotechnol* 8:48.

Ganeshpurkar A, Saluja A (2019) The pharmacological potential of hesperidin. *Indian J Biochem Biophys* 56:287–300.

Garcia-Bennett A, Nees M, Fadeel B (2011) In search of the Holy Grail: Folate-targeted nanoparticles for cancer therapy. *Biochem Pharmacol* 81:976–984.

Garg A, Garg S, Zaneveld LJ, Singla AK (2001) Chemistry and pharmacology of the citrus bioflavonoid hesperidin. *Phyther Res* 15:655–669

Geng Y, Wu Z, Buist-Homan M, Blokzijl H, Moshage H (2020) Hesperetin protects against palmitate-induced cellular toxicity via induction of GRP78 in hepatocytes. *Toxicol Appl Pharmacol* 404:115183.

Gera M, Sharma N, Ghosh M, Huynh DL, Lee SJ, Min T, Kwon T, Jeong DK (2017) Nanoformulations of curcumin: An emerging paradigm for improved remedial application. *Oncotarget* 8:66680–66698.

Goldberg JE, Schwertfeger KL (2010) Proinflammatory cytokines in breast cancer: Mechanisms of action and potential targets for therapeutics. *Curr Drug Targets* 11:1133–1146.

Grivennikov SI, Greten FR, Karin M (2010) Immunity, inflammation, and cancer. *Cell* 140:883–899.

Guo X, Li K, Guo A, Li E (2020) Intestinal absorption and distribution of naringin, hesperidin, and their metabolites in mice. *J Funct Foods* 74:104158.

Haddadi GH, Rezaeyan A, Mosleh-Shirazi MA, Hosseinzadeh M, Fardid R, Najafi M, Salajegheh A (2017) Hesperidin as radioprotector against radiation-induced lung damage in rat: A histopathological study. *J Med Phys* 42:25–32.

Haghmorad D, Mahmoudi MB, Salehipour Z, Jalayer Z, Abbas A, Rastin M, Kokhaei P, Mahmoudi M (2016) Hesperidin ameliorates immunological outcome and reduces neuroinflammation. *J Neuroimmunol* 302:23–33.

Haghmorad D, Mahmoudi MB, Salehipour Z, Jalayer Z, Momtazi Brojeni AA, Rastin M, Kokhaei P, Mahmoudi M (2017) Hesperidin ameliorates immunological outcome and reduces neuroinflammation in the mouse model of multiple sclerosis. *J Neuroimmunol* 302:23–33.

Hanedan B, Ozkaraca M, Kirbas A, Kandemir FM, Aktas MS, Kilic K, Comakli S, Kucukler S, Bilgili A (2018) Investigation of the effects of hesperidin and chrysin on renal injury induced by colistin in rats. *Biomed Pharmacother* 108:1607–1616.

Hemanth Kumar B, Dinesh Kumar B, Diwan PV (2017) Hesperidin, a citrus flavonoid, protects against l-methionine-induced hyperhomocysteinemia by abrogation of oxidative stress, endothelial dysfunction and neurotoxicity in Wistar rats. *Pharm Biol* 55:146–155.

Hijiya H, Miyake T (1997). U.S. Patent No. 5,652,124. Washington, DC: U.S. Patent and Trademark Office.

Ikemura M, Sasaki Y, Giddings JC, Yamamoto J (2012) Preventive effects of hesperidin, glucosyl hesperidin and naringin on hypertension and cerebral thrombosis in stroke-prone spontaneously hypertensive rats. *Phyther Res* 26:1272–1277.

Ikram M, Muhammad T, Rehman SU, Khan A, Jo MG, Ali T, Kim MO (2019) Hesperetin confers neuroprotection by regulating Nrf2/TLR4/NF-κB signaling in an Aβ Mouse model. *Mol Neurobiol* 56:6293–6309.

Jahan ST, Sadat SMA, Walliser M, Haddadi A (2017) Targeted therapeutic nanoparticles: An immense promise to fight against cancer. *J Drug Deliv* 2017:9090325.

Jung UJ, Kim DW, Kim S, Kwon JY (2018) Beneficial effects of hesperetin in a mouse model of temporal lobe epilepsy. *J Med Food* 21:1306–1309.

Kamaraj S, Anandakumar P, Jagan S, Ramakrishnan G, Devaki T (2011) Hesperidin attenuates mitochondrial dysfunction during benzo(a)pyrene-induced lung carcinogenesis in mice. *Fundam Clin Pharmacol* 25:91–98.

Karimi N, Monfared AS, Haddadi GH, Soleymani A, Mohammadi E (2017) Radioprotective effect of hesperidin on reducing oxidative stress in the lens tissue of rats. *Int J Pharm Investig* 7:149–154.

Kesh S, Kannan RR, Sivaji K, Balakrishnan A (2020) Hesperidin downregulates kinases lrrk2 and gsk3beta in a 6-OHDA induced Parkinson's disease model. *Neurosci Lett* 740:135426.

Kilic K, Sedat M, Serkan S, Fatih Y, Kandemir M, Sitki M (2018) The amendatory effect of hesperidin and thymol in

allergic rhinitis: An ovalbumin-induced rat model. *Eur Arch Oto-Rhino-Laryngology* 8:249–258.
Köksoy FN, Yankol Y, Şen Oran E, Őzkan Gürdal S, Yüksel M, Akyildiz Iğdem A, Yildirim Yazgan N, Soybir GR (2013) Preventive effects of enoxaparin and hesperidin in cerulein-induced acute pancreatitis in rats. *Turkish J Gastroenterol Off J Turkish Soc Gastroenterol* 24:495–501.
Kumar A, Lalitha S, Mishra J (2013) Epilepsy & behavior possible nitric oxide mechanism in the protective effect of hesperidin against pentylenetetrazole (PTZ)-induced kindling and associated cognitive dysfunction in mice. *Epilepsy Behav* 29:103–111.
Kumar A, Lalitha S, Mishra J (2014) Hesperidin potentiates the neuroprotective effects of diazepam and gabapentin against pentylenetetrazole-induced convulsions in mice: Possible behavioral, biochemical and mitochondrial alterations. *Indian J Exp Biol* 46:309–316.
Kumar R, Akhtar F, Ibrahim S (2020) Hesperidin attenuates altered redox homeostasis in an experimental hyperlipidaemic model of rat. *Clin Exp Pharmacol Physiol* 47:571–582.
Kundu A, Padhi S, Behera A, Hasnain MS, Nayak AK (2022) Tumor targeting strategies by chitosan-based nanocarriers. In: Hasnain MS, Beg S, Nayak AK (Eds.) *Chitosan in Biomedical Applications*. Academic Press, London, pp. 163–188.
Lawrence T, Gilroy DW, Colville-Nash PR, Willoughby DA (2001) Possible new role for NF-kappaB in the resolution of inflammation. *Nat Med* 7:1291–1297.
Lee J, Lee J, Kim M, Kim JH (2018) Fermented extraction of citrus unshiu peel inhibits viability and migration of human pancreatic cancers. *J Med Food* 21:5–12.
Lee K-H, Yeh M-H, Kao S-T, Hung C-M, Liu C-J, Huang Y-Y, Yeh C-C (2010) The inhibitory effect of hesperidin on tumour cell invasiveness occurs via suppression of activator protein 1 and nuclear factor-kappaB in human hepatocellular carcinoma cells. *Toxicol Lett* 194:42–49.
Li X, Xie X, Zhang L, Meng Y, Li N, Wang M (2019a) Hesperidin inhibits keratinocyte proliferation and imiquimod-induced psoriasis-like dermatitis via the IRS-1/ERK1/2 pathway. *Life Sci* 219:311–321.
Li X, Xie X, Zhang L, Meng Y, Li N, Wang M, Zhai C, Liu Z, Di T, Zhang L, Li P (2019b) Hesperidin inhibits keratinocyte proliferation and imiquimod-induced psoriasis-like dermatitis via the IRS-1/ERK1/2 pathway. *Life Sci* 219:311–321.
Liu J, Ma X (2020) Modulating effect of graphene oxide loaded hesperidin nanocomposite on the 1,2-dimethylhydrazine provoked colon carcinogenesis in rats via inhibiting the iNOS and COX-2 pathways. *Arab J Chem* 13:6708–6723.
Lorzadeh E, Ramezani-Jolfaie N, Mohammadi M, Khoshbakht Y, Salehi-Abargouei A (2019) The effect of hesperidin supplementation on inflammatory markers in human adults: A systematic review and meta-analysis of randomized controlled clinical trials. *Chem Biol Interact* 307:8–15.
Mahato K, Nagpal S, Shah MA, Srivastava A, Maurya PK, Roy S, Jaiswal A, Singh R, Chandra P (2019) Gold nanoparticle surface engineering strategies and their applications in biomedicine and diagnostics. *3 Biotech* 9:57.
Mantovani A (2010) Molecular pathways linking inflammation and cancer. *Curr Mol Med* 10:369–373.
Mary Lazer L, Sadhasivam B, Palaniyandi K, Muthuswamy T, Ramachandran I, Balakrishnan A, Pathak S, Narayan S, Ramalingam S (2018) Chitosan-based nanoformulation enhances the anticancer efficacy of hesperetin. *Int J Biol Macromol* 107:1988–1998.
Mehrdad I, Ramin R, Hamideh P, Ali Roohbakhsh FS (2015) Protective effects of flavonoids against microbes and toxins: The cases of hesperidin and Hesperetin. *Life Sci* 137:125–132.
Melekoglu R, Ciftci O, Eraslan S, Cetin A, Basak N (2018) The beneficial effects of nerolidol and hesperidin on surgically induced endometriosis in a rat model. *Gynecol Endocrinol Off J Int Soc Gynecol Endocrinol* 34:975–980.
Miguel J, Oliveira PFD, Santos C, Fernandes E (2019) Therapeutic potential of hesperidin and its aglycone hesperetin: Cell cycle regulation and apoptosis induction in cancer models. *Phytomedicine* 73:152887.
Minami T, Aird WC (2005) Endothelial cell gene regulation. *Trends Cardiovasc Med* 15:174–184.
Mota L, César B, Bordignon L, Nathália L, Mariano B, Mariott M, Boeing T, Carolina A, Longo B, Cechinel-filho V, Souza PD, Faloni S, Andrade D (2019) Hesperidin, a citrus flavanone glycoside, accelerates the gastric healing process of acetic acid-induced ulcer in rats. *Chem Biol Interact* 308:45–50.
Musa AE, Omyan G, Esmaely F, Shabeeb D (2019) Radioprotective effect of hesperidin: A systematic review. *Medicina (B Aires)* 55:1–12.
Naz H, Tarique M, Ahamad S, Alajmi MF, Hussain A, Rehman MT, Luqman S, Hassan MI (2019) Hesperidin-CAMKIV interaction and its impact on cell proliferation and apoptosis in the human hepatic carcinoma and neuroblastoma cells. *J Cell Biochem* 120:15119–15130.
Nizamutdinova IT, Jeong JJ, Xu GH, Lee S-H, Kang SS, Kim YS, Chang KC, Kim HJ (2008) Hesperidin, hesperidin methyl chalone and phellopterin from Poncirus trifoliata (Rutaceae) differentially regulate the expression of adhesion molecules in tumor necrosis factor-alpha-stimulated human umbilical vein endothelial cells. *Int Immunopharmacol* 8:670–678.
Ohara T, Muroyama K, Yamamoto Y, Murosaki S (2016) Oral intake of a combination of glucosyl hesperidin and caffeine elicits an anti-obesity effect in healthy, moderately obese subjects: a randomized double-blind placebo-controlled trial. *Nutr J* 15(1):6.
Ota-kontani A, Hirata H, Ogura M, Tsuchiya Y (2019) Comprehensive analysis of mechanism underlying hypouricemic effect of glucosyl hesperidin. *Biochem Biophys Res Commun* 521(4):861–867.
Padhi S, Behera A (2020) Nanotechnology based targeting strategies for the delivery of Camptothecin. In: Saneja A, Panda AK, Lichtfouse E, (Eds.) *Sustainable Agriculture Reviews 44: Pharmaceutical Technology for Natural Products Delivery Vol. 2 Impact of Nanotechnology*. Cham, Springer International Publishing, pp. 243–272.
Padhi S, Behera A (2021) Advanced drug delivery systems in the treatment of ovarian cancer. In: Dua K, Mehta M, de Jesus Andreoli Pinto T, Pont LG, Williams KA, Rathbone MJ (Eds.) *Advanced Drug Delivery Systems in the Management of Cancer*. Academic Press, London, pp. 127–139.
Padhi S, Behera A (2022) Cellular internalization and toxicity of polymeric nanoparticles. In: Padhi S, Behera A, Lichtfouse E (Eds.) *Polymeric Nanoparticles for Treating Solid Tumours*. Environmental Chemistry for a Sustainable World, vol 71. Springer, Cham, pp. 473–488.
Padhi S, Behera A, Hasnain MS, Nayak AK (2022) Chitosan-based drug delivery systems in cancer therapeutics. In: Hasnain MS, Beg S, Nayak AK (Eds.) *Chitosan in Drug Delivery*. Academic Press, London, pp. 159–193.
Padhi S, Dash M, Behera A (2021) Nanophytochemicals for the treatment of type II diabetes mellitus: A review. *Environ Chem Lett* 19:4349–4373.

Padhi S, Mirza MA, Verma D, Khuroo T, Panda AK, Talegaonkar S, Khar RK, Iqbal Z (2016) Revisiting the nanoformulation design approach for effective delivery of Topotecan in its stable form: An appraisal of its in vitro behavior and tumour amelioration potential. *Drug Deliv* 23(8):2827–2837.

Padhi S, Nayak AK, Behera A (2020) Type II diabetes mellitus: A review on recent drug-based therapeutics. *Biomed Pharmacother* 131:110708.

Palit S, Kar S, Sharma G, Das PK (2015) Hesperetin induces apoptosis in breast carcinoma by triggering accumulation of ROS and activation of ASK1/JNK pathway. *J Cell Physiol* 230:1729–1739.

Panda S, Kar A (2014) Antithyroid effects of naringin, hesperidin and rutin in L-T4 induced hyperthyroid rats: Possible mediation through 50 DI activity. *Pharmacol Rep* 66:1092–1099.

Pandey P, Sayyed U, Tiwari RK, Siddiqui MH, Pathak N, Bajpai P (2019) Hesperidin induces ROS-mediated apoptosis along with cell cycle arrest at G2/M phase in human gall bladder carcinoma. *Nutr Cancer* 71:676–687.

Parhiz H, Roohbakhsh A, Soltani F, Rezaee R, Iranshahi M (2015) Antioxidant and anti-inflammatory properties of the citrus flavonoids hesperidin and Hesperetin: An updated review of their molecular mechanisms and experimental models. *Phytother Res* 29:323–331.

Park WS, Park MS, Kang SW, Jin SA, Jeon Y, Hwang J, Kim SK (2019) Hesperidin shows protective effects on renal function in ischemia-induced acute kidney injury (Sprague-Dawley rats). *Transplant Proc* 51:2838–2841.

Patnaik S, Gorain B, Padhi S, Choudhury H, Gabr GA, Md S, Kumar Mishra D, Kesharwani P (2021) Recent update of toxicity aspects of nanoparticulate systems for drug delivery. *Eur J Pharm Biopharm* 161:100–119.

Pradeep K, Ko KC, Choi MH, Kang JA, Chung YJ, Park SH (2012) Protective effect of hesperidin, a citrus flavanoglycone, against γ-radiation-induced tissue damage in Sprague-Dawley rats. *J Med Food* 15:419–427.

Pradhan SP, Sahoo S, Behera A, Sahoo R, Sahu PK (2022) Memory amelioration by hesperidin conjugated gold nanoparticles in diabetes induced cognitive impaired rats. *J Drug Deliv Sci Technol* 69:103145.

Pritha B, Amiya P, Rajiv K, Sakthi PP (2020) Quercetin loaded folate targeted plasmonic silver nanoparticles for light activated chemo-photothermal therapy of DMBA induced breast cancer in Sprague Dawley rats. *RSC Adv* 10:31961–31978.

Purushothaman BK, Uma Maheswari P, Meera Sheriffa Begum KM (2020) Magnetic casein-$CaFe_2O_4$ nanohybrid carrier conjugated with progesterone for enhanced cytotoxicity of citrus peel derived hesperidin drug towards breast and ovarian cancer. *Int J Biol Macromol* 151:293–304.

Qi W, Lin C, Fan K, Chen Z, Liu L (2019) Hesperidin inhibits synovial cell inflammation and macrophage polarization through suppression of the PI3K/AKT pathway in complete Freund's adjuvant-induced arthritis in mice. *Chem Biol Interact* 306:19–28.

Rajendran P, Chen Y-F, Chen Y-F, Chung L-C, Tamilselvi S, Shen C-Y, Day CH, Chen R-J, Viswanadha VP, Kuo W-W, Huang C-Y (2018) The multifaceted link between inflammation and human diseases. *J Cell Physiol* 233:6458–6471.

Roohbakhsh A, Parhiz H, Soltani F, Rezaee R, Iranshahi M (2015) Molecular mechanisms behind the biological effects of hesperidin and hesperetin for the prevention of cancer and cardiovascular diseases. *Life Sci* 124:64–74.

Sahu BD, Kuncha M, Sindhura GJ, Sistla R (2013) Hesperidin attenuates cisplatin-induced acute renal injury by decreasing oxidative stress, inflammation and DNA damage. *Phytomedicine* 20:453–460.

Saiprasad G, Chitra P, Manikandan R, Sudhandiran G (2014) Hesperidin induces apoptosis and triggers autophagic markers through inhibition of Aurora-A mediated phosphoinositide-3-kinase/Akt/mammalian target of rapamycin and glycogen synthase kinase-3 beta signalling cascades in experimental colon carcinogenesis. *Eur J Cancer* 50:2489–2507.

Samie A, Sedaghat R, Baluchnejadmojarad T, Roghani M (2018) Hesperetin, a citrus flavonoid, attenuates testicular damage in diabetic rats via inhibition of oxidative stress, inflammation, and apoptosis. *Life Sci* 210:132–139.

Shahbazi M-A, Herranz B, Santos HA (2012) Nanostructured porous Si-based nanoparticles for targeted drug delivery. *Biomatter* 2:296–312.

Shahbazi R, Cheraghpour M, Homayounfar R, Nazari M, Nasrollahzadeh J, Davoodi SH (2018) Hesperidin inhibits insulin-induced phosphoinositide 3-kinase/Akt activation in human pre-B cell line NALM-6. *J Cancer Res Ther* 14:503–508.

Shanmugam MK, Sethi G (2013) Role of epigenetics in inflammation-associated diseases. *Subcell Biochem* 61:627–657.

Shanmugam MK, Warrier S, Kumar AP, Sethi G, Arfuso F (2017) Potential role of natural compounds as antiangiogenic agents in cancer. *Curr Vasc Pharmacol* 15:503–519.

Shokoohi M, Khaki A, Shoorei H, Khaki AA, Moghimian M, Abtahi-Eivary S-H (2020) Hesperidin attenuated apoptotic-related genes in the testicle of a male rat model of varicocele. *Andrology* 8:249–258.

Siveen KS, Ahn KS, Ong TH, Shanmugam MK, Li F, Yap WN, Kumar AP, Fong CW, Tergaonkar V, Hui KM, Sethi G (2014) Y-Tocotrienol inhibits angiogenesis-dependent growth of human hepatocellular carcinoma through abrogation of AKT/mTOR pathway in an orthotopic mouse model. *Oncotarget* 5:1897–1911.

Subramani PA, Panati K, Narala VR (2017) Curcumin nanotechnologies and its anticancer activity. *Nutr Cancer* 69:381–393.

Sulaiman GM, Waheeb HM, Jabir MS, Khazaal SH, Dewir YH, Naidoo Y (2020) Hesperidin loaded on gold nanoparticles as a drug delivery system for a successful biocompatible, anti-cancer, anti-inflammatory and phagocytosis inducer model. *Sci Rep* 10:9362.

Sun Y, Chen J, Shen L, Zhou J, Wang C (2017) Anti-atherosclerotic effect of hesperidin in LDLr –/– mice and its possible mechanism. *Eur J Pharmacol* 815:109–117.

Sutradhar KB, Amin ML (2014) Nanotechnology in cancer drug delivery and selective targeting. *ISRN Nanotechnol* 2014:939378.

Tabeshpour J, Hosseinzadeh H, Hashemzaei M, Karimi G (2020) A review of the hepatoprotective effects of hesperidin, a flavanon glycoside in citrus fruits, against natural and chemical toxicities. *DARU J Pharm Sci* 28:305–317.

Thenmozhi AJ, Raja T, Raja W, Janakiraman U, Essa MM, Thenmozhi AJ, Raja T, Raja W (2016) Oxidative stress and apoptosis against aluminium chloride induced rat model of Alzheimer's disease hesperidin ameliorates cognitive dysfunction, oxidative stress and apoptosis against aluminium chloride induced rat model of Alzheimer's disease. *Nutr Neurosci* 20(6):360–368.

Tonga GY, Moyano DF, Kim CS, Rotello VM (2014) Inorganic nanoparticles for therapeutic delivery: Trials, tribulations and promise. *Curr Opin Colloid Interface Sci* 19:49–55.

Uma Maheswari P, Muthappa R, Bindhya KP, Meera Sheriffa Begum KM (2021) Evaluation of folic acid functionalized

BSA-CaFe_2O_4 nanohybrid carrier for the controlled delivery of natural cytotoxic drugs hesperidin and eugenol. *J Drug Deliv Sci Technol* 61:102105.

Umar S, Kumar A, Sajad M, Zargan J (2012) Hesperidin inhibits collagen-induced arthritis possibly through suppression of free radical load and reduction in neutrophil activation and infiltration. *Rheumatol Int* 33:657–663.

Vijayalakshmi S, Mariadoss AVA, Ramachandran V, Shalini V, Agilan B, Sangeetha CC, Balu P, Kotakadi VS, Karthikkumar V, Ernest D (2019) Polydatin encapsulated poly[lactic-co-glycolic acid] nanoformulation counteract the 7,12-dimethylbenz[a] anthracene mediated experimental carcinogenesis through the inhibition of cell proliferation. *Antioxidants* 8(9), 375.

Villaverde G, Baeza A (2019) Targeting strategies for improving the efficacy of nanomedicine in oncology. *Beilstein J Nanotechnol* 10:168–181.

Wang Y, Yu H, Zhang J, Gao J, Ge X, Lou G (2015) Hesperidin inhibits HeLa cell proliferation through apoptosis mediated by endoplasmic reticulum stress pathways and cell cycle arrest. *BMC Cancer* 15:682.

Wei Q-Y, He K-M, Chen J-L, Xu Y-M, Lau ATY (2019) Phytofabrication of nanoparticles as novel drugs for anticancer applications. *Molecules* 24(23):4246.

Wu D, Zhang J, Wang J, Li J, Liao F, Dong W (2016) Hesperetin induces apoptosis of esophageal cancer cells via mitochondrial pathway mediated by the increased intracellular reactive oxygen species. *Tumour Biol J Int Soc Oncodevelopmental Biol Med* 37:3451–3459.

Wunpathe C, Potue P, Maneesai P (2018) Hesperidin suppresses renin-angiotensin system mediated NOX_2 over-expression and sympathoexcitation in 2K-1C hypertensive rats. *Am J Chin Med* 46:1–17.

Xia R, Sheng X, Xu X, Yu C, Lu H (2018a) Hesperidin induces apoptosis and G0/G1 arrest in human non-small cell lung cancer A549 cells. *Int J Mol Med* 41:464–472.

Xia R, Xu G, Huang Y, Sheng X, Xu X, Lu H (2018b) Hesperidin suppresses the migration and invasion of non-small cell lung cancer cells by inhibiting the SDF-1/CXCR-4 pathway. *Life Sci* 201:111–120.

Xiong H (2019) Hesperidin: A therapeutic agent for obesity. *Drug Des Devel Ther* 13:3855–3866.

Yamada M, Tanabe F, Arai N, Mitsuzumi H, Kubota M, Chaen H, Kibata M (2014) Bioavailability of glucosyl hesperidin in rats bioavailability of glucosyl hesperidin in rats. *Biosci Biotechnol Biochem* 70:1386–1394.

Yang Z, Yang H, Dong X, Pu M, Ji F (2020) Hesperidin loaded Zn^{2+}@SA/PCT nanocomposites inhibit the proliferation and induce the apoptosis in colon cancer cells (HCT116) through the enhancement of pro-apoptotic protein expressions. *J Photochem Photobiol B* 204:111767.

Yeh M-H, Kao S-T, Hung C-M, Liu C-J, Lee K-H, Yeh C-C (2009) Hesperidin inhibited acetaldehyde-induced matrix metalloproteinase-9 gene expression in human hepatocellular carcinoma cells. *Toxicol Lett* 184:204–210.

Yin H, Liao L FJ (2014) Enhanced permeability and retention (EPR) effect based tumor targeting: The concept, application and prospect. *JSM Clin Oncol Res* 2:1010.

Zarubin T, Han J (2005) Activation and signalling of the p38 MAP kinase pathway. *Cell Res* 15:11–18.

Zhang J, Song J, Wu D, Wang J, Dong W (2015a) Hesperetin induces the apoptosis of hepatocellular carcinoma cells via mitochondrial pathway mediated by the increased intracellular reactive oxygen species, ATP and calcium. *Med Oncol* 32:101.

Zhang J, Wu D, Vikash, Song J, Wang J, Yi J, Dong W (2015b) Hesperetin induces the apoptosis of gastric cancer cells via activating mitochondrial pathway by increasing reactive oxygen species. *Dig Dis Sci* 60:2985–2995.

Zhao X, Liu J, Feng L, Ge S, Yang S, Chen C, Li X, Peng L, Mu Y, Wang Y, Gu D, Guo Y, Lin G, Deng B, Cheng Z, Cai D (2018) Antiangiogenic effects of Qingdu granule on breast cancer through inhibiting NFAT signaling pathway. *J Ethnopharmacol* 222:261–269.

Zhao Y, Jiao Y, Wang L (2020) Hesperidin methyl chalcone alleviates spinal tuberculosis in New Zealand white rabbits by suppressing immune responses. *J Spinal Cord Med* 43(4):532–539.

Zhu Y, Zhu M, Lance P (2012) iNOS signaling interacts with COX-2 pathway in colonic fibroblasts. *Exp Cell Res* 318:2116–2127.

11 Nanoenabled Systems for the Delivery of Thymoquinone as an Anticancer Agent

Puja Das, Sayantan Ghosh, Subhashree Mohapatra, and Bismita Nayak
Immunology and Molecular Medicine Laboratory, Department of Life Science, National Institute of Technology, Rourkela, India

11.1 INTRODUCTION

Since time immemorial, an inseparable bond between plants and humanity has existed. Consumption of a wide variety of vegetables, fruits, spices, and herbs has proved its worth by preventing various diseases due to their therapeutic properties, and they are found to contain multiple bioactive compounds (Anas Ahmad et al., 2019). *Nigella sativa* (also known as black seed, black cumin, and kalonji), which belongs to the Ranunculaceae family, is well known for its medicinal properties; majorly cultivated across the Mediterranean region and Western Asian countries (India, Afghanistan, Pakistan) (Rathore et al., 2020).

The therapeutic activity of *N. sativa* is attributed to its various bioactive components, such as thymoquinone, alkaloids, monoterpenes, saponins, and so forth. Among these, thymoquinone(2-Isopropyl-5-methyl-1, 4-benzoquinone; Figure 11.1) is the prime bioactive component of *N.sativa* seeds, which forms 30%–48% of its total composition (Aftab Ahmad et al., 2013). It was first isolated by El-Dakhakhny in 1963 from black cumin seeds using thin-layer chromatography (Rathore et al., 2020). Thymoquinone is known to treat various diseases due to its several health-beneficial properties like anticancer, anti-inflammatory, analgesic, antipyretic, antimicrobial, antioxidant, and hypoglycaemic properties (El-Far et al., 2018). Within normal tissues, thymoquinone acts as a vigorous antioxidant reagent, downregulating lipid peroxidation and synthesizing superoxide radicals.

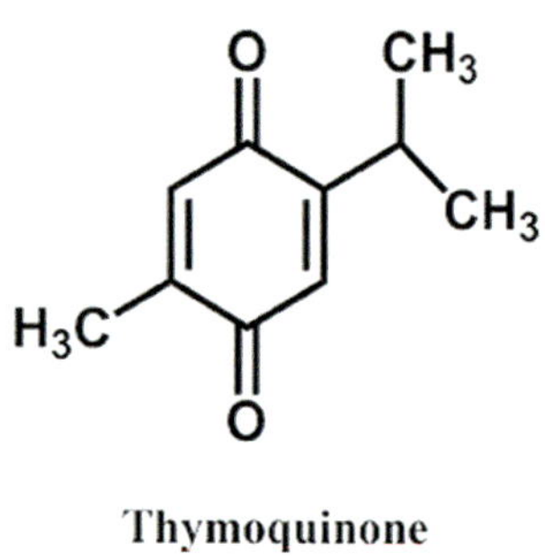

FIGURE 11.1 Chemical structure of thymoquinone.

Moreover, it upregulates the activities of numerous antioxidant enzymes like glutathione *S*-transferase, catalase, reduced glutathione, superoxide dismutase, and quinone reductase (Rathore et al., 2020).

The anticancer property of the phytomolecule is eminently known against human carcinomas with negligible toxicity in normal cells. It has been proven in various *in vitro* cell lines, such as breast (MDA-MB-231, MCF-7), colorectal (LoVo), hepatic (HepG2), cervical cancer (HeLa), glioma (U-87), colon (HCT-116), etc. (Khan et al., 2017). The antitumour property of thymoquinone is attributed to its capability to target several molecular pathways and cellular mechanisms like cell proliferation, apoptosis induction, DNA synthesis, reactive oxygen species production, angiogenesis prevention, immunomodulation, and cellular metastasis (Khan et al., 2017).

Even though thymoquinone shows promising antineoplastic properties, its clinical application is very much limited owing to its poor bioavailability, low solubility, high hydrophobicity, poor stability in body fluids, and sensitivity to pH and light (Rathore et al., 2020). Further, its capability to bind to plasma proteins prevents it from reaching the targeted tumour sites (El-Far et al., 2018). Nanotechnology is an emerging interdisciplinary field that is pivotal in cancer treatment. Cancer nanotechnology is evolving as unique multidisciplinary research, interrelating the disciplines of engineering, chemistry, biology, and medicine. It is predicted to have a substantial role in cancer detection and diagnosis (Padhi et al., 2018, Behera and Padhi, 2022). The medical application of nanotechnology (Nanomedicine) has a promising potential for transforming cancer therapeutics by designing and constructing effective biocompatible drug delivery systems (Misra et al., 2010).

Consequently, research on thymoquinone-nanoformulation could serve as a new dimension for overcoming the limitations of free thymoquinone mentioned above. Nanoenabled delivery of thymoquinone improves its bioavailability and targeting efficacy, protects it from nonspecific binding, and reduces its renal clearance, thereby promoting its prolonged pharmacological effect (Ballout et al., 2018). Further, nanoparticles can effortlessly penetrate the biomembranes, facilitating the sustained release

DOI: 10.1201/9781003611516-14

of thymoquinone in the targeted sites and thereby improving its biodistribution. Moreover, the site-specific targeting efficacy of these nanoparticles could be enhanced by modifying them with various ligands specific to numerous hallmarked surface receptors of cancer cells, which are generally absent in normal cells. Many thymoquinone nanoformulations have been studied extensively against colon, cervical, lung, and breast cancer, including polymeric, liposomal, nanostructured lipid carriers, etc. Thus, this book chapter summarises thymoquinone nanoformulations, their properties, and therapeutic efficacy against different cancers.

11.2 MECHANISMS OF ACTION OF THYMOQUINONE AS AN ANTICANCER AGENT

The anticancer activity of the multifunctional thymoquinone is attributed to various underlying mechanisms such as preferential oxidant-antioxidant activity, alteration in DNA configuration and synthesis, interference with the carcinogenic signalling pathway, epigenetic modulation, and immunomodulation (Figure 11.2).

11.2.1 Antioxidant Activity of Thymoquinone

Studies in numerous cancer models, like hepatic, colon, and prostate cancers, have shown that the antioxidant property of thymoquinone is responsible for its chemotherapeutic activities. Another theory suggested that at a lower concentration, thymoquinone exhibits antioxidant activity, whereas at a higher concentration, it imparts pro-oxidant properties (Zubair et al., 2013). Also, thymoquinone stimulates the activation of several cytoprotective antioxidant enzymes like catalase, superoxide dismutase, glutathione, reduced glutathione, peroxidases, glutathione-*S*-transferases, and glutathione reductase, which thereby counteract the effects of oxidative stress-mediated malignancies (Jrah Harzallah et al., 2012; Woo et al., 2012).

11.2.2 Thymoquinone Hampers DNA Structural Conformation and Its Synthesis

Thymoquinone alters the DNA structure and hinders its synthesis. As reported in an earlier study, thymoquinone targets the intracellular copper closely associated with guanine in chromatin, thereby leading to reactive oxygen species-induced nucleic acid damage and cancer cell death (Zubair et al., 2013). Apart from this, thymoquinone was also found to obstruct the synthesis of DNA, viability, and propagation of several cancer cells, such as prostate cancer (DU145, PC-3, and LNCaP) cells, but not non-cancerous prostate epithelial cells (BPH-1) (Kaseb et al., 2007).

11.2.3 Thymoquinone Targets Cancer Signalling Pathways

Thymoquinone targets various carcinogenic signalling pathway and their signalling molecule. One of the major targets of thymoquinone is the signal transducer and

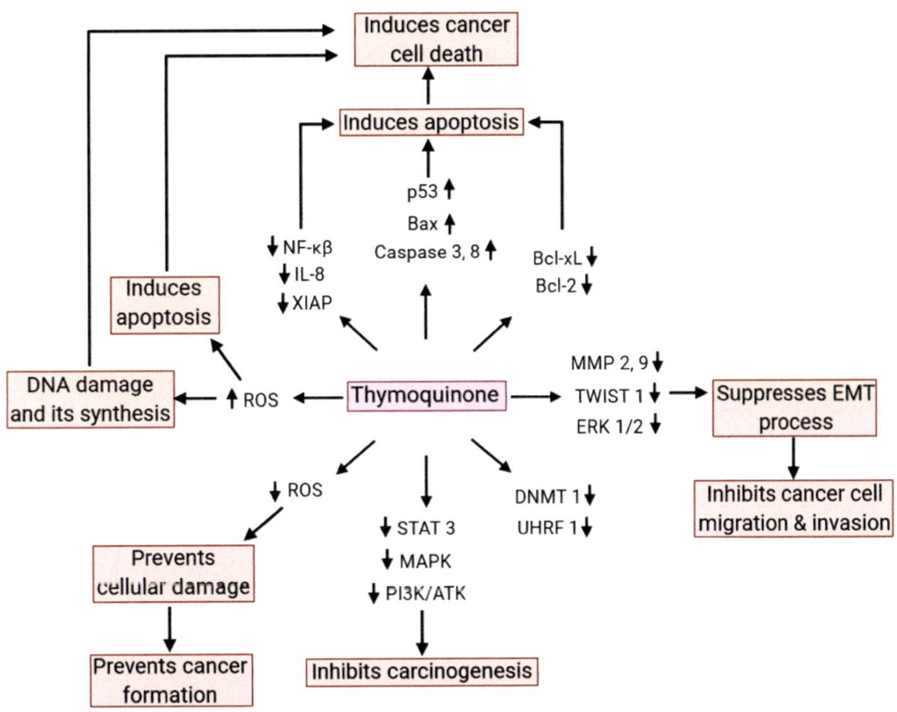

FIGURE 11.2 Anticancer mechanisms of thymoquinone [The anticancer activity of thymoquinone includes induction of apoptosis or cancer cell death, suppression of epithelial-mesenchymal transition (EMT), inhibition of metastasis by inhibiting cancer cell migration and invasion, control of reactive oxygen species (ROS) which inhibits the DNA synthesis and controls the cancer cell growth, and inhibition of carcinogenesis by maintaining different signalling systems].

activator of the transcription-3 pathway. Another research reported that thymoquinone inhibits both the constitutive and interleukin-6-inducible signal transducer and activator of transcription-3 phosphorylation by blocking the activation of Janus kinase 2 and cellular sarcoma kinase. Apart from this, thymoquinone also downregulates signal transducer and activator of the transcription-3-regulated proteins, such as B-cell lymphoma-extra large, B-cell lymphoma 2, cyclin D1, vascular endothelial growth factor, survivin, and myeloid cell leukaemia sequence 1, resulting in enhanced apoptosis and inhibition of cancer cell growth (Li et al., 2010). Notably, certain studies have reported that thymoquinone alters the phosphoinositide-3-kinase–protein kinase B signalling pathway in breast cancer cell lines (T47D and MDA-MB-468), thereby resulting in G-1 arrest and increased apoptosis (Rajput et al., 2013). Interestingly, research carried out in human colon cancer cell line (HCT116) suggests that thymoquinone facilitates apoptosis, which is correlated with increased expression of B-cell lymphoma 2-associated protein x, decreased expression of B-cell lymphoma 2, the stimulation of caspases-3, -7, and -9, and the cleavage of poly-(ADP-ribose) polymerase (Kundu et al., 2014).

11.2.4 Immunomodulatory Effects of Thymoquinone

Another crucial mechanism of thymoquinone is its immunomodulatory activity. Numerous reports suggest that thymoquinone inhibits the activation of tumour necrosis factor-α-stimulated nuclear factor kappa-light-chain- enhancer of activated B cells. Subsequently, it also downregulates the expression of nuclear factor kappa-light-chain-enhancer of activated B cells regulated antiapoptotic proteins, like B-cell lymphoma-extra large, B-cell lymphoma 2, X-linked inhibitor of apoptosis protein, inhibitor of apoptosis proteins-1, inhibitor of apoptosis proteins-2; angiogenic proteins like vascular endothelial growth factor and matrix metalloproteinase-9; and surviving proliferative proteins, like cyclin D1, cyclooxygenase-2, and cellular myelocytomatosis oncogene (Sethi et al., 2008). Another study states that when hepatic carcinoma cells (HepG2) undergo thymoquinone treatment, it showed enhanced expression of proapoptotic B-cell lymphoma-extra small and tumour-necrosis factor-related apoptosis-inducing ligand death receptors, suppression of interleukin-8, and nuclear factor kappa B, reduced expression of antiapoptotic gene B-cell lymphoma 2, thereby inducing apoptosis (Ashour et al., 2014). Also, thymoquinone treatment in cholangiocarcinoma cell lines (HuCCT1 and TFK-1) suppresses the expression of phosphoinositide-3-kinase–protein kinase B and nuclear factor kappa B and other related proteins, such as vascular endothelial growth factor, cyclooxygenase-2, X-linked inhibitor of apoptosis protein, B-cell lymphoma 2, p65, and phosphorylated - protein kinase B (Xu et al., 2014). Consequently, exposure of colorectal cancer cells to thymoquinone alters the conformation and scaffold function of the P21-activated kinases-1 oncogene. It changes the rapidly accelerated fibrosarcoma/mitogen-activated protein kinase/extracellular signal-regulated kinase 1 and 2 pathway (El-Baba et al., 2014).

11.2.5 Effects of Thymoquinone on Metastasis

Thymoquinone treatment is also known to cause a significant downregulation of focal adhesion kinase, which is correlated with the suppression of matrix metalloproteinase-2 and matrix metalloproteinase-9 secretion and extracellular signal-regulated kinase phosphorylation, thereby suppressing cell migration and invasion (Kolli-Bouhafs et al., 2012). Numerous evidence suggest that the application of thymoquinone to breast cancer cell lines suppresses the twist-related protein 1 promoter activity and its expression, thereby leading to the suppression of epithelial-mesenchymal transition-mediated cancer cell invasion, migration, and metastasis (Khan et al., 2015). Thymoquinone also hampers the mammalian target of rapamycin activity and phosphoinositide-3-kinase–protein kinase B signalling pathway in bladder cancer (Iskender et al., 2016a, b).

11.2.6 Epigenetic Modulation by Thymoquinone

Lastly, thymoquinone is known to mediate its anticancer effects *via* the epigenetic modulation of certain specific genes. Accordingly, thymoquinone induces p73 expression, which thereby downregulates the expression of the antiapoptotic epigenetic modulator known as ubiquitin-like-containing PHD and ring finger domains (UHRF1) and its DNA methyltransferase 1 (Alhosin et al., 2010; Abusnina et al., 2011). And since the UHRF1-DNA methyltransferase 1-histone deacetylase complex plays a significant role in epigenetic regulation, the influence of thymoquinone on UHRF1 and p73 provides an epigenetic mechanism through which thymoquinone modulates the regulation of various pro- and antiapoptotic proteins (Qadi et al., 2019). Interestingly, it has been studied that the treatment of Jurkat cells with thymoquinone leads to the upregulation of several tumour suppressor genes (DLC1, ST7, FOXO6, PPARG, CYP1B1, SALL4, TET2, and DDIT3) along with numerous proapoptotic genes (GNG3, RASL11B, BIK, RASD1, and BAD) *via* epigenetic modulation (Qadi et al., 2019).

11.3 NANOENABLED TARGETING STRATEGIES FOR THYMOQUINONE

Even though thymoquinone is a potentially promising candidate for anticancer therapy, its clinical application is restricted due to its poor physicochemical properties, including low solubility, hydrophobicity, and bioavailability. Further, its sensitivity to light and pH limits its access to tumour sites, thereby obstructing its clinical translation (Ballout et al., 2018). Nanotechnology plays a substantial role in overcoming the abovementioned limitations, encouraging the clinical application of thymoquinone. Encapsulation of thymoquinone in various nanoformulations could aid its effective delivery, bioavailability, targeting efficiency and prevent its nonspecific binding (Rathore et al., 2020). Targeting tumour tissue via drug-encapsulated nanoparticles can be achieved by passive and active targeting.

11.3.1 Passive Targeting Strategies

In passive targeting, nanoparticles exploit the enhanced permeability and retention effect and utilize the leaky vasculature for its preferential accumulation in the tumour microhabitat (Figure 11.3). The enhanced permeability and retention effect is a distinctive phenomenon of solid tumours attributed to their abnormal anatomy and physiology absent in normal tissues (Padhi and Behera, 2022; Behera et al., 2023; Patnaik et al., 2021). For instance, high vascular density due to enhanced angiogenesis, preferential extravasation, retention of nanoparticles, and large fenestration ranging from 300 to 400 nm in size between endothelial cells in neoplastic blood vessels. Moreover, the highly proliferating nature of cancer cells causes dysfunctional lymphangiogenesis and compacted lymphatic vessels, thereby leading to compromised lymphatic drainage. To summarise, the enhanced permeability of cancer vasculature facilitates the entry of nanomedicines into the interstitial space of cancer cells. In contrast, the impaired lymphatic filtration facilitates its retention inside, promoting effective drug delivery into tumour sites (Fang et al., 2011; Nehoff et al., 2014).

Accordingly, an effective nano-drug delivery system comprising poly lactic-co-glycolic acid nanoparticles was formulated by Soni et al. (2015) for the co-delivery of paclitaxel-thymoquinone in breast cancer cell line (MCF-7). They showed that the designed formulation showed enhanced efficacy compared to free drugs alone. Similarly, Ganea et al. (2010) studied that thymoquinone-encapsulated poly lactic-co-glycolic acid nanoparticles exhibited excellent antioxidant and antitumour properties as compared to free thymoquinone in breast cancer cells (MDA-MB-231). Moreover, Ibrahim et al. (2020) observed that thymoquinone-encapsulated poly lactic-co-glycolic acid nanoparticles showed a sustained release and enhanced accumulation due to the enhanced permeability and retention effect in melanoma cancer cells (A375).

In another research, Fathy et al. (2020) fabricated TQ loaded iron oxide nanocarriers for combined chemo-photothermal chemotherapy. The synthesized nanocarrier enhanced the bioavailability of the loaded thymoquinone when treated against highly aggressive breast cancer cell lines (MDA-MB-231). Interestingly, when thymoquinone was co-delivered with doxorubicin *via* cockle shell-derived aragonite calcium carbonate nanoparticles, it resulted in enhanced apoptosis, reduced cell migration, and invasion in breast cancer cell line, as studied by Ibiyeye and coworkers (2019). Further, Sunoqrot et al. (2020) fabricated thymoquinone-encapsulated methoxy poly(ethylene glycol)-b-poly(caprolactone) copolymer nanoparticles for the effective delivery of thymoquinone against breast (MCF-7), pancreatic (PANC-1), and colon (Caco-2) cancer cell lines. Their results revealed that the formulation increased the bioavailability of thymoquinone compared to free thymoquinone.

Kumar et al. (2020) synthesized thymoquinone- encapsulated porous polyvinylpyrrolidone conjugated iron oxide nanoparticles to effectively deliver thymoquinone against triple-negative breast cancer cells (MDA-MB-231). It was observed that the synthesized thymoquinone- polyvinylpyrrolidone - iron oxide nanoparticles induced apoptosis-related

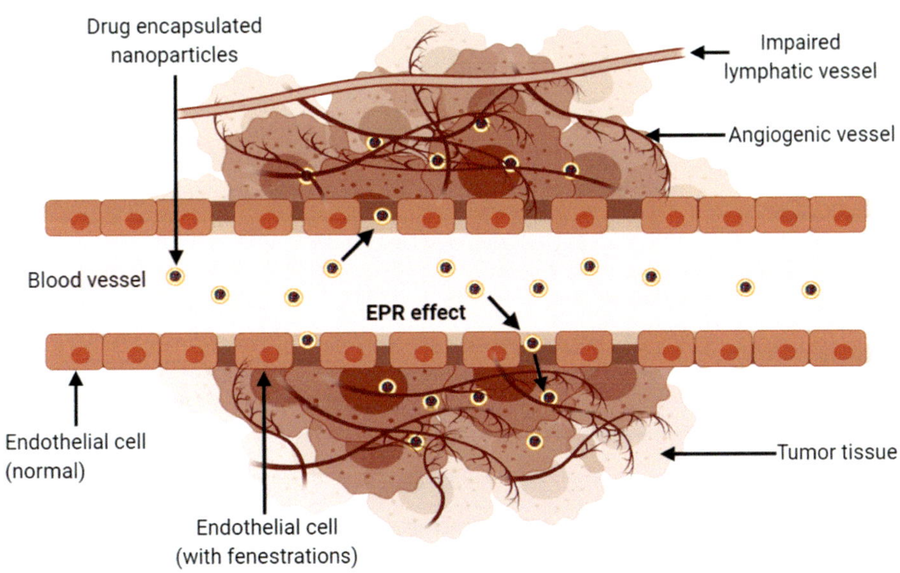

FIGURE 11.3 Schematic illustration of passive targeting into tumour tissue [Nanoencapsulated drugs enter through the leaky vasculature and angiogenic vessels of the cancer cells and exhibit anticancer activity by enhanced permeation and retention (EPR) effect].

cell death in triple-negative breast cancer cells *via* pH-dependent sustained release of thymoquinone. Goel and Mishra (2019) studied that when breast (MCF-7) and cervical cancer (HeLa) cell lines were treated with thymoquinone-loaded mesoporous silica nanoparticles, it exerted its cytotoxic effects *via* the generation of reactive oxygen species, which in turn stimulates apoptosis in tumour cells while inhibiting their metastatic properties.

Ahmad et al. (2020) studied the anticancer properties of thymoquinone-encapsulated poly lactic-co-glycolic acid-poly (ethylene glycol) nanoparticles in breast (MCF-7 and UACC 732) cancer cell lines. They concluded that these particles tend to show selective cytotoxicity towards breast cancer cell subtypes, wherein the UACC 732 cell line was relatively more sensitive toward thymoquinone-nanoformulations than the MCF-7 tamoxifen-resistant cells. Furthermore, Fakhoury et al. also demonstrated that thymoquinone nanoparticles exhibited anticancer activity against breast cancer cell lines (MDA-MB-231 and MCF-7), thereby suggesting a newer approach for the application of thymoquinone as an anticancer molecule (Fakhoury et al., 2016). Notably, Ng et al. (2015) fabricated a thymoquinone-loaded nanostructured lipid carrier and demonstrated its cytotoxicity on breast cancer (MDA-MB-231 and MCF-7) and cervical cancer (HeLa and SiHa) cell lines. They showed that the formulation exhibited antiproliferative activity towards all the cell lines in a dose-dependent manner, which showed higher cytotoxicity towards MDA-MB-231 cells.

Zafar and coworkers (2020) designed and successfully synthesized 1,2-distearoyl-sn-glycero-3-phos phoethanolamine with conjugated methoxyl poly(ethylene glycol)-vitamin E-tocopheryl polyethylene glycol succinate-lipid nanocapsule for co-delivery of docetaxel and thymoquinone in breast cancer cell line (MCF-7 & MDA-MB-231). Their formulated nanoparticles were reported to exhibit sustained drug release and enhanced chemotherapeutic efficacy against the breast cancer cell line. Moreover, Alhakamy et al. (2020) reported that thymoquinone-loaded soy-phospholipid-based phytosomes exhibit antitumour properties against the human lung cancer cell line (A549). Their findings revealed that the formulation showed a prominent increase in dose-dependent cytotoxicity, induced apoptosis, and enhanced cell necrosis *via* activation of caspase-3, reactive oxygen species-mediated stress in the cell line.

In another work, Abu-Dahab et al. (2013) constructed thymoquinone-cyclodextrin nanoparticles and demonstrated their cytotoxic effects against human breast adenocarcinoma cell lines (MCF7). Their report suggested that thymoquinone-loaded cyclodextrin nanoparticles could become an effective approach to improve thymoquinone's solubility, enhance its antiproliferative activity, and lower toxic effects against normal cells. Ong et al. (2018) designed and fabricated thymoquinone-encapsulated nanostructured lipid carriers to deliver thymoquinone in 4T1 tumour-bearing female BALB/c mice. Their study reported that applying nanostructured lipid carriers as a drug delivery vehicle has elevated thymoquinone's therapeutic index,

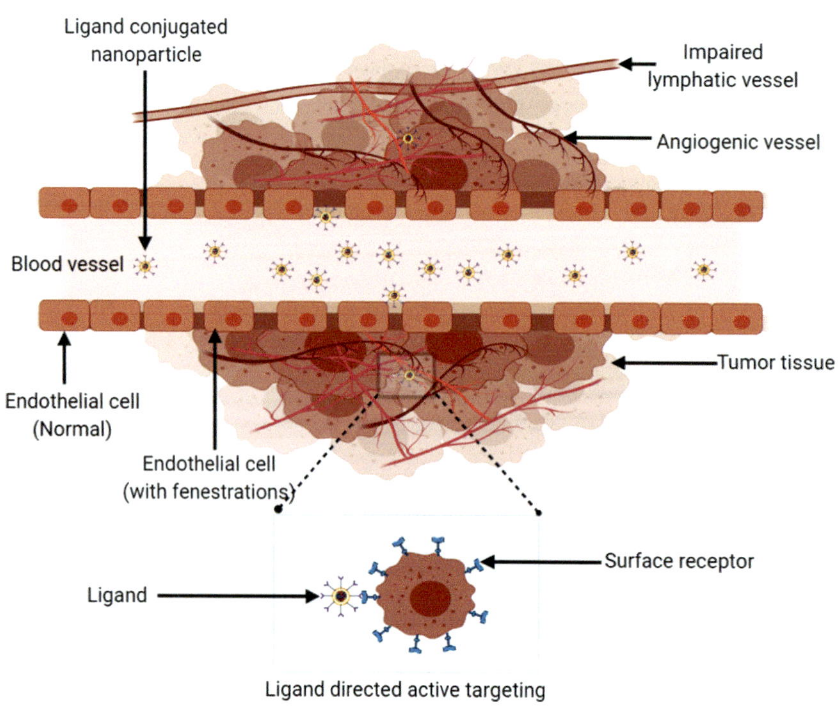

FIGURE 11.4 Schematic illustration of active targeting into tumour tissue [In active targeting, the drug moieties are conjugated with ligands that are specific to the overexpressed receptors on cancer cells so that the nanoformulations can specifically bind to the surface receptors and internalize the drugs into the cancer cells by receptor-mediated endocytosis].

thereby improving mice's survival rate. Also, the formulation was reported to modulate Bcl-2 and caspase-8 of the intrinsic apoptotic pathway, thereby inducing apoptosis.

Bhattacharya and his coworkers reported the delivery of thymoquinone *via* PEGylated-thymoquinone-nanoparticles; it exhibited superior targeting efficiency against breast cancer cell lines (MCF-7 and HBL-100). Moreover, the drug-loaded nanoparticles inhibited cell migration and showed less toxicity towards normal cells than free thymoquinone alone (Bhattacharya et al., 2015). In another study, Shahein et al. (2019) synthesized thymoquinone-encapsulated mesoporous silica core-shell nanoformulations for pH-dependent delivery of thymoquinone to glioma cells. The *in-vitro* assays with cancer and normal cells showed that the formulated nanoparticles exhibited improved anticancer efficacy compared to free thymoquinone.

11.3.2 Active Targeting Strategies

Even though the enhanced permeation and retention effect promotes the access of nanoformulations inside the tumour microenvironment, it does not aid in target-specific drug delivery. This limitation of passive targeting could be bridged by modifying the nanocarriers with ligands of tumour-specific receptors (Bahrami et al., 2017; Padhi et al., 2016; Kundu et al., 2022). This approach is termed active targeting (Figure 11.4), which has several merits, such as improved drug efficacy, lesser nonspecific toxicity, enhanced therapeutic index, etc (Kumari et al., 2016). Regarding the active targeting of thymoquinone, Upadhyay et al. (2019) synthesized transferrin decorated thymoquinone-encapsulated polyethylene glycol - poly(lactide-*co*-glycolide) nanoparticles for targeted delivery of thymoquinone against non-small-cell-lung-carcinoma (A549 cell line). The constructed NPs showed enhanced internalization and increased antitumour activity *via* modulation of micro-RNA-34a and micro-RNA-16.

In another approach, Singh et al. (2018) fabricated a novel method for the targeted delivery of thymoquinone. In their study, they used starch to synthesize the thymoquinone-nanoparticle formulation as a core which is subsequently coated with poly (ε-caprolactone)/poly(ethylene glycol) copolymer functionalized with anti-human neurotensin receptor 1 – monoclonal antibody, which facilitated the binding to specific human neurotensin receptor 1 present on non-small cell lung cancer (MRC-5, NCI-H1975, and NCI-H1563). These nanoparticles showed rapid internalization and enhanced anticancer activities *via* upregulation of proapoptotic markers and downregulation of antiapoptotic markers

Notably, Khattabi et al. (2017) formulated folic acid-decorated silica nanoparticles ligated with a long polymer and carboxymethyl-β-cyclodextrin to co-deliver melatonin and thymoquinone. The formulation was more toxic to cervical cancer (HeLa) cells when compared to unconjugated nanoparticles and the free drug, as confirmed by the *in vitro* cell viability assay. Similarly, İnce et al. (2020) synthesized folic acid-decorated chitosan nanoparticles for targeted delivery of thymoquinone to ovarian cancer cells. The folic acid-modified nanoparticles exhibited enhanced and specific internalization of thymoquinone compared to free thymoquinone in ovarian cancer (SKOV3) cell lines. Interestingly, Bhattacharya et al. (2020) prepared hyaluronic acid ligated mixed pluronic P123 and F127 copolymer nanoparticles for target-specific delivery of thymoquinone. Their designed nanoparticles inhibited metastasis and angiogenesis simultaneously in the breast cancer cell line (MDA-MB-231). Hence, site-specific targeted nanoformulations could be a better platform for effectively delivering the potential anticancer molecule, thymoquinone. Nazeer and George (2017) formulated thymoquinone-encapsulated colon-targeting pH-dependent microcapsules for treating colorectal cancer. They demonstrated that there is a very low release of thymoquinone at a lower pH (pH 1.2). In contrast, around neutral pH (pH=6.8 and 7.4), the formulation initially showed burst release and gradually exhibited a constant release rate.

11.4 REPORT ON CLINICAL TRIALS AND PATENTS OF THYMOQUINONE

Though thymoquinone is a potential anticancer phytomolecule, its application in healthcare is restricted due to its hydrophobicity, poor bioavailability, poor membrane permeability, pH- and light-sensitive nature. Due to the above-mentioned pitfalls, there are negligible reported studies on the antitumour chemotherapeutic efficacy of thymoquinone in humans (Odeh et al., 2012). Interestingly, Al-Amri and Bamosa (2009) investigated the potential effectiveness of oral thymoquinone as a chemotherapeutic molecule in cancer patients. Weekly safety assessments were conducted during treatment, and no significant symptoms were demonstrated. Although tumour markers were found to be reduced in these cancer patients, the decrease was not more than 25% of baseline levels. Hence, no anticancer effects were reported by tumour markers or imaging. Therefore, this phase I study concluded that thymoquinone could be safely consumed at a dose ranging from 75 to 2,600 mg/day with barely minimum toxicities or therapeutic side effects being observed. Subsequently, a phase II clinical trial was carried out to assess the impact of thymoquinone on oral cancer lesions, which at present is currently registered by the U.S government; however, the experimental outcome of this study is yet to be declared (https://clinicaltrials.gov). Notably, several research studies have been made to overcome these physicochemical downsides of thymoquinone and its adverse consequences. Accordingly, some evidence suggests that thymoquinone, in combination with other chemotherapeutic drugs, can impart enhanced chemotherapeutic activities, which could be a potentially effective approach for future clinical investigation of thymoquinone.

11.5 FUTURE PERSPECTIVES

Thymoquinone is a well-known natural product with great potential to cure several diseases. Being a medicinal phytochemical, it has grabbed the attention of several researchers due to its easy availability, safety, and efficacy. Thymoquinone has been extensively studied as a promising anticancer candidate in targeting several forms of aggressive cancers by inhibiting mechanisms related to cell cycle regulation, apoptosis, tumour metastasis, and immunomodulation. Despite the beneficial properties of thymoquinone, its clinical application is still very far from reality due to its poor biodistribution, low solubility, sensitivity to light, and pH. The application of nanotechnology as a drug delivery system could bridge the abovementioned pitfalls of thymoquinone. Hence, taking into consideration the multifunctional nature of thymoquinone, various mechanisms involved in its anticancer activity, low dose requirement, synergistic effects with other conventional chemotherapeutics, improved anticancer efficacy of nanoparticles, enhanced biodistribution, thymoquinone-nanoformulations are a relatively more deserving candidate to reach clinical studies as compared to free thymoquinone. Numerous parameters should be fulfilled to facilitate their clinical transition, including minimum side effects, effective size range, and so forth. Apart from these, developing an efficient yet sustainable process for the large-scale synthesis of nanoparticles is of prime importance. Subsequently, Human trials should be conducted to evaluate these formulations' toxicological effects and establish their superiority over free thymoquinone. Also, it is highly essential to assess the synergistic effects and efficacy of these formulations with conventional chemotherapeutics to attain better-grade modalities to treat cancer with minimum side effects.

11.6 CONCLUSION

In previous years, numerous pharmaceutical studies on the multifunctional thymoquinone have claimed its promising chemotherapeutic potential against multiple cancer cells. However, it has not reached clinical application due to its poor physicochemical properties, low solubility, low bioavailability, and sensitivity to pH and light. A wide array of novel thymoquinone-nanoformulations has been designed and developed to overcome these hurdles. This novel drug delivery system enhances thymoquinone's therapeutic efficacy and decreases its required dosage. Even though thymoquinone nanoparticles have superior efficiency relative to the free drug, thymoquinone-encapsulated nanoformulations have not been clinically tried yet. Hence, to encourage their clinical transition, human trials should be conducted to evaluate the toxicity of these nanoformulations and thereby determine their chemotherapeutic efficacy. Also, it is crucial to assess the effectiveness of these formulations in combination with conventional chemotherapy to attain an effective approach to treating cancer with minimum toxicity. Nevertheless, this nanocarrier-based drug delivery system could be an imminent approach to turning the clinical application of thymoquinone into reality.

REFERENCES

Abu-Dahab, R., Odeh, F., Ismail, S. I., Azzam, H., & Al Bawab, A. (2013). Preparation, characterization and antiproliferative activity of thymoquinone-β-cyclodextrin self assembling nanoparticles. *Pharmazie*, 68(12), 939–944.

Abusnina, A., Alhosin, M., Keravis, T., Muller, C. D., Fuhrmann, G., Bronner, C., & Lugnier, C. (2011). Down-regulation of cyclic nucleotide phosphodiesterase PDE1A is the key event of p73 and UHRF1 deregulation in thymoquinone-induced acute lymphoblastic leukemia cell apoptosis. *Cellular Signalling*, 23(1), 152–160.

Ahmad, A., Husain, A., Mujeeb, M., Khan, S. A., Najmi, A. K., Siddique, N. A., … & Anwar, F. (2013). A review on therapeutic potential of Nigella sativa: A miracle herb. *Asian Pacific Journal of Tropical Biomedicine*, 3(5), 337–352.

Ahmad, A., Mishra, R. K., Vyawahare, A., Kumar, A., Rehman, M. U., Qamar, W., … & Khan, R. (2019). Thymoquinone (2-Isopropyl-5-methyl-1, 4-benzoquinone) as a chemopreventive/anticancer agent: Chemistry and biological effects. *Saudi Pharmaceutical Journal*, 27(8), 1113–1126.

Ahmad, R., Kaus, N. H. M., & Hamid, S. (2020). Synthesis and characterisation of PLGA-PEG thymoquinone nanoparticles and its cytotoxicity effects in tamoxifen-resistant breast cancer cells. *Advances in Experimental Medicine and Biology*, 1292, 65–82.

Al-Amri, A. M., & Bamosa, A. O. (2009). Phase I safety and clinical activity study of thymoquinone in patients with advanced refractory malignant disease. *Shiraz E-Medical Journal*, 10(3), 107–11.

Alhakamy, N. A., Badr-Eldin, S. M., Fahmy, U. A., Alruwaili, N. K., Awan, Z. A., Caruso, G., Alfaleh, M. A., Alaofi, A. L., Arif, F. O., Ahmed, O. A. A., & Alghaith, A. F. (2020). Thymoquinone-loaded soy-phospholipid-based phytosomes exhibit anticancer potential against human lung cancer cells. *Pharmaceutics*, 12(8), 1–17.

Alhosin, M., Abusnina, A., Achour, M., Sharif, T., Muller, C., Peluso, J., … & Fuhrmann, G. (2010). Induction of apoptosis by thymoquinone in lymphoblastic leukemia Jurkat cells is mediated by a p73-dependent pathway which targets the epigenetic integrator UHRF1. *Biochemical Pharmacology*, 79(9), 1251–1260.

Ashour, A. E., Abd-Allah, A. R., Korashy, H. M., Attia, S. M., Alzahrani, A. Z., Saquib, Q., … & Rishi, A. K. (2014). Thymoquinone suppression of the human hepatocellular carcinoma cell growth involves inhibition of IL-8 expression, elevated levels of TRAIL receptors, oxidative stress and apoptosis. *Molecular and Cellular Biochemistry*, 389, 85–98.

Bahrami, B., Hojjat-Farsangi, M., Mohammadi, H., Anvari, E., Ghalamfarsa, G., Yousefi, M., & Jadidi-Niaragh, F. (2017). Nanoparticles and targeted drug delivery in cancer therapy. *Immunology Letters*, 190, 64–83.

Ballout, F., Habli, Z., Rahal, O. N., Fatfat, M., & Gali-Muhtasib, H. (2018). Thymoquinone-based nanotechnology for cancer therapy: Promises and challenges. *Drug Discovery Today*, 23(5), 1089–1098.

Behera, A., & Padhi, S. (2022). pH-Sensitive polymeric nanoparticles for cancer treatment. In: Padhi, S., Behera, A., & Lichtfouse, E. (Eds.) *Polymeric Nanoparticles for the Treatment of Solid Tumours*. Environmental Chemistry for a Sustainable World, vol 71. Springer, Cham, pp. 401–425.

Behera, A., Padhi, S., & Nayak, A. K. (2023). Engineered liposomes as drug delivery and imaging agents. In: Ray, S., & Nayak, A. K. (Eds.) *Design and Applications of Theranostic Nanomedicines*. Woodhead Publishing, Cambridge, MA, pp. 75–108.

Bhattacharya, S., Ahir, M., Patra, P., Mukherjee, S., Ghosh, S., Mazumdar, M., Chattopadhyay, S., Das, T., Chattopadhyay, D., & Adhikary, A. (2015). PEGylated- thymoquinone-nanoparticle mediated retardation of breast cancer cell migration by deregulation of cytoskeletal actin polymerization through miR-34a. *Biomaterials*, 51, 91–107.

Bhattacharya, S., Ghosh, A., Maiti, S., Ahir, M., Debnath, G. H., Gupta, P., Bhattacharjee, M., Ghosh, S., Chattopadhyay, S., Mukherjee, P., & Adhikary, A. (2020). Delivery of thymoquinone through hyaluronic acid-decorated mixed Pluronic® nanoparticles to attenuate angiogenesis and metastasis of triple-negative breast cancer. *Journal of Controlled Release*, 322, 357–374.

El-Baba, C., Mahadevan, V., Fahlbusch, F. B., Rau, T. T., Gali-Muhtasib, H., & Schneider-Stock, R. (2014). Thymoquinone-induced conformational changes of PAK1 interrupt prosurvival MEK-ERK signaling in colorectal cancer. *Molecular Cancer*, 13, 1–14.

El-Far, A. H., Al Jaouni, S. K., Li, W., & Mousa, S. A. (2018). Protective roles of thymoquinone nanoformulations: Potential nanonutraceuticals in human diseases. *Nutrients*, 10(10), 1369.

Fakhoury, I., Saad, W., Bouhadir, K., Nygren, P., Schneider-Stock, R., & Gali-Muhtasib, H. (2016). Uptake, delivery, and anticancer activity of thymoquinone nanoparticles in breast cancer cells. *Journal of Nanoparticle Research*, 18(7), 1–16.

Fang, J., Nakamura, H., & Maeda, H. (2011). The EPR effect: Unique features of tumor blood vessels for drug delivery, factors involved, and limitations and augmentation of the effect. *Advanced Drug Delivery Reviews*, 63(3), 136–151.

Fathy, M. M. (2020). Multifunctional thymoquinone-capped iron oxide nanoparticles for combined chemo-photothermal therapy of cancer. *Journal of Superconductivity and Novel Magnetism*, 33(7), 2125–2131.

Ganea, G. M., Fakayode, S. O., Losso, J. N., Van Nostrum, C. F., Sabliov, C. M., & Warner, I. M. (2010). Delivery of phytochemical thymoquinone using molecular micelle modified poly (D,L lactide-co-glycolide)(PLGA) nanoparticles. *Nanotechnology*, 21(28), 285104.

Goel, S., & Mishra, P. (2019). Thymoquinone loaded mesoporous silica nanoparticles retard cell invasion and enhance in vitro cytotoxicity due to ROS mediated apoptosis in HeLa and MCF-7 cell lines. *Materials Science and Engineering C*, 104, 109881.

Ibiyeye, K. M., Nordin, N., Ajat, M., & Zuki, A. B. Z. (2019). Ultrastructural changes and antitumour effects of doxorubicin/thymoquinone-loaded $CaCO_3$ nanoparticles on breast cancer cell line. *Frontiers in Oncology*, 9, 1–14.

Ibrahim, W. N., Rosli, L. M. B. M., & Doolaanea, A. A. (2020). Formulation, cellular uptake and cytotoxicity of thymoquinone-loaded PLGA nanoparticles in malignant melanoma cancer cells. *International Journal of Nanomedicine*, 15, 8059–8074.

İnce, İ., Yıldırım, Y., Güler, G., Medine, E. İ., Ballıca, G., Kuşdemir, B. C., & Göker, E. (2020). Synthesis and characterization of folic acid-chitosan nanoparticles loaded with thymoquinone to target ovarian cancer cells. *Journal of Radioanalytical and Nuclear Chemistry*, 324(1), 71–85.

Iskender, B., Izgi, K., & Canatan, H. (2016a). Novel anticancer agent myrtucommulone-A and thymoquinone abrogate epithelial–mesenchymal transition in cancer cells mainly through the inhibition of PI3K/AKT signalling axis. *Molecular and Cellular Biochemistry*, 416, 71–84.

Iskender, B., Izgi, K., Hizar, E., Jauch, J., Arslanhan, A., Yuksek, E. H., & Canatan, H. (2016b). Inhibition of epithelial-mesenchymal transition in bladder cancer cells via modulation of mTOR signalling. *Tumour Biology*, 37, 8281–8291.

Jrah Harzallah, H., Grayaa, R., Kharoubi, W., Maaloul, A., Hammami, M., & Mahjoub, T. (2012). Thymoquinone, the Nigella sativa bioactive compound, prevents circulatory oxidative stress caused by 1,2-dimethylhydrazine in erythrocyte during colon postinitiation carcinogenesis. *Oxidative Medicine and Cellular Longevity*, 2012, 854065.

Kaseb, A. O., Chinnakannu, K., Chen, D., Sivanandam, A., Tejwani, S., Menon, M., … & Reddy, G. P. V. (2007). Androgen receptor–and E2F-1–targeted thymoquinone therapy for hormone-refractory prostate cancer. *Cancer Research*, 67(16), 7782–7788.

Khan, M. A., Tania, M., Fu, S., & Fu, J. (2017). Thymoquinone, as an anticancer molecule: From basic research to clinical investigation. *Oncotarget*, 8(31), 51907–51919.

Khan, M. A., Tania, M., Wei, C., Mei, Z., Fu, S., Cheng, J., … & Fu, J. (2015). Thymoquinone inhibits cancer metastasis by downregulating TWIST1 expression to reduce epithelial to mesenchymal transition. *Oncotarget*, 6(23), 19580.

Khattabi, A. M., Talib, W. H., & Alqdeimat, D. A. (2017). A targeted drug delivery system of anticancer agents based on folic acid-cyclodextrin-long polymer functionalized silica nanoparticles. *Journal of Drug Delivery Science and Technology*, 41, 367–374.

Kolli-Bouhafs, K., Boukhari, A., Abusnina, A., Velot, E., Gies, J. P., Lugnier, C., & Rondé, P. (2012). Thymoquinone reduces migration and invasion of human glioblastoma cells associated with FAK, MMP-2 and MMP-9 down-regulation. *Investigational New Drugs*, 30, 2121–2131.

Kumar, S. R., Thangam, R., Vivek, R., Srinivasan, S., & Ponpandian, N. (2020). Synergetic effects of thymoquinone-loaded porous PVPylated Fe_3O_4 nanostructures for efficient pH-dependent drug release and anticancer potential against triple-negative cancer cells. *Nanoscale Advances*, *2*(8), 3209–3221.

Kumari, P., Ghosh, B., & Biswas, S. (2016). Nanocarriers for cancer-targeted drug delivery. *Journal of Drug Targeting*, 24(3), 179–191.

Kundu, A., Padhi, S., Behera, A., Hasnain, M. S., & Nayak, A. K. (2022). Tumour targeting strategies by chitosan-based nanocarriers. In: Hasnain, M. S., Beg, S., & Nayak, A. K. (Eds.) *Chitosan in Biomedical Applications*. Academic Press, London, pp. 163–188.

Kundu, J., Choi, B. Y., Jeong, C. H., Kundu, J. K., & Chun, K. S. (2014). Thymoquinone induces apoptosis in human colon cancer HCT116 cells through inactivation of STAT3 by blocking JAK2-and Src-mediated phosphorylation of EGF receptor tyrosine kinase. *Oncology Reports*, 32(2), 821–828.

Li, F., Rajendran, P., & Sethi, G. (2010). Thymoquinone inhibits proliferation, induces apoptosis and chemosensitizes human multiple myeloma cells through suppression of signal transducer and activator of transcription 3 activation pathway. *British Journal of Pharmacology*, 161(3), 541–554.

Misra, R., Acharya, S., & Sahoo, S. K. (2010). Cancer nanotechnology: Application of nanotechnology in cancer therapy. *Drug Discovery Today*, 15(19–20), 842–850.

Nazeer, A., & George, N. (2017). Formulation and evaluation of colon targeted pH dependent microcapsules of thymoquinone for colorectal cancer. *Saudi Journal of Medical and Pharmaceutical Sciences*, 4929, 1168–1179.

Nehoff, H., Parayath, N. N., Domanovitch, L., Taurin, S., & Greish, K. (2014). Nanomedicine for drug targeting: Strategies beyond the enhanced permeability and retention effect. *International Journal of Nanomedicine*, 9, 2539–2555.

Ng, W. K., Saiful Yazan, L., Yap, L. H., Wan Nor Hafiza, W. A. G., How, C. W., & Abdullah, R. (2015). Thymoquinone-loaded nanostructured lipid carrier exhibited cytotoxicity towards breast cancer cell lines (MDA-MB-231 and MCF-7) and cervical cancer cell lines (HeLa and SiHa). *BioMed Research International*, 2015, 263131.

Odeh, F., Ismail, S. I., Abu-Dahab, R., Mahmoud, I. S., & Al Bawab, A. (2012). Thymoquinone in liposomes: A study of loading efficiency and biological activity towards breast cancer. *Drug Delivery*, 19(8), 371–377.

Ong, Y. S., Saiful Yazan, L., Ng, W. K., Abdullah, R., Mustapha, N. M., Sapuan, S., … & Zakarial Ansar, F. H. (2018). Thymoquinone loaded in nanostructured lipid carrier showed enhanced anticancer activity in 4T1 tumor-bearing mice. *Nanomedicine*, 13(13), 1567–1582.

Padhi, S., & Behera, A. (2022). Cellular internalization and toxicity of polymeric nanoparticles. In: Padhi, S., Behera, A., & Lichtfouse, E. (Eds.) *Polymeric Nanoparticles for Treating Solid Tumours*. Environmental Chemistry for a Sustainable World, vol 71. Springer, Cham, pp. 473–488.

Padhi, S., Kapoor, R., Verma, D., Panda, A. K., & Iqbal, Z. (2018). Formulation and optimization of topotecan nanoparticles: In vitro characterization, cytotoxicity, cellular uptake and pharmacokinetic outcomes. *Journal of Photochemistry and Photobiology B: Biology*, 183, 222–232.

Padhi, S., Mirza, M. A., Verma, D., Khuroo, T., Panda, A. K., Talegaonkar, S., Khar, R. K., & Iqbal, Z. (2016). Revisiting the nanoformulation design approach for effective delivery of Topotecan in its stable form: An appraisal of its in vitro behavior and tumour amelioration potential. *Drug Delivery*, 23(8), 2827–2837.

Patnaik, S., Gorain, B., Padhi, S., Choudhury, H., Gabr, G. A., Md, S., Kumar Mishra, D., & Kesharwani, P. (2021). Recent update of toxicity aspects of nanoparticulate systems for drug delivery. *European Journal of Pharmaceutics and Biopharmaceutics*, 161, 100–119.

Qadi, S. A., Hassan, M. A., Sheikh, R. A., Baothman, O. A., Zamzami, M. A., Choudhry, H., … & Alhosin, M. (2019). Thymoquinone-induced reactivation of tumor suppressor genes in cancer cells involves epigenetic mechanisms. *Epigenetics Insights*, 12, 2516865719839011.

Rajput, S., Kumar, B. P., Dey, K. K., Pal, I., Parekh, A., & Mandal, M. (2013). Molecular targeting of Akt by thymoquinone promotes G1 arrest through translation inhibition of cyclin D1 and induces apoptosis in breast cancer cells. *Life Sciences*, 93(21), 783–790.

Rathore, C., Rathbone, M. J., Chellappan, D. K., Tambuwala, M. M., Pinto, T. D. J. A., Dureja, H., … & Negi, P. (2020). Nanocarriers: More than tour de force for thymoquinone. *Expert Opinion on Drug Delivery*, 17(4), 479–494.

Sethi, G., Ahn, K. S., & Aggarwal, B. B. (2008). Targeting nuclear factor-κB activation pathway by thymoquinone: Role in suppression of antiapoptotic gene products and enhancement of apoptosis. *Molecular Cancer Research*, 6(6), 1059–1070.

Shahein, S. A., Aboul-Enein, A. M., Higazy, I. M., Abou-Elella, F., Lojkowski, W., Ahmed, E. R., Mousa, S. A., & Abouaitah, K. (2019). Targeted anticancer potential against glioma cells of thymoquinone delivered by mesoporous silica core-shell nanoformulations with pH-dependent release. *International Journal of Nanomedicine*, 14, 5503–5526.

Singh, S. K., Lillard, J. W., & Singh, R. (2018). Abstract 4665: Neurotensin receptor-based targeted delivery of thymoquinone for non-small cell lung cancer. *Cancer Research*, 78, 4665–4665.

Soni, P., Kaur, J., & Tikoo, K. (2015). Dual drug-loaded paclitaxel–thymoquinone nanoparticles for effective breast cancer therapy. *Journal of Nanoparticle Research*, 17(1), 1–12.

Sunoqrot, S., Alfaraj, M., Hammad, A. M., Kasabri, V., Shalabi, D., Deeb, A. A., Ibrahim, L. H., Shnewer, K., & Yousef, I. (2020). Development of a thymoquinone polymeric anticancer nanomedicine through optimization of polymer molecular weight and nanoparticle architecture. *Pharmaceutics*, 12(9), 1–16.

Upadhyay, P., Sarker, S., Ghosh, A., Gupta, P., Das, S., Ahir, M., … & Adhikary, A. (2019). Transferrin-decorated thymoquinone-loaded PEG-PLGA nanoparticles exhibit anticarcinogenic effect in non-small cell lung carcinoma via the modulation of miR-34a and miR-16. *Biomaterials Science*, 7(10), 4325–4344.

Woo, C. C., Kumar, A. P., Sethi, G., & Tan, K. H. B. (2012). Thymoquinone: Potential cure for inflammatory disorders and cancer. *Biochemical Pharmacology*, 83(4), 443–451.

Xu, D., Ma, Y., Zhao, B., Li, S., Zhang, Y., Pan, S., … & Jiang, H. (2014). Thymoquinone induces G2/M arrest, inactivates PI3K/Akt and nuclear factor-κB pathways in human cholangiocarcinomas both in vitro and in vivo. *Oncology Reports*, 31(5), 2063–2070.

Zafar, S., Akhter, S., Garg, N., Selvapandiyan, A., Kumar Jain, G., & Ahmad, F. J. (2020). Co-encapsulation of docetaxel and thymoquinone in mPEG-DSPE-vitamin E TPGS-lipid nanocapsules for breast cancer therapy: Formulation optimization and implications on cellular and in vivo toxicity. *European Journal of Pharmaceutics and Biopharmaceutics*, 148(December 2019), 10–26.

Zubair, H., Khan, H. Y., Sohail, A., Azim, S., Ullah, M. F., Ahmad, A., … & Hadi, S. M. (2013). Redox cycling of endogenous copper by thymoquinone leads to ROS-mediated DNA breakage and consequent cell death: Putative anticancer mechanism of antioxidants. *Cell Death & Disease*, 4(6), e660–e660.

12 Nanoenabled Delivery Systems for Epigallocatechin Gallate as an Anticancer Agent

Sweta Priyadarshini Pradhan[1], *Santwana Padhi*[2], *Anindita Behera*[3], *and Pratap Kumar Sahu*[1]
[1]School of Pharmaceutical Sciences, Siksha O Anusandhan (Deemed to be University), Bhubaneswar, India
[2]Scientific Communication Manager, Publication Practice Counsel, USA
[3]Amity Institute of Pharmacy, Major Arterial Road, Action Area II Rajarhat, New Town Kolkata, India

12.1 INTRODUCTION

Epigallocatechin gallate, commonly called green tea, is obtained from *Camellia sinesis* and is extensively consumed in East Asia and associated with various health benefits (Afzal et al. 2015; Chowdhury et al. 2016). Firstly, the green tea plant was found in China as a common tea beverage (Graham 1992). Green tea is mainly composed of 26% fibres, 5% vitamins and minerals, 15% proteins, 2%–7% lipids, 1%–2% secondary metabolites as pigments, 30%–40% polyphenols of which 80% contains flavonoids and 3%–4% methylxanthines (Graham 1992; Cabrera et al. 2003, 2006). This composition may also vary according to the geographical location, such as soil, climate, season, agricultural practices, properties of plant-like variety, the position of the harvested shoot, age of leaf, etc. (Cabrera et al. 2006).

They commonly possess antioxidant properties due to the presence of several polyphenolic compounds, such as catechins (Graham 1992; Cirillo et al. 2016; Shikha et al. 2024). Green tea contains different polyphenols and is classified into three groups, i.e., flavanols, flavones, and flavonols (Botten et al. 2015). Flavanols differ from flavonols by lacking the presence of a double bond at the 2 and 3 positions and oxygen molecule at the 4th position in the C-ring (Hollman and Arts 2000). From these three groups, flavanols are the most potent and comprise a majority of elements of green tea with catechins of about one-third of the total dry leaf weight and distributed into four molecules such as (–)-epicatechin, (–)-epicatechin gallate, (–)-epigallocatechin and (–)-epigallocatechin gallate (Figure 12.1) (Graham 1992; Ito et al. 2023). Of these four molecules, epigallocatechin gallate is the most pharmacologically active compound with various therapeutic properties and content and is reported to contain about 65% of catechin (Islam 2012).

Epigallocatechin gallate is chiefly obtained from green tea. Also, it exists in small amounts in different varieties of fruits such as strawberries, cranberries, kiwis, cherries, blackberries, peaches, apples, avocados, pecans, hazelnuts, pistachios, etc. Epigallocatechin gallate has a complex structure of flavanol as a core element with gallocatechol group and gallate ester (Botten et al. 2015). These two gallocatechol groups are responsible for the potent antioxidant and chelating properties of epigallocatechin gallate and act directly by scavenging the free radicals from the environment with higher efficiency (Nanjo et al. 1996; Braicu et al. 2013a,b). Due to the presence of various polyphenols, they have recently been commonly used for the prevention and treatment of cancer and other vascular disorders, neurodegenerative disorder, obesity, diabetes, and other health concerns (Crespy and Williamson 2004; Zaveri 2006; Singh et al. 2015; Kciuk et al. 2023; Wang et al. 2023). It also acts as a powerful anti-inflammatory (Farzaei et al. 2015; Hsu 2015), antioxidant, antibacterial, and antiviral agent and is capable of altering the pathways, thereby changing the metabolism of lipids (Reygaert 2014; Legeay et al. 2015; Yang et al. 2016).

Although epigallocatechin gallate has various potential therapeutic properties, it still has some limitations, such as low stability and poor bioavailability (Dube et al. 2010; Dai et al. 2020). The bioavailability of oral epigallocatechin gallate is relatively poor, and it is reported that only 0.15 μM reaches the plasma concentration following the consumption of two cups of green tea (Natarajan et al. 2019). Some results also proposed that epigallocatechin gallate is prone to oxidative degradation when exposed to high temperatures and neutral or weak alkali conditions, and the decomposition rate is enhanced with the increase in temperature and oxidative concentration. Epigallocatechin gallate is also degraded under intestinal pH, which is less than 10%. In addition, the stability of epigallocatechin gallate is also affected by the effect of intestinal microorganisms *in vitro* and *in vivo* (Wang et al. 2008; Alavi et al. 2023). Epigallocatechin gallate is mainly hydrolysed into epigallocatechin and gallic acid in the presence of intestinal

DOI: 10.1201/9781003611516-15

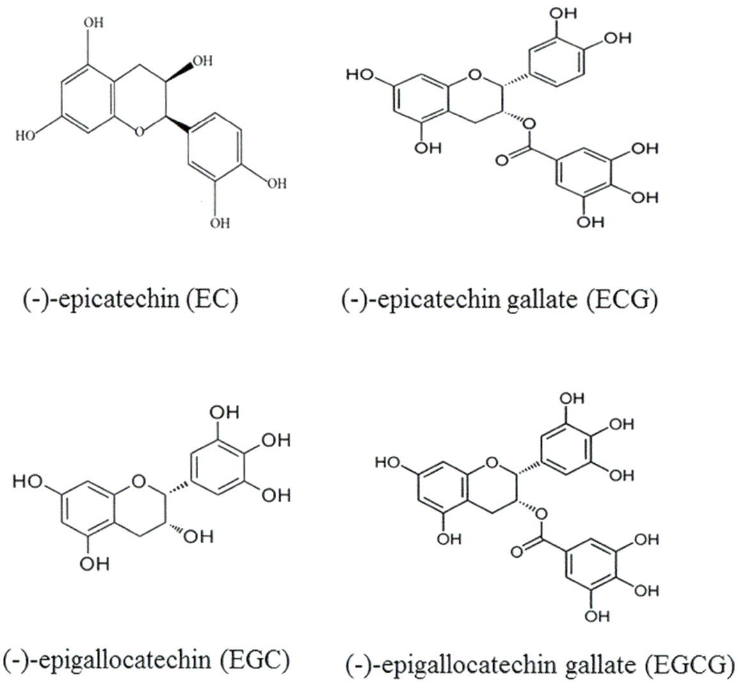

FIGURE 12.1 Different catechins present in green tea.

microorganisms and undergoes intestinal microbial metabolism before its absorption. So, due to its chemical instability, it degrades before the absorption process, which readily affects the rate of bioavailability. The blood-brain barrier permeability of epigallocatechin gallate is also decreased due to its metabolism process due to conversion to 5-(3′,5′-dihydroxy phenyl)-γ-valerolactone, and a lack of hydrophobicity can also be the reason for its lower rate of absorption (Pervin et al. 2017, 2019).

Cancer is mainly defined as an abnormal outgrowth or proliferation of cells, either caused by inheritance or caused by inheritance factors. The appropriate mechanism behind this disease is unknown, which now becomes the greatest burden in society due to increased mortality and morbidity rates (Padhi et al. 2016; Kundu et al. 2022; Sun et al. 2024). However, it is mainly caused due to epigenetic and nucleosome modification, histone production, DNA methylation, etc. (Hassanpour and Dehghani 2017). The optimal use of conventional chemotherapy is widely used to prevent and treat cancer. Still, it is involved with wide adverse events and toxicity due to its administration of overdoses, as safer doses failed to produce the desired pharmacological effect against cancer (Falzone et al. 2018). In recent years, the nanotechnology domain has evolved, garnered wide attention from researchers around the globe (Padhi et al. 2020; Behera et al. 2020a, b; Sahoo et al. 2022). Nanotechnology has several advantages and has been rapidly included in cancer-specific drug delivery systems due to its enhanced bioavailability rate and reduced systemic adverse effects resulting in greater drug efficiency and efficacy (Patra et al. 2018; Behera and Padhi 2022; Padhi et al. 2018). Various nanoformulations, including metal nanoparticles, lipid nanoparticles, polymeric nanoparticles, mesoporous silica nanostructures, and protein nano-assemblies, can serve as the target carriers for delivering epigallocatechin gallate to its site of action by increasing their efficiency, which is highly dependent on the morphological and surface characteristics of nanovehicles (Din et al. 2017).

Moreover, nano derivatives enhance the physiological stability and cellular uptake phenomena of different drug targets by improving their chemical structure, dimensions, surface chemistry, and electrostatics of the drug-loaded nanoparticles (Sutradhar and Amin 2014; Jain and Jain 2016; Padhi and Behera 2022; Behera et al. 2023; Patnaik et al. 2021). This chapter mainly focuses on the applications of epigallocatechin gallate and its health benefits regarding treatment and prevention as an anticancer agent. In addition, the advantages of nanoformulations of epigallocatechin gallate are that they are delivered directly to the target site by passive and active targeting strategies.

12.2 MECHANISMS OF ACTION OF EPIGALLOCATECHIN GALLATE AS AN ANTICANCER DRUG

The mechanism of epigallocatechin gallate against cancer includes inhibition and modulation of biological pathways such as blockade of nuclear factor kappa B, cancer cell migration and invasion factor, autophagy, caspase dependant and independent apoptosis mechanism, lysosomal membrane permeabilization induced cell death, enhancement of pro-antioxidant enzymes like superoxide dismutase, catalase, glutathione transferase thereby resulting the decreased activity of oxidative stress (Figure 12.2). Cancer mainly

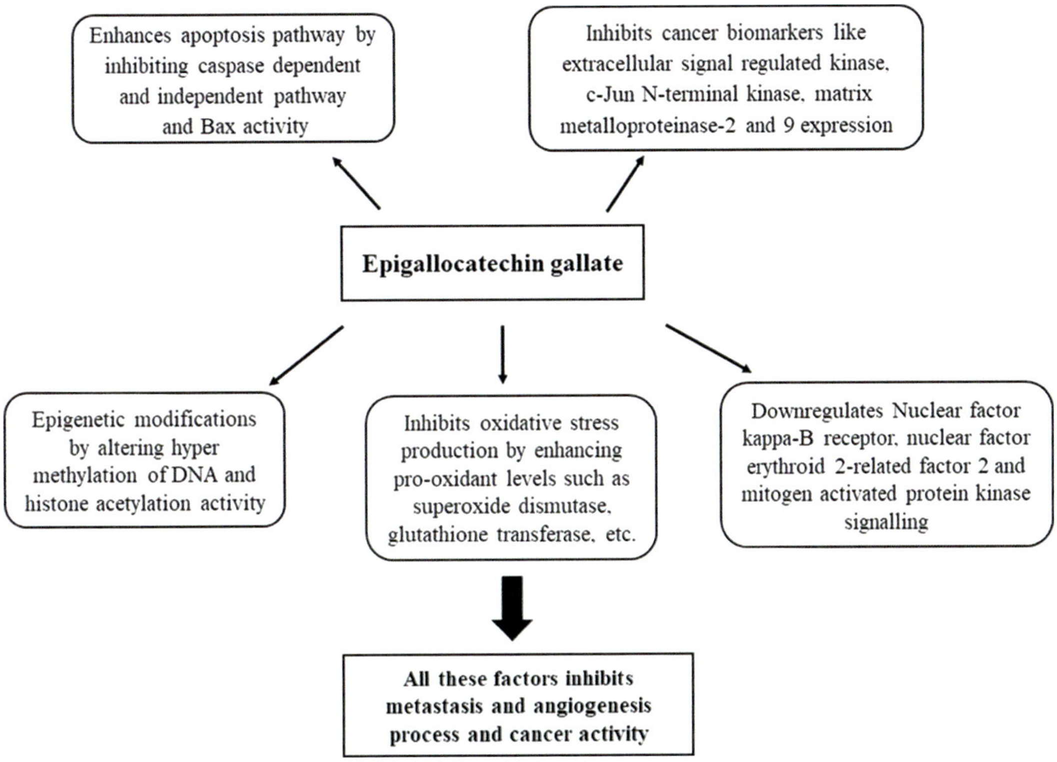

FIGURE 12.2 Schematic representation of different mechanisms of epigallocatechin gallate for inhibition of cancer activity. This figure explains that epigallocatechin gallate enhances the apoptosis pathway, prevents cancer biomarkers, combats oxidative stress pathways, modulates epigenetic pathways, and inhibits metastasis and angiogenesis expressions.

occurs due to the initiation and development of epigenetic alteration in different cells, including DNA acetylation and methylation. Epigallocatechin gallate significantly inhibits the tumorigenic activities of the lungs, prostate, and oral digestive tract. Epigallocatechin gallate hinders the nitrosamine (4-methylnitrosaamino-1,3-pyridyl-1-butanone), which is responsible for inducing lung tumorigenesis by blocking the action of 8-hydroxydeoxyguanosine and formation of antioxidant functions (Xu et al. 1992). It also inhibits dimethyl arsinic acid or cisplatin-induced lung tumourigenesis and diethyl nitrosamine-induced liver tumorigenesis by blocking the insulin-like growth factor signalling in diabetic and obese C57BL/KsJ-db/db mice model (An et al. 2008; Shimizu et al. 2011).

In addition, oral administration of epigallocatechin gallate also decreases the upregulation of the apoptosis mechanism and inhibits the development stage of prostate cancer cells in a xenograft model (Khan et al. 2014). Growth of tumour cells mainly occurs through different mechanisms like the process of angiogenesis as this provides the nutrients to the tumour cells, upregulation of vascular endothelial growth factor, epidermal growth factor receptor, and platelet-derived growth factor receptor factors, which are the different angiogenic factor and readily provides oxygen and nutrients to the tumour cells. Epigallocatechin gallate significantly inhibits cancer growth and angiogenesis development, reduces vascular endothelial growth factor (Jung et al. 2001; Shimizu et al. 2010), epidermal growth factor receptor (Adachi et al. 2009), and epidermal growth factor expression (Ahn et al. 1999) and enhances the cytotoxic T-lymphocyte infiltration into the tumour, which has a direct impact on the proliferation of cancer cells (Braicu et al. 2013a,b).

Epigallocatechin gallate moderates epidermal growth factor receptor signalling in numerous pathways, i.e., by downregulating the autophosphorylation process of epidermal growth factor receptor, initiating epidermal growth factor receptor internalization into endosomes and modulating the membrane lipid organization thereby inhibiting the binding of epidermal growth factor to epidermal growth factor receptor (Liang et al. 1997; Adachi et al. 2009, 2008). Migration and invasion of cancer cells are also the leading factors in cancer development and can be the target for anticancer therapy (Lim et al. 2008). Epigallocatechin gallate obstructs the hepatocyte growth factor, mainly involving the induction of matrix metalloproteinase 9 and stimulation of urokinase-type plasminogen activator to degrade metastasis and invasion in hypopharyngeal tumour cells. Epigallocatechin gallate also decreases the expression of matrix metalloproteinase-2 and 9 by suppressing the activity of hepatocyte growth factor-tyrosine kinase signalling (Koh et al. 2011; Kwak et al. 2011). Epigallocatechin gallate also modifies the small guanosine triphosphate proteins (Rho and Rac) and vasodilator-enthused phosphoprotein expression in squamous cell carcinoma and breast cancer, respectively, which are the signalling factors of the cellular migration and invasion process (Hwang et al. 2013). It also significantly inhibits the invasion of medulloblastoma cells effectively on collagen tissues by activating β1 integrin factor, which effectually controls the proliferation of cancer cells (Pilorget et al. 2003; Kushima et al. 2009).

Apoptosis also plays an important role in the elimination of cancer cells. Epigallocatechin gallate significantly activates apoptosis by downregulating the antiapoptotic family, i.e., B-cell leukaemia/lymphoma-2 protein. It increases the proapoptotic concentration of B-cell leukaemia/lymphoma-2-associated X protein in breast carcinoma cells (Roy et al. 2005), hepatoma cells (Nishikawa et al. 2006), prostate (Hastak et al. 2005) and ovarian carcinoma cells (Chen et al. 2013), nasopharyngeal carcinoma cells and bladder carcinoma cells (Qin et al. 2007). Cancer cell-mediated p53 suppressor cell also acts as a key strategy for diminishing the cycle of tumourigenesis by apoptosis, regulating cell cycle, DNA repair, and checkpoint activation. Epigallocatechin gallate increases the stabilization and activity of p53-mediated apoptosis, thereby stimulating the p53-mediated nonsteroidal anti-inflammatory drugs activated gene-1 expression, which possesses anti-tumourigenic and proapoptotic effects (Kang et al. 2013). Moreover, epigallocatechin gallate also regulates the p53 expression by phosphorylating p53 at the serine residue (Hastak et al. 2003). A study attributes the inhibition of class 1 histone deacetylases, causing acetylation of p53 by epigallocatechin gallate, which increases the p53 transcriptional activity, resulting in the anticancer effect (Thakur et al. 2012).

The generation of reactive oxygen species is a critical factor in the induction of cancer action. Epigallocatechin gallate directly scavenges the activity of reactive oxygen species. The antioxidant activity of epigallocatechin gallate occurs due to a reaction with superoxide (O^{2-}), which activates oxidation and scavenges the free radicals OH and O^{2-} efficiently (Shi et al. 2000; Severino et al. 2009). Moreover, epigallocatechin gallate decreases cell proliferation, induces cell apoptosis, and inhibits the 12-*O*-tetradecanoylphorbol-13-acetate, facilitating oxidative stress (Bhimani et al. 1993). It is responsible for blocking the adhesion and invasion properties of hepatoma cells exhibiting a potent antioxidant effect (Zhang et al. 2000). It also markedly induces nuclear factor erythroid 2-related factor 2 protein expression of antioxidant enzymes, which diminishes tumour cell growth and metastasis (Yuan et al. 2007; Min and Kwon 2014). Nuclear factor kappa B is the regulator of gene expression, which mainly acts in the progression and development of different stages of cancer, i.e., migration, invasion, proliferation, and apoptosis. Epigallocatechin gallate induces the reduction of nuclear factor kappa B factor by decreasing the activity of B-cell leukaemia/lymphoma-2, downregulating the effect of cyclooxygenase 2, inhibiting invasion through diminished matrix metalloproteinase-9 expression, and also blocks migration of various cancer cells (Hastak et al. 2005; Peng et al. 2006; Qin et al. 2012; Rahmani et al. 2015).

Furthermore, epigallocatechin gallate significantly blocks the production of vascular endothelial growth factor resulting in the action of antiproliferative and antiangiogenic activities (Mirossay et al. 2018). Downregulation of extracellular signal-regulated kinase phosphorylation and c-Jun N-terminal kinase inhibits the effect of matrix metalloproteinase-2 and 9, thereby suppressing the cell growth and proliferation and inducing apoptosis, which controls and invades the cancer activity (Kim et al. 2005; Lim and Cha 2011; Khoi et al. 2013). Epigenetic alteration includes hypermethylation of DNA and the histone acetylation process, which changes the genetic functions, leading to malignant cellular production. This also limits the binding of transcriptional factors to the promotors, producing a protein with a silencing factor and a gene mediated by an enzyme called DNA methyltransferase. Epigallocatechin gallate is a powerful inhibitor of DNA methyltransferase, which directly inhibits the interaction or catalytic site of DNA methyltransferase (Lee et al. 2005). It significantly decreases methylation-mediated DNA activity, reduces tumour production, suppresses O6-methyl guanine methyltransferase, downregulates retinoic acid receptor-β and reverses mismatch DNA repair gene, resulting in reduced cell proliferation and differentiation of cell colony (Fang et al. 2003).

Moreover, it also significantly enhances histone acetylation by increasing histone H3 and H4. It inhibits androgen-mediated transcription factors, leading to the tumour suppressive effect and degrading cell growth and colony with enhanced apoptosis (Nandakumar et al. 2011). Epigallocatechin gallate also significantly inhibits androgen receptor-mediated transcription and cell growth, reversing its acetylation process and inhibiting prostate cancer (Lee et al. 2012). It also modulates the Smad signalling process in lung carcinoma (Ko et al. 2013).

12.3 TARGETING STRATEGIES OF EPIGALLOCATECHIN GALLATE AS AN ANTICANCER AGENT

12.3.1 Passive Targeting Strategy of Epigallocatechin Gallate

The nanomaterials-based anticancer drugs are generally administered by intravenous or intratumoral routes due to their better absorption rate than oral administration (Zhao and Castranova 2011). Vascular tissues are generally disrupted at tumour sites, enabling nanocarriers for better adsorption (Maeda et al. 2013). The gap between the tumour vasculature in the endothelial cells ranges from about 200–2,000 nm, depending on the tumour type, localization, and environment. Moreover, due to the deprived lymphatic function, nanoparticles are not hastily cleared and accumulate in the interstitial tumour cells (Jain and Stylianopoulos 2010; Padhi and Behera 2020). This phenomenon is known as the enhanced permeability and retention effect, which is the fundamental principle of passive targeting (Hobbs et al. 1998). This phenomenon of nanoparticle-loaded drug accumulation in passive targeting requires prolonged systemic circulation to reach the site for its drug delivery to elicit the therapeutic action. The nanocarrier accumulation mainly depends on several physicochemical parameters such as size, morphology, surface charge, and surface chemistry (Bertrand et al. 2014; Verma et al. 2017). The extent of nanocarriers should have a smaller particle size than the cut-off proportion in the neovasculature for better permeation. The permeation and bioavailability factor of nanoparticles were also affected by blood perfusions, immunological methods like renal clearance, phagocytosis, and biomolecule interaction (Rosenblum et al. 2018; Padhi et al. 2022a,b; Padhi and Behera 2021).

Chavva et al. (2019) demonstrated a comparative study of epigallocatechin gallate, epigallocatechin gallate gold nanoparticles, and epigallocatechin gallate-citrate stabilized gold nanoparticles for their efficacies against various established cancer cell lines. *In-vitro* studies in four human cancer cell lines, i.e., malignant melanoma (A375SM), human pancreatic cancer (MIA PaCa), prostate cancer (PC3), and resistant breast cancer (MDA-MB-231) were analysed by water-soluble tetrazolium salt assay. It was found that epigallocatechin gallate gold nanoparticles at a dose of 200 µg/mL produced significant dose-dependent growth inhibition of cancer cells as compared to free epigallocatechin gallate and epigallocatechin gallate-citrate stabilized gold nanoparticles with higher cellular uptake both by normal and cancerous cells with sustainable release. Epigallocatechin gallate gold nanoparticles also enhanced apoptosis by blocking prostate and breast cancer cells by 48% and 34.7%, respectively. The epigallocatechin gallate gold nanoparticles reduced antiapoptotic proteins such as B-cell leukaemia/lymphoma-2 protein and B-cell lymphoma-extra-large and stimulated expression of B-cell leukaemia/lymphoma-2-associated X protein, cleaved caspase 3 in resistant breast cancer cells, and cleaved caspase 7 in prostate cells. Moreover, epigallocatechin gallate gold nanoparticles showed excellent suppression of nuclear factor kappa B transcriptional activity in both breast and prostate cell lines by 73.7% and 65.5%, respectively, than epigallocatechin gallate and epigallocatechin gallate-citrate stabilized gold nanoparticles.

In a study, Hsieh et al. (2011) designed gold nanoparticles loaded with epigallocatechin gallate by the ultrasonication method and evaluated their anticancer effect against bladder cancer both in *in-vitro* and *in-vivo* models. It was observed that the *in-vitro* model of epigallocatechin gallate gold nanoparticles significantly induced higher levels of cytotoxicity in bladder cancer cells (MBT-2) without causing any possible damage to normal Vero cells. It also induced cell apoptosis by activating caspases 3 and 7, triggering intrinsic apoptotic pathways. In *in-vivo* studies, C3H/HeN mice induced with MBT-2 cells were effectively reversed by oral formulation of epigallocatechin gallate gold nanoparticles than non-encapsulated epigallocatechin gallate. Further, studies were extended and observed that gold nanoparticles encapsulating epigallocatechin gallate produced 4.91 times better efficiency than free/non-encapsulated epigallocatechin gallate in lowering the level of apoptosis in murine melanoma cells (B16F10) in an *in-vitro* model. Whereas in the *in-vivo* model, intratumoral injection of epigallocatechin gallate gold nanoparticles showed 1.66 times more potency than free epigallocatechin gallate in tumour growth reduction in the mouse melanoma model when compared with a control group with higher biocompatibility with producing lower damage to red blood cells.

Sanna et al. and Nune et al. also synthesized epigallocatechin gallate gold nanoparticles that possessed higher stability in simulated biological fluid and significantly induced apoptosis and antioxidant properties in neuroblastoma (SH-SY5Y-CFP-DEVD-YFP) cells in a dose-dependent concentration after 72 hours of exposure (Sanna et al. 2014; Nune et al. 2009).

Radhakrishnan et al. (2016) developed epigallocatechin gallate-loaded solid lipid nanoparticles using a double emulsification method to improve the rate of bioavailability and stability. Encapsulated epigallocatechin gallate-loaded solid lipid nanoparticles caused 8.1-fold increased cytotoxicity against breast cancer cells (MDA-MB-231) and 3.8 times increased activity against prostate cancer cells (DU-145) with cell apoptosis of 32.7% and 24.9%, respectively in *in-vitro* models. However, encapsulated epigallocatechin gallate-loaded solid lipid nanoparticles exhibited better activity with increased stability and bioavailability than non-encapsulated epigallocatechin gallate.

Khan et al. (2014) and Siddiqui et al. (2014) encapsulated epigallocatechin gallate with chitosan nanoparticles and evaluated them against human melanoma (Mel 928) cells and prostate (22Rv1) cancer cell tumour xenograft model, respectively. The polymeric nanoparticles showed slow release at simulated gastric pH and faster release at simulated intestinal pH. Treatment of chitosan-loaded epigallocatechin gallate in both studies significantly inhibited the tumour cell growth by inducing poly (ADP-ribose) polymerase cleavage, increasing expression of apoptotic protein B-cell leukaemia/lymphoma 2 protein-associated X protein with concomitant decreasing that of antiapoptotic, i.e., B-cell leukaemia/lymphoma-2-protein expression, activating different caspase dependant pathways, reducing Ki-67 protein involved in the regulation of cell cycle and cell proliferation and proliferating cell nuclear antigen and downregulating the activities of vascular endothelial growth factor, the cluster of differentiation 31, cell proliferation, and angiogenesis.

Mukherjee et al. (2015) synthesized the epigallocatechin gallate-loaded gold nanoparticles. Also, they revealed results in his study that epigallocatechin gallate-conjugated gold nanoparticles increased antioxidant activity, cellular internalization, and uptake and produced more toxicity in cancer cells than free epigallocatechin gallate. Epigallocatechin gallate-loaded gold nanoparticles (20 µg/mL) caused significant cellular death in Ehrlich's ascites carcinoma cells effectively with increased 30% activity than non-encapsulated epigallocatechin gallate. It also induces cell apoptosis, inhibits activation of nuclear factor kappa B, and protects hepatocyte cells from undesirable damage (Mukherjee et al. 2015).

Chen et al. (2014) also reported that epigallocatechin gallate-loaded gold nanoparticles possess *in-vitro* and cytotoxicity and *in-vivo* anticancer activity. It improved the anticancer effect by targeting murine melanoma cells and produced 4.91-fold increased activity than free epigallocatechin gallate. It induced cell apoptosis, significantly decreased the haemoglobin release from red blood cells in the hemolysis test, and increased anticancer activity.

12.3.2 Active Targeting Strategy of Epigallocatechin Gallate

Active targeting or ligand-mediated targeting of anticancer drug delivery systems increases the efficacy of the drug-loaded nanocarriers and makes it more specific to reach the target site with improved bioavailability (Galvin et al. 2012; Bertrand et al. 2014). Active targeting of nano-drug delivery system interacts with the tumour site through several molecular interactions such as charge-based interactions, receptor-ligand interactions, and simplified motif-based interactions with substrate or enzyme molecules (Varki 2007; Padhi et al. 2022a,b). Active targeting is achieved when a nano-formulated drug is conjugated with a ligand system like proteins, nucleic acids, carbohydrates, antibodies, and small organic molecules like vitamins for the overexpressed receptor on the surface of tumour cells or simply by adsorption phenomenon (Weissleder et al. 2005; Verma and Stellacci 2010; Mout et al. 2012). The ligand attached to drug moieties undergoes a receptor-mediated endocytosis process in the cell membrane, which enhances drug uptake into the tumour cells (Weissleder et al. 2005). Several studies also revealed that the nanosystem method with ligand attachment possesses better nanocarrier internalization and enhanced therapeutic activities with higher efficacy when compared to passive targeting systems (Behera and Padhi 2020).

Sanna et al. (2017) encapsulated epigallocatechin gallate nanoparticles to target and bind to the prostate-specific membrane antigen, a transmembrane protein overexpressed in prostate cancer. This study evaluated that epigallocatechin gallate-loaded nanoparticles conjugated with pseudo-mimetic dipeptide exhibited a tenfold enhancement of antiangiogenic and proapoptotic activity against prostate cancer in both *in-vitro* and *in-vivo* models.

Hajipour et al. (2018) designed conjugated epigallocatechin gallate nanoparticles loaded with nano lipid carriers and evaluated their potency against breast cancer. It was reported that arginyl-glycyl-aspartic acid conjugated epigallocatechin gallate has significantly enhanced cellular uptake and accumulation to the targeted site, better apoptotic activity, arrested cell cycle, reduced tumour growth, and downregulates cell adhesion and invasion by inhibiting integrin alpha V and integrin beta 3-integrin receptor resulting in higher anticancer activity. Moreover, doxorubicin and epigallocatechin gallate coloaded nano liquid carrier produced a more synergistic effect than epigallocatechin gallate-loaded nano liquid carrier-arginyl-glycyl-aspartic acid by inhibiting *P*-glycoprotein.

Liang et al. (2014) developed folate-conjugated chitosan-coated epigallocatechin gallate nanoparticles. The nanoformulation produced a sustained release of epigallocatechin gallate at different pH values. It significantly inhibited tumour cells' growth and proliferation effectively. Ding et al. (2015) designed a drug delivery system with a colloidal mesoporous silica target ligand system conjugated with peptides and epigallocatechin gallate to increase stability and target specificity for cancer activity. It produced greater tumour inhibition activity, approximately 89.66%, without causing any damage to normal cancer cells. The nanocarrier system promoted efficacy against breast cancer cells by inducing cellular apoptosis by blocking antiapoptotic B-cell leukaemia/lymphoma-2-protein expression and displaying better biocompatibility rates.

Yang et al. (2017) designed a new nanoformulation of epigallocatechin gallate with 3-mercapto-1-hexanol and β-lactoglobulin nano complexes to improve anticancer activity. Nanovehicles of modified β-lactoglobulin, 3-mercapto-1-hexanol, and epigallocatechin gallate produced greater activity in stability and sustained release in both *in-vitro* and *in-vivo* models effectively than free epigallocatechin gallate. It inhibits hepatocarcinoma (Hep G2), human melanoma cell line (A375), and oesophagal carcinoma cells (TE-1) tumour cell proliferation at 60.44%, 65.90%, and 32.88%, respectively, inducing anticancer activity. The nano complex at the highest dose of 100 mg/kg also significantly inhibited implanted human melanoma (A375) growth at 57.78% without causing any toxic effect and was twofold more effective than free epigallocatechin gallate.

Wu et al. (2017) also reported that the nanocarrier epigallocatechin gallate-β-lactoglobulin improved the antitumour activity of epigallocatechin gallate by inducing cell apoptosis by protein kinase B-mediated phosphorylation and blocking B-cell leukaemia/lymphoma-2-protein expression, inhibiting proliferation of human oesophagal carcinoma (TE-1) cells and human malignant melanoma (A375) cells at 63.7%, and 69.0% respectively.

Shafiei et al. (2015) incorporated epigallocatechin gallate with calcium/aluminium nitrate layered double hydroxide nanoparticles and evaluated its efficacy against human prostate (PC3) cancer cell lines in an *in-vitro* model. It has a relatively smaller particle size with sustained release behaviour at pH 7.45 and 4.25. It effectively increased apoptotic cell activity, inhibiting tumour cell growth and proliferation. It increased activity fivefold over non-encapsulated epigallocatechin gallate with longer duration activity than a physical mixture of epigallocatechin gallate and layered double hydroxide.

Rocha et al. (2011) investigated epigallocatechin gallate-loaded carbohydrate nanoparticles composed of maltodextrin and gum arabica. The nanoparticles significantly reduced the cell viability of prostate cancer cells (Du 145) effectively by increasing caspase-3 activity and inducing cell apoptosis. The results were greater than epigallocatechin gallate at lower concentrations.

Ramadass et al. (2015) proposed a liposomal codelivery system of paclitaxel and epigallocatechin gallate for anticancer activity against the breast cancer cell line (MDA-MB-231). The result showed a synergistic effect in enhancing cell apoptosis by inhibiting cell viability and invasion and increased caspase-dependent effect with decreased matrix metalloproteinase expression. The anticancer effects were greater than the individual effects of epigallocatechin gallate and paclitaxel.

Karahaliloğlu et al. (2018) also developed a dual combination of anticancer therapy for more effective and efficient activity of epigallocatechin gallate-conjugated curcumin and α-tocopheryl succinate loaded polystyrene-polysoyaoil-diethanol amine nanoparticles. The nanoparticles showed enhanced bioavailability and entrapment efficiency of dual curcumin and α-tocopheryl succinate in epigallocatechin gallate, found to be 68% and 80%, respectively. It inhibited tumour growth with a synergistic effect against breast cancer cells compared to each drug's effect and reduced systemic toxicity.

Chen et al. (2020) designed an epigallocatechin gallate drug delivery system with cisplatin-loaded gelatin nanoparticles. The particle size and zeta potential were evaluated by dynamic light scattering, and the morphological structure was determined by transmission electron microscopy. The gelatin-loaded cisplatin and epigallocatechin gallate showed significantly increased antitumour activity by inhibiting the human adenocarcinoma cell lines (A549) within 7 days with approximately 24% cell death and also overcoming limitations of cisplatin, thereby resulting in a synergistic effect.

Zhou et al. (2016) reported that functionalization enhanced the uptake of nanoparticles to cancer cells by the receptor-mediated endocytosis process, thereby exerting higher therapeutic activities. In this research, it was found that ruthenium nanoparticles loaded with luminescent ruthenium complexes with epigallocatechin gallate had a better target on laminin receptor 67-kDa, which was overexpressed in hepatocellular (HCC) cancer cells. The epigallocatechin gallate-ruthenium nanoparticles possess high selectivity for cancer cells and enhanced stability. Ruthenium complex-loaded epigallocatechin gallate ruthenium nanoparticles significantly inhibited the laminin or 67LR blocking antibody. Moreover, analysis of TEM and confocal microscopic images showed the accumulation of ruthenium complex-loaded epigallocatechin gallate ruthenium nanoparticles in the cytoplasm of hepatocarcinoma cell lines (SMMC-7721). Functionalized epigallocatechin gallate nanoparticles also exhibited anticancer effects with reactive oxygen species scavenging activity, and proapoptotic and anti-invasive effects against hepatocarcinoma cells. *In-vivo* studies also suggested epigallocatechin gallate encapsulated with ruthenium nanoparticles possesses high anti-tumorigenic and antiangiogenic effects in nude mice.

Hu et al. (2014) synthesized an epigallocatechin gallate-loaded genipin crosslinked casein phosphopeptide-chitosan nanoparticles. The nanoparticles showed improved stability rate, high biocompatibility, and adjustable release profile in the gastrointestinal tract. It possessed strong cytotoxicity activity against human gastrointestinal cancer cells (BGC823) and human liver cancer cells (HepG2) (Hu et al. 2014).

12.4 CONCLUSION AND FUTURE PERSPECTIVES

Much evidence of catechin, especially the epigallocatechin gallate, contributes several health benefits in the prevention and treatment of disorders such as cardiovascular disease, Parkinson's disease, Alzheimer's disease, obesity, and also cancer. Nevertheless, tea catechin is not relatively used in the clinical setting because of its poor oral bioavailability and limited stability. In recent years, many studies have been reported regarding the stability and bioavailability of epigallocatechin gallate. However, issues like less intestinal stability of oral epigallocatechin gallate still need to be addressed. Therefore, further studies are required to analyse the pharmacokinetic models, i.e., absorption, distribution, metabolism, and excretion process of epigallocatechin gallate in animal models.

Nanotechnologies are being researched and implemented to enhance the pharmacokinetics and pharmacodynamics of chemopreventive mediators to prevent cancer. In the above studies, different forms of nanoformulations such as polymeric nanoparticles, metal nanoparticles, ligand-based, and liposomal nanocarriers are observed to be used as the delivery system of epigallocatechin gallate with small particle size and higher encapsulation efficiency. The results showed that the nanoparticles promoted increased circulatory flow in the blood, enhanced cell internalization, and significantly inhibited the advancement and proliferation of tumour cells both in *in-vitro* and *in-vivo* animal models. Surface modification methods were also derived to improve intestinal absorption, maintain sustained drug-release properties of epigallocatechin gallate, and increase its cellular uptake mechanism. Moreover, the further use of targeting ligands, mainly known as the active process, also improves cancer cell specificity and is more prominent in inhibiting tumour cells than the passive process. The various nanoenabled formulations of epigallocatechin gallate highlight the importance of epigallocatechin gallate as an anticancer agent in upcoming clinical studies. However, many studies reported that combining epigallocatechin gallate with different cytotoxic drugs produces a synergistic effect that mainly enhances the effectiveness of the drugs with decreased dose concentrations and adverse side effects and finally improves the quality of life of cancer patients. Thus, further studies on clinical trials must be employed on epigallocatechin gallate shortly as it holds promising anticancer activity with antioxidant properties.

REFERENCES

Adachi S, Nagao T, To S, Joe AK, Shimizu M, Matsushima-Nishiwaki R, Kozawa O, Moriwaki H, Maxfield FR, Weinstein IB (2008) (−)-Epigallocatechin gallate causes internalization of the epidermal growth factor receptor in human colon cancer cells. *Carcinogenesis* 29, 1986–1993.

Adachi S, Shimizu M, Shirakami Y, Yamauchi J, Natsume H, Matsushima-Nishiwaki R, To S, Weinstein IB, Moriwaki H, Kozawa O (2009) (−)-Epigallocatechin gallate downregulates EGF receptor via phosphorylation at Ser1046/1047 by p38 MAPK in colon cancer cells. *Carcinogenesis* 30, 1544–1552.

Afzal M, Safer AM, Menon M (2015) Green tea polyphenols and their potential role in health and disease. *Inflammopharmacology* 23(4), 151–161.

Ahn HY, Hadizadeh KR, Seul C, Yun YP, Vetter H, Sachinidis, A (1999) Epigallocathechin-3 gallate selectively inhibits the PDGF-BB–induced intracellular signaling transduction pathway in vascular smooth muscle cells and inhibits transformation of sis-transfected NIH 3T3 fibroblasts and human glioblastoma cells (A172). *Mol Biol Cell* 10, 1093–1104.

Alavi M, Hamblin MR, Aghaie E, Mousavi SAR, Hajimolaali M (2023) Antibacterial and antioxidant activity of catechin, gallic acid, and epigallocatechin-3-gallate: Focus on nanoformulations. *Cell Mol Biomed Rep* 3(2), 62–72.

An Y, Li Z, Wang S, Wang Z (2008) Inhibition of (-) epigallocatechin gallate on dimethylarsinic acid promoting lung tumourigenesis through the induction of oxidative stress in mice. *Wei Sheng yan jiu= J Hyg Res* 37, 748–750.

Behera A, Mittu B, Padhi S, Patra N, Singh J (2020b) Bimetallic nanoparticles: Green synthesis, applications, and future perspectives. In: Abd-Elsalam KA (Ed.) *Multifunctional Hybrid Nanomaterials for Sustainable Agri-food and Ecosystems*. Elsevier, London, pp. 639–682.

Behera A, Mittu B, Padhi S, Singh A (2020a) Antimicrobial efficacy of essential oil nanoemulsions. In: Dhull SB, Chawla P, Kaushik R (Eds.) *Nanotechnological Approaches in Food Microbiology*. CRC Press, Boca Raton, FL, pp. 293–309.

Behera A, Padhi S (2020) Passive and active targeting strategies for the delivery of the camptothecin anticancer drug: A review. *Environ Chem Lett* 18(5), 1557–1567.

Behera A, Padhi S (2022) pH-Sensitive polymeric nanoparticles for cancer treatment. In: Padhi S, Behera A, Lichtfouse E (Eds.) *Polymeric Nanoparticles for the Treatment of Solid Tumours*. Environmental Chemistry for a Sustainable World, vol 71. Springer, Cham, pp. 401–425.

Behera A, Padhi S, Nayak AK (2023) Engineered liposomes as drug delivery and imaging agents. In: Ray, S, Nayak AK (Eds.) *Design and Applications of Theranostic Nanomedicines*. Woodhead Publishing, Cambridge, MA, pp. 75–108.

Bertrand N, Wu J, Xu X, Kamaly N, Farokhzad OC (2014) Cancer nanotechnology: The impact of passive and active targeting in the era of modern cancer biology. *Adv Drug Deliv Rev* 66, 2–25.

Bhimani RS, Troll W, Grunberger D, Frenkel K (1993) Inhibition of oxidative stress in HeLa cells by chemopreventive agents. *Cancer Res* 53, 4528–4533.

Botten D, Fugallo G, Fraternali F, Molteni C (2015) Structural properties of green tea catechins. *J Physic Chem B* 119, 12860–12867. https://doi.org/10.1021/acs.jpcb.5b08737

Braicu C, Gherman CD, Irimie A, Berindan-Neagoe I (2013b) Epigallocatechin-3-Gallate inhibits cell proliferation and migratory behaviour of triple negative breast cancer cells. *J Nanosci Nanotechnol* 13, 632–637.

Braicu C, Ladomery MR, Chedea VS, Irimie A, Berindan-Neagoe I (2013a) The relationship between the structure and biological actions of green tea catechins. *Food Chem* 141, 3282–3289.

Cabrera C, Artacho R, Giménez R (2006) Beneficial effects of green tea: A review. *J Am Coll Nutr* 25, 79–99.

Cabrera C, Gimenez R, Lopez CM (2003) Determination of antioxidant activity of tea components. *J Agric Food Chem* 31, 4427–4435.

Chavva SR, Deshmukh SK, Kanchanapally R, Tyagi N, Coym JW, Singh AP, Singh S (2019) Epigallocatechin gallate-gold nanoparticles exhibit superior antitumour activity compared to conventional gold nanoparticles: Potential synergistic interactions. *Nanomaterials* 9, 396.

Chen CC, Hsieh DS, Huang KJ, Chan YL, Hong PD, Yeh MK, Wu CJ (2014) Improving anticancer efficacy of (−)-epigallocatechin-3-gallate gold nanoparticles in murine B16F10 melanoma cells. *Drug Des Devel Ther* 8, 459.

Chen H, Landen CN, Li Y, Alvarez RD, Tollefsbol TO (2013) Epigallocatechin gallate and sulforaphane combination treatment induce apoptosis in paclitaxel-resistant ovarian cancer cells through hTERT and Bcl-2 down-regulation. *Exp Cell Res* 319, 697–706.

Chen YJ, Wang ZW, Lu TL, Gomez CB, Fang HW, Wei Y, Tseng CL (2020) The synergistic anticancer effect of dual drug-(Cisplatin/Epigallocatechin gallate) loaded gelatin nanoparticles for lung cancer treatment. *J Nanomater* 2020, 1–15.

Chowdhury A, Sarkar J, Chakraborti T, Pramanik PK, Chakraborti S (2016) Protective role of epigallocatechin-3-gallate in health and disease: A perspective. *Biomed Pharmacother* 78, 50–59.

Cirillo G, Curcio M, Vittorio O, Iemma F, Restuccia D, Spizzirri UG, Puoci F, Picci N (2016) Polyphenol conjugates and human health: A perspective review. *Crit Rev Food Sci Nutr* 56(2), 326–337.

Crespy V, Williamson G (2004) A review of the health effects of green tea catechins in in vivo animal models. *J Nutr* 134, 3431S–3440S.

Dai W, Ruan C, Zhang Y, Wang J, Han J, Shao Z, Sun Y, Liang J (2020) Bioavailability enhancement of epigallocatechin gallate by structural modification and nano-delivery: A review. *J Funct Foods* 65, 103732.

Ding J, Yao J, Xue J, Li R, Bao B, Jiang L, Zhu JJ, He Z (2015) Tumour-homing cell-penetrating peptide linked to colloidal mesoporous silica encapsulated (-)-epigallocatechin-3-gallate as drug delivery system for breast cancer therapy in vivo. *ACS Appl Mater Interfaces* 7, 18145–18155.

Dube A, Nicolazzo JA, Larson I (2010) Chitosan nanoparticles enhance the intestinal absorption of the green tea catechins (+)-catechin and (−)-epigallocatechin gallate. *Eur J Pharma Sci* 41, 219–225.

Falzone L, Salomone S, Libra M (2018) Evolution of cancer pharmacological treatments at the turn of the third millennium. *Front Pharmacol* 9, 1300.

Fang MZ, Wang Y, Ai N, Hou Z, Sun Y, Lu H, Welsh W, Yang CS (2003) Tea polyphenol (−)-epigallocatechin-3-gallate inhibits DNA methyltransferase and reactivates methylation-silenced genes in cancer cell lines. *Cancer Res* 63, 7563–7570.

Farzaei HM, Rahimi R, Abdollahi M (2015) The role of dietary polyphenols in the management of inflammatory bowel disease. *Curr Pharm Biotechnol* 16, 196–210.

Galvin P, Thompson D, Ryan KB, McCarthy A, Moore AC, Burke CS, Dyson M, MacCraith BD, Gun'ko YK, Byrne MT,

Volkov Y (2012) Nanoparticle-based drug delivery: Case studies for cancer and cardiovascular applications. *Cell Mol Life Sci* 69, 389–404.

Graham HN (1992) Green tea composition, consumption, and polyphenol chemistry. *Prev Med* 21, 334–350.

Hajipour H, Hamishehkar H, Nazari Soltan Ahmad S, Barghi S, Maroufi NF, Taheri RA (2018) Improved anticancer effects of epigallocatechin gallate using RGD-containing nanostructured lipid carriers. *Artif Cells Nanomed Biotechnol* 46, 283–292.

Hassanpour SH, Dehghani M (2017) Review of cancer from perspective of molecular. *J Cancer Res Prac* 4, 127–129.

Hastak K, Agarwal MK, Mukhtar H, Agarwal ML (2005) Ablation of either p21 or Bax prevents p53-dependent apoptosis induced by green tea polyphenol epigallocatechin-3-gallate. *FASEB J* 19, 1–19.

Hastak K, Gupta S, Ahmad N, Agarwal MK, Agarwal ML, Mukhtar H (2003) Role of p53 and NF-κB in epigallocatechin-3-gallate-induced apoptosis of LNCaP cells. *Oncogene* 22, 4851–4859.

Hobbs SK, Monsky WL, Yuan F, Roberts WG, Griffith L, Torchilin VP, Jain RK (1998) Regulation of transport pathways in tumour vessels: Role of tumour type and microenvironment. *Proc Natl Acad Sci USA* 95, 4607–4612.

Hollman PC, Arts IC (2000) Flavonols, flavones and flavanols–nature, occurrence and dietary burden. *J Sci Food Agric* 80, 1081–1093.

Hsieh DS, Wang H, Tan SW, Huang YH, Tsai CY, Yeh MK, Wu CJ (2011) The treatment of bladder cancer in a mouse model by epigallocatechin-3-gallate-gold nanoparticles. *Biomaterials* 32, 7633–7640.

Hsu S (2015) Compounds derived from epigallocatechin-3-gallate (epigallocatechin gallate) as a novel approach to the prevention of viral infections. *Inflamm Allergy Drug Targets* 14(1), 13–18.

Hu B, Xie M, Zhang C, Zeng X (2014) Genipin-structured peptide–polysaccharide nanoparticles with significantly improved resistance to harsh gastrointestinal environments and their potential for oral delivery of polyphenols. *J Agric Food Chem* 62, 12443–12452.

Hwang YS, Park KK, Chung WY (2013) Epigallocatechin-3 gallate inhibits cancer invasion by repressing functional invadopodium formation in oral squamous cell carcinoma. *Eur J Pharmacol* 715, 286–295.

Islam MA (2012) Cardiovascular effects of green tea catechins: Progress and promise. *Recent Pat Cardiovasc Drug Discov* 7(2), 88–99.

Ito H, Shoji Y, Matsumoto KI, Fukuhara K, Nakanishi I (2023) Anti-cancer effect of a planar catechin analog through the decrease in mitochondrial membrane potential. *ACS Med Chem Lett* 14(10), 1478–1481.

Jain A, Jain SK (2016) Optimization of chitosan nanoparticles for colon tumours using experimental design methodology. *Artif Cells Nanomed Biotechnol* 44, 1917–1926.

Jain RK, Stylianopoulos T (2010) Delivering nanomedicine to solid tumours. *Nat Rev Clin Oncol* 7, 653.

Jung YD, Kim MS, Shin BA, Chay KO, Ahn BW, Liu W, Bucana CD, Gallick GE, Ellis LM (2001) epigallocatechin gallate, a major component of green tea, inhibits tumour growth by inhibiting VEGF induction in human colon carcinoma cells. *Br J Cancer* 84, 844–850.

Kang SU, Lee BS, Lee SH, Baek SJ, Shin YS, Kim CH (2013) Expression of NSAID-activated gene-1 by epigallocatechin gallate in head and neck cancer: Involvement of ATM-dependent p53 expression. *J Nutr Biochem* 24, 986–999.

Karahaliloğlu Z, Kilicay E, Alpaslan P, Hazer B, BakiDenkbas E (2018) Enhanced antitumour activity of epigallocatechin gallate–conjugated dual-drug-loaded polystyrene–polysoyaoil–diethanol amine nanoparticles for breast cancer therapy. *J Bioactiv Compat Polym* 33, 38–62.

Kciuk M, Alam M, Ali N, Rashid S, Głowacka P, Sundaraj R, Celik I, Yahya EB, Dubey A, Zerroug E, Kontek R (2023) Epigallocatechin-3-gallate therapeutic potential in cancer: Mechanism of action and clinical implications. *Molecules* 28(13), 5246.

Khan N, Bharali DJ, Adhami VM, Siddiqui IA, Cui H, Shabana SM, Mousa SA, Mukhtar H (2014) Oral administration of naturally occurring chitosan-based nanoformulated green tea polyphenol epigallocatechin gallate effectively inhibits prostate cancer cell growth in a xenograft model. *Carcinogenesis* 35, 415–423.

Khoi PN, Park JS, Kim JH, Xia Y, Kim NH, Kim KK, Jung YD (2013) (-)-Epigallocatechin-3-gallate blocks nicotine-induced matrix metalloproteinase-9 expression and invasiveness via suppression of NF-κB and AP-1 in endothelial cells. *Int J Oncol* 43(3), 868–876.

Kim M, Murakami A, Kawabata K, Ohigashi H (2005) (−)-Epigallocatechin-3-gallate promotes pro-matrix metalloproteinase-7 production via activation of the JNK1/2 pathway in HT-29 human colorectal cancer cells. *Carcinogenesis* 26, 1553–1562.

Ko H, So Y, Jeon H, Jeong MH, Choi HK, Ryu SH, Lee SW, Yoon HG, Choi KC (2013) TGF-β1-induced epithelial–mesenchymal transition and acetylation of Smad2 and Smad3 are negatively regulated by epigallocatechin gallate in human A549 lung cancer cells. *Cancer Lett* 335, 205–213.

Koh YW, Choi EC, Kang SU, Hwang HS, Lee MH, Pyun J, Park R, Lee Y, Kim CH (2011) Green tea (−)-epigallocatechin-3-gallate inhibits HGF-induced progression in oral cavity cancer through suppression of HGF/c-Met. *J NutrBiochem* 22, 1074–1083.

Kundu A, Padhi S, Behera A, Hasnain MS, Nayak AK (2022) Tumour targeting strategies by chitosan-based nanocarriers. In: Hasnain MS, Beg S, Nayak AK (Eds.) *Chitosan in Biomedical Applications*. Academic Press, London, pp. 163–188.

Kushima Y, Iida K, Nagaoka Y, Kawaratani Y, Shirahama T, Sakaguchi M, Baba K, Hara Y, Uesato S (2009) Inhibitory effect of (−)-epigallocatechin and (−)-epigallocatechin gallate against heregulin β1-induced migration/invasion of the MCF-7 breast carcinoma cell line. *Biol Pharm Bull* 32, 899–904.

Kwak IH, Shin YH, Kim M, Cha HY, Nam HJ, Lee BS, Chaudhary SC, Pai KS, Lee JH (2011) Epigallocatechin-3-gallate inhibits its paracrine and autocrine hepatocyte growth factor/scatter factor-induced tumour cell migration and invasion. *Exp Mol Med* 43, 111–120.

Lee WJ, Shim JY, Zhu BT (2005) Mechanisms for the inhibition of DNA methyltransferases by tea catechins and bioflavonoids. *Mol Pharmacol* 68, 1018–1030.

Lee YH, Kwak J, Choi HK, Choi KC, Kim S, Lee J, Jun W, Park HJ, Yoon HG (2012) Epigallocatechin gallate suppresses prostate cancer cell growth modulating acetylation of androgen receptor by anti-histone acetyltransferase activity. *Int J Mol Med* 30, 69–74.

Legeay S, Rodier M, Fillon L, Faure S, Clere N (2015) Epigallocatechin gallate: A review of its beneficial properties to prevent metabolic syndrome. *Nutrients* 7, 5443–5468.

Liang J, Cao L, Zhang L, Wan XC (2014) Preparation, characterization, and in vitro antitumour activity of folate conjugated

chitosan coated epigallocatechin gallate nanoparticles. *Food Sci Biotechnol* 23, 569–575.
Liang YC, Lin-Shiau SY, Chen CF, Lin JK (1997) Suppression of extracellular signals and cell proliferation through EGF receptor binding by (–)-epigallocatechin gallate in human A431 epidermoid carcinoma cells. *J Cell Biochem* 67, 55–65.
Lim YC, Cha YY (2011) Epigallocatechin-3-gallate induces growth inhibition and apoptosis of human anaplastic thyroid carcinoma cells through suppression of EGFR/ERK pathway and cyclin B1/CDK1 complex. *J Surg Oncol* 104, 776–780.
Lim YC, Park HY, Hwang HS, Kang SU, Pyun JH, Lee MH, Choi EC, Kim CH (2008) (–)-Epigallocatechin-3-gallate (epigallocatechin gallate) inhibits HGF-induced invasion and metastasis in hypopharyngeal carcinoma cells. *Cancer Lett* 271, 140–152.
Maeda H, Nakamura H, Fang J (2013) The EPR effect for macromolecular drug delivery to solid tumours: Improvement of tumour uptake, lowering of systemic toxicity, and distinct tumour imaging in vivo. *Advan Drug Deliv Rev* 65, 71–79.
Min KJ, Kwon TK (2014) Anticancer effects and molecular mechanisms of epigallocatechin-3-gallate. *Integ Med Res* 3, 16–24.
Mirossay L, Varinská L, Mojžiš J (2018) Antiangiogenic effect of flavonoids and chalcones: An update. *Int J Mol Sci* 19, 27.
Mout R, Moyano DF, Rana S, Rotello VM (2012) Surface functionalization of nanoparticles for nanomedicine. *Chem Soc Rev* 41, 2539–2544.
Mukherjee S, Ghosh S, Das DK, Chakraborty P, Choudhury S, Gupta P, Adhikary A, Dey S, Chattopadhyay S (2015) Gold-conjugated green tea nanoparticles for enhanced antitumour activities and hepatoprotection: Synthesis, characterization and in vitro evaluation. *J Nutr Biochem* 26, 1283–1297.
Nandakumar V, Vaid M, Katiyar SK (2011) (–)-Epigallocatechin-3-gallate reactivates silenced tumour suppressor genes, Cip1/p21 and p16 INK4a, by reducing DNA methylation and increasing histones acetylation in human skin cancer cells. *Carcinogenesis* 32, 537–544.
Nanjo F, Goto K, Seto R, Suzuki M, Sakai M, Hara Y (1996) Scavenging effects of tea catechins and their derivatives on 1,1-diphenyl-2-picrylhydrazyl radical. *Free Radic Biol Med* 21, 895–902.
Natarajan SB, Chandran SP, Vinukonda A (2019) Green tea catechin loaded nanodelivery systems for the treatment of pandemic diseases. *Asian J Pharm Clin Res* 12, 1–7.
Nishikawa T, Nakajima T, Moriguchi M, Jo M, Sekoguchi S, Ishii M, Takashima H, Katagishi T, Kimura H, Minami M, Itoh Y (2006) A green tea polyphenol, epigalocatechin-3-gallate, induces apoptosis of human hepatocellular carcinoma, possibly through inhibition of Bcl-2 family proteins. *J Hepatol* 44, 1074–1082.
Nune SK, Chanda N, Shukla R, Katti K, Kulkarni RR, Thilakavathy S, Mekapothula S, Kannan R, Katti KV (2009) Green nanotechnology from tea: Phytochemicals in tea as building blocks for production of biocompatible gold nanoparticles. *J Mater Chem* 19, 2912–2920.
Padhi S, Behera A (2020) Nanotechnology based targeting strategies for the delivery of camptothecin. In: Saneja A, Panda A, Lichtfouse E (Eds.) *Sustainable Agriculture Reviews*, vol 44. Springer, Cham, pp. 243–272.
Padhi S, Behera A (2021) Advanced drug delivery systems in the treatment of ovarian cancer. In: Dua K, Mehta M, de Jesus Andreoli Pinto T, Pont LG, Williams KA, Rathbone MJ (Eds.) *Advanced Drug Delivery Systems in the Management of Cancer*. Academic Press, London, pp. 127–139.
Padhi S, Behera A (2022) Cellular internalization and toxicity of polymeric nanoparticles. In: Padhi S, Behera A, Lichtfouse E (Eds.) *Polymeric Nanoparticles for Treating Solid Tumours*. Environmental Chemistry for a Sustainable World, vol 71. Springer, Cham, pp. 473–488.
Padhi S, Behera A, Hasnain MS, Nayak AK (2022a) Chitosan-based drug delivery systems in cancer therapeutics. In: Hasnain MS, Beg S, Nayak AK (Eds.) *Chitosan in Drug Delivery*. Academic Press, London, pp. 159–193.
Padhi S, Behera A, Hasnain MS, Nayak AK (2022b) Uses of chitosan in drug delivery. In: Hasnain MS, Beg S, Nayak AK (Eds.) *Chitosan in Biomedical Applications*. Academic Press, London, pp. 139–162.
Padhi S, Kapoor R, Verma D, Panda AK, Iqbal Z (2018) Formulation and optimization of topotecan nanoparticles: In vitro characterization, cytotoxicity, cellular uptake and pharmacokinetic outcomes. *J Photochem Photobiol B Biol* 183, 222–232.
Padhi S, Mirza MA, Verma D, Khuroo T, Panda AK, Talegaonkar S, Khar RK, Iqbal Z (2016) Revisiting the nanoformulation design approach for effective delivery of Topotecan in its stable form: An appraisal of its in vitro behavior and tumour amelioration potential. *Drug Deliv* 23(8), 2827–2837.
Padhi S, Nayak AK, Behera A (2020) Type II diabetes mellitus: A review on recent drug-based therapeutics. *Biomed Pharmacother* 131, 110708.
Patnaik S, Gorain B, Padhi S, Choudhury H, Gabr GA, Md S, Kumar Mishra D, Kesharwani P (2021) Recent update of toxicity aspects of nanoparticulate systems for drug delivery. *Eur J Pharm Biopharm* 161, 100–119.
Patra JK, Das G, Fraceto LF, Campos EV, del Pilar Rodriguez-Torres M, Acosta-Torres LS, Diaz-Torres LA, Grillo R, Swamy MK, Sharma S, Habtemariam S (2018) Nano based drug delivery systems: Recent developments and future prospects. *J Nanobiotechnol* 16, 1–33.
Peng G, Dixon DA, Muga SJ, Smith TJ, Wargovich MJ (2006) Green tea polyphenol (–)-epigallocatechin-3-gallate inhibits cyclooxygenase-2 expression in colon carcinogenesis. *Mol Carcinog* 45, 309–319.
Pervin M, Unno K, Nakagawa A, Takahashi Y, Iguchi K, Yamamoto H, Hoshino M, Hara A, Takagaki A, Nanjo F, Minami A (2017) Blood brain barrier permeability of (–)-epigallocatechin gallate, its proliferation-enhancing activity of human neuroblastoma SH-SY5Y cells, and its preventive effect on age-related cognitive dysfunction in mice. *Biochem Biophys Rep* 9, 180–186.
Pervin M, Unno K, Takagaki A, Isemura M, Nakamura Y (2019) Function of green tea catechins in the brain: Epigallocatechin gallate and its metabolites. *Int J Mol Sci* 20, 3630.
Pilorget A, Berthet V, Luis J, Moghrabi A, Annabi B, Béliveau R (2003) Medulloblastoma cell invasion is inhibited by green tea (–) epigallocatechin-3-gallate. *J Cell Biochem* 90, 745–755.
Qin J, Wang Y, Bai Y, Yang K, Mao Q, Lin Y, Kong D, Zheng X, Xie L (2012) Epigallocatechin-3-gallate inhibits bladder cancer cell invasion via suppression of NF-κB-mediated matrix metalloproteinase-9 expression. *Mol Med Rep* 6, 1040–1044.
Qin J, Xie LP, Zheng XY, Wang YB, Bai Y, Shen HF, Li LC, Dahiya R (2007) A component of green tea, (–)-epigallocatechin-3-gallate, promotes apoptosis in T24 human bladder cancer cells via modulation of the PI3K/Akt pathway and Bcl-2 family proteins. *Biochem Biophys Res Commun* 354, 852–857.

Radhakrishnan R, Kulhari H, Pooja D, Gudem S, Bhargava S, Shukla R, Sistla R (2016) Encapsulation of biophenolic phytochemical EPIGALLOCATECHIN GALLATE within lipid nanoparticles enhances its stability and cytotoxicity against cancer. *Chem Phys Lipids* 198, 51–60.

Rahmani AH, Allemailem KS, Aly SM, Khan MA (2015) Implications of green tea and its constituents in the prevention of cancer via the modulation of cell signalling pathway. *BioMed Res Int* 2015, 925640.

Ramadass SK, Anantharaman NV, Subramanian S, Sivasubramanian S, Madhan B (2015) Paclitaxel/epigallocatechin gallate coloaded liposome: A synergistic delivery to control the invasiveness of MDA-MB-231 breast cancer cells. *Colloids Surf B Biointerfaces* 125, 65–72.

Reygaert WC (2014) The antimicrobial possibilities of green tea. *Front Microbiol* 5, 434.

Rocha S, Generalov R, Pereira MD, Peres I, Juzenas P, Coelho MA (2011) Epigallocatechin gallate-loaded polysaccharide nanoparticles for prostate cancer chemoprevention. *Nanomedicine* 6(1), 79–87.

Rosenblum D, Joshi N, Tao W, Karp JM, Peer D (2018) Progress and challenges towards targeted delivery of cancer therapeutics. *Nat Commun* 9, 1–12.

Roy AM, Baliga MS, Katiyar SK (2005) Epigallocatechin-3-gallate induces apoptosis in estrogen receptor–negative human breast carcinoma cells via modulation in protein expression of p53 and Bax and caspase-3 activation. *Mol Cancer Ther* 4, 81–90.

Sahoo P, Dey J, Mahapatra SR, Ghosh A, Jaiswal A, Padhi S, Prabhuswamimath SC, Misra N, Suar M (2022) Nanotechnology and COVID-19 convergence: Toward new planetary health interventions against the pandemic. *OMICS J Integr Biol* 26(9), 473–488.

Sanna V, Pala N, Dessì G, Manconi P, Mariani A, Dedola S, Rassu M, Crosio C, Iaccarino C, Sechi M (2014) Single-step green synthesis and characterization of gold-conjugated polyphenol nanoparticles with antioxidant and biological activities. *Int J Nanomed* 9, 4935.

Sanna V, Singh CK, Jashari R, Adhami VM, Chamcheu JC, Rady I, Sechi M, Mukhtar H, Siddiqui IA (2017) Targeted nanoparticles encapsulating (−)-epigallocatechin-3-gallate for prostate cancer prevention and therapy. *Sci Rep* 7, 1–5.

Severino JF, Goodman BA, Kay CW, Stolze K, Tunega D, Reichenauer TG, Pirker KF (2009) Free radicals generated during oxidation of green tea polyphenols: Electron paramagnetic resonance spectroscopy combined with density functional theory calculations. *Free RadicBiol Med* 46, 1076–1088.

Shafiei SS, Solati-Hashjin M, Samadikuchaksaraei A, Kalantarinejad R, Asadi-Eydivand M, Osman NA (2015) Epigallocatechin gallate/layered double hydroxide nanohybrids: Preparation, characterization, and in vitro antitumour study. *PLoS One* 10, e0136530.

Shi X, Ye J, Leonard S, Ding M, Vallyathan V, Castranova V, Rojanasakul Y, Dong Z (2000) Antioxidant properties of (-)-epicatechin-3-gallate and its inhibition of Cr(VI)-induced DNA damage and Cr(IV)-or TPA-stimulated NF-κB activation. *Mol Cell Biochem* 206(1), 125–132.

Shikha D, Singh A, Rangra NK, Monga V, Bhatia R (2024) Insights to therapeutic potentials, pharmaceutical formulations, chemistry and analytical methods of catechin. *Phytochem Rev* 23, 1–42.

Shimizu M, Sakai H, Shirakami Y, Yasuda Y, Kubota M, Terakura D, Baba A, Ohno T, Hara Y, Tanaka T, Moriwaki H (2011) Preventive effects of (−)-epigallocatechin gallate on diethylnitrosamine-induced liver tumourigenesis in obese and diabetic C57BL/KsJ-db/db mice. *Cancer Prev Res* 4, 396–403.

Shimizu M, Shirakami Y, Sakai H, Yasuda Y, Kubota M, Adachi S, Tsurumi H, Hara Y, Moriwaki H (2010) (−)-Epigallocatechin gallate inhibits growth and activation of the VEGF/VEGFR axis in human colorectal cancer cells. *Chem Biol Interact* 185, 247–252.

Siddiqui IA, Bharali DJ, Nihal M, Adhami VM, Khan N, Chamcheu JC, Khan MI, Shabana S, Mousa SA, Mukhtar H (2014) Excellent antiproliferative and proapoptotic effects of (−)-epigallocatechin-3-gallate encapsulated in chitosan nanoparticles on human melanoma cell growth both in vitro and in vivo. *Nanomed Nanotechnol Biol Med* 10, 1619–1626.

Singh G, Kaur H, Harikumar SL (2015) Pleiotropic effects of green tea: An overview. *Int J Pharm Phytopharmacol Res* 4(4), 223–226.

Sun W, Yang Y, Wang C, Liu M, Wang J, Qiao S, Jiang P, Sun C, Jiang S (2024) Epigallocatechin-3-gallate at the nanoscale: A new strategy for cancer treatment. *Pharm Biol* 62(1), 676–690.

Sutradhar KB, Amin M (2014) Nanotechnology in cancer drug delivery and selective targeting. *Int Sch Res Notices* 2014, 939378.

Thakur VS, Gupta K, Gupta S (2012) Green tea polyphenols cause cell cycle arrest and apoptosis in prostate cancer cells by suppressing class I histone deacetylases. *Carcinogenesis* 33, 377–384.

ud Din F, Aman W, Ullah I, Qureshi OS, Mustapha O, Shafique S, Zeb A (2017) Effective use of nanocarriers as drug delivery systems for the treatment of selected tumours. *Int J Nanomed* 12, 7291.

Varki A (2007) Glycan-based interactions involving vertebrate sialic-acid-recognizing proteins. *Nature* 446, 1023–1029.

Verma A, Stellacci F (2010) Effect of surface properties on nanoparticle–cell interactions. *Small* 6, 12–21.

Verma D, Thakur PS, Padhi S, Khuroo T, Talegaonkar S, Iqbal Z (2017) Design expert assisted nanoformulation design for codelivery of Topotecan and thymoquinone: Optimization, in vitro characterization, and stability assessment. *J Mol Liquids* 242, 382–394.

Wang C, Bai M, Sun Z, Yao N, Zhang A, Guo S, Asemi Z (2023) Epigallocatechin-3-gallate and cancer: Focus on the role of microRNAs. *Cancer Cell Int* 3(1), 241.

Wang R, Zhou W, Jiang X (2008) Reaction kinetics of degradation and epimerization of epigallocatechin gallate (epigallocatechin gallate) in aqueous system over a wide temperature range. *J Agric Food Chem* 56(8), 2694–2701.

Weissleder R, Kelly K, Sun EY, Shtatland T, Josephson L (2005) Cell-specific targeting of nanoparticles by multivalent attachment of small molecules. *Nat Biotechnol* 23, 1418–1423.

Wu M, Jin J, Jin P, Xu Y, Yin J, Qin D, Wang K, Du Q (2017) Epigallocatechin gallate-β-lactoglobulin nanoparticles improve the antitumour activity of epigallocatechin gallate for inducing cancer cell apoptosis. *J Funct Foods* 39, 257–263.

Xu Y, Ho CT, Amin SG, Han C, Chung FL (1992) Inhibition of tobacco-specific nitrosamine-induced lung tumourigenesis in A/J mice by green tea and its major polyphenol as antioxidants. *Cancer Res* 52(14), 3875–3879.

Yang CS, Zhang J, Zhang L, Huang J, Wang Y (2016) Mechanisms of body weight reduction and metabolic syndrome alleviation by tea. *Mol Nutr Food Res* 60, 160–174.

Yang Y, Jin P, Zhang X, Ravichandran N, Ying H, Yu C, Ying H, Xu Y, Yin J, Wang K, Wu M (2017) New epigallocatechin gallate (epigallocatechin gallate) nanocomplexes co-assembled with 3-mercapto-1-hexanol and β-lactoglobulin for improvement of antitumour activity. *J Biomed Nanotechnol* 13, 805–814.

Yuan JH, Li, YQ, Yang XY (2007) Inhibition of epigallocatechin gallate on orthotopic colon cancer by upregulating the Nrf2-UGT1A signal pathway in nude mice. *Pharmacology* 80, 269–278.

Zaveri NT (2006) Green tea and its polyphenolic catechins: Medicinal uses in cancer and noncancer applications. *Life Sci* 78, 2073–2080.

Zhang G, Miura Y, Yagasaki K (2000) Suppression of adhesion and invasion of hepatoma cells in culture by tea compounds through antioxidative activity. *Cancer Lett* 159, 169–173.

Zhao J, Castranova V (2011) Toxicology of nanomaterials used in nanomedicine. *J Toxicol Environ Health* 14, 593–632.

Zhou Y, Yu Q, Qin X, Bhavsar D, Yang L, Chen Q, Zheng W, Chen L, Liu J (2016) Improving the anticancer efficacy of laminin receptor-specific therapeutic ruthenium nanoparticles (RuBB-loaded epigallocatechin gallate-RuNPs) via ROS-dependent apoptosis in SMMC-7721 cells. *ACS Appl Mater Interfaces* 8, 15000–15012.

13 Nanoenabled Systems for the Delivery of Quercetin as an Anticancer Agent

Rahul Lotan Shirole[1], *Vijayeta Pralhad Nage*[2], *Nitin Lotan Shirole*[3], *and Sushil Vilas Chawre*[2]
[1]Department of Pharmacology, DCS's A. R. A. College of Pharmacy, Dhule, India
[2]Department of Kayachikitsa, Government Ayurvedic College, Nagpur, India
[3]Department of Pharmaceutical Chemistry, DCS's A. R. A. College of Pharmacy, Dhule, India

13.1 INTRODUCTION

Quercetin or 2-(3,4-dihydroxy phenyl)-3,5,7-trihydroxy 4*H*-chromen-4-one is (Figure 13.1) a polyphenolic flavonoid that occurs commonly in various fruits and vegetables like capers, lovage, dill, cilantro, onions, apples, and berries as in cranberries and chokeberries. Quercetin has a wide range of pharmacological activities, including antioxidant, anti-inflammatory, antidiabetic, antihypertensive, immune-modulator, antiviral, gastro-protective, and anti-allergic activities (Rauf et al., 2018; Aghababaei & Hadidi, 2023), making it useful in the treatment of cancers of the liver, breast, lungs, prostate, colon, and cervical. Quercetin's anticancer properties are mediated by various strategies, including cellular signalling, inhibiting carcinogen-initiating enzymes, and acting on cellular receptors and proteins (Rauf et al., 2018; Gibellini et al., 2011).

Chemotherapeutics have significant limits in cancer treatment due to their poor solubility and bioavailability, resulting in lower drug concentrations at the target tumour location and multiple unfavourable effects on normal cells (Nam et al., 2016). Because of its hydrophobic nature (Cai et al., 2013), quercetin has a low solubility and bioavailability (Figure 13.2). It also suffers a more enzymatic breakdown in the gastrointestinal tract, resulting in lower bioavailability and shorter plasma circulation duration (Khushnud and Mousa, 2013; Wang et al., 2013d).

Quercetin has been reported to treat a variety of cancers. However, the appended limitations curtail its potential clinical usage (Park et al., 2005; Azeem et al., 2023). The nanoenabled drug delivery method increases quercetin loading at the target location, enhancing therapeutic efficacy and lowering toxicity. It also allows sustained drug release and the ability to be administered via various routes. The cytotoxicity of anticancer drugs frequently affects healthy tissues, resulting in significant systemic toxicity and unpleasant drug responses. Because of the stated issue and considering the advantages of nanoenabled systems, targeting approaches such as passive and active targeting are employed to deliver the drug at the targeted tumour site (Byrne et al., 2008; Joshi et al., 2023).

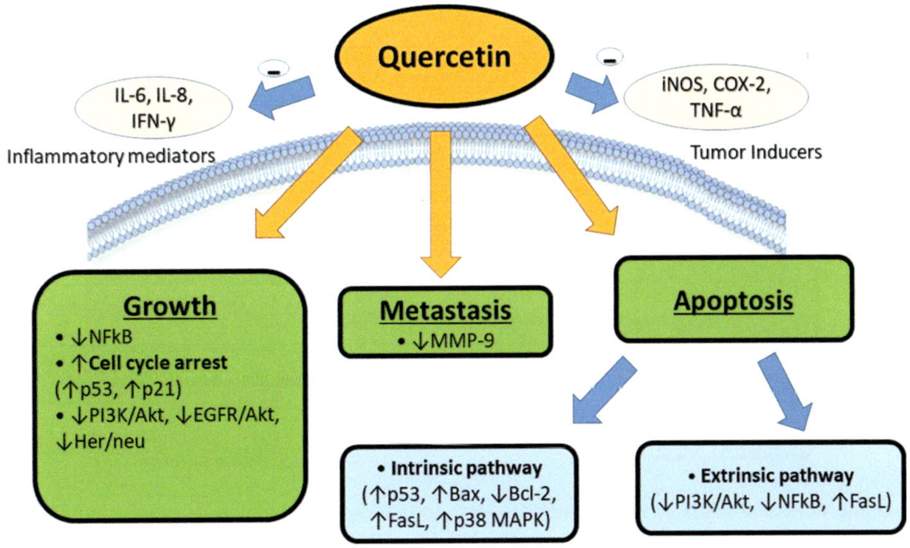

FIGURE 13.1 Chemical structure of quercetin [2-(3,4-dihydroxy phenyl)-3,5,7-trihydroxy-4*H*-chromen-4-one].

DOI: 10.1201/9781003611516-16

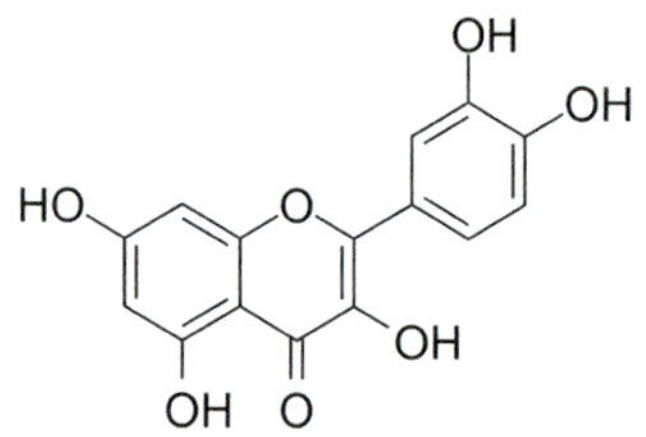

FIGURE 13.2 Limitations of quercetin as an anticancer drug.

Normal and malignant cells have varied vasculature. Compared to normal cells with a tight endothelium of blood vessels, cancerous cells are more heterogeneous, have higher vascular density, are larger, and have higher density more permeable and leaky due to gap sizes of 100 nm to 2 µm. Extravasation mediated by excessive generation of vascular mediators causes a rise in the accumulation of drugs with a high molecular weight in solid tumours. Endocytosis causes drug-containing nanoenabled carriers to concentrate more within the tumour interstitium, resulting in higher drug concentrations for antitumour activity (Byrne et al., 2008; Parveen and Sahoo, 2008; Brannon-Peppas and Blanchette, 2004). In addition, different nanoenabled active targeting systems commonly utilize ligands such as proteins, antibodies, nucleic acid aptamers or small molecules to target the overexpressed receptors on tumour surfaces, resulting in greater drug concentration at the tumour site (Sakurai et al., 2015; Muller et al., 2016; Shahbaz et al., 2023).

To overcome the poor solubility and bioavailability of quercetin, various quercetin nanoformulations have been reported utilizing different nanocarriers such as liposomes (Caddeo et al., 2016a, 2016b), solid lipid nanoparticles (Chen et al., 2018), polymeric nanoparticles (Wang et al., 2013c, 2013b, 2013a), micelles (Moussa et al., 2014), nanoemulsions (Ojeda-Serna et al., 2019), metallic nanoparticles (Balakrishnan et al., 2017; Srisa-Nga et al., 2019; Yuan et al., 2017) and nanocapsules (Omwenga et al., 2018).

13.2 MECHANISM OF ACTION OF QUERCETIN AS ANTICANCER AGENT

Quercetin shows anticancer activity by modulating various cellular processes and preventing metastasis and cancerous cell proliferation (Figure 13.3) (Shu et al., 2010; Bishayee, 2009). It interacts with phosphoinositide-3-kinase–protein kinase B, human epidermal growth factor receptor 2/neu, Wnt/-catenin and epidermal growth factor receptor signalling pathways. It also inhibits the mammalian target of rapamycin, controlling the autophagy and biosynthesis of proteins and the cell growth pathways (Bruning, 2013). The spread of cancer from one organ or part to another within a body through metastasis. Quercetin prevents the metastasis of breast cancer cells by inhibiting matrix metalloproteinases-9 in breast (MCF-7) and epidermal cells (JB6 P+) (Lin et al., 2008; Hwang et al., 2009). Programmed cell death is called apoptosis, essential for controlling the physiological processes controlled by intrinsic (mitochondria-mediated) or extrinsic (mediated by signals from other cells). Apoptosis dysregulation can lead to cancer progression (Lowe and Lin, 2000). It increases apoptosis against various cancers such as human promyelocytic leukaemia cell line (HL-60) (Yuan et al., 2012), murine mammary cancer (4T1) (Kim et al., 2013), colon cancer (SW-480) (Shan et al., 2009) and epidermoid tumour cells (A431) (Huang et al., 1999).

13.3 NANOENABLED TARGETING STRATEGIES FOR THE QUERCETIN

13.3.1 Passive Targeting Strategy for the Quercetin

The delivery of the drug in passive tumour targeting is based on the "enhanced permeability and retention effect" in tumour tissues instead of normal tissues. Commonly used passive-targeting drug delivery systems include liposomes, microemulsions, and nanoparticles; upon administration, the drug enters the tissue through gaps in the endothelium due to the weak difference of enhanced permeability and

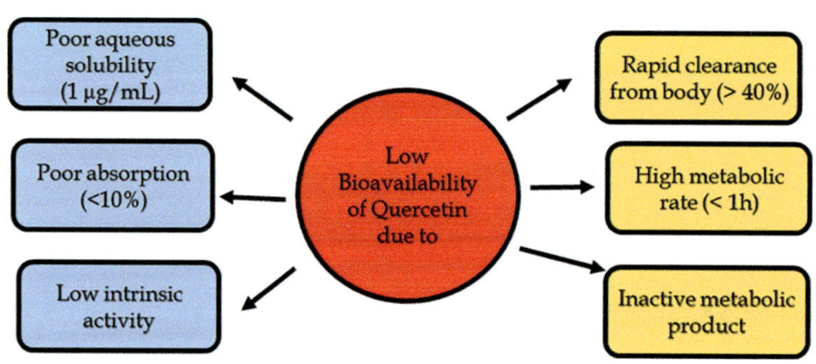

FIGURE 13.3 Cellular processes modulated by quercetin [It regulates various metabolic processes for exhibiting anticancer activity. It inhibits tumour inducers like inducible nitric oxide synthase (iNOS), cyclooxygenase-2 (COX-2), tumour necrosis factor-α (TNF-α), inflammatory mediators like interleukin 6 and 8 (IL-6, IL-8), and interferon-γ. Quercetin regulates tumour growth and metastasis by inducing apoptotic factors through intrinsic and extrinsic pathways, inhibiting the uncontrolled cell cycle (↓ downregulation and ↑ upregulation).

retention effect between tumour tissues and peripheral tissues (Li et al., 2020).

13.3.1.1 Liposomes

Goniotaki et al. (2004) encapsulated quercetin, rutin, isoscutellarein, and isoscutellarein diglycoside into liposomes containing egg phosphatidylcholine by the film hydration method and tested against human cancer cell line. The results showed that the thermal and dynamic effects depend on the flavonoid structure and its interaction with model membranes. In an *in vitro* cytotoxic study, quercetin, rutin, isoscutellarein, and isoscutellarein diglycoside, along with their liposomal formulations, showed reduction of cancer cell growth in human glioblastoma cells (SF267), breast cells (MCF7), and human lung carcinoma (H460) cell lines.

Tammela et al. (2004) studied the permeability and membrane affinity of quercetin/luteolin and alkyl gallates on colon cancer cells (Caco-2). Both quercetin and luteolin showed a greater extent of hydroxylation with no transport across the cell monolayer with higher partition coefficients at 7.1 and 7.5 µM, respectively. Yuan et al. (2006a, 2006b) prepared PEGylated quercetin-loaded nanoliposomes and evaluated their antitumour activity *in vitro* and *in vivo* studies on C57BL/6N mice bearing Lewis lung cancer (LL/2) and BALB/c mice bearing colon adenocarcinoma (CT26) and hepatoma carcinoma cells (H22). Pegylated liposomal quercetin showed enhanced apoptosis *in vitro* and dose-dependent tumour growth inhibition in *in-vivo* studies. There was an increased quercetin accumulation, extended release, and prolongation of residual time in tumour-bearing mice.

Ghosh et al. (2012) prepared galactosylated quercetin-loaded liposomes and tested its effectiveness against hepatocarcinoma in Swiss albino rats. Intravenous administration of galactosylated quercetin (8.98 µM/kg) once a week for 16 weeks prevented the diethylnitrosamine-induced hepatocarcinoma and oxidative damage in rat liver. Ghosh et al. (2010) evaluated the effect of quercetin liposomes on arsenic-induced acute liver toxicity. The single dose of quercetin liposomes (2.71 mg/kg b.wt) injection to rats protected the arsenic-induced hepatic oxidative stress. It inhibited fibrosis-associated upregulation of transforming Growth Factor-β and cytochrome-C release in the cytoplasm.

Wong and Chiu (2010) entrapped quercetin and vincristine into a liposomal delivery to increase the effectiveness against estrogen-receptor-negative breast cancer treatment. A combination of liposomal formulation (quercetin: vincristine; 1:2) showed a synergistic effect in *in-vitro* (3-[4,5-dimethylthiazol-2-yl]-2,5 diphenyl tetrazolium bromide) assay with a combination index of 0.113 and the dose reduction value of 115 at ED_{50} of vincristine, suggesting the formulation is useful for optimal anticancer effect.

Ghosh et al. (2011a, 2011b) studied the effect of quercetin-loaded liposomes with an arsenic chelator (*meso*-2,3-dimercaptosuccinic acid) against arsenic-induced oxidative stress and mitochondrial dysfunction in liver and brain in rats. Co-encapsulation of quercetin and *meso*-2,3-dimercaptosuccinic acid into liposomes showed a synergistic effect in reducing the liver and brain arsenic levels from 9.5 and 4.8 to 2.2 and 1.5 µg/g, respectively by significantly inhibiting cytochrome-c expression in liver and brain. Wong and Chiu (2011) entrapped vincristine and quercetin (1:2) in liposomes for the treatment of breast cancer. Quercetin-vincristine nanoliposomes showed a synergistic effect at two-thirds of the maximum tolerated dose of vincristine in human breast tumour xenograft (JIMT-1) by increasing antitumour efficiency as compared to both free drugs without significant body weight loss in the animals.

Yang et al. (2011) studied the effect of radiofrequency ablation combined with quercetin and doxorubicin liposomes to enhance antitumour activity by modulation of heat shock protein in a rat tumour model. The results showed increased coagulation mass with quercetin combination compared to radiofrequency or quercetin alone. Further adjuvant doxorubicin to quercetin-radiofrequency-doxorubicin showed much greater increases in tumour coagulation for better antitumour actions by suppressing heat shock protein production to increase apoptosis and radiofrequency ablation-induced tumour destruction.

Wang et al. (2012a, 2012b) studied the effect of quercetin liposomes on apoptosis in glioma cells (C6). The results of quercetin liposomes showed dose and time-dependent caspase-independent induction of type III (necrotic) programmed cell death of glioma cells, whereas free quercetin had no effect. Yang et al. (2012) studied the combined effect of radiofrequency ablation and quercetin liposomes intravenously alone or with liposomal doxorubicin to assess the rat model's tumour growth and endpoint survival. The results showed that radiofrequency ablation and intravenous quercetin liposomes significantly decreased the tumour growth rates and increased animal endpoint survival. The addition of doxorubicin to quercetin liposomes further enhanced ablation efficiency.

He et al. (2013) explored the effect of quercetin liposomes for the improvement of sensitivity to thermotherapy and thermos-chemotherapy using tumour *in vitro* studies and *in vivo* tumour-bearing peritoneal carcinomatosis and subcutaneous mouse model of colon cancer. The results suggested the upregulation of heat shock protein 70 (especially Hsp70) and increased apoptosis-induced hyperthermia and thermo-chemotherapy, suggesting the potential use of quercetin liposomes for cancer treatment. (Hu et al., 2013) prepared quercetin liposomes to enhance the distribution through human skin to get an anti-ultraviolet B effect on human epidermal keratinocytes. The results revealed that quercetin liposomes enhanced the permeation more than 3.8-fold compared to free quercetin and reduced the reactive oxygen species and malondialdehyde levels. Thus, it helps to reduce oedema and inflammation, suggesting the potential use against skin cancer.

Liu et al. (2013) investigated the role of quercetin liposomes against radiation-induced acute pneumonitis in a murine C57BL/6J mice model. The results revealed the protective effect of quercetin liposomes against radiation-induced acute pneumonitis by diminishing pulmonary oxidative damage and improving lung fibrosis. Wang et al. (2013a, 2013b, 2013c) studied the effect of quercetin liposomes on glioma cell (C6) death on a molecular basis. The results revealed that quercetin liposomes induced death in glioma cells due to modulation of the signalling cross point in the mitochondrial pathway and the Janus kinase 2/signal transducer and activator of transcription 3 signalling pathway, suggesting the potential use against chemical-resistant glioma.

Zheng et al. (2014a) showed that quercetin liposomes doped with neutralizing antibody dose-dependently inhibited human oesophagal cancer cell (Ec9706) growth and proliferation. Zheng et al. (2014b) investigated the CD133/CD44 expression in esophageal carcinomas (Eca109 and Ec9706 cells). Quercetin liposomes downregulated the biomarkers nuclear factor kappa B, p65, histone deacetylase 1, cyclin D1, and upregulated caspase-3. Dabbagh-Bazarbachi et al. (2014) studied the combined liposomal formulation of quercetin, epigallocatechin gallate, and clioquinol in mouse hepatocarcinoma (HeLa 1–6) cells. They found a rapid upsurge of FluoZin-3 (Zn-specific detector) fluorescence inside the nanoliposomes. Thus, it suggests the role of quercetin, epigallocatechin gallate, and clioquinol on lipid membranes.

Moussa et al. (2014) investigated the effects of various drug-loaded liposomes and micelles on their distribution and effectiveness in managing radiofrequency ablation. The results revealed greater deposition and deeper penetration in liver cancer using beads of size 20 nm. Micelles of doxorubicin and quercetin were more effective in suppressing periablational hypoxia-inducible factor-1α or heat shock protein 70 rim thickness than liposomal formulations. Radiofrequency ablation, doxorubicin liposomes, and quercetin liposomes showed better tumour penetration, prolonged drug release, and tumour inhibition.

Caddeo et al. (2016a) studied the quercetin and resveratrol co-encapsulated liposomes to treat skin cancer potentially. The results showed that the formulation showed better penetration of both drugs in fibroblast and the synergistic anti-reactive oxygen species activity, reducing the 12-*O*-tetra decanoyl phorbol 13-acetate-induced inflammation. Thus, it could be useful for the management of skin cancer. Caddeo et al. (2016b) co-encapsulated pH-sensitive quercetin and resveratrol liposomes using succinyl-chitosan polymer for oral administration. The results revealed no alteration of internal characteristics of liposome bilayers with succinyl-chitosan polymer, suggesting the potential use for treating inflammatory conditions and oxidative stress related to the intestine. Hu et al. (2016) encapsulated quercetin and temozolomide nanoliposomes using novel polymer 1,2-distearoyl-sn-glycero-3-phosphoethanolamine-*N*-[methoxy (polyethylene glycol)-2000] for brain cancer. The results suggested that quercetin-temozolomide nanoliposomes increased the drug loading inside the liposomes, sustained release, enhanced circulation time, and lesser drug clearance. Thus, it was more effective in reducing the growth of glioma cells (C6) than free temozolomide or quercetin.

Using the lipid hydration method, Ravichandiran et al. (2017) developed curcumin nanoliposomes with or without quercetin. They investigated their toxicity *in-vitro* and efficacy *in-vivo* in Dalton's ascites lymphoma in mice models. The results suggested that the quercetin loading in curcumin liposomes enhanced the cytotoxicity *in vitro* in human colon cancer cell lines (HT-29 and HCT-15) and *in vivo* increased the lifespan of tumour-bearing mice.

Rodriguez et al. (2019) prepared and characterized quercetin-loaded liposomes using rice bran and assessed their antioxidant and antiangiogenic properties. The liposomes showed an enhanced % entrapment efficiency of 84.9% with a size of 157.33 nm and enhanced free antioxidant and anti-angiogenic actions of quercetin. Dos Santos et al. (2018) prepared liposomes of *Rapanea ferruginea* leaf extracts (entailing quercetin flavonoids), which showed antioxidant activity in 2,2-Diphenyl-1-picrylhydrazyl and thiobarbituric acid reactive substance assay. The quercetin-containing extract liposomes in the (3-(4,5-dimethyl thiazolyl-2)-2,5-diphenyltetrazolium bromide) assay showed decreased viability of glioma cells (C6) *in vitro* up to 70.8%. In contrast, the free-form extract did up to 40%, suggesting better antioxidant and antitumour actions than free *R. ferruginea* leaf extracts. Zhou et al. (2018) prepared quercetin-loaded PEGylated liposomes using short peptides with amino acid arginine-glycine-aspartic acid to treat lung cancer. The RGD-modified liposomes showed an enhanced encapsulation ratio and drug loading of 89.2% and 9.2%, respectively. There was a slow release of quercetin *in vitro* and greater tumour targeting in tumour-bearing A549 mice.

Kruszewski et al. (2018) showed that quercetin nanoliposomes in an alkaline pH medium altered the surface charge for glioblastoma cells (LN-18 and LN-229) as per micro-electrophoretic mobility calculations. The quercetin moves inside the cells and hinders the propagation of glioblastoma cells. Zuo et al. (2019) prepared the liposomal formulation of 7-*O*-geranyl and quercetin combined with survivin small interfering RNA or interleukin-10 small interfering RNA for treating breast cancer. The results showed that in breast cancer cells (MFC-7), 7-*O*-geranyl and quercetin and survivin small interfering RNA-containing liposomes showed antiproliferative and pro-apoptosis due to intracellular reduction of the amount of surviving and increased caspase-7 levels, suggesting the potential use in the treatment of breast cancer.

Riaz et al. (2019) targeted the transferrin-7 surface receptor, which is overexpressed and restricted in tumours using quercetin-loaded liposomes. The transferrin-7-quercetin nanoliposomes showed threefold increased cytotoxicity in an *in vitro* study due to increased apoptosis and inhibition of *S*-phase in the tumour cell cycle. Pulmonary administration of transferrin-7-quercetin nanoliposomes significantly increased the lifespan in orthotopic lung tumour-bearing mice, suggesting that quercetin-filled phosphatidylcholine liposomes and injection of mesenchymal stem cells can efficiently regenerate the infra-red rays induced vascular dysfunction (Soloviev and Kizub, 2019). Additionally, it includes RNA interface techniques directed to various protein kinase C isoforms. Yu et al. (2019) formulated the quercetin and doxorubicin liposomes for treating multi-drug-resistant cancer. The suppression of P-glycoprotein transmembrane protein mediated the action. The results revealed the enhancement of plasma doxorubicin levels with extended half-life. Quercetin/doxorubicin-nanoliposomes enhanced the cytotoxic effect in doxorubicin-resistant human leukaemia cells (HL-60/ADR), and breast cancer cells (MCF-7/ADR) were increased ~4.81 and 3.21 fold, respectively. In an *in vivo* study, quercetin/doxorubicin-nanoliposomes decreased tumour size and growth.

13.3.1.2 Nanoparticles

Khoee and Rahmatolahzadeh (2012) synthesized folate nanoparticles based on self-assembled brush-like poly (lactic-co-glycolic acid)/polyethylene glycol/2-Aminoethyl methacrylamide copolymer for targeting cancer using the nanoprecipitation method. The nanoparticles prepared showed size <100 nm in dynamic light scattering analysis and low polydispersity in size at different pHs (Bose et al., 2013) formulated topical delivery quercetin from solid lipid-based nanosystem using probe sonication method. The nanosystem showed physical stability for up to 8 weeks at 2°C–8°C. In an *in vitro* release study, the nanosystem showed biphasic release of quercetin, with an initial spurt release followed by sustained release for up to 24 hours. In *in vitro* permeation studies using full-thickness human skin, the nanosystem showed more quercetin accumulation within the skin than the control formulation co-encapsulated tamoxifen and quercetin in polymeric nanoparticles for antitumour efficacy. In the 2,2-Diphenyl-1-picrylhydrazyl assay, the quercetin-loaded freeze-dried nanoparticles retained the antioxidant activity. In breast cancer cells (MCF-7), the tamoxifen-quercetin polymeric nanoparticles showed greater cellular uptake, cytotoxicity, and nuclear co-localization. The uptake studies of nanoformulation in colon cancer cells (Caco-2) showed increased oral bioavailability compared to free drugs. At the same time, higher cellular uptake revealed its potential for oral delivery, which was well corroborated with in-vivo pharmacokinetics, suggesting ~ five and ~threefold increase in oral bioavailability as compared to the free tamoxifen citrate and free quercetin, respectively. In 7,12–dimethylbenz(*a*)anthracene-induced breast cancer model in female Sprague-Dawley rats, the formulations showed higher tumour suppression on multiple administrations by reducing the angiogenesis through reduction of biomarkers like matrix metalloproteinases-2 and 9. The formulations were found safe, with no hepatotoxicity in safety studies. Sahu (2013) formulated quercetin-loaded nanoparticles using ethyl cellulose as a polymer for the treatment of skin cancer. They found entrapment efficiency of 51%–54% with a particle size range of 228 nm and increased drug loading > 34%. In *ex vivo* studies, the quercetin-loaded nanoparticles showed an increase in drug release from nanoparticles and drug retained in skin for longer duration, reducing the dose and frequency of drug administration. Verma et al. (2013) synthesized and characterized surface-coated iron oxide magnetic nanoparticles with poly(lactic-co-glycolic acid) for drug delivery by nebulization. The developed nanocarriers were biocompatible when tested for cytotoxicity on cultured human lung epithelial cells up to 100 µg/mL. In *in-vitro* studies on human lung carcinoma cell line (A549), the quercetin-loaded poly (lactic-co-glycolic acid)-magnetic nanoparticles aerosol formation significantly reduced the number of viable lung carcinoma cells as compared to the nebulization method or by direct pipetting.

Jain et al. (2013) fabricated the quercetin-loaded GeluPearl lipid nanoparticles with or without oil, with a particle size less than 380 nm with more than 90% entrapment efficiency and sustained drug release *in-vitro* for cancer treatment. In C57BL/6 tumour-bearing mice, the formulations reduced the flank tumour volume over 22 days. GeluPearl lipid nanoparticles enhanced lung colonization, and antimetastatic activity was observed against melanoma cells (B16F10) in C57BL/6 mice compared to free quercetin. Kumar et al. (2014) formulated quercetin-conjugated dextran-coated iron oxide nanoparticles using the nanoprecipitation method for breast cancer therapy. In 3-[4,5-dimethylthiazol-2-yl]-2,5 diphenyl tetrazolium bromide assay, the quercetin-loaded nanoparticles showed significant cytotoxicity compared to pure quercetin. Mittal et al. (2014) synthesized bimetallic silver and selenium nanoparticles (Ag/Se Nanoparticles) with quercetin and gallic acid. The nanoparticles showed good in vitro antioxidant and antimicrobial activity (against gram-positive and gram-negative bacteria). In an in-vitro study on Dalton lymphoma cells, the Ag/Se nanoparticles showed excellent antitumour activity, reducing the cell viability by 80% at 50 µg/mL. Varshosaz et al. (2014) prepared quercetin-loaded sterol-containing solid lipid nanoparticles for the treatment of hepatocellular carcinoma by the emulsification solvent evaporation method. In 3-[4,5-dimethylthiazol-2-yl]-2,5 diphenyl tetrazolium bromide assay on hepatocellular cancer cells (HepG2), quercetin-solid lipid nanoparticles were found to be more effective and enhanced accumulation was observed in cells as compared to free quercetin.

Zafar et al. (2014) prepared sterically stabilized polymeric nanoparticles of mitoxantrone dihydrochloride (MTO) with sodium deoxycholate (SDC) MTO-SDC-complex along with an efflux transporter (*P*-glycoprotein/breast cancer resistance protein) inhibitor to reduce the drug resistance in breast cancer. The nanoparticles showed reduced IC_{50} values on MCF7, A2780(p), and A2780(ADR) cells *in-vitro*. The nanoparticles also showed increased residence time in blood with decreased macrophage uptake, thus prolonging release from the formulation. In vitro studies on breast cancer cells (MCF7), the hyaluronic acid-coated nanoparticles showed enhanced cellular uptake, suggesting their targeting potential. The hyaluronic acid-coated nanoparticles and *P*-glycoprotein/breast cancer resistance protein showed better effectiveness for targeted chemotherapy in CD44 overexpressing multidrug-resistant breast cancer.

Bishayee et al. (2015) prepared poly(lactic-co-glycolic acid) - loaded gold nanoparticles precipitated with quercetin and tested them against hepatocarcinoma cells. The nanoparticles induced conformational changes in DNA and affected cell cycle control by modulating various proteins related to epigenetic modifications. In an in vitro cytotoxicity on hepatocellular carcinoma cells (HepG2), the nanoparticles inhibited histone deacetylases to control cell proliferation and arrest the cell cycle at the sub-G stage by restraining the action of proteins p21WAF, cdk1, and pAkt. The nanoparticles also promoted apoptosis in hepatocellular carcinoma cells by stimulating the p53-ROS crosstalk to induce epigenetic modifications, thereby reducing cell multiplication.

David et al. (2015) prepared quercetin and 5-fluorouracil-loaded chitosan nanoparticles with enhanced % entrapment efficiency (quercetin 95% and 5-fluorouracil 75%) with the formulation ratio of quercetin: 5-fluorouracil: chitosan (1:2:3). The nanoparticles showed enhanced toxicity against pancreatic cancer cells suggesting the potential use in the treatment of pancreatic cancer.

Du et al. (2015) prepared quercetin-loaded self-aggregates based on *O*-carboxymethyl chitosan-cholic acid conjugates. They assessed the bioavailability when altered with glycyrrhetinic acid in the nanocarrier fabrication. The quercetin nanoconjugate showed acidic pH selectivity. At pH 5.7 and 6.5, it released quercetin-*O*-carboxymethyl chitosan-cholic acid and quercetin-glycyrrhetinic acid - *O*-carboxymethyl chitosan-cholic acid. The cellular toxicity *in vitro* and higher cell apoptosis *in vivo* were observed with quercetin-glycyrrhetinic acid - *O*-carboxymethyl chitosan-cholic acid due to enhanced retention time. Pandey et al. (2015) prepared quercetin-loaded poly(lactic acid) nanoparticles using an emulsified nanoprecipitation method and tested them on human breast cancer cells. Quercetin-loaded poly(lactic acid) nanoparticles showed sustained release of quercetin from the polymer matrix because of delayed diffusion and a stronger interface. In an *in vitro* cytotoxicity study, quercetin-loaded poly(lactic acid) nanoparticles killed ~50% of breast cancer cells, suggesting the potential use in the treatment of breast cancer.

Sharma et al. (2015) prepared methoxy poly (ethylene glycol)-poly (lactide) nanoparticles encapsulated quercetin nanoparticles for the treatment of breast cancer. In an in-vitro study, polymeric quercetin nanoparticles induced apoptosis in a triple-negative breast cancer cell line (MDA-MB-231) with more quercetin in cell lysate than free quercetin. An *in-vivo* study involving a xenograft mice model for breast cancer polymeric quercetin nanoparticles inhibited tumour growth. Also, the terminal deoxynucleotidyl transferase deoxyuridine triphosphate nick end labelling assay exhibited more apoptotic cells in polymeric quercetin nanoparticles-treated tumours than free quercetin at the same dose.

Balakrishnan et al. (2016) prepared quercetin-conjugated gold nanoparticles for breast cancer. They assessed mesenchymal transition, angiogenesis, and invasiveness using scratch wound-healing assay, chick embryo angiogenesis assay and 7,12-dimethylbenz[*a*]-anthracene-induced mammary carcinoma in Sprague-Dawley rats, respectively. There was a significant decrease in protein expression of vimentin, *N*-cadherin, snail, slug, twist, matrix metalloproteinases-2, matrix metalloproteinases-9, *p*-epidermal growth factor receptor, vascular endothelial growth factor receptors-2, *p*-phosphatidylinositol-3 kinase, protein kinase B, and *p*-glycogen synthase kinase, and increased E-cadherin protein expression in response post-quercetin-gold nanoparticles treatment. Quercetin gold nanoparticles inhibited the cell migration and invasion of breast cancer cells (MCF-7 and MDA-MB-231) more effectively than free quercetin. Treatment with quercetin-gold nanoparticles of human umbilical vein endothelial cells exhibited reduced cell viability and new blood vessel formation. Quercetin gold nanoparticles showed excellent *in vitro* and *in vivo* antiangiogenic activity and suppressed tumour growth in 7,12-dimethylbenz[*a*]-anthracene-induced mammary carcinoma in Sprague-Dawley rats more effectively than free quercetin.

Fatma et al. (2016) prepared poly(D,L-lactic-co-glycolic acid)-conjugated biodegradable nanoparticles with etoposide and quercetin for oral delivery in treating cancer. In 3-[4,5-dimethylthiazol-2-yl]-2,5 diphenyl tetrazolium bromide assay, etoposide-loaded nanoparticles and poly(D,L-lactic-co-glycolic acid)-etoposide/quercetin-nanoparticles showed nine and 11-fold reduction in IC_{50} value, respectively, as compared to free drugs. Poly(D,L-lactic-co-glycolic acid)-etoposide/quercetin-nanoparticles in pharmacokinetic studies on rats revealed 2.4-fold enhanced bioavailability of etoposide but not quercetin.

Guan et al. (2016) prepared quercetin-loaded poly(D,L-lactic-co-glycolic acid)-tocopheryl polyethylene glycol 1000 succinate nanoparticles for the treatment of liver cancer and quercetin/coumarin-6-loaded poly(D,L-lactic-co-glycolic acid)-tocopheryl polyethylene glycol 1000 succinate nanoparticles with coumarin-6 as a fluorescent marker. Both showed significant activity for cellular uptake by liver cancer cells (HepG2 and HCa-F) and enhanced apoptosis of HepG2 cells. *In-vivo* studies on solid tumour-bearing mouse models showed that quercetin-loaded poly(D,L-lactic-co-glycolic acid)-tocopheryl polyethylene glycol 1000 succinate nanoparticles suppress tumour growth by 59%, thus used as an intravenous dosage form for the treatment of liver cancer with improved liver targeting.

Lou et al. (2016) showed that quercetin gold nanoparticles induced autophagy and apoptosis in human neuroglioma cells through protein kinase B/extracellular signal-regulated kinase/caspase-3 signalling pathway. Treatment with quercetin gold nanoparticles in flow cytometry revealed enhanced apoptotic cells compared to the untreated group. Furthermore, quercetin gold nanoparticles downregulated activated phosphatidylinositol 3-kinase/protein kinase B and B-cell leukaemia 2 expression in human neuroglioma cells. Also, the expression of microtubule-associated protein 1A/1B-light chain 3, extracellular signal-regulated kinase, cytoplasm p53 and cleaved caspase-3 mediated poly (ADP-ribose) polymerases were positively correlated with the concentration of quercetin gold nanoparticles. Pre-treatment with quercetin gold nanoparticles downregulated *p*-mammalian target of rapamycin and Gα-interacting protein dose-dependently.

Luo et al. (2016) prepared quercetin gold nanoparticles to treat cervical cancer. They found the effect by inducing apoptosis, autophagy, and anti-proliferation via Janus kinase 2 suppression in cervical cancer cell lines. Results suggested that Janus kinase 2 is expressed more in cancer cell lines than in normal cervical cells. Also, autophagy and apoptosis were observed in Janus kinase 2-inhibited cancer cells through activating caspase-3 and inhibiting cyclin-d1 and mammalian target of rapamycin regulated by signal transducer and activator of transcription 3/5 and phosphatidylinositide 3-kinase/protein kinases B signalling pathway. Thus, quercetin gold nanoparticles repressed cervical cancer by Janus kinase 2 suppression.

Lv et al. (2016) co-encapsulated quercetin and doxorubicin biotin-conjugated polyethylene glycol-polycaprolactone nanoparticles for treating drug-resistant breast cancer. The results showed that the doxorubicin-quercetin-biotin-polyethylene glycol-polycaprolactone nanoparticles had more effective cellular uptake in breast cancer cells (MCF-7/ADR cells) than free drugs. Mainly, it reduced the drug resistance both *in vivo* and *in vitro*. Martirosyan et al. (2016) studied the effect of quercetin silver nanoparticles on treating intestinal cancer. The results showed that treatment with quercetin silver nanoparticles reduced the cell viability, reduced thiol groups induced oxidative/nitrosative stress and lipid peroxidation, and increased the activity of various antioxidant enzymes. In an *in-vitro* colon cancer cell (Caco-2) culture as model of the human intestinal mucosa, quercetin silver nanoparticles showed anti-inflammatory effect due to downregulation of interleukin-8 levels. Thus, quercetin-loaded silver nanoparticles reduction of cytotoxicity, and oxidative stress and recovered the metabolic activity of Caco-2 cells. It also helps reduce the dangerous activities of quercetin silver nanoparticles *per se* on intestinal cells.

Saha et al. (2016) synthesized quercetin-loaded poly(D,L-lactic-co-glycolic acid) nanoparticles of adriamycin and mitoxantrone by surface modification using bovine serum albumin or histones for the cancer treatment of multi-drug-resistant cancer. The nanoformulations showed extended release up to 20 days with an early burst release in an in vitro study. Surface modification with histone yielded greater drug loading in quercetin nanoparticles than albumin. Using such quercetin nanoparticles may overcome the problem of developing multi-drug resistance to cancerous cells.

Suksiriworapong et al. (2016) studied the comparison of different hydrophobic chain lengths of poly (ε-caprola ctone)-co-D-α-tocopheryl-poly (ethylene glycol) 1000 succinate nanoparticles to increase quercetin loading in breast cancer cells (SKBR3). The results showed that longer poly (ε-caprolactone) chains exhibited enhanced crystallinity than shorter chains and increased the drug loading by increasing the particle size, which simultaneously retard the quercetin release from nanoparticles. Poly(ε-caprolact one)-co-D-α-tocopheryl-poly (ethylene glycol) 1000 succinate nanoparticles enhanced the cytotoxicity of quercetin by 2.9 times against breast cancer cells as compared to free quercetin, suggesting their use for the delivery of quercetin to breast cancer cells.

Zhu et al. (2016) prepared quercetin-loaded poly(D,L-lactic-co-glycolic acid)-D-α-tocopheryl-poly (ethylene glycol) 1000 succinate for skin cancer. In vitro studies suggest the protective effect of polymeric nanoparticles of quercetin to reduce ultraviolet-B irradiation by inhibiting cyclooxygenase-2 up-expression and - activation in the skin cancer cell line (HaCaT). Quercetin-poly(D,L-lactic-co-glycolic acid)-D-α-tocopheryl-poly (ethylene glycol) 1000 succinate nanoparticles showed increased penetration through the epidermis and reached the dermis in tumour xenograft mouse model, suggesting their use for external application targeted skin cancers.

Malekzadeh et al. (2017) prepared quercetin-loaded magnetic nanoparticles coated with polycitric acid and polyethylene glycol and conjugated with folic acid to target magnetic resonance imaging and cancer therapy. Quercetin-polycitric acid - polyethylene glycol-folic acid nanoparticles showed enhanced quercetin loading and release from nanoparticles as well as increased cellular uptake by folate receptor overexpressing cervical cancer cells (HeLa) as compared to polycitric acid - polyethylene glycol-quercetin magnetic nanoparticles. Quercetin- polycitric acid - polyethylene glycol-folic acid nanoparticles also showed increased cytotoxicity, and magnetic resonance imaging signals were increased concentration-dependently, suggesting their use for drug targeting in cancer chemotherapy and magnetic resonance imaging contrast agent.

Ren et al. (2017) investigated the antitumour activity of quercetin nanoparticles in liver cancer cells by inhibiting proliferation and inducing apoptosis. The results confirmed the actions due to the inactivation of the caspase/cytochrome-C pathway, inhibiting AP-2β/human telomerase reverse transcriptase, nuclear factor kappa B/cyclooxygenase-2 and hindering protein kinase B/extracellular signal-regulated protein kinases 1 and 2 signalling pathways. Thus, it suggests using quercetin nanoparticles as a potential approach for targeting liver cancer.

Xing et al. (2017) prepared quercetin nanoparticles conjugating with triphenylphosphine, which was further modified by polyethylene glycol to get a pH-responsive quercetin-polyethylene glycol - triphenylphosphine nanoparticles. The nanoparticles targeted mitochondrial apoptosis with reactive oxygen species. They activated the caspase cascade by releasing Cytochrome-C from mitochondria to cytosol, thus suggesting triphenylphosphine-based nanoparticles for cancer therapy. Xu et al. (2017) prepared quercetin-loaded gold nanoparticles and assessed the anti-inflammatory and anti-oxidative stress effect on endotoxemia-induced kidney injury in C57BL/6 mice. Results showed that quercetin gold nanoparticles inhibited the upregulated nuclear factor kappa B (inflammatory) kelch-Like ECH-associated protein 1/nuclear factor (erythroid-derived 2)-like 2 protein (oxidative stress) pathways by high-fat diet intake in mice, eventually lessening podocyte injury. It suggested the use of quercetin gold nanoparticles in metabolic disorders and to prevent kidney injury, insulin resistance and pro-inflammatory cytokine production by inhibition of Toll-like receptor 4/nuclear factor kappa B and oxidative stress, further increasing superoxide dismutase activity.

Yuan et al. (2017) prepared quercetin-loaded graphene oxide-silver nanoparticle nanocomposites to treat neuroblastoma. Results showed that quercetin-graphene oxide-silver nanoparticles exhibited greater cytotoxicity against human neuroblastoma cancer cells (SH-SY5Y) due to a decrease in cell viability, cell proliferation, mitochondrial membrane potential, mitochondria number, anti-apoptotic genes expression and increased leakage of lactate dehydrogenase, reactive oxygen species production, pro-apoptotic genes expression. Quercetin-graphene oxide-silver nanoparticles caused increased caspase-9/3-dependent apoptosis via DNA fragmentation and accumulation of autophagosomes and autophagic vacuoles in neuroblastoma cancer cells to enhance cytotoxicity.

Yang et al. (2017) prepared quercetin/cadmium selenide/zinc sulfide nanoparticles with a diameter <10 nm. Quercetin/cadmium selenide/zinc sulfide nanoparticles showed two to six fold increased cytotoxicity in 3-[4,5-dimethylthiazol-2-yl]-2,5 diphenyl tetrazolium bromide assay compared to free quercetin. Additionally, in using real-time cell electronic sensing studies of gastric cancer cell line (BGsC-823), the quercetin/cadmium selenide/zinc sulfide nanoparticles exhibit an effect on migration and proliferation of cells, thus showing good anticancer activity.

Zhu et al. (2017) co-delivered vincristine and quercetin using lipid-polymeric nanocarriers (vincristine/quercetin-lipid-polymeric nanocarriers) for the treatment of non-Hodgkin's lymphoma. They investigated them in vitro cell culture models and a tumour xenograft mouse model. The results suggested the sustained release profile of both drugs and the best antitumour efficacy for treating chemoresistant lymphoma. Baksi et al. (2018) prepared quercetin-loaded chitosan nanoparticles with enhanced % entrapment efficiency and sustained drug release activity. Quercetin chitosan nanoparticles reduced IC_{50} compared to free quercetin in an *in vitro* cytotoxicity study. In tumour xenograft mice with lung cancer cells (A549) and triple-negative breast cancer cells (MDA MB 468), Quercetin chitosan nanoparticles showed reduced tumour volume and increased serum antioxidant enzymes, suggesting the potential use of quercetin chitosan nanoparticles in targeted cancer therapy.

Yamina et al. (2018) prepared quercetin-loaded polyethylene glycol-graft-halloysite nanotubes for the treatment of cancer to enhance the loading capacity and controlled release. Quercetin-halloysite-*g*-polyethylene glycol was decorated with carbon quantum dots for fluorescent properties and biotin for precise targeting of tumour tissue and higher cellular uptake. The prepared nanoparticles showed superior quercetin loading capacity, *in-vitro* sustained quercetin release and low toxicity in cervical cancer cells (HeLa) and higher antitumour activity in 3-[4,5-dimethylth iazol-2-yl]-2,5-diphenyl tetrazolium bromide assay as compared to free quercetin.

13.3.1.3 Nanoemulsion

Schwingel et al. (2014) investigated the effects of resveratrol, rutin, quercetin, and quercetin nanoemulsion on oxaliplatin-induced hepatotoxicity and neurotoxicity in BALB/c mice. The results showed that treatment with resveratrol, rutin, quercetin, and quercetin nanoemulsion ameliorated the oxaliplatin-induced mechanical allodynia by reducing the immunopositivity for caspase-3, an apoptosis marker. Quercetin and quercetin nanoemulsion inhibited the myeloperoxidase activity; thus, the neutrophil infiltration in oxaliplatin-induced hepato and neurotoxicity.

Jain et al. (2014) prepared a quercetin-loaded self-nanoemulsifying drug delivery system. *In vitro* studies on breast cancer cells (MCF-7), quercetin-self-nanoemulsifying drug delivery showed dose-dependent cytotoxicity and antioxidant activity against 7,12-dimethylbenz[*a*]-anthracene-induced breast tumours, thereby increasing the latency to tumour growth as compared to free quercetin. Normalization of levels of tumour angiogenesis markers (matrix metalloproteinase-2, matrix metalloproteinase-9, tumour necrosis factor-α, and interleukin-6) further confirmed the finding. Quercetin-self-nano emulsifying drug delivery showed pro-oxidant activity at 100 mg/kg in the same model and exhibited significantly higher therapeutic anticancer efficacy. It was found to be safe without hampering the hepatic enzymes.

Pangeni et al. (2018) developed nanoemulsion (w/o/w) co-delivery of quercetin and pemetrexed by ion-pairing complex of pemetrexed with *N*α-deoxycholyl-L-lysyl-meth ylester, which merged with quercetin nanoemulsion. Oral administration of nanoemulsion enhanced the bioavailability by 4.51 and 23.9 fold of pemetrexed and quercetin, respectively. It exhibited a synergistic inhibitory effect on lung cancer cells (A549) proliferation and migration in tumour-bearing mice.

Arbain et al. (2018) formulated the quercetin nanoemulsion using palm oil ester for pulmonary delivery in lung cancer treatment. The results showed that the aerosol output, median mass aerodynamic diameter, and aerosol rate of 99.31%, 4.25 μm and 0.19 g/mL, respectively, suggesting the potential drug delivery for lung cancer. Arbain et al. (2019) developed the quercetin nanoemulsion (O/W) using palm oil ester or ricinoleic acid as an oil phase as a lung delivery system. The emulsion showed enhanced bioavailability and prolonged release of quercetin and induced the cytotoxicity of lung cancer cells (A549), sparing the normal cells. Ojeda-Serna et al. (2019) developed organogel-based nanoemulsions loaded with three different flavones: betulin, curcumin, and quercetin, vegetable oil (oil phase) and studied the *in vitro* digestion (oral, gastric, and intestinal phase), lipolysis assays, permeability, and bioaccessibility on colon cancer (Caco-2) cells. There was an increased bioaccessibility of quercetin, while curcumin and betulin did not improve. The permeability was improved with betulin but not with curcumin.

13.3.1.4 Micelles

Gao et al. (2012) improved the solubility of quercetin in water by encapsulating it in biodegradable monomethoxypoly (ethylene glycol)-poly(ε-caprolactone) micelles for the treatment of ovarian cancer. In an *in vitro* study, the formulation showed a dose-dependent reduction of the growth of ovarian cancer cells (A2780S). In an *in vivo* study involving xenograft A2780S ovarian tumours, the intravenous administration of monomethoxypoly (ethylene glycol)-poly(ε-caprolactone) micelles showed increased apoptosis and decreased angiogenesis compared to free quercetin.

Xu et al. (2015) prepared quercetin micelles for enhancing the solubility and bioavailability of quercetin using biodegradable monomethoxypoly (ethylene glycol)-poly(ε-caprolactone) for the treatment of colon cancer. The formulation improved cell apoptosis and inhibited angiogenesis and cell proliferation, suggesting the potential use against colon cancer. Wang et al. (2016) prepared the quercetin-loaded freeze-dried nanomicelles for enhancing absorption and increasing efficiency in the management of glioma. The results showed that treatment with quercetin freeze-dried nanomicelles increased intracellular uptake *in vitro* in colon cancer (Caco-2) cells and in vivo increased penetration of blood-brain barrier and increased cytotoxicity on glioma-bearing mice.

Zhao et al. (2016) prepared the quercetin nanomicelles using 1,2 distearoyl-sn-glycero-3-phosphatidylethanolamin e-polyethene glycol 1000 to manage prostate cancer. The results suggested that quercetin nanomicelles enhanced water solubility up to 45-fold, with reduced IC_{50} value in vitro compared to free quercetin. Quercetin nanomicelles were more effective in inhibiting cell proliferation and induced more apoptosis of human androgen-independent PC-3 cells, suggesting that the quercetin nanomicelles possess superior anticancer activity against prostate cancer. Patidar et al. (2017) investigated the effect of glucose on micellization using amphiphilic copolymer Tetronic® 1307 in water and formed quercetin and curcumin micelles. The results showed enhanced solubilization, controlled release kinetics, and cytotoxicity, suggesting the potential use of quercetin/curcumin-loaded Tetronic® 1307 (T1307) micelles to manage cancer.

Rich et al. (2017) studied the upregulation of quercetin in colon adenocarcinoma (CaCo-2/TC7) cells with and without blended nanomicelles. The results showed that the nanomicelles subdued the transportation of quercetin into the cells. The segregation of quercetin into simple bile salt micelles was less. For mixed micelles, it was suppressed by increasing bile salt concentration, which suggests the possibility of low bioavailability of quercetin.

Sandhu et al. (2017) prepared quercetin micelles with surface-modified salicylic acid to manage skin cancer effectively. The quercetin-salicylic acid nanomicelles showed enhanced permeation, permeability and cellular uptake *in vitro*. The quercetin-salicylic acid nanomicelles increased apoptosis, suggesting their potential use for skin cancer treatment. Ahmad et al. (2018) prepared quercetin nanomicelles and investigated their role in hepatic radiofrequency ablation heating parameters and initiation of heat shock proteins in decreasing remote tumour growth. The results showed that lesser radiofrequency heating leads to bigger tumours, while higher radiofrequency heating decreases tumour growth. Also, hepatic radiofrequency ablation at lower doses possesses more periablational heat shock proteins 70 than anti-heat shock proteins 70 quercetin nanomicelles.

Mu et al. (2019) formulated multifunctional quercetin-conjugated chitosan nanomicelles for oral delivery of doxorubicin. The conjugate thus prepared showed *P*-glycoprotein inhibition and permeation enhancement of doxorubicin, thereby promoting twofold higher cellular uptake of doxorubicin than free doxorubicin. *In vitro* studies on colon cancer cells (Caco-2), quercetin-conjugated chitosan nanomicelles also lowered the transepithelial electrical resistance value with tenfold higher level of DOX, suggesting the potential role of quercetin-conjugated chitosan nanomicelles for the oral delivery of insoluble anticancer drugs.

13.3.1.5 Dextran/Chitin Conjugates

Tabassum et al. (2013) synthesized quercetin-Cu(II)-Sn2(IV) and Zn(II)-Sn2(IV)1 for potential treatment against cancer. The plasmid Bolivar Rodriguez322 DNA cleavage by quercetin complexes 1 and 2 was observed via oxidative and hydrolytic pathways. *In vitro,* the antioxidant activity of complex 1 for superoxide dismutase showed IC_{50} of less than 2.27 μM. Complex 1 in an *in vivo* cytotoxicity in human carcinoma cell lines revealed more tumour inhibitory activity via Topo-I.

Oliver et al. (2018) prepared a dextran-quercetin conjugate with potential anticancer properties. Water-soluble quercetin is obtained by conjugating quercetin to an aldehyde functionalized dextran by a hydrochloric acid-catalysed condensation reaction. Compared to free quercetin, improved stability and antioxidant activity have been observed for the synthesized conjugate. Singh et al. (2018a) synthesized quercetin conjugate using chitin-glucan-aldehyde, which showed antioxidant activity. In *in-vitro* studies, the formulation showed excellent cytotoxicity on macrophage cancer cell lines (J774), but no effect was observed in peripheral blood mononuclear cells. Singh et al. (2018b) synthesized curcumin-loaded conjugate using chitin-glucan with quercetin to treat cancer. *In-vitro* studies on peripheral blood mononuclear cells and cytotoxicity tests on macrophage cancer cell line (J774) revealed excellent cytotoxicity and biocompatibility, respectively. The conjugate showed much higher antioxidant activity than free quercetin.

13.3.1.6 Cubosomes, Phytosomes, and Niosome-Based Nanoformulations

Murgia et al. (2013) reported monoolein-doped quercetin cubosomes with dansyl and fluorescein (fluorescent probes) to enhance the encapsulation efficiency. The results showed that the quercetin cubosomes showed significant anticancer activity *in vitro* against fibroblast cells (3T3), suggesting the potential use of fluorescent cubosomes in cancer treatment. Minaei et al. (2016) investigated the role of quercetin-phytosomes in adjunct with doxorubicin for apoptosis of breast cancer. The results showed that quercetin-doxorubicin-phytosomes inhibited the proliferation in 3-[4,5-dimethylthiazol-2-yl]-2,5 diphenyl tetrazolium bromide assay and significantly enhanced the apoptosis of breast cancer cells (MCF-7).

Karthivashan et al. (2016) prepared the phyto-cubosomes from the ethyl acetate fraction of *Moringa oleifera* leaves using a sequential loading technique. The ethyl acetate fraction was quantified for quercetin, kaempferol content, and apigenin. Phyto-cubosomes showed excellent release kinetics from the phyto-cubosome indirectly in 2,2-Diphenyl-1-picrylhydrazyl assay with no toxicity on human hepatoma cell line when assessed with 3-(4,5-dimethylthiazol-2-yl)-2,5-diphenyltetrazolium bromide assay. Thus, it suggests using a sequential loading technique in multi-flavonoid nanodelivery for sustained release and antioxidant properties.

Tavano et al. (2016) entrapped quercetin, doxorubicin, and curcumin in a niosomal formulation for the treatment of breast cancer. The results showed that the niosomal formulations have an increased bioavailability and drug loading at the tumour site and excellent anticancer activity in vitro, suggesting their use in breast cancer treatment. Babazadeh et al. (2018) reviewed the promising use of phytosomes complexes in chemotherapy treatment. Phytosomal formulations enhance the anticancer properties of phytoconstituents. Hemati et al. (2019) entrapped the doxorubicin and quercetin in PEGylated niosomes and small snooping/silencing RNA to enhance the anticancer activity against various cancers like stomach, prostrate, human foreskin fibroblast, and breast cancer. PEGylated – doxorubicin/quercetin niosomes exhibited greater anticancer potential than the blank formulations.

13.3.1.7 Solid Lipid Nanoparticle-Based Nanoformulation

Bose et al. (2013) prepared quercetin solid lipid nanoparticles to treat skin cancer. The results of *in vitro* investigation revealed initial biphasic burst release of quercetin from solid lipid nanoparticles followed by sustained release up to 24 hours. The skin permeation studies showed more accumulation of quercetin in epidermal layers than in control samples, which protects against ultraviolet radiation-induced cell necrosis of epidermal cells. Varshosaz et al. (2014) entrapped quercetin in solid lipid nanoparticles using cholesterol, stigmasterol, and stigmastanol to increase the permeability and target hepatocellular carcinoma cells (HepG-2). The results of 3-(4,5-dimethylthiazol-2-yl)-2,5-diphenyltetrazolium bromide assay showed that the quercetin in cholesterol solid lipid nanoparticles was able to reduce the IC_{50} value by 6 times and 2 times less as compared to free quercetin and phytosterol solid lipid nanoparticles, respectively and showed more drug accumulation in hepatocellular carcinoma cells (HepG2) suggesting the potential use in hepatocellular carcinoma.

Chen et al. (2018) formulated redox-responsive novel quercetin-solid lipid nanoparticles based on disulfide (S–S) bonds and conjugation with glyceryl caprylate caprate. The quercetin-solid lipid nanoparticles showed greater entrapment of paclitaxel due to aromatic rings and selective rapid drug release due to the breakdown of S–S bonds only in tumour cells as compared to normal cells.

13.3.1.8 Nanocapsule-Based Nanoformulations

Ghosh et al. (2009) investigated the effect of quercetin nanocapsules on arsenic-induced brain and liver oxidative stress in rats. The results showed that the quercetin-nanocapsules ameliorated the reactive oxygen species-mediated oxidative stress and improved the arsenic-induced membrane microviscosities in the brain and liver. Ghosh et al. (2010) investigated the effect of quercetin nanocapsules on arsenic-induced hepatic oxidative stress, fibrosis, gene expression, and transforming growth factor-beta levels in rats. The results showed that the single oral administration of quercetin nanocapsules 0.5 mL ameliorated the arsenic-induced hepatic oxidative stress, fibrosis, and gene expression in the liver, suggesting the potential use in liver cancer and fibrosis. Ghosh et al. (2011a, 2011b) formulated quercetin nanocapsules dimercaptosuccinic acid nanocapsules either *per se* or in adjunct with poly(lactide-co-glycolide) and investigated their action on arsenic-*induced* toxicity in rats. The results showed that the quercetin/dimercapto succinic acid- poly(lactide-co-glycolide) nanocapsules showed protection against arsenic-induced toxicity by suppressing the reactive oxygen species and stabilising the mitochondrial activity in the brain and liver.

Ghosh et al. (2012) studied the effect of quercetin nanocapsules against diethylnitrosamine-induced liver cancer in rats. The results showed that quercetin nanocapsules suppressed oxidative stress and enhanced the levels of glutathione-*S*-transferase and cytochrome-C expression in the liver, thereby preventing liver cancer.

Ding et al. (2014) entrapped quercetin into folate nanocapsules using the phase inversion method, which showed prolonged release of quercetin in rats. *In-vitro* studies on cervical cancer cells (HeLa) and resistant breast cancer cells (MCF-7/MDR) showed significant antiproliferative actions and enhanced cellular uptake of quercetin as compared to free quercetin. Also, it inhibited the tumour growth *in vivo* in hepatocellular carcinoma (H22) tumour-bearing mice. Mandal et al. (2014) studied the effect of quercetin nanocapsules on diethylnitrosamine-induced hepatocarcinogenesis in Swiss rats by estimating an inflammatory mediator matrix metalloproteinase-13. The results showed that the quercetin nanocapsules ameliorated the reactive oxygen species-induced oxidative stress, tumour necrosis factor-α, and interleukin-6, initiation of matrix metalloproteinase-13. Thus, it was found to be effective against diethylnitrosamine-induced hepatocarcinogenesis.

Hatahet et al. (2017) enhanced the water solubility of quercetin more than 5,000-fold by entrapment of quercetin in lipid nanocapsules. The lipid-quercetin nanocapsules did not modify the antioxidant activity of quercetin in the 2,2-Diphenyl-1-picrylhydrazyl assay, but were found to be effective in tamm-horsfall protein-1, used to test leukaemia cell lines from hydrogen peroxide-induced stress, suggesting the potential use against skin cancer. Hatahet et al. (2018) developed and studied quercetin liposomes, quercetin nanocapsules, and quercetin-smart Crystals® for quercetin delivery through the skin. The results showed no toxicity in skin cancer cells (HaCaT), further confirmed by the excellent antioxidant ability of these formulations on tamm-horsfall protein-1 cells. There was more deposition of quercetin in the epidermis by quercetin-smart Crystals® and quercetin nanocapsules, suggesting their use in skin treatments. Omwenga et al. (2018) studied the effect of free quercetin and baicalein and after conjugation with chitosan nanocapsules on biofilm establishment, cytotoxicity effect and quorum sensing on Madin-Darby canine kidney (MDCK-7) cells. Both free-form drugs showed minor inhibition of quorum sensing response, biofilm establishment, and higher concentration cytotoxicity against mammalian cells, though their nanoenabled delivery system showed cytoprotective actions.

13.3.1.9 Miscellaneous

Gismondi et al. (2015) entrapped quercetin and ciproten in nanodiamonds and investigated the effect of surface interaction by different methods (oxidized, wet chemical reduced, and plasma reduced). The results showed that quercetin nanodiamonds enhanced the antiproliferative actions and cytotoxic actions of quercetin and ciproten on cervical (HeLa) and murine (B16F10) tumour cells as compared to free drugs.

Gupta et al. (2015) prepared quercetin-conjugated poly-β-amino ester nanogels, which enhanced quercetin aqueous solubility in the presence of polyethylene glycol, more controlled and prolonged release. Quercetin poly-β-amino ester nanogel enhanced antioxidant activity for a longer period, thus suggesting its use in cancer management.

Cirillo et al. (2013) developed a three-functional nanocomposite of quercetin to treat cancer by radical polymerization of methacrylic acid around carbon nanotubes. The nanocomposite showed better antioxidant properties in 2,2-Diphenyl-1-picrylhydrazyl and 2,2′-azino-bis(3-ethylbenzothiazoline-6-sulfonic acid) assay as compared to free quercetin and no *in-vivo* toxicity in cell viability assay.

Lakshmi et al. (2019) prepared ruthenium-quercetin-conjugated nanoclusters to treat cancer. The ruthenium-quercetin nanoclusters showed anticancer properties on non-small cell lung cancer (A549) but no significant cytotoxicity activity in human dermal fibroblast (HDFa) cell lines. For the first time, Lakshmi and Kim (2019) prepared quercetin-mediated gold nanoclusters for the treatment of lung cancer. Studies using bio-atomic force microscopy and confocal imaging of lung cancer cell lines (A549) revealed high anticancer activity with minimal toxicity on fibroblast cells (HDFa).

Lockhart et al. (2015) entrapped tamoxifen and quercetin in polyester nanosponge particles. To assess the effect on metabolism and drug release kinetics in gastrointestinal fluids, tamoxifen-quercetin nanosponge was administered at concentrations of nanosponge-4% and nanosponge-8%. The metabolism of tamoxifen was decreased by twofold for nanosponge-4% and threefold for nanosponge-8%, respectively. In contrast, the metabolism of quercetin was reduced by threefold in nanosponge-4% and fourfold in nanosponge-8%. The cytotoxicity results supported this and showed similar anticancer effectiveness of tamoxifen-quercetin nanosponge.

Dos Santos et al. (2019) prepared quercetin asolectin-based magnetoliposomes for contrast and anticancer purposes. The interaction of quercetin with certain magnetoliposome regions significantly reduced the efficacy of magnetoliposomes encapsulated with maghemite nanoparticles. Also, there was a reduction in magnetic capacity with the incorporation of quercetin without affecting the superparamagnetic activity of magnetoliposomes. In the presence of quercetin, the lipid peroxidation was reduced by eightfold *in vitro*, and cell viability of glioma cells (C6) was reduced by 45%. Thus, quercetin-loaded magnetoliposomes could be used in contrast in cancer management.

George et al. (2019) formulated quercetin-loaded chitosan-cellulose hydrogel with green zinc oxide nanoparticles for anticancer activity. Quercetin-loaded zinc oxide nanoparticles showed a remarkable increase in drug loading compared to free quercetin at pH 5, which is suitable for anticancer applications. The formulation showed excellent biocompatibility and anticancer activity *in-vitro* using murine fibroblast cells (L929) and human skin carcinoma cell lines (A431).

13.3.2 Active Targeting Strategy for the Quercetin

Passive targeting drug delivery has few clinical applications due to a lack of selectivity and affinity, which causes toxicity and side effects in normal tissues. On the other hand, active targeting is designed by using the differences in receptors and antigens expressed on the surface of normal and cancer cells. The ligands have a high affinity to bind to specific receptors, thus increasing the drug loading at the cancer site, sparing the normal tissue. In active targeting, vectors are commonly proteins, peptides, small molecules, and aptamers; examples include folic acid, transferrin, and arginyl-glycyl-aspartic acid peptide (Li et al., 2020).

El-Gogary et al. (2014) prepared the quercetin nanocapsules for active targeting tumours with folic acid doped with poly(lactic-co-glycolic acid). The results showed that the folic acid-targeted nanocapsules showed selective uptake and enhanced cytotoxicity in cervical cancer cells (HeLa); thus, they could be used for active targeting of the tumours. Jiang et al. (2018) actively targeted quercetin using the co-delivery of quercetin with copper sulfide nanoparticle using zeolitic imidazolate framework-8 for photothermal therapy and stabilized by folic acid–bovine serum albumin to get quercetin- copper sulfide - zeolitic imidazolate framework-8 - folic acid–bovine serum albumin. The results of near-infrared fluorescent imaging showed an enhanced accumulation of quercetin nanoparticles inside the cells due to folic acid receptor-mediated endocytosis. The nanoparticles and photothermal therapy showed the synergistic antitumour effect *in-vitro* and *in-vivo* studies, suggesting multifunctional drug delivery in cancer management.

Gupta et al. (2016) actively targeted quercetin using mannose-quercetin-poly(DL-lactide-co-glycolic acid) and folic acid-quercetin-poly(DL-lactide-co-glycolic acid) nanoparticles for the treatment of skin cancer. The results showed that the nanoparticles excelled in the *in vitro* release profile and targeted anticancer potential in the 3-(4,5-dimethylthiazol-2-yl)-2,5-diphenyltetrazolium bromide assay on two different cell lines of human skin cancer (HaCaT & A431).

Shitole et al. (2020) entrapped quercetin and docetaxel in nanocapsules. They conjugated luteinizing hormone-releasing hormone ligand to poly-lactide-co-glycolide using polyethylene glycol as a spacer for targeting prostate cancer. Due to polyethylene glycol coating, the quercetin-polyethylene glycol-docetaxel nanocapsules showed sustained in-vitro release and serum stability. Luteinizing hormone-releasing hormone-targeted quercetin-polyethylene glycol-docetaxel nanocapsules bestowed enhanced cellular uptake in *in-vitro* studies on prostate cancer cell lines (PC-3 and LNCaP) and higher tumour accumulation *in-vivo.* Also, it showed higher caspase-3 activity and enhanced cell inhibition activity of the combinatorial docetaxel-quercetin compared to free and nontargeted quercetin-polyethylene glycol-docetaxel nanocapsules for the treatment of prostate cancer.

Liu et al. (2020) entrapped quercetin and paclitaxel in chondroitin sulfate-coated mesoporous silica nanoparticles. Results showed that quercetin-paclitaxel-chondroitin sulfate mesoporous silica nanoparticles showed active targeting to CD44-receptor, down-regulating *P*-glycoprotein expression, decreasing the IC_{50} value and having higher apoptosis rate *in vitro* in resistant breast cancer cells (MCF-7/ADR) as compared to free drugs. Thus, quercetin-paclitaxel-chondroitin sulfate mesoporous silica nanoparticles showed excellent tumour-targeting ability, longer tumour retention, and enhanced antitumour actions without showing adverse reactions on normal cells suggesting potential to use for the treatment of breast cancer.

Riaz et al. (2019) prepared actively targeted transferrin T7 receptors for quercetin delivery via liposomes to treat lung cancer. The results showed that the T7-quercetin liposomes enhanced target specificity in lung cancer cells (A549) and deeper penetration in the tumour, thus enhancing tumour growth inhibition. Pulmonary delivery of T7-quercetin liposomes showed increased cytotoxicity and apoptosis *in vivo* in tumour-bearing mice. Wang et al. (2021) prepared an active targeted chitosan-bonded cetuximab-paclitaxel and cetuximab-quercetin nanoparticles for the management of non-small cell lung cancer resistant to paclitaxel. The results showed that cetuximab-paclitaxel nanoparticles/cetuximab-quercetin nanoparticles showed synergistic cytotoxic action in lung cancer cells (A549 and A549/taxol), showing enhanced pharmacological effect of paclitaxel-based therapy and reversed the resistance to paclitaxel by inhibiting the phosphorylation of protein kinase B and extracellular signal-regulated protein kinases.

13.4 REPORTS ON CLINICAL TRIALS AND PATENTS OF THE QUERCETIN

Despite the promising effect of quercetin nanoformulations in preclinical studies, no single formulation is in clinical trials. The list of clinical trials performed on quercetin as a dietary supplement for cancer prevention is mentioned in Table 13.1.

TABLE 13.1
List of Clinical Trials of Quercetin

Disease	Purpose	Phase	Status	Design	No of Participants	Study Start	Study End
Fanconi anaemia and squamous cell carcinoma	Prevention	Phase II	Recruiting	Single group	55	8 May 2018	September, 2023
Lung carcinoma Tobacco-related carcinoma	Prevention	Not applicable	Recruiting	Randomized crossover	42	25 October 2019	31 December 2022
Adenocarcinoma of prostate	Prevention	Phase I	Active not recruiting	Randomized parallel design double-blind	31	1 January 2014	31 January 2021

Based on a search of Clinicaltrials.gov (https://clinicaltrials.gov/ct2/results?term=quercetin&cond=Cancer+AND+%22Carcinoma%22).

13.4.1 Patents Based on Quercetin Nanoformulations

CN105106117A: Wang Xiangtao Di Jing Hong Jingyi Liu Yingying invented the quercetin nanoparticles using tocopheryl polyethylene glycol succinate as a stabilising agent with a ratio of 1:20 to 10:1. The quercetin-tocopheryl polyethylene glycol succinate nanoparticles had high drug loading and small particle size and slow release effect *in-vitro*. Thus, quercetin-tocopheryl polyethylene glycol succinate nanoparticles enhance cancer therapy drugs to the lethal effect of tumour cells and can be used as adjuvant reversible drug resistance of tumours.

CN102697795A: Sun Yiyi Zhong Ling Lin Dong Yin Jia copolymerised pharmorubicin and quercetin with poly-lactide-co-glycolide nanoparticles to reverse pharmorubicin-resistant cancer. There was an increase in entrapment efficiency and drug loading. Also, epichlorohydrin-quercetin nanoparticles changed the tolerance to tumour cells with lesser toxicity on normal cells.

Numerous clinical trials of quercetin nanoenabled formulations, patented for their anticancer potential, are mentioned below (Table 13.2).

13.5 FUTURE PERSPECTIVES

From the studies described in the present chapter, the effectiveness of quercetin nanoenabled formulations in cancer management is very clear. Very few clinical trials of quercetin are ongoing. The nanoenabled drug delivery systems offer the huge potential of quercetin to the pharmaceutical industry in cancer management. Therefore, there is a need to undertake randomized clinical trials of the nanoenabled drug delivery systems for the clinical development of quercetin.

13.6 CONCLUSION

The present chapter explained various active and passive drug delivery systems of quercetin like liposomes, nanoparticles, nanoemulsions, micelles, cubosomes, phytosomes, niosomes, solid lipid nanoparticles and active drug delivery systems like nanodelivery systems conjugated with folic acid and luteinizing hormone-releasing hormone ligand to enhance the oral bioavailability, stability, increase safety, targeted drug delivery and to achieve greater tumour inhibitory actions using various in vitro cell lines and in vivo tumour-bearing animals. Few clinical trials are ongoing for quercetin. Thus, there is huge potential for nanoenabled drug delivery systems of quercetin in clinical use for cancer management.

REFERENCES

Aghababaei F, Hadidi M (2023) Recent advances in potential health benefits of quercetin. *Pharmaceuticals* 16(7), 1020.

Ahmad N, Ahmad R, Naqvi AA, Alam MA, Ashafaq M, Abdur Rub R, Ahmad FJ (2018) Intranasal delivery of quercetin-loaded mucoadhesive nanoemulsion for treatment of

TABLE 13.2
List of Patented Quercetin Nanoenabled Formulations

Patent No	Date	Inventor	Formulation	Purpose
1020160149181	27.12.2016	Lee Sang-soo	Methoxy poly (ethylene glycol)-poly(DL-lactic acid)-quercetin nanoparticles	Breast cancer
1020160057716	24.05.2016	Lee, Sang Soolee, Sang Soo	Methoxy poly (ethylene glycol)-poly(DL-lactic acid)-quercetin nanoparticles	Breast cancer
WO/2019/071229	11.04.2019	Nezami, Mohammed, Amin	Nanoformulations	Neoplasms
20210015787	21.01.2021	Mohammed Amin Nezami	Nanoformulations	Neoplasms
WO/2019/099750	23.05.2019	NEZAMI, Mohammed Amin	Nanoformulations	Neoplasms
WO/2018/172942	27.09.2018	Huang, Leaf	Lipid nanoparticles	Chemotherapeutic
WO/2018/157081	30.08.2018	Nezami, Mohammed Amin	Poly(lactic-co-glycolic acid)-methoxy poly(ethylene glycol) nanoparticles	Neoplasms
WO/2020/219628	29.10.2020	Huang, Leaf	Micelles	Colorectal cancer
WO/2018/170457	20.09.2018	Nezami, Mohammed Amin	Poly(lactic-co-glycolic acid)-methoxy poly(ethylene glycol) nanoparticles	Neoplasms
109999002	12.07.2019	Jiang Longguang	Nanoparticles	Breast cancer
110876805	13.03.2020	Wang Lei	Nanoparticles	Skin cancer
106265597	04.01.2017	Feng Chao	Nanoparticles	Cancer
109288819	01.02.2019	Wu Hong	Nanoparticles	Colon cancer
3691632	12.08.2020	Nezami Mohammed Amin	Nanoparticles	Cancer
3710434	23.09.2020	Nezami Mohammed Amin	Nanoparticles	Cancer

Based on a search patentscope.wipo.int/search (https://patentscope.wipo.int/search/en/result.jsf?_vid=P11-KNVLGG-41575).

cerebral ischaemia. *Artificial Cells, Nanomedicine, and Biotechnology* 46(4), 717–729.

Arbain NH, Salim N, Masoumi HRF, Wong TW, Basri M, Abdul Rahman MB (2019) In vitro evaluation of the inhalable quercetin loaded nanoemulsion for pulmonary delivery. *Drug Delivery and Translational Research* 9(2), 497–507.

Arbain NH, Salim N, Wui WT, Basri M, Rahman MBA (2018) Optimization of quercetin loaded palm oil ester based nanoemulsion formulation for pulmonary delivery. *Journal of Oleo Science* 67(8), 933–940.

Azeem M, Hanif M, Mahmood K, Ameer N, Chughtai FRS, Abid U (2023) An insight into anticancer, antioxidant, antimicrobial, antidiabetic and anti-inflammatory effects of quercetin: A review. *Polymer Bulletin* 80(1), 241–262.

Babazadeh A, Zeinali M, Hamishehkar H (2018) Nano-phytosome: A developing platform for herbal anti-cancer agents in cancer therapy. *Current Drug Targets* 19(2), 170–180.

Baksi R, Singh DP, Borse SP, Rana R, Sharma V, Nivsarkar M (2018) *In vitro* and *in vivo* anticancer efficacy potential of Quercetin loaded polymeric nanoparticles. *Biomedicine & Pharmacotherapy* 106, 1513–1526.

Balakrishnan S, Bhat FA, Raja Singh P, Mukherjee S, Elumalai P, Das S, Patra CR, Arunakaran J (2016) Gold nanoparticle-conjugated quercetin inhibits epithelialmesenchymal transition, angiogenesis and invasiveness via EGFR/VEGFR2 mediated pathway in breast cancer. *Cell Proliferation* 49(6), 678–697.

Balakrishnan S, Mukherjee S, Das S, Bhat FA, Raja Singh P, Patra CR, Arunakaran J (2017) Gold nanoparticles-conjugated quercetin induces apoptosis via inhibition of EGFR/PI3K/Akt-mediated pathway in breast cancer cell lines (MCF-7 and MDA-MB-231). *Cell Biochemistry and Function* 35(4), 217–231.

Bishayee A (2009) Cancer prevention and treatment with resveratrol: From rodent studies to clinical trials. *Cancer Prevention Research* 2(5), 409–418.

Bishayee K, Khuda-Bukhsh AR, Huh S-O (2015) PLGA-loaded gold-nanoparticles precipitated with quercetin downregulate HDAC-Akt activities controlling proliferation and activate p53-ROS crosstalk to induce apoptosis in hepatocarcinoma cells. *Molecules and Cells* 38(6), 518.

Bose S, Du Y, Takhistov P, Michniak-Kohn B (2013) Formulation optimization and topical delivery of quercetin from solid lipid based nanosystems. *International Journal of Pharmaceutics* 441(1–2), 56–66.

Brannon-Peppas L, Blanchette JO (2004) Nanoparticle and targeted systems for cancer therapy. *Advanced Drug Delivery Reviews* 56(11), 1649–1659.

Bruning A (2013) Inhibition of mTOR signaling by quercetin in cancer treatment and prevention. *Anti-Cancer Agents in Medicinal Chemistry* 13(7), 1025–1031. doi:10.2174/1871520611313139990114.

Byrne JD, Betancourt T, Brannon-Peppas L (2008) Active targeting schemes for nanoparticle systems in cancer therapeutics. *Advanced Drug Delivery Reviews* 60(15), 1615–1626.

Caddeo C, DÃez-Sales O, Pons R, Carbone C, Ennas G, Puglisi G, Fadda AM, Manconi M (2016a) Cross-linked chitosan/liposome hybrid system for the intestinal delivery of quercetin. *Journal of Colloid and Interface Science* 461, 69–78.

Caddeo C, Nacher A, Vassallo A, Armentano MF, Pons R, Fernandez-Busquets X, Carbone C, Valenti D, Fadda AM, Manconi M (2016b) Effect of quercetin and resveratrol co-incorporated in liposomes against inflammatory/oxidative response associated with skin cancer. *International Journal of Pharmaceutics* 513(1–2), 153–163.

Cai X, Fang Z, Dou J, Yu A, Zhai G (2013) Bioavailability of quercetin: Problems and promises. *Current Medicinal Chemistry* 20(20), 2572–2582.

Chen SQ, Wang C, Tao S, Wang YX, Hu FQ, Yuan H (2018) Rational design of redox-responsive and P-gp-inhibitory lipid nanoparticles with high entrapment of paclitaxel for tumour therapy. *Advanced Healthcare Materials* 7(17), 1800485.

Cirillo G, Vittorio O, Hampel S, Iemma F, Parchi P, Cecchini M, Puoci F, Picci N (2013) Quercetin nanocomposite as novel anticancer therapeutic: Improved efficiency and reduced toxicity. *European Journal of Pharmaceutical Sciences* 49(3), 359–365.

Dabbagh-Bazarbachi H, Clergeaud G, Quesada IM, Ortiz M, O'Sullivan CK, Fernandez-Larrea JB (2014) Zinc ionophore activity of quercetin and epigallocatechin-gallate: From Hepa 1–6 cells to a liposome model. *Journal of Agricultural and Food Chemistry* 62(32), 8085–8093.

David KI, Jaidev LR, Sethuraman S, Krishnan UM (2015) Dual drug loaded chitosan nanoparticles-sugar-coated arsenal against pancreatic cancer. *Colloids and Surfaces B: Biointerfaces* 135, 689–698.

Ding B, Chen P, Kong Y, Zhai Y, Pang X, Dou J, Zhai G (2014) Preparation and evaluation of folate-modified lipid nanocapsules for quercetin delivery. *Journal of Drug Targeting* 22(1), 67–75.

dos Santos DM, Rocha CVJ, da Silveira EF, Marinho MAG, Rodrigues MR, Silva NO, da Silva Ferreira A, de Moura NF, Darelli GJS, Braganhol E (2018) In vitro anti/pro-oxidant activities of R. ferruginea extract and its effect on glioma cell viability: Correlation with phenolic compound content and effects on membrane dynamics. *The Journal of Membrane Biology* 251(2), 247–261.

Dos Santos SC, Silva NO, Junior JoBd SE, Marinho MAG, Borges ZV, Branco Nl BC, Faita FL, Soares BM, Horn AP, Parize AL (2019) Molecular interactions and physico-chemical characterization of quercetin-loaded magnetoliposomes. *Chemistry and Physics of Lipids* 218, 22–33.

Du H, Liu M, Yang X, Zhai G (2015) The role of glycyrrhetinic acid modification on preparation and evaluation of quercetin-loaded chitosan-based self-aggregates. *Journal of Colloid and Interface Science* 460, 87–96.

El-Gogary RI, Rubio N, Wang JT-W, Al-Jamal WT, Bourgognon M, Kafa H, Naeem M, Klippstein R, Abbate V, Leroux F (2014) Polyethylene glycol conjugated polymeric nanocapsules for targeted delivery of quercetin to folate-expressing cancer cells in vitro and in vivo. *ACS Nano* 8(2), 1384–1401.

Fatma S, Talegaonkar S, Iqbal Z, Panda AK, Negi LM, Goswami DG, Tariq M (2016) Novel flavonoid-based biodegradable nanoparticles for effective oral delivery of etoposide by P-glycoprotein modulation: An in vitro, ex vivo and in vivo investigations. *Drug Delivery* 23(2), 500–511.

Gao X, Wang B, Wei X, Men K, Zheng F, Zhou Y, Zheng Y, Gou M, Huang M, Guo G (2012) Anticancer effect and mechanism of polymer micelle-encapsulated quercetin on ovarian cancer. *Nanoscale* 4(22), 7021–7030.

George D, Maheswari PU, Begum KMMS (2019) Synergic formulation of onion peel quercetin loaded chitosan-cellulose hydrogel with green zinc oxide nanoparticles towards controlled release, biocompatibility, antimicrobial and anticancer activity. *International Journal of Biological Macromolecules* 132, 784–794.

Ghosh A, Ghosh D, Sarkar S, Mandal AK, Choudhury ST, Das N (2012) Anticarcinogenic activity of nanoencapsulated quercetin in combating diethylnitrosamine-induced hepatocarcinoma in rats. *European Journal of Cancer Prevention* 21(1), 32–41.

Ghosh A, Mandal AK, Sarkar S, Das N (2011a) Hepatoprotective and neuroprotective activity of liposomal quercetin in combating chronic arsenic induced oxidative damage in liver and brain of rats. *Drug Delivery* 18(6), 451–459.

Ghosh A, Mandal AK, Sarkar S, Panda S, Das N (2009) Nanoencapsulation of quercetin enhances its dietary efficacy in combating arsenic-induced oxidative damage in liver and brain of rats. *Life Sciences* 84(3–4), 75–80.

Ghosh D, Ghosh S, Sarkar S, Ghosh A, Das N, Saha KD, Mandal AK (2010) Quercetin in vesicular delivery systems: Evaluation in combating arsenic-induced acute liver toxicity associated gene expression in rat model. *Chemico-Biological Interactions* 186(1), 61–71.

Ghosh S, Dungdung SR, Chowdhury ST, Mandal AK, Sarkar S, Ghosh D, Das N (2011b) Encapsulation of the flavonoid quercetin with an arsenic chelator into nanocapsules enables the simultaneous delivery of hydrophobic and hydrophilic drugs with a synergistic effect against chronic arsenic accumulation and oxidative stress. *Free Radical Biology and Medicine* 51(10), 1893–1902.

Gibellini L, Pinti M, Nasi M, Montagna JP, De Biasi S, Roat E, Bertoncelli L, Cooper EL, Cossarizza A (2011) Quercetin and cancer chemoprevention. *Evidence-Based Complementary and Alternative Medicine* 15, 591356.

Gismondi A, Reina G, Orlanducci S, Mizzoni F, Gay S, Terranova ML, Canini A (2015) Nanodiamonds coupled with plant bioactive metabolites: A nanotech approach for cancer therapy. *Biomaterials* 38, 22–35.

Goniotaki M, Hatziantoniou S, Dimas K, Wagner M, Demetzos C (2004) Encapsulation of naturally occurring flavonoids into liposomes: Physicochemical properties and biological activity against human cancer cell lines. *Journal of Pharmacy and Pharmacology* 56(10), 1217–1224.

Guan X, Gao M, Xu H, Zhang C, Liu H, Lv L, Deng S, Gao D, Tian Y (2016) Quercetin-loaded poly (lactic-co-glycolic acid)-d-α-tocopheryl polyethylene glycol 1000 succinate nanoparticles for the targeted treatment of liver cancer. *Drug Delivery* 23(9), 3307–3318.

Gupta A, Kaur CD, Saraf S, Saraf S (2016) Formulation, characterization, and evaluation of ligand-conjugated biodegradable quercetin nanoparticles for active targeting. *Artificial Cells, Nanomedicine, and Biotechnology* 44(3), 960–970.

Gupta P, Authimoolam SP, Hilt JZ, Dziubla TD (2015) Quercetin conjugated poly (Î²-amino esters) nanogels for the treatment of cellular oxidative stress. *Acta Biomaterialia* 27, 194–204.

Hatahet T, Morille M, Hommoss A, Devoisselle JM, Muller RH, Begu S (2018) Liposomes, lipid nanocapsules and smartCrystals®: A comparative study for an effective quercetin delivery to the skin. *International Journal of Pharmaceutics* 542(1–2), 176–185.

Hatahet T, Morille M, Shamseddin A, Aubert-Pouessel A, Devoisselle JM, Begu S (2017) Dermal quercetin lipid nanocapsules: Influence of the formulation on antioxidant activity and cellular protection against hydrogen peroxide. *International Journal of Pharmaceutics* 518(1–2), 167–176.

He B, Wang X, Shi H-S, Xiao W-J, Zhang J, Mu B, Mao Y-Q, Wang W, Wang Y-S (2013) Quercetin liposome sensitizes colon carcinoma to thermotherapy and thermochemotherapy in mice models. *Integrative Cancer Therapies* 12(3), 264–270.

Hemati M, Haghiralsadat F, Yazdian F, Jafari F, Moradi A, Malekpour-Dehkordi Z (2019) Development and characterization of a novel cationic PEGylated niosome-encapsulated forms of doxorubicin, quercetin and siRNA for the treatment of cancer by using combination therapy. *Artificial Cells, Nanomedicine, and Biotechnology* 47(1), 1295–1311.

Hu F, Bu YZ, Liang R, Duan RM, Wang S, Han RM, Wang P, Ai XC, Zhang JP, Skibsted LH (2013) Quercetin and daidzein β-apo-14′-carotenoic acid esters as membrane antioxidants. *Free Radical Research* 47(5), 413–421.

Hu J, Wang J, Wang G, Yao Z, Dang X (2016) Pharmacokinetics and antitumour efficacy of DSPE-PEG2000 polymeric liposomes loaded with quercetin and temozolomide: Analysis of their effectiveness in enhancing the chemosensitization of drug-resistant glioma cells. *International Journal of Molecular Medicine* 37(3), 690–702.

Huang YT, Hwang JJ, Lee PP, Ke FC, Huang JH, Huang CJ, Kandaswami C, Middleton Jr E, Lee MT (1999) Effects of luteolin and quercetin, inhibitors of tyrosine kinase, on cell growth and metastasis- associated properties in A431 cells overexpressing epidermal growth factor receptor. *British Journal of Pharmacology* 128(5), 999–1010.

Hwang MK, Song NR, Kang NJ, Lee KW, Lee HJ (2009) Activation of phosphatidylinositol 3-kinase is required for tumour necrosis factor-alpha-induced upregulation of matrix metalloproteinase-9: Its direct inhibition by quercetin. *The International Journal of Biochemistry & Cell Biology* 41, 1592–1600.

Jain AK, Thanki K, Jain S (2013) Co-encapsulation of tamoxifen and quercetin in polymeric nanoparticles: Implications on oral bioavailability, antitumour efficacy, and drug-induced toxicity. *Molecular Pharmaceutics* 10(9), 3459–3474.

Jain AK, Thanki K, Jain S (2014) Novel self-nanoemulsifying formulation of quercetin: Implications of pro-oxidant activity on the anticancer efficacy. *Nanomedicine* 10(5), 959–969.

Jain AS, Shah SM, Nagarsenker MS, Nikam Y, Gude RP, Steiniger F, Thamm J, Fahr A (2013) Lipid colloidal carriers for improvement of anticancer activity of orally delivered quercetin: Formulation, characterization and establishing *in vitro-in vivo* advantage. *Journal of Biomedical Nanotechnology* 9(7), 1230–1240.

Jiang W, Zhang H, Wu J, Zhai G, Li Z, Luan Y, Garg S (2018) CuS@MOF-based well-designed quercetin delivery system for chemo-photothermal therapy. *ACS Applied Materials & Interfaces* 10(40), 34513–34523.

Joshi H, Gupta DS, Kaur G, Singh T, Ramniwas S, Sak K, Aggarwal D, Chhabra RS, Gupta M, Saini AK, Tuli HS (2023) Nanoformulations of quercetin for controlled delivery: A review of preclinical anticancer studies. *Naunyn-Schmiedeberg's Archives of Pharmacology*, 396(12), 3443–3458.

Karthivashan G, Masarudin MJ, Kura AU, Abas F, Fakurazi S (2016) Optimization, formulation, and characterization of multiflavonoids-loaded flavanosome by bulk or sequential technique. *International Journal of Nanomedicine* 11, 3417.

Khoee S, Rahmatolahzadeh R (2012) Synthesis and characterization of pH-responsive and folated nanoparticles based on self-assembled brush-like PLGA/PEG/AEMA copolymer with targeted cancer therapy properties: A comprehensive kinetic study. *European Journal of Medicinal Chemistry* 50, 416–427.

Khushnud T, Mousa SA (2013) Potential role of naturally derived polyphenols and their nanotechnology delivery in cancer. *Molecular Biotechnology* 55(1), 78–86.

Kim H, Seo E-M, Sharma AR, Ganbold B, Park J, Sharma G, Kang Y-H, Song D-K, Lee S-S, Nam J-S (2013) Regulation of Wnt signaling activity for growth suppression induced by quercetin in 4T1 murine mammary cancer cells. *International Journal of Oncology* 43(4), 1319–1325.

Kruszewski M, Kusaczuk M, Kotynska J, Gal M, Kretowski R, Cechowska-Pasko M, Naumowicz M (2018) The effect of quercetin on the electrical properties of model lipid membranes and human glioblastoma cells. *Bioelectrochemistry* 124, 133–141.

Kumar SR, Priyatharshni S, Babu VN, Mangalaraj D, Viswanathan C, Kannan S, Ponpandian N (2014) Quercetin conjugated superparamagnetic magnetite nanoparticles for in-vitro analysis of breast cancer cell lines for chemotherapy applications. *Journal of Colloid and Interface Science* 436, 234–242.

Lakshmi BA, Bae JY, An JH, Kim S (2019) Nanoclusters prepared from ruthenium(II) and quercetin for fluorometric detection of cobalt(II), and a method for screening their anticancer drug activity. *Mikrochim Acta* 186(8), 539.

Lakshmi BA, Kim S (2019) Quercetin mediated gold nanoclusters explored as a dual functional nanomaterial in anticancer and bio-imaging disciplines. *Colloids and Surfaces B: Biointerfaces* 178, 230–237.

Li J, Zhao J, Tan T, Liu M, Zeng Z, Zeng Y, Zhang L, Fu C, Chen D, Xie T (2020) Nanoparticle drug delivery system for glioma and its efficacy improvement strategies: A comprehensive review. *International Journal of Nanomedicine* 15, 2563–2582.

Lin CW, Hou W-C, Shen S-C, Juan S-H, Ko C-H, Wang L-M, Chen Y-C (2008) Quercetin inhibition of tumour invasion via suppressing PKC δ/ERK/AP-1-dependent matrix metalloproteinase-9 activation in breast carcinoma cells. *Carcinogenesis* 29, 1807–1815.

Liu H, Xue JX, Li X, Ao R, Lu Y (2013) Quercetin liposomes protect against radiation-induced pulmonary injury in a murine model. *Oncology Letters* 6(2), 453–459.

Liu M, Fu M, Yang X, Jia G, Shi X, Ji J, Liu X, Zhai G (2020) Paclitaxel and quercetin co-loaded functional mesoporous silica nanoparticles overcoming multidrug resistance in breast cancer. *Colloids and Surfaces B: Biointerfaces* 196, 111284.

Lockhart JN, Stevens DM, Beezer DB, Kravitz A, Harth E (2015) Dual drug delivery of tamoxifen and quercetin: Regulated metabolism for anticancer treatment with nanosponges. *Journal of Controlled Release* 220, 751–757.

Lou M, Zhang L-N, Ji P-G, Feng F-Q, Liu J-H, Yang C, Li B-F, Wang L (2016) Quercetin nanoparticles induced autophagy and apoptosis through AKT/ERK/Caspase-3 signaling pathway in human neuroglioma cells: In vitro and in vivo. *Biomedicine & Pharmacotherapy* 84, 1–9.

Lowe SW, Lin AW (2000) Apoptosis in cancer. *Carcinogenesis*, 21(3), 485–495.

Luo C-L, Liu Y-Q, Wang P, Song C-H, Wang K-J, Dai L-P, Zhang J-Y, Ye H (2016) The effect of quercetin nanoparticle on cervical cancer progression by inducing apoptosis, autophagy and anti-proliferation via JAK2 suppression. *Biomedicine & Pharmacotherapy* 82, 595–605.

Lv L, Liu C, Chen C, Yu X, Chen G, Shi Y, Qin F, Ou J, Qiu K, Li G (2016) Quercetin and doxorubicin co-encapsulated biotin receptor-targeting nanoparticles for minimizing drug resistance in breast cancer. *Oncotarget* 7(22), 32184.

Malekzadeh AM, Ramazani A, Rezaei SJT, Niknejad H (2017) Design and construction of multifunctional hyperbranched polymers coated magnetite nanoparticles for both targeting magnetic resonance imaging and cancer therapy. *Journal of Colloid and Interface Science* 490, 64–73.

Mandal AK, Ghosh D, Sarkar S, Ghosh A, Swarnakar S, Das N (2014) Nanocapsulated quercetin downregulates rat hepatic MMP-13 and controls diethylnitrosamine-induced carcinoma. *Nanomedicine* 9(15), 2323–2337.

Martirosyan A, Grintzalis K, Polet M, Laloux L, Schneider Y-J (2016) Tuning the inflammatory response to silver nanoparticles via quercetin in Caco-2 (co-) cultures as model of the human intestinal mucosa. *Toxicology Letters* 253, 36–45.

Minaei A, Sabzichi M, Ramezani F, Hamishehkar H, Samadi N (2016) Co-delivery with nano-quercetin enhances doxorubicin-mediated cytotoxicity against MCF-7 cells. *Molecular Biology Reports* 43(2), 99–105.

Mittal AK, Kumar S, Banerjee UC (2014) Quercetin and gallic acid mediated synthesis of bimetallic (silver and selenium) nanoparticles and their antitumour and antimicrobial potential. *Journal of Colloid and Interface Science* 431, 194–199.

Moussa M, Goldberg SN, Kumar G, Sawant RR, Levchenko T, Torchilin VP, Ahmed M (2014) Nanodrug-enhanced radiofrequency tumour ablation: Effect of micellar or liposomal carrier on drug delivery and treatment efficacy. *PLoS One* 9(8), e102727.

Mu Y, Fu Y, Li J, Yu X, Li Y, Wang Y, Wu X, Zhang K, Kong M, Feng C, Chen X (2019) Multifunctional quercetin conjugated chitosan nano-micelles with P-gp inhibition and permeation enhancement of anticancer drug. *Carbohydrate Polymers* 203, 10–18.

Muller J, Martins A, Csabi J, Fenyvesi F, Konczol Ã, Hunyadi A, Balogh GT (2016) BBB penetration-targeting physicochemical lead selection: Ecdysteroids as chemo-sensitizers against CNS tumours. *European Journal of Pharmaceutical Sciences* 96, 571–577.

Murgia S, Bonacchi S, Falchi AM, Lampis S, Lippolis V, Meli V, Monduzzi M, Prodi L, Schmidt J, Talmon Y (2013) Drug-loaded fluorescent cubosomes: Versatile nanoparticles for potential theranostic applications. *Langmuir* 29(22), 6673–6679.

Nam J-S, Sharma AR, Nguyen LT, Chakraborty C, Sharma G, Lee S-S (2016) Application of bioactive quercetin in oncotherapy: From nutrition to nanomedicine. *Molecules* 21(1), 108.

Ojeda-Serna IE, Rocha-Guzmán NE, Gallegos-Infante JA, Cháirez-Ramírez MH, Rosas-Flores W, Pérez-Martínez JD, Moreno-Jiménez MR, González-Laredo RF (2019) Water-in-oil organogel based emulsions as a tool for increasing bioaccessibility and cell permeability of poorly water-soluble nutraceuticals. *Food Research International* 120, 415–424.

Oliver S, Yee E, Kavallaris M, Vittorio O, Boyer C (2018) Water soluble antioxidant dextran-quercetin conjugate with potential anticancer properties. *Macromolecular Bioscience* 18(4), e1700239.

Omwenga EO, Hensel A, Shitandi A, Goycoolea FM (2018) Chitosan nanoencapsulation of flavonoids enhances their quorum sensing and biofilm formation inhibitory activities against an E. coli top 10 biosensor. *Colloids and Surfaces B: Biointerfaces* 164, 125–133.

Pandey SK, Patel DK, Thakur R, Mishra DP, Maiti P, Haldar C (2015) Anti-cancer evaluation of quercetin embedded PLA nanoparticles synthesized by emulsified nanoprecipitation. *International Journal of Biological Macromolecules* 75, 521–529.

Pangeni R, Panthi VK, Yoon I-S, Park JW (2018) Preparation, characterization, and in vivo evaluation of an oral multiple nanoemulsive system for co-delivery of pemetrexed and quercetin. *Pharmaceutics* 10(3), 158.

Park CH, Chang JY, Hahm ER, Park S, Kim H-K, Yang CH (2005) Quercetin, a potent inhibitor against β-catenin/Tcf signaling in SW480 colon cancer cells. *Biochemical and Biophysical Research Communications* 328(1), 227–234.

Parveen S, Sahoo SK (2008) Polymeric nanoparticles for cancer therapy. *Journal of Drug Targeting* 16(2), 108–123.

Patidar P, Pillai SA, Sheth U, Bahadur P, Bahadur A (2017) Glucose triggered enhanced solubilisation, release and cytotoxicity of poorly water soluble anti-cancer drugs from T1307 micelles. *Journal of Biotechnology* 254, 43–50.

Rauf A, Imran M, Khan IA, Ur-Rehman M, Gilani SA, Mehmood Z, Mubarak MS (2018) Anticancer potential of quercetin: A comprehensive review. *Phytotherapy Research* 32(11), 2109–2130.

Ravichandiran V, Masilamani K, Senthilnathan B, Maheshwaran A, Wui Wong T, Roy P (2017) Quercetin-decorated curcumin liposome design for cancer therapy: In-vitro and in-vivo studies. *Current Drug Delivery* 14(8), 1053–1059.

Ren K-W, Li Y-H, Wu G, Ren J-Z, Lu H-B, Li Z-M, Han X-W (2017) Quercetin nanoparticles display antitumour activity via proliferation inhibition and apoptosis induction in liver cancer cells. *International Journal of Oncology* 50(4), 1299–1311.

Riaz MK, Zhang X, Wong KH, Chen H, Liu Q, Chen X, Zhang G, Lu A, Yang Z (2019) Pulmonary delivery of transferrin receptors targeting peptide surface-functionalized liposomes augments the chemotherapeutic effect of quercetin in lung cancer therapy. *International Journal of Nanomedicine* 14, 2879.

Rich GT, Buchweitz M, Winterbone MS, Kroon PA, Wilde PJ (2017) Towards an understanding of the low bioavailability of quercetin: A study of its interaction with intestinal lipids. *Nutrients* 9(2), 111.

Rodriguez EB, Almeda RA, Vidallon MLP, Reyes CT (2019) Enhanced bioactivity and efficient delivery of quercetin through nanoliposomal encapsulation using rice bran phospholipids. *Journal of the Science of Food and Agriculture* 99(4), 1980–1989.

Saha C, Kaushik A, Das A, Pal S, Majumder D (2016) Anthracycline drugs on modified surface of quercetin-loaded polymer nanoparticles: A dual drug delivery model for cancer treatment. *Plos One* 11(5), e0155710.

Sahu S (2013) Biocompatible nanoparticles for sustained topical delivery of anticancer phytoconstituent quercetin. *Pakistan Journal of Biological Sciences* 16(13), 601–609.

Sakurai Y, Kajimoto K, Hatakeyama H, Harashima H (2015) Advances in an active and passive targeting to tumour and adipose tissues. *Expert Opinion on Drug Delivery* 12(1), 41–52.

Sandhu PS, Kumar R, Katare OP, Singh B (2017) Surface-tailored nanomixed micelles containing quercetinsalicylic acid physical complex for enhanced cellular and in vivo activities: A quality by design perspective. *Nanomedicine* 12(11), 1281–1303.

Schwingel TE, Klein CP, Nicoletti NF, Dora CL, Hadrich G, Bica CG, Lopes TG, da Silva VD, Morrone FB (2014) Effects of the compounds resveratrol, rutin, quercetin, and quercetin nanoemulsion on oxaliplatin-induced hepatotoxicity and neurotoxicity in mice. *Naunyn-Schmiedeberg's Archives of Pharmacology* 387(9), 837–848.

Shahbaz M, Naeem H, Momal U, Imran M, Alsagaby SA, Al Abdulmonem W, Abdulmonem WA, Waqar AB, El-Ghorab AH, Ghoneim MM, Abdelgawad MA, Shaker ME, Umar M, Hussain M, Kumar R, Al Jbawi E (2023). Anticancer and apoptosis inducing potential of quercetin against a wide range of human malignancies. *International Journal of Food Properties* 26(1), 2590–2626.

Shan B-E, Wang M-X, Li R-Q (2009) Quercetin inhibit human SW480 colon cancer growth in association with inhibition of cyclin D1 and survivin expression through Wnt/Î²-catenin signaling pathway. *Cancer Investigation* 27(6), 604–612.

Sharma G, Park J, Sharma AR, Jung J-S, Kim H, Chakraborty C, Song D-K, Lee S-S, Nam J-S (2015) Methoxy poly (ethylene glycol)-poly (lactide) nanoparticles encapsulating quercetin act as an effective anticancer agent by inducing apoptosis in breast cancer. *Pharmaceutical Research* 32(2), 723–735.

Shitole AA, Sharma N, Giram P, Khandwekar A, Baruah M, Garnaik, B, Koratkar S (2020) LHRH-conjugated, PEGylated, poly-lactide-co-glycolide nanocapsules for targeted delivery of combinational chemotherapeutic drugs Docetaxel and Quercetin for prostate cancer. *Materials Science and Engineering: C*, 114, 111035.

Shu L, Cheung K-L, Khor TO, Chen C, Kong A-N (2010) Phytochemicals: Cancer chemoprevention and suppression of tumour onset and metastasis. *Cancer and Metastasis Reviews* 29(3), 483–502.

Singh A, Dutta PK, Kumar H, Kureel AK, Rai AK (2018a) Synthesis of chitin-glucan-aldehyde-quercetin conjugate and evaluation of anticancer and antioxidant activities. *Carbohydrate Polymers* 193, 99–107.

Singh A, Lavkush, Kureel AK, Dutta PK, Kumar S, Rai AK (2018b) Curcumin loaded chitin-glucan quercetin conjugate: Synthesis, characterization, antioxidant, in vitro release study, and anticancer activity. *International Journal of Biological Macromolecules* 110, 234–244.

Soloviev AI, Kizub IV (2019) Mechanisms of vascular dysfunction evoked by ionizing radiation and possible targets for its pharmacological correction. *Biochemical Pharmacology* 159, 121–139.

Srisa-Nga K, Mankhetkorn S, Okonogi S, Khonkarn R (2019) Delivery of superparamagnetic polymeric micelles loaded with quercetin to hepatocellular carcinoma cells. *Journal of Pharmaceutical Sciences* 108(2), 996–1006.

Suksiriworapong J, Phoca K, Ngamsom S, Sripha K, Moongkarndi P, Junyaprasert VB (2016) Comparison of poly(ε-caprolactone) chain lengths of poly(ε-caprolacto ne)-co-d-α-tocopheryl-poly(ethylene glycol) 1000 succinate nanoparticles for enhancement of quercetin delivery to SKBR3 breast cancer cells. *European Journal of Pharmaceutics and Biopharmaceutics* 101, 15–24.

Tabassum S, Zaki M, Afzal M, Arjmand F (2013) New modulated design and synthesis of quercetin-Cu(II)/Zn(II)-Sn2(IV) scaffold as anticancer agents: In vitro DNA binding profile, DNA cleavage pathway and Topo-I activity. *Dalton Transactions* 42(27), 10029–10041.

Tammela P, Laitinen L, Galkin A, Wennberg T, Heczko R, Vuorela H, Slotte JP, Vuorela P (2004) Permeability characteristics and membrane affinity of flavonoids and alkyl gallates in Caco-2 cells and in phospholipid vesicles. *Archives of Biochemistry and Biophysics* 425(2), 193–199.

Tavano L, Mauro L, Naimo GD, Bruno L, Picci N, Andò S, Muzzalupo R (2016) Further evolution of multifunctional niosomes based on pluronic surfactant: Dual active targeting and drug combination properties. *Langmuir* 32(35), 8926–8933

Varshosaz J, Jafarian A, Salehi G, Zolfaghari B (2014) Comparing different sterol containing solid lipid nanoparticles for targeted delivery of quercetin in hepatocellular carcinoma. *Journal of Liposome Research* 24(3), 191–203.

Verma NK, Crosbie-Staunton K, Satti A, Gallagher S, Ryan KB, Doody T, McAtamney C, MacLoughlin R, Galvin P, Burke CS (2013) Magnetic core-shell nanoparticles for drug delivery by nebulization. *Journal of Nanobiotechnology* 11(1), 1.

Wang BL, Gao X, Men K, Qiu J, Yang B, Gou ML, Huang MJ, Huang N, Qian ZY, Zhao X (2012a) Treating acute cystitis with biodegradable micelle-encapsulated quercetin. *International Journal of Nanomedicine* 7, 2239.

Wang G, Wang J, Luo J, Wang L, Chen X, Zhang L, Jiang S (2013a) PEG2000-DPSE-coated quercetin nanoparticles remarkably enhanced anticancer effects through induced programed cell death on C6 glioma cells. *Journal of Biomedical Materials Research Part A* 101(11), 3076–3085.

Wang G, Wang J-J, Chen X-L, Du L, Li F (2016) Quercetin-loaded freeze-dried nanomicelles: Improving absorption and anti-glioma efficiency in vitro and in vivo. *Journal of Controlled Release* 235, 276–290.

Wang G, Wang JJ, Chen XL, Du SM, Li DS, Pei ZJ, Lan H, Wu LB (2013b) The JAK2/STAT3 and mitochondrial pathways are essential for quercetin nanoliposome-induced C6 glioma cell death. *Cell Death & Disease* 4(8), e746–e746.

Wang G, Wang JJ, Yang GY, Du SM, Zeng N, Li DS, Li RM, Chen JY, Feng JB, Yuan SH (2012b) Effects of quercetin nanoliposomes on C6 glioma cells through induction of type III programmed cell death. *International Journal of Nanomedicine* 7, 271.

Wang Y, Yu H, Wang S, Gai C, Cui X, Xu Z, Li W, Zhang W (2021) Targeted delivery of quercetin by nanoparticles based on chitosan sensitizing paclitaxel-resistant lung cancer cells to paclitaxel. *Materials Science and Engineering: C*, 119, 111442.

Wang Q, Bao Y, Ahire J, Chao Y (2013c) Co-encapsulation of biodegradable nanoparticles with silicon quantum dots and quercetin for monitored delivery. *Advanced Healthcare Materials* 2(3), 459–466.

Wang S, Zhang J, Chen M, Wang Y (2013d) Delivering flavonoids into solid tumours using nanotechnologies. *Expert Opinion on Drug Delivery* 10(10), 1411–1428.

Wong M-Y, Chiu GNC (2010) Simultaneous liposomal delivery of quercetin and vincristine for enhanced estrogen-receptor-negative breast cancer treatment. *Anti-Cancer Drugs* 21(4), 401–410.

Wong M-Y, Chiu GNC (2011) Liposome formulation of co-encapsulated vincristine and quercetin enhanced antitumour activity in a trastuzumab-insensitive breast tumour xenograft model. *Nanomedicine: Nanotechnology, Biology and Medicine* 7(6), 834–840.

Xing L, Lyu J-Y, Yang Y, Cui P-F, Gu L-Q, Qiao J-B, He Y-J, Zhang T-Q, Sun M, Lu J-J (2017) pH-Responsive de-PEGylated nanoparticles based on triphenylphosphine-quercetin-self-assemblies for mitochondria-targeted cancer therapy. *Chemical Communications* 53(62), 8790–8793.

Xu G, Shi H, Ren L, Gou H, Gong D, Gao X, Huang N (2015) Enhancing the anti-colon cancer activity of quercetin by self-assembled micelles. *International Journal of Nanomedicine* 10, 2051.

Xu M-X, Wang M, Yang W-W (2017) Gold-quercetin nanoparticles prevent metabolic endotoxemia-induced kidney injury by regulating TLR4/NF-κβ signaling and Nrf2 pathway in high fat diet fed mice. *International Journal of Nanomedicine* 12, 327.

Yamina AM, Fizir M, Itatahine A, He H, Dramou P (2018) Preparation of multifunctional PEG-graft-halloysite nanotubes for controlled drug release, tumour cell targeting, and bio-imaging. *Colloids and Surfaces B: Biointerfaces* 170, 322–329.

Yang W, Ahmed M, Tasawwar B, Levchenko T, Sawant RR, Collins M, Signoretti S, Torchilin V, Goldberg SN (2011) Radiofrequency ablation combined with liposomal quercetin to increase tumour destruction by modulation of heat shock protein production in a small animal model. *International Journal of Hyperthermia* 27(6), 527–538.

Yang W, Ahmed M, Tasawwar B, Levchenko T, Sawant RR, Torchilin V, Goldberg SN (2012) Combination radiofrequency (RF) ablation and IV liposomal heat shock protein suppression: Reduced tumour growth and increased animal endpoint survival in a small animal tumour model. *Journal of Controlled Release* 160(2), 239–244.

Yu J, Chen H, Jiang L, Wang J, Dai J, Wang J (2019) Codelivery of adriamycin and P-gp inhibitor quercetin using PEGylated liposomes to overcome cancer drug resistance. *Journal of Pharmaceutical Sciences* 108(5), 1788–1799.

Yuan Y-G, Wang Y-H, Xing H-H, Gurunathan S (2017) Quercetin-mediated synthesis of graphene oxide- silver nanoparticle nanocomposites: A suitable alternative nanotherapy for neuroblastoma. *International Journal of Nanomedicine* 12, 5819.

Yuan Z, Long C, Junming T, Qihuan L, Youshun Z, Chan Z (2012) Quercetin-induced apoptosis of HL-60 cells by reducing PI3K/Akt. *Molecular Biology Reports* 39(7), 7785–7793.

Yuan Z-P, Chen L-J, Fan L-Y, Tang M-H, Yang G-L, Yang H-S, Du X-B, Wang G-Q, Yao W-X, Zhao Q-M (2006a) Liposomal quercetin efficiently suppresses growth of solid tumours in murine models. *Clinical Cancer Research* 12(10), 3193–3199.

Yuan Z-P, Chen L-J, Wei Y-Q, Fan L-Y, Tang M-H, Yang G-L (2006b) Nanoliposomal quercetin inhibits formation of malignant ascites of hepatocellular carcinoma. *Ai zheng Aizheng Chinese Journal of Cancer* 25(8), 941–945.

Yang X, Zhang W, Zhao Z, Li N, Mou Z, Sun, D, Cai Y, Wang W, Lin Y (2017) Quercetin loading CdSe/ZnS nanoparticles as efficient antibacterial and anticancer materials. *Journal of Inorganic Biochemistry*, 167, 36–48.

Zafar S, Negi LM, Verma AK, Kumar V, Tyagi A, Singh P, Iqbal Z, Talegaonkar S (2014) Sterically stabilized polymeric nanoparticles with a combinatorial approach for multi drug resistant cancer: In vitro and in vivo investigations. *International Journal of Pharmaceutics* 477(1–2), 454–468.

Zhao J, Liu J, Wei T, Ma X, Cheng Q, Huo S, Zhang C, Zhang Y, Duan X, Liang X-J (2016) Quercetin-loaded nanomicelles to circumvent human castration-resistant prostate cancer in vitro and in vivo. *Nanoscale* 8(9), 5126–5138.

Zheng N-G, Mo S-J, Li J-P, Wu J-L (2014a) Anti-CSC effects in human esophageal squamous cell carcinomas and Eca109/9706 cells induced by nanoliposomal quercetin alone or combined with CD 133 antiserum. *Asian Pacific Journal of Cancer Prevention* 15(20), 8679–8684.

Zheng N-G, Wang J-L, Yang S-L, Wu J-L (2014b) Aberrant epigenetic alteration in Eca9706 cells modulated by nanoliposomal quercetin combined with butyrate mediated via epigenetic-NF-Î°B signaling. *Asian Pacific Journal of Cancer Prevention* 15(11), 4539–4543.

Zhou X, Liu H-Y, Zhao H, Wang T (2018) RGD-modified nanoliposomes containing quercetin for lung cancer targeted treatment. *OncoTargets and Therapy* 11, 5397.

Zhu B, Yu L, Yue Q (2017) Co-delivery of vincristine and quercetin by nanocarriers for lymphoma combination chemotherapy. *Biomedicine & Pharmacotherapy* 91, 287–294.

Zhu X, Zeng X, Zhang X, Cao W, Wang Y, Chen H, Wang T, Tsai H-I, Zhang R, Chang D (2016) The effects of quercetin-loaded PLGA-TPGS nanoparticles on ultraviolet B-induced skin damages in vivo. *Nanomedicine: Nanotechnology, Biology and Medicine* 12(3), 623–632.

Zuo J, Jiang Y, Zhang E, Chen Y, Liang Z, Zhu J, Zhao Y, Xu H, Liu G, Liu J (2019) Synergistic effects of 7-O-geranylquercetin and siRNAs on the treatment of human breast cancer. *Life Sciences* 227, 145–152.

Section IV

Phenolic Phytomolecules in Cancer Chemotherapy

14 Nanoenabled Systems for the Delivery of Curcumin as an Anticancer Agent

Saswati Mishra[1], *Sweta Priyadarshini Pradhan*[2], *and Anindita Behera*[3]
[1]School of Allied Health Sciences, Malla Reddy University, India
[2]School of Pharmaceutical Sciences, Siksha O Anusandhan (Deemed to be University), Bhubaneswar, India
[3]Amity Institute of Pharmacy, Amity University, Major Arterial Road, Action Area II, Rajarhat, New Town, Kolkata, India

14.1 INTRODUCTION

Cancer is a life-threatening disease, creating havoc worldwide in the public health sector. More than 609,000 fatalities in the United States alone were attributed to cancer in 2018, according to reported research (Torre et al., 2018). The prevalence of cancer and its mortality rate have not decreased over the past 30 years despite the advancements in cancer treatment (Gupta et al., 2013). Changes at the cellular and molecular level influence the pathogenesis and development of cancer. Hence, a deeper insight into the associated factors is anticipated to help with cancer treatment and prevention (Umar et al., 2012).

By preventing the growth, progression, and metastasis of cancer cells, several approaches for the reversal of cancer may be implemented. However, owing to its elevated treatment costs, cancer therapy is frequently disregarded in this cutthroat society(Martin et al., 2013; Falzone et al., 2018). Cancer treatment methods that are extensively and often utilized include chemotherapy and radiotherapy (Abbas and Rehman, 2018). However, the main disadvantage of these cancer therapeutic medications is that cancer cells develop resistance to them and cause a variety of side effects that prevent patients from receiving additional chemotherapy and radiation treatment (Wang et al., 2019b). A handful of synthetic drugs are available for cancer therapy, in addition to those extracted from plant derivatives, which possess extensive therapeutic activities against cancer. These plants include *Betula alba*, *Curcuma longa*, *Taxus brevifolia*, *Catharanthus roseus*, and *Erythroxylum previllei* (Gupta et al., 2017). Curcumin, found in turmeric, was first isolated in pure crystalline form in 1870 (Goel et al., 2008; Alibeiki et al., 2017).

Curcumin [diferuloylmethane or 1,7-bis(4-hydroxy-3-methoxy phenyl)-2,6-heptadine-3,5-dione] is considered the principal curcuminoid species of a rhizomatous perennial herbaceous plant belonging to Zingiberaceae family (Chanda and Ramachandra, 2019). Curcumin and its derivatives possess immense pharmacological activities like anticancer, anti-inflammatory, antioxidant, anti-mutagenic, and antibiotic properties, which are effectively used for the treatment of Alzheimer's disease, rheumatoid arthritis, cancer, diabetes, depression, wound healing, etc. (Nagahama et al., 2016). These characteristics result from the structure-activity relationships, which aim to enhance curcumin's physicochemical and biological properties (Aggarwal et al., 2014).

Curcumin's primary strategy for displaying its distinct anticancer capabilities includes activating apoptosis, preventing tumour cell proliferation, and blocking tumour cell invasion by overpowering various cellular signalling pathways (Kunnumakkara et al., 2017). By demonstrating its target effect against numerous cancer cell lines, other publications demonstrated curcumin's remarkable anticancer activity on additional body areas, including the breast, head, neck, lungs, prostate, brain, etc. (Anand et al., 2008).

Despite the benefits of curcumin described above, its in vivo and clinical applications are limited because of its poor water solubility profile, which further slows the rate of oral bioavailability and poses significant stability issues. The amount of less bioavailable curcumin causes an increase in metabolism and faecal excretion in addition to a decrease in the rate of absorption (Barry et al., 2009; Tsukamoto et al., 2014). Additionally, curcumin's hydrophobic nature makes it less capable of being taken up by cells, which tends to cause tight bonds with lipid membranes (fatty acyl chains) through hydrophobic interactions and hydrogen binding. This results in less bioavailability of the said drug inside cancer cells (Koo et al., 2015; Wong et al., 2019; Kuzminska et al., 2024).

As a result, various nanoformulations are developed to improve curcumin's effectiveness to get beyond the restrictions above (Torchilin, 2009; Lee et al., 2014; Rahiman et al., 2024). Several nanotechnology-based drug delivery systems, such as nanoemulsions, liposomes, dendrimers, micelles, nanogels, polymeric nanoparticles, solid lipid nanoparticles, magnetic nanoparticles, and cyclodextrin complexes, have shown increased therapeutic efficacy with higher target-specificity (Bose et al., 2015; Yallapu et al., 2015; Su and Kang, 2020). Innovative nanotechnologies like dual (thermal and magnetic) responsive nanoparticles are designed to produce a synergistic effect by delivering

DOI: 10.1201/9781003611516-18

curcumin at the target site and increasing the therapeutic efficacy of the curcumin (Ayubi et al., 2019). Novel nanotechnology-mediated drug delivery strategies, such as intranasal administration, can improve the rate of bioavailability, potency, and efficacy of curcumin (Hu et al., 2015). Hence, it may be stated that nanotechnology plays an important role in delivering the drug by improving its rate of bioavailability at the target site. Therefore, curcumin-loaded nanoparticles may be a potential therapeutic strategy for treating and preventing cancer (Gera et al., 2017; Patra et al., 2018).

The mechanism of action of curcumin against cancer, active and passive cancer targeting techniques, and associated adverse effects are all outlined in this chapter. The chapter also covers several curcumin-loaded nanotools that exhibit significant anticancer action and have improved bioavailability, target selectivity, and efficacy.

14.2 MECHANISM OF ACTION OF CURCUMIN AS AN ANTICANCER AGENT

Curcumin, alone or in combination, exhibits potential chemotherapeutic and chemopreventive anticancer properties with reported uses against several cancers, including prostate, breast, colorectal, pancreatic, lung, and oral. Curcumin exhibits its chemopreventive activity by promoting tumour apoptosis, inhibiting cell proliferation, anti-angiogenesis, anti-metastasis, scavenging reactive oxygen species, and decreasing inflammatory molecules in cancer (Wilken et al., 2011; Rahmani et al., 2014). Curcumin efficiently inhibits different cell signalling pathways which initiate the development and progression of cancer, i.e., nuclear factor kappa B, activated protein-1, signal transducer and activator of transcription-3, and epidermal growth response-1 (Wang et al., 2019a).

These transcriptional factors generally remain upregulated in tumour cells, which assists in cell proliferation, angiogenesis, and the development of tumours (Rahmani et al., 2014; Wang et al., 2019a). Curcumin significantly downregulates the nuclear factor kappa B pathways, which elevates the expression of various apoptotic linked proteins such as cyclin D1, B-cell lymphoma-extra-large, and B-cell lymphoma 2 and enzymes, i.e., matrix metalloproteinase-9 and cyclooxygenase-2, thereby promoting stimulation of cell proliferation, cell cycle arrest, and apoptosis leading to cell death (Chen et al., 2016). Curcumin also modulates the activated protein-1 and signal transducer and activator of transcription-3 expression, which helps in the cell proliferation stage of cancer (Sultana et al., 2021). The cancer progression depends on the epidermal growth factor receptor and is mainly associated with solid tumour cells (Sultana et al., 2021). Curcumin is responsible for the downregulation of epidermal growth factor receptor factor, which is mediated by epidermal growth response-1 action in colon cancer cells (HT29 and Caco2) which, thereby, produces a reduction in the cancer cell growth, metastasis, and invasion (Panda et al., 2017; Sultana et al., 2021).

Furthermore, curcumin is known to target the apoptosis process and elevate the tumour suppressor gene (p53) level, which causes apoptosis and cell death at the G2 phase. This further leads to the activation of B-cell lymphoma extra-large and B-cell lymphoma 2 (anti-apoptotic gene) and Bcl-2-associated X protein (pro-apoptotic gene), which inhibits cancer and tumour cells (Pfeffer and Singh, 2018; Talib et al., 2018). Curcumin inhibits endothelial nitric oxide synthase activity and its derivatives, nitric oxide, which plays a vital role in cell proliferation-inducing cancer. It was also reported that curcumin significantly inhibits angiogenic factors like interleukin-8, fibroblast growth factor, vascular endothelial growth factor, vascular endothelial growth factor receptor-2, and epidermal growth factor, which mainly induce angiogenesis, resulting in the formation of cancer (Chen et al., 2016; Sultana et al., 2021).

Several reports have also demonstrated that curcumin possesses both anti-inflammatory and antioxidant defence mechanisms responsible for the development and progression of multiple cancer cells (Wilken et al., 2011; Amalraj et al., 2017). Despite the extensive anticancer properties of curcumin, its preclinical and clinical application is restricted because of its lower oral bioavailability. In such a scenario, curcumin nanoformulation can tackle the said challenges and improve the delivery of curcumin to the desired tumour site for better therapeutic outcomes.

14.3 NANOENABLED TARGETING STRATEGIES FOR CURCUMIN AS AN ANTICANCER AGENT

The treatment strategy for cancer includes the targeted delivery of anticancer drugs to the tumour's target site. Conventional drug delivery systems have limitations, including lesser bioavailability due to differential pharmacokinetic parameters. The drugs administered through traditional dosage forms undergo first-pass metabolism, bind to plasma protein, and are excreted through the kidneys. Reduced circulation time, increased systemic toxicity, and the emergence of drug resistance render chemotherapy for cancer treatment more difficult and limit the effectiveness of anticancer drugs. As a result, passive and active targeting is employed to deliver anticancer medications to specific tumour sites. The uncontrolled growth of cancer cells causes angiogenesis and extravasation of blood vessels. The extravasation leads to the development of irregular blood vessels, facilitating the transit and accumulation of anticancer medications, resulting in enhanced permeation and retention effect (Jain and Stylianopoulos, 2010; Padhi and Behera, 2020).

In contrast, active targeting of chemotherapeutic drugs exhibits enhanced drug efficacy by targeted delivery of the active moieties to the overexpressed receptors on the cancer cells. In active targeting, the anticancer drugs are conjugated with ligands that selectively bind with the overexpressed receptors, and the drug molecules are taken up by receptor-mediated endocytosis. The conjugating ligands may be peptides, antibodies, enzymes, aptamers, folate, transferrin, biotin, hyaluronic acids, etc. The ligated anticancer drugs elicit anticancer activity in response to internal and external stimuli in the tumour microenvironment. The pH, temperature, redox potential, and magnetic field-sensitive nanoformulations elicit the delivery of anticancer drugs by active targeting (Behera and Padhi, 2020). The various publications on the targeted delivery of curcumin as an anticancer agent, both passive and active, are covered in the following section.

14.3.1 Passive Targeting Strategy of Curcumin

Kumari et al. (2021) designed nanoparticulate systems containing the polymer, polyglycerol malic acid, and dodecanedioic acid along with curcumin in the ratios of 7:3 and 6:4 and evaluated the anticancer potency against breast cancer cell lines (MCF-7 and MDA-MB-231). The in vitro anticancer assay of polymeric curcumin nanoparticles (7:3 and 6:4) in both the cell lines had inhibitory concentrations (50%) of 40 and 33.6 µM in MCF-7 cancer cells, respectively. In contrast, the values were 43.4 and 30.5 µM for MDA-MB-231 cancer cells. The percentage of cell migration mobility by the starch assay was observed at 39.1% and 45.9% at 48 hours by the curcumin nanoparticle treatment (6:4 and 7:3) in MDA-MB-231 cell lines, whereas 21.04% and 17.09% in MCF-7 treated cancer cells. Different staining assays with ethidium bromide, 4′,6-diamidino-2-phenylindole, and acridine orange revealed nuclear anomalies and apoptotic features, and further western blotting assay showcased overexpression of caspase-9, which confirmed the induction of apoptosis.

Chen et al. (2020a) designed a curcumin nanoformulation by conjugating a nano-micelle to a food-derived hydrophilic hydroxyethyl starch by an acid-ester linkage. The nano-micelle of curcumin possessed desirable drug loading efficiency and higher colloidal stability. It increased solubility more than a thousand times that of free curcumin with an acid-responsive release factor. The *in vitro* study of cytotoxicity assay and radical scavenging activity confirmed the potential anticancer, cytocompatibility, and antioxidant activity of curcumin nano-micelle with improved bioavailability.

Due to curcumin's reduced water stability and solubility, Duse et al. (2019) nano-encapsulated it in biodegradable poly (lactic-co-glycolic acid). The therapeutic efficacy of the nanoformulation was evaluated in human ovarian adenocarcinoma cells (SKOV-3). It was found that the curcumin nanoparticles showed enhanced reactive oxygen species inhibition compared to free curcumin. The curcumin nanoparticles caused tumour cell death due to the inhibitory effect of reactive oxygen species. Additionally, higher amounts of polymeric nanoparticles of curcumin exhibited strong apoptotic effects on tumour cells compared to free curcumin.

In another study, curcumin loaded in chitosan nanoparticles was evaluated to increase the bioavailability and solubility of curcumin. Curcumin loaded in nanoparticles was further investigated for its binding interaction using *in-silico* molecular docking studies. The formulated curcumin nanoparticles exhibited a fourfold increase in cytotoxicity than free curcumin. Furthermore, the cellular uptake, apoptotic effect, cell cycle blockage, DNA damage, and decreased reactive oxygen species levels confirmed the potential anticancer activity of curcumin nanoparticles (Yadav et al., 2018).

Wang et al. (2018) discovered that curcumin-loaded solid lipid nanoparticles elicited improved therapeutic efficacy against breast cancer. *In vitro* studies revealed that curcumin-loaded solid lipid nanoparticles showed greater uptake efficiency and stronger cytotoxicity activity against breast cancer cells (SKBR3) than free curcumin. Moreover, curcumin-loaded solid lipid nanoparticles induced a higher apoptosis rate (36.7%), promoting anti-apoptotic factors such as B-cell lymphoma protein 2-associated X/B-cell lymphoma protein 2. It reduced the cyclin-dependent kinase 4 and cyclin D1 expression. It also significantly inhibited cell migration property at the G1 phase and increased cell death by inducing apoptosis, indicating that it has superior therapeutic potential against breast cancer cells.

A novel drug delivery system, like a gold quantum cluster, was developed by Khandelwal et al. (2018) to entrap curcumin. The formulated gold quantum cluster concealed the poor solubility profile of curcumin. The anticancer efficacy of the curcumin-gold quantum cluster, glutathione-gold quantum cluster, and curcumin gold nanoparticles was evaluated in vitro against the normal cell (NIH 3T3) and breast cancer cell line (MCF7). The research study indicated increased cytotoxicity of curcumin-gold quantum clusters compared to curcumin-gold nanoparticles and glutathione-gold quantum clusters. The gold quantum cluster retained curcumin's cytotoxic properties, and the increased activity was associated with higher drug loading and smaller size compared to the curcumin gold nanoparticles and glutathione-gold quantum cluster.

Pectin-curcumin conjugates that could self-assemble into water-soluble nano-micelles in an aqueous solution were formulated (Bai et al., 2017). Transmission electron microscopy study revealed that these nano-micelles were spherical or ellipsoidal and aggregated with an approximate size range of 70–190 nm. Curcumin's stability was enhanced, and its antioxidant capability was maintained in the nano-micelle solution. The 3-(4,5-dimethylthiazol-2-yl)-2,5-diphenyltetrazolium bromide] test employing the cell lines such as hepatic cancer cell line (HepG2), cervical cancer cell line (HeLa), breast cancer cell line (MCF-7), and human normal kidney cell line (293A) revealed the fact that as a comparison to native curcumin, the curcumin of pectin-curcumin conjugates had a more pronounced inhibitory effect on cancer cells with a lower cytotoxicity on normal cells.

Azandeh et al. (2017) investigated the toxic effect and the anticancer activity of curcumin and curcumin-loaded poly lactic-coglycolic acid nanoparticles on prostate cancer cells

(PC3). The anticancer activity was evaluated by cell viability and proliferation activity study by the (3-[4,5-dimethylthiazol-2-yl]-2,5-diphenyl tetrazolium bromide) assays. It exhibited a significant reduction in tumour cell growth upon treatment with curcumin nanoparticles than native curcumin when used at a concentration of 25 μg/mL for a period of 48 hours. Secondly, a nuclear morphology study revealed that chromatin condensation increased the curcumin polymeric nanoparticles' apoptotic index. The number of prostate cancer cells that underwent apoptosis was substantially higher in the group treated with curcumin nanoparticles ($p<0.01$) than in the group treated with pure curcumin when analysed by annexin V-fluorescein isothiocyanate/propidium iodide assay. The relevant studies proved that curcumin polymeric nanoparticles have more potential as adjuvant therapy for prostate cancer.

Rompicharla et al. (2017) reported improved anticancer efficacy of curcumin-loaded solid lipid nanoparticles against *in vitro* breast cancer (MDA-MB-231) cells. The curcumin-loaded solid lipid nanoparticles exhibited superior cytotoxicity and cellular uptake activity compared to free curcumin. Furthermore, curcumin-conjugated solid lipid nanoparticles significantly enhanced the apoptosis rate compared to free curcumin (63.87% ± 3.11% vs 55.4% ± 1.80%).

In another study, cell growth inhibition in a human breast cancer cell line (MDA-MB-231) was studied employing native curcumin and curcumin capped with copper nanoparticles. In the chorioallantoic membrane assay, native curcumin was more anti-angiogenic than the curcumin-capped copper nanoparticles. The inhibition of pro-angiogenic cytokines like tumour necrosis factor-α, insulin-like growth factor, vascular endothelial growth factor, interleukin, fibroblast growth factor receptor, transforming growth factor-β, epidermal growth factor, and leptin by native curcumin and curcumin capped with copper nanoparticles showed selective inhibition of the said tumour cells. In the cell migration assay, it was observed that curcumin and the curcumin-capped copper nanoparticles considerably decreased the migration activity of breast cancer cells in a dose-dependent fashion. In the (3-[4,5-dimethylthiazol-2-yl]-2,5-diphenyl tetrazolium bromide) assay, curcumin significantly reduced cellular proliferation. In contrast, curcumin-capped copper nanoparticles presented increased cell proliferation activity. Hence, the curcumin-capped copper nanoparticles were not found to be potent anticancer agents compared to native curcumin (Kamble et al., 2016).

Cai et al. (2016) demonstrated the anticancer potential of curcumin micelles employing an amphiphilic copolymer (Pluronic F68). The pluronic F68-curcumin nano-micelle significantly produced higher cytotoxicity efficiency against human breast cancer cells (MDA-MB-231), having the inhibitory concentration (50%) value 1.94-fold lower than free curcumin, specifying the enhanced anticancer efficacy of the nanoformulation. Moreover, prevention with pluronic F68-curcumin nano-micelles at a dose level of 8 μg/mL increased the apoptosis rate to 56.63%. In contrast, the apoptotic cell percentage was reduced to 3.7-fold in the free curcumin-treated group. Thus, it may be apt to state that pluronic F68-curcumin nano-micelles exhibited a higher anticancer activity than free curcumin.

Curcumin-loaded Eudragit E 100 nanoparticles were developed using the emulsification-diffusion-evaporation technique (Chaurasia et al., 2016). Curcumin-loaded Eudragit E 100 nanoparticles demonstrated a satisfactory percentage of entrapment efficiency and nanometric particle size. The formulated nanoparticles showed a 19-fold decrease in 50% cell growth inhibition values. When curcumin is taken orally as curcumin-loaded eudragit E 100 nanoparticles as opposed to pure curcumin, there is a notable improvement in the oral bioavailability of curcumin, as evidenced by a ~91-fold increase in Cmax and ~95-fold increase in AUC_{0-12h}. Tumour volume, body weight, and survival rate increased significantly in the in vivo anticancer trial using CENPs compared to pure curcumin. It was evident from the data that the produced polymeric nanoparticles have significant promise for enhancing the antitumour efficacy and bioavailability of hydrophobic chemotherapeutics.

Curcumin-entrapped nanoparticles with varying sizes coated in polyvinyl alcohol were developed. The curcumin nanoparticles' particle size, encapsulation effectiveness, stability, cytotoxicity, and cellular uptake were evaluated using several cancer cell lines. Confocal microscopy showed that the bigger curcumin nanoparticles (200 nm) were only found in the cytoplasm, while the smaller curcumin nanoparticles (28 nm) were found subcellular in both the nucleus and cytoplasm (Lee et al., 2015)

The therapeutic potential for novel poly(lactic-co-glycolic acid)-curcumin nanoparticles for the treatment of prostate cancer was investigated in a study. The findings showed that prostate cancer cells effectively internalize poly(lactic-co-glycolic acid)-curcumin nanoparticles, which then deliver biologically active curcumin into the cytosolic compartment of the cells for optimal therapeutic action. The results of western blot, clonogenic, and cell proliferation 3-(4,5-dimethylthiazol-2-yl)-5-(3-carboxy methoxyphenyl)-2-(4-sulfophenyl)-2*H*-tetrazolium tests demonstrate that poly(lactic-co-glycolic acid)-curcumin nanoparticles are more effective than free curcumin in preventing prostate cancer cells from proliferating and forming colonies. In xenograft mice, poly(lactic-co-glycolic acid)-curcumin nanoparticles exhibited better tumour regression than curcumin. Subsequent studies indicated that poly(lactic-co-glycolic acid)-curcumin nanoparticles suppressed the expression of androgen and nuclear β-catenin in cells and tumour xenograft tissues. In prostate cancer, the results indicated that poly(lactic-co-glycolic acid)-curcumin nanoparticles can substantially accumulate and show better anticancer efficacy (Yallapu et al., 2014).

The apolipoprotein-E3-mediated curcumin-loaded poly(butyl)cyanoacrylate nanoparticles were prepared using the anionic polymerization method (Mulik et al., 2012). *In-vitro* cell culture studies revealed that apolipoprotein-E3 mediated curcumin-loaded poly(butyl) cyanoacrylate nanoparticles were more effective against neuroblastoma cells (SH-SY5Y) than curcumin solution and curcumin-loaded poly(butyl)cyanoacrylate nanoparticles. Curcumin indicated increased activity against SH-SY5Y neuroblastoma cells using curcumin-loaded poly(butyl)cyanoacrylate nanoparticles when compared to curcumin solution and curcumin-loaded poly(butyl)cyanoacrylate nanoparticles, with apoptosis being the underlying cause, according to flow cytometry studies. The investigation results showed that curcumin-loaded poly(butyl)cyanoacrylate nanoparticles have great potential to become a successful treatment method for brain cancer.

14.3.2 Active Targeting Strategy of Curcumin

Chen et al. (2020b) designed a delivery system utilizing platinum-based curcumin complex loaded in pH and redox-dual sensitive nanoparticles evaluated against non-small cell lung cancer cells. The nanoparticles exhibited enhanced intracellular curcumin release, posing a synergistic anticancer activity by inhibiting the phosphoinositide-3-kinase–protein kinase B/Ak strain transforming signalling pathway. The matrix metallopeptidase 2 and vascular endothelial growth factor receptor 2 were inhibited, which resulted in the arrest of the metastatic growth of the tumour cells.

Bian and Guo (2020) designed pH-responsive lactosylated nanoparticles for the co-delivery of curcumin and sorafenib, which were conjugated in a ratio of 1:1 (w/w). The nanoparticles were evaluated in the human hepatic carcinoma cells (HepG2) and hepatocellular tumour xenograft model. The *in-vitro* study revealed the synergistic effect of curcumin and sorafenib for tumour inhibition. The lactosylated-sorafenib-curcumin nanoparticles inhibited tumour cell growth up to 77.4%, with the lowest tumour volume of $239 \pm 14\,mm^3$. It significantly inhibited the proliferation, migration, and viability of cancer cell lines even at smaller doses, proving its potential in tumour cell inhibition activity.

Ghaffari et al. (2020) designed a curcumin nanocomposite using zinc oxide nanoparticles functionalized with *N*-succinyl chitosan for a pH-sensitive drug delivery system. The curcumin-conjugated nanoparticles possessed higher anticancer activity when compared to free curcumin against breast cancer cells (MDA-MB-231) using Annexin-V fluorescein isothiocyanate/propidium iodide and (3-[4,5-dimethylthiazol-2-yl]-2,5 diphenyl tetrazolium bromide) assay by inducing apoptosis. The nanoparticles didn't show any cytotoxicity on normal cells (HEK293).

Pooresmaeil and Namazi (2020) formulated a porous MIL-88 (iron) using pre-synthesized graphene quantum dots and loaded curcumin in the fabricated nanohybrid. The graphene quantum dots@MIL-88 (iron) were coated with chitosan by green coating to increase the biocompatibility of the nanohybrid. The fabricated nanohybrid exhibited a 38.3% release of curcumin at a pH of 5.0 after 72 hours of incubation against breast cancer cells (MDA-MB 231). The cytotoxicity of the nanohybrid was assessed by (3-[4,5-dimethylthiazol-2-yl]-2,5 diphenyl tetrazolium bromide) assay against breast cancer cells (MDA-MB-231), and the cell viability was found to be 62% and 50% for curcumin@graphene quantum dots@MIL-88(iron) and chitosan/curcumin@graphene quantum dots@MIL-88(iron), respectively. Hence, the nanohybrid acted as a biocompatible carrier of anticancer drugs exhibiting controlled release.

Kundu et al. (2019) designed a pH-responsive nanohybrid of curcumin conjugated with phenylboronic acid–zinc oxide nanoparticles. Curcumin-loaded zinc oxide nanoparticles targeted the overexpressed sialic acid through phenylboronic acid on the breast cancer cell membranes. The nanohybrid developed oxidative stress and mitochondrial damage to induce apoptosis. The nanohybrid inhibited tumour cell growth with fewer side effects and no systemic toxicity.

Mahalunkar et al. (2019) designed a sustained-release nanoformulation using folic acid conjugated with curcumin to compose folate-curcumin-loaded gold-polyvinyl pyrrolidine nanoparticles. This pH-responsive nanoconjugate was characterized and optimized for its target delivery approach and evaluated against *in vivo* breast cancer activity. The nanoconjugates efficiently exhibited anticancer effects at a reduced concentration in only progesterone/estrogen receptors in negative cell lines and not in positive cell lines. However, it did not produce cytotoxicity in mouse fibroblast and human breast epithelial cells. The nanoparticles also inhibited cell migration properties and possessed great anticancer efficacy in the *in vivo* studies. Thus, using folate to target cancer cells using folate-curcumin-loaded gold-polyvinyl pyrrolidine nanoparticles is a potential treatment approach for breast cancer without targeting common cells.

Amani et al. (2019) designed a magnetic nanohybrid containing ferric oxide modified by alginate and conjugated to bovine serum albumin and poly((3-acrylamidopropyl) trimethylammonium chloride). The encapsulation of curcumin in the nanohybrid was 95% by electrostatic interaction between negatively charged alginate/ferric oxide and positively charged bovine serum albumin/poly((3-acrylamidopropyl) trimethylammonium chloride). The (3-[4,5-dimethylthiazol-2-yl]-2,5 diphenyl tetrazolium bromide) assay revealed that the nanohybrid had no cytotoxicity activity. In contrast, the curcumin-encapsulated nanohybrid exhibited enhanced anticancer activity in breast cancer cells (MCF-7) and inhibition of cell proliferation in a dose-dependent manner.

Biodegradable poly(D,L-lactide-co-glycolic acid) microspheres containing magnetic iron oxide nanoparticles were synthesized and loaded with curcumin. However, pH-sensitive drug-release profiles resulted in the polymeric magnetic microspheres with a cumulative curcumin-releasing efficiency of 73.25%. The cytotoxic effect was evaluated using (3-[4,5-dimethylthiazol-2-yl]-2,5-diphenyl tetrazolium bromide) assay against human cervical cancer cell lines (HeLa). The *in vitro* cytotoxicity effect of polymeric magnetic microspheres was significantly greater than that of free microspheres and curcumin. The corresponding value of inhibitory concentration (50%) values of curcumin, curcumin-polymeric magnetic microspheres, and polymeric magnetic microspheres was found to be 10.3±0.560, 1.0±0.054, and 4.2±0.214 μg/mL, which signifies that the curcumin loaded with microspheres exhibited greater potential cytotoxic value when compared with free curcumin. Again, the cytotoxic effect was measured via the rate of apoptosis or cell death using fluorescent microscopic analysis with acridine orange/ethidium bromide stains, and the curcumin-polymeric magnetic microspheres possessed stronger apoptosis ability and anticancer activity than curcumin and polymeric magnetic microspheres (Ayyanaar et al., 2019).

Ahmadi Nasab et al. (2018) formulated pH-sensitive curcumin-encapsulated chitosan nanoparticles in a mesoporous silica nanocarrier system. The nano delivery system enhanced curcumin's solubility, bioavailability, and anticancer activity in glioblastoma cancer cell lines (U87MG). *In vitro,* the release study of the mesoporous silica nanosystem exhibited a controlled release of curcumin in the acidic microenvironment of cancer cells.

Xie et al. (2018) formulated pH-sensitive prodrug nanoformulation to co-deliver curcumin and methotrexate for combinational and multi-drug therapy for cancer treatment. Methotrexate acted as an anticancer agent and folate targeting ligand for overexpressed receptors in breast (MCF-7) and cervical (HeLa) cancer cells. Methotrexate was conjugated with 1,2-distearoyl-sn-glycero-3-phosphoethanolamine-*N*-[aldehyde(polyethyleneglycol)-2000] by Schiff's reaction. The obtained conjugated prodrug was loaded with curcumin in the core. The acid-labile nanoformulation released both drugs by folate receptor-mediated endocytosis. Moreover, the conjugated prodrug nanoformulation exhibited higher *in vitro* cytotoxicity activity and *in vivo* antitumour properties than individual free drugs and combination with free drugs towards cancer cell lines and the cervical tumour-bearing mice model.

Yuan et al. (2018) formulated polymeric nanoparticles with [monomethoxy (polyethylene glycol)-*b*-*P* (L-glutamic acid)] for co-delivery of doxorubicin and curcumin to treat tumour cells and breast cancer stem cells. The nanoparticles exhibited sustained release, among which curcumin was released faster than doxorubicin at a pH of 7.4. In an *in-vivo* study, it was found that nanoparticles were effective in lowering tumour growth. Cancer stem line growth declined from 39.9% to 6.82% in the control group after the curcumin and doxorubicin combination was administered as a nanoformulation. It was found effective in breast cancer treatment.

Using poly-L-lysine, Yang et al. (2018) developed a curcumin polymeric nanoparticle system employing the dialysis method. The *in-vitro* tests of the nanocomposite revealed that the changes in size and charge of the nanoconjugate at lower pH improved curcumin's cell uptake into human hepatoma (Hep3B) cells via the endocytosis process. It was observed that the curcumin nanoconjugate significantly decreased the tumour growth in the treated group of animals and possessed remarkable antitumour activity without any evidence of toxicity when compared to the control group. The results also showed that the nanoconjugate had an extensive circulation time.

Sauraj et al. (2018) prepared novel pH-responsive prodrug nanoparticles employing xylan-curcumin conjugate to enhance curcumin's therapeutic efficacy. The xylan-curcumin nanoparticles were assessed for anticancer efficacy against human colon cancer cells (HT-29, HCT-15). The xylan-curcumin nanoparticles exhibited a significant cytotoxicity effect as compared to free curcumin. The results showed that xylan-curcumin nanoparticles possessed 2.8- and 1.28-fold lower inhibitory concentration (50%) values than free curcumin in colon cancer cells, HT-29 and HCT-15 cell lines, respectively, indicating the increased anticancer effect of xylan-curcumin nanoparticles.

Cui et al. (2017) designed transferrin-conjugated nanoparticles to co-deliver curcumin and doxorubicin against breast cancer cells (MCF–7). A pH-responsive prodrug system of transferrin-poly (ethylene glycol)-curcumin was prepared, and doxorubicin was incorporated into the said system. The transferrin-poly (ethylene glycol)-curcumin-doxorubicin nanoparticles were evaluated for anticancer activity in breast cancer cells (MCF-7) and MCF-7 cells bearing mice. The nanoparticles showed more targeted drug delivery with enhanced cytotoxicity along with retardation of tumour growth by 83.5% as compared to transferrin-poly (ethylene glycol)-curcumin nanoparticles (68.4%), doxorubicin liposomes (53.2%), and curcumin-doxorubicin mixed injection (37.9%), respectively.

Baek and Cho (2017) conjugated copolymer *N*-carboxymethyl chitosan with curcumin. The said conjugate was formulated as solid lipid nanoparticles (NCC-SLN) to enhance the release rate of curcumin in the acidic medium and thereby improve its bioavailability. The nanoparticles released 40.4%±5.4% of curcumin at 2 hours, while chitosan-conjugated curcumin in solid lipid nanoparticles released 6.7%±1.5% at 2 hours. In the cytotoxicity study, the cytotoxic effect of curcumin in solid lipid nanoparticles and *N*-carboxymethyl chitosan-conjugated curcumin solid lipid nanoparticles was studied against breast cancer cells (MDA-MD-231). In (3-[4,5-dimethylthiazol-2-yl]-2,5 diphenyl tetrazolium bromide) assay, the cytotoxicity effect of curcumin-loaded solid lipid nanoparticles and *N*-carboxymethyl chitosan-conjugated curcumin solid lipid nanoparticles exhibited enhanced, controlled, and sustained release in the acidic environment of cancer cells. However, the oral bioavailability and lymphatic uptake of *N*-carboxymethyl chitosan-conjugated curcumin solid lipid nanoparticles were 9.5- and 6.3-fold higher than free curcumin.

Zhang et al. (2016) formulated prodrug nanoparticles using doxorubicin, curcumin, and polyethylene glycol to co-deliver curcumin and doxorubicin to treat cancer. From the *in vitro* studies, curcumin-doxorubicin polyethyleneglycol nanoparticles exhibited increased anticancer activity compared to free curcumin, the combination of curcumin and doxorubicin, and free doxorubicin. The in vivo study of the nano-prodrug demonstrated high antitumour activity, targeted drug delivery, and tumour growth inhibition.

When multiple drugs work simultaneously on the target, they exhibit enhanced therapeutic efficacy, notably in chemotherapy, making combination therapy with various drugs more effective than single or monotherapy. However, a synergistic interaction between pH-sensitive nanoparticles and anti-angiogenic activity reveals strong anticancer capabilities. Hence, a copolymer, amphiphilic poly(-amino ester), was used to develop a nanoparticle delivery system that co-entrapped the anti-angiogenic drug, curcumin and the pro-apoptotic drug, doxorubicin. Doxorubicin and curcumin nanoparticle co-delivery system showed improved cellular uptake and increased anticancer activities. To synergistically prevent human hepatocellular carcinoma growth, doxorubicin and curcumin were incorporated in pH-sensitive nanoparticles. The obtained nanoparticles were evaluated against human liver cancerous cells (SMMC 7221) and human umbilical vein endothelial cells (HUVECs). By lowering mitochondrial membrane potential, the doxorubicin-curcumin nanoparticles greatly increased the degree of apoptosis in human liver malignant cells. Additionally, it demonstrated anti-angiogenic properties, inhibiting the growth, migration, invasion, and tube formation of HUVECs in both in vivo and in vitro models. This was accomplished by modulating the vascular endothelial growth factor pathway. Nevertheless, doxorubicin-curcumin nanoparticles outperformed free curcumin in anticancer activity in human liver and umbilical vein endothelial cancer cells (Zhang et al., 2017).

Mulik et al. (2010) formulated transferrin-conjugated curcumin solid lipid nanoparticles to enhance curcumin's photostability and anticancer effect against breast cancer cells (MCF-7). The *in vitro* release study of the transferrin-curcumin-solid lipid nanoparticles confirmed the encapsulation of curcumin (amorphous state) in the solid lipid nanoparticles, which prevented the photodegradation of curcumin and exhibited sustained drug release for a protracted period. Transferrin-curcumin-solid lipid nanoparticles demonstrated improved cytotoxicity and cellular uptake compared to curcumin solid lipid nanoparticles and pure curcumin. Compared to free curcumin and curcumin solid lipid nanoparticles, the transferrin-targeted nanoformulation showed significantly higher anti-proliferative effects at low doses of 3 and 9 µM, indicating improved cellular uptake.

14.4 REPORTS ON CLINICAL TRIALS AND PATENTS OF CURCUMIN

Curcumin and its nanoformulations have been studied extensively in several animal models and cell culture assays. Numerous studies have also examined the effectiveness and safety of curcumin when used alone or in combination with other chemotherapeutic drugs to treat various cancers. The pharmacokinetic analysis, safety profile, and efficacy of curcumin in different cancer cells remain the core objectives of the clinical investigations. However, using curcumin or its combination with other chemotherapeutic agents against cancer shows interesting results, which have led to many studies on curcumin and its nano derivatives. The first clinical trial on curcumin was conducted by Kuttan et al. (1987) for treating cancer patients with external lesions. Curcumin has been employed as an anticancer drug ever since, and numerous reports have found it to have encouraging and effective results. Curcumin was said to have anticancer properties in a variety of neoplastic conditions, including gastric, pancreatic, colon, and myeloma cancer. Clinical research has recently concentrated on ovarian, lung, and breast cancer. Here are a few current research summaries on curcumin's use in treating and preventing various cancer types (Tables 14.1 and 14.2). The patents for curcumin nanoformulations used alone or in conjunction with other chemotherapeutic drugs to treat multiple tumours are listed in Table 14.3.

TABLE 14.1

Different Clinical Trial Studies of Curcumin and Its Treatment Strategy to Treat Cancer

Type of Cancer	Type of Study	No. of Patients	Dose	Outcome of the Study	References
Benign prostatic hyperplasia	Pilot, product evaluation registry study	61	1 g/day (24 weeks)	Decreased clinical and subclinical episodes of urinary block and infections and improved quality of patient's life	Ledda et al. (2012)
Chronic myeloid leukaemia	Randomized controlled study	50	5 g of turmeric powder with imatinib (400 mg/kg) three times per day for 6 weeks	Reduced nitric oxide levels significantly	Ghalaut et al. (2012)
Advanced and metastatic breast cancer	Phase I dose escalation trial	14	0.5–8 g/day with 100 mg/m^2 docetaxel for 7 days	No progression in further cancer was found, few patients responded, some side effects were observed, and the vascular endothelial growth factor levels	Bayet-Robert et al. (2010)

(Continued)

TABLE 14.1 (*Continued*)
Different Clinical Trial Studies of Curcumin and Its Treatment Strategy to Treat Cancer

Type of Cancer	Type of Study	No. of Patients	Dose	Outcome of the Study	References
Colorectal cancer	Dose escalation pilot study	15	40–200 mg/day for 29 days	Decreased prostaglandin E2 levels in a dose-dependent manner	Plummer et al. (2001)
	Phase I clinical trial study	126	360 mg three times a day for 10–30 days	Inhibited tumour necrosis factor-alpha levels, enhanced apoptotic tumour cells and their pathways, and increased p53 expression in the colorectal tissue	He et al. (2011)
	Clinical pilot study	26	2.35 g/day for 14 days	Safe and well-tolerated, sustained biologically active levels of curcumin were found in the colon tissues	Irving et al. (2013)
Colorectal neoplasia	Phase IIa clinical trial study	44	2 and 4 g/day for 1 month	Well-tolerated and decreased numbers of aberrant crypt foci with the dose level of 4 g	Carroll et al. (2011)
Head and neck squamous cell carcinoma cancer	Pilot study	21	1 g/day	Decreased inhibitor of nuclear factor kappa-β kinase activity in the salivary cells, which in turn reduces the expression of different cytokines	Kim et al. (2011)
Solid tumours	Randomized, controlled trial study	80	180 mg/day for 8 weeks	Improved life quality and decreased the expression of inflammatory mediators	Panahi et al. (2014)
Prostate cancer	Randomized double-blind controlled study	85	100 and 40 mg soy isoflavones for 6 months	Decreased prostate-specific antigens levels in patients with ≥10 μg/mL and androgen receptor expression	Ide et al. (2010)
	Double-blinded, randomized, placebo-controlled trial study	40	3 g/day for 3 months	Reduced the levels of prostate-specific antigens	Hejazi et al. (2016)
Pancreatic cancer	Phase I clinical study	16	200–400 mg/day for 9 months	Safe, improved quality of patient's life, easily bioavailable, but no changes in nuclear factor kappa-β expression and inflammatory response	Kanai et al. (2013)
	Phase II clinical trial	21	8,000 mg/day with gemcitabine for 14 days	Time to tumour progression was 1–12 months with a high frequency of side effects	Epelbaum et al. (2010)
	Phase I/II clinical trial	17	8,000 mg/day for 8 weeks	Safe and well-tolerated	Kanai et al. (2011)
Intestinal adenoma	Double-blinded randomized trial study	44	1,500 mg two times for 12 months	No significant clinical response was observed, with fewer side effects	Cruz-Correa et al. (2018)
Colorectal liver metastases	Phase I dose-escalation study	12	5 μM daily for 2 weeks with 5 μM 5-fluorouracil and 2 μM oxaliplatin	Safe and well-tolerated, enhanced anti-proliferative and anti-apoptotic activity, declined stem-cell associated markers expression such as high aldehyde dehydrogenase activity and CD133	James et al. (2015)

14.5 CONCLUSION AND FUTURE PERSPECTIVES

Curcumin is known to be a promising natural drug and is known to be effective against cancer, whether used alone or in combination with conventional chemotherapy treatment regimens. In both in vitro and in vivo studies, it has demonstrated significant anticancer benefits in a variety of malignancies, including prostate, breast, colon, colorectal, head, neck, and pancreatic cancer. Clinical trial investigations assessed its safety and effectiveness in human subjects, either alone or in combination with anticancer drugs. It possesses anticancer activity by targeting multiple pathways, such as inhibiting various

TABLE 14.2
Different Clinical Trial Studies of Curcumin and Its Updated Status at the Present

Type of Cancer	Type of Study	Identifier No.	Title	Dose	No. of Participants	Recruitment Status
Breast cancer	Randomized controlled trial	NCT01042938	Curcumin for the prevention of radiation-induced dermatitis in breast cancer patients	2 g three times daily for 4–7 weeks	35	Phase II completed
	Randomized clinical trial	NCT01740323	Phase II study of curcumin vs. placebo for chemotherapy-treated breast cancer patients undergoing radiotherapy	500 mg twice per day	30	Phase II completed
	Randomized pilot phase II trial	NCT02556632	Prophylactic topical agents in reducing radiation-induced dermatitis in patients with non-inflammatory breast cancer		191	Phase II completed
Cervical cancer	Phase II clinical trial study	NCT02017353	Effect of curcumin addition to standard treatment on tumour-induced inflammation in endometrial carcinoma	2 g/day for 2 weeks	7	Phase II completed
	Randomized clinical trial	NCT02944578	Biomolecular effects of topical curcumin in high-grade squamous intraepithelial lesion cervical neoplasia	2 g once a week for 12 weeks	40	Phase II ongoing
Pancreatic cancer	Interventional clinical trial	NCT00094445	Phase II trial of curcumin in patients with advanced pancreatic cancer	8 g/day for 8 weeks	50	Phase II completed
Lung cancer	Clinical trial study	NCT02321293	A phase 1 open-label prospective cohort trial of curcumin plus tyrosine kinase inhibitors for epidermal growth factor receptor-mutant advanced non-small cell lung cancer	80 mg daily for 8 weeks	20	Phase I ongoing
Colon cancer	Phase I clinical trial study	NCT00027495	Phase I pharmacokinetic trial of curcuminoids administered in a capsule formulation			Phase I completed
	Phase I clinical trial study	NCT02724202	A pilot feasibility study of curcumin in combination with 5-fluorouracil for patients with 5-fluorouracil-resistant metastatic colon cancer	500 mg twice per day for 2 weeks	13	Phase I ongoing
Prostate cancer	Randomized trial study	NCT01917890	Radiation therapy with or without curcumin supplements in treating patients with prostate cancer	3 g	40	Completed
Head and neck cancer	Phase I clinical trial study	NCT01160302	An exploratory biomarker trial of the food substances curcumin C3 complex® in subjects with newly diagnosed head and neck squamous cell carcinoma	4 g twice daily for 21–28 days	33	Early Phase I completed
Gastric cancer	Randomized clinical trial study	NCT02782949	Curcumin in preventing gastric cancer in patients with chronic atrophic gastritis or gastric intestinal metaplasia	Two times a day for 180 days	100	Phase II ongoing
Lymphoma	Phase II clinical trial study	NCT02100423	A Phase II study of curcumin and vitamin D in previously untreated patients with early-stage chronic lymphocytic leukaemia or small lymphocytic lymphoma	Per oral daily for 1–28 days	35	Phase II completed

cytokine and enzyme-related pathways and interfering with the metastasis and cell proliferation properties by inducing apoptosis. However, it is restricted from being used in clinical settings because of its poor bioavailability and low solubility characteristics. Different nanoformulations have been developed to overcome these challenges, including nanoparticles, liposomes, dendrimers, nanospheres, solid lipid nanoparticles, and micelles. Appropriate clinical trial studies are needed to fully realize the potential of various nano delivery systems entrapping curcumin to demonstrate its therapeutic efficacy in preventing and treating multiple malignancies. Numerous studies on curcumin and its nanoformulations are still in the preclinical stage.

TABLE 14.3
Details of Patents for Curcumin and Its Nanoformulations for the Treatment of Cancer

Name of the Drug (s)	Patent Number	Type of Nanoparticle	Outcomes	References
Camptothecin and curcuminoids	WO202008WO202008702A1	Nanocrystals	Decrease anticancer and antitumour activity	Xiao and Sedlarik (2020)
Curcumin, lutein, hesperidin, apigenin, ajoene etc.	WO2020249383A1	Nanoparticle using poly(lactic-co-glycolic acid), polyethyleneglycol, polyvinyl alcohol	Anticancer effect	Reymond et al. (2020)
Curcumin	IN201711026345A	Nanoparticle	It efficiently treats hepatocellular carcinoma by decreasing tumour volume and rate of apoptosis	Katare et al. (2019)
Curcumin	IN201811027482A	Nanoparticle	Antitumour effect	Sharma et al. (2019)
Curcumin and erlotinib	US20210369861A1	Nanoparticle	Anticancer activity	Xu (2021)
Curcumin, rutin and ginger	WO2021232072A1	Solid lipid nanoparticle	Inhibits cancer and tumour cells effectively with improvement in liver function	Van et al. (2021)
Curcumin, quercetin, fisetin, equol, resveratrol	WO2019210189A1	Nanoparticle	Anticancer and antitumour activity	Zhang (2019)
Curcumin and doxorubicin	IN201841019915A	Nanoparticle	Anticancer effect	Gupta et al. (2019)

REFERENCES

Abbas, Z., & Rehman, S. (2018). An overview of cancer treatment modalities. *Neoplasm*, 1, 139–157.

Aggarwal, B. B., Deb, L., & Prasad, S. (2014). Curcumin differs from tetrahydro curcumin for molecular targets, signalling pathways and cellular responses. *Molecules*, 20(1), 185–205.

Ahmadi Nasab, N., Hassani Kumleh, H., Beygzadeh, M., Teimourian, S., & Kazemzad, M. (2018). Delivery of curcumin by pH-responsive chitosan mesoporous silica nanoparticles for cancer treatment. *Artificial Cells, Nanomedicine, and Biotechnology*, 46(1), 75–81.

Alibeiki, F., Jafari, N., Karimi, M., & Peeri Dogaheh, H. (2017). Potent anticancer effects of less polar Curcumin analogues on gastric adenocarcinoma and esophageal squamous cell carcinoma cells. *Scientific Reports*, 7(1), 1–9.

Amalraj, A., Pius, A., Gopi, S., & Gopi, S. (2017). Biological activities of curcuminoids, other biomolecules from turmeric and their derivatives: A review. *Journal of Traditional and Complementary Medicine*, 7(2), 205–233.

Amani, S., Mohamadnia, Z., & Mahdavi, A. (2019). pH-responsive hybrid magnetic polyelectrolyte complex based on alginate/BSA as efficient nanocarrier for curcumin encapsulation and delivery. *International Journal of Biological Macromolecules*, 141, 1258–1270.

Anand, P., Sundaram, C., Jhurani, S., Kunnumakkara, A. B., & Aggarwal, B. B. (2008). Curcumin and cancer: An "old-age" disease with an "age-old" solution. *Cancer Letters*, 267(1), 133–164.

Ayubi, M., Karimi, M., Abdpour, S., Rostamizadeh, K., Parsa, M., Zamani, M., & Saedi, A. (2019). Magnetic nanoparticles decorated with PEGylated curcumin as dual targeted drug delivery: Synthesis, toxicity and biocompatibility study. *Materials Science and Engineering: C*, 104, 109810.

Ayyanaar, S., Kesavan, M. P., Sivaraman, G., Maddiboyina, B., Annaraj, J., Rajesh, J., & Rajagopal, G. (2019). A novel curcumin-loaded PLGA micromagnetic composite system for controlled and pH-responsive drug delivery. *Colloids and Surfaces A: Physicochemical and Engineering Aspects*, 573, 188–195.

Azandeh, S. S., Abbaspour, M., Khodadadi, A., Khorsandi, L., Orazizadeh, M., & Heidari-Moghadam, A. (2017). Anticancer activity of curcumin-loaded PLGA nanoparticles on PC3 prostate cancer cells. *Iranian Journal of Pharmaceutical Research: IJPR*, 16(3), 868.

Baek, J. S., & Cho, C. W. (2017). Surface modification of solid lipid nanoparticles for oral delivery of curcumin: Improvement of bioavailability through enhanced cellular uptake, and lymphatic uptake. *European Journal of Pharmaceutics and Biopharmaceutics*, 117, 132–140.

Bai, F., Diao, J., Wang, Y., Sun, S., Zhang, H., Liu, Y., Wang, Y., & Cao, J. (2017). A new water-soluble nano-micelle through self-assembly pectin-curcumin conjugates: Preparation, characterization and anticancer activity evaluation. *Journal of Agricultural and Food Chemistry*, 65(32), 6840–6847.

Barry, J., Fritz, M., Brender, J. R., Smith, P. E., Lee, D. K., & Ramamoorthy, A. (2009). Determining the effects of lipophilic drugs on membrane structure by solid-state NMR spectroscopy: The case of the antioxidant curcumin. *Journal of the American Chemical Society*, 131(12), 4490–4498.

Bayet-Robert, M., Kwiatowski, F., Leheurteur, M., Gachon, F., Planchat, E., Abrial, C., Durando, X., Barthomeuf, C., & Chollet, P. (2010). Phase I dose escalation trial of docetaxel plus curcumin in patients with advanced and metastatic breast cancer. *Cancer Biology & Therapy*, 9(1), 8–14.

Behera, A., & Padhi, S. (2020). Passive and active targeting strategies for the delivery of the camptothecin anticancer drug: A review. *Environmental Chemistry Letters*, 18(5), 1557–1567.

Bian, Y., & Guo, D. (2020). Targeted therapy for hepatocellular carcinoma: Co-delivery of sorafenib and curcumin using lactosylated pH-responsive nanoparticles. *Drug Design, Development and Therapy*, 14, 647.

Bose, S., Panda, A. K., Mukherjee, S., & Sa, G. (2015). Curcumin and tumour immune-editing: Resurrecting the immune system. *Cell Division*, 10, 6.

Cai, Y., Sun, Z., Fang, X., Fang, X., Xiao, F., Wang, Y., & Chen, M. (2016). Synthesis, characterization and anticancer activity of Pluronic F68–curcumin conjugate micelles. *Drug Delivery*, 23(7), 2587–2595.

Carroll, R. E., Benya, R. V., Turgeon, D. K., Vareed, S., Neuman, M., Rodriguez, L., Kakarala, M., Carpenter, P. M., McLaren, C., Meyskens, F. L., & Brenner, D. E. (2011). Phase IIa clinical trial of curcumin for the prevention of colorectal neoplasia. *Cancer Prevention Research*, 4(3), 354–364.

Chanda, S., & Ramachandra, T. V. (2019). Phytochemical and pharmacological importance of turmeric (*Curcuma longa*): A review. *Research & Reviews: A Journal of Pharmacology*, 9(1), 16–23.

Chaurasia, S., Chaubey, P., Patel, R. R., Kumar, N., & Mishra, B. (2016). Curcumin-polymeric nanoparticles against colon-26 tumour-bearing mice: Cytotoxicity, pharmacokinetic and anticancer efficacy studies. *Drug Development and Industrial Pharmacy*, 42(5), 694–700.

Chen, J., He, Z. M., Wang, F. L., Zhang, Z. S., Liu, X. Z., Zhai, D. D., & Chen, W. D. (2016). Curcumin and its promise as an anticancer drug: An analysis of its anticancer and antifungal effects in cancer and associated complications from invasive fungal infections. *European Journal of Pharmacology*, 772, 33–42.

Chen, S., Wu, J., Tang, Q., Xu, C., Huang, Y., Huang, D., Luo, F., Yuanzi, W., Yan, F., Weng, Z., & Wang, S. (2020a). Nano-micelles based on hydroxyethyl starch-curcumin conjugates for improved stability, antioxidant and anticancer activity of curcumin. *Carbohydrate Polymers*, 228, 115398.

Chen, Y., Chen, C., Zhang, X., He, C., Zhao, P., Li, M., Fan, T., Yan, R., Lu, Y., Khan, M. W., Sarfraz, M., Ma, X., Yang, T., & Xiang, G. (2020b). Platinum complexes of curcumin delivered by dual-responsive polymeric nanoparticles improve chemotherapeutic efficacy based on the enhanced anti-metastasis activity and reduce side effects. *Acta Pharmaceutica Sinica B*, 10(6), 1106–1121.

Cruz-Correa, M., Hylind, L. M., Marrero, J. H., Zahurak, M. L., Murray-Stewart, T., Casero Jr, R. A., Montgomery, E. A., Iacobuzio-Donahue, C., Brosens, L. A., Offerhaus, G. J., Umar, A., Rodriguez, L. M., & Giardiello, F. M. (2018). Efficacy and safety of curcumin in treatment of intestinal adenomas in patients with familial adenomatous polyposis. *Gastroenterology*, 155(3), 668–673.

Cui, T., Zhang, S., & Sun, H. (2017). Co-delivery of doxorubicin and pH-sensitive curcumin prodrug by transferrin-targeted nanoparticles for breast cancer treatment. *Oncology Reports*, 37(2), 1253–1260.

Duse, L., Agel, M. R., Pinnapireddy, S. R., Schäfer, J., Selo, M. A., Ehrhardt, C., & Bakowsky, U. (2019). Photodynamic therapy of ovarian carcinoma cells with curcumin-loaded biodegradable polymeric nanoparticles. *Pharmaceutics*, 11(6), 282.

Epelbaum, R., Schaffer, M., Vizel, B., Badmaev, V., & Bar-Sela, G. (2010). Curcumin and gemcitabine in patients with advanced pancreatic cancer. *Nutrition and Cancer*, 62(8), 1137–1141.

Falzone, L., Salomone, S., & Libra, M. (2018). Evolution of cancer pharmacological treatments at the turn of the third millennium. *Frontiers in Pharmacology*, 9, 1300.

Gera, M., Sharma, N., Ghosh, M., Huynh, D. L., Lee, S. J., Min, T., & Jeong, D. K. (2017). Nanoformulations of curcumin: An emerging paradigm for improved remedial application. *Oncotarget*, 8(39), 66680.

Ghaffari, S. B., Sarrafzadeh, M. H., Salami, M., & Khorramizadeh, M. R. (2020). A pH-sensitive delivery system based on N-succinyl chitosan-ZnO nanoparticles for improving antibacterial and anticancer activities of curcumin. *International Journal of Biological Macromolecules*, 151, 428–440.

Ghalaut, V. S., Sangwan, L., Dahiya, K., Ghalaut, P., Dhankhar, R., & Saharan, R. (2012). Effect of imatinib therapy with and without turmeric powder on nitric oxide levels in chronic myeloid leukemia. *Journal of Oncology Pharmacy Practice*, 18(2), 186–190.

Goel, A., Kunnumakkara, A. B., & Aggarwal, B. B. (2008). Curcumin as "Curecumin": From kitchen to clinic. *Biochemical Pharmacology*, 75(4), 787–809.

Gupta, A. P., Khan, S., Manzoor, M. M., Yadav, A. K., Sharma, G., Anand, R., & Gupta, S. (2017). Anticancer curcumin: Natural analogues and structure-activity relationship. *Studies in Natural Products Chemistry*, 54, 355–401.

Gupta, A. P., Pandotra, P., Sharma, R., Kushwaha, M., & Gupta, S. (2013). Marine resource: A promising future for anticancer drugs. *Studies in Natural Products Chemistry*, 40, 229–325.

Gupta, P. K., Anjanetulu, K., Gupta, S., Chakraborty, D., & Verma, R. S. (2019). Metal-free polyester based nano-drug carrier. Patent IN201841019915A.

He, Z. Y., Shi, C. B., Wen, H., Li, F. L., Wang, B. L., & Wang, J. (2011). Upregulation of p53 expression in patients with colorectal cancer by administration of curcumin. *Cancer Investigation*, 29(3), 208–213.

Hejazi, J., Rastmanesh, R., Taleban, F. A., Molana, S. H., Hejazi, E., Ehtejab, G., & Hara, N. (2016). Effect of curcumin supplementation during radiotherapy on oxidative status of patients with prostate cancer: A double blinded, randomized, placebo-controlled study. *Nutrition and Cancer*, 68(1), 77–85.

Hu, L., Kong, D., Hu, Q., Gao, N., & Pang, S. (2015). Evaluation of high-performance curcumin nanocrystals for pulmonary drug delivery both in vitro and in vivo. *Nanoscale Research Letters*, 10(1), 1–9.

Ide, H., Tokiwa, S., Sakamaki, K., Nishio, K., Isotani, S., Muto, S., Hama, T., Masuda, H., & Horie, S. (2010). Combined inhibitory effects of soy isoflavones and curcumin on the production of prostate-specific antigen. *The Prostate*, 70(10), 1127–1133.

Irving, G. R., Howells, L. M., Sale, S., Kralj-Hans, I., Atkin, W. S., Clark, S. K., Britton, R. G., Jones, D. J. L., Scott, E. N., Berry, D. P., Hemingway, D., Miller, A.S., Brown, K., Gescher, A. J., & Steward, W. P. (2013). Prolonged biologically active colonic tissue levels of curcumin achieved after oral administration: A clinical pilot study including assessment of patient acceptability. *Cancer Prevention Research*, 6(2), 119–128.

Jain, R. K., & Stylianopoulos, T. (2010). Delivering nanomedicine to solid tumors. *Nature Reviews Clinical Oncology*, 7(11), 653–664.

James, M. I., Iwuji, C., Irving, G., Karmokar, A., Higgins, J. A., Griffin-Teal, N., Thomas, A., Greaves, P., Cai, H., Patel, S. R., Morgan, B., & Brown, K. (2015). Curcumin inhibits cancer stem cell phenotypes in ex vivo models of colorectal liver metastases, and is clinically safe and tolerable in combination with FOLFOX chemotherapy. *Cancer Letters*, 364(2), 135–141.

Kamble, S., Utage, B., Mogle, P., Kamble, R., Hese, S., Dawane, B., & Gacche, R. (2016). Evaluation of curcumin capped copper nanoparticles as possible inhibitors of human breast cancer cells and angiogenesis: A comparative study with native curcumin. *AAPS PharmSciTech*, 17(5), 1030–1041.

Kanai, M., Otsuka, Y., Otsuka, K., Sato, M., Nishimura, T., Mori, Y., Kawaguchi, M., Hatano, E., Kodama, Y., Matsumoto, S., & Shibata, H. (2013). A phase I study investigating the safety and pharmacokinetics of highly bioavailable curcumin (Theracurmin®) in cancer patients. *Cancer Chemotherapy and Pharmacology*, 71(6), 1521–1530.

Kanai, M., Yoshimura, K., Asada, M., Imaizumi, A., Suzuki, C., Matsumoto, S., Nishimura, T., Mori, Y., Kawaguchi, Y., Yazumi, S., Chiba, T., Guha, S., & Aggarwal, B. B. (2011). A phase I/II study of gemcitabine-based chemotherapy plus curcumin for patients with gemcitabine-resistant pancreatic cancer. *Cancer Chemotherapy and Pharmacology*, 68(1), 157–164.

Katare, D. P., Saxena, S., & Dhiman, N. (2019). A novel composition for the treatment of hepatocellular carcinoma. Patent IN201711026345A.

Khandelwal, P., Alam, A., Choksi, A., Chattopadhyay, S., & Poddar, P. (2018). Retention of anticancer activity of curcumin after conjugation with fluorescent gold quantum clusters: An in vitro and in vivo xenograft study. *ACS Omega*, 3(5), 4776–4785.

Kim, S. G., Veena, M. S., Basak, S. K., Han, E., Tajima, T., Gjertson, D. W., Starr, J., Eidelman, O., Pollard, H. B., Srivastava, M., & Wang, M. B. (2011). Curcumin treatment suppresses IKKβ kinase activity of salivary cells of patients with head and neck cancer: A pilot study. *Clinical Cancer Research*, 17(18), 5953–5961.

Koo, H. J., Shin, S., Choi, J. Y., Lee, K. H., Kim, B. T., & Choe, Y. S. (2015). Introduction of methyl groups at C2 and C6 positions enhances the anti-angiogenesis activity of curcumin. *Scientific Reports*, 5(1), 1–12.

Kumari, M., Sharma, N., Manchanda, R., Gupta, N., Syed, A., Bahkali, A. H., & Nimesh, S. (2021). PGMD/curcumin nanoparticles for the treatment of breast cancer. *Scientific Reports*, 11(1), 1–17.

Kundu, M., Sadhukhan, P., Ghosh, N., Chatterjee, S., Manna, P., Das, J., & Sil, P. C. (2019). pH-Responsive and targeted delivery of curcumin via phenylboronic acid-functionalized ZnO nanoparticles for breast cancer therapy. *Journal of Advanced Research*, 18, 161–172.

Kunnumakkara, A. B., Bordoloi, D., Padmavathi, G., Monisha, J., Roy, N. K., Prasad, S., & Aggarwal, B. B. (2017). Curcumin, the golden nutraceutical: Multitargeting for multiple chronic diseases. *British Journal of Pharmacology*, 174(11), 1325–1348.

Kuttan, R., Sudheeran, P. C., & Josph, C. D. (1987). Turmeric and curcumin as topical agents in cancer therapy. *Tumori Journal*, 73(1), 29-31.

Kuzminska, J., Szyk, P., Mlynarczyk, D. T., Bakun, P., Muszalska-Kolos, I., Dettlaff, K., Sobczak, A., Goslinski, T., & Jelinska, A. (2024). Curcumin derivatives in medicinal chemistry: Potential applications in cancer treatment. *Molecules*, 29(22), 5321.

Ledda, A., Belcaro, G., Dugall, M., Luzzi, R., Scoccianti, M., Togni, S., Appendino, G., & Ciammaichella, G. (2012). Meriva®, a lecithinized curcumin delivery system, in the control of benign prostatic hyperplasia: A pilot, product evaluation registry study. *Panminerva Medica*, 54(4), 17.

Lee, W. H., Bebawy, M., Loo, C. Y., Luk, F., Mason, R. S., & Rohanizadeh, R. (2015). Fabrication of curcumin micellar nanoparticles with enhanced anticancer activity. *Journal of Biomedical Nanotechnology*, 11(6), 1093–1105.

Lee, W. H., Loo, C. Y., Young, P. M., Traini, D., Mason, R. S., & Rohanizadeh, R. (2014). Recent advances in curcumin nanoformulation for cancer therapy. *Expert Opinion on Drug Delivery*, 11(8), 1183–1201.

Mahalunkar, S., Yadav, A. S., Gorain, M., Pawar, V., Braathen, R., Weiss, S., Bogen, B., Gosavi, S. W., & Kundu, G. C. (2019). Functional design of pH-responsive folate-targeted polymer-coated gold nanoparticles for drug delivery and in vivo therapy in breast cancer. *International Journal of Nanomedicine*, 14, 8285.

Martin, T. A., Ye, L., Sanders, A. J., Lane, J., & Jiang, W. G. (2013). Cancer invasion and metastasis: Molecular and cellular perspective. In: Jandial, R. (Ed.) *Metastatic Cancer Clinical and Biological Perspectives*. Landes Bioscience, Austin, TX, pp. 135–168.

Mulik, R. S., Mönkkönen, J., Juvonen, R. O., Mahadik, K. R., & Paradkar, A. R. (2012). ApoE3 mediated polymeric nanoparticles containing curcumin: Apoptosis induced in vitro anticancer activity against neuroblastoma cells. *International Journal of Pharmaceutics*, 437(1–2), 29–41.

Mulik, R. S., Mönkkönen, J., Juvonen, R. O., Mahadik, K. R., & Paradkar, A. R. (2010). Transferrin mediated solid lipid nanoparticles containing curcumin: Enhanced in vitro anticancer activity by induction of apoptosis. *International Journal of Pharmaceutics*, 398(1–2), 190–203.

Nagahama, K., Utsumi, T., Kumano, T., Maekawa, S., Oyama, N., & Kawakami, J. (2016). Discovery of a new function of curcumin which enhances its anticancer therapeutic potency. *Scientific Reports*, 6(1), 1–14.

Padhi, S., & Behera, A. (2020). Nanotechnology based targeting strategies for the delivery of Camptothecin. In: *Sustainable Agriculture Reviews 44: Pharmaceutical Technology for Natural Products Delivery Vol. 2 Impact of Nanotechnology*. Springer nature, Switzerland, Edited by Ankit Saneja, Amulya K. Panda, Eric Lichtfouse. pp. 243–272.

Panahi, Y., Saadat, A., Beiraghdar, F., & Sahebkar, A. (2014). Adjuvant therapy with bioavailability-boosted curcuminoids suppresses systemic inflammation and improves quality of life in patients with solid tumours: A randomized double-blind placebo-controlled trial. *Phytotherapy Research*, 28(10), 1461–1467.

Panda, A. K., Chakraborty, D., Sarkar, I., Khan, T., & Sa, G. (2017). New insights into therapeutic activity and anticancer properties of curcumin. *Journal of Experimental Pharmacology*, 9, 31.

Patra, J. K., Das, G., Fraceto, L. F., Campos, E. V. R., Rodriguez-Torres, M. D. P., Acosta-Torres, L. S., & Shin, H. S. (2018). Nano based drug delivery systems: Recent developments and future prospects. *Journal of Nanobiotechnology*, 16(1), 1–33.

Pfeffer, C. M., & Singh, A. T. (2018). Apoptosis: A target for anticancer therapy. *International Journal of Molecular Sciences*, 19(2), 448.

Plummer, S. M., Hill, K. A., Festing, M. F., Steward, W. P., Gescher, A. J., & Sharma, R. A. (2001). Clinical development of leukocyte cyclooxygenase 2 activity as a systemic biomarker for cancer chemopreventive agents. *Cancer Epidemiology and Prevention Biomarkers*, 10(12), 1295–1299.

Pooresmaeil, M., & Namazi, H. (2020). Facile preparation of pH-sensitive chitosan microspheres for delivery of curcumin; characterization, drug release kinetics and evaluation of anticancer activity. *International Journal of Biological Macromolecules*, 162, 501–511.

Rahiman, N., Kesharwani, P., Karav, S., & Sahebkar, A. (2024). Curcumin-based nanofibers: A promising approach for cancer therapy. *Pathology-Research and Practice*, 266, 155791.

Rahmani, A. H., Al Zohairy, M. A., Aly, S. M., & Khan, M. A. (2014). Curcumin: A potential candidate in prevention of cancer via modulation of molecular pathways. *BioMed Research International*, 2014, 761608.

Reymond, M. A., Schneider, M., Linder, A., & Sahoo, R. (2020). Active substance delivery system, preferably anticancer agent included in nanoparticles. Patent WO2020249383A1.

Rompicharla, S. V. K., Bhatt, H., Shah, A., Komanduri, N., Vijayasarathy, D., Ghosh, B., & Biswas, S. (2017). Formulation optimization, characterization, and evaluation of in vitro cytotoxic potential of curcumin loaded solid lipid nanoparticles for improved anticancer activity. *Chemistry and Physics of Lipids*, 208, 10–18.

Sauraj, Kumar, S. U., Kumar, V., Priyadarshi, R., Gopinath, P., & Negi, Y. S. (2018). pH-responsive prodrug nanoparticles based on xylan-curcumin conjugate for the efficient delivery of curcumin in cancer therapy. *Carbohydrate Polymers*, 188, 252–259.

Sharma, P. K., Kaushik, Y., & Alam, M. A. (2019). Antitumour nanoparticle formulation. Patent IN201811027482A.

Su, S., & M Kang, P. (2020). Recent advances in nanocarrier-assisted therapeutics delivery systems. *Pharmaceutics*, 12(9), 837.

Sultana, S., Munir, N., Mahmood, Z., Riaz, M., Akram, M., Rebezov, M., Kuderinova, N., Moldabayeva, Z., Rauf, A., & Rengasamy, K. R. (2021). Molecular targets for the management of cancer using Curcuma longa Linn. phytoconstituents: A review. *Biomedicine & Pharmacotherapy*, 135, 111078.

Talib, W. H., Al-Hadid, S. A., Ali, M. B. W., Al-Yasari, I. H., & Abd Ali, M. R. (2018). Role of curcumin in regulating p53 in breast cancer: An overview of the mechanism of action. *Breast Cancer: Targets and Therapy*, 10, 207.

Torchilin, V. (2009). Multifunctional and stimuli-sensitive pharmaceutical nanocarriers. *European Journal of Pharmaceutics and Biopharmaceutics*, 71(3), 431–444.

Torre, L. A., Trabert, B., DeSantis, C. E., Miller, K. D., Samimi, G., Runowicz, C. D., Gaudet, M. M., Jemal, A., & Siegel, R. L. (2018). Ovarian cancer statistics, 2018. *CA: A Cancer Journal for Clinicians*, 68(4), 284–296.

Tsukamoto, M., Kuroda, K., Ramamoorthy, A., & Yasuhara, K. (2014). Modulation of raft domains in a lipid bilayer by boundary-active curcumin. *Chemical Communications*, 50(26), 3427–3430.

Umar, A., Dunn, B. K., & Greenwald, P. (2012). Future directions in cancer prevention. *Nature Reviews Cancer*, 12(12), 835–848.

Van, N. A., Nghia, N. D., Hung, K. D., & Binh, N. T. N. (2021). Tricomponent solid lipid nano-composition comprising curcumin, ginger oleoresin, rutin having a cancer-cell-killing property and method of preparing the same. Patent WO2021232072A1.

Wang, M., Jiang, S., Zhou, L., Yu, F., Ding, H., Li, P., Zhou, M., & Wang, K. (2019a). Potential mechanisms of action of curcumin for cancer prevention: Focus on cellular signaling pathways and miRNAs. *International Journal of Biological Sciences*, 15(6), 1200.

Wang, W., Chen, T., Xu, H., Ren, B., Cheng, X., Qi, R., Liu, H., Wang, Y., Lei, Y., Chen, S., Yang, Q., & Chen, C. (2018). Curcumin-loaded solid lipid nanoparticles enhanced anticancer efficiency in breast cancer. *Molecules*, 23(7), 1578.

Wang, X., Zhang, H., & Chen, X. (2019b). Drug resistance and combating drug resistance in cancer. *Cancer Drug Resistance (Alhambra, Calif.)*, 2, 141.

Wilken, R., Veena, M. S., Wang, M. B., & Srivatsan, E. S. (2011). Curcumin: A review of anticancer properties and therapeutic activity in head and neck squamous cell carcinoma. *Molecular Cancer*, 10(1), 1–19.

Wong, K. E., Ngai, S. C., Chan, K. G., Lee, L. H., Goh, B. H., & Chuah, L. H. (2019). Curcumin nanoformulations for colorectal cancer: A review. *Frontiers in Pharmacology*, 10, 152.

Xiao, H., & Sedlarik, V. (2020). Methods of making nanocrystals with enhanced biological availability and formulation for such nanocrystals preparation for use in anticancer therapy. Patent WO2020088702A1.

Xie, J., Fan, Z., Li, Y., Zhang, Y., Yu, F., Su, G., Xie, L., & Hou, Z. (2018). Design of pH-sensitive methotrexate prodrug-targeted curcumin nanoparticles for efficient dual-drug delivery and combination cancer therapy. *International Journal of Nanomedicine*, 13, 1381.

Xu, P. (2021). Carrier-free curcumin nanoparticle for EGFR positive cancer therapy. Patent US20210369861A1.

Yadav, P., Bandyopadhyay, A., Chakraborty, A., & Sarkar, K. (2018). Enhancement of anticancer activity and drug delivery of chitosan-curcumin nanoparticle via molecular docking and simulation analysis. *Carbohydrate Polymers*, 182, 188–198.

Yallapu, M. M., Khan, S., Maher, D. M., Ebeling, M. C., Sundram, V., Chauhan, N., Ganju, A., Balakrishna, S., Gupta, B. K., Zafar, N., Jaggi, M., & Chauhan, S. C. (2014). Anticancer activity of curcumin loaded nanoparticles in prostate cancer. *Biomaterials*, 35(30), 8635–8648.

Yallapu, M. M., Nagesh, P. K., Jaggi, M., & Chauhan, S. C. (2015). Therapeutic applications of curcumin nanoformulations. *The AAPS Journal*, 17, 1341–1356.

Yang, D. H., Kim, H. J., Park, K., Kim, J. K., & Chun, H. J. (2018). Preparation of poly-l-lysine-based nanoparticles with pH-sensitive release of curcumin for targeted imaging and therapy of liver cancer in vitro and in vivo. *Drug Delivery*, 25(1), 950–960.

Yuan, J. D., ZhuGe, D. L., Tong, M. Q., Lin, M. T., Xu, X. F., Tang, X., & Xu, H. L. (2018). pH-Sensitive polymeric nanoparticles of mPEG-PLGA-PGlu with hybrid core for simultaneous encapsulation of curcumin and doxorubicin to kill the heterogeneous tumour cells in breast cancer. *Artificial Cells, Nanomedicine, and Biotechnology*, 46(sup1), 302–313.

Zhang, H.-G. (2019). Edible plant exosome-like nanovectors for vaccination. Patent WO2019210189A1.

Zhang, J., Li, J., Shi, Z., Yang, Y., Xie, X., Lee, S. M., Wang, Y., Leong, K. W., & Chen, M. (2017). pH-Sensitive polymeric nanoparticles for co-delivery of doxorubicin and curcumin to treat cancer via enhanced pro-apoptotic and anti-angiogenic activities. *Acta Biomaterialia*, 58, 349–364.

Zhang, Y., Yang, C., Wang, W., Liu, J., Liu, Q., Huang, F., Chu, L., Gao, H., Li, C., Kong, D., Liu, Q., & Liu, J. (2016). Co-delivery of doxorubicin and curcumin by pH-sensitive prodrug nanoparticle for combination therapy of cancer. *Scientific Reports*, 6(1), 1–12.

15 Nanoenabled Systems for the Delivery of Resveratrol as an Anticancer Agent

Santwana Palai
College of Veterinary Science & Animal Husbandry, Odisha University of Agriculture and Technology, Bhubaneswar, India

15.1 INTRODUCTION

Natural drugs continuously serve as precious novel chemical entities for healing purposes owing to their immense structural variety. Anti-microbial, anticancer, antidiabetic, and antihypertensive drug research is increasing as the spread of drug resistance is becoming more widespread. More than half of anticancer drugs are derived from secondary metabolites of the plant. Resveratrol, an antioxidant, cardioprotective, and anti-inflammatory compound found in jackfruit, grapes, legumes, berries, and peanuts, offers promising relief for a variety of cancers, including those of the head, thyroid, neck, stomach, pancreas, colon, breast, prostate, melanoma squamous, and ovary (Figure 15.1) (Shukla and Singh 2011). The antitumour effect of resveratrol is due to inhibiting different signalling pathways like hypoxia-inducible factor-1α and induction of apoptosis through p53 activation, establishing it as a therapeutic molecule for cancer therapy (Whitlock and Baek 2012). However, the vast therapeutic potential of resveratrol against cancer is hindered by its poor pharmacokinetic properties. To tackle the said challenge, resveratrol has been entrapped in various novel nano drug delivery systems (Saraf, 2010). These nanoformulations of resveratrol enable them to target cells specifically, lowering their dose requirements and off-site toxicities. The solubility, stability, half-life, and undesirable effects of resveratrol in normal cells and tissues are all overcome by its nanoformulations, which concentrate the drug at the targeted site (Coimbra et al., 2011). With enhanced absorption, resveratrol nanoformulations have demonstrated superior anticancer and anti-inflammatory benefits (Pavan et al. 2016).

Resveratrol is a secondary metabolite found in various dietary sources, principally in fruits (e.g., berries and grapes). Resveratrol is a stilbene class of polyphenols having the chemical name 3,5,4′-trihydroxystilbene or 5-(2-(4-hydroxyphenyl) vinyl) benzene-1,3-diol with molecular formula $C_{14}H_{12}O_3$ and molecular weight of 228.247 g/mol (Figure 15.1). The structure of resveratrol consists of two aromatic rings connected by a bridge (-methylene) and three phenolic hydroxyls (–OH) groups (Figure 15.1). The methylene bond, hydroxyl groups, and aromatic rings of resveratrol modify it into various derivatives, like trans-3,4′,5-trimethoxy stilbene, 4,4′-dihydroxy-trans-stilbene, pterostilbene, monoalkoxy, and dialkoxy derivatives having versatile therapeutic effects. Resveratrol exhibits antioxidant, antibacterial, anti-inflammatory, and antiobesity effects. It is reported to possess potential antitumour activity in different types of cancers (Wang et al., 2017b).

Resveratrol was first isolated from the roots of white hellebore (Veratrum grandiflorum) in the year 1940 (Varoni et al. 2016). As moderate red wine (high in resveratrol) consumption improved the cardiovascular health of individuals with high-fat diets in the early 1990s, resveratrol was recognized as a significant drug molecule (Renaud and de Lorgeril 1992). Other than the skin of red grapes, in varying amounts, resveratrol can be found in edible natural foods such as raspberries, pistachios, mulberries, peanuts, blueberries, cranberries, etc. Resveratrol naturally occurs in two isomeric forms: cis-resveratrol and trans-resveratrol. The trans-isomer is more versatile and has better potential as an antioxidant and anticancer agent (Singh and Pai, 2014). Upon exposure to artificial UV radiation or sunlight, rapid isomerisation, i.e., conversion of trans-resveratrol to cis-resveratrol and vice versa, is triggered, making it unstable and photosensitive (Vian et al., 2005). Resveratrol is a hydrophobic compound soluble in solvents like dimethyl sulfoxide and ethanol because of the presence of three –OH groups of the aromatic rings. The metabolism of resveratrol is fast, with a very short plasma half-life of 8–14 minutes, and resveratrol plasma concentrations are low even with larger doses of resveratrol given. Low water solubility and short half-life lead to low bioavailability, limiting resveratrol applications in the biomedical field (Guo et al. 2024).

Nanotechnology has recently become a recognized scientific approach for avoiding the adverse consequences of conventional anticancer therapies, including chemotherapy (Padhi et al. 2016; Kundu et al. 2022; Behera and Padhi 2022). The revised theory relies on the controlled and targeted delivery of medications via nano-based carriers. In general, polymeric and lipid-based nanoparticles, with particle sizes ranging from 10 to 1,000 nm, are among the most often utilized nanocarriers for the delivery of anticancer drugs, with 10–200 nm being the ideal particle size for the said use (Padhi et al. 2018). Typically, nanoparticles are classified into two types based on their composition (Behera et al., 2020).

DOI: 10.1201/9781003611516-19

FIGURE 15.1 Chemical structure and biological sources of resveratrol.

i. Inorganic nanoparticles, which are carbon-based, metal-based, and silica-based
ii. Organic nanoparticles comprise lipids (liposomes, solid lipid nanoparticles, nanostructured lipid carriers, polymers, and proteins).

For the development of nanotherapeutics for cancer, the characteristics of nanoparticles in terms of zeta potential, architecture, and particle size should be taken into consideration since they affect encapsulation for regulated drug delivery to malignant cells (Padhi and Behera 2022; Behera et al. 2023; Hartshorn et al. 2018).

Resveratrol undergoes extensive metabolism due to its low water solubility, resulting in low bioavailability and preventing optimal delivery to the target tissue. Encapsulation or conjugation with nanosized structures protects and maintains the structural integrity of polyphenolic compounds while increasing their water solubility. Along with improving physicochemical properties, nanotechnology-based carriers deliver polyphenols to target cells, increasing bioavailability and, as a result, bioactivity. The improvement in the drug delivery systems in cancer treatment can be used for modulating the release of the drug, protecting the encapsulated bioactive in the biological environment with better physicochemical effects (Padhi and Behera 2020; Kundu et al. 2022). Nanocarriers in drug delivery improve drug solubility, biodistribution, and bioavailability, thereby improving therapeutic effects with lower drug doses (Padhi et al. 2021). Thus, nanocarriers represent a better approach to improving traditional and plant-derived anticancer drug efficacy (Patnaik et al. 2021; Behera and Padhi 2022).

This chapter aims to provide a concise assessment of the anticancer potential of resveratrol when it is encapsulated or conjugated with nanocarriers, aiding in the more effective transport of resveratrol.

15.2 MECHANISM OF ACTION OF RESVERATROL AS AN ANTICANCER AGENT

Resveratrol, a nutritional phytochemical, regulates the expression of transcription factors essential for carcinogenesis, such as carcinogen activation, detoxification, proinflammatory-mediated signalling pathways (cell proliferation, invasion, angiogenesis, metastasis), arresting growth, and apoptosis (Jang et al. 1997). It participates in different mechanisms, either by preserving the functional status of cells, extending cellular life span, or triggering cell death of physiologically unbalanced cells (Fulda 2010). Resveratrol influences signal transduction pathways linked to tumorigenesis by altering essential transcription factors and carcinogenesis (Elshaer et al. 2018). Resveratrol affects the transcriptional machinery, causing essential regulatory proteins to be transactivated, acting as a signalling pathway modulator. Resveratrol's anticancer effects, therefore, result from a combined alteration of tumorigenesis and carcinogenesis in vitro and in vivo (Figure 15.2). The following section provides a full explanation of the said mechanism.

Resveratrol's chemopreventive effect is related to its ability to regulate inflammation, eliminate reactive oxygen species, neutralize pro-carcinogens, and stimulate calorie management. Resveratrol inhibits the nucleotide-binding domain leucine-rich repeat-containing family pyrin domain containing three inflammasomes, which include the apoptosis-associated speck-like protein, nucleotide-binding domain leucine-rich repeat-containing family pyrin domain containing three protein, and procaspase-1. Resveratrol inhibits the nucleotide-binding domain leucine-rich repeat-containing family pyrin domain containing three inflammasomes by upregulation of sirtuin, which causes downregulation of the nucleotide-binding domain leucine-rich repeat-containing family pyrin domain containing three inflammasomes by making autophagy via phosphatidylinositol-3-kinase/mammalian target of rapamycin path inhibition (Misawa et al. 2015). Resveratrol activates the mitogen-activated protein kinase phosphatase-1 pathway, suppressing inflammation by blocking the mitogen-activated protein kinases/nuclear factor-kappa B. It stimulates the nuclear factor erythroid 2-related factor 2 gene expression. Resveratrol reduces oxidative stress and slows the evolution of periodontitis by activating the sirtuin/adenosine monophosphate-activated protein kinase and nuclear factor erythroid 2-related factor 2/antioxidant defence pathways. Nuclear factor erythroid 2-related factor 2 overturns inflammation by attaching to interleukin-6 and interleukin-1β, like pro-inflammatory cytokine genes, constraining their macrophage expression (Zarebavani et al. 2023).

Resveratrol inhibits cyclooxygenase by reducing cyclooxygenase-2 expression in lipopolysaccharide-treated colorectal adenocarcinoma cells (Caco-2), decreasing prostaglandin E2 synthesis. Furthermore, resveratrol inhibits *N*-nitroso methyl benzylamine-induced oesophageal carcinogenesis in Fischer 344 rats via downregulating cyclooxygenase-2. The suppression of nuclear factor-kappa B by resveratrol is linked to the downregulation of cyclooxygenase-2. Nuclear factor-kappa B regulates the transcription of pro-inflammatory mediators like cyclooxygenase-2, critical in p53-dependent apoptosis in cancer cells (Cianciulli et al. 2012).

Resveratrol regulates drug-metabolising enzymes in human and rat liver cells by suppressing cytochrome P450 3A4, cytochrome P450 1A, and cytochrome P450 2E1. Furthermore, it has a marginal inhibitory effect on cytochrome P450 3A4 and a modest inhibitory effect on cytochrome P450 2C19. Furthermore, in BALB/c mice,

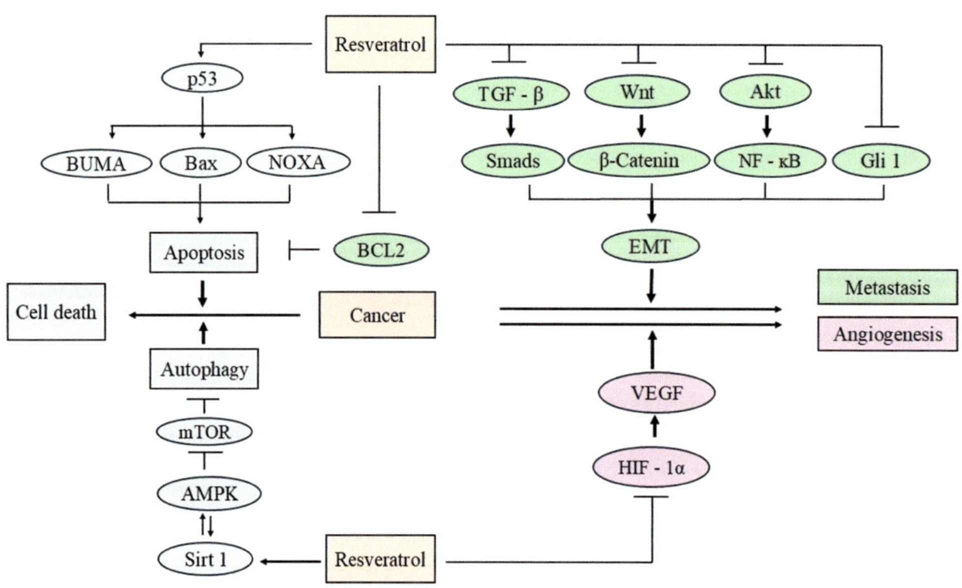

FIGURE 15.2 Anticancer mechanism of resveratrol inducing apoptosis, inhibiting the epithelial to mesenchymal transition, arresting the metastasis, and restricting angiogenesis.

resveratrol inhibits cytochrome P450 1A1 enzymatic activity induced by benzo[*a*]pyrene, as well as the generation of benzo[*a*]pyrene diol epoxide-DNA adducts (Aggarwal et al. 2004).

Phytoestrogens can intermingle with the oestrogen receptors (α and β) and inhibit the actions of estradiol due to their structural similarities. Resveratrol works as both oestrogen-agonist and antagonist in the absence of 17 beta-estradiol but as an anti-oestrogen in the presence of 17 beta-estradiol. Resveratrol has a high affinity for both estrogen receptor-α and oestrogen receptor-β. Oestrogen receptor-α promotes cell proliferation and is increased in prostate cancer cell lines, while oestrogen receptor-β protects against cancer cell proliferation (Sakamoto et al. 2010).

Resveratrol prevents estradiol-induced downregulation of nuclear factor erythroid 2-related factor 2 downstream antioxidant and detoxifying genes in estradiol-induced mammary tumours, decreasing tumour incidence and increasing tumour latency (Elshaer et al. 2018).

Resveratrol promotes autophagy by inhibiting the mammalian target of rapamycin via the sirtuin/adenosine monophosphate-activated protein kinase pathway. It causes cell death in a variety of malignancies by triggering autophagy. It promoted lung cancer (A549) cell death by autophagic cell death via the Ca^{2+}/adenosine monophosphate-activated protein kinase-mammalian target of the rapamycin signalling pathway. It stimulates the production of autophagic proteins such as Atg5, Atg12, and beclin-1, which causes autophagy and death in cisplatin-resistant human oral squamous cell carcinoma (CAR) cells. In T-cell acute lymphoblastic leukaemia (T-ALL cells), an aggressive malignant neoplasm of the bone marrow triggered both autophagy and apoptosis in a dose- and time-dependent manner (Wang et al. 2018).

It promotes apoptosis in many cancer cell types in different ways. It activates several apoptotic pathways. It causes both p53-dependent and p53-independent cell cycle arrest and apoptosis. Resveratrol encourages apoptosis in mouse skin cancer via activating p53 and increasing the expression of proapoptotic bcl-2-like protein 4 while lessening the antiapoptotic protein bcl-2. Resveratrol and its methylated derivatives activate p53, causing apoptosis in rat and human glioma cells. It induces p53 activity, which increases apoptosis in the human breast cancer cell line (MCF-7 cells). It increases the sensitivity of tumour cells to apoptosis by meddling with various signal transduction pathways regulating p53 activation (Almatroodi et al. 2022). Resveratrol activates p53 via extracellular signal-regulated protein kinases and p38 in the mouse epidermal cell line (JB6 cells). It makes hepatocellular (HepG2) cells more susceptible to cell death by boosting extracellular signal-regulated kinase activity and decreasing cyclin D1, AKt, and Pak1 expression and activity. Resveratrol also causes apoptosis in bladder cancer (T24 and 5,637 cells) by inhibiting microRNA-21 expression, which decreases Akt activity and decreases bcl-2 (Elshaer et al. 2018).

The p53-independent apoptosis of resveratrol in human nasopharyngeal cancer cells is intermediated by the N-terminal truncated isoform of p63 downregulation. Resveratrol alters the bcl-2-like protein 4/bcl2 ratio, leading to the mitochondrial transmembrane potential loss and cytochrome C release in activating caspases 3 and 9, resulting in apoptosis. High resveratrol stimulates bcl-2-like protein 4 expression, whereas low doses trigger bcl-2-like protein 4 translocation and mitochondria-mediated death. This demonstrates the resveratrol mechanism causing mitochondrial transmembrane potential loss (Almatroodi et al., 2022).

Resveratrol can reduce angiogenesis by regulating vascular endothelial growth factor. Hypoxia-inducible factor 1 downregulation overpowers vascular endothelial growth factor expression in osteosarcoma and human ovarian cancer cells. It suppresses Akt- and mitogen-activated protein kinase-driven hypoxia-inducible factor 1 baseline expression and insulin-like growth factor-1-induced hypoxia-inducible factor 1 induction and promotes hypoxia-inducible factor 1proteasomal degradation (Tino et al. 2016).

It suppresses colorectal, liver, and lung metastases by downregulating the transforming growth factor-β/smads pathway, which enhances E-cadherin expression while suppressing vimentin expression. Resveratrol inhibits the metastatic capacity of epithelioid carcinoma-attached pancreatic cancer cells (Panc-1) *in vitro* by altering epithelial-mesenchymal transition-related factors (N-cadherin, E-cadherin, vimentin, and matrix metalloproteinase-2,9) via the phosphoinositide-3-kinase–protein kinase B/nuclear factor-kappa B signalling path (Cheng et al. 2018).

The "Warburg effect," a notably altered glucose metabolism, is where cancer cells demonstrate noticeably increased glucose uptake rates and glycolysis, transforming pyruvate produced during glycolysis into lactate. Resveratrol is reported to counteract the Warburg effect in colon cancer cells by boosting the pyruvate dehydrogenase complex's activity. Resveratrol prevents microRNA-21-mediated glycolysis in the tumour stroma, which prevents glycolysis in pancreatic stellate cells (Brockmueller et al. 2021).

Resveratrol also inhibits the growth of non-small cell lung cancer by reducing glycolysis and targeting the adenosine-monophosphate-activated protein kinase/mammalian target of the rapamycin signalling pathway. In contrast, hexokinase II-mediated glycolysis regulation inhibits ovarian cancer cells' growth by inducing death (Brockmueller et al. 2021).

15.3 MOLECULAR TARGETS IN CANCER FOR RESVERATROL

The main mechanisms involved in the anticancer activity of resveratrol are induction of apoptosis, antiproliferation, and anti-inflammation, indicating its chemotherapeutic and chemopreventive potential. Resveratrol behaves as a proapoptotic agent. Caspase pathways regulate apoptosis induction in normal cells. These mechanisms are blocked by inhibiting important regulatory tumour suppressor genes in cancer cells. Resveratrol boosted caspase activation and lowered the expression of heat shock protein family anti-apoptotic proteins (Santos et al. 2019).

Resveratrol also targets the nuclear transcription factor (p53), a tumour suppressor gene that prevents tumour growth and controls several pathways. Through the mitogen-activated protein kinase pathway, resveratrol activates the p53 protein in breast and prostate cancer cells, leading to apoptosis and cell cycle arrest. Resveratrol induces apoptosis by interacting with protein kinase B and dephosphorylating the tumour suppressor protein (Whitlock and Baek 2012).

Resveratrol and its metabolites prevent DNA synthesis, arrest the cell cycle, and interfere with cancer cell proliferation. Resveratrol modulates the cell cycle and progresses apoptosis. The antioxidant activity exhibited by resveratrol is related to its chemopreventive activity of resveratrol by reducing the damage due to reactive oxygen species generation in cancer cells, resulting in apoptosis.

Resveratrol's antiproliferative effects are brought about by a disruption in signalling activity, which slows the spread of cancer. Nuclear factor-kappa B is suppressed by resveratrol. As a result, antiapoptotic proteins such as X-linked inhibitors of apoptosis protein, vascular endothelial growth factor, B-cell lymphoma-extra-large protein, and B-cell leukaemia/lymphoma 2 protein are downregulated. In particular, downregulation of B-cell leukaemia/lymphoma 2 protein expression prevents the release of cytochrome C from the mitochondrial intermembranous region, promoting invasion (Varoni et al. 2016).

The interaction of resveratrol with the growth factor heregulin-beta 1 activates a torrent of signalling pathways showing matrix metalloproteinase expression, especially matrix metalloproteinase-9. Resveratrol inhibits the heregulin-beta 1 growth factor pathway and matrix metalloproteinase-9 expressions, which decreases the proliferative effect in breast cancer cells (Xiong et al. 2016).

15.4 NANOENABLED TARGETING STRATEGIES FOR RESVERATROL AS AN ANTICANCER AGENT

Targeting techniques enable nanocarriers' active or passive transport to the intended site. Passive transport improves permeability and retention. An aberrant expansion of blood vessels in the tumour is linked to this increased permeability and retention effect, leading to significant molecular extravasation (Padhi et al. 2022; Khuroo et al. 2014; Ali et al. 2024). The passive targeting of anticancer medications exhibits off-target cytotoxicity by acting on normal, healthy cells since it lacks cancer cell selectivity. On the other hand, active targeting of anticancer medications is accomplished by combining them with receptor-specific ligands that can bind to overexpressed receptors on tumour cells (Verma et al. 2017). Some specific receptors are overexpressed in the cancer cells. Accordingly, ligands like transferrin, hyaluronic acid, antibodies, aptamers, and peptides are conjugated with the nano drug delivery systems to enhance the target-specific binding. This may increase bioavailability, therapeutic efficacy, and reduce non-specific toxicity. The active targeted nano delivery systems enter the cancer cells by receptor-mediated endocytosis and release the load in the tumour microenvironment (Behera and Padhi 2020; Padhi and Behera 2021).

15.4.1 Passive Targeting of Resveratrol-Loaded Nano Delivery Systems

Sudha et al. (2020) reported the anticancer efficacy of polymeric nanoparticles entrapping resveratrol in human colon cancer (COLO205-luc) in mice's xenograft and orthotopic implantation models. The study investigated the antiproliferative and antiangiogenic activity of poly lactic-co-glycolic acid – polyethylene glycol chitosan-coated polymeric nanoparticles of resveratrol. The polymeric nanoparticles of resveratrol exhibited enhanced anticancer activity by reducing the growth of tumour and haemoglobin content, reflecting the higher antitumour and antiangiogenic efficacy with increased bioavailability compared to free resveratrol.

Guo et al. (2019) co-delivered resveratrol and docetaxel in methoxyl poly(ethylene glycol)-poly(D,L-lactide) copolymer-based polymeric nanocarrier system and evaluated them in human breast adenocarcinoma cell line (MCF-7 cells). The cytotoxic efficacy of individual drugs, resveratrol and docetaxel, was determined in terms of IC_{50} value, which was found to be 23.0 and 10.4 µg/mL, respectively. The polymeric micelles combined had the IC_{50} value of 4.8 µg/mL in the breast cancer cell lines. The combination ratio of 1:1 for resveratrol and docetaxel exhibited a synergistic, prolonged release profile with increased cytotoxicity in the *in-vitro* study. The intravenous administration of the combination showed a 1.6 and threefold increase in the area under the curve for resveratrol and docetaxel compared to the individual drugs. Hence, the combination was proved to be an efficient anticancer lead for breast cancer therapy.

Feng et al. (2017) prepared resveratrol-loaded lipid-core-nanocapsules and reported their anticancer activity in colon cancer cells. The resveratrol-loaded lipid-core-nanocapsules exhibited a controlled and sustained release up to ~70% by 48 hours of the study period. An excellent cellular uptake was seen for resveratrol-loaded lipid-core-nanocapsules in colon cancer cells (HT29 cells). Encapsulation of resveratrol in nanocapsules displayed a higher anticancer effect than free resveratrol. A remarkable cell apoptosis of ~36% was shown in the colon cancer cells in the presence of resveratrol-loaded lipid-core-nanocapsules.

The lignin-based magnetic nanoparticles entrapping resveratrol for drug delivery were designed for passive targeting. In contrast to free drugs, the hybrid nanoparticles showed enhanced resveratrol stability, accumulation, and anticancer efficacy. The hybrid nanoparticles demonstrated better drug loading capacity and good biocompatibility. The composition of nanoparticles was optimized to form green lignin nanoparticles in an efficient technique for the delivery of insoluble drugs, contributing to environmental and economic sustainability (Dai et al. 2017).

Wang et al. (2017b) designed resveratrol-loaded solid lipid nanoparticles and evaluated them in breast cancer (MDA-MB-231) cells. Nanoformulation exhibited a superior antiproliferative effect on cancer cells, inhibiting invasion and migration of cancer cells and inducing cell cycle G0/G1 arrest, hence inhibiting the process of metastasis. The western blot study revealed the upregulation of Bax/Bcl-2 genes while downregulating the expression of cyclinD1 and c-Myc.

Temozolomide exhibits resistance to prolonged chemotherapy in malignant glioma. Resveratrol is known to be chemosensitive. To improve antitumour efficacy, temozolomide and resveratrol were loaded into nanoparticles with methoxy poly (ethylene glycol)-poly epsilon-caprolactone as drug carriers. Temozolomide/resveratrol-coloaded methoxy poly(ethylene glycol)-poly epsilon-caprolactone nanoparticles were evaluated for their antitumour activity against glioma cells in a xenograft model system (Xu et al. 2017). The co-encapsulation improved the in vitro stability of temozolomide-resveratrol-loaded nanoparticles and showed increased uptake by glioma cells through increased cell apoptosis. The coloaded temozolomide/resveratrol-coloaded nanoparticles demonstrated enhanced cytotoxicity effects, increasing the expression of downstream apoptotic proteins; hence, they can be used in glioma chemotherapy.

Wang et al. (2016) designed a co-delivery of bicalutamide-loaded PEGylated resveratrol and studied the cytotoxicity efficacy in breast (MCF-7) and cervical (HeLa) cancer cells by (3-(4,5-dimethylthazolk-2-yl)-2,5-diphenyl tetrazolium bromide) assay. The PEGylated resveratrol with bicalutamide exhibited synergistic anticancer activity in the cancer cells. Confocal laser scanning microscopy and flow cytometry analyses demonstrated more efficient internalisation of PEGylated resveratrol-bicalutamide conjugates into cells.

Huang et al. (2016) prepared a biopolymer-based core-shell type resveratrol polymeric nanoparticles with zein and pectin and further studied its efficacy against hepatocellular carcinoma (Bel-7402) cells. The IC_{50} value for the nanoformulation was found to be 17.6 µg/mL, whereas the said value for free resveratrol was noted to be 25.6 µg/mL. The biopolymeric nanoformulation was found to exhibit superior antioxidant and anticancer activity.

Resveratrol loaded onto polyethylene glycol–polylactic acid polymeric nanoparticles for increased solubility and stability demonstrated superior anticancer properties in CT26 colon cancer cells in vitro and CT26 tumour-bearing mice in vivo. Positron emission tomography with 18F-fluorodexoyglucose was used to monitor the anticancer effect. The effect of resveratrol on cancer cells was tested using resveratrol polymeric nanoparticles, which caused a comparable reduction in CT26 cell 18F-fluorodexoyglucose uptake and production of reactive oxygen species. As free resveratrol has a shorter plasma half-life owing to faster in vivo glucuronide and sulphate metabolism, increased uptake of resveratrol polymeric nanoparticles by tumour cells was demonstrated by 18F-fluorodexoyglucose (Jung et al. 2015). Micro-positron emission tomography-computed tomography scans showed that 18F-fluorodexoyglucose - nanoparticle-treated animals significantly reduced tumour growth.

Mohanty et al. (2014) synthesized resveratrol gold nanoparticles by green synthesis and loaded doxorubicin and assessed the anticancer efficacy against human glioma carcinoma cell lines (LN 229) by (3-(4,5-dimethylthazolk-2-yl)-2,5-diphenyl tetrazolium bromide) assay. The resveratrol gold nanoparticles were stable in physiological pH and were found to be biocompatible. The doxorubicin-loaded resveratrol gold nanoparticles inhibited the growth of glioblastoma cells in a dose-dependent manner, and the IC_{50} values for doxorubicin-loaded resveratrol gold nanoparticles and free doxorubicin were found to be 4 and 6 µg/mL, respectively. The study concluded that the increase in doxorubicin-loaded resveratrol gold nanoparticle efficacy was due to higher cellular uptake.

Green-synthesized nanoparticles are also gaining popularity as effective drug delivery systems for brain tumours. Based on *in-vitro* (C6 glioma cell line) and *in-vivo* (brain-implanted C6 cells) models of glioblastoma, resveratrol-loaded lipid-core nanocapsules lowered the viability of C6 glioma cells better than free resveratrol. Resveratrol-loaded lipid-core nanocapsules not cytotoxic to healthy neural cells showed cancer cell selectivity. Annexin-fluorescein isothiocyanate/propidium iodide assay revealed the apoptotic activity of resveratrol-loaded lipid-core nanocapsules in the G1 stages of the cell cycle. Resveratrol-loaded lipid-core nanocapsules therapy reduced tumour size and malignancy traits in brain-implanted with C6 tumour cells more significantly than free resveratrol. With its antiglioma activity, resveratrol nanoformulation can serve as a new chemotherapeutic treatment modality for glioma therapy (Figueiro et al. 2013; Ohara and Ohyama 2013).

Karthikeyan et al. (2013) encapsulated resveratrol in gelatin nanoparticles and studied its anticancer efficacy in lung cancer cells (NCI-H460). The cross-linking of resveratrol with gelatin was achieved by glutaraldehyde through Schiff base reaction and hydrogen bond interaction. *In-vitro,* drug release kinetic study showed an initial burst release followed by a slow and sustained release of resveratrol from the nanoformulation. The resveratrol gelatin nanoparticles displayed an enhanced cellular uptake with higher antiproliferative efficacy than free resveratrol in the lung cancer cells. The resveratrol nanoformulation generated more reactive oxygen species, leading to DNA damage and apoptosis. Erythrocyte aggregation assay showed no toxicity associated with the nanoformulation. The resveratrol gelatin nanoparticles had higher bioavailability and a longer half-life. Hence, it may be offered as a suitable delivery system for cancer therapy.

Glioblastoma is a type of malignant brain tumour that has a poor prognosis. The effectiveness of transferrin-modified polyethyleneglycol-polylactic acid nanoparticles coupled with resveratrol on the viability of glioma cells (C6 and U87 cells) was determined by the 3-(4,5-dimethylthazolk-2-yl)-2,5-diphenyl tetrazolium bromide test. Due to improved cellular absorption of the transferrin-resveratrol-polymeric nanoparticles by glioma cells, the anticancer activity of transferrin-resveratrol-polymeric nanoparticles demonstrated a greater in vitro cytotoxicity against the glioma cells in a C6 glioma-bearing rat model than free resveratrol. Resveratrol nanoformulation significantly reduced the tumour volume compared to free resveratrol, increasing the life of C6 glioma-bearing rats. Thus, transferrin-resveratrol-polymeric nanoparticles proved to have efficient potential against glioma therapy (Guo et al. 2013).

15.4.2 Active Targeting of Resveratrol-Loaded Nano Delivery Systems

Jangid et al. (2020) synthesized resveratrol-loaded amphiphilic bioconjugate with pluronic F68, polysaccharide inulin, and a lipid stearic acid. The nanomicelles containing resveratrol showed high drug entrapment efficiency, controlled drug release, and protected the said drug during storage. The nanoformulation exhibited a pH-responsive drug release in colorectal cell lines (HCT 116 cells), induced apoptosis, and altered the mitochondrial membrane potential with enhanced therapeutic efficacy.

Zhao et al. (2020) designed a co-delivery of resveratrol and paclitaxel in bovine serum albumin nanoparticles. They assessed their anticancer efficacy in murine mammary carcinoma (4T1) cells, human lung adenocarcinoma (A549) and paclitaxel-resistant human lung adenocarcinoma (A549/PTX). The co-delivery of both drugs enhanced the cellular internalisation with a synergistic action, and the in vitro study showcased cytotoxicity in multidrug-resistant cancer cell lines. The hybrid nanoparticles exhibited superior anticancer efficacy than the individual drugs. The in-silico study revealed the lower binding energy of bovine serum albumin-paclitaxel than bovine serum albumin-resveratrol, resulting in high encapsulation efficiency and sustained release of the entrapped drugs.

A thermosensitive hydrogel was developed by Wen et al. (2020) for intraperitoneal delivery as part of targeted therapy for patients with advanced hepatocellular carcinoma. Resveratrol and cisplatin microspheres were formulated using thermosensitive pluronic F127 hydrogel. The hepatocellular carcinoma cell lines (H22 cells) used in the in-vitro investigation displayed apoptosis and cell cycle arrest at the G1 phase. The in vivo research in mice with advanced hepatocellular carcinoma and ascites demonstrated a reduction in ascites, an antiproliferative effect with constrained micro angiogenesis. It prolonged the survival of the treated mice. Resveratrol and cisplatin intraperitoneal chemotherapy were safe and effective, with reduced toxicity to other organs when evaluated by immunohistochemical staining.

Mekonnen et al. (2020) designed a co-delivery system entrapping doxorubicin and resveratrol in an injectable form, which formed an in situ sodium deoxycholate hydrogel at the tumour extracellular microenvironment pH. To assess the release kinetics of the nanoformulation, generation 4.5 poly(amidoamine) dendrimer of doxorubicin was entrapped into the resveratrol hydrogel. The synergistic anticancer activity was observed in the cervical cell lines (HeLa cells). In vitro cytotoxicity study exhibited negligible cytotoxicity in cervical (HeLa), resistant breast cancer (MDA-MB-231), and skin (HaCaT) cells. The release kinetic study showcased a rapid release of resveratrol and sustained release of doxorubicin at pH 6.5, followed by a single intratumoral injection. The treatment suppressed the tumour growth during the 28 days of treatment. The hydrogel delivered the dual drug combination safely and efficiently in response to the tumour's extracellular pH environment.

Xiang et al. (2019) prepared an IR780 iodide and titanium disulfide nanosheet and intended to deliver resveratrol targeting mitochondria. The resveratrol-loaded nanosheets were stable and biocompatible and, hence, were targeted for photothermal-triggered tumour chemotherapy. On exposure to near-infrared radiation, the nanosheets released the drug by generating heat. The mitochondria-specific ligand IR780 favoured the accumulation of resveratrol in mitochondria. To attain the chemotherapeutic effect, it reduced the mitochondrial membrane potential and upregulated the intrinsic apoptotic factors such as cytochrome C with the initiation of caspase cascades. The nanosheets exhibited high antitumour efficacy with no significant cytotoxicity.

Wang et al. (2017a) developed a resveratrol-loaded chitosan liposome system coated with gold nanoshells. The nanoformulation had a broad near-infrared absorbance, and the system was targeted for photothermal therapy for cervical cancer (HeLa cells). The nanoformulation exhibited superior anticancer activity in response to thermal and pH stimuli in tumour cells. The liposomes coated with gold nanoshells responded to near-infrared laser irradiation and showed an increase in cellular uptake, exhibiting 57.6% drug release at pH = 5.0 compared to physiological pH = 7.4 (20.5%). Comparing the individual chemotherapeutic strategies, the dual mechanism of resveratrol liposomes exhibited higher anticancer efficacy.

Resveratrol-loaded human serum albumin nanoparticles were formulated and prepared for highly active targeted pancreatic tumour treatment by conjugating arginine–glycine–aspartate tripeptide through a polyethylene glycol "bridge". The coating with human serum albumin and polyethylene glycol augmented the *in-vitro* biocompatibility of resveratrol. Confocal fluorescence images implicated that human serum albumin-tripeptide-resveratrol nanoparticles had a better cellular uptake than human serum albumin nanoparticles, indicating the tripeptide-mediated effect. Human serum albumin-tripeptide-resveratrol nanoparticles improved blood circulation of resveratrol. It was inferred from the study that human serum albumin-tripeptide-resveratrol nanoparticles had high *in vivo* biocompatibility and enhanced tumour growth suppression effect with no systemic toxicity (Geng et al. 2017).

Resveratrol–bovine serum albumin nanoparticles exhibited improved antiproliferative effects in primary ovarian cancer cells (SKOV3). Resveratrol–bovine serum albumin nanoparticles prevented tumour growth by causing necrosis and promoting cell apoptosis in cancer cells. Terminal deoxynucleotidyl transferase dUTP nick end labelling staining indicated that resveratrol–bovine serum albumin nanoparticles-treated tumour-bearing mice were more sensitive to higher dosages of resveratrol–bovine serum albumin nanoparticles through necrosis and/or other death pathways. Cells of the treated group exhibited apoptosis and necrosis, and levels of expression of caspase-3, caspase-9, and cytochrome C proteins were significantly higher in all treated groups. The levels of these protein expressions in tumour tissue were correlated with tumour growth suppression by resveratrol–bovine serum albumin nanoparticles (Guo et al. 2010).

15.5 CONCLUSION

Significant advancements in the development of herbal drug delivery systems enabled these novel carriers to be precisely delivered as needed by the body during therapy, channelling the bioactive herbal drug entity to the intended site. Because of its pleiotropic action, resveratrol has a broad spectrum of bioactivities and has excellent potential as an anticancer agent. Nanoparticles are a viable technical approach for overcoming physical, chemical, and metabolic barriers for efficient resveratrol delivery. Developing nanocarriers such as polymeric nanoparticles, nanocapsules, liposomes, solid lipid nanoparticles, phytosomes, and nanoemulsions in phyto-oriented research can be advantageous for herbal drugs owing to the increase in solubility, pharmacological activity, enhanced bioavailability, and decrease in toxicity. These nanoparticles can release drugs in response to a pH or enzymatic reaction, allowing the visualisation of nanoparticle intake and helping in diagnosis and treatments. So, novel drug delivery systems for herbal drugs such as resveratrol have a broad and bright future in treating various types of cancer. More detailed preclinical and clinical testing is needed to establish its potential as a better therapeutic and diagnostic agent for cancer treatment.

REFERENCES

Aggarwal BB, Bhardwaj A, Aggarwal RS, Seeram NP, Shishodia S, Takada Y. Role of resveratrol in prevention and therapy of cancer: preclinical and clinical studies. *Anticancer Research*. 2004 Sep 1;24(5A):2783–2840.

Ali M, Benfante V, Di Raimondo D, Salvaggio G, Tuttolomondo A, Comelli A. Recent developments in nanoparticle formulations for resveratrol encapsulation as an anticancer agent. *Pharmaceuticals*. 2024;17(1):126.

Almatroodi SA, Alsahli MA, Aljohani ASM, Alhumaydhi FA, Babiker AY, Khan AA, Rahmani AH. Potential therapeutic targets of resveratrol, a plant polyphenol, and its role in the therapy of various types of cancer. *Molecules*. 2022 Apr 21;27(9):2665.

Behera A, Mittu B, Padhi S, Patra N, Singh J. Bimetallic nanoparticles: green synthesis, applications, and future perspectives. In: Abd-Elsalam KA (Ed.) *Multifunctional Hybrid Nanomaterials for Sustainable Agri-food and Ecosystems*. Elsevier, Amsterdam, 2020; pp. 639–682.

Behera A, Padhi S, Nayak AK. Engineered liposomes as drug delivery and imaging agents. In: Ray S, Nayak AK (Eds.) *Design and Applications of Theranostic Nanomedicines*. Woodhead Publishing, Cambridge, MA, 2023; pp. 75–108.

Behera A, Padhi S. Passive and active targeting strategies for the delivery of the camptothecin anticancer drug: a review. *Environmental Chemistry Letters*. 2020;18(5):1557–1567.

Behera A, Padhi S. pH-Sensitive polymeric nanoparticles for cancer treatment. In: Padhi S, Behera A, Lichtfouse, E. (Eds.) *Polymeric Nanoparticles for the Treatment of Solid Tumours*. Environmental Chemistry for a Sustainable World, vol 71. Springer, Cham, 2022; pp. 401–425.

Brockmueller A, Sameri S, Liskova A, Zhai K, Varghese E, Samuel SM, Büsselberg D, Kubatka P, Shakibaei M. Resveratrol's anticancer effects through the modulation of tumour glucose metabolism. *Cancers*. 2021 Jan 7;13(2):188.

Cheng L, Yan B, Chen K, Jiang Z, Zhou C, Cao J, Qian W, Li J, Sun L, Ma J, Ma Q. Resveratrol-induced downregulation of NAF-1 enhances the sensitivity of pancreatic cancer cells to gemcitabine via the ROS/Nrf2 signaling pathways. *Oxidative Medicine and Cellular Longevity*. 2018 Oct;2018:9482018.

Cianciulli A, Calvello R, Cavallo P, Dragone T, Carofiglio V, Panaro MA. Modulation of NF-κB activation by resveratrol in LPS treated human intestinal cells results in downregulation of PGE2 production and COX-2 expression. *Toxicology in Vitro*. 2012 Oct 1;26(7):1122–1128.

Coimbra M, Isacchi B, van Bloois L, Torano JS, Ket A, Wu X, Broere F, Metselaar JM, Rijcken CJ, Storm G, Bilia R, Schiffelers RM. Improving solubility and chemical stability of natural compounds for medicinal use by incorporation into liposomes. *International Journal of Pharmaceutics*. 2011;416:433–442.

Dai L, Liu R, Hu LQ, Zou ZF, Si CL. Lignin nanoparticle as a novel green carrier for the efficient delivery of resveratrol. *ACS Sustainable Chemistry & Engineering*. 2017;5(9):8241–8249.

Elshaer M, Chen Y, Wang XJ, Tang X. Resveratrol: an overview of its anticancer mechanisms. *Life Sciences*. 2018;207:340–349.

Feng M, Zhong LX, Zhan ZY, Huang ZH, Xiong JP. Enhanced antitumour efficacy of resveratrol-loaded nanocapsules in colon cancer cells: physicochemical and biological characterisation. *European Review for Medical and Pharmacological Sciences*. 2017;21(2):375–382.

Figueiro F, Bernardi A, Frozza RL, Terroso T, Zanotto-Filho A, Jandrey EH, Moreira JCF, Salbego CG, Edelweiss MI, Pohlmann AR, Guterres SS, Battastini AMO. Resveratrol-loaded lipid-core nanocapsules treatment reduces in vitro and in vivo glioma growth. *Journal of Biomedical Nanotechnology*. 2013;9(3):516–526.

Fulda, S. Resveratrol and derivatives for the prevention and treatment of cancer. *Drug Discovery Today*. 2010;15(17–18):757–765.

Geng T, Zhao X, Ma M, Zhu G, Yin L. Resveratrol-loaded albumin nanoparticles with prolonged blood circulation and improved biocompatibility for highly effective targeted pancreatic tumour therapy. *Nanoscale Research Letters*. 2017;12(1):1–10.

Guo L, Peng Y, Yao J, Sui L, Gu A, Wang J. Anticancer activity and molecular mechanism of resveratrol–Bovine serum albumin nanoparticles on subcutaneously implanted human primary ovarian carcinoma cells in Nude mice. *Cancer Biotherapy & Radiopharmaceuticals*. 2010;25(4):471–477.

Guo W, Li A, Jia Z, Yuan Y, Dai H, Li H. Transferrin modified PEG-PLA-resveratrol conjugates: in vitro and in vivo studies for glioma. *European Journal of Pharmacology*. 2013;718(1–3):41–47.

Guo X, Zhao Z, Chen D, Qiao M, Wan F, Cun D, Sun Y, Yang M. Co-delivery of resveratrol and docetaxel via polymeric micelles to improve the treatment of drug-resistant tumours. *Asian Journal of Pharmaceutical Sciences*. 2019;14(1):78–85.

Guo Y, Kitts DD, Dolati D, Pratap-Singh A, Singh A. Enhancing resveratrol bioavailability and intestinal uptake using an oil-based blueberry extract formulated with chitosan/PEG containing nanoparticles. *Food Hydrocolloids*. 2024;156:110373.

Hartshorn CM, Bradbury MS, Lanza GM, Nel AE, Rao J, Wang AZ, Wiesner UB, Yang L, Grodzinski P. Nanotechnology strategies to advance outcomes in clinical cancer care. *ACS Nano*. 2018;12(1):24–43.

Huang X, Dai Y, Cai J, Zhong N, Xiao H, McClements DJ, Hu K. Resveratrol encapsulation in core-shell biopolymer nanoparticles: impact on antioxidant and anticancer activities. *Food Hydrocolloids*. 2016;64:157–165.

Jang M, Cai L, Udeani GO, Slowing KV, Thomas CF, Beecher CW, Fong HH, Farnsworth NR, Kinghorn AD, Mehta RG, Moon RC, Pezzuto JM. Cancer chemopreventive activity of resveratrol, a natural product derived from grapes. *Science*. 1997;275(5297):218–220.

Jangid AK, Patel K, Jain P, Patel S, Gupta N, Pooja D, Kulhari H. Inulin-pluronic-stearic acid based double folded nanomicelles for pH responsive delivery of resveratrol. *Carbohydrate Polymers*. 2020;247:116730.

Jung KH, Lee JH, Park JW, Quach CHT, Moon SH, Cho YS, Lee KH. Resveratrol-loaded polymeric nanoparticles suppress glucose metabolism and tumour growth in vitro and in vivo. *International Journal of Pharmaceutics*. 2015;478(1):251–257.

Karthikeyan S, Prasad NR, Ganamani A, Balamurugan E. Anticancer activity of resveratrol-loaded gelatin nanoparticles on NCI-H460 non-small cell lung cancer cells. *Biomedicine & Preventive Nutrition*, 2013;3(1):64–73.

Khuroo T, Verma D, Talegaonkar S, Padhi S, Panda AK, Iqbal Z. Topotecan–tamoxifen duple PLGA polymeric nanoparticles: investigation of in vitro, in vivo and cellular uptake potential. *International Journal of Pharmaceutics*, 2014;473(1–2):384–394.

Kundu A, Padhi S, Behera A, Hasnain MS, Nayak AK. Tumour targeting strategies by chitosan-based nanocarriers. In: Hasnain MS, Beg S, Nayak AK (Eds.) *Chitosan in Biomedical Applications*. Academic Press, London, 2022; pp. 163–188.

Mekonnen TW, Andrgie AT, Darge HF, Birhan, YS, Hanurry EY, Chou HY, Lai J-Y, Tsai H-C, Yang JM, Chang YH. Bioinspired composite, pH-responsive sodium deoxycholate hydrogel and generation 4.5 poly(amidoamine) dendrimer improves cancer treatment efficacy via doxorubicin and resveratrol co-delivery. *Pharmaceutics*. 2020;12(11):1069.

Misawa T, Saitoh T, Kozaki T, Park S, Takahama M, Akira S. Resveratrol inhibits the acetylated α-tubulin-mediated assembly of the NLRP3-inflammasome. *International Immunology*. 2015 Sep 1;27(9):425–434.

Mohanty RK, Thennarasu S, Mandal AB. Resveratrol stabilised gold nanoparticles enable surface loading of doxorubicin and anticancer activity. *Colloids and Surfaces B: Biointerfaces*. 2014;114:138–143. https://doi.org/10.1016/j.colsurfb.2013.09.057

Ohara M, Ohyama Y. Delivery and application of dietary polyphenols to target organs, tissues and intracellular organelles. *Current Drug Metabolism*. 2014;15(1):37–47.

Padhi S, Behera A, Hasnain MS, Nayak AK. Chitosan-based drug delivery systems in cancer therapeutics. In: Hasnain MS, Beg S, Nayak AK (Eds.) *Chitosan in Drug Delivery*, Academic Press, London, 2022; pp. 159–193.

Padhi S, Behera A. Advanced drug delivery systems in the treatment of ovarian cancer. In: Dua K, Mehta M, de Jesus Andreoli Pinto T, Pont LG, Williams KA, Rathbone MJ (Eds.) *Advanced Drug Delivery Systems in the Management of Cancer*. Academic Press, London, 2021; pp. 127–139.

Padhi S, Behera A. Cellular internalization and toxicity of polymeric nanoparticles. In: Padhi S, Behera A, Lichtfouse E (Eds.) *Polymeric Nanoparticles for Treating Solid Tumours*. Environmental Chemistry for a Sustainable World, vol 71. Springer, Cham, 2022; pp. 473–488.

Padhi S, Behera A. Nanotechnology based targeting strategies for the delivery of Camptothecin. In: Saneja A, Panda AK, Lichtfouse E (Eds.) *Sustainable Agriculture Reviews 44: Pharmaceutical Technology for Natural Products Delivery Vol. 2 Impact of Nanotechnology*. Springer, Cham, 2020; pp. 243–272.

Padhi S, Dash M, Behera, A. Nanophytochemicals for the treatment of type II diabetes mellitus: a review. *Environmental Chemistry Letters*. 2021;19:4349–4373.

Padhi S, Kapoor R, Verma D, Panda AK, Iqbal, Z. Formulation and optimisation of topotecan nanoparticles: in vitro characterisation, cytotoxicity, cellular uptake and pharmacokinetic outcomes. *Journal of Photochemistry and Photobiology B: Biology*. 2018;183:222–232.

Padhi S, Mirza MA, Verma D, Khuroo T, Panda AK, Talegaonkar S, Khar RK, Iqbal, Z. Revisiting the nanoformulation design approach for effective delivery of Topotecan in its stable form: an appraisal of its in vitro behavior and tumour amelioration potential. *Drug Delivery*. 2016;23(8):2827–2837.

Patnaik S, Gorain B, Padhi S, Choudhury H, Gabr GA, Md S, Kumar Mishra D, Kesharwani, P. Recent update of toxicity aspects of nanoparticulate systems for drug delivery. *European Journal of Pharmaceutics and Biopharmaceutics*. 2021;161:100–119.

Pavan AR, Silva GDBD, Jornada DH, Chiba DE, Fernandes GFDS, Man Chin C, Dos Santos JL. Unraveling the anticancer effect of curcumin and resveratrol. *Nutrients*. 2016;8(11):628.

Renaud S, de Lorgeril, M. Wine, alcohol, platelets, and the French paradox for coronary heart disease. *The Lancet*. 1992;339(8808):1523–1526.

Sakamoto T, Horiguchi H, Oguma E, Kayama F. Effects of diverse dietary phytoestrogens on cell growth, cell cycle and apoptosis in estrogen-receptor-positive breast cancer cells. *The Journal of Nutritional Biochemistry*. 2010 Sep 1;21(9):856–864.

Santos AC, Pereira I, Magalhães M, Pereira-Silva M, Caldas M, Ferreira L, Figueiras A, Ribeiro AJ, Veiga F. Targeting cancer via resveratrol-loaded nanoparticles administration: focusing on in vivo evidence. *The AAPS Journal*. 2019;21(4):1–16.

Saraf S. Applications of novel drug delivery system for herbal formulations. *Fitoterapia*. 2010;81(7):680–689.

Shukla Y, Singh R. Resveratrol and cellular mechanisms of cancer prevention. *Annals of the New York Academy of Sciences*. 2011;1215(1):1–8.

Singh G, Pai RS. Optimised PLGA nanoparticle platform for orally dosed trans-resveratrol with enhanced bioavailability potential. *Expert Opinion on Drug Delivery*. 2014;11:647–659.

Sudha T, El-Far AH, Mousa DS, Mousa SA. Resveratrol and its nanoformulation attenuate growth and the angiogenesis of xenograft and orthotopic colon cancer models. *Molecules*. 2020;25(6):1412.

Tino AB, Chitcholtan K, Sykes PH, Garrill A. Resveratrol and acetyl-resveratrol modulate activity of VEGF and IL-8 in ovarian cancer cell aggregates via attenuation of the NF-κB protein. *Journal of Ovarian Research*. 2016 Dec;9(1):1–2.

Varoni EM, Lo Faro AF, Sharifi-Rad J, Iriti M. Anticancer molecular mechanisms of resveratrol. *Frontiers in Nutrition*. 2016;3:8.

Verma D, Thakur PS, Padhi S, Khuroo T, Talegaonkar S, Iqbal, Z. Design expert assisted nanoformulation design for co-delivery of topotecan and thymoquinone: optimisation, in vitro characterisation and stability assessment. *Journal of Molecular Liquids*. 2017;242:382–394.

Vian MA, Tomao V, Gallet S, Coulomb PO, Lacombe JM. Simple and rapid method for cis- and trans-resveratrol and piceid isomers determination in wine by high-performance liquid chromatography using chromolith columns. *Journal of Chromatography A*. 2005;1085(2):224–229.

Wang J, Li J, Cao N, Li Z, Han J, Li L. Resveratrol, an activator of SIRT1, induces protective autophagy in non-small-cell lung cancer via inhibiting Akt/mTOR and activating p38-MAPK. *OncoTargets and Therapy*. 2018 Nov 2;11:7777–7786.

Wang M, Liu Y, Zhang X, Luo L, Li L, Xing S, Yuchu H, Weiwei C, Zhu R, Gao D. Gold nanoshell coated thermo-pH dual responsive liposomes for resveratrol delivery and chemo-photothermal synergistic cancer therapy. *Journal of Materials Chemistry B*. 2017a;5(11):2161–2171.

Wang W, Zhang L, Chen T, Guo W, Bao X, Wang D, Li Y, Wang Y, Chen S, Tang B, Yang Q, Chen C. Anticancer effects of resveratrol-loaded solid lipid nanoparticles on human breast cancer cells. *Molecules*. 2017b;22(11):1814.

Wang W, Zhang L, Le Y, Chen JF, Wang J, Yun J. Synergistic effect of PEGylated resveratrol on delivery of anticancer drugs. *International Journal of Pharmaceutics*. 2016;498(1–2):134–141.

Wen Q, Zhang Y, Luo J, Xiong K, Lu Y, Wu Z, Wang BQ, Wu J, Chen Y, Fu S. Therapeutic efficacy of thermosensitive Pluronic hydrogel for co-delivery of resveratrol microspheres and cisplatin in the treatment of liver cancer ascites. *International Journal of Pharmaceutics*. 2020;582:119334.

Whitlock NC, Baek SJ. The anticancer effects of resveratrol: modulation of transcription factors. *Nutrition and Cancer*. 2012;64(4):493–502.

Xiang S, Zhang K, Yang G, Gao D, Zeng C, He M. Mitochondria-targeted and resveratrol-loaded dual-function titanium disulfide nanosheets for photothermal-triggered tumour chemotherapy. *Nanoscale Research Letters*. 2019;14(1):1–10.

Xiong W, Yin A, Mao X, Zhang W, Huang H, Zhang X. Resveratrol suppresses human glioblastoma cell migration and invasion via activation of RhoA/ROCK signaling pathway. *Oncology Letters*. 2016;11(1):484–90.

Xu H, Jia, F, Singh PK, Ruan S, Zhang H, Li X. Synergistic anti-glioma effect of a coloaded nano-drug delivery system. *International Journal of Nanomedicine*. 2017;12:29–40.

Zarebavani M, Baghaei Naeini F, Farahvash A, Moradi F, Dashti N. Resveratrol attenuates chronic social isolation stress-induced affective disorders: involvement of NF-κB/NLRP3 axis. *Journal of Biochemical and Molecular Toxicology*. 2023;13:e23311.

Zhao Y, Cai C, Liu M, Zhao Y, Wu Y, Fan Z, Ding Z, Zangh H, Wang Z, Han J. Drug-binding albumins forming stabilised nanoparticles for co-delivery of paclitaxel and resveratrol: in vitro/in vivo evaluation and binding properties investigation. *International Journal of Biological Macromolecules*. 2020;153:873–882.

16 Nanoenabled Systems for the Delivery of Cannabinoids as Anticancer Agents

Plachikkattu Parambil Akhila, Basheer Aaliya, and Kappat Valiyapeediyekkal Sunooj
Department of Food Science and Technology, Pondicherry University, Puducherry, India

16.1 INTRODUCTION

Cancer is one of the world's most deadly diseases that affects people of all ages, genders, and socio-economic statuses, posing a significant challenge to the scientific community (Zafar et al., 2020; Padhi et al., 2016; Kundu et al., 2022). It is characterized by unregulated cell differentiation and destruction (Guindon and Hohmann, 2011). External (tobacco, infectious species, radiation, chemicals, and others) or internal (inherited mutations, metabolic mutations, hormones, and others) factors, or sometimes hereditary factors, may trigger cancer (Guindon and Hohmann, 2011). According to the World Health Organisation Survey, cancer accounts for over 70% of deaths in underdeveloped and developing nations (Sadhu et al., 2020). The tumour cells respond to treatment in various ways, making it difficult to predict the therapeutic outcome for each patient (Greish et al., 2018). Conventional therapies have adverse effects, and a low survival rate, so less invasive but successful cancer therapies are often desirable (Sadhu et al., 2020; Behera and Padhi, 2022; Padhi et al., 2018). Traditional chemotherapy has a significant flaw because it fails to discriminate between diseased and non-cancerous cells, resulting in organ damage and low survival rates.

Plants have long been used in traditional therapy to treat several diseases, including cancer, cardiovascular disease, inflammatory disease, metabolic syndrome, and parasitic and viral diseases (Singh et al., 2020). Cannabis has been used for medicinal purposes for thousands of years. The first recorded evidence of cannabis use in traditional medicine can be found in Chinese Pharmacopoeia (Pisanti and Bifulco, 2019). Cannabis was reported to treat rheumatoid pain, malaria, constipation, gynaecological disorders, and beriberi in Chinese medicine. Other countries like Egypt, India, Iran, Japan, and Arab nations have uncovered the use of cannabis in ancient herbal medicine (Almogi-Hazan and Or, 2020).

In the second half of the twentieth century, basic research helped to reveal cannabinoid receptors and the endocannabinoid system. The main focus of research in the cannabinoid sector is the possible use of cannabinoids to treat various disorders (De La Ossa et al., 2013). Because of their analgesic and anti-emetic properties, cannabinoid-based drugs are used as palliative medicines in cancer patients (Almogi-Hazan and Or, 2020). Cannabinoids' anticancer effects have been the subject of intense scientific research over the past few years. (Dariš et al., 2019). The discovery of the potential utility of cannabinoids for targeting tumours was an important development in their use in cancer care. The discovery of cannabinoid receptors and a better understanding of how psychoactive agents interact with these receptors can be used to treat cancer.

Cannabinoids can arrest tumour growth, induce cell death, and restrict cell proliferation in various cells, including lymphocytes, neurons, and transformed neural and non-neural cells (Guzmán et al., 2001). Other antitumourigenic mechanisms of cannabinoids are now being discovered, including their capacity to inhibit tumour neovascularisation, cancer cell adhesion, invasion, migration, and metastasis (Massi et al., 2013). However, unfavourable physical chemistry, such as high instability, low bioavailability, oily-resin composition, and poor solubility in water, is the most challenging obstacle to clinical application (Martín-Banderas et al., 2014). Epidiolex® is a non-toxic cannabinoid (Cannabidiol) derived from the cannabis plant that has been licensed by the United States Food and Drug Administration for the management of severe myoclonic epilepsy of infancy, Lennox-Gastaut syndrome, and epileptic seizures in children aged two and up (Onaivi et al., 2020). Sativex®, a sublingual spray made by combining Δ^9-tetrahydrocannabinol and cannabidiol, has been approved in Spain, the United Kingdom, New Zealand, and Canada. However, Sativex® may irritate the buccal cavity due to propylene glycol and ethanol added as solubilizing agents (De La Ossa et al., 2012). Hence, we undoubtedly need a better formulation of cannabinoids with fewer side effects and more bioavailability for administration. The difficulty in efficiently administering cannabinoids is also a significant barrier to developing new therapeutic applications for these compounds (De La Ossa et al., 2012).

Delivering cannabis and cannabinoid medications more effectively for therapeutic purposes is made possible by the application of nanotechnology. Indeed, the increased interest in cannabis and cannabinoids as nano-delivery systems is accompanied by opportunities and challenges in the formulation of cannabis products, as well as concerns about quality control and safety. This ensures that cannabinoids can be loaded into nanoparticles, effectively delivering

DOI: 10.1201/9781003611516-20

these phytomolecules to diseased sites. At the same time, nanotechnology has paved the way for cancer detection and treatment (Ngwa et al., 2017). Nanotechnology in medicine also enhances the early detection and diagnosis of disease and control of the biodistribution of the drug with high efficacy (Onaivi et al., 2020). Many pharmaceutical drugs that utilize nanocarriers are already being used in clinical trials to treat cancer. As cannabis nanoscience evolves, nanotechnology can allow targeted delivery of cannabinoid medicines, elevating their use in cancer treatment. This chapter summarises the mechanism of cannabinoid action against the cancer cell, nanotechnological advances in the field of effective delivery of the phytomolecule and major clinical trials happening in the therapeutic area of cannabinoid.

16.2 ANTICANCER ACTIVITY OF CANNABINOIDS

Cannabinoids are active aromatic hydrocarbon compounds naturally occurring in the plant *Cannabis sativa*. Based on the source, cannabinoids are classified into three categories: phytocannabinoids (plant-derived cannabinoids), endocannabinoids (produced by the human body), and synthetic cannabinoids (synthesized artificially) (Figure 16.1) (Kovalchuk and Kovalchuk, 2020). Even though all the classical cannabinoids are formed from various sources, they show structural similarity, which is exploited in the medical field (Kovalchuk and Kovalchuk, 2020). More than 480 types of cannabinoids have been identified so far; around 110 are phytocannabinoids. Cannabinoids interact with cells by G-protein coupled receptors, such as cannabinoid receptor types 1 and 2. The two types of cannabinoid receptors differ by their location. Cannabinoid receptor type 1 receptors are more general and widespread in the body, primarily located in the central nervous system and, to a lesser extent, in extra-neural tissues and peripheral nerve terminals. Cannabinoid receptor type 2 receptors are found in most body parts but are particularly abundant in the spleen, bone marrow, lymph nodes, and appendix. Cannabinoid receptor type 2 receptors are also present in the CNS, often in microglia and neurons (Kovalchuk and Kovalchuk, 2020). Δ^9-Tetrahydrocannabinol is the most potent psychoactive agent, first isolated in 1964. Cannabinol, cannabidiol, cannabivarins, cannabigerol, and cannabinoid acids are further therapeutically relevant derivatives (Bruni et al., 2018).

Antitumour properties of cannabis compounds were first suggested over 35 years ago. However, the potential of using the endocannabinoid system to treat cancer has recently drawn attention (Pisanti et al., 2013). *N*-Arachidonoylethanolamide and 2-arachidonoyl glycerol are the proteins involved in the synthesis, transport, and degradation of endocannabinoids, and the receptors by which their signals are transferred, altogether are referred to as the endocannabinoid system (Velasco et al., 2016). The endocannabinoid system helps to maintain the homeostasis of the cell. The endocannabinoid system regulates appetite, sensation, thermoregulation, muscle function, mood, pain and inflammation, intraocular pressure, motivation/reward, and memory (Nikan et al., 2016). Modulation of the ECS has been proposed as a possible therapeutic target for a variety of diseases, including neurodegeneration, gastrointestinal, obesity/metabolic syndrome, diabetes, psychiatric disorders, cardiovascular, cachexia, liver, skin, pain, cancer, and chemotherapy-induced vomiting and nausea (Nikan et al., 2016). When cannabinoids interact with a cell, they have various effects, including reducing oxygen and nutrient delivery, preventing cell proliferation, suppressing angiogenesis, promoting autophagy, interrupting the cell cycle, and triggering cell death (Pisanti et al., 2013). The most common methods of their activity against tumour cells are the activation of apoptosis and the inhibition of tumour cell proliferation (Velasco et al., 2012). Table 16.1 highlights contributions in the area of cannabinoids in various cancer treatments.

16.2.1 Mechanisms of Anticancer Activity of Cannabinoids

The antineoplastic activity of cannabinoids has been demonstrated in various animal and human cancer cells (Bogdanović et al., 2017). These chemicals are hypothesized to affect tumour growth by multiple mechanisms, the most prevalent of which is the stimulation of cancer cell death (Apoptosis) and the suppression of cancer cell proliferation (Bogdanović et al., 2017). The action of cannabinoids is associated with the stimulation of the endocannabinoid

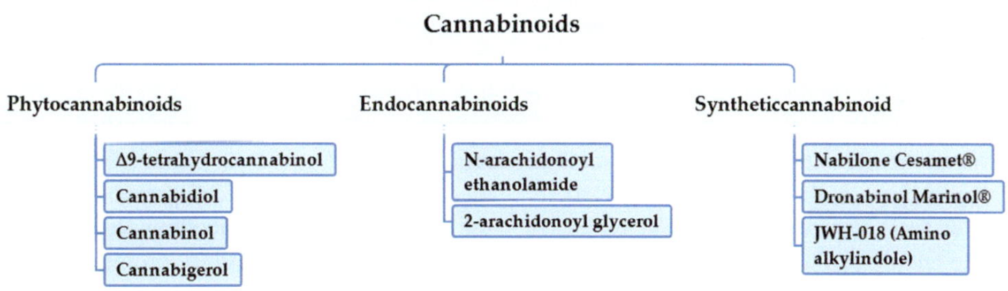

FIGURE 16.1 Classification of cannabinoids based on origin [Based on source, cannabinoids are classified into phytocannabinoids, endocannabinoids, and synthetic cannabinoids. Again, each class is subdivided according to its chemical nature].

TABLE 16.1
Studies on Cannabinoids in Cancer Treatment

Cannabinoid	Type of Cancer	Type of Study	Major Findings	References
Δ^9 – tetrahydrocannabinol + Cannabidiol at a 1:1 ratio	Glioblastoma multiforme	*In vivo*	Reduced the growth of orthotopic xenografts	López-Valero et al. (2018)
Synthetic cannabidiol	Breast cancer and glioma	Clinical	Reduction in circulating tumour cells in some patients reduction in tumour size	Kenyon et al. (2018)
Δ^9-tetrahydrocannabinol + Cannabidiol at a 1:1 ratio	v-rapidly accelerated fibrosarcoma murine sarcoma viral oncogene homolog B1/neuroblastoma renin-angiotensin-system viral oncogene homolog wildtype tumours	*In vivo*	Non-canonical autophagy-mediated apoptosis of melanoma cells	Armstrong et al. (2015)
Cannabidiol	Malignant melanoma	*In vivo*	Significant decrease in tumour size with better quality of life	Simmerman et al. (2019)
Δ^9-tetrahydrocannabinol	Skin cancer	*In vivo*, in vivo	Inhibited cancer cell growth through antagonistic effects *in vivo* study	Glodde et al. (2015)
Cannabidiol	Cervical cancer	*In vitro*	Prevent cell growth and induce cell death	Lukhele and Motadi (2016)
Δ^9-tetrahydrocannabinol, cannabidiol	Leukaemia	*In vitro*	Combination treatment showed more effectiveness toward the human myeloid cancer (HL60) cell line. Inhibitory concentrations at 50% for cannabidiol and Δ^9-tetrahydrocannabinol are 8 and 13 μM, respectively	Scott et al. (2017)
Cannabidiol	Colorectal cancer	*In vitro*	induced apoptosis in a Noxa-and-reactive oxygen species-dependent manner	Jeong et al. (2019b)
Cannabidiol	Gastric cancer	*In vitro*	Induced apoptosis by suppressing X-linked inhibitor apoptosis	Jeong et al. (2019a)

system in the cell. When the endocannabinoid system is activated, it triggers several signalling pathways that affect cellular function, including cyclic adenosine monophosphate activation, sodium potassium pump modulation, and attenuation of members of the mitogen-activated protein kinase family (p38, extracellular signal-regulated kinase-1 and 2, and c-Jun N-terminal kinase). Studies have revealed that prognosis is often related to enhanced activity of endocannabinoid signalling pathways and increased expression of endocannabinoid receptors in many cancer types (Turgeman and Bar-Sela, 2019).

16.2.2 Apoptosis by Cannabinoid Drug

Cannabinoid-related drugs induce apoptosis, autophagy, or both as programmed cell death mechanisms (Pisanti et al., 2013). According to the existing literature, autophagy suppresses tumour formation in the initial stages of tumour progression. However, in the advanced stage of cancer, autophagy provides energy to the cancer cell due to its increased demand for fuel for growth. This makes it easier for the tumour cell to endure cytotoxic treatments and death (Armstrong et al., 2015; Bogdanović et al., 2017). Research studies on melanoma, glioma, pancreatic, and hepatic tumour cells verified that the triggering of autophagy by cannabinoids contributes to cell death (Armstrong et al., 2015; Galluzzi et al., 2015). The pro-apoptotic effects of cannabinoids and anticancer properties are likely to be explained by several mechanisms (Bogdanović et al., 2017). Many apoptotic effects are related to cannabinoid receptors and mitogen-activated protein kinase pathways (p38 mitogen-activated protein kinase, c-Jun N-terminal kinase, and extracellular signal-regulated protein kinases 1 and 2) activation or inhibition of the phosphoinositide-3-kinase–protein kinase B/Akt pathway, which triggers the activation of pro-apoptotic and inactivation of anti-apoptotic proteins, respectively (Pisanti et al., 2013).

Cannabinoid receptors get activated by binding Δ^9-tetrahydrocannabinol on cannabinoid receptor type 1 or 2, triggering the sphingolipids (ceramide) synthesis through de novo pathways. This causes endoplasmic reticulum stress-related signalling to be stimulated, promoting autophagy (Bogdanović et al., 2017). The endoplasmic reticulum stress-related signalling pathway upregulates the transcriptional nuclear protein 1(P8) and its effector, the pseudo-kinase tribbles homolog 3. The pseudo-kinase tribbles homolog 3 inhibits protein kinase B/mammalian target of the rapamycin, promoting autophagy-mediated apoptosis (Armstrong et al., 2015; Velasco et al., 2016). During stress responses, pseudo-kinase tribbles homolog 3 is a decision-maker in controlling cell survival and death. The pseudo-kinase tribbles homolog three produced in melanoma cells by activating Δ^9-tetrahydrocannabinol may lead to apoptosis (Velasco et al., 2016). While autophagy is primarily a cytoprotective mechanism, its activation may also result in cell death. Research showed that in cannabinoid-activated cell death, autophagy is upstream of apoptosis (Velasco et al., 2016).

Cannabidiol is a cannabinoid originating from plants with a low affinity for cannabinoid receptors. Cannabidiol shows anticancerous activity by apoptosis via enhanced production of reactive oxygen species. Cannabidiol has also been shown to reduce mitochondrial oxidative metabolism and modulate the extracellular signal-regulated kinase and reactive oxygen species pathways associated with apoptosis initiation in glioma cells (Bogdanović et al., 2017). Furthermore, Mcallister et al. (2015) supported cannabidiol-modulated signalling patterns related to extracellular signal-regulated kinase and reactive oxygen species through their experiments. Singer et al. (2015), in their *in-vivo* study, reported the first proof of anticancerous activity of cannabidiol-dependent generation of reactive oxygen species.

16.2.3 Inhibition of Migration, Invasion, Angiogenesis Invasion, and Metastasis by Cannabinoids

The role of cannabinoids in cancer treatment by different mechanisms, such as invasion, angiogenesis, and migration, was explained in several published reports (Laezza et al., 2020). These mechanisms are triggered through the activation of receptors by various agonists (e.g., 2-arachidonoyl glycerol, Δ^9-tetrahydrocannabinol, *N*-arachidonoylethanolamide, etc.) when agonists come in contact with the cannabinoid receptor type 1 and type 2 receptors, which inhibit the ras homolog gene family, member A-focal adhesion kinase-proto-oncogene tyrosine-protein kinase Src axis. As a result, placental growth factor, vascular endothelial growth factor, and angiopoietin-2 get downregulated (Fonseca et al., 2017). Furthermore, cannabinoids trigger the release of tissue inhibitors of matrix metalloproteinases-1 that act as endogenous inhibitors of matrix metalloproteinase 2, thereby inhibiting invasion and angiogenesis (Laezza et al., 2020). The family of inhibitors of DNA binding acts in cancer growth and invasion. It can inhibit the basic helix-loop-helix transcriptional factors (Ramer and Hinz, 2016). Cannabidiol treatment in breast cancer cells shows inhibition of tumour invasion and proliferation by decreasing inhibitors of DNA-1 expression (McAllister et al., 2007).

16.3 TYPES OF TARGETED DRUG DELIVERY OF CANNABINOIDS

Nanocarriers can cross the biological barriers inside the body and reach the desired tumour cell (Padhi and Behera, 2022; Behera et al., 2023). After getting into the cancer cell, the nanocarrier should deliver the drug to the target cell and produce cytotoxicity. Nanoparticle drug delivery to the cancer tissue through systematic circulation is done by passive and active targeting (Figure 16.2) (Kumari et al., 2016; Behera and Padhi, 2020).

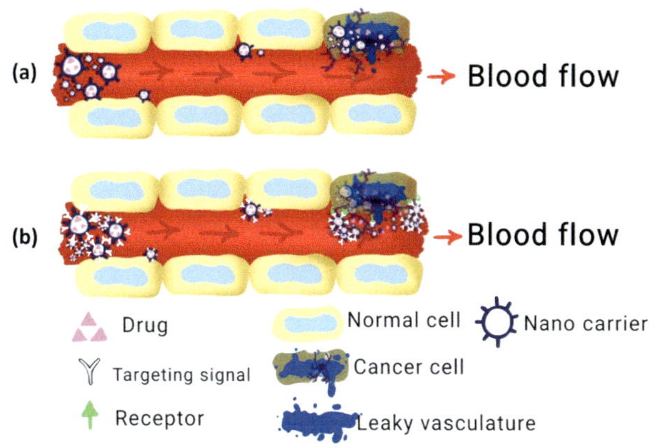

FIGURE 16.2 Schematic presentation of drug delivery to cancer cells using nanocarriers. (a) Passive drug delivery, (b) active drug delivery.

16.3.1 Passive Targeting

Passive targeting can be defined as the accumulation of the therapeutic drug in the target cells based on the pharmacological or physicochemical factors of the diseases (Khodabandehloo et al., 2016; Padhi and Behera, 2020; Kundu et al., 2022). Cellular changes in the cancer tissue have been utilized in the passive targeting of nanoparticle delivery (Wicki et al., 2015). In cancer treatment, nanoparticles accumulate on the surface of tumour cells due to their enhanced permeability and retention response in the tissue (Khodabandehloo et al., 2016). The cancer cells are metabolically highly active. Hence, to meet the higher metabolic demand of the cells, they undergo neovascularization. Consequently, endothelial cells become damaged, with numerous pores showing enlarged gap junctions. This phenomenon of the cancer cell is called the enhanced permeability and retention effect, i.e., incomplete endothelial linings, abnormal basement membranes, and leaky vessels of the tumour tissues, allowing the entry of nanoparticles passively through space. This results in the tumour region's high therapeutic drug concentration (Rahman et al., 2020). As a result, drugs act on specific cells, and healthy tissue is protected (Verma et al., 2018; Patnaik et al., 2021; Behera and Padhi, 2022). The extended circulation time of the nanocarriers ensures the success of the passive targeting method by increasing the possibility of passing the target site (Rahman et al., 2020). However, usually, the nanoparticles have a short life span due to the self-defence mechanism (mononuclear phagocytic system) inside the body. Modifying the nanoparticles' surface with a suitable hydrophilic polymer (e.g., polyethylene glycol) helps them to survive and increases the circulation time inside the body. The properties of polyethylene glycol, such as the availability of terminal primary hydroxyl groups for derivatization, chemical inertness of the polymer backbone, and low immunogenicity and antigenicity, make it suitable for nanoparticle modification (Rahman et al., 2020). Passive targeting is not based on ligand-specificity. However, the efficiency of the target cell can be increased through the combined effect of active and passive drug targeting (Wicki et al., 2015).

Passive targeting of anticancer drugs occurs through the leaky vasculature of cancer cells and delivers the drugs to elicit the anticancer activity through enhanced permeation and retention effect. Whereas the active targeting of anticancer drugs involves the interaction of overexpressed receptors of cancer cells with the ligand-conjugated drug molecules, which elicit the anticancer activity through internalization into the cells via endocytosis in response to stimuli like changes in the pH, redox potential, and temperature in the tumour microenvironment.

16.3.2 Active Targeting

Active targeting involves adding high-affinity ligands on the surface of a nanocarrier, and ligands bind specifically to the target cell (Wicki et al., 2015). Through active targeting, the drug enters the cells at the molecular level (Anarjan, 2019). Overexpressed receptors have been seen on the cancer cell's surface. Drug-loaded nanoparticles can be effectively delivered to the targeted cells using a suitable ligand that is specific to bind to surface receptors. Active targeting helps to deliver therapeutic drugs to the cell, whose cellular absorption requires endocytosis, fusion, or other processes (Rahman et al., 2020). Vitamins (folic acid), peptides (arginine–glycine–aspartic acid), proteins (antibody or antibody fragments), glycol protein (transferrin), oligonucleotide, and nucleic acid (aptamers) are ligands used in nanoparticles for therapeutic purposes in human (Wicki et al., 2015; Felice et al., 2014). The active targeting approach increases the drug's therapeutic index by increasing its efficacy while lowering non-specific toxicity (Kumari et al., 2016). Delivering therapeutic drugs in pH-sensitive nanocarriers is another form of active targeting (e.g., liposomes).

Oxygen depletion in the tumour cell uses the glycolytic pathway to get extra energy, making the microenvironment acidic. Exploiting this property, a therapeutic drug with pH-sensitive liposomes can stabilize at physiological pH and release the drug to the cancer cells in response to an acidic environment. This will retain the healthy tissue and act specifically on the tumour cells (Khodabandehloo et al., 2016). Even though passive targeting aids in the efficient delivery of drugs to the targeted tissue, it cannot further enhance the uptake of the drug into the cell. It is possible through ligand-receptor interaction through active targeting (Rahman et al., 2020). A large number of the receptors are overexpressed in the cancer cell compared to a normal cell. Vascular endothelial growth factor receptor, human epidermal receptor-2, transferrin receptor, avb3-Integrin, vascular cell adhesion molecules, etc., are some of the overexpressed receptors in different cancer cell types. Cytotoxic response through this receptor can be achieved by either blocking the regular role of the receptor or by delivering the therapeutic drugs to the cancer cell (Kumari et al., 2016). Specific interaction between tumour cell receptors and the nanocarrier surface ligand may facilitate nanoparticle internalization by triggering receptor-mediated endocytosis (Rahman et al., 2020). Increased protein absorption and immunogenicity are the significant drawbacks of ligand utilization in drug delivery to humans. Therefore, clinical studies in active targeted therapeutic nano-delivery of a drug are limited, and only a few are under advanced clinical trials (Felice et al., 2014).

16.4 NANOTECHNOLOGY-BASED APPROACHES OF CANNABINOIDS FOR CANCER TREATMENT

Nanotechnology is a multidisciplinary research area involving chemistry, physics, biology, and medicine (Verma et al., 2018). Nanoparticles have a variety of distinguishing characteristics due to their small size and physical properties (Singh et al., 2017). Nanoparticle sizes ranging from 1 to 100 nm have been extensively studied and widely exploited by researchers in many areas, including early detection, accurate diagnosis, and drug delivery of various diseases (Verma et al., 2018; Mokoena et al., 2019; Rana and Sharma, 2019; Padhi et al., 2020). Different nanoparticle-based therapeutic products have recently been used due to their solubility, ease of release characteristics, biocompatibility, and enhanced half-life (Mokoena et al., 2019). The small size helps to deliver drugs and fluorescent or contrast agents into the cell or onto the surface without disturbing its function. These nanoparticles can penetrate the tissue or enter the blood circulation without forming a clot or aggregation of platelets (Kushwaha et al., 2012; Padhi and Behera, 2020).

Nanocarriers for targeted, regulated, and sustained drug delivery, such as solid lipid nanoparticles, liposomes, nanoemulsions, micelles, dendrimers, and biodegradable polymer nanoparticles, are typical and offer alternatives to conventional drug delivery approaches (Bruni et al., 2018). The use of a nanocarrier for drug delivery could provide several benefits, including protection from hostile conditions (such as the acidic pH of the digestive tract), optimal delivery of lipophilic medicines at high doses, and selective and regulated delivery to enable exact dosing to a specific tissue over time (Bruni et al., 2018). The benefits and advantages of nanotechnology are due to many features and interactions unique to the nanoscale structure (Enrico, 2019; Padhi et al., 2020). However, limitations include low renal clearance, less stability during blood circulation (Bruni et al., 2018). The development of nanocarriers for treating neurological diseases, use as antimicrobial agents, cardiovascular diseases, and cancer, with systemic administration, has received a lot of research effort (Bruni et al., 2018). Injection moulding, nanoimprinting, and soft lithography are some of the techniques for producing polymeric nanocarriers with complex patterns (Kushwaha et al., 2012). Nanotechnology paved the way for theranostic medicine, allowing therapeutic and diagnostic strategies to implement in malignant gliomas (Wang et al., 2016). Lipids (liposomes), polymers (polymeric carriers, dendrimers, or micelles), solid lipid carriers, gold carriers, nanotubes magnetic carriers, and viruses are used as carriers in drug delivery systems (Conti et al., 2006). Natural polymers such as gelatin, dextran, and chitosan are nanocarriers in drug delivery, showing high encapsulating efficiency (Conti et al., 2006).

The limitations faced during the drug delivery of cannabinoids due to their high lipophilicity can be overcome through nanotechnology (Bruni et al., 2018). The advantage of nanocarriers is exploited in cannabinoid delivery for distinct therapeutic purposes (Aparicio-Blanco et al., 2019). Table 16.2 represents recent studies on cannabinoid nano-delivery in cancer treatment. The data from the preclinical research indicate that hemp oil nanoemulsions and nanoparticles are a useful alternative to the existing formulation technologies and can be used for oral delivery of drugs with bioavailability concerns (Yin et al., 2017). Durán-Lobato et al. (2016) successfully developed an orally delivered cannabinoid-loaded lipid-nanoparticle through solvent-emulsion evaporation, and optimization was done by in-vitro interaction with fibroblast (NIH 3T3), human embryonic kidney (HEK 293T), and colon cancer (Caco-2) cells. Further, a clinical study proved the effectiveness and bioavailability of the cannabinoid-loaded pro-nanolipo-sphere formulations. The pharmacokinetic evaluation was done using nine healthy volunteers. The study demonstrated a threefold and fourfold increase in maximum concentration for Δ^9-tetrahydrocannabinol and cannabidiol compared to Sativex (Cherniakov et al., 2017b). Atsmon et al. (2018) conducted a similar study. Cannabinoids have created a lot of interest in cancer research. Many researchers aim to develop cannabinoid-based nanoparticles to increase anticancerous efficacy, including the progress of polymeric nanoparticles (Fraguas-Sánchez et al., 2020).

Breast cancer treatment with cannabinoid nanoparticles showed translational potential by targeting the endocannabinoid system in the cell (Onaivi et al., 2020). Aparicio-Blanco et al. (2019) studied the efficacy of cannabidiol-loaded lipid nanocapsules in the glioblastoma cell line (U373MG) using the *in vitro* method. The work revealed that nanolipid carriers encapsulating cannabidiol drug are a promising dually-targeted candidate with better-targeting properties for intravenous treatment of glioma. In recent years, liposomal, micellar, and nanosized formulations have been used in systematic formulation (Bruni et al., 2018). A list of different nanoformulations of cannabinoids is represented in Table 16.1.

16.4.1 Polymeric Carriers

Polymeric carriers are natural or synthetic polymers in which a drug can be entrapped, adsorbed, or covalently attached, forming nanospheres or nanocapsules (Gupta and Kumar, 2012). Biodegradable and biocompatible synthetic polymers such as polyglycolic acid, poly lactic acid, polylactide-co-glycolide, poly methyl methacrylate, and poly ε-caprolactone have been approved by the United States Food and Drug Administration for the administration in human beings (Gupta and Kumar, 2012). The polymer, polyglycolic acid, and polylactide-co-glycolide can be hydrolysed into simple monomers, such as glycolic acid and lactic acid, which are excreted from the body through normal metabolic pathways. Incorporating drugs during the preparation time of polymeric nanocarrier shows higher entrapment efficiency than adsorption on preformed nanoparticles. The drug is released by the degradation of the polymer followed by desorption, diffusion, or erosion (Gupta and Kumar, 2012). Cannabinoids are commonly

TABLE 16.2
Studies on Cannabinoid-Loaded Nanocarriers in Cancer Therapy

Cannabinoid	Type of Cancer	Type of Study	Nanocarrier Used	Major Findings	References
Δ^9-THC	Lung cancer	In vitro and in vivo	Poly(D,L-lactide-co-glycolide)	Showed cytotoxic effect	Martín-Banderas et al. (2015)
Cannabinoid	Lung cancer	In vivo	Gold nanoparticle	Inhalation route offers more promise for targeting tumour cells	Ngwa et al. (2017)
WIN55,212-2 (WIN)	Triple-negative breast cancer	In vivo	Micelle	Favour drug tumour accumulation minimized adverse psychoactive effects	Greish et al. (2018)
Cannabidiol	Glioma cells	*In vitro*	Lipid nanocapsules	3.4-fold increase in drug delivery to the target cell	Aparicio-Blanco et al. (2019)
Cannabidiol	Ovarian cancer	In vitro and in vivo	PLGA	Onset of apoptosis tumour growth inhibition	Fraguas-Sánchez et al. (2020)

delivered using formulations based mostly on polymer poly-lactic-co-glycolic acid (Fathordoobady et al., 2019).

Preclinical trials for evaluating cannabinoid-loaded poly-lactic-co-glycolic acid nanocarriers for neuropathic pain treatment showed long-lasting pain relief up to 11 days after a single oral dose (Berrocoso et al., 2017). Durán-Lobato et al. (2014) investigated the physicochemical properties of surface-modified and non-modified poly-lactic-co-glycolic acid nanoparticles, including mean particle size, drug loading, and zeta potential, as in vitro release cytotoxicity mucoadhesion, and uptake in the colon cancer (Caco-2) cell line, and *in vivo* biodistribution. The type of modifier affected the mean size, zeta potential values, and spherical form of nanoparticles. The particles with a mean size of 253–344 nm and a spherical shape showed higher release rates when vitamin E and lecithin were used. According to cell culture studies, chitosan-poly-lactic-co-glycolic acid nanoparticles had the highest oral absorption and uptake value. The study effectively delivered cannabinoid derivative (CB13) into the cell's interior by combining the surface-modified polymers. *In vitro* and *in vivo* studies showed that Δ^9-tetrahydrocannabinol-loaded poly-lactic-co-glycolic acid nanoparticles were effective against lung cancer (Martín-Banderas et al., 2015). When the nanoformulations were tested among human embryo lung fibroblastic cells, murine lung adenocarcinoma cells, and human lung cancer cells showed substantial cell viability differences (Martín-Banderas et al., 2015). Durán-Lobato et al. (2015) examined polyethylene glycol and chitosan-encapsulated lipid and poly-lactic-co-glycolic acid nanoparticles for therapeutic oral delivery of cannabinoid derivative (CB13). The study revealed that adding chitosan and polyethylene glycol to poly-lactic-co-glycolic acid substantially increased zeta potential values. Spherical particles with different core-shell configurations were also formed using lipid and poly-lactic-co-glycolic acid. The study showed that the release profile of poly-lactic-co-glycolic acid nanoparticles was significantly longer in contrast to lipid-based nanoparticles.

16.4.2 Nano-Liposomes and Nanocapsules

Liposomes are spherical lipid colloids derived from synthetic or natural phospholipid dispersion in water. It has been widely used as a biocompatible and biodegradable drug carrier to decrease the toxicity of chemotherapeutics and increase the drug's efficacy (Majidinia et al., 2020). Liposomes have been used to deliver vaccines, anti-human immunodeficiency virus, anticancer drugs, toxoids, and genes. Surface modification of the carriers by attaching polyethylene glycol, poly-*N*-vinylpyrrolidones, and dextran to the lipid bilayer increases their blood circulation time (Gupta and Kumar, 2012). In liposomes, the bilayer lipid is surrounded by the central aqueous space (Verma et al., 2018). The carrier can load hydrophilic drugs in lipid layers and hydrophobic drugs in the aqueous core (Majidinia et al., 2020). Myocet® and Doxil® were the first liposome-based cancer treatments approved by the Food and Drug Administration (Khodabandehloo et al., 2016).

Nanocapsules are colloidal drug delivery systems consisting of an oily or aqueous core coated with a thin polymer matrix. Nanocapsules have most recently sparked a great interest in producing controlled-release delivery systems due to the availability of biocompatible and biodegradable polymers. Dispersed polymer nanocapsules can be used as nanosized drug delivery systems to ensure controlled release and efficient drug targeting. The structure and composition of the outer coating and type of surfactant are the critical determinants of dispersion stability, release and degradation properties, and the primary physiological response of the nanoparticles. Another significant criterion is capsule size, with radii ranging from 100 to 500 nm, which is considered optimal (Gupta and Kumar, 2012). Cannabidiol-loaded lipid nanocapsules appear to be promising dually-targeted candidates for intravenous glioma therapy. Aparicio-Blanco et al. (2019) studied cannabidiol delivery using lipid nanocapsules against cancer cell lines (glioblastoma). The findings revealed that the nanocapsule's size significantly impacts the amount of cannabidiol released. The inhibitory concentration (50%) value of lipid nanocapsules with a diameter of 20 nm was decreased by 3.0 times compared to lipid nanocapsules with a diameter of 50 nm.

16.4.3 Micelles

Micelles/polymeric micelles are amphiphilic copolymers with a hydrophobic core and a hydrophilic shell in which the hydrophobic core carries the drug (Mokoena et al., 2019). The size of micelles varies from 20 to 80 nm. Polymeric micelles can carry drugs at the core, up to 20%–30% of their weight. They are most commonly used in phytochemical delivery (Manickam et al., 2021). The shell protects the drug from interactions with serum proteins and untargeted cells by stabilizing micelles in dispersion and releasing the drug by diffusion after reaching the target cell (Gupta and Kumar, 2012). It emerges as a powerful tool in drug delivery for cancer treatment due to its *in-vivo* stability, tunable size, efficiency in solubilizing insoluble anticancer drugs in water, and increased retention capacity of the drug in blood circulation (Priya et al., 2017). A preclinical study on triple-negative breast cancer using cannabinoid-loaded nano micellar formulation reduced the tumour growth to a greater degree with minimal psychoactive side effects than the free drug. The report opens a new therapeutic strategy to treat triple-negative breast cancer (Greish et al., 2018).

16.4.4 Solid Lipid Nanoparticles and Nanostructured Lipid Carriers

Solid lipid nanoparticles can be synthesized by replacing oil with solid lipids in an oil and water emulsion (Khodabandehloo et al., 2016). These nanocarriers are suitable for delivering small molecular-weight drug payloads (Majidinia et al., 2020). Fatty acids, waxes, highly purified triglycerides, or complex glyceride mixtures are commonly used as lipids in solid lipid nanoparticles. The limitations, such as drug expulsion during phase modification and limited loading capacity, reduce its practical applications (Gupta and Kumar, 2012). Nanostructured lipid carriers are the second generation of lipid-based nanocarriers comprising a combination of solid and liquid lipids. Nanostructured lipid carriers were designed to solve the limitations of solid lipid nanoparticles (Ghasemiyeh and Mohammadi-Samani, 2018). Due to imperfect crystal structure, nanostructured lipid carriers show higher loading capacity than solid lipid nanoparticles. Nanostructured lipid carriers use the enhanced permeation and retention effect to achieve passive tumour targeting (Selvamuthukumar and Velmurugan, 2012). The blood-brain barrier limits drug penetration into the brain, resulting in poor prognosis and healing process for brain-related diseases. Although delivering drugs to the brain is difficult, nanostructured lipid carriers overcome these obstacles (Mulvihill et al., 2020). Esposito et al. (2016) validated and designed cannabinoid-encapsulated nanostructured lipid carriers to optimize nanoparticle recovery and drug encapsulation efficiency. The study revealed that adapting the nanostructured lipid carriers development protocol made it possible to efficiently nano-encapsulate cannabinoids, resulting in a formulation suitable for clinical and preclinical studies. Furthermore, long-term physicochemical stability and nanostructured lipid carriers carried Δ^9-tetrahydrocannabinol composition optimization seem feasible (Hommoss et al., 2017).

16.4.5 Inorganic Nanoparticles

Due to their optoelectrical properties, the inorganic nanoparticles have remarkably impacted drug delivery in cancer treatment. Their properties, such as wide availability, site-specific delivery, rich functionality, strong biocompatibility, and regulated drug release, attract drug delivery (Priya et al., 2017). Gold, silver, copper, aluminium, and silica are used to prepare inorganic nanoparticles. Gold nanoparticles are the most extensively used and studied drug delivery method (Gupta and Kumar, 2012). Connor et al. (2005) showed the nontoxic nature of gold carriers at the cellular level in several human cell lines. Their size, shape, functional group, and charge of gold nanoparticles significantly affected their ability to accumulate on various immune cells (Surendran et al., 2018). Gold nanoparticles are remarkably effective contrast agents in cancer diagnosis and photo-dermal cancer treatment (Gupta and Kumar, 2012).

16.4.6 Self-Emulsifying Drug-Delivery Systems

Self-emulsifying drug-delivery systems enhance the oral bioavailability of poorly water-soluble drugs known as pro-nanolipospheres (Sarpal et al., 2010; Chen and Rogers, 2019). These systems can easily be dissolved in intestinal fluid and rapidly get absorbed by the lymphatic system in the body. Self-emulsifying drug-delivery systems are formed by the isotropic mixture of natural or synthetic oil with surfactant and co-solvent. The large surface area and small size enhance the oral bioavailability of the molecule (Sarpal et al., 2010; Priya et al., 2017). The lipophilic substrate is entrapped inside the triglyceride core, resulting in aqueous solubility.

In a pharmacokinetic experiment in a rat model, Cherniakov et al. (2017b) discovered that adding a solubility enhancer like piperine resulted in a six-fold increase in cannabidiol bioavailability and a 9.3-fold increase in Δ^9-tetrahydrocannabinol bioavailability. Cherniakov et al. (2017a) performed a clinical trial with healthy participants to evaluate the plasma exposure and bioavailability of the oral Δ^9-tetrahydrocannabinol-cannabidiol-piperine-pro-nanolipospheres formulation to the Sativex® spray (buccal spray). Compared to Sativex®, a single oral administration of the piperine- pro-nanolipospheres formulation resulted in a 1.5-fold increase in area under the curve and a threefold increase in maximum plasma concentration for Δ^9-tetrahydrocannabinol. Cannabidiol was found to have a fourfold increase in overall observed plasma concentration and a 2.2-fold rise in area under the curve. The excipients' "generally recognized as safe" status allowed for the completion of the recorded clinical study and is an advantage in regulatory compliance and future clinical trials in disease models. The study showed that pro-nanolipospheres provide the benefits of an oral formulation while being adapted to the cannabinoids' unique lipophilicity and metabolism characteristics. Because of these characteristics, it is the best choice for a standardized oral cannabinoid formulation.

16.5 CLINICAL STUDIES INVOLVING CANNABINOIDS

In vitro and *in vivo* studies have shown that cannabinoids have potent anticancer activity in various animal cancer cells (Bogdanović et al., 2017). However, studies on human anticancer efficacy are still under development (Turgeman and Bar-Sela, 2019). The first clinical trial evaluated the antitumoural effect of cannabinoids in glioblastoma patients in 2006. This pilot phase 1 study took nine patients and administered Δ^9-tetrahydrocannabinol intratumourally. The study assessed the safety of intracranial administration of the drug and its effect on the cancer cells. The result showed possible anti-proliferative action and a good safety profile for Δ^9-tetrahydrocannabinol administration in patients without any psychoactive effect (Guzmán et al., 2006). Based on preclinical evidence of Δ^9-tetrahydrocannabinol: cannabidiol combination treatment for a cancer cell, a clinical trial is carried out using 27 adult patients with glioblastoma in two-part, part A and part B. Part A was an open-label phase that assessed the adverse effect and safety of the patient who was treated with Sativex (Δ^9-tetrahydrocannabinol: cannabidiol combination 27 mg/mL:25 mg/mL) in combination with dose-intense temozolomide. Part B was a randomized, double-blind, and placebo-controlled trial. Pharmacokinetic examination of temozolomide and its metabolites, clinical laboratory tests, adverse reactions, and vital signs were used to evaluate Sativex's efficacy and safety compared to placebo. The survival rate of the patients was assessed in part B. The drug was sprayed orally (maximum of twelve sprays/day) with dose-intense temozolomide for 365 days. Part A study observed the most common problems in patients, like dizziness, fatigue, vomiting, nausea, and headache.

In part 2, after one year's study, 44% of placebo-treated patients ($p=0.042$) and 83% of Nabiximols survived. This clinical study concluded that Sativex with personalized dosing showed acceptable safety and tolerability (Part 1- NCT01812603; Part 2- NCT01812616) (ClinicalTrials.gov; Twelves et al., 2021). Another clinical study based on the effect of cannabinoid administration in cancer patients is in progress. A tightly controlled randomization study is carried out with 30 patients (adults) who are ready to take cannabinoids for their cancer treatment and related symptoms. This is a method of patient-oriented prospective observational study, and the first phase of the study was expected to be completed on September 20, 2021 (NCT03617692) (ClinicalTrials.gov). A more recent study on the safety and efficacy of inhaled therapeutic-based medication (PPP011) is under investigation. In this trial, 334 participants with advanced cancer were observed for 12 weeks. The physical functioning and cachexia progression of the patients will be evaluated, and the work is under phase 3 (NCT04001010) (ClinicalTrials.gov).

Another clinical phase 1 trial using a synthetic analogue of Δ^9-tetrahydrocannabinol (dronabinol) was conducted to investigate the safety, preliminary antitumour activity, drug penetration capacity to the central nervous system, and recommended phase II dose of patients with brain cancer. Patients (number - 26) were subjected to systemic exposure to dose concentration (at 2, 4, 8, 16, 24, 28, and 36 mg/kg) once a week. The minimum phase II dose of dexanabinol was identified at 28 mg/kg from the study. The result showed that a considerable amount of drugs were present in the cerebrospinal fluid, which ensured the penetration capacity of the drug and the possibility of reaching the tumour cells. However, the study did not show the antitumour activity in patients; only 21% experienced stable disease (Juarez et al., 2021). A case study was conducted on a 14-year-old lymphoblastic leukaemia patient who was administered cannabinoid extracts orally for 78 days. Clinical research provides dose-dependent disease control (Singh and Bali, 2013).

Furthermore, a combination of cannabidiol, Δ^9-tetrahydrocannabinol with temozolomide, and radiotherapy is being tested in glioblastoma patients. The study was designed to determine the maximum tolerable dose of Δ^9-tetrahydrocannabinol-cannabidiol, find the adverse effects, antitumour activity of the combination treatment, and overall survival with a time-bound of 12 months and has not yet reported results (Clinical Trials number: NCT03529448) (ClinicalTrials.gov). Taha et al. (2019) evaluated the impact of tumour response on patients with highly advanced cancer. A one-year retrospective observational study with 140 cancer patients discovered that the use of cannabis in the immunotherapeutic procedure reduced response rate. At the same time, it did not affect the overall survival or progression-free survival. There seemed to be no correlation between the composition of cannabis and response rate. Kenyon et al. (2018) conducted an objective clinical response study on cancer patients to assess the effect of synthetic cannabinoids. One hundred nineteen solid cancer patients were taken for this study over four years. The result showed the safe use of cannabinoids, and clinical response to this treatment was observed in 92% of patients. The work revealed the reduction of circulating tumour cells; in some cases, tumour size reduction was also seen in repeated scanning. The study concluded that synthetic cannabidiol could treat breast cancer and glioma. More research is required in preclinical and clinical trials to monitor the effects of cannabinoids alone or in combination with conventional anticancer strategies. The efficacy of the targeted drug can be improved by a nanoparticle delivery system. This area has to be explored in cancer treatment by suitable preclinical and clinical trials (Bogdanović et al., 2017).

16.6 CONCLUSION AND FUTURE PERSPECTIVES

The main problem with current anticancer research is that it is limited to cell lines and animal experiments, which prevents it from drawing any firm conclusions or extrapolating to human cancer. Systematic analysis has contributed to the increasing quantity of knowledge surrounding the use of cannabis in oncology. Cannabinoids can play a role in cancer treatment in the future. Because of their selectivity and effectiveness, cannabinoids have been demonstrated in clinical research to be potential therapeutic therapy antecedents. It is anticipated that the use of cannabinoids in therapeutic settings will become more appropriate as we gain a better knowledge of the mechanisms of action behind their impact. In academic cannabis studies, there are some potential challenges and limitations. High uncertainty and lack of standardization in trial construction and drug formulation make a comprehensive investigation of research on cannabis and related drugs difficult. Overwhelming evidence from positive *in-vitro* and *in-vivo* trials of cannabinoids' therapeutic translation is hampered by psychoactive side effects. As a result, cannabinoid-based treatments or methods to avoid these psychotic side effects are desperately required.

Cancer nanotechnology is growing as a tool for early detection, diagnosis, and treatment. It can result in a gradual transition in clinical practice towards a life-saving strategy. Some nanotechnology-based formulations are already on the market, and many more are being researched and evaluated in clinical trials. *In vitro* and *in vivo*, cannabinoid nanomedicines display high drug-loading ability, cellular absorption, and therapeutic impact. They will address issues such as cannabinoid drugs' poor water solubility, a significant barrier to effective cancer care. Cannabinoid drug delivery through nanotechnology is a positive step towards saving people's lives worldwide, and the potential is enormous.

ACKNOWLEDGEMENTS

The authors thank Dr Shijin A, veterinary surgeon, Government of Kerala, and Ms Jiya Fathima KV, KV No: 1, JIPMER, Puducherry.

REFERENCES

Almogi-Hazan O, Or R (2020) Cannabis, the endocannabinoid system and immunity: The journey from the bedside to the bench and back. *Int J Mol Sci* 21:1–17.

Anarjan SF (2019) Active targeting drug delivery nanocarriers: Ligands. *Nano-Struct Nano-Objects* 19:100370.

Aparicio-Blanco J, Sebastián V, Benoit JP, Torres-Suárez AI (2019) Lipid nanocapsules decorated and loaded with cannabidiol as targeted prolonged release carriers for glioma therapy: In vitro screening of critical parameters. *Eur J Pharm Biopharm* 134:126–137.

Armstrong JL, Hill DS, McKee CS, Hernandez-Tiedra S, Lorente M, Lopez-Valero I, Anagnostou ME, Babatunde F, Corazzari M, Redfern CPF, Velasco G, Lovat PE (2015) Exploiting cannabinoid-induced cytotoxic autophagy to drive melanoma cell death. *J Invest Dermatol* 135:1629–1637.

Atsmon J, Cherniakov I, Izgelov D, Hoffman A, Domb AJ, Deutsch L, Deutsch F, Heffetz D, Sacks H (2018) PTL401, a new formulation based on pro-nano dispersion technology, improves oral cannabinoids bioavailability in healthy volunteers. *J Pharm Sci* 107:1423–1429.

Behera A, Padhi S (2020) Passive and active targeting strategies for the delivery of the camptothecin anticancer drug: A review. *Environ Chem Lett* 18(5):1557–1567.

Behera A, Padhi S (2022) pH-Sensitive polymeric nanoparticles for cancer treatment. In: Padhi S, Behera A, Lichtfouse E (Eds.) *Polymeric Nanoparticles for the Treatment of Solid Tumours*. Environmental Chemistry for a Sustainable World, vol 71. Springer, Cham, pp. 401–425.

Behera A, Padhi S, Nayak AK (2023) Engineered liposomes as drug delivery and imaging agents. In: Ray S, Nayak AK (Eds.) *Design and Applications of Theranostic Nanomedicines*. Woodhead Publishing, Cambridge, MA, pp. 75–108.

Berrocoso E, Rey-Brea R, Fernández-Arévalo M, Micó JA, Martín-Banderas L (2017) Single oral dose of cannabinoid derivate loaded PLGA nanocarriers relieves neuropathic pain for eleven days. *Nanomed Nanotechnol Biol Med* 13:2623–2632.

Bogdanović V, Mrdjanović J, Borišev I (2017) A review of the therapeutic antitumour potential of cannabinoids. *J Altern Complement Med* 23:831–836.

Bruni N, Pepa CD, Oliaro-Bosso S, Pessione E, Gastaldi D, Dosio F (2018) Cannabinoid delivery systems for pain and inflammation treatment. *Molecules* 23:2478.

Chen PX, Rogers MA (2019) Opportunities and challenges in developing orally administered cannabis edibles. *Curr Opin Food Sci* 28:7–13.

Cherniakov I, Izgelov D, Barasch D, Davidson E, Domb AJ, Hoffman A (2017a) Piperine-pro-nanolipospheres as a novel oral delivery system of cannabinoids: Pharmacokinetic evaluation in healthy volunteers in comparison to buccal spray administration. *J Control Release* 266:1–7.

Cherniakov I, Izgelov D, Domb AJ, Hoffman A (2017b) The effect of Pro NanoLipospheres (PNL) formulation containing natural absorption enhancers on the oral bioavailability of delta-9-tetrahydrocannabinol (THC) and cannabidiol (CANNABIDIOL) in a rat model. *Eur J Pharm Sci* 109:21–30.

ClinicalTrials.gov[internet] National Library of Medicine (US). 2013 March 18, Identifier-NCT01812603, A Safety Study of Sativex in Combination With Dose-intense Temozolomide in Patients with Recurrent Glioblastoma, 2021 February 5, Available. In: https://www.clinicaltrials.gov/ct2/show/NCT01812603.

ClinicalTrials.gov[internet] National Library of Medicine (US). 2013 March 18, Identifier-NCT01812616, A Safety Study of Sativex Compared with Placebo (Both With Dose-intense Temozolomide) in Recurrent Glioblastoma Patients. 2016 August 11, Available. In: https://www.clinicaltrials.gov/ct2/show/NCT01812616.

ClinicalTrials.gov[internet] National Library of Medicine (US). 2018 August 6, Identifier-NCT03617692, Cannabis Use in Cancer Patients. Available. In: https://clinicaltrials.gov/ct2/show/NCT03617692.

ClinicalTrials.gov[internet] National Library of Medicine (US). 2019 June 27, Identifier-NCT04001010, Safety and Efficacy of Inhaled Synthetic THC/CANNABIDIOL for Improving Physical Functioning and for Modulating Cachexia Progression in Patients with Advanced Cancer and Assciated Cachexia (S. In: https://clinicaltrials.gov/ct2/show/NCT04001010.

ClinicalTrials.gov[internet] Spain: National Library of Medicine (US). 2018 May 18, Identifier-NCT03529448, TN-TC11G (THC+CANNABIDIOL) Combination with Temozolomide and Radiotherapy in Patients with Newly-diagnosed Glioblastoma (GEINOCANN); December 16, 2020. Available. In: https://clinicaltrials.gov/ct2/show/NCT03529448#contact.

Connor EE, Mwamuka J, Gole A, Murphy CJ, Wyatt MD (2005) Gold nanoparticles are taken up by human cells but do not cause acute cytotoxicity. *Small* 1:325–327.

Conti M, Tazzari V, Baccini C, Pertici G, Serino LP, De Giorgi U (2006) Anticancer drug delivery with nanoparticles. *In Vivo* 20:697–702.

Dariš B, Verboten MT, Knez Ž, Ferk P (2019) Cannabinoids in cancer treatment: Therapeutic potential and legislation. *Bosn J Basic Med Sci* 19:14–23.

De La Ossa DHP, Ligresti A, Gil-Alegre ME, Aberturas MR, Molpeceres J, Di Marzo V, Torres Suárez AI (2012) Poly-ε-caprolactone microspheres as a drug delivery system for cannabinoid administration: Development, characterisation and in vitro evaluation of their antitumoural efficacy. *J Control Release* 161:927–932.

De La Ossa DHP, Lorente M, Gil-Alegre ME, Torres S, García-Taboada E, Aberturas M del R, Molpeceres J, Velasco G, Torres-Suárez AI (2013) Local delivery of cannabinoid-loaded microparticles inhibits tumour growth in a murine xenograft model of glioblastoma multiforme. *PLoS One* 8:2–9.

Durán-Lobato M, Martín-Banderas L, Gonçalves LMD, Fernández-Arévalo M, Almeida AJ (2015) Comparative study of chitosan- and PEG-coated lipid and PLGA nanoparticles as oral delivery systems for cannabinoids. *J Nanoparticle Res* 17:1–17.

Durán-Lobato M, Martín-Banderas L, Lopes R, Gonçalves LMD, Fernández-Arévalo M, Almeida AJ (2016) Lipid nanoparticles as an emerging platform for cannabinoid delivery: Physicochemical optimisation and biocompatibility. *Drug Dev Ind Pharm* 42:190–198.

Durán-Lobato M, Muñoz-Rubio I, Holgado MÁ, Álvarez-Fuentes J, Fernández-Arévalo M, Martín-Banderas L (2014) Enhanced cellular uptake and biodistribution of a synthetic cannabinoid loaded in surface-modified poly(lactic-co-glycolic acid) nanoparticles. *J Biomed Nanotechnol* 10:1068–1079.

Enrico C (2019) *Nanotechnology-Based Drug Delivery of Natural Compounds and Phytochemicals for the Treatment of Cancer and Other Diseases*, 1st edn. Elsevier BV.

Esposito E, Drechsler M, Cortesi R, Nastruzzi C (2016) Encapsulation of cannabinoid drugs in nanostructured lipid carriers. *Eur J Pharm Biopharm* 102:87–91.

Fathordoobady F, Singh A, Kitts DD, Pratap Singh A (2019) Hemp (*Cannabis sativa* L.) extract: Antimicrobial properties, methods of extraction, and potential oral delivery. *Food Rev Int* 35:664–684.

Felice B, Prabhakaran MP, Rodríguez AP, Ramakrishna S (2014) Drug delivery vehicles on a nano-engineering perspective. *Mater Sci Eng C* 41:178–195.

Fonseca BM, Teixeira NA, Correia-da-Silva G (2017) Cannabinoids as modulators of cell death: Clinical applications and future directions. *Reviews of Physiology, Biochemistry and Pharmacology* 173:63–88.

Fraguas-Sánchez AI, Torres-Suárez AI, Cohen M, Delie F, Bastida-Ruiz D, Yart L, Martin-Sabroso C, Fernández-Carballido A (2020) PLGA nanoparticles for the intraperitoneal administration of CANNABIDIOL in the treatment of ovarian cancer: In vitro and in Ovo assessment. *Pharmaceutics* 12:1–19.

Galluzzi L, Pietrocola F, Bravo-San Pedro JM, Amaravadi RK, Baehrecke EH, Cecconi F, Codogno P, Debnath J, Gewirtz DA, Karantza V, Kimmelman A, Kumar S, Levine B, Maiuri MC, Martin SJ, Penninger J, Piacentini M, Rubinsztein DC, Simon H, Simonsen A, Thorburn AM, Velasco G, Ryan KM, Kroemer G (2015) Autophagy in malignant transformation and cancer progression. *EMBO J* 34:856–880.

Ghasemiyeh P, Mohammadi-Samani S (2018) Solid lipid nanoparticles and nanostructured lipid carriers as novel drug delivery systems: Applications, advantages and disadvantages. *Res Pharm Sci* 13:288–303.

Glodde N, Jakobs M, Bald T, Tüting T, Gaffal E (2015) Differential role of cannabinoids in the pathogenesis of skin cancer. *Life Sci* 138:35–40.

Greish K, Mathur A, Al Zahrani R, Elkaissi S, Al Jishi M, Nazzal O, Taha S, Pittalà V, Taurin S (2018) Synthetic cannabinoids nano-micelles for the management of triple negative breast cancer. *J Control Release* 291:184–195.

Guindon J, Hohmann AG (2011) The endocannabinoid system and cancer: Therapeutic implication. *Br J Pharmacol* 163:1447–1463.

Gupta S, Kumar P (2012) Drug delivery using nanocarriers: Indian perspective. *Proc Natl Acad Sci India Sect B Biol Sci* 82:167–206.

Guzmán M, Duarte MJ, Blázquez C, Ravina J, Rosa MC, Galve-Roperh I, Sánchez C, Velasco G, González-Feria L (2006) A pilot clinical study of Δ9-tetrahydrocannabinol in patients with recurrent glioblastoma multiforme. *Br J Cancer* 95:197–203.

Guzmán M, Sánchez C, Galve-Roperh I (2001) Control of the cell survival/death decision by cannabinoids. *J Mol Med* 78:613–625.

Hommoss G, Pyo SM, Müller RH (2017) Mucoadhesive tetrahydrocannabinol-loaded NLC: Formulation optimisation and long-term physicochemical stability. *Eur J Pharm Biopharm* 117:408–417.

Jeong S, Jo MJ, Yun HK, Kim DY, Kim BR, Kim JL, Park SH, Na YJ, Jeong YA, Kim BG, Ashktorab H, Smoot DT, Heo JY, Han J, Lee SI, Kim HD, Kim DH, Oh SC, Lee D (2019a) Cannabidiol promotes apoptosis via regulation of XIAP/Smac in gastric cancer. *Cell Death Dis* 10:1–13.

Jeong S, Yun HK, Jeong YA, Jo MJ, Kang SH, Kim JL, Kim DY, Park SH, Kim BR, Na YJ, Lee SI, Kim HD, Kim DH, Oh SC, Lee DH (2019b) Cannabidiol-induced apoptosis is mediated by activation of Noxa in human colorectal cancer cells. *Cancer Lett* 447:12–23.

Juarez TM, Piccioni D, Rose L, Nguyen A, Brown B, Kesari S (2021) Phase I dose-escalation, safety, and CNS pharmacokinetic study of dexanabinol in patients with brain cancer. *Neuro-Oncol Adv* 3:1–9.

Kenyon J, Liu W, Dalgleish A (2018) Report of objective clinical responses of cancer patients to pharmaceutical-grade synthetic cannabidiol. *Anticancer Res* 38:5831–5835.

Khodabandehloo H, Zahednasab H, Hafez AA (2016) Nanocarriers usage for drug delivery in cancer therapy. *Int J Cancer Manag* 9(2):e3966.

Kovalchuk O, Kovalchuk I (2020) Cannabinoids as anticancer therapeutic agents. *Cell Cycle* 19:961–989.

Kumari P, Ghosh B, Biswas S (2016) Nanocarriers for cancer-targeted drug delivery. *J Drug Target* 24:179–191.

Kundu A, Padhi S, Behera A, Hasnain MS, Nayak AK (2022) Tumour targeting strategies by chitosan-based nanocarriers. In: Hasnain MS, Beg S, Nayak AK (Eds.) *Chitosan in Biomedical Applications*, Academic Press, London, pp. 163–188.

Kushwaha SKS, Rastogi A, Rai AK, Singh S (2012) Novel drug delivery system for anticancer drug: A review. *Int J PharmTech Res* 4:542–553.

Laezza C, Pagano C, Navarra G, Pastorino O, Proto MC, Fiore D, Piscopo C, Gazzerro P, Bifulco M (2020) The endocannabinoid system: A target for cancer treatment. *Int J Mol Sci Rev* 21:747.

López-Valero I, Saiz-Ladera C, Torres S, Hernández-Tiedra S, García-Taboada E, Rodríguez-Fornés F, Barba M, Dávila D, Salvador-Tormo N, Guzmán M, Sepúlveda JM, Sánchez-Gómez P, Lorente M, Velasco G (2018) Targeting glioma initiating cells with A combined therapy of cannabinoids and temozolomide. *Biochem Pharmacol* 157:266–274.

Lukhele ST, Motadi LR (2016) Cannabidiol rather than Cannabis sativa extracts inhibit cell growth and induce apoptosis in cervical cancer cells. *BMC Complement Altern Med* 16:1–16.

Majidinia M, Mirza-Aghazadeh-Attari M, Rahimi M, Mihanfar A, Karimian A, Safa A, Yousefi B (2020) Overcoming multidrug resistance in cancer: Recent progress in nanotechnology and new horizons. *IUBMB Life* 72:855–871.

Manickam V, Sundar V, Panchangam RL, Amiti KAS, Tamizhselvi R (2021) Nanotechnology in delivery and targeting of phytochemicals. In: Yata VK, Ranjan S, Dasgupta N, Lichtfouse E (Eds.) *Nanopharmaceuticals: Principles and Applications.* Environmental Chemistry for a Sustainable World, vol. 2. Springer, Cham, pp. 211–264.

Martín-Banderas L, Muñoz-Rubio I, Álvarez-Fuentes J, Durán-Lobato M, Arias JL, Holgado MÁ, Fernández-Arévalo M (2014) Engineering of Δ^9-tetrahydrocannabinol delivery systems based on surface modified-PLGA nanoplatforms. *Colloids Surf B Biointerfaces* 123:114–122.

Martín-Banderas L, Muñoz-Rubio I, Prados J, Álvarez-Fuentes J, Calderón-Montaño JM, López-Lázaro M, Arias JL, Leiva MC, Holgado MA, Fernández-Arévalo M (2015) In vitro and in vivo evaluation of Δ^9-tetrahidrocannabinol/PLGA nanoparticles for cancer chemotherapy. *Int J Pharm* 487:205–212.

Massi P, Solinas M, Cinquina V, Parolaro D (2013) Cannabidiol as potential anticancer drug. *Br J Clin Pharmacol* 75:303–312.

McAllister SD, Christian RT, Horowitz MP, Garcia A, Desprez PY (2007) Cannabidiol as a novel inhibitor of Id-1 gene expression in aggressive breast cancer cells. *Mol Cancer Ther* 6:2921–2927.

Mcallister SD, Soroceanu L, Desprez P (2015) The antitumour activity of plant-derived non-psychoactive cannabinoids. *J neuroimmune Pharmacol* 10:255–267.

Mokoena DR, George BP, Abrahamse H (2019) Enhancing breast cancer treatment using a combination of cannabidiol and gold nanoparticles for photodynamic therapy. *Int J Mol Sci* 20(19):4771.

Mulvihill JJ, Cunnane EM, Ross AM, Duskey JT, Tosi G, Grabrucker AM (2020) Drug delivery across the blood-brain barrier: Recent advances in the use of nanocarriers. *Nanomedicine* 15:205–214.

Ngwa W, Kumar R, Moreau M, Dabney R, Herman A (2017) Nanoparticle drones to target lung cancer with radiosensitisers and cannabinoids. *Front Oncol* 7:1–7.

Nikan M, Nabavi SM, Manayi A (2016) Ligands for cannabinoid receptors, promising anticancer agents. *Life Sci* 146:124–130.

Onaivi ES, Singh Chauhan BP, Sharma V (2020) Challenges of cannabinoid delivery: How can nanomedicine help? *Nanomedicine* 15:2023–2028.

Padhi S, Behera A (2020) Nanotechnology based targeting strategies for the delivery of Camptothecin. In: Saneja A, Panda AK, Lichtfouse E (Eds.) *Sustainable Agriculture Reviews 44: Pharmaceutical Technology for Natural Products Delivery Vol. 2 Impact of Nanotechnology*. Springer, Cham, pp. 243–272.

Padhi S, Behera A (2022) Cellular internalization and toxicity of polymeric nanoparticles. In: Padhi S, Behera A, Lichtfouse E (Eds.) *Polymeric Nanoparticles for Treating Solid Tumours*. Environmental Chemistry for a Sustainable World, vol 71. Springer, Cham, pp. 473–488.

Padhi S, Kapoor R, Verma D, Panda AK, Iqbal Z (2018) Formulation and optimisation of topotecan nanoparticles: In vitro characterisation, cytotoxicity, cellular uptake and pharmacokinetic outcomes. *J Photochem Photobiol B Biol* 183:222–232.

Padhi S, Mirza MA, Verma D, Khuroo T, Panda AK, Talegaonkar S, Khar RK, Iqbal Z (2016) Revisiting the nanoformulation design approach for effective delivery of Topotecan in its stable form: An appraisal of its in vitro behavior and tumour amelioration potential. *Drug Deliv* 23(8):2827–2837.

Padhi S, Nayak AK, Behera A (2020) Type II diabetes mellitus: A review on recent drug-based therapeutics. *Biomed Pharmacother* 131:110708

Patnaik S, Gorain B, Padhi S, Choudhury H, Gabr GA, Md S, Kumar Mishra D, Kesharwani P (2021) Recent update of toxicity aspects of nanoparticulate systems for drug delivery. *Eur J Pharm Biopharm* 161:100–119.

Pisanti S, Bifulco M (2019) Medical Cannabis: A plurimillennial history of an evergreen. *J Cell Physiol* 234:8342–8351.

Pisanti S, Picardi P, Alessandro AD, Laezza C, Bifulco M (2013) The endocannabinoid signaling system in cancer. *Trends Pharmacol Sci* 34:273–282.

Priya LB, Baskaran R, Padma VV (2017) Phytonanoconjugates in oral medicine. In: Andronescu E, Grumezescu AM (Eds.) *Nanostructures for Oral Medicine*. Elsevier Inc., Amsterdam, pp. 639–668.

Rahman MA, Yusuf M, Alshammari T, Faiyazuddin M (2020) Nanopharmaceuticals: In relevance to drug delivery and targeting. In: Yata V, Ranjan S, Dasgupta N, Lichtfouse E (Eds.) *Nanopharmaceuticals: Principles and Applications*. Springer, Cham, pp. 77–112.

Ramer R, Hinz B (2016) Antitumourigenic targets of cannabinoids: Current status and implications. *Expert Opin Ther Targets* 20:1219–1235.

Rana V, Sharma R (2019) *Recent Advances in Development of Nano Drug Delivery*. Elsevier Inc., pp. 93–131.

Sadhu PK, Rajput A, Seth AK, Dash DK, Shah NV, Kumari M, Patel S (2020) A combined approach of gold nanoparticles with cannabinoids for the treatment of cancer: A review. *Int J Pharm Res* 12:393–405.

Sarpal K, Pawar Y, Bansal A (2010) Self-emulsifying drug delivery systems: A strategy to improve oral bioavailability. Current Res Inform Pharm Sci 11:42–49.

Scott KA, Dalgleish AG, Liu WM (2017) Anticancer effects of phytocannabinoids used with chemotherapy in leukaemia cells can be improved by altering the sequence of their administration. *Int J Oncol* 51:369–377.

Selvamuthukumar S, Velmurugan R (2012) Nanostructured lipid carriers: A potential drug carrier for cancer chemotherapy. *Lipids Health Dis* 11:1–8.

Simmerman E, Qin X, Yu JC, Baban B (2019) Cannabinoids as a potential new and novel treatment for melanoma: A pilot study in a murine model. *J Surg Res* 235:210–215.

Singer E, Judkins J, Salomonis N, Matlaf L, Soteropoulos P, McAllister S, Soroceanu L (2015) Reactive oxygen species-mediated therapeutic response and resistance in glioblastoma. *Cell Death Dis* 6:e1601–e1611.

Singh K, Jamshidi N, Zomer R, Piva TJ, Mantri N (2020) Cannabinoids and prostate cancer: A systematic review of animal studies. *Int J Mol Sci* 21:1–15.

Singh R, Kumari P, & Kumar, S. (2017). Nanotechnology for enhanced bioactivity of bioactive phytomolecules. In: Grumezescu AM (Ed.) *Nutrient Delivery*. Academic Press, London, pp. 413–456.

Singh Y, Bali C (2013) Cannabis extract treatment for terminal acute lymphoblastic leukemia with a Philadelphia chromosome mutation. *Case Rep Oncol* 6:585–592.

Surendran SP, Moon MJ, Park R, Jeong YY (2018) Bioactive nanoparticles for cancer immunotherapy. *Int J Mol Sci* 19:1–18.

Taha T, Meiri D, Talhamy S, Wollner M, Peer A, Bar-Sela G (2019) Cannabis impacts tumour response rate to nivolumab in patients with advanced malignancies. *Oncologist* 24:549–554.

Turgeman I, Bar-Sela G (2019) Cannabis for cancer–illusion or the tip of an iceberg: A review of the evidence for the use of cannabis and synthetic cannabinoids in oncology. *Expert Opin Investig Drugs* 28:285–296.

Twelves C, Sabel M, Checketts D, Miller S, Tayo B, Jove M, Brazil L, Short SC (2021) A phase 1b randomised, placebo-controlled trial of nabiximols cannabinoid oromucosal spray with temozolomide in patients with recurrent glioblastoma. *Br J Cancer* 124:1379–1387.

Velasco G, Hernández-Tiedra S, Dávila D, Lorente M (2016) The use of cannabinoids as anticancer agents. *Prog Neuro-Psychopharmacol Biol Psychiatry* 64:259–266.

Velasco G, Sánchez C, Guzmán M (2012) Towards the use of cannabinoids as antitumour agents. *Nat Rev Cancer* 12:436–444.

Verma M, Sheoran P, Chaudhury A (2018) Application of nanotechnology for cancer treatment. In: Gahlawat S, Duhan J, Salar R, Siwach P, Kumar SKP (Eds.) *Advances in Animal Biotechnology and its Applications*. Springer, Singapore, pp. 161–178.

Wang L, Mori W, Cheng R, Yui J, Hatori A, Ma L, Zhang Y, Rotstein BH, Fujinaga M, Shimoda Y, Yamasaki T, Xie L, Nagai Y, Minamimoto T, Higuchi M, Vasdev N, Zhang MR, Liang SH (2016) Synthesis and preclinical evaluation of sulfonamidobased [11C-Carbonyl]-carbamates and ureas for imaging monoacylglycerol lipase. *Theranostics* 6:1145–1159.

Wicki A, Witzigmann D, Balasubramanian V, Huwyler J (2015) Nanomedicine in cancer therapy: Challenges, opportunities, and clinical applications. *J Control Release* 200:138–157.

Yin J, Xiang C, Wang P, Yin Y, Hou Y (2017) Biocompatible nanoemulsions based on hemp oil and less surfactants for oral delivery of baicalein with enhanced bioavailability. *Int J Nanomedicine* 12:2923–2931.

Zafar S, Kumar Gaurav J, Ahmad FJ (2020) Nanomedicine approaches for the delivery of herbal anticancer drugs. In: Rahman M, Beg S, Kumar V, Ahmad FJ (Eds.) *Nanomedicine for Bioactives*. Springer Nature, Singapore, pp. 201–229.

Section V

Other Natural Drugs in Cancer Chemotherapy

17 Nanoenabled Systems for the Delivery of Anthracyclines as Anticancer Agents

Bharti Mittu[1] *and Abida Bhat*[2]
[1]National Institute of Pharmaceutical Education and Research (NIPER), S.A.S. Nagar, Mohali, Chandigarh, India
[2]Department of Immunology and Molecular Medicine, Sher-e-Kashmir Institute of Medical Sciences, Kashmir Medical Sciences, Kashmir, India

17.1 INTRODUCTION

Cancer is caused by gene defects, which cause uncontrolled cell division and growth. The alteration in cell replication because of gene mutation, chromosome translocation, and disabled apoptosis results in carcinoma (Padhi et al., 2016; Kundu et al., 2022). It is expected that by 2030, 21.4 million new cancer cases and 13.2 million cancer deaths will become a global burden (Bray et al., 2012). Treatment of cancer is still restricted due to poor drug delivery, poor drug retention at tumour sites, resistance to many drugs, considerable harm to normal cells, and other major after-effects (Padhi et al., 2018; Padhi et al., 2020; Behera et al., 2020a, 2020b). Nano-delivery systems have advantages over conventional delivery systems, such as specificity, stability, directed release, high payload, enhanced solubility, and bioavailability (Davis et al., 2008; Hassan et al., 2021; Mohapatra et al., 2020; Pradhan et al., 2023a; Behera et al., 2023a; Pradhan et al., 2024).

Anthracyclines are drugs of bacterial origin. This class of drug is extracted from the *Streptomyces* bacterium. Anthracyclines are used to treat different types of cancers such as bladder, lungs, stomach, thyroid, ovary, uterus, leukaemias, lymphomas, osteogenic sarcoma, Ewing's sarcoma, soft tissue sarcoma, and neuroblastoma. Chemotherapy is the most effective treatment, which delivers anticancer drugs systemically to quench the uncontrolled proliferation of cancerous cells (Behera and Padhi, 2022). Anthracyclines and their derivatives are potent chemotherapeutic drugs. Anthracyclines function by inhibiting DNA replication and transcription in cancer cells. By blocking topoisomerase II, which inhibits DNA repair and reduces protein synthesis by binding to nuclear DNA, this molecule can enhance the formation of reactive oxygen species inside a cell and impede DNA repair (Chen et al., 2013). The cell cycle is blocked during the S phase of mitosis due to the inhibition of topoisomerase II and intercalation in DNA. Thus, in rapidly proliferating cancer cells, inhibition of topoisomerase II is a prime functional target for its antitumour action (Lyu et al., 2006). The first discovered anthracycline compound is 'daunorubicin' with a trade name 'daunomycin', isolated from *Streptomyces peucetius*, a mutant strain of the actinobacteria (discovered by Arcamone and Di Marco). In 1969, the anthracycline derivative, 'doxorubicin', was discovered. The World Health Organization has frequently identified doxorubicin and daunomycin as the two anthracyclines as the most potent and broadly effective chemotherapy drugs (Cassinelli, 2016). Epirubicin, idarubicin, aclarubicin, pirarubicin, mitoxantrone, and valrubicin are other clinically significant anthracycline derivatives used to treat various malignancies. The chemical structures of all these derivatives are shown in Figure 17.1 (Shaul et al., 2013).

Although anthracycline derivatives are effective in chemotherapy, there are still some substantial drawbacks to their anticancer therapies. Multi-drug resistance is one of such limitations. Anthracyclines do not cross the blood-brain barrier and also act as one of the substrates for multi-drug resistance protein-1 (P-glycoprotein 1). So, this group of cytotoxic drugs did not emerge as suitable for treating malignant gliomas (Eramo et al., 2006). They show decreased activity or expression of type II topoisomerase, DNA damage by enzymes, overexpressed superoxide dismutase enzyme, mutation of p53, and cell defences emerge against oxidative stress. They interfere with the apoptotic signalling pathways, which prevent cell death. An altered membrane carrier system is the principal mechanism and primary cause of anthracycline resistance. Multiple drug resistance developments appear to be the main limitation of their clinical applications (Ma and Mumper, 2013). Another major limitation of the clinical use of anthracyclines is severe cardiotoxicity (Cardinale et al., 2020). Anthracycline-induced myocardial toxicity event initiates from cell injury and is followed by left ventricular ejection fraction. This event, if not detected and rectified early, will progressively lead to the chronic condition of congestive heart failure (Cardinale et al., 2020).

Nano-drug delivery methods are advanced technology to overcome severe cardiotoxicity and multi-drug resistance. By increasing permeability through biological membranes, intervening molecular interactions, and detecting molecular changes in solid tumours, nano-delivery systems are promising carriers to enhance the anticancer

DOI: 10.1201/9781003611516-22

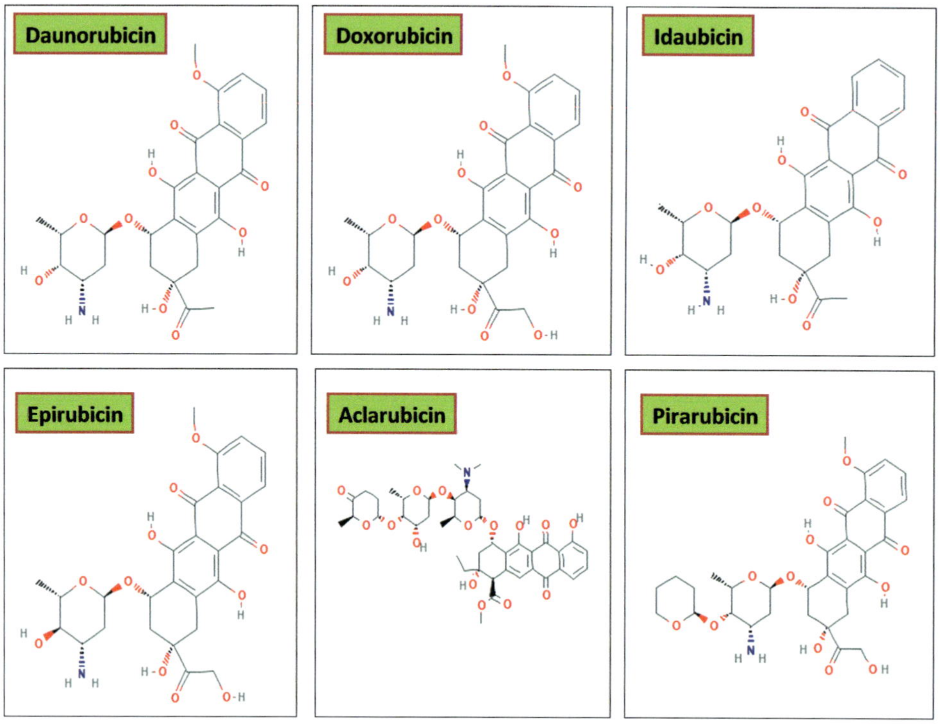

FIGURE 17.1 Chemical structures of various derivatives of anthracyclines.

activity of anthracyclines. This eventually minimizes systemic exposure and non-specific toxicity. Nanotechnology in cancer adds advantages like improvement in the diagnosis and treatment of the said disease (Padhi and Behera, 2022; Behera et al., 2023b). Research studies are conducted to evaluate potential nanotechnological methods for cancer therapeutics that will minimize the limitations of conventional ones. Nanoparticles as carriers possess properties like large surface area, optical activity, magnetic activity, and biological role compared to macroparticles. For cancer treatment, nano-drug delivery systems include polymeric micelles, liposomes, dendrimers, nanocapsules, nanospheres, and nanotubes (Patnaik et al., 2021). Doxil (liposomal doxorubicin) and Abraxane (albumin-bound paclitaxel) are the two commercialized nano-drug formulations that are currently utilized to treat cancer (Ventola, 2017).

Anthracyclines and their derivatives are a very efficient chemotherapeutic class of drugs for treating various cancers. Nano-delivery systems of anthracyclines and their analogues provide a method to overcome multi-drug resistance in cancer treatment. To date, many nano-based systems containing anthracyclins have been developed and reported to be efficient in overcoming multi-drug resistance in both experimental studies (*in-vitro* and *in-vivo*), and some even entered clinical trials such as *N*-(2-hydroxypropyl) methacrylamide conjugate doxorubicin) and PEGylated liposomal doxorubicin (Eramo et al., 2006). The targeted therapeutic approaches include agents delivered at sufficient concentration and prolonged retention at the targeted site. Nanomedicine has generated NPs with a safer and more effective therapeutic potential to achieve this property. These NPs are prone to aggregation and protein opsonization (binding to nanoparticle surface as a tag for immune system recognition) when circulating in the bloodstream. Opsonized nanoparticles are cleared by the body's clearance system, which involves phagocytosis or filtration in the liver, spleen, and kidneys. One of the major issues with this clearance mechanism is the rapid and non-specific removal of these compounds from the bloodstream, limiting their bioavailability at the target site. Masking of nanoparticle surface with polymers like polyethylene glycol, carbohydrates, acetyl groups, or protein moieties (arginine-glycine-aspartate peptide, albumin) can increase their retention time in the body (Shreffler et al., 2019). However, these surface modifications are prone to alter their recognition ability for targeted delivery. The illustrations of various anthracycline-based nanoparticles used to overcome multiple drug resistance and other limitations are as follows.

17.1.1 Liposomes

Doxorubicin, as a liposomal formulation, exhibited higher therapeutic efficacy and was able to overcome multi-drug resistance. The study on empty liposomes demonstrated the affinity of binding to P-glycoproteins, which facilitates the uptake of drugs resistant to cancer cells by altering the cells' plasma membrane properties (Hu et al., 1995). To combat multi-drug resistance, liposomal doxorubicin's multifaceted mechanism employs a variety of mechanisms. In both *in-vitro* and *in-vivo*, the liposomal doxorubicin demonstrated improved chemotherapeutic efficiency against resistant cells (Chen et al., 2010).

17.1.2 Solid-Lipid Nanoparticles

Doxorubicin-encapsulated solid-lipid nanoparticles have been designed by many scientists in their research for their ample benefits in chemotherapy. The doxorubicin-encapsulated solid-lipid nanoparticles enhanced the cellular uptake by up to 20-fold and increased the apoptotic cell death compared to free doxorubicin. Subedi et al. developed solid-lipid nanoparticles of doxorubicin with glyceryl caprate (capmul MCM C10) as the lipid core, curdlan as shell material, and polyethylene glycol 660 as the surfactant (Subedi et al., 2009). This core-shell encapsulated doxorubicin showed more potential to overcome multi-drug resistance. Anionic ion pairing agents like sodium taurodeoxycholate and sodium tetradecyl sulfate were applied to enhance the drug entrapment of doxorubicin. Thus, it exorbitantly improved the antitumour efficiency compared to the free doxorubicin molecule. Dong et al. (2009) studied the mechanisms of the doxorubicin solid-lipid nanoparticles in overcoming multi-drug resistance, and it was concluded that the multi-drug resistance reversal of solid-lipid nanoparticles may be due to the P-gp inhibition by Brij 78, tocofersolan, and ATP depletion. However, solid-lipid nanoparticles possess some limitations, which can be overcome by incorporating liquid lipids into solid-lipid nanoparticles to increase the drug loading capacity, burst drug release behaviour, and potentially cause drug explosion upon storage. Thus, nanostructured lipid carriers enhance the imperfections of solid-lipid nanoparticles and achieve more efficient drug delivery (Müller et al., 2000).

17.1.3 Gold Nanoparticles

These are most effective among the drug carriers as they are biocompatible, smaller, more permeable, and stable. Anticancer drugs can be associated with the nanoparticles by ionic/covalent bonding or physical adsorption (Pradhan et al., 2022, 2023b; Sa et al., 2023). Zhang et al. conjugated doxorubicin with gold nanoparticles via an amide bond to form doxorubicin–gold nanoparticles with a particle size of 2.7 nm. It was suggested that the ability of nanoparticles to overcome multi-drug resistance is related to the cell membrane properties and not the nucleus, as it does not interact with DNA molecules. These particles were non-toxic, about 20-fold greater cytotoxic, and faster in action than free doxorubicin, respectively (Zhang et al., 2010). A pH-responsive polymeric doxorubicin (polyethylene glycol) was conjugated to gold nanoparticles, which showed significantly enhanced intracellular drug uptake and dramatically increased cytotoxicity compared to free doxorubicin in resistant cells (Wang et al., 2011).

17.1.4 Polymeric Nanoparticles

The biodegradable poly-isobutyl-cyano-acrylate polymer was employed to entrap doxorubicin to form nanoparticles, demonstrating the complete reversal of drug resistance *in vitro*. The polymeric nanoparticles conjugated to doxorubicin entered the cells through the endocytosis process, followed by the interaction with DNA by releasing doxorubicin close to the nuclear membrane. The overexpressed receptor-specific polymeric doxorubicin led to more permeability of doxorubicin inside the membrane (Alavi and Varma, 2020). Soma et al. (2000) have prepared poly(alkyl-cyanoacrylate) nanoparticles encapsulating doxorubicin and cyclosporine-A, which showed inhibition in cell growth as compared to the other formulation combinations of both drugs in solution or particle form or with drug alone in resistant cells. Shieh et al. (2011) developed a more complex nanoparticle encapsulating the chemotherapeutic agent, doxorubicin, and a photosensitizer (4-armed porphyrin-polylactide) into porphyrin polylactide particles with tocopherol polyethylene glycol succinate published another report. The investigation concluded that the synergistic effect had increased doxorubicin delivery in resistant cells, effectively overcoming multi-drug resistance.

17.2 MECHANISM OF ACTION OF ANTHRACYCLINE DERIVATIVES AS AN ANTICANCER AGENT

Anthracyclines are well known for their cytostatic and cytotoxic activity, which involves free radical formation, lipid peroxidation, and membrane interaction. However, its antitumour activity in solid and leukaemia cancers is attributed to its interaction with DNA or DNA-topoisomerase II complex that alters DNA replication and transcription (Alavi and Varma, 2020). The pharmacological role of anthracyclines and their analogues as chemotherapeutic drugs is mainly as DNA topoisomerase inhibitors (Bayles et al., 2023).

Anthracyclines and their analogues, such as doxorubicin, have topoisomerase II-related effects at cellular levels. DNA damage demonstrated by the dose-response curves of doxorubicin and its analogues indicates their dosage-dependent levels of DNA damage up to threshold concentrations, as at increased concentrations, DNA breakage levels dropped to control levels (Marinello et al., 2018). Structural modifications to anticancer compounds that enhance their efficacy in cancer treatment mainly focus on their controlled site selection in cancer cells (Aung et al., 2017). The main structural characteristic of anthracyclines is an anthraquinone ring, which intercalates within base pairs of the DNA perpendicular to the long axis of the double helix (Figure 17.2). The stability of this complex depends on daunosamine sugar alignment in the minor groove through hydrogen bonding to one of the rings as an anchor arrangement. A single positive charge on daunorubicin is responsible for binding (Kizek et al., 2012).

For cancer cells, the maintenance of telomeres is an essential factor for their growth. The sensitivity of cancer cells to chemotherapy is greatly enhanced when the function of telomeres is affected by the inhibition of telomerase

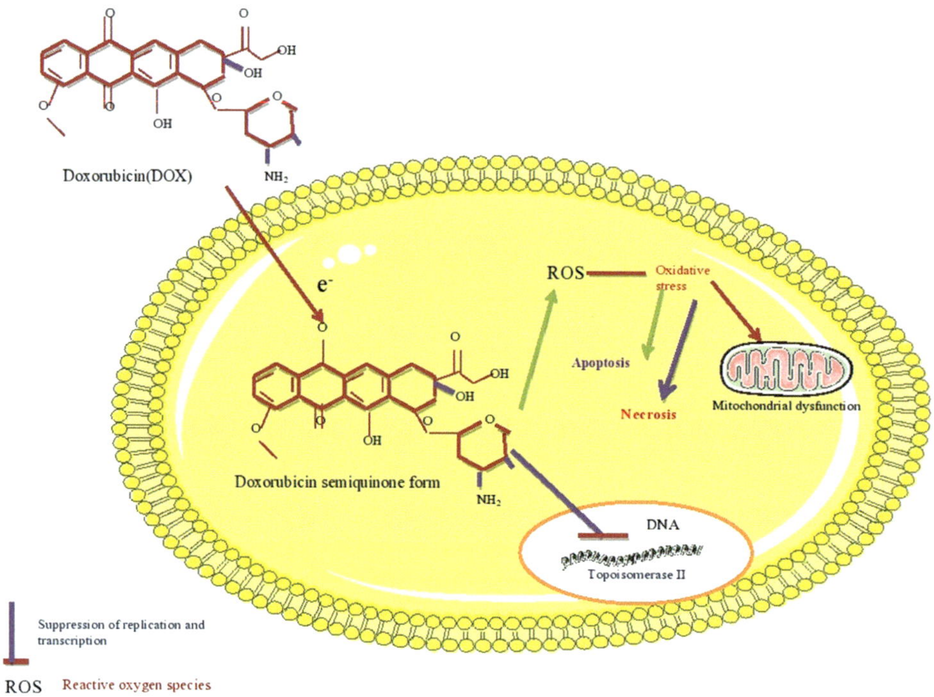

FIGURE 17.2 The mechanism of doxorubicin involves the reversible conversion from anthraquinone to a semiquinone form inside the cancer cell, which acts by intercalating to DNA with disruption of topoisomerase II that breaks DNA and also the synthesis of reactive oxygen species leads to the generation of other processes inside the cell that show cardiotoxicity.

enzymes or telomeric proteins. "PIN2 interacting protein 1" is a microtubule-binding protein whose primary function is to mediate the association of telomerase with telomeres in cancer cells. The over-expression of PIN2 interacting protein 1 is related to decreased telomerase activity and alteration to the telomerase/telomere recruitment pathway; for example, PIN2 interacting protein 1 silencing leads to telomere shortening, suppression of tumour growth, and making telomerase-positive cancer cells more sensitive to the cytotoxic effects of DNA damaging agents (Yu et al., 2020). Chemotherapeutic anthracyclines function by causing acute telomere dysfunction by inducing downregulation of PIN2 interacting protein 1, which contributes to the killing of cancer cells (Zhang et al., 2012). Anthracyclines mediate cancer cell death mainly by acting on type II topoisomerase enzymes. Another gene, TP53, is a tumour suppressor that senses DNA damage responses due to the interaction of doxorubicin in the apoptosis pathway (Borišev et al., 2018). Doxorubicin interacts with lymphoblastoid cell lines through the p53-mediated apoptotic pathways (Shaaban et al., 2014). Doxorubicin also influences negatively charged actin polymerization by inhibiting filament growth with drug concentration (stoichiometric and sub-stoichiometric).

Multi-drug resistance involves a common modification, i.e. amplification of the P-glycoprotein, which affects chemotherapy. Using an *in-vitro* system, it has been seen that the inhibitor of the proteasome, i.e. MG-132 (carbobenzoxy-leu-leu-leucinal), leads to intracellular accumulation of anthracyclines, inhibiting P-glycoprotein function. At additional levels, P-glycoprotein and topoisomerase II proteasome-dependent degradation in the drug-resistance state is blocked by the presence of proteasome inhibitors (Fekete et al., 2005). The formation of anthracycline antibiotics such as doxorubicin or daunorubicin secondary alcohol metabolites (doxorubicinol and daunorubicinol) is documented to be more cardiotoxic than their parent compounds inside the cell, which are catalysed by two cytosolic enzymes, i.e. carbonyl reductase and aldoketo reductase. But in resistant cancer cells, the carbonyl reductase and aldoketo reductase inhibitors have cardioprotective and radio-sensitizing properties (Piska et al., 2017). Secondary alcohol metabolites formed, which have been implicated in anthracycline-induced cardiotoxicity, involve a two-electron hydroxylation reduction of parent anthracycline at the C-13 carbonyl group.

This Phase I biotransformation of anthracycline is mediated by a diverse family of nicotinamide adenine dinucleotide phosphate-dependent carbonyl and aldo-keto reductase enzymes in cytosol together known as carbonyl-reducing enzymes that drive the formation of doxorubicinol, epirubicinol, and daunorubicinol (Piska et al., 2017). The latest anthracycline analogues, such as WP 631 or WP 744, are reported to be potential in overcoming multi-drug resistance. The binding affinity of WP 631 to DNA is more than that of the first-generation anthracyclines. Experimental studies showed that WP 631 is more efficient than its parent compounds, i.e. doxorubicin or daunorubicin, in different types of cancer cells such as Jurkat T lymphocytes, human non-small cell lung carcinoma cells (H1299), chronic myelogenous leukaemia cells (K562) and colon carcinoma cells (HCT116). First and second-generation anthracyclines and their analogues show cytotoxic effects in solid tumours and leukaemic cancer cells. The process of cancer cell death is mainly contributed by apoptosis. Doxorubicin and WP 631 showed increased annexin – V - fluorescein isothiocyanate binding over time (24–48 hours) in ovarian cancer cell lines (OV-90, SKOV-3) with apoptotic cells after 48 hours of incubation (31.8% for WP 631 and 30.6% for doxorubicin), binding of annexin V results in increased membrane permeability (Gajek et al., 2014).

17.3 NANOENABLED TARGETING STRATEGIES FOR THE ANTHRACYCLINE DERIVATIVES

Nanotechnology has made a great revolution in the targeted delivery of anticancer drugs. The nanoparticles can be designed through various modifications, such as changing their size, shape, chemical and physical properties, and so forth, to programme them to target the desired cells. They can target the neoplastic cells through active or passive targeting (Behera and Padhi, 2020). Combining therapeutic agents entrapped within nanoparticles is a promising approach that exhibits unique physicochemical and biological properties suitable for targeting and transporting these molecules to their target locations in the body (Jahan et al., 2017; Verma et al., 2017). This targeted strategy increases their bioavailability in cells and tissues; therefore, low doses can be used for maximum therapeutic efficacy, which improves their therapeutic index. To overcome limitations like multiple drug resistance and cardiotoxicity, nano-delivery systems have been developed and evaluated in vitro and *in vivo* for the proficiency of nanoenabled drug delivery systems.

17.3.1 Passive Targeting

Nanomaterials combined with anticancer drugs have overcome the physiological barriers to their delivery deeply at the tumour site (Padhi et al., 2022a; Padhi and Behera, 2021). The leaky vasculature in the cancer cells allows the entry of nano-delivery systems to cancer cells, and the nanoformulations exhibit anticancer activity by enhanced retention and permeation effect (Figure 17.3) (Padhi and Behera, 2020; Maeda et al., 2000, 2013). Different nanoformulations showing anticancer activity by passive targeting are discussed in the following section.

Kanwal et al. (2019) designed a polymeric nanoparticle loaded with quaternary ammonium palmitoyl glycol chitosan. The polymeric nanoparticle was evaluated for its release pattern and the uptake by cells and organs like the heart, liver, and cancer cells. The polymeric nanoparticle entrapping doxorubicin exhibited a sustained release for 48 hours, and fluorescence microscopy showcased an enhanced cellular uptake of the nanoparticles in human rhabdomyosarcoma cells compared to free doxorubicin. The IC_{50} value for in vitro cytotoxicity of doxorubicin polymeric nanoparticles was found to be 1.7 ± 0.404 µmol as compared to free doxorubicin (3.0 ± 0.968 µmol). Optical

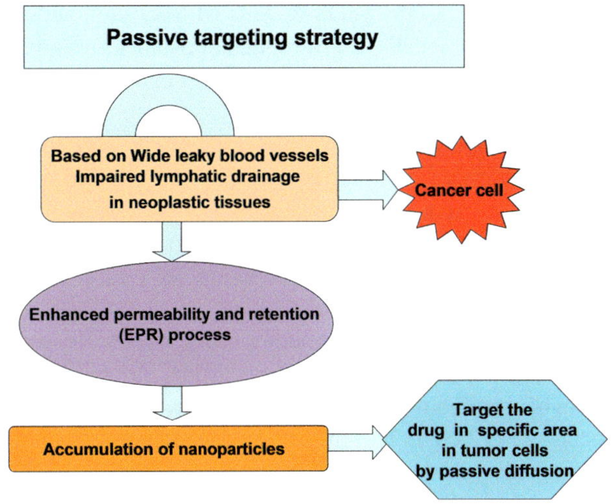

FIGURE 17.3 Passive targeting strategies for targeted delivery of anticancer drugs [Leaky vasculature of cancer cells allows the passage, distribution, retention, and bioavailability of nanoformulations containing anticancer drugs and exhibits enhanced permeation and retention effect].

imaging revealed lower accumulation of the nanoformulation in the heart and liver and higher accumulation in skin tumour xenograft cells.

Liang et al. (2019) designed poly(ethylene glycol)-β-poly(L-lactide-co-glycolide) based polymeric nanoparticles entrapping idarubicin and evaluated its anticancer potential against acute myeloid leukaemia. The in vivo study of idarubicin polymeric nanoparticles enhanced the survival of MLL-AF9-induced murine leukaemia mice. In contrast, the in vitro human leukaemia cell lines study on HL60 and U937 exhibited a two to fourfold increase in the inhibition of cancer cell growth.

Yang et al. (2018) designed a polymeric nanoparticle of doxorubicin by conjugating it with dichloroacetate, which self-assembled with a small amount of 1,2-distearoyl-sn-glycero-3-phosphoethanolamine-poly(ethylene glycol). The polymeric nanoparticles were examined against murine hepatic carcinoma cells (H22) and murine melanoma cells (B16F10), and the nanoformulation exhibited 15 times higher efficacy than free doxorubicin. Sun et al. (2013) designed eight types of bioreducible polyethylene glycol-g-poly-amide amine copolymers graft copolymeric micelles entrapping doxorubicin and evaluated their anticancer efficacy against breast cancer cells (4T1 tumour-bearing mice). Encapsulation of doxorubicin in micelles improved their bioavailability and exhibited better anticancer effects in breast cancer cells. The micelles prevented organ damage and metastasis.

Ding et al. (2019) designed silver-capped epirubicin nanoparticles and evaluated the anticancer activity against hepatocellular cancer cell lines (Hep G2). The silver nanoparticles of epirubicin exhibited antitumour activity with an IC_{50} value of 1.92 µg/mL.

Haghiralsadat et al. (2020) evaluated the efficacy and potency of liposomal doxorubicin in an osteosarcoma cell line (MG-63, U2-OS, SaOS-2, SaOS-LM7) and human primary osteoblast cells. The designed liposomes entrapped 84% doxorubicin and exhibited enhanced cell death compared to free doxorubicin.

Gold nanoparticles entrapping doxorubicin exhibited higher latency to tumour growth in chemical-induced fibrosarcoma. At a dose of 2.5 kg/mg, doxorubicin–gold nanoparticles suppressed the tumour growth by 81% compared to free doxorubicin, showing 48% suppression (Dhamecha et al., 2016).

Kong et al. (2020) designed a small molecule cell-penetrating peptide (RPV) conjugated dual drug delivery system of epirubicin and dioscin against non-small cell lung cancer (A549 cells). The liposomal formulation inhibited vasculogenic mimicry channel formation, tumour angiogenesis, migration, and invasion. The conjugation enhanced the circulation time, increased tumour cell uptake, and inhibited tumour angiogenesis.

Chen et al. (2016) reported the anticancer activity of surface-modified poly(D,L-lactide-co-glycolide) nanoparticles with biotinylated chitosan entrapping epirubicin. The modified nanoparticles showed a controlled release of epirubicin for 48 hours with increased cytotoxicity against human breast cancer (MCF-7) cells. The modified nanoparticles accumulated at the tumour site and exhibited excellent antitumour activity.

17.3.2 Active Targeting

Active targeting can overcome the limitations of passive targeting like off-site toxicity and fast clearance. The nanoformulations are designed with ligands specific to the overexpressed receptors on the cancer cells (Padhi et al., 2022b). These nanoformulations are stimulus-sensitive in response to cancer cells' temperature, pH, enzymes/proteins, and microenvironment conditions. The efficacy of anticancer drugs by active targeting is more than passive targeting as the cellular uptake by receptor-mediated endocytosis and delivery of anticancer drugs in the cancer cells increases the bioavailability (Behera and Padhi, 2020).

Ansari et al. (2020) designed polyethylene glycol and folic acid-conjugated hydroxyapatite nanoparticles entrapping epirubicin. The folic acid-conjugated polyethylene glycol nanoparticles exhibited a 2.5-fold enhanced drug release in acidic pH, and the efficacy for tumour growth inhibition was more for folic acid-conjugated nanoparticles than for free epirubicin.

Liao et al. (2018) prepared intracellular pH-sensitive nanoparticles by conjugating hyaluronic acid with doxorubicin through a hydrazone linkage. The nanoparticles showed a pH-responsive burst release in the first 6 hours at pH 5.0 in human cervical cancer cells (HeLa cells). The drug got internalized through the high affinity of hyaluronic acid towards CD44 receptors in the cancer cells; hence, the bioavailability of doxorubicin inside the cells increased.

Jiang et al. (2018) designed peptide-conjugated PEGylated liposomes of doxorubicin targeting fibronectin overexpressed on tumour vessels and tumour stroma walls, mostly in metastatic sites. The peptide Cys-Arg-Glu-Lys-Ala (CREKA) modified liposomal doxorubicin was evaluated against metastatic breast tumours *in vivo* and *in vitro*. The nanoformulation exhibited reduced metastasis in the lungs.

Zhang et al. (2022) designed a urokinase plasminogen activator receptor-targeting U11 peptide-conjugated pH-sensitive, dual drugs co-encapsulated nanoparticles containing doxorubicin and curcumin against doxorubicin-resistant lung cancer cells (A549/ADR cells). The dual drug entrapped peptide-conjugated nanoparticles exhibited a pH-responsive drug delivery in cancer cells at pH = 5.0 with a synergistic effect.

Kolahkaj et al. (2019) conjugated a monoclonal antibody to polymeric nanoparticles of epirubicin. The poly(lactic-co-glycolic acid) nanoparticles of epirubicin were conjugated with therapeutic monoclonal antibody trastuzumab and evaluated against the human epidermal growth factor receptor 2 (HER2) positive and negative breast cancer cells. Enhanced cytotoxicity and cellular uptake of antibody-conjugated polymeric nanoparticles of epirubicin were observed in HER2-positive cancer cells than negative cells.

Hyun et al. (2021) prepared an engineered cyclodextrin-based polyethylene glycol-folic acid-conjugated doxorubicin and reported the pH-sensitive anticancer effect in breast cancer cells (MCF7 cells). The drug was delivered through receptor-mediated endocytosis at pH = 5.5 and decreased the tumour volume within 14 days without systemic toxicity and cardiotoxicity.

Di-wen et al. (2016) designed peptide-conjugated polymeric nanoparticles of epirubicin targeting liver cancer cells (HepG2). The polymeric nanoparticles were conjugated with a linear peptide LFC131, which has high specificity to the overexpressed CXCR4 receptors in liver cancer cells. The peptide-conjugated poly(lactic-co-glycolic acid) nanoparticles of epirubicin showed a threefold enhanced uptake by cancer cells at a pH of 5.5. The pH-responsive sustained drug delivery of epirubicin inhibited the growth of liver tumours in situ and reduced systemic toxicity.

Soe et al. (2019) designed transferrin-conjugated polymeric nanoparticles of doxorubicin against doxorubicin-resistant breast tumours. The polymeric nanoparticles containing poloxamer 407 (F127) and 123 (P123) were conjugated with doxorubicin and combined with transferrin, then evaluated against breast cancer (OVCAR-3, MDA-MB-231, and MDA-MB-231) cells. The transferrin-targeted nanoparticles were efficacious against doxorubicin-sensitive and resistant breast tumours.

Jalalian et al. (2018) designed an aptamer (5TR1) coated selenium nanoparticles entrapping polymeric epirubicin loaded with DNA aptamer NAS-24. They evaluated the anticancer efficacy against hepatocellular cancer cells (HepG2), breast cancer cells (MCF-7), and murine colon cancer cells (C26). The complex reduced the off-target toxicity and cell viability and reduced tumour growth compared to free epirubicin.

Xia et al. (2019) prepared galactose-modified selenium nanoparticles of doxorubicin and reported the antitumour activity of the nanoparticles in hepatocellular carcinoma. The cellular uptake via clathrin-mediated endocytosis induced apoptosis via caspase signalling and proteins of the BCL 2 family. The galactose-modified selenium nanoparticles showed no organ toxicity to the heart, liver, spleen, and kidney and exhibited more efficacy than passive targeting selenium nanoparticles of doxorubicin. The same scientific group reported the folic acid-modified selenium nanoparticles of doxorubicin against human cervical cancer cells (HeLa cells) against lung cancer cells (A549 cells). The folic acid overexpressed receptors on cervical cancer cells showed increased uptake of modified selenium nanoparticles of doxorubicin by clathrin-mediated endocytosis and showed significant antitumour activity (Xia et al., 2019).

Karimi et al. (2018) designed a human serum albumin nanocapsule containing drug-modified cobalt ferrite ($CoFe_2O_4$) magnetic nanoparticle entrapping epirubicin and evaluated its anticancer efficacy against breast cancer cells (MCF-7 cells). The slow and controlled release of the magnetic nanoparticles exhibited the biocompatibility of the drug delivery system.

Verma et al. (2016) designed temperature and pH-sensitive polymeric nanogels of doxorubicin. The nanogel was based on poly(*N*-isopropyl acrylamide-co-2-acrylamide-2-methyl-1-propanesulfonate-co-1-propene-2-3-dicarboxylate [poly-(NIPAAm-IA-AMPS)] cross-linked by ethylene glycol dimethacrylate. These optimized nanogels were further coupled to *N,O*-carboxymethyl chitosan by coupling reaction. The nanogel was internalized easily into breast cancer cells (MCF-7 and MDA-MB231) compared to human breast epithelial cells (MCF10A cells). The nanogels caused intracellular DNA damage and arrested the cell cycle, leading to mitochondrial-mediated apoptosis in MCF-7 cells.

17.4 REPORTS ON CLINICAL TRIALS AND PATENTS OF THE ANTHRACYCLINE DERIVATIVES FOR ANTICANCER ACTIVITY

Many clinical and scientific studies have reported on promising approaches of passive/active targeting of cancer cells and their successful adaptation into clinical therapeutics. The first nano-drug approved by the Food and Drug Administration is doxorubicin (PEGylated liposome entrapped doxorubicin) targeted against ovarian cancer and human immunodeficiency virus-related kaposi sarcoma tumour (Duggan and Keating, 2011). Encapsulated lipid-doxorubicin material was developed to boost the drug's pharmacokinetics with superior biodistribution and release (Shi et al., 2017; Navya et al., 2019). Another FDA-approved nanoformulation consisting of liposomal encapsulated CPX-351 Vyxeos™ (cytarabine–daunorubicin) combination has shown an overall survival of about 9.6 months compared to previous survival of 6.0 months for the drug (free form) in patients with high-risk acute myeloid leukaemia (Lancet et al., 2016). Many such formulations have been approved (Shi et al., 2017), creating new approaches toward cancer therapeutics.

The development of more sophisticated nanomaterials (micelles, protein-based nanoparticles, inorganic and metallic particles) utilized in clinical trials is trending in the direction of using biomaterials that the Food and Drug Administration has approved in liposomal, polymeric, and nanocrystal formulations. Food and Drug Administration approves the route of application of nanomedicines in the drug form, devices, or biologicals. Nanoparticles' efficiency, toxicity, and physicochemical properties undergo investigational new drug application for a clinical trial. Investigational new drug is approved for clinical trial as per the safety and efficacy of new nanomedicines. The number of investigational new drug-approved nanomedicines granted to undertake clinical trials has increased steadily since 2007 (Fatemeh et al., 2022). Table 17.1 lists the significant anthracycline-derived nanoformulations that have received clinical approval and are expected to become more popular chemotherapeutics in the upcoming years. Though nanoformulations have reduced the limitations of traditional drug delivery systems, they still have challenges and limitations that will open a path for the future area of research (Farjadian et al., 2019). Figure 17.4 illustrates the different structurally modified anthracycline analogues developed for enhanced anticancer activity.

17.5 FUTURE PROSPECTS

Anthracyclines are the most effective anticancer drugs. They play an accredited role in the treatment of many neoplastic diseases. However, some limitations add to its limited usage as a chemotherapeutic drug. These compounds exhibit multiple cellular effects, including DNA damage mediated by the inhibition of topoisomerase II, and the generation of reactive oxygen species involved in regulating many biological functions and causing multi-drug resistance. The risk of cardiotoxicity is considered the most serious adverse effect. This outlines the potential of nanotechnology to avoid some of the problems connected with conventional pharmaceutical formulations. Considerable efforts have been made to modify various nanomaterials for cancer treatment with high specificity, sensitivity, and efficacy. Engineered nanoparticles complexed with ligands efficiently target tumour cells and deliver encapsulated payloads successfully. In addition, tailor-made nanomaterials constructed for improved drug loading, enhanced half-life, restricted release, and selective distribution by changing their composition, size, morphology, and surface chemistry can add to their therapeutic options. Polymeric nanoparticles, metal nanoparticles, carbon-based nanomaterials, liposomes, and dendrimers have all been developed as smart drug delivery systems for cancer treatment, and they exhibit improved pharmacokinetic and pharmacodynamic profiles compared to conventional formulations.

TABLE 17.1

Chemotherapeutic Anthracycline-Based Nanoformulations Approved by the Food and Drug Administration and European Medicines Agency

Brand Name	Type of Cancer	Year of FDA Approval	Composition
Doxil (Caelyx)	Ovarian cancer, breast cancer, human immunodeficiency virus-associated Kaposi sarcoma and multiple myeloma	November 1995	PEGylated liposomal doxorubicin
DaunoXome (Galen)	Human immunodeficiency virus-related Kaposi sarcoma	April 1996	Non-PEGylated liposomal daunorubicin
Myocet (Teva)	Metastatic breast cancer	European Medicines Agency approval in 2000 in Europe and Canada	Non-PEGylated liposomal doxorubicin
Lipo-Dox	Ovarian cancer and breast cancer	Food and Drug Administration approval in 2012	Liposomal doxorubicin
VYXEOS (CPX-351)	Acute myeloid leukaemia	Food and Drug Administration approval in 2017 European Medicines Agency approval in 2018	Combination of daunorubicin and cytarabine liposome in a 1:5 molar ratio

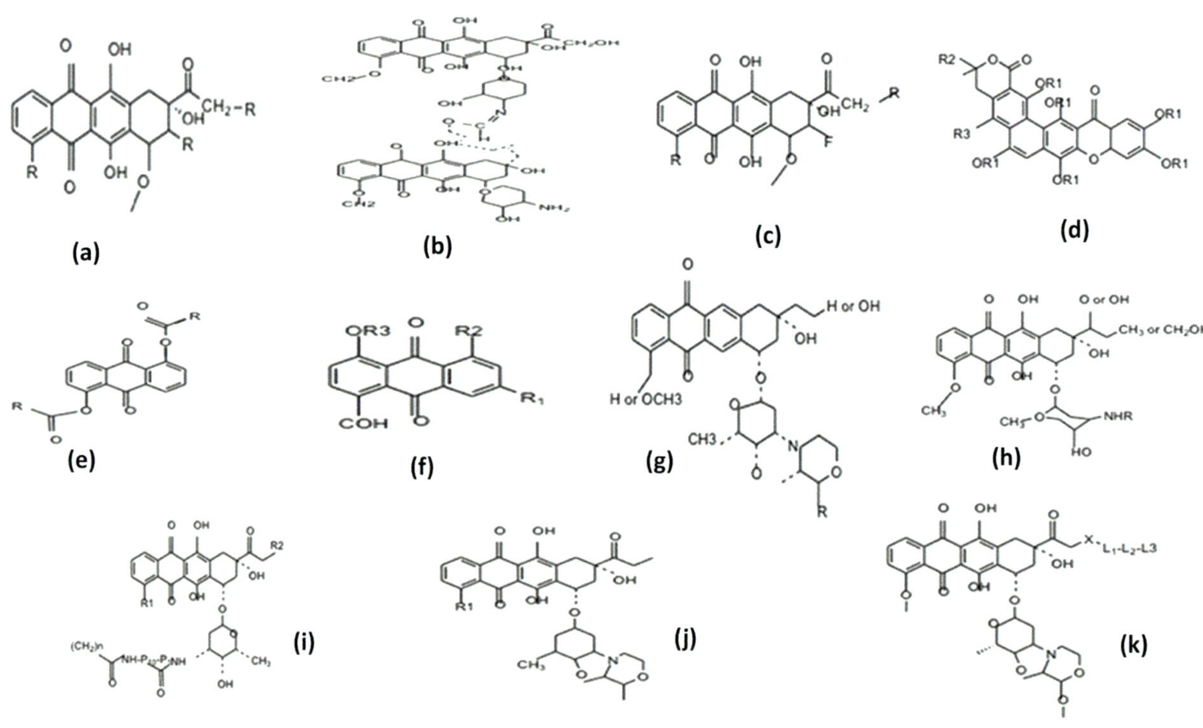

FIGURE 17.4 Different structurally modified anthracycline analogues developed for enhanced anticancer activity.

For future perspectives, efforts are being made to generate new anthracycline analogues as nanoformulations. For this, the chemical modification of parent anthracyclines that would favour the pharmacokinetic and pharmacodynamic properties is a bigger challenge. New, already present derivatives proved to be better-tolerated drugs than parent anthracyclines, such as epirubicin (4-epidoxorubicin), idarubicin, and 4-deoxy-4-idodoxorubicin. With nanotechnology, liposomal anthracycline analogues are being formulated, leading to target and tissue-specific clinical applications with reduced side effects. A nanotechnology-based combinatorial drug delivery system is becoming a popular therapeutic approach for generating synergistic anticancer effects with reduced drug-related toxicity and multi-drug resistance. Diverse drug carriers for the design and characterization of nanoparticles are being developed in the field of cancer due to molecular heterogeneity and complexity. To effectively localize these medications at the region of interest, focus is placed more on non-targeted (passive) and targeted (active) therapeutic nanoparticles.

17.6 CONCLUSION

Anthracyclines are part of the class of chemotherapy agents that have been highly effective in treating a variety of cancers, including ovarian, breast, lymph, and hepatocellular carcinoma. Structural modifications of their parent compounds are developed to create new anthracycline analogues with improved characteristics including sub-cellular localization, cellular absorption, and target affinity for cancer cells. Due to their capacity to produce reactive oxygen species, anthracyclines have a major side effect: dose-dependent cardiotoxicity. Potential cardioprotective analogues with improved antitumour activities are being developed due to a better understanding of the processes involved in its mode of action and the production of its structurally changed forms. Nanoformulations like liposomes entrapping anthracyclines, dexrazoxane (bisdioxopiperazine chelate form with iron), and morpholinyl anthracyclines have overcome the cardiotoxicity and free radical formation to reduce the toxicity associated with this class of compounds. Nanoparticles as a drug delivery approach have proved to be the most potent approach in reducing multi-drug resistance by enhancing its cytotoxicity effect. A benefit of nano vehicles is that they can easily pass through leaky blood arteries for targeted binding with tumour-specific proteins because they are 100–1,000 times smaller than cancer cells. Surface charge, PEGylation, target, and stimulus responses are used in nanoformulation procedures to generate liposomal and polymeric nanoparticles that distribute anthracycline drugs, including daunorubicin, doxorubicin, epirubicin, and idarubicin effectively. A framework of physical, chemical, and functional descriptions of nanoparticles being investigated for cancer diagnostics and treatments makes up the current advancement in nanoparticle Ontology (NPO). Characteristics of nanoparticles, such as small size, improved solubility, surface functionalization ability, and multifunctional approach, have opened new research methods in cancer therapeutics. Nanotechnology, with promising benefits and diverse applications, deals with issues such as accurate target delivery, thermodynamic instability, and the high production cost that still needs to be dealt with for its worldwide use.

REFERENCES

Alavi M, Varma RS (2020). Overview of novel strategies for the delivery of anthracyclines to cancer cells by liposomal and polymeric nanoformulations. *International Journal of Biological Macromolecules*, 164, 2197–2203.

Ansari L, Derakhshi M, Bagheri E, Shahtahmassebi N, Malaekeh-Nikouei B (2020). Folate conjugation improved uptake and targeting of porous hydroxyapatite nanoparticles containing epirubicin to cancer cells pharmaceutical development and technology. *Pharmaceutical Development and Technology*, 25, 5.

Aung TN, Qu Z, Kortschak RD, Adelson DL (2017). Understanding the effectiveness of natural compound mixtures in cancer through their molecular mode of action. *International Journal of Molecular Sciences*, 18(3), 656.

Bayles CE, Hale DE, Konieczny A, Anderson VD, Richardson CR, Brown KV, et al. (2023). Upcycling the anthracyclines: new mechanisms of action, toxicology, and pharmacology. *Toxicology and Applied Pharmacology*, 459, 116362.

Behera A, Mittu B, Padhi S, Patra N, Singh J (2020b). Bimetallic nanoparticles: green synthesis, applications, and future perspectives. In: Abd-Elsalam KA (Ed.) *Multifunctional Hybrid Nanomaterials for Sustainable Agri-food and Ecosystems*. Elsevier, Amsterdam, pp. 639–682.

Behera A, Mittu B, Padhi S, Singh A (2020a). Antimicrobial efficacy of essential oil nanoemulsions. In: Dhull SB, Chawla P, Kaushik R (Eds.) *Nanotechnological Approaches in Food Microbiology*, CRC Press, Boca Raton, FL, pp. 293–309.

Behera A, Padhi S (2020). Passive and active targeting strategies for the delivery of the camptothecin anticancer drug: a review. *Environmental Chemistry Letters*, 18(5), 1557–1567.

Behera A, Padhi S (2022). pH-Sensitive polymeric nanoparticles for cancer treatment. In: Padhi S, Behera A, Lichtfouse E (Eds.) *Polymeric Nanoparticles for the Treatment of Solid Tumours*. Environmental Chemistry for a Sustainable World, vol 71. Springer, Cham, pp. 401–425.

Behera A, Padhi S, Nayak AK (2023b). Engineered liposomes as drug delivery and imaging agents. In: Ray S, Nayak AK (Eds.) *Design and Applications of Theranostic Nanomedicines*. Woodhead Publishing, Cambridge, MA, pp. 75–108.

Behera A, Sa N, Pradhan SP, Swain S, Sahu PK (2023a). Metal nanoparticles in Alzheimer's disease. *Journal of Alzheimer's Disease Reports*, 7(1), 791–810.

Borišev I, Mrđanovic J, Petrovic D, Seke M, Jović D, Srđenović B, et al. (2018). Nanoformulations of doxorubicin: how far have we come and where do we go from here? *Nanotechnology*, 29(33), 332002.

Bray F, Jemal A, Grey N, Ferlay J, Forman D (2012). Global cancer transitions according to the Human Development Index (2008–2030): a population-based study. *The Lancet Oncology*, 13(8), 790–801.

Cardinale D, Iacopo F, Cipolla CM (2020). Cardiotoxicity of anthracyclines. *Frontiers in Cardiovascular Medicine*, 7, 26.

Cassinelli, G. (2016). The roots of modern oncology: from discovery of new antitumor anthracyclines to their clinical use. *Tumori Journal*, 102(3), 226-235.

Chen H, Xie LQ, Qin J, Jia Y, Cai X, Nan WB, Yang W, Lv F, Zhang QQ (2016). Surface modification of PLGA nanoparticles with biotinylated chitosan for the sustained in vitro release and the enhanced cytotoxicity of epirubicin. *Colloids and Surfaces B: Biointerfaces*, 138, 1–9.

Chen SH, Chan NL, Hsieh TS (2013). New mechanistic and functional insights into DNA topoisomerases. *Annual Review of Biochemistry*, 82, 139–170.

Chen Y, Bathula SR, Li J, Huang L. (2010) Multifunctional nanoparticles delivering small interfering RNA and doxorubicin overcome drug resistance in cancer. *Journal of Biological Chemistry*, 285, 22639–22650.

Davis ME, Chen Z, Shin DM (2008). Nanoparticle therapeutics: an emerging treatment modality for cancer. *Nature Reviews Drug discovery*, 7(9), 771–782.

Dhamecha D, Jalalpure S, Jadhav K, Jagwani S, Chavan R (2016). Doxorubicin loaded gold nanoparticles: implication of passive targeting on anticancer efficacy. *Pharmacological Research*, 113, 547–556.

Ding J, Chen G, Chen G, Guo M (2019). One-pot synthesis of epirubicin-capped silver nanoparticles and their anticancer activity against Hep G2 cells. *Pharmaceutics*, 11(3), 123.

Di-Wen S, Pan G-Z, Hao L, Zhang J, Xue Q-Z, Wang P, Yuan Q-Z (2016). Improved antitumour activity of epirubicin-loaded CXCR4-targeted polymeric nanoparticles in liver cancers. *International Journal of Pharmaceutics*, 500(1–2), 54–61.

Dong X, Mattingly CA, Tseng MT, Cho MJ, Liu Y, Adams VR, Mumper RJ (2009). Doxorubicin and paclitaxel-loaded lipid-based nanoparticles overcome multi-drug resistance by inhibiting P-glycoprotein and depleting ATP. *Cancer Research*, 69, 3918–3926.

Duggan ST, Keating GM (2011). Pegylated liposomal doxorubicin: a review of its use in metastatic breast cancer, ovarian cancer, multiple myeloma and AIDS-related Kaposi's sarcoma. *Drugs*, 71, 2531–2558.

Eramo A, Ricci-Vitiani L, Zeuner A, Pallini R, Lotti F, Sette G, et al. (2006). Chemotherapy resistance of glioblastoma stem cells. *Cell Death & Differentiation*, 13(7), 1238–1241.

Farjadian F, Ghasemi A, Gohari O, Roointan A, Karimi M, Hamblin MR (2019). Nanopharmaceuticals and nanomedicines currently on the market: challenges and opportunities. *Nanomedicine*, 14(1), 93–126.

Fatemeh SM, Maryam T, Mehdi P (2022). Overcoming drug resistance in breast cancer using quercetin nanodrugs. *Journal of Advanced Biomedical Sciences*, 12(3), 292–301.

Fekete MR, McBride WH, Pajonk F (2005). Anthracyclines, proteasome activity and multi-drug-resistance. *BMC Cancer*, 5(1), 1–7.

Gajek A, Denel M, Bukowska B, Rogalska A, Marczak A (2014). Pro-apoptotic activity of new analog of anthracyclines–WP 631 in advanced ovarian cancer cell line. *Toxicology in Vitro*, 28(2), 273–281.

Haghiralsadat BF, Naderinezhad S, Amoabediny G, Montazeri F, Zandieh Doulabi B (2020). Evaluation of the effects of surface charge on cytotoxicity of liposomal Doxorubicin on bone cancer cell line (Osteosarcoma). *Daneshvar Medicine*, 25(6), 19–26.

Hassan N, Firdaus S, Padhi S, Ali A, Iqbal Z (2021). Investigating natural antibiofilm components: a new therapeutic perspective against candidal vulvovaginitis. *Medical Hypotheses*, 148, 110515.

Hu YP, Henrytoulme N, Robert J (1995) Failure of liposomal encapsulation of doxorubicin to circumvent multi-drug resistance in an in vitro model of rat glioblastoma cells. *European Journal of Cancer*, 31A, 389–394.

Hyun H, Park MH, Jo G, Lee BY, Choi JW, Chun HJ, et al. (2021). Injectable glycol chitosan hydrogel containing folic acid-functionalized cyclodextrin-paclitaxel complex for breast cancer therapy. *Nanomaterials*, 11(2), 317.

Jahan ST, Sadat, SMA, Walliser, M, Haddadi A (2017) Targeted therapeutic nanoparticles: an immense promise to fight against cancer. *Journal of Drug Delivery*, 2017, 9090325.

Jalalian SH, Ramezani M, Abnous K, Taghdisi SM (2018). Targeted co-delivery of epirubicin and nas-24 aptamer to cancer cells using selenium nanoparticles for enhancing tumour response in vitro and in vivo. *Cancer Letters*, 416, 87–93.

Jiang K, Song X, Yang L, Li L, Wan Z, Sun X, Gong T, Lin Q, Zhang Z (2018). Enhanced antitumour and anti-metastasis efficacy against aggressive breast cancer with a fibronectin-targeting liposomal doxorubicin. *Journal of Controlled Release*, 271, 21–30.

Kanwal U, Bukhari NI, Rana NF, Rehman M, Hussain K, Abbas N, Mehmood A, Raza A (2019). Doxorubicin-loaded quaternary ammonium palmitoyl glycol chitosan polymeric nanoformulation: uptake by cells and organs. *International Journal of Nanomedicine*, 14, 1–15.

Karimi F, Fallah Shojaei A, Tabatabaeian K, Karimi-Maleh H, Shakeri S (2018). HSA loaded with $CoFe_2O_4$/mnps as a high-efficiency carrier for epirubicin anticancer drug delivery. *IET Nanobiotechnology*, 12(3), 336–342.

Kizek R, Adam V, Hrabeta J, Eckschlager T, Smutny S, Burda JV, et al. (2012). Anthracyclines and ellipticines as DNA-damaging anticancer drugs: recent advances. *Pharmacology & Therapeutics*, 133(1), 26–39.

Kolahkaj FF, Derakhshandeh K, Khaleseh F, Azandaryani AH, Mansouri K, Khazaei M (2019). Active targeting carrier for breast cancer treatment: monoclonal antibody conjugated epirubicin loaded nanoparticle. *Journal of Drug Delivery Science and Technology*, 53, 1773–2247.

Kong L, Zhang SM, Chu JH, Liu XZ, Zhang L, He SY, et al. (2020). Tumor microenvironmental responsive liposomes simultaneously encapsulating biological and chemotherapeutic drugs for enhancing antitumor efficacy of NSCLC. *International Journal of Nanomedicine*, 15, 6451–6468.

Kundu A, Padhi S, Behera A, Hasnain MS, Nayak AK (2022). Tumour targeting strategies by chitosan-based nanocarriers. In: Hasnain MS, Beg S, Nayak AK (Eds.) *Chitosan in Biomedical Applications*. Academic Press, London, pp. 163–188.

Lancet JE, Uy GL, Cortes JE, Newell LF, Lin TL, Ritchie EK, et al. (2016). Final results of a phase III randomized trial of CPX-351 versus 7+3 in older patients with newly diagnosed high risk (secondary) AML. *Journal of Clinical Oncology*, 34(15), 7000.

Liang B, Li N, Zhang S, Qi A, Feng J, Jing W, Shi C, Ma Z, Gao S (2019). Idarubicin-loaded methoxy poly(ethylene glycol)-b-poly(L-Lactide-co-glycolide) nanoparticles for enhancing cellular uptake and promoting antileukemia activity. *International Journal of Nanomedicine*, 14, 543–556.

Liao JH, Zheng H, Fei ZM, Lu B, Zheng H, Li D, Xiong, Yi Y (2018). Tumour-targeting and pH-responsive nanoparticles from hyaluronic acid for the enhanced delivery of doxorubicin. *International Journal of Biological Macromolecules*, 113, 737–747.

Lyu YL, Lin CP, Azarova AM, Cai L, Wang JC, Liu LF (2006). Role of topoisomerase IIβ in the expression of developmentally regulated genes. *Molecular and Cellular Biology*, 26(21), 7929–7941.

Ma P, Mumper RJ (2013). Paclitaxel nano-delivery systems: a comprehensive review. *Journal of Nanomedicine & Nanotechnology*, 4(2), 1000164.

Maeda H, Nakamura H, Fang J (2013). The EPR effect for macromolecular drug delivery to solid tumours: improvement of tumour uptake, lowering of systemic toxicity, and distinct tumour imaging in vivo. *Advanced Drug Delivery Reviews*, 65(1), 71–79.

Maeda H, Wu J, Sawa T, Matsumura Y, Hori K (2000). Tumour vascular permeability and the EPR effect in macromolecular therapeutics: a review. *Journal of Controlled Release*, 65, 271–284.

Marinello J, Delcuratolo M, Capranico G (2018). Anthracyclines as topoisomerase II poisons: from early studies to new perspectives. *International Journal of Molecular Sciences*, 19(11), 3480.

Mohapatra S, Iqbal Z, Ahmad S, Kohli K, Farooq U, Padhi S, et al. (2020). Menopausal remediation and quality of life (QoL) improvement: insights and perspectives. *Endocrine, Metabolic & Immune Disorders-Drug Targets (Formerly Current Drug Targets-Immune, Endocrine & Metabolic Disorders)*, 20(10), 1624–1636.

Müller RH, Mäder K, Gohla S (2000). Solid lipid nanoparticles (SLN) for controlled drug delivery: a review of the state of the art. *European Journal of Pharmaceutics and Biopharmaceutics*, 50(1), 161–177.

Navya PN, Kaphle A, Srinivas SP, Bhargava SK, Rotello VM, Daima HK (2019). Current trends and challenges in cancer management and therapy using designer nanomaterials. *Nano Convergence*, 6(1), 1–30.

Padhi S, Behera A (2021). Advanced drug delivery systems in the treatment of ovarian cancer. In: Dua K, Mehta M, de Jesus Andreoli Pinto T, Pont LG, Williams KA, Rathbone MJ (Eds.) *Advanced Drug Delivery Systems in the Management of Cancer*. Academic Press, London, pp. 127–139.

Padhi S, Behera A (2022). Cellular internalization and toxicity of polymeric nanoparticles. In: Padhi S, Behera A, Lichtfouse E (Eds.) *Polymeric Nanoparticles for Treating Solid Tumours*. Environmental Chemistry for a Sustainable World, vol 71. Springer, Cham, pp. 473–488.

Padhi S, Behera A, Hasnain MS, Nayak AK (2022a). Chitosan-based drug delivery systems in cancer therapeutics. In: Hasnain MS, Beg S, Nayak AK (Eds.) *Chitosan in Drug Delivery*. Academic Press, London, pp. 159–193.

Padhi S, Behera A, Hasnain MS, Nayak AK (2022b). Uses of chitosan in drug delivery. In: Hasnain MS, Beg S, Nayak AK (Eds.) *Chitosan in Drug Delivery*. Academic Press, London, pp. 139–162.

Padhi S, Behera, A (2020) Nanotechnology based targeting strategies for the delivery of Camptothecin. In: Ankit S, Amulya PK, Eric L (Eds.) *Pharmaceutical Technology for Natural Products Delivery, Impact of Nanotechnology*. Springer Nature, Switzerland, pp. 243–272.

Padhi S, Kapoor R, Verma D, Panda AK, Iqbal Z (2018). Formulation and optimization of topotecan nanoparticles: in vitro characterization, cytotoxicity, cellular uptake and pharmacokinetic outcomes. *Journal of Photochemistry and Photobiology B: Biology*, 183, 222–232.

Padhi S, Mirza MA, Verma D, Khuroo T, Panda AK, Talegaonkar S, Khar RK, Iqbal Z (2016). Revisiting the nanoformulation design approach for effective delivery of Topotecan in its stable form: an appraisal of its in vitro behavior and tumour amelioration potential. *Drug Delivery*, 23(8), 2827–2837.

Padhi S, Nayak AK, Behera A (2020). Type II diabetes mellitus: a review on recent drug-based therapeutics. *Biomedicine & Pharmacotherapy*, 131, 110708

Patnaik S, Gorain B, Padhi S, Choudhury H, Gabr GA, Md S, Kumar Mishra D, Kesharwani P (2021). Recent update of toxicity aspects of nanoparticulate systems for drug delivery. *European Journal of Pharmaceutics and Biopharmaceutics*, 161, 100–119.

Piska, K., Koczurkiewicz, P., Bucki, A., Wójcik-Pszczoła, K., Kołaczkowski, M., & Pękala, E. (2017). Metabolic carbonyl reduction of anthracyclines—role in cardiotoxicity and cancer resistance. Reducing enzymes as putative targets for novel cardioprotective and chemosensitizing agents. *Investigational New Drugs*, 35(3), 375–385.

Pradhan SP, Sahoo S, Behera A, Sahoo R, Sahu PK (2022). Memory amelioration by hesperidin conjugated gold nanoparticles in diabetes induced cognitive impaired rats. *Journal of Drug Delivery Science and Technology*, 69, 103145.

Pradhan SP, Sahu PK, Behera A (2023a). New insights toward molecular and nanotechnological approaches to antidiabetic agents for Alzheimer's disease. *Molecular and Cellular Biochemistry*, 478, 1–24.

Pradhan SP, Tejaswani P, Behera A, Sahu PK (2024). Phytomolecules from conventional to nano form: next-generation approach for Parkinson's disease. *Ageing Research Reviews*, 93, 102136.

Pradhan SP, Tejaswani P, Sa N, Behera A, Sahoo RK, Sahu PK (2023b). Mechanistic study of gold nanoparticles of Vildagliptin and Vitamin E in diabetic cognitive impairment. *Journal of Drug Delivery Science and Technology*, 84, 104508.

Sa N, Tejaswani P, Pradhan SP, Alkhayer KA, Behera A, Sahu PK (2023). Antidiabetic and antioxidant effect of magnetic and noble metal nanoparticles of Clitoria ternatea. *Journal of Drug Delivery Science and Technology*, 84, 104521.

Shaaban S, Negm A, Ibrahim EE, Elrazak AA (2014). Chemotherapeutic agents for the treatment of hepatocellular carcinoma: efficacy and mode of action. *Oncology Reviews*, 8(1), 246.

Shaul P, Frenkel M, Goldstein EB, Mittelman L, Grunwald A, Ebenstein Y, et al. (2013). The structure of anthracycline derivatives determines their subcellular localization and cytotoxic activity. *ACS Medicinal Chemistry Letters*, 4(3), 323–328.

Shi J, Kantoff PW, Wooster R, Farokhzad OC (2017). Cancer nanomedicine: progress, challenges and opportunities. *Nature Reviews Cancer*, 17(1), 20–37.

Shieh MJ, Hsu CY, Huang LY, Chen HY, Huang FH, Lai PS. (2011) Reversal of doxorubicin-resistance by multifunctional nanoparticles in MCF-7/ADR cells. *Journal of Control Release*, 152, 418–425.

Shreffler JW, Pullan JE, Dailey KM, Mallik S, Brooks AE (2019) Overcoming hurdles in nanoparticle clinical translation: the influence of experimental design and surface modification. *International Journal of Molecular Sciences*, 20, 6056.

Soe ZC, Kwon JB, Thapa RK, Ou W, Nguyen HT, Gautam M, Oh KT, Choi H-G, Ku SK, Yong CS, Kim JO (2019). Transferrin-conjugated polymeric nanoparticle for receptor-mediated delivery of doxorubicin in doxorubicin-resistant breast cancer cells. *Pharmaceutics*, 11(2), 63.

Soma CE, Dubernet C, Bentolila D, Benita S, Couvreur P (2000). Reversion of multidrug resistance by co-encapsulation of doxorubicin and cyclosporin A in polyalkylcyanoacrylate nanoparticles. *Biomaterials*, 21(1), 1–7.

Subedi RK, Kang KW, Choi HK (2009). Preparation and characterization of solid lipid nanoparticles loaded with doxorubicin. *European Journal of Pharmaceutical Sciences*, 37, 508–513.

Sun, Y., Zou, W., Bian, S., Huang, Y., Tan, Y., Liang, J., Fan, Y., & Zhang, X. (2013). Bioreducible PAA-g-PEG graft micelles with high doxorubicin loading for targeted antitumor effect against mouse breast carcinoma. *Biomaterials*, 34(28), 6818–6828.

Ventola CL (2017). Progress in nanomedicine: approved and investigational nanodrugs. *Pharmacy and Therapeutics*, 42(12), 742.

Verma D, Thakur PS, Padhi S, Khuroo T, Talegaonkar S, Iqbal Z (2017). Design expert assisted nanoformulation design for co-delivery of Topotecan and thymoquinone: optimization, in vitro characterization and stability assessment. *Journal of Molecular Liquids*, 242, 382–394.

Verma NK, Purohit MP, Equbal D, Dhiman N, Singh A, Kar AK, et al. (2016). Targeted smart pH and thermoresponsive N,O-carboxymethyl chitosan conjugated nanogels for enhanced therapeutic efficacy of doxorubicin in MCF-7 breast cancer cells. *Bioconjugate Chemistry*, 27(11), 2605–2619.

Wang F, Wang YC, Dou S, Xiong MH, Sun TM, Wang J (2011). Doxorubicin-tethered responsive gold nanoparticles facilitate intracellular drug delivery for overcoming multi-drug resistance in cancer cells. *ACS Nano*, 5, 3679–3692.

Xia Y, Zhong J, Zhao M, Tang Y, Han N, Hua L, Xu T, Wang C, Zhu B (2019). Galactose-modified selenium nanoparticles for targeted delivery of doxorubicinto hepatocellular carcinoma. *Drug Delivery*, 1, 1–11.

Yang C, Wu T, Qin Y, Qi Y, Sun Y, Kong M, Jiang X, Qin X, Shen Y, Zhang Z (2018). A facile doxorubicin-dichloroacetate conjugate nanomedicine with high drug loading for safe drug delivery. *International Journal of Nanomedicine*, 13, 1281–1293.

Yu C, Chen F, Wang X, Cai Z, Yang M, Zhong Q, et al. (2020). Pin2 telomeric repeat factor 1-interacting telomerase inhibitor 1 (PinX1) inhibits nasopharyngeal cancer cell stemness: implication for cancer progression and therapeutic targeting. *Journal of Experimental & Clinical Cancer Research*, 39, 1–13.

Zhang B, Qian D, Ma HH, Jin R, Yang PX, Cai MY, et al. (2012). Anthracyclines disrupt telomere maintenance by telomerase through inducing PinX1 ubiquitination and degradation. *Oncogene*, 31(1), 1–12.

Zhang X, Chibli H, Mielke R, Nadeau J. (2010). Ultrasmall gold-doxorubicin conjugates rapidly kill apoptosis-resistant cancer cells. *Bioconjugate Chemistry*, 22, 235–243.

Zhang Y, Li T, Hu Y, Chen J, He Y, Gao X, Zhang Y (2022). Co-delivery of doxorubicin and curcumin via cRGD-peptide modified PEG-PLA self-assembly nanomicelles for lung cancer therapy. *Chinese Chemical Letters*, 33(5), 2507–2511.

18 Nanoenabled Systems for the Delivery of Different Essential Oils as Anticancer Agents

Muhammed Navaf, Plachikkattu Parambil Akhila, and Kappat Valiyapeediyekkal Sunooj
Department of Food Science and Technology, Pondicherry University, Puducherry, India

18.1 INTRODUCTION

One of the growing health problems and the leading cause of mortality worldwide is cancer (Padhi et al., 2016). Cancer is characterized as uncontrolled cell growth in response to mutation, regardless of the traditional cell division rules. Various cancers are reported nowadays, including lung, colon, breast, liver, stomach, etc. Different physical and biochemical agents, such as ultraviolet radiation, ionizing radiation, infections by viruses, bacteria, and other reactive oxygen species, may act as cancer-causing agents. The most effective approach to prevent carcinogenesis is to interfere with cell cycle modulation steps like initiation, propagation, and termination (Padhi et al., 2018).

Conventionally, chemotherapy and radiotherapy are used for cancer treatment. These are not fully effective and have side effects and environmental impacts. However, natural bioactive compounds possess cancer prevention properties with unique advantages like low toxicity and fewer side effects. The compounds obtained from medicinal plants are excellent in anticancer activities. The anticancer activity of bioactive compounds involves activating cellular systems, such as antioxidant enzymes, detoxifying activity, antiproliferative action, and anti-cell growth signalling pathway by arresting the cell cycle (Kundu et al., 2022; Behera and Padhi, 2022).

Essential oils undergo more research among different bioactive compounds due to their wide range of anticancer properties. Essential oils represent a complex blend of plant-borne volatile compounds. It has excellent potential for antioxidant, antimutagenic, and antiproliferative activity against various cancerous cells. However, the significant challenges of using essential oils in cancer therapy are the low biocompatibility, stability, and poor solubility and absorption in biological tissue with a non-specific target delivery (Sharma et al., 2022).

Nanotechnology is applying a material or substance on a nanometer scale with specific controlled physical and chemical properties (Padhi and Behera, 2022a; Behera et al., 2023a; Pradhan et al. 2024a). Nanosystems provide the same formulation for both hydrophilic and lipophilic medicines. Drug delivery using nanosystems protects from several *in vitro* and *in vivo* threats, improves treatment effectiveness, and regulates drug release in the target cell (Behera et al., 2023b; Pradhan et al. 2024b). Nanotechnology in drug delivery plays a significant role in delivering essential oils to cancer cells. It guides the essential oils to the target cells (Padhi et al., 2021; Sahoo et al., 2022). Hence, nano-delivery systems help overcome the challenges of using essential oils in cancer therapy (Alabrahim et al., 2024).

18.2 ESSENTIAL OILS

Volatile oils and ethereal oils are also terms used to describe essential oils. These are secondary metabolites found in lower concentrations in different plant organs, the epicarp of citrus fruits being the most abundant. In other words, essential oils are volatile, complex, aromatic compounds synthesized from aromatic plant secretory cells. However, "essential oil" describes plants' liquids and highly volatile compounds with a distinct aroma. Essential oils are hydrophobic and are mixtures of different aromatic compounds present in glands of various plant parts, like leaves, flowers, fruits, etc., of aromatic plants. Unlike fixed oils, essential oils are known as essences due to their characteristic volatility and pleasant and intense aroma (Blowman et al., 2018).

Climate, soil composition, plant part, age, vegetative cycle stage, and other factors influence essential oils' chemical composition and biological activities. The extraction method plays a vital role in the quality of essential oils. False or inappropriate extraction methods of essential oils will affect the chemical constituents and their biological functions. Hydrodistillation, steam distillation, solvent extraction, supercritical carbon dioxide, pressure hot water, expression, and microwave are all methods for extracting volatile oils. Hydro distillation is one of the most popular extraction techniques for industrial and medical applications (Bakkali et al., 2008; Weisany et al., 2022).

Researchers identified over 300 essential oil compounds, classified into different groups. The main groups of compounds present in essential oils are terpenes and terpenoids, aromatic (phenolic) compounds, and aliphatic compounds present in trace amounts (Ousaid et al., 2024).

DOI: 10.1201/9781003611516-23

18.2.1 Terpene and Terpenoids

Terpenes have a simple structure that follows a general rule. The carbon skeleton of terpenes is built up by 2-methylbutane, also known as isoprene units. Terpenes are categorized into monoterpenes, sesquiterpenes, diterpenes, and tetraterpenes, depending on the number of isoprene units in the carbon skeleton. Monoterpenes are made up of two isoprene units ($C_{10}H_{16}$). It occurs linear or cyclic in structure, including β-myrcene, limonene, α-phellandrene, etc. Sesquiterpenes contain three isoprene units ($C_{15}H_{24}$), including zingiberene, humulene, β-bisabolene, etc. Diterpenes consist of four isoprene units with a formula of $C_{20}H_{32}$, which include cembrene C, kaurene, camphorene, etc. Tetraterpenes are a class of carotenoids comprised of eight isoprene units. It has a general formula of $C_{40}H_{64}$ (Lai et al., 2022).

Terpenoids are compounds with one or more functional groups obtained from terpenes. Terpenoids include oxygenated monoterpenes such as camphor, carveol, limonene oxide and oxygenated sesquiterpenes such as caryophyllene oxide, humulene epoxide, and bisabolene oxide. The sources and the anticancer mechanism of some terpenes and terpenoids are given in Table 18.1.

18.2.2 Aromatic Compounds

Aromatic compounds are less plentiful than terpenes and are phenylpropane derivatives. Two classes of aromatic compounds are nuclear-substituted and benzene-substituted compounds, in which the substituents are directly attached to the side chain of the benzene ring (Bayala et al., 2014).

18.3 MECHANISM OF ANTICANCER ACTIVITY OF ESSENTIAL OILS

18.3.1 Antioxidant Action

Numerous studies have concentrated on antioxidant activity since oxidation harms various biological tissues and results in several diseases, including cancer, liver disorders, inflammation, etc. Researchers reported the antioxidant activity of thyme essential oils, clove, eucalyptus leaves, chamomile oil, etc. As a result, natural antioxidants found in essential oils can treat several diseases. Essential oils from *Salvia cryptantha, Salvia multicaulis, and Curcuma zedoaria* have excellent antioxidant activity (Parham et al., 2020). Many of the volatile compounds found in essential oils are also potent antioxidants. In eukaryotic cells, the superoxide anions and hydrogen peroxide produce hydroxyl radicals, capable of causing potential damage to mitochondrial DNA (Bolisetty and Jaimes, 2013). It hinders the electron transport protein expression and results in the accumulation of reactive oxygen species. The free radical produced by the damaged mitochondrial membrane combines with essential oils to create reactive phenoxyl radicals, which combine with reactive oxygen species and prevent proliferation, as illustrated in Figure 18.1 (Ancuceanu et al., 2024; Jayakodi et al., 2024).

TABLE 18.1
Source and Anticancer Activity of Some Terpenes and Terpenoids

Monoterpene	Source	Anticancer Activity	References
Paessential oilsniflorin	*Paeonia lactiflora*	It inhibited the nuclear factor kappa-B pathway and down-regulated anti-apoptotic gene bcl-2, thus up-regulating pro-apoptotic genes caspase-3 and bax.	Wang et al. (2012)
Vincosamide – *N*-oxide, d isodihydro aminocadambine	*Anthocephalus cadamba*	As a result of apoptosis, both compounds showed substantial chromatin fragmentation within nucleus cells.	Mishra et al. (2018)
Linalool		It induces apoptosis, cell cycle arrest, and modulation of the phosphoinositide-3-kinase–protein kinase B signalling pathway.	Pan and Zhang (2019)
Geraniol		Causes mitochondrial apoptosis by increasing Bax expression while decreasing bcl-2 expression.	Qi et al. (2018)
α-tThujone	*Thuja occidentalis* L	It accelerates oxidative stress and autophagy in tumour cells	Pudełek et al. (2019)
Citral	*Cymbopogon citratus*	Citral inhibited cancer cells' ability to clone and changed the shape of cancer cells. Suppress lipogenesis.	Balusamy et al. (2020)
lactones	Asteraceae family	In cancer cells, it disrupts cellular redox equilibrium and induces oxidative stress.	Gach et al. (2015)
Zingiberene	*Zingiber officinale*	In colorectal cancer (HT-29) cells, it activates autophagy-related caspases. In addition, it suppresses the mammalian target of rapamycin/ phosphoinositide-3-kinase/protein kinase B signalling pathway in colorectal cancer cells.	Chen et al. (2019)
Ursolic acid	Betulautilis	Inactivate extrinsic apoptotic pathway. It was responsible for the production of intracellular reactive oxygen species and the disruption of mitochondrial membrane potential.	Mishra et al. (2016)

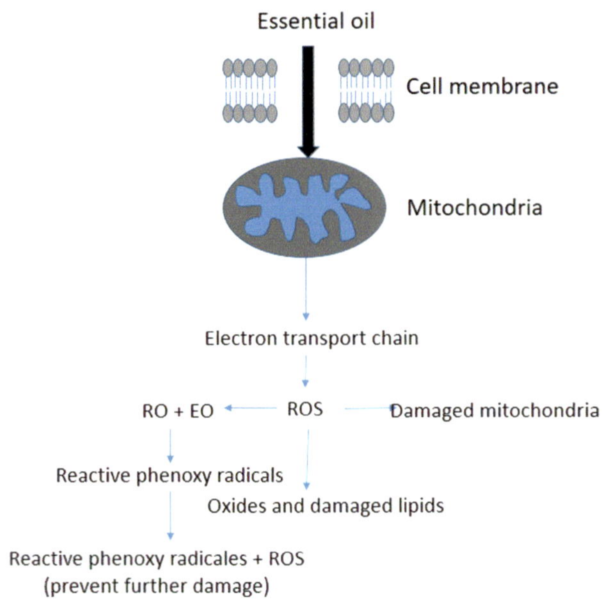

FIGURE 18.1 Antioxidant activity of essential oils [The mechanism of antioxidant activity of essential oils by inhibiting the generation of reactive oxygen species, thereby preventing the damage of mitochondria and lipids].

18.3.2 Antimutagenic Action

The antimutagenic activity of essential oils works by different mechanisms, summarized in Figure 18.2. Antimutagenic compounds work by either activating or inhibiting error-free or error-prone DNA repair. By interfering with pro-mutagens or pro-carcinogen xenobiotic biotransformation, essential oils constituents such as α-terpinene, α-terpineol, 1,8-cineole, D-limonene, camphor, citronellal, and citral modulate hepatic monooxygenase activity (Odin, 1997; Agrawal et al., 2024).

18.3.3 Antiproliferative Action

Antiproliferative action on human erythroleukemic cells (K562) was reported for essential oils like Satureja montana and Mentha arvensis. Eugenol, a constituent in many essential oils, also exhibited antiproliferative action against various cancer cells. Citrus aurantifolia showed apoptosis-mediated proliferation prevention in human colon cancer cells. Similarly, lime volatile oil may also help to prevent colon cancer. Human colon cancer was controlled by 78% at a concentration of 100 mg/mL volatile lime oil. The essential oil from *Thymus vulgaris, Carum copticum*, containing carvacrol as a primary compound, activated the apoptosis in a metastatic breast cancer cell by decreasing mitochondrial potential and increasing the release of cytochrome C. It also reduced the bcl-2/bax ratio, increasing caspase activity, poly(ADP-ribose) polymerase cleavage, and DNA fragmentation. Similarly, Astrodaucus orientalis essential oil also shows antiproliferative action in the breast cancer cell lines (T47D) (Edris, 2007; Dash et al., 2024).

18.3.4 Enhancement of Immune Function and Surveillance

Aromatherapy is a highly effective technique for improving the body's immune function, and it has a significant role in improving immune function. It uses essential oils through different media and works in many ways. It regulates adrenal gland hormone secretion, alleviates stress, supports and stimulates the lymph to expel toxins, immune-boosting cell development, and degradation of harmful microorganisms. Aromatherapy with citrus limonum essential oils affects positive mood and boosts the release of norepinephrine (Agarwal et al., 2022).

Inflammation is a complex reaction of the body to toxic substances that results from the release of soluble mediators such as reactive oxygen species, lipid mediators, proteases, and cytokines by an accumulation of cells. It is a fundamental component of several pathologies and reacts to an injury caused by noxious physical or chemical stimuli. Macrophage cells also control the initiation and propagation of various inflammatory cells caused by septic shock and chronic inflammatory conditions (Hamidzadeh et al., 2017).

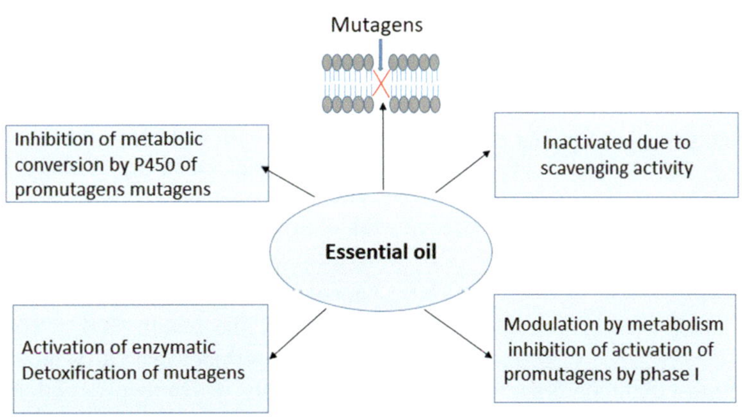

FIGURE 18.2 Mechanisms of antimutagenic activity of essential oils [Essential oils inhibit the metabolic conversion of promutagens and modulate the activation of promutagens. Essential oils activate the enzymatic detoxification of mutagens, and promutagens are inactivated by the scavenging activity of essential oils].

The essential oil isolated from Curcuma heyneana inhibits Inducible nitric oxide synthase, cyclooxygenase-2, and pro-inflammatory cytokines in lipopolysaccharide-stimulated murine macrophages by down-regulating the nuclear factor kappa-B pathway. Similarly, eugenol has antitumour necrosis factor activity. Another study found that essential oils from Cryptomeria japonica inhibited nuclear factor kappa-B activation caused by lipopolysaccharide by increasing the transcription of many cytokines involved in acute inflammatory responses (Miguel, 2010).

18.3.5 Enzyme Induction and Enhancing Detoxification

Essential oils-containing sulphur compounds work as an anticarcinogenic agent and affect procarcinogen metabolism. The enhanced rate of detoxification by these compounds is due to an increase in phase II enzymes like glutathione *S*-transferase, uridine 5′-diphosphate-glucuronyltransferase, and quinine reductase, as well as a decrease in phase I enzymes like cytochrome P450. The essential oils constituent of allium-containing organosulphur shows anticarcinogenic properties. Allylsulphides compound present in Allium sativum had shown an increased phase - enzyme-like glutathione *S*-transferase, uridine 5′-diphosphate-glucuronyltransferase, and epoxide hydrolase and inhibition to cytochrome P450 2E1 (Ho et al., 2010). Similarly, in animal models, Rosmarinus officinalis essential oils act as an anticarcinogenic agent against various structurally varied compounds. The hepatic glutathione *S*-transferase and reduced nicotinamide adenine dinucleotide phosphate, increase functions of quinine reductase in animals fed with rosemary extracts (Nagoor Meeran et al., 2017).

Citral (3,7-dimethyl-2,6-octadienal) is an unsaturated aldehyde found in many essential oils, and it induces the amino terpene of glutathione *S*-transferase class π. The total glutathione *S*-transferase activity in rat hepatocyte cells was increased upon citral treatment. According to the researchers, the induction of glutathione *S*-transferase class π is primarily due to the action of trans citral and an aldehyde group conjugated with a trans double bond. The improved glutathione *S*-transferase class π activity of citral can be used as a cancer chemopreventive agent against inflammation-related carcinogens like skin and colon cancer (Li et al., 2018; Gao et al., 2024).

18.4 EFFECT OF ESSENTIAL OILS ON VARIOUS CANCER CELLS

18.4.1 Colon Cancer

Colon cancer is one of the most common types of cancer, and it has a great possibility of recurrence and metastasis, and it has acquired resistance to conventional chemotherapy (Kaur and Kaur, 2015). α-Humulene, valencene, β-caryophyllene oxide, and trans-nerolidol are sesquiterpenes found in essential oils of several fruits and herbs that have an antiproliferative effect against colon cancer cell lines (Caco-2 and SW-620) (Ambrož et al., 2019). An essential oil obtained from *lebanese sage* showed antitumour activities. However, the combined application of linalyl acetate, terpineol, and camphor has a significant role in the growth of colon cancer cell lines (HCT-116) ($p53^{+/+}$ and $p53^{-/-}$) (Itani et al., 2008). Similarly, *Illicium verum* essential oils have a cytotoxic effect against colon cancer cells (HCT 116) and human colorectal adenocarcinoma cell lines (HT 29). It is due to the combined impact of free radical scavenging, induction of apoptosis, and tumour metastasis inhibition (Muhammad Asif et al., 2016).

Carvacrol is a significant compound in thyme essential oils and is effective against colon cancer cells (HCT 116 and LoVo). It reduces the proliferation, migration, and invasion of these cells (Fan et al., 2015). Extract from *Rosa canina* exhibited cytotoxicity against colon cancer cells (WiDr), and it reduced the cell viability and increased the number of late and early apoptotic cells (Turan et al., 2018). The plant's essential oil is used as a folk medicine like *Tageteserecta* L, *Tetradenia Riparia* Codd, *Bidens Sulphurea* Sch. Bip. and *Foeniculum vulgare* Mill show cytotoxicity against human colon carcinoma (HT29) (de Oliveira et al., 2015). Similarly, mastic essential oils from *Pistacia lentiscus* var. Chia shows antiproliferative action against human and murine colon cancer cell lines (Spyridopoulou et al., 2017). The *in-vitro* studies of essential oils, including nerolidol, thymol, geraniol, methyl isoeugenol, eugenol, and linalool, showed renoprotective effects on colonic epithelial cells (HT-29) (Thapa et al., 2019).

18.4.2 Oral Cancer

Essential oils containing aqueous and dry extract from *Ocimum sanctum* showed anti-cancer activity in oral squamous cancer cell lines (Ca9–22) (Luke et al., 2020). Caffeic acid phenethyl ester inhibits human oral cancer cell proliferation, survival, metastasis, epidermal growth factor receptor, cycloxygenase-2 activity, phosphoinositide-3-kinase–protein kinase B signalling, and S-phase kinase-associated protein 2, as well as inducing cell cycle arrest and apoptosis activity (Kuo et al., 2015). Essential oils of *Cinnamomum cassia* showed anticancer activity in human oral squamous carcinoma cell lines (HSC-3) by impacting apoptosis and G2/M cell cycle arrest by causing DNA damage and oxidative stress, as well as the stimulation of mitochondria-mediated signalling (Chang et al., 2017). Similarly, *Thymus caramanica's* essential oils have potential antiproliferative properties on squamous cell carcinoma of the mouth (KB cells) (Fekrazad et al., 2017). All allyl isothiocyanate kills human cisplatin-resistant oral cancer cells, which reduces cell viability. It inhibits protein kinase B/mammalian target of rapamycin proliferation signalling and promotes mitochondrial-dependent oral cancer cell apoptosis by causing DNA fragmentation and increased caspase 3 and caspase 9 activity (Chang et al., 2021).

18.4.3 Breast Cancer

Nigella sativa L. essential oils inhibit the growth of breast cancer cells (MCF-7). In the targeted cells, it causes cell membrane blebbing, cytoplasmic vacuolation, chromatin marginalization, and nucleus fragmentation. (Periasamy et al., 2016a). *Foeniculum vulgare* essential oil's anticancer activity was also evaluated *in vitro* against human breast cancer cells (MDA-MB) (Akhbari et al., 2019). *Ocimum canum Sims* essential oil also showed cytotoxicity and DNA fragmentation in breast cancer cells (MCF-7) (Tamil Selvi et al., 2015).

Sesquiterpene β-bisabolene opaline has a selective cytotoxic activity for human breast cancer cells (Yessential oils et al., 2016). Potential cytotoxic activity of *Achillea fragrantissima* essential oil constituents like caryophyllene oxide, 1-terpinene-4-Ol, viridiflorol, and guaiacol (Mouchira, 2017), and *Hedyosmum spruce* Solms essential oil on human breast cancer cells (MCF7) is reported (Guerrini et al., 2016). α-Pinene β-myrcene present in the essential oil of *Juniperus oxycedrus* shows antiproliferative effects and activates the mitochondrial pathway of apoptosis in estrogen receptor-positive breast cancer cells (El-Abid et al., 2019). Expression of mutant p53 in hormone-receptor-negative and human epidermal growth factor receptor 2-negative cells is decreased by curcumol from *Curcuma wenyujin* (Huang et al., 2017).

Piper cubeba essential oil exhibited cytotoxic activity against different types of breast cancer cells (MCF-7, MDA-MB-468, MDA-MB-231) and fibroblast cells (L929), which inhibit cell growth and appear to have induced apoptosis (Graidist et al., 2015). Similarly, the essential oil of *Juniperus excels*, showing cytotoxic activity against human breast cancer cell lines (MCF-7, MDA-MB-231, and T-47D) (Khanavi et al., 2019).

18.4.4 Prostate Cancer

Linalool, a key component of several essential oils, has an effective dose- and time-dependent cytotoxicity on prostate cancer cells (DU145), causing sub-G1 phase growth arrest, resulting in an increase in the sub-G0/G1 cell population (Sun et al., 2015). Essential oils from *Nepeta cataria* and *Rosa canina* have shown potent cytotoxic activity against the human prostate cell line (PC3 and DU-145) (Emami et al., 2016; Kilinc et al., 2020). Similarly, *in vitro,* antitumour action against human prostate cell lines was reported in *Lavandula angustifolia* essential oils (Zhao et al., 2017). Caryophyllene oxide, a primary compound in essential oils, reduces the viability of human prostate cancer cells, enhances the apoptosis process, and enhances ROS production (Russo et al., 2018). α-Pinene inhibits human prostate cancer cells' growth, cell cycle arrest, and apoptosis (Zhao et al., 2018).

18.4.5 Leukemia

Capillin, a constituent of essential oil from *Artemisia capillaries* thumb flower, has a potential apoptotic effect against human leukaemia cells (HL-60) (Masuda et al., 2015a). Similarly, García et al. observed significant antiproliferative action of *Lippia alba* essential oil in human leukaemia cells (K562) (García et al., 2017). Henisol, a sesquiterpenoid from *Atractylodes lancea* rhizome, exhibits apoptosis-induced activity in HL-60 human leukemia cells (Masuda et al., 2015b). Linalool obtained from Plantaginaceae has significant cytotoxicity against human leukaemia cells (U937) (M. Y. Chang et al., 2015). Likewise, essential oils from *Juniperus excels* fruit, *Juniperus oxycedrus*, *Cedrus libani,* and *Pinus pinea* woods showed cytotoxicity against human leukaemia cells (CCRF-CEM) (Saab et al., 2012). The essential oils from garlic produce reactive oxygen species, potentially affecting the induction of apoptosis and differentiation of human promyelocytic leukaemia cell line HL-60 cells (Agassi et al., 2020).

18.4.6 Lung Cancer

Essential oil from *Nepeta deflersiana* shows potential *in-vitro* cytotoxicity in lung cancer cell lines (A-549) (Al-Oqail et al., 2015). Luteolin is the major constituent in the essential oils of *Dracocephalum kotschyi and* has cytotoxicity against lung cancer cell lines (Calu-6 and Mehr-80) (Sani et al., 2017). Similarly, essential oils from *Thymus vulgaris* show cytotoxic effects against non-small-cell lung cancer cell lines (H460), and it modulates the release of interleukin-1 beta, interleukin-8, and down-regulated nuclear factor kappa-B p65 and nuclear factor kappa-B p52 proteins (Oliviero et al., 2016). The essential oil from *Cymbopogon* has been reported to retards the proliferation and survival of small cell lung cancer cells (Maruoka et al., 2018). *Lavandula dentate* essential oil shows potential cytotoxicity and a significant reduction in cell viability in the lung cancer cell line (Calu-3) (Justus et al., 2019).

18.4.7 Melanoma

Camphene, a constituent of the essential oil of *Piper cernuum,* has cytotoxicity against murine melanoma cells (B16F10-Nex 2). The cytotoxic activity of camphene is due to the induced apoptosis by increasing endoplasmic reticulum stress, which releases Ca^{2+} along with high mobility group box 1 and calreticulin, loss of mitochondrial membrane potential, and improves the regulation of caspase-3 activity (Girola et al., 2015). Carvacrol shows cytotoxicity to human melanoma cancer cells (A375). It inhibits cell growth by cleavage of poly (ADP-ribose) polymerases and reduces bromodomain-2 gene expression (Govindaraju & Arulselvi, 2018). Essential oils from *Eryngium campestre* and *Eryngium amethystium* (Cianfaglione et al., 2017) and *Liriodendron tulipifera L* (Quassinti et al., 2019) also have cytotoxic activity against human melanoma cancer cells (A375). Similarly, cytotoxic activity against HT144 melanoma and M14 melanoma cells was reported in *Callistemon viminalis* essential oil (de Oliveira et al., 2015) and *Salvia verbenaca* essential oil (Russo et al., 2015), respectively. Essential oils from *Matricaria recutita* L. and *Calendula officinalis* L. show cytotoxic activity against human melanoma cell lines (SK-MEL-2) (Sak et al., 2017). Frankincense (Boswellia) essential oil has a potential dose-dependent reduction in FM94 human melanoma cells' cell viability. It induces cleavage of caspase 3, 9, and poly (ADP-ribose) polymerases (Hakkim et al., 2019). *Pituranthos tortuosus* essential oil induces apoptosis and inhibits migration and invasion of melanoma cancer cells (B16F10) (Krifa et al., 2016).

18.5 NANO-ENABLED TARGETING OF ESSENTIAL OILS

A perfect cancer therapeutic material would be selective and able to distinguish between cancer and healthy cells. It must be accomplished through effective anticancer drug delivery into cancer cells (Padhi and Behera, 2020; Kundu et al., 2022). Nanocarriers are substances with nanometer-scale dimensions ranging from 1 to 100 nm (Hassan et al., 2021). It is made up of particles that are in one or more dimensions. Nanocarriers must be

- stable as long as they reach the target cell
- capable of escape from the reticuloendothelial system clearance,
- accumulate in cells through irregular tumour vasculature,
- capable of penetrating the interstitial tumour fluid at high pressure,
- capable of reaching the active site and exclusively interacting with the targeted cells

Nanocarriers can carry different drugs and protect them from degradation (Padhi and Behera, 2022b; Satpathy et al., 2024). Also, nano delivery of drugs facilitates the uniformity in pharmacokinetics and biodistribution of medicines (Padhi et al., 2022a). A single nanocarrier can carry different drug molecules, and the release kinetics of various drug molecules are controlled by internal and external stimuli in the cell (Maldonado et al., 2013; Padhi et al., 2020).

18.5.1 Passive and Active Targeting

Targeted therapeutic substance delivery has some benefits, including protecting normal cells from toxicity, reducing dose-limiting adverse effects, and combating cancerous cells resistant to drugs (Padhi and Behera, 2021). Many therapies, including cancer therapy, target the nucleus as a primary target.

18.5.1.1 Passive Targeting

Under certain conditions, such as in a tumour, the changes in physicochemical and pharmacological factors facilitate new vessel formation or the expansion of existing vessels. These vascular system changes facilitate the permeation of the nanosystem into tumour stroma, leading to the deposition of drug or drug nanocarriers at a specific site. Solid tumour cells show more favourable conditions for the accumulation of drugs in macromolecular size and colloidal size, including liposomes, polymeric-drug conjugates, and polymeric nanoparticles (Padhi et al., 2022b). The encapsulation of phytomolecules in nanocarrier enhances pharmacokinetics, selectivity, and decreases side effects. With the faulty lymphatic drainage of fast-growing tumours, the vascular system enlarges, and vascular permeability increases, resulting in improved permeability and retention of the nanosystem in the cell. This type of tumour-targeting without a legend for a tissue or target binding is passive targeting (Patel and Patel, 2019; Padhi and Behera, 2020).

Compared to the normal cells, the enhanced permeability and retention effect in the cancerous cell increases drug specificity and a 20%–30% increase in drug delivery efficiency. The tumour microenvironment and intrinsic factors such as angiogenesis and lymphangiogenesis, perivascular tumour development, stromal response density, and intratumour pressure influence enhanced permeation and retention effect. The increased rate of permeation of nanocarriers through the breakage of enlarged vessels in tumour cells leads to high interstitial pressure, which inhibits the accumulation of drug carriers in cancer cells. Nanoparticles crossing the vasculature to the tumour cells will be hindered by the presence of interstitial tumour milieu, which acts as a profound infiltration to tumour tissue (Attia et al., 2019).

Another passive targeting technique is tumour-activated prodrug therapy, which uses the specific tumour microenvironment in the nanocarrier system. Cancerous cells that expand uncontrollably have a higher metabolic rate and consume more oxygen and nutrients. The lack of the required amount of oxygen and nutrients causes cancerous cells to produce energy by glycolysis, resulting in an acidic milieu. Also, the cancerous cells secrete enzymes like matrix metalloproteinases. The anticancer drug loaded in a nanocarrier system will remain inert until it reaches the target site. Anticancer drugs are released at the target site by hydrolysis of the association between the nanocarrier and the drug due to enzymes' action or changes in the tumour site's pH (Patel and Patel, 2019).

18.5.1.2 Active Targeting

Active targeting is necessary for the successful and effective delivery of drugs to the target site. The nanocarrier or legends are explicitly maintained and taken up by cancerous cells in the active targeting strategy, also known as legend-mediated targeting. It can be achieved by altering or conjugating the nanocarrier surface with ligands binding to receptors over-expressed onto the tumour cells (Attia et al., 2019; Behera and Padhi, 2020). The affinity between the carrier and the tumour cell surface will be enhanced due to these alterations on the nanocarrier surface, resulting in increased drug penetration rates. This attraction may be due to physical adsorption by forming a covalent bond between them. The conjugation procedure, particle size and shape, legend density, surface hydrophobicity, and surface, nanocarrier, and legend charges influence the active targeting of the biomolecules. Conjugating nanocarrier with ligands improves its therapeutic efficiency, reducing the toxic effects on the cells. Nanocarriers conjugated with ligands will be absorbed by receptor-mediated endocytosis and thus result in intercellular drug release. Nanocarriers such as solid lipid nanocarriers, nanostructured lipid carriers, and liposomes can produce phytomolecules in active targeting (Shreya et al., 2019).

18.5.1.2.1 Features of Nano Delivery of Essential Oils

- Protection of essential oils in both extra and intracellular levels
- Sustained and controlled delivery of the essential oils in target cells
- Increased penetration rate and depth due to the nano size of the particles
- Increased rate of cellular uptake and subcellular activity

Numerous nanocarriers are used to deliver essential oils successfully and are broadly classified into polymer-based nanocarriers and lipid-based nanocarriers.

18.6 NANO DELIVERY OF ESSENTIAL OILS FOR ANTICANCER ACTIVITY

Several studies have been conducted on the suitable delivery of essential oils for cancer care. Table 18.2 enlists a range of nanocarriers used to deliver some essential oils.

18.6.1 Polymer-Based Nano Delivery

Nanocapsules and nanospheres are two types of polymeric nanoparticles. Nanocapsules are oily and consist of a wall and a core made of a polymer. However, nanospheres are the colloidal system that entraps the therapeutic substance within a matrix. The polymers used as nanocarriers are natural or synthetic in origin and are highly biodegradable and biocompatible.

Synthetic polymers used in phytochemical nanodelivery include poly cyanoacrylate alkyl esters, polyvinyl alcohol, polylactic acid, polyglycolic acid, and polylactic glycolic acid. Polylactic glycolic acids include polysaccharides, and proteins are natural polymers used in nanodelivery. Polysaccharides are of plant origin (pectin, cellulose, starch, arabic gum, alginate, etc.) and microbial or animal origin (xanthan gum and chitosan). Nanoparticles prepared from polysaccharides have unique hydrophilic characteristics. Hence, it is an acceptable carrier for delivering essential oils for therapeutic purposes.

TABLE 18.2
Nanocarriers Used for the Delivery of Some Essential Oils

Essential Oil	Nanocarrier	Cancerous Cell	References
Linalool	Gold nanoparticles, conjugate the complex with CALNN peptide	Breast cancer cells (MCF-7)	Jabir et al. (2019)
Isofuranodiene	Nanostructured liquid crystalline particles	Breast cancer cells (MDA-MB 231)	Pisani et al. (2020)
Rosemary essential oil	Nanoliposomes	Breast cancer cells (MCF7)	Salari and Salari (2019)
3β,6β,16β-tri-hydroxilup-20(29)-ene	Polycaprolactone (PCL) polymer	Colon carcinoma line (HCT116), prostate (PC3), glioblastoma (SNB19)	Silva-Filho et al. (2020)
*Saussurealappa*Essential Oil	Polymethyl methacrylate based nanocapsules	A549 cell line	Lammari et al. (2021)
Saccocalyxsatureioides essential oil	Nanoemulsions by high-pressure homogenization	Liver cancer cells (Hep-G2)	Aouf et al. (2020)
Zataria Multiflora Essential Oil	Chitosan nanoparticles	Breast cancer cells (DA-MB-231)	Salehi et al. (2020)
Turmeric oil, lemongrass oil	Chitosan–alginate nanocapsules	A549 cell	Natrajan et al. (2015)
Jasmine oil	Pectin/chitosan composite nanoparticles	Breast cancer cells (MCF-7)	Attallah et al. (2020)
Curcumin	Poly lactide-co-glycolide-co-polyacrylic acid nano-encapsulation	Pancreatic cell (PANC-1)	Jadid et al. (2021)
Cardamom oil	Chitosan nano-particles	Liver cancer cell line (Hep G2)	Jamil et al. (2016)
Origanumglandulosum oil	Nanoemulsion	Liver cancer cell line (Hep G2)	Ali et al. (2020)
Nigella sativa oil	Nanoemulsion	Breast cancer cells (MCF 7)	Periasamy et al. (2016b)

Albumin, gelatine, soy proteins, and casein are widely exploited proteins to prepare nanostructured carrier substances (Bilia et al., 2014).

18.6.2 Nanoencapsulation of Essential Oils

The use of essential oils and their derivative compounds is challenging because of their volatility and instability in the presence of light, air, moisture, and heat. Processing, transportation, storage, and consumption will degrade thermal and oxidative labile essential oils. These products may cause harm by forming toxic substances. Nanoencapsulation has been introduced to overcome these limitations. The nanoencapsulation of essential oils is an approach to enhance the bioavailability and bioactivity of essential oils through high cellular uptake and controlled release of essential oils. Nanoprecipitation, also known as solvent displacement or interfacial deposition, is a widely used technique for nanoencapsulation (Padhi et al., 2022c).

18.6.2.1 Polylactic (Co-Glycolic Acid)-Based Nanoencapsulation

Polylactic (co-glycolic acid) is a polylactic acid and polyglycolic acid copolymer. It is one of the best biodegradables and biocompatible materials used in drug delivery systems and is approved by the Food and Drug Administration as a drug carrier. Polylactic (co-glycolic acid) has been extensively studied drug delivery systems, showing an exciting activity as a drug carrier vehicle.

Trachyspermum ammi seed essential oils loaded polylactic (co-glycolic acid) nanoparticles were synthesized by evaporation and sonication-based nano encapsulation. It induced selective apoptosis and potent antioxidant activity in colorectal cancer cells (HT-29) (Almnhawy et al., 2020). Garcinol, found in *Garcinia cambogia,* has promising therapeutic activity due to its potential anti-cancer and antiproliferative activity in a broad range of cancer cell lines. However, its low solubility reduces its therapeutic application. Garcinol was encapsulated in polylactic (co-glycolic acid) nanoparticles synthesized by the nanoprecipitation technique to overcome this limitation. And it showed enhanced anticancer activity against hepatocellular (HepG2), melanoma (B16F10), and human epithelial carcinoma (KB) cell lines (Mir et al., 2017).

Curcumin has anticancer properties against various cancers. Due to its low dissolution rate and poor bioavailability, it has minimal clinical application. Its bioavailability and solubility were increased by nanodelivery by using a polylactic (co-glycolic acid) prepared nanoprecipitation method (Meena et al., 2017). In breast cancer cell lines (MDA-MB 231, MCF-10A, and MCF-7), Ahmed et al. (2019) found that *Callistemon citrinus* essential oil's nanoencapsulation with polylactic (co-glycolic acid) increased aqueous solubility, cellular absorption, passive cellular efflux, and gastrointestinal stability.

18.6.2.2 Polycaprolactone-Based Nanoencapsulation

Polycaprolactone is a biodegradable synthetic polymer formed by the ring-opening polymerization of caprolactone. Eugenol is the principal constituent of many essential oils, and its storage stability is improved by polycaprolactone nanoencapsulation and reduced light-induced oxidation (Falcão et al., 2015). The pentacyclic triterpene 3,6,16-trihydroxylup–20(29)-ene, which comes from the plant *Combretum leprosum*, is poisonous to human cancer cell lines. However, the low water solubility of this compound reduces its bioavailability. Silva-Filho et al. (2020) encapsulated the 3β,6β,16β-trihydroxylup-20(29)-ene with polycaprolactone and improved the water solubility and bioavailability of the same. The encapsulated nanoformulation was assayed against the prostate (PC3), human colon carcinoma line (HCT116), glioblastoma (SNB19), and the normal epithelial embryo kidney cell line (Hek293T). The nanoformulation was more selective toward prostate and glioblastoma cell lines.

18.6.2.3 Poly(Methyl Methacrylate) Based Nanoencapsulation

Poly(methyl methacrylate) is a synthetic polymer with potential application in drug delivery systems. Poly(methyl methacrylate) nanoparticles have greater biocompatibility and cell permeability, are non-toxic, non-irritant, have a controlled drug release profile, and are well known for their mucoadhesive properties. Lammari et al. found that encapsulating *Saussurea lappa* essential oils in poly (methyl methacrylate) nanoparticles increased its bioavailability and pharmacological activity (Lammari et al., 2021).

18.6.2.4 Chitosan-Based Nanoencapsulation

Chitosan is a biodegradable, biocompatible polymer approved for wound dressing applications and considered safe for human consumption. It is widely used as a polymeric nanocarrier for drug administration. Due to the positive surface charge and the mucoadhesive property, the chitosan and chitosan-derived nanocarrier enable mucosal surface adhesion and sustained release of phytochemicals. Chitosan nanoparticles have several applications in nonparenteral drug delivery (Mohammed et al., 2017).

Morinda citrifolia essential oil was encapsulated with chitosan nanoparticles, forming a coating on the surface of the essential oil, which enhanced the biocompatibility and reduced the toxic effect of nanoformulation. Nano-chitosan *Morinda citrifolia, an* essential oils complex, showed 54% inhibition at 40 μg/mL in lung cancer cells (A459) (Rajivgandhi et al., 2020). Attallah et al. (2020) encapsulated jasmine essential oils in chitosan/pectin nanoparticles as an encapsulating material. It increased the solubility and reduced the sensitivity problems using the green ionotropic gelation method. Chitosan/pectin nano delivery shows a 13-fold increase in potential anti-cancer activity.

Turmeric oil and lemongrass oil have a wide range of therapeutic applications. However, the low stability, higher volatility, and less solubility in water hinder its use in new formulations. Natrajan et al. (2015) successfully encapsulated these oils with chitosan and alginate nanoparticles, enhancing solubility and stability. 3-(4,5-Dimethylthiazol-2-yl)-2,5-diphenyltetrazolium bromide assay on lung cancer cells (A549) revealed the antiproliferative activity of the nanoformulation.

Cyperus articulatus essential oils-chitosan nanoformulation's bioactivity and physicochemical properties have shown promising applications in cancer treatment (Kavaz et al., 2019). Likewise, nanoencapsulation with chitosan nanoparticles enhanced cumin essential oils' thermal stability and biocompatibility. The cytotoxicity study of loaded nanoparticles exhibited good biocompatibility, and at higher doses, the nanoparticles showed slight prevention effects on the brain and breast tumours (Amir et al., 2020).

18.6.2.5 Cellulose-Based Nanoencapsulation

Cellulose is a naturally abundant biopolymer from plants and various living organisms, including fungi, algae, and bacteria. The polymerization of β-D-glucopyranose monomeric units forms cellulose, and the presence of three hydroxyl groups provides higher functionality. It is a non-toxic, flexible, highly biodegradable, and biocompatible polymer. Cellulose is widely applied in drug delivery systems, nanofibers, biosensors, etc. The cationic modification of the cellulose backbone can enhance the interaction with bioactive compounds and the adhesion properties of cellulose nanoparticles (Shi et al., 2024).

Curcumin is a hydrophobic, bioactive polyphenolic compound, and many studies have confirmed its biological significance. However, the limited stability and solubility, quick elimination, and faster metabolic activity of curcumin restrict its availability in the target cells. Asabuwa Ngwabebhoh et al. (2018) formulated a nanoemulsion by homogenization of medium-chain triglyceride and tween 80 as an organic phase and containing aminated nanocellulose particles as a hydrophilic phase, resulting in the formation of fine droplets of essential oil coated by cellulose nanoparticles by the rapid diffusion of surfactant from the hydrophobic phase into the hydrophilic phase. Pickering emulsion encapsulation based on nanocellulose has greater encapsulation effectiveness and longer drug release, increasing curcumin bioavailability at the target site (Mitbumrung and Matsukawa, 2024).

18.6.3 Inclusion Complex with Cyclodextrin

A drug's bioavailability and therapeutic index depend much on its aqueous solubility. Essential oils are commonly complexed with cyclodextrin molecules to enhance solubility. Cyclodextrins are produced by the enzymatic conversion of a starch molecule consisting of α-1,4 glycosidic linked glucose units. These can form a complex with hydrophobic molecules like essential oils, which will enhance the solubility of the hydrophobic molecule. The increased solubility is due to the characteristic hydrophobic internal and hydrophilic external structures of inclusion complexes.

Carvacrol is a phenolic monoterpene present in thyme oil and *origanum* essential oils. It has great interest due to its broad anticancer activities. Trindadea et al. (2019) encapsulated carvacrol in β-cyclodextrin by ultrasound and freeze-drying methods to improve the solubility of carvacrol. Higher encapsulation capacity was reported in the freeze-drying (81.2%) method over the ultrasound treatment (34.02%). In addition, freeze-dried inclusion complexed cells have increased dose-dependent cytotoxic activity in 2D and 3D culture systems.

Another method of carvacrol encapsulation with cyclodextrin is the slurry complex method. Compared to the physical mixing, the slurry, the complex method, will produce complexity with increased stability and solubility. The particle size of the complex by the slurry complex method is much lower than that of the physical mixing method. The cyclodextrin inclusion of carvacrol forms a stable complex with enhanced solubility, bioavailability, and antiproliferative activity (Falcão et al., 2015).

Santana et al. (2020) encapsulated *Eugenia brejoensis* essential oils with β-cyclodextrin using the co-precipitation method, and the amount of essential oils encapsulated in the inclusion complex was equal to 4.7 μg of essential oils/mg. They reported that the 50% cytotoxic concentration (CC_{50}) value of the *Eugenia brejoensis* essential oils β-cyclodextrin complex is four times more than the free *Eugenia brejoensis* essential oils and is highly selective and efficient in cervical cancer cells. Similarly, various concentrations of 4:96, 12:88, and 16:84 (w/w) of concentrated orange oil are also encapsulated with β-cyclodextrin by co-precipitation, and a higher recovery and biological activity were observed in a proportion of 16:84 (w/w) (Torres-Alvarez et al., 2020).

The essential oils compounds menthol, menthone, pulegone, and eucalyptol from peppermint (*Mentha piperita*) were complexed with α-cyclodextrin, β-cyclodextrin, γ-cyclodextrin, hydroxypropyl-β-cyclodextrin, randomly methylated-β-cyclodextrin, a low methylated-β-cyclodextrin, and cross-linked-β-cyclodextrin. Multiple headspace extraction experiments studied the controlled release of essential oil compounds from cyclodextrin complexes. Among the 1:1 inclusion complexes of all polymers, the *β*-cyclodextrin complex showed improved retention ability and better release efficiency of essential oil constituents (Falcão et al., 2015).

18.6.4 Nanoemulsion-Based Delivery of Essential Oils

Emulsions are soft materials consisting of droplets of one immiscible liquid dispersed in another liquid. The droplet size of a nanoemulsion ranges from 50 to 200 nm (Behera et al., 2020). It is a nanocarrier with an amorphous, lipophilic, and negatively charged surface. Nanoemulsions are also known as mini, ultrafine, and submicron emulsions.

Nanoemulsions are kinetically stable, isotropic heterogeneous systems with a translucent appearance and are prepared by droplet dispersion of one liquid. Nanoemulsions have long-term physical stability without exhibiting any visible changes. The enhanced stability of these nanoemulsions is due to the steric stabilization happening in the presence of a nonionic surfactant or polymer material. The nanoemulsion has several advantages over microemulsions. Nanoemulsion requires less surfactant and gives uniform surface coating and good wetting properties. The emulsification process requires high energy, like sonication, and the property of the nanoformulation depends on the essential oils/surfactant combination and their ratio (Campolo et al., 2020).

Emulsification procedures include high-pressure homogenization, microfluidization, phase inversion, spontaneous emulsification, solvent evaporation, and ultrasonication. In recent years, widespread investigation has been conducted on the effective delivery of therapeutic agents in nanoemulsion. Nanoemulsion formulations of essential oils will increase the drug retention time in the cells with minimum side effects. It also shows a higher penetration rate, enhanced bioavailability, prolonged retention time, and minimal drug loss during cell lysis (Joyce Nirmala et al., 2019).

The ultrasonic emulsification method is simple, cheap, high energy, clean, quick, and aseptic. It evolved as a highly effective method for overcoming viscous resistance by applying a powerful shear force to minimize droplet size using high-energy ultrasound. The emulsification by ultrasonic involves two mechanisms. First, the eruption of the oil phase into the water phase in the form of droplets is caused by the formation of an acoustic field that generates a combination of interfacial waves and instability. Second, cavitation near the interphase causes low-frequency ultrasonic waves to collapse the droplets. The increased drug retention and bioavailability of nanoemulsion are due to the smaller particle size (Periasamy et al., 2016a).

Nirmala et al. (2020) formulated celery seed oil nanoemulsion with Tween 80 using ultrasonication. The IC_{50} concentration of the formulation was 1.4 µL/mL. It significantly reduced cell proliferation by inhibiting anchorage-independent cell growth, and it induced cell apoptosis of cancer cells (Nirmala et al., 2020). Likewise, eugenol nanoemulsion with tween 20 and tween 80 exhibited cytotoxicity against thyroid cancer cell line (HTH-7) and Hek-293 cancer cell lines (Joyce Nirmala et al., 2019). Periasamy et al. (2016a) formulated *Nigella sativa* L. essential oil nanoemulsion with a nonionic surfactant, polysorbate 80, and water by ultrasonication. He observed that the droplet size and turbidity depended on surfactant concentration. It showed excellent apoptosis activity in MCF-7 cancer cells, decreased cell viability, and affected the nuclear morphology in a time and dose-dependent manner.

Abedinpour et al. (2021) reported the anticancer activity of *Mentha piperita* essential oils formulated by a spontaneous approach with Tween 20 and ethanol. Nanoemulsion with a droplet size of 136 ± 2 nm exhibited significant anticancer activity in human breast cancer cell lines (MCF-7, MDA-MB-468, and MDA-MB-231). Similarly, enhanced tissue-specific cytotoxicity against human breast cancer cells (MCF7) was reported in sour cherry pit oil nanoemulsion formulated by mixing oil with polysorbate 20, polysorbate 80, polyethylene glycol as a co-solvent, and water. It showed a significant caspase-3 up-regulation and increased Sub G1 peaks against nanoemulsion dose in breast cancer cells (MCF7) (Maragheh et al., 2019).

The apoptotic activity of essential oils of *Arachis hypogaea* was studied by formulating nanoemulsion with Tween 80 and Tween 20 using an ultrasonic wave frequency of 20 kHz. It has characteristic tissue-dependent cytotoxic activity in the lung cancer cell line (A549) (Fazelifar et al., 2020). Khan et al. (2018) optimized nanoemulsion with varying concentrations of carvacrol and polysorbate 80. The emulsion caused the activation of apoptosis regulators like Jun N-terminal kinase, B-cell leukaemia/lymphoma 2 protein-associated X protein, and B-cell leukaemia/lymphoma 2 protein and cytochrome C release and activation of the caspase cascade by enhanced reactive oxygen species production in lung cancer cells (A549). Similarly, *Citrus aurantium* L. bloom essential oils nanoemulsion with a particle size of 76.9 nm showed cytotoxicity in lung cancer cells (A549) with the IC_{50} value of 152 µg/mL. Additionally, it induced the overexpression of caspase-3 and activated apoptotic cell death (Navaei Shoorvarzi et al., 2020). *Citrus lemon* essential oil nanoemulsion with Tween 80 and Tween 20 has significant apoptotic activity in lung cancer cells (A549). Additionally, it inhibited the chick embryo chorioallantoic membrane angiogenesis (Yousefian Rad et al., 2020). *Carum carvi* essential oils nanoemulsion with a droplet size of 34.6 nm showed higher stability and enhanced anticancer activity in the human colon cancer cell lines (HT-29). A 12.5 and 25 µg/mL concentration induced the caspase-3 enzyme (Khatamian et al., 2019).

18.6.5 Liposome-Based Nanocarriers

Liposomes are commonly used to produce a wide range of therapeutic agents. It consists of lipid vacuoles with two phospho-lipid layers and is self-assembled colloidal particles with spherical shapes. It forms spontaneously with a curved lipid bilayer, which traps the solvent. Liposomes have the same supermolecular lipid organization as the natural membranes of living cells. Liposomes have considerable significance in drug delivery due to their biocompatibility and intercellular delivery targeting. They can hold hydrophilic and hydrophobic molecules in the aqueous core and into a lipid bilayer. Besides, liposome-based delivery enhances the biopharmaceutical functions of encapsulated essential oils by increasing the release time, reducing drug clearance, and improving tumour accumulation.

The liposome's characteristic structure allows it to

1. encapsulate phytochemical compounds having different physicochemical properties,
2. enhance the stability of the phytochemicals by preventing their destabilization,
3. reduce the side effects by lowering the dosage supply.

Based on the lipid bilayer's structure and size, liposomes are grouped into multilamellar vesicles, large unilamellar vesicles, short unilamellar vesicles, and multivesicular vesicles. These vesicular carriers have a significant role in the delivery of drugs through the oral route. The therapeutic efficiency of liposomes is affected by liposome composition and phospho-lipid bilayer fluidity and stability, and it also affects the physicochemical characteristics of essential oils (Shreya et al., 2019).

Nanoliposomes have great applications for delivering essential oils due to the improved solubility, stability, and increased absorption of lyophobic essential oils. Its enhanced cellular interaction results in increased therapeutic efficiency of essential oils. The cationic liposomes have preferential activity on the angiogenic endothelium of tumours and stop vascular functions. This ultimately limits the growth of tumour cells and metastasis.

Liposome-based nano delivery of curcumin enhances its bioavailability and stability. Saengkrit et al. (2014) encapsulated curcumin in liposomes modified by didodecyldimethylammonium bromide, cholesterol, and a nonionic surfactant. Didodecyldimethylammonium bromide in liposomes containing curcumin shows a potential anti-cancer activity. Didodecyldimethylammonium bromide enhances the cellular internalization of curcumin-loaded liposome nanoparticles. Cholesterol in liposomes increases the lipid bilayer's fluidity and curcumin entrapment.

The surface modifications of liposomes with polyethylene glycol enhance their stability and circulation properties. The incorporation of polyethylene glycol reduces positive surface charges significantly. The mechanism behind it is that the hydrated polyethylene glycol shell forms an aqueous layer on nanoparticles. It affects the interaction of liposomes with serum proteins, reducing opposition, and uptake of liposome macrophages. Oleanolic acid is a triterpenoid found in many plants with low solubility and bioavailability. Gao et al. (2012) formulated oleanolic acid nanoliposomes modified with polyethylene glycol. The higher stability of polyethylene glycol-modified nanoliposomes containing oleanolic acids enhances the longevity of essential oils in circulation. It has a slow release in the target centre due to the stability of essential oils-loaded liposomes.

Eugenol is the primary constituent of clove essential oil, and its stability was improved by encapsulating it in liposomes using saturated and unsaturated soybean phospholipids in combination with cholesterol using the ethanol injection method. The liposomal delivery protected eugenol from UV-induced degradation and provided nanometric oligolamellar and spherical-shaped vesicles (Sebaaly et al., 2015).

18.6.6 Functionalized Nanoparticles

The conventional way of nano delivery of anticancer drugs using various nanoparticles, including polymeric nanoparticles like poly lactic-co-glycolic acid, chitosan nanoparticles, cellulose nanoparticles, nanoemulsion, etc., has limitations like lack of long-term drug release. Over 90% of encapsulated drugs get released within a few hours. The limited delivery of the drug is due to the lower stability of the nanocarrier. The inorganic nanoparticles, such as mesoporous silica, magnetic, etc., have better stability, giving long-term drug release.

Hydroxyapatite is a primary inorganic compound. It shows excellent biocompatibility compared to other inorganic molecules and has a wide therapeutic application. AbouAitah et al. (2020) created a piperine-loaded hydroxyapatite nanocarrier coated with gum arabic and folic acid conjugation on the surface. It shows excellent targeting and selectivity. It enables the prolonged release of piperin and complete inhibition of monolayer colon cancer cells (HCT116).

Shetty et al. (2021) delivered cinnamaldehyde to breast cancer cells using magnetic iron oxide nanoparticles functionalized with fluorescein isothiocyanate and folic acid. Folic acid conjugation enhanced the absorption of nanoparticles in both the cytoplasm and the nucleus. This nanoformulation improved the apoptosis rate and increased the caspase gene expression.

Uma Maheswari et al. (2020) used a nanoformulation with folic acid conjugated bovine serum albumin-calcium ferrite to distribute hesperidin and eugenol to MCF-7 cells. The nanoformulations demonstrated encapsulation efficiency of 62.94% and 85.58% for hesperidin and eugenol, respectively. Superparamagnetic calcium ferrite nanoparticles in the nanoformulation allowed for more efficient loading of hesperidin and eugenol and faster drug release when exposed to an external magnetic field. The drug carrier systems were biocompatible with normal subcutaneous areolar adipose cells (L929). The increased cytotoxicity of hesperidin and eugenol in cell viability experiments on breast cancer cell lines (MCF-7), with IC_{50} values reduced by 20–30 fold. They resulted from a specific interaction with receptors on breast cancer cells (MCF-7).

Similarly, the solubility and stability of curcumin were enhanced by loading in zinc oxide nanoparticles covalently conjugated with 3-mercaptopropionic acid. And it did not affect the anti-cancer activity of curcumin. Besides, the functionalization of curcumin-zinc oxide nanocomposite with 3-mercaptopropionic acid enhances the cytotoxicity of curcumin against tumour cells. It may also have applications in cancer treatment, including chemotherapy, photodynamic therapy, and imaging (Ghaffari et al., 2017).

18.6.7 Solid Lipid Nanoparticles

To build a new method of colloidal drug delivery, solid lipid nanoparticles combine the properties of liposomes and polymeric nanoparticles. As a result, it has the stability of a solid matrix, the biocompatibility of a lipid carrier, and the ability to stay solid at room and body temperatures. Due to several advantages, solid lipid nanoparticles can be a good nanocarrier for delivering essential oils. Essential oils are mostly hydrophobic. It shows good miscibility in the matrix and has higher entrapment efficiency. Solid lipid nanoparticles protect the ESSENTIAL OILSs from the degradation of active compounds, decreasing sensitivity to light, air, and temperature and reducing volatilization.

A hot, high-pressure homogenization technique and a high-shear dispersion method are often used to make solid lipid nanoparticles. The aqueous phase is added to the oil phase with continuous stirring. The high-pressure homogenization employed particle size reduction. Another form of size reduction is using an ultrasound probe in preformed emulsions.

Curcumin-loaded solid lipid nanoparticles were blended with liquid-liquid Sefsol 218 to minimize their size and stabilize them, improving their chemical stability and dispersibility in aqueous media. It displayed long-term curcumin release, increased cellular intake, and inhibited cell development (Sun et al., 2013). Similarly, Mulik et al. developed transferrin-conjugated solid lipid nanoparticles using the hot homogenization method. The nanoformulation of curcumin with transferrin-conjugated solid lipid nanoparticles is resistant to photostability and improves drug delivery and anticancer activity against SH-SY5Y neuroblastoma cells. The ROS production and the cellular uptake of curcumin are enhanced by nanoformulation with transferrin-loaded solid lipid nanoparticles (Mulik et al., 2012).

The anti-cancer activity of curcumin-loaded solid lipid nanoparticles against SKBR3 breast cancer cells was studied. Nano formulation consists of uniformly sized particles (40 nm) with well-defined spherical shapes and a negative charge on the surface. Drug loading and encapsulating efficiencies were 23.38% and 72.47%, respectively. It has a high level of cytotoxicity against SKBR3 breast cancer cells. It may also increase the ratio of Bax to Bcl-2 while reducing the expression of cyclin D1 and CDK4 (Wang et al., 2018). Yang et al. studied the chitosan-mediated zedoary turmeric oil-loaded solid lipid nanoparticles, revealed a threefold increase in the release time and the targeting delivery to the cell also improved (Yang et al., 2020).

Linalool successfully loaded into lipids like myristyl myristate, cetyl esters, and cetyl palmitate. A controlled drug release was achieved with a 100 nm-sized lipid particle. The nano lipid carrier exhibited good incorporation of linalool, and weak interaction with components was formed. Linalool incorporation also changes the thermal and structural properties of free lipid nanoparticles. A 1.0 mM or higher concentration of linalool exhibited enhanced cytotoxic activity in human lung (A549) and liver cell lines (HepG2), respectively (Rodenak-Kladniew et al., 2017).

Researchers loaded monoterpenes such as alpha-pinene, citral, geraniol, and limonene to glycerol monostearate solid lipid nanoparticles by high-pressure homogenization. Geraniol shows 43.28% encapsulation efficiency. However, the encapsulation efficiency of alpha-pinene, citral, and limonene was recorded as 4.48%, 4.34%, and 5.58%, respectively. A lower rate of drug release was observed in limonene solid lipid nanoparticles. Sixteen percent of alpha-pinene, 6% of citral, and 3.5% of geraniol were released during the same period (Rodenak-Kladniew et al., 2017).

18.6.8 Nanostructured Lipid Carrier

Nanostructured lipid carriers are a novel pharmaceutical formulation of lipid nanoparticles, and they have several advantages over solid lipid nanoparticles. It consists of physiological and biocompatible lipids, surfactants, and co-surfactants. Incorporating oil into a solid lipid nanocarrier causes imperfections and leads to the degradation of crystalline alignment, preventing drug leakage and endowing a high drug load. Nanostructured lipid carriers are gaining much attention among researchers. The use of these nanocarriers in the delivery of both hydrophilic and lipophilic drugs is highly beneficial. (Chauhan et al., 2020).

Oral bio efficiency of citral was increased by delivery of citral in nanostructured lipid carrier. According to *in vitro* cytotoxic activity on MDA-MB-231 cells, nanostructured lipid carriers formulated with citral inhibited cell proliferation significantly more than free citral. Citral, in combination with a nanostructured lipid carrier, accelerates the depolarization of the mitochondrial membrane, resulting in apoptosis. This formulation induces the regulation of pro-apoptotic genes among the Bcl-2 family, a well-known gene to mediate cell apoptosis (Nordin et al., 2019). Hosseinpour et al. used a nanostructured lipid carrier to carry zerumbone to a human mammary gland adenocarcinoma (MDA-MB-231) cell line. It was discovered that a zerumbone-loaded nanostructured lipid carrier with an IC50 value of 6.01 g/mL inhibits MDA-MB-231 cell proliferation and stops the cell cycle in the G2/M phase. The induced apoptotic activity of zerumbone is due to the activation of caspase-3 and caspase-9 and the release of cytochrome C. It also causes anti-apoptotic Bcl-2, Bcl-xL, and proliferating cell nuclear protein to be down-regulated when pro-apoptotic Bax protein is up-regulated (Hosseinpour et al., 2014). Zerumbone-loaded nanostructured lipid particles similarly function in the human lymphoblastic leukaemia (Jurkat) cell line (Rahman et al., 2014).

In 4T1 cancer cell lines, citral delivered in a nanostructured lipid carrier triggered superior downward regulation of inducible nitric oxide synthase, matrix metallopeptidase 9, eotaxin, fundamental fibroblast growth factor, vascular endothelial growth factor, and macrophage colony-stimulating factor (Nordin et al., 2020).

18.7 FUTURE PERSPECTIVES

The effective delivery of essential oils in cancer treatment via nanotechnology is a growing topic. Numerous studies have been done in this field in the past few years. The precise nano delivery of essential oils via different nanocarriers to cancer cells, such as lung, colon, breast, etc., will continue to be a focus of study and development in the future.

The nanoparticles used for the various studies are not uniform in their size. Some of them are in the submicron level size. More research must be conducted on the nano delivery of phytomolecules with consistently uniform size, drug loading, and release properties. Understanding the impact of nano delivery of essential oils is still limited, and these attributes to the field are to be exploited in further research. Many important markers of disease tissues, including critical biological characteristics that allow absolute targeting without disrupting normal cellular functions, are unknown, and these are key areas for further research, such as the use of nanotechnology in essential oils delivery targets on the molecular level application. Hence, the advancements in understanding the molecular level of cancer cells will lead to nano-delivery of essential oils to the target molecules.

18.8 CONCLUSION

Essential oils are a complex blend of plant-derived volatile oils. These are well known for their biological functions, and it has significant anticancer activity. However, their physicochemical properties, like the high volatility, lower solubility, etc., hinder their potential applications in medical treatment. Nanocarriers are the practical methods that help deliver the essential oil to the target site, which helps overcome the barriers to using essential oils with improved stability and solubility. Various kinds of nanocarriers are used to successfully deliver essential oils into the target cell, including polymer-based nanocarriers such as poly(lactic-co-glycolic acid)-based nano-encapsulation, polycaprolactone-based nanoencapsulation, chitosan-based nanoencapsulation, etc., and lipid-based nanocarriers such as liposome, solid lipid nanocarrier, and nanostructured lipid carriers. Delivering essential oils by nanocarrier will enhance efficacy and stability and reduce side effects.

REFERENCES

Abedinpour, N., Ghanbariasad, A., Taghinezhad, A., & Osanloo, M. (2021). Preparation of nanoemulsions of *Mentha piperita* essential oil and investigation of their cytotoxic effect on human breast cancer lines. *BioNanoScience,* 11, 428–436

AbouAitah, K., Stefanek, A., Higazy, I. M., Janczewska, M., Swiderska-Sroda, A., Chodara, A., Wojnarowicz, J., Szałaj, U., Shahein, S. A., Aboul-Enein, A. M., Elella, F. A., Gierlotka, S., Ciach, T., & Lojkowski, W. (2020). Effective targeting of colon cancer cells with piperine natural anticancer prodrug using functionalized clusters of hydroxyapatite nanoparticles. *Pharmaceutics*, 12(1), 12010070.

Agarwal, P., Sebghatollahi, Z., Kamal, M., Dhyani, A., Shrivastava, A., Singh, K. K., … & Baek, K. H. (2022). Citrus essential oils in aromatherapy: Therapeutic effects and mechanisms. *Antioxidants*, 11(12), 2374.

Agassi, S. F. T., Yeh, T. M., Chang, C. D., Hsu, J. L., & Shih, W. L. (2020). Potentiation of differentiation and apoptosis in a human promyelocytic leukemia cell line by garlic essential oil and its organosulfur compounds. *Anticancer Research*, 40(11), 6345–6354.

Agrawal, M., Singhal, M., Nayak, A. K., & Bhatt, S. (2024). An update of anticancer application of essential oils. In: Singh, S., Chaurasia, P. K., & Bharati, S. L. (Eds.) *Pharmacological Aspects of Essential Oils*. CRC Press, Boca Raton, FL, pp. 265–275.

Ahmed, R., Tariq, M., Ahmad, I. S., Fouly, H., Fakhar-I-Abbas, Hasan, A., & Kushad, M. (2019). Poly(lactic-co-glycolic acid) nanoparticles loaded with *Callistemon citrinus* phenolics exhibited anticancer properties against three breast cancer cell lines. *Journal of Food Quality*, 2019, 1–12.

Akhbari, M., Kord, R., Jafari Nodooshan, S., & Hamedi, S. (2019). Analysis and evaluation of the antimicrobial and anticancer activities of the essential oil isolated from Foeniculum vulgare from Hamedan, Iran. *Natural Product Research*, 33(11), 1629–1632.

Alabrahim, O. A. A., Lababidi, J. M., Fritzsche, W., & Azzazy, H. M. E. S. (2024). Beyond aromatherapy: Can essential oil loaded nanocarriers revolutionize cancer treatment? *Nanoscale Advances*, 6(22), 5511–5562.

Ali, H., Al-Khalifa, A. R., Aouf, A., Boukhebti, H., & Farouk, A. (2020). Effect of nanoencapsulation on volatile constituents, and antioxidant and anticancer activities of Algerian *Origanum glandulosum* Desf. essential oil. *Scientific Reports*, 10(1), 1–9.

Almnhawy, M., Jebur, M., Alhajamee, M., Marai, K., & Tabrizi, M. H. (2020). PLGA-based nano-encapsulation of Trachyspermum ammi seed essential oil (TSESSENTIAL OILS-PNP) as a safe, natural, efficient, anticancer compound in human HT-29 colon cancer cell line. *Nutrition and Cancer*, 73, 1–13.

Al-Oqail, M., Al-Sheddi, E., Siddiqui, M., Musarrat, J., Al-Khedhairy, A., & Farshori, N. (2015). Anticancer activity of chloroform extract and sub-fractions of nepeta deflersiana on human breast and lung cancer cells: An in vitro cytotoxicity assessment. *Pharmacognosy Magazine*, 11(44), 598–604.

Ambrož, M., Šmatová, M., Šadibolová, M., Pospíšilová, E., Hadravská, P., Kašparová, M., Skarková, V. H., Králová, V., & Skálová, L. (2019). Sesquiterpenes α-humulene and β-caryophyllene oxide enhance the efficacy of 5-fluorouracil and oxaliplatin in colon cancer cells. *Acta Pharmaceutica*, 69(1), 121–128.

Amiri, A., Mousakhani-Ganjeh, A., AmiriYi, Z., Guo, G., Singh, A. P., & Kenari, R. E. (2020). Fabrication of cumin loaded-chitosan particles: Characterized by molecular, morphological, thermal, antioxidant and anticancer properties as well as its utilization in food system. *Food Chemistry*, 310, 1258212.

Ancuceanu, R., Anghel, A. I., Hovaneţ, M. V., Ciobanu, A. M., Lascu, B. E., & Dinu, M. (2024). Antioxidant activity of essential oils from Pinaceae species. *Antioxidants*, 13(3), 286.

Aouf, A., Ali, H., Al-Khalifa, A. R., Mahmoud, K. F., & Farouk, A. (2020). Influence of nanoencapsulation using high-pressure homogenization on the volatile constituents and anticancer and antioxidant activities of algerian *Saccocalyx satureioides* Coss. et Durieu. *Molecules*, 25(20), 1–11.

Asabuwa Ngwabebhoh, F., Ilkar Erdagi, S., & Yildiz, U. (2018). Pickering emulsions stabilized nanocellulosic-based nanoparticles for coumarin and curcumin nanoencapsulations: In vitro release, anticancer and antimicrobial activities. *Carbohydrate Polymers*, 201, 317–328.

Asif, M., Salem Yehya, A. H., Ahmed Al-Mansoub, M., Revadigar, V., Ezzat, M. O., Ahamed, M. B. K., Oon, C. E., Murugaiyah, V., Majid, A. S. A., & Majid, A. M. S. A. (2016). Anticancer attributes of *Illicium verum* essential oils against colon cancer. *South African Journal of Botany*, 103, 156–161.

Attallah, O. A., Shetta, A., Elshishiny, F., & Mamdouh, W. (2020). Essential oil loaded pectin/chitosan nanoparticles preparation and optimization: Via Box-Behnken design against MCF-7 breast cancer cell lines. *RSC Advances*, 10(15), 8703–8708.

Attia, M. F., Anton, N., Wallyn, J., Omran, Z., & Vandamme, T. F. (2019). An overview of active and passive targeting strategies to improve the nanocarriers efficiency to tumour sites. *Journal of Pharmacy and Pharmacology*, 71(8), 1185–1198.

Bakkali, F., Averbeck, S., Averbeck, D., & Idaomar, M. (2008). Biological effects of essential oils: A review. *Food and Chemical Toxicology*, 46(2), 446–475.

Balusamy, S. R., Perumalsamy, H., Veerappan, K., Huq, M. A., Rajeshkumar, S., Lakshmi, T., & Kim, Y. J. (2020). Citral induced apoptosis through modulation of key genes involved in fatty acid biosynthesis in human prostate cancer cells: In silico and in vitro study. *BioMed Research International*, 2020, 6040727.

Bayala, B., Bassole, I. H., Scifo, R., Gnoula, C., Morel, L., Lobaccaro, J. M. A., & Simpore, J. (2014). Anticancer activity of essential oils and their chemical components: A review. *American Journal of Cancer Research*, 4(6), 591.

Behera, A., & Padhi, S. (2020). Passive and active targeting strategies for the delivery of the camptothecin anticancer drug: A review. *Environmental Chemistry Letters*, 18(5), 1557–1567.

Behera, A., & Padhi, S. (2022). pH-Sensitive polymeric nanoparticles for cancer treatment. In: Padhi, S., Behera, A., & Lichtfouse, E. (Eds.) *Polymeric Nanoparticles for the Treatment of Solid Tumours*. Environmental Chemistry for a Sustainable World, vol 71. Springer, Cham, pp. 401–425.

Behera, A., Mittu, B., Padhi, S., & Singh, A. (2020). Antimicrobial efficacy of essential oil nanoemulsions. In: Dhull, S. B., Chawla, P., & Kaushik, R. (Eds.) *Nanotechnological Approaches in Food Microbiology*. CRC Press, Boca Raton, FL, pp. 293–309.

Behera, A., Padhi, S., & Nayak, A. K. (2023a). Engineered liposomes as drug delivery and imaging agents. In: Ray, S., & Nayak, A. K. (Eds.) *Design and Applications of Theranostic Nanomedicines*. Woodhead Publishing, Cambridge, MA, pp. 75–108.

Behera, A., Sa, N., Pradhan, S. P., Swain, S., & Sahu, P. K. (2023b). Metal nanoparticles in Alzheimer's disease. *Journal of Alzheimer's Disease Reports*, 7(1), 791–810.

Bilia, A. R., Guccione, C., Isacchi, B., Righeschi, C., Firenzuoli, F., & Bergonzi, M. C. (2014). Essential oils loaded in nanosystems: A developing strategy for a successful therapeutic approach. *Evidence-Based Complementary and Alternative Medicine*, 2014, 651593.

Blowman, K., Magalhães, M., Lemos, M. F. L., Cabral, C., & Pires, I. M. (2018). Anticancer properties of essential oils and other natural products. *Evidence-Based Complementary and Alternative Medicine*, 2018, 3149362.

Bolisetty, S., & Jaimes, E. A. (2013). Mitochondria and reactive oxygen species: Physiology and pathophysiology. *International Journal of Molecular Sciences*, 14(3), 6306–6344.

Campolo, O., Giunti, G., Laigle, M., Michel, T., & Palmeri, V. (2020). Essential oil-based nanoemulsions: Effect of different surfactants, sonication and plant species on physicochemical characteristics. *Industrial Crops and Products*, 157, 1112935.

Chang, M. Y., Shieh, D. E., Chen, C. C., Yeh, C. S., & Dong, H. P. (2015). Linalool induces cell cycle arrest and apoptosis in leukemia cells and cervical cancer cells through CDKIs. *International Journal of Molecular Sciences*, 16(12), 28169–28179.

Chang, P.-Y., Tsai, F.-J., Bau, D.-T., Hsu, Y.-M., Yang, J.-S., Tu, M.-G., & Chiang, S. L. (2021). Potential effects of allyl isothiocyanate on inhibiting cellular proliferation and inducing apoptotic pathway in human cisplatin-resistant oral cancer cells. *Journal of the Formosan Medical Association*, 120, 515–523.

Chang, W.-L., Cheng, F.-C., Wang, S.-P., Chou, S.-T., & Shih, Y. (2017). Cinnamomum cassia essential oil and its major constituent cinnamaldehyde induced cell cycle arrest and apoptosis in human oral squamous cell carcinoma HSC-3 cells. *Environmental Toxicology*, 32(2), 456–468.

Chauhan, I., Yasir, M., Verma, M., & Singh, A. P. (2020). Nanostructured lipid carriers: A groundbreaking approach for transdermal drug delivery. *Advanced Pharmaceutical Bulletin*, 10(2), 150–165.

Chen, H., Tang, X., Liu, T., Jing, L., & Wu, J. (2019). Zingiberene inhibits in vitro and in vivo human colon cancer cell growth via autophagy induction, suppression of PI3K/AKT/mTOR pathway and caspase 2 deactivation. *Journal of BUON*, 24(4), 1470–1475.

Cianfaglione, K., Blomme, E. E., Quassinti, L., Bramucci, M., Lupidi, G., Dall'Acqua, S., & Maggi, F. (2017). Cytotoxic essential oils from Eryngium campestre and *Eryngium amethystinum* (Apiaceae) growing in Central Italy. *Chemistry and Biodiversity*, 14(7), 1–11.

Dash, K. T., Sahoo, A., Ray, A., Jena, S., & Nayak, S. (2024). Assessment of *Piper nigrum* essential oil from Eastern India: Insights into chemical profiling, antimicrobial and antiproliferative potential. *Journal of Biologically Active Products from Nature*, 14(6), 719–734.

de Oliveira, C. M., Graças Cardoso, M. das, Ionta, M., Soares, M. G., Andrade Santiago, J. de, da Silva, G. Á. F., Teixeira, M. L., Selvati Rezende, D. A. de C., de Souza, R. V., Soares, L. I., Nelson, D. L., & Sousa Carvalho, M. S. (2015). Chemical characterization and in vitro antitumour activity of the essential oils from the leaves and flowers of *Callistemon viminalis*. *American Journal of Plant Sciences*, 06(16), 2664–2671.

de Oliveira, P. F., Alves, J. M., Damasceno, J. L., Oliveira, R. A. M., Júnior Dias, H., Crotti, A. E. M., & Tavares, D. C. (2015). Cytotoxicity screening of essential oils in cancer cell lines. *Revista Brasileira de Farmacognosia*, 25(2), 183–188.

Edris, A. E. (2007). Pharmaceutical and therapeutic potentials of essential oils and their individual volatile constituents: A review. *Phytotherapy Research: An International Journal Devoted to Pharmacological and Toxicological Evaluation of Natural Product Derivatives*, 21(4), 308–323.

El-Abid, H., Amaral, C., Cunha, S. C., Augusto, T. V., Fernandes, J. O., Correia-da-Silva, G., Teixeira, N., & Moumni, M. (2019). Chemical composition and anti-cancer properties of *Juniperus oxycedrus* L. essential oils on estrogen receptor-positive breast cancer cells. *Journal of Functional Foods*, 59, 261–271.

Emami, S. A., Asili, J., HosseinNia, S., Yazdian-Robati, R., Sahranavard, M., & Tayarani-Najaran, Z. (2016). Growth inhibition and apoptosis induction of essential oils and extracts of *Nepeta cataria* L. on human prostatic and breast cancer cell lines. *Asian Pacific Journal of Cancer Prevention*, 17, 125–130.

Falcão, D. Q., Mourão, S. C., Araujo, J. L. de, Pereira, P. A. K., Cardoso, A. C. A., Almeida, K. B. de, Zibetti, F. M., & Lima, B. G. (2015). Challenges in development of essential oil nanodelivery systems and future prospects. In: J. Naik (Ed.) *Nano-Based Drug Delivery* (1st ed.). IAPC Publishing, Zagreb, Croatia, pp. 559–575.

Fan, K., Li, X., Cao, Y., Qi, H., Li, L., Zhang, Q., & Sun, H. (2015). Carvacrol inhibits proliferation and induces apoptosis in human colon cancer cells. *Anti-Cancer Drugs*, 26(8), 813–823.

Fazelifar, P., Tabrizi, M. H., & Rafiee, A. (2020). The *Arachis hypogaea* essential oil nanoemulsion as an efficient safe apoptosis inducer in human lung cancer cells (A549). *Nutrition and Cancer*, 5581, 1–9.

Fekrazad, R., Afzali, M., Pasban-Aliabadi, H., Esmaeili-Mahani, S., Aminizadeh, M., & Mostafavi, A. (2017). Cytotoxic effect of *Thymus caramanicus* Jalas on human oral epidermoid carcinoma KB cells. *Brazilian Dental Journal*, 28(1), 72–77.

Gach, K., Długosz, A., & Janecka, A. (2015). The role of oxidative stress in anticancer activity of sesquiterpene lactones. *Naunyn-Schmiedeberg's Archives of Pharmacology*, 388(5), 477–486.

Gao, D., Tang, S., & Tong, Q. (2012). Oleanolic acid liposomes with polyethylene glycol modification: Promising antitumour drug delivery. *International Journal of Nanomedicine*, 7, 3517–3526.

Gao, W., Hua, X., Liao, S., Xiahou, Z., Yang, H., Hu, L., & Chi, Y. (2024). Newly synthesized citral derivatives serve as novel inhibitor in HepG2 cells. *ChemistryOpen*, 14, e202400112.

García, L. T., Leal, A. F., Moreno, É. M., Stashenko, E. E., & Arteaga, H. J. (2017). Differential anti-proliferative effect on K562 leukemia cells of Lippia alba (Verbenaceae) essential oils produced under diverse growing, collection and extraction conditions. *Industrial Crops and Products*, 96, 140–148.

Ghaffari, S. B., Sarrafzadeh, M. H., Fakhroueian, Z., Shahriari, S., & Khorramizadeh, M. R. (2017). Functionalization of ZnO nanoparticles by 3-mercaptopropionic acid for aquessential oilsus curcumin delivery: Synthesis, characterization, and anticancer assessment. *Materials Science and Engineering C*, 79, 465–472.

Girola, N., Figueiredo, C. R., Farias, C. F., Azevedo, R. A., Ferreira, A. K., Teixeira, S. F., Capello, T. M., Martins, E. G. A., Matsuo, A. L., Travassos, L. R., & Lago, J. H. G. (2015). Camphene isolated from essential oil of Piper cernuum (Piperaceae) induces intrinsic apoptosis in melanoma cells and displays antitumour activity in vivo. *Biochemical and Biophysical Research Communications*, 467(4), 928–934.

Govindaraju, S., & Arulselvi, P. I. (2018). Characterization of Coleus aromaticus essential oil and its major constituent carvacrol for in vitro antidiabetic and antiproliferative activities. *Journal of Herbs, Spices and Medicinal Plants*, 24(1), 37–51.

Graidist, P., Martla, M., & Sukpondma, Y. (2015). Cytotoxic activity of Piper cubeba extract in breast cancer cell lines. *Nutrients*, 7(4), 2707–2718.

Guerrini, A., Sacchetti, G., Grandini, A., Spagnoletti, A., Asanza, M., & Scalvenzi, L. (2016). Cytotoxic effect and TLC bioautography-guided approach to detect health properties of Amazonian *Hedyosmum sprucei* essential oil. *Evidence-Based Complementary and Alternative Medicine*, 2016, 1–8.

Hakkim, F. L., Bakshi, H. A., Khan, S., Nasef, M., Farzand, R., Sam, S., Rashan, L., Al-Baloshi, M. S., Hasson, S. S. A. A., Al Jabri, A., McCarron, P. A., & Tambuwala, M. M. (2019). Frankincense essential oil suppresses melanoma cancer through down regulation of Bcl-2/Bax cascade signaling and ameliorates heptotoxicity via phase I and II drug metabolizing enzymes. *Oncotarget*, 10(37), 3472–3490.

Hamidzadeh, K., Christensen, S. M., Dalby, E., Chandrasekaran, P., & Mosser, D. M. (2017). Macrophages and the recovery from acute and chronic inflammation. *Annual Review of Physiology*, 79, 567–592.

Hanušová, V., Caltová, K., Svobodová, H., Ambrož, M., Skarka, A., Murínová, N., Králová, V., Tomšík, P., & Skálová, L. (2017). The effects of β-caryophyllene oxide and trans-nerolidol on the efficacy of doxorubicin in breast cancer cells and breast tumour-bearing mice. *Biomedicine & Pharmacotherapy*, 95, 828–836.

Hassan, N., Firdaus, S., Padhi, S., Ali, A., & Iqbal, Z. (2021). Investigating natural antibiofilm components: A new therapeutic perspective against candidal vulvovaginitis. *Medical Hypotheses*, 148, 110515.

Ho, B. E., Shen, D. D., McCune, J. S., Bui, T., Risler, L., Yang, Z., & Ho, R. J. (2010). Effects of garlic on cytochromes P450 2C9-and 3A4-mediated drug metabolism in human hepatocytes. *Scientia Pharmaceutica*, 78(3), 473–482.

Hosseinpour, M., Abdul, A. B., Rahman, H. S., Rasedee, A., Yeap, S. K., Ahmadi, N., Othman, H. H., & Chartrand, M. S. (2014). Comparison of apoptotic inducing effect of zerumbone and zerumbone-loaded nanostructured lipid carrier on human mammary adenocarcinoma MDA-MB-231 cell line. *Journal of Nanomaterials*, 2014, 742738.

Huang, L., Li, A., Liao, G., Yang, F., Yang, J., Chen, X., & Jiang, X. (2017). Curcumol triggers apoptosis of p53 mutant triple-negative human breast cancer MDA-MB 231 cells via activation of p73 and PUMA. *Oncology Letters*, 14(1), 1080–1088.

Itani, W. S., El-Banna, S. H., Hassan, S. B., Larsson, R. L., Bazarbachi, A., & Gali-Muhtasib, H. U. (2008). Anti colon cancer components from lebanese sage (*Salvia libanotica*) essential oil: Mechanistic basis. *Cancer Biology and Therapy*, 7(11), 1765–1773.

Jabir, M. S., Taha, A. A., Sahib, U. I., Taqi, Z. J., Al-Shammari, A. M., & Salman, A. S. (2019). Novel of nano delivery system for Linalool loaded on gold nanoparticles conjugated with CALNN peptide for application in drug uptake and induction of cell death on breast cancer cell line. *Materials Science and Engineering C*, 94, 949–964.

Jadid, M. F. S., Shademan, B., Chavoshi, R., Seyyedsani, N., Aghaei, E., Taheri, E., Goleij, P., Hajazimian, S., Karamad, V., Behroozi, J., Sabet, M. N., Isazadeh, A., & Baradaran, B. (2021). Enhanced anticancer potency of hydroxytyrosol and curcumin by PLGA-PAA nano-encapsulation on PANC-1 pancreatic cancer cell line. *Environmental Toxicology*, 36, 1–12.

Jamil, B., Abbasi, R., Abbasi, S., Imran, M., Khan, S. U., Ihsan, A., Javed, S., & Bokhari, H. (2016). Encapsulation of cardamom essential oil in chitosan nano-composites: In-vitro efficacy on antibiotic-resistant bacterial pathogens and cytotoxicity studies. *Frontiers in Microbiology*, 7(OCT), 1–10.

Jayakodi, Y., Thiviya, P., Gamage, A., Evon, P., Madhujith, T., & Merah, O. (2024). Antioxidant activity of essential oils extracted from Apiaceae family plants. *Agrochemicals*, 3(1), 57–69.

Joyce Nirmala, M., Durai, L., Gopakumar, V., & Nagarajan, R. (2019). Anticancer and antibacterial effects of a clove bud essential oil-based nanoscale emulsion system. *International Journal of Nanomedicine*, 14, 6439–6450.

Justus, B., Kanunfre, C. C., Budel, J. M., de Faria, M. F., Raman, V., de Paula, J. P., & Farago, P. V. (2019). New insights into the mechanisms of French lavender essential oil on non-small-cell lung cancer cell growth. *Industrial Crops and Products*, 136, 28–36.

Kaur, J., & Kaur, G. (2015). An insight into the role of citrus bioactives in modulation of colon cancer. *Journal of Functional Foods*, 13, 239–261.

Kavaz, D., Idris, M., & Onyebuchi, C. (2019). Physiochemical characterization, antioxidative, anticancer cellsproliferation and food pathogens antibacterial activity of chitosannanoparticles loaded with *Cyperus articulates* rhizome essential oils. *International Journal of Biological Macromolecules*, 123, 837–845.

Khan, I., Bahuguna, A., Kumar, P., Bajpai, V. K., & Kang, S. C. (2018). In vitro and in vivo antitumour potential of carvacrol nanoemulsion against human lung adenocarcinoma A549 cells via mitochondrial mediated apoptosis. *Scientific Reports*, 8(1), 712–714.

Khanavi, M., Enayati, A., Ardekani, M. R. S., Akbarzadeh, T., Razkenari, E. K., & Eftekhari, M. (2019). Cytotoxic activity of juniperus excelsa m. Bieb. leaves essential oil in breast cancer cell lines. *Research Journal of Pharmacognosy*, 6(2), 1–7.

Khatamian, N., Homayouni Tabrizi, M., Ardalan, P., Yadamani, S., & Darchini Maragheh, A. (2019). Synthesis of *Carum Carvi* essential oil nanoemulsion, the cytotoxic effect, and expression of caspase 3 gene. *Journal of Food Biochemistry*, 43(8), 1–8.

Kilinc, K., Demir, S., Turan, I., Mentese, A., Orem, A., Sonmez, M., & Aliyazicioglu, Y. (2020). Rosa canina extract has antiproliferative and proapoptotic effects on human lung and prostate cancer cells. *Nutrition and Cancer*, 72(2), 273–282.

Krifa, M., El Meshri, S. E., Bentouati, N., Pizzi, A., Sick, E., Chekir-Ghedira, L., & Rondé, P. (2016). In vitro and in vivo anti-melanoma effects of Pituranthos tortuosus essential oil via inhibition of FAK and Src activities. *Journal of Cellular Biochemistry*, 117(5), 1167–1175.

Kundu, A., Padhi, S., Behera, A., Hasnain, M. S., & Nayak, A. K. (2022). Tumour targeting strategies by chitosan-based nanocarriers. In: Hasnain, M. S., Beg, S., & Nayak, A. K. (Eds.) *Chitosan in Biomedical Applications*. Academic Press, London, pp. 163–188.

Kuo, Y. Y., Jim, W. T., Su, L. C., Chung, C. J., Lin, C. Y., Huo, C., Tseng, J. C., Huang, S. H., Lai, C. J., Chen, B. C., Wang, B. J., Chan, T. M., Lin, H. P., Chang, W. S. W., Chang, C. R., & Chuu, C. P. (2015). Caffeic acid phenethyl ester is a potential therapeutic agent for oral cancer. *International Journal of Molecular Sciences*, 16(5), 10748–10766.

Lai Shi Min, S., Liew, S. Y., Chear, N. J. Y., Goh, B. H., Tan, W. N., & Khaw, K. Y. (2022). Plant terpenoids as the promising source of cholinesterase inhibitors for anti-AD therapy. *Biology*, 11(2), 307.

Lammari, N., Demautis, T., Louaer, O., Meniai, A. H., Casabianca, H., Bensouici, C., Devouassoux, G., Fessi, H., Bentaher, A., & Elaissari, A. (2021). Nanocapsules containing Saussurea lappa essential oil: Formulation, characterization, antidiabetic, anti-cholinesterase and anti-inflammatory potentials. *International Journal of Pharmaceutics*, 593, 120138.

Li, C. C., Yu, H. F., Chang, C. H., Liu, Y. T., & Yao, H. T. (2018). Effects of lemongrass oil and citral on hepatic drug-metabolizing enzymes, oxidative stress, and acetaminophen toxicity in rats. *Journal of Food and Drug Analysis*, 26(1), 432–438.

Luke, A. M., Patnaik, R., Kuriadom, S. T., Jaber, M., & Mathew, S. (2020). An in vitro study of *Ocimum sanctum* as a chemotherapeutic agent on oral cancer cell-line. *Saudi Journal of Biological Sciences*, 28, 887–890.

Maldonado, C. R., Salassa, L., Gomez-Blanco, N., & Mareque-Rivas, J. C. (2013). Nano-functionalization of metal complexes for molecular imaging and anticancer therapy. *Coordination Chemistry Reviews*, 257(19–20), 2668–2688.

Maragheh, A. D., Tabrizi, M. H., Karimi, E., Seyedi, S. M. R., & Khatamian, N. (2019). Producing the sour cherry pit oil nanoemulsion and evaluation of its anti-cancer effects on both breast cancer murine model and MCF-7 cell line. *Journal of Microencapsulation*, 36(4), 399–409.

Maruoka, T., Kitanaka, A., Kubota, Y., Yamaoka, G., Kameda, T., Imataki, O., Dobashi, H., Bandoh, S., Kadowaki, N., & Tanaka, T. (2018). Lemongrass essential oil and citral inhibit Src/Stat3 activity and suppress the proliferation/survival of small-cell lung cancer cells, alone or in combination with chemotherapeutic agents. *International Journal of Oncology*, 52(5), 1738–1748.

Masuda, Y., Asada, K., Satoh, R., Takada, K., & Kitajima, J. (2015a). Capillin, a major constituent of *Artemisia capillaris* Thunb. flower essential oil, induces apoptosis through the mitochondrial pathway in human leukemia HL-60 cells. *Phytomedicine*, 22(5), 545–552.

Masuda, Y., Kadokura, T., Ishii, M., Takada, K., & Kitajima, J. (2015b). Hinesol, a compound isolated from the essential oils of *Atractylodes lancea* rhizome, inhibits cell growth and induces apoptosis in human leukemia HL-60 cells. *Journal of Natural Medicines*, 69(3), 332–339.

Meena, R., Kumar, S., kumar, R., Gaharwar, U. S., & Rajamani, P. (2017). PLGA-CTAB curcumin nanoparticles: Fabrication, characterization and molecular basis of anticancer activity in triple negative breast cancer cell lines (MDA-MB-231 cells). *Biomedicine and Pharmacotherapy*, 94, 944–954.

Miguel, M. G. (2010). Antioxidant and anti-inflammatory activities of essential oils: A short review. *Molecules*, 15(12), 9252–9287.

Mir, M., Ahmed, N., & Rehman, A. U. (2017). Recent applications of PLGA based nanostructures in drug delivery. *Colloids and Surfaces B: Biointerfaces*, 159, 217–231.

Mishra, D. P., Khan, M. A., Yadav, D. K., Rawat, A. K., Singh, R. K., Ahamad, T., Hussain, M. K., Saquib, M., & Khan, M. F. (2018). Monoterpene indole alkaloids from Anthocephalus cadamba fruits exhibiting anticancer activity in human lung cancer cell line H1299. *Chemistry Select*, 3(29), 8468–8472.

Mishra, T., Arya, R. K., Meena, S., Joshi, P., Pal, M., Meena, B., Upreti, D. K., Rana, T. S., & Datta, D. (2016). Isolation, characterization and anticancer potential of cytotoxic triterpenes from Betula utilis bark. *PLoS ONE*, 11(7), 1–14.

Mitbumrung, W., & Matsukawa, S. (2024). Cellulose nanocrystal stabilized pickering nanoemulsions and their coalescence stability studied by fluorescence microscopy. *Food Hydrocolloids*, 149, 109594.

Mohammed, M. A., Syeda, J. T. M., Wasan, K. M., & Wasan, E. K. (2017). An overview of chitosan nanoparticles and its application in non-parenteral drug delivery. *Pharmaceutics*, 9, 1–26.

Mouchira, A. C. (2017). Chemical composition and anticancer activity of *Achillea fragrantissima* (Forssk.) Sch. Bip. (Asteraceae) essential oil from Egypt. *Journal of Pharmacognosy and Phytotherapy*, 9(1), 1–5.

Mulik, R. S., Mönkkönen, J., Juvonen, R. O., Mahadik, K. R., & Paradkar, A. R. (2012). Apoptosis-induced anticancer effect of transferrin-conjugated solid lipid nanoparticles of curcumin. *Cancer Nano*, 3, 65–81.

Nagoor Meeran, M. F., Javed, H., Al Taee, H., Azimullah, S., & Ojha, S. K. (2017). Pharmacological properties and molecular mechanisms of thymol: Prospects for its therapeutic potential and pharmaceutical development. *Frontiers in Pharmacology*, 8, 380.

Natrajan, D., Srinivasan, S., Sundar, K., & Ravindran, A. (2015). Formulation of essential oil-loaded chitosan–alginate nanocapsules. *Journal of Food and Drug Analysis*, 23(3), 560–568.

Navaei Shoorvarzi, S., Shahraki, F., Shafaei, N., Karimi, E., & Oskoueian, E. (2020). Citrus aurantium L. bloom essential oil nanoemulsion: Synthesis, characterization, cytotoxicity, and its potential health impacts on mice. *Journal of Food Biochemistry*, 44(5), 1–11.

Nirmala, M. J., Durai, L., Gopakumar, V., & Nagarajan, R. (2020). Preparation of celery essential oil-based nanoemulsion by ultrasonication and evaluation of its potential anticancer and antibacterial activity. *International Journal of Nanomedicine*, 15, 7651–7666.

Nordin, N., Yeap, S. K., Rahman, H. S., & Zamberi, N. R. (2020). Antitumour and anti-metastatic effects of citral-loaded nanostructured lipid carrier in 4T1-induced breast cancer mouse model. *Molecules*, 25, 1–17.

Nordin, N., Yeap, S. K., Rahman, H. S., Zamberi, N. R., Abu, N., Mohamad, N. E., How, C. W., Masarudin, M. J., Abdullah, R., & Alitheen, N. B. (2019). In vitro cytotoxicity and anticancer effects of citral nanostructured lipid carrier on MDA MBA-231 human breast cancer cells. *Scientific Reports*, 9(1), 1–19.

Odin, A. P. (1997). Vitamins as antimutagens: Advantages and some possible mechanisms of antimutagenic action. *Mutation Research/Reviews in Mutation Research*, 386(1), 39–67.

Oliviero, M., Romilde, I., Beatrice, M. M., Mattessential oils, V., Giovanna, N., Consuelo, A., Claudio, C., Giorgio, S., Maggi, F., & Massimo, N. (2016). Evaluations of thyme extract effects in human normal bronchial and tracheal epithelial cell lines and in human lung cancer cell line. *Chemico-Biological Interactions*, 256, 125–133.

Ousaid, F. E. Z., Guenaou, I., Irahal, I. N., Lahlou, F. A., Errami, A., Si, L. A., Abdou-Allah, F., Ridaoui, K., Zouheir, Y., Hmimid, F., & Bourhim, N. (2024). The chemical composition, in vitro, and in silico studies of Lavandula mairei essential oil. *Scientific African*, 23, e02062.

Padhi, S., & Behera, A. (2020). Nanotechnology based targeting strategies for the delivery of Camptothecin. In: Ankit, S., Panda Amulya, K., & Eric, L. (Eds.) *Pharmaceutical Technology for Natural Products Delivery, Impact of Nanotechnology*. Springer Nature, Switzerland, pp. 243–272.

Padhi, S., & Behera, A. (2021). Advanced drug delivery systems in the treatment of ovarian cancer. In: Dua, K., Mehta, M., de Jesus Andreoli Pinto, T., Pont, L. G., Williams, K. A., & Rathbone, M. J. (Eds.) *Advanced Drug Delivery Systems in the Management of Cancer*. Academic Press, London, pp. 127–139.

Padhi, S., & Behera, A. (2022a). Cellular internalization and toxicity of polymeric nanoparticles. In: Padhi, S., Behera, A., & Lichtfouse, E. (Eds.) *Polymeric Nanoparticles for Treating Solid Tumours*. Environmental Chemistry for a Sustainable World, vol 71. Springer, Cham, pp. 473–488.

Padhi, S., & Behera, A. (2022b). Biosynthesis of silver nanoparticles: Synthesis, mechanism, and characterization. In: Abd-Elsalam, K. A., Periakaruppan, R., & Rajeshkumar, S. (Eds.) *Agri-Waste and Microbes for Production of Sustainable Nanomaterials*. Elsevier, Amsterdam, pp. 397–440.

Padhi, S., Azharuddin, M., Behera, A., Zakir, F., Mirza, M. A., Chyad, A. A., Iqbal, Z., & Mansoor, S. (2022a). Nanocarriers as delivery tool for COVID-19 drugs. In: Egbuna, C. (Ed.) *Coronavirus Drug Discovery*. Elsevier, Amsterdam, pp. 293–332.

Padhi, S., Behera, A., Hasnain, M. S., & Nayak, A. K. (2022b). Uses of chitosan in drug delivery. In: Hasnain, M. S., Beg, S., & Nayak, A. K. (Eds.) *Chitosan in Drug Delivery*. Academic Press, London, pp. 139–162.

Padhi, S., Dash, M., & Behera, A. (2021). Nanophytochemicals for the treatment of type II diabetes mellitus: A review. *Environmental Chemistry Letters*, 19, 4349–4373.

Padhi, S., Kapoor, R., Verma, D., Panda, A. K., Iqbal, Z. (2018). Formulation and optimization of topotecan nanoparticles: In vitro characterization, cytotoxicity, cellular uptake and pharmacokinetic outcomes. *Journal of Photochemistry and Photobiology B: Biology*, 183, 222–232.

Padhi, S., Mirza, M. A., Verma, D., Khuroo, T., Panda, A. K., Talegaonkar, S., Khar, R. K., & Iqbal, Z. (2016). Revisiting the nanoformulation design approach for effective delivery of Topotecan in its stable form: An appraisal of its in vitro behavior and tumour amelioration potential. *Drug Delivery*, 23(8), 2827–2837.

Padhi, S., Nayak, A. K., & Behera, A. (2020). Type II diabetes mellitus: A review on recent drug-based therapeutics. *Biomedicine & Pharmacotherapy*, 131, 110708.

Padhi, S., Pradhan, S. P., & Behera, A. (2022c). Methods to formulate polymeric nanoparticles. In: Padhi, S., Behera, A., & Lichtfouse, E. (Eds.) *Polymeric Nanoparticles for the Treatment of Solid Tumors*. Springer International Publishing, Cham, pp. 51–74.

Pan, W., & Zhang, G. (2019). Linalool monoterpene exerts potent antitumour effects in OECM 1 human oral cancer cells by inducing sub-G1 cell cycle arrest, loss of mitochondrial membrane potential and inhibition of PI3K/AKT biochemical pathway. *Journal of BUON*, 24(1), 323–328.

Parham, S., Kharazi, A. Z., Bakhsheshi-Rad, H. R., Nur, H., Ismail, A. F., Sharif, S., Ramakrishna, S., & Berto, F. (2020). Antioxidant, antimicrobial and antiviral properties of herbal materials. *Antioxidants*, 9(12), 1309.

Patel, J. K., & Patel, A. P. (2019). Passive targeting of nanoparticles to cancer. In: Pathak, Y. V. (Ed.) *Surface Modification of Nanoparticles for Targeted Drug Delivery*. Springer, Cham, pp. 125–143.

Periasamy, V. S., Athinarayanan, J., & Alshatwi, A. A. (2016a). Anticancer activity of an ultrasonic nanoemulsion formulation of Nigella sativa L. essential oil on human breast cancer cells. *Ultrasonics Sonochemistry*, 31, 449–455.

Periasamy, V. S., Athinarayanan, J., & Alshatwi, A. A. (2016b). Anticancer activity of an ultrasonic nanoemulsion formulation of Nigella sativa L. essential oil on human breast cancer cells. *Ultrasonics Sonochemistry*, 31, 449–455.

Pisani, M., Quassinti, L., Bramucci, M., Galassi, R., Maggi, F., Rossi, B., Damin, A., Carloni, P., & Astolfi, P. (2020). Nanostructured liquid crystalline particles as delivery vectors for isofuranodiene: Characterization and in-vitro anticancer activity. *Colloids and Surfaces B: Biointerfaces*,192, 111050.

Pradhan, S. P., Behera, A., & Sahu, P. K. (2024b). Effect of selenium nanoparticles conjugated vildagliptin on cognitive dysfunction associated with diabetes mellitus. *Journal of Drug Delivery Science and Technology*, 98, 105907.

Pradhan, S. P., Tejaswani, P., Behera, A., & Sahu, P. K. (2024a). Phytomolecules from conventional to nano form: Next-generation approach for Parkinson's disease. *Ageing Research Reviews*, 93, 102136.

Pudełek, M., Catapano, J., Kochanowski, P., Mrowiec, K., Janik-Olchawa, N., Czyż, J., & Ryszawy, D. (2019). Therapeutic potential of monoterpene α-thujone, the main compound of Thuja occidentalis L. essential oil, against malignant glioblastoma multiforme cells in vitro. *Fitoterapia*, 134(February), 172–181.

Qi, F., Yan, Q., Zheng, Z., Liu, J., Chen, Y., & Zhang, G. (2018). Geraniol and geranyl acetate induce potent anticancer effects in colon cancer Colo-205 cells by inducing apoptosis, DNA damage and cell cycle arrest. *Journal of BUON*, 23(2), 346–352.

Quassinti, L., Maggi, F., Ortolani, F., Lupidi, G., Petrelli, D., Vitali, L. A., Miano, A., & Bramucci, M. (2019). Exploring new applications of tulip tree (*Liriodendron tulipifera* L.): Leaf essential oil as apoptotic agent for human glioblastoma. *Environmental Science and Pollution Research*, 26(29), 30485–30497.

Rahman, H. S., Rasedee, A., Abdul, A. B., Zeenathul, N. A., Othman, H. H., Yeap, S. K., How, C. W., & Wan Nor Hafiza, W. A. G. (2014). Zerumbone-loaded nanostructured lipid carrier induces G2/M cell cycle arrest and apoptosis via mitochondrial pathway in a human lymphoblastic leukemia cell line. *International Journal of Nanomedicine*, 9(1), 527–538.

Rajivgandhi, G., Saravanan, K., Ramachandran, G., Li, J. L., Yin, L., Quero, F., Alharbi, N. S., Kadaikunnan, S., Khaled, J. M., Manoharan, N., & Li, W. J. (2020). Enhanced anti-cancer activity of chitosan loaded Morinda citrifolia essential oil against A549 human lung cancer cells. *International Journal of Biological Macromolecules*, 164, 4010–4021.

Rodenak-Kladniew, B., Islan, G. A., Bravo, M. G. de, Durán, N., & Castro, G. R. (2017). Design, characterization and in vitro evaluation of linalool-loaded solid lipid nanoparticles as potent tool in cancer therapy. *Colloids and Surfaces B: Biointerfaces*, 154, 123–133.

Russo, A., Cardile, V., Graziano, A. C. E., Avola, R., Bruno, M., & Rigano, D. (2018). Involvement of Bax and Bcl-2 in induction of apoptosis by essential oils of three Lebanese salvia species in human prostate cancer cells. *International Journal of Molecular Sciences*, 19(1), 292.

Russo, A., Cardile, V., Graziano, A. C. E., Formisano, C., Rigano, D., Canzoneri, M., Bruno, M., & Senatore, F. (2015). Comparison of essential oil components and in vitro anticancer activity in wild and cultivated Salvia verbenaca. *Natural Product Research*, 29(17), 1630–1640.

Saab, A. M., Guerrini, A., Sacchetti, G., Maietti, S., Zeino, M., Arend, J., Gambari, R., Bernardi, F., & Efferth, T. (2012). Phytochemical analysis and cytotoxicity towards multidrug-resistant leukemia cells of essential oils derived from lebanese medicinal plants. *Planta Medica*, 78(18), 1927–1931.

Saengkrit, N., Saesoo, S., Srinuanchai, W., Phunpee, S., & Ruktanonchai, U. R. (2014). Influence of curcumin-loaded cationic liposome on anticancer activity for cervical cancer therapy. *Colloids and Surfaces B: Biointerfaces*, 114, 349–356.

Sahoo, P., Dey, J., Mahapatra, S. R., Ghosh, A., Jaiswal, A., Padhi, S., Prabhuswamimath, S. C., Misra, N., & Suar, M. (2022). Nanotechnology and COVID-19 convergence: Toward new planetary health interventions against the pandemic. *OMICS: A Journal of Integrative Biology*, 26(9), 473–488.

Sak, K., Nguyen, T. H., Ho, V. D., Do, T. T., & Raal, A. (2017). Cytotoxic effect of chamomile (*Matricaria recutita*) and marigold (*Calendula officinalis*) extracts on human melanoma SK-MEL-2 and epidermoid carcinoma KB cells. *Cogent Medicine*, 4(1), 1–7.

Salari, S., & Salari, R. (2019). Nanoliposomal system of rosemary essential oil made by specific human cell phospholipids and evaluation of its anti-cancer properties. *Applied Nanoscience*, 9(8), 2085–2089.

Salehi, F., Behboudi, H., Kavoosi, G., & Ardestani, S. K. (2020). Incorporation of *Zataria multiflora* essential oil into chitosan biopolymernanoparticles: A nanoemulsion based delivery system to improve thein-vitro efficacy, stability and anticancer activity of ZESSENTIAL OILS against breastcancer cells. *International Journal of Biological Macromolecules*, 143, 382–392.

Sani, T. A., Mohammadpour, E., Mohammadi, A., Memariani, T., Yazdi, M. V., Rezaee, R., Calina, D., Docea, A. O., Goumenou, M., Etemad, L., & Shahsavand, S. (2017). Cytotoxic and apoptogenic properties of *Dracocephalum kotschyi* aerial part different fractions on calu-6 and mehr-80 lung cancer cell lines. *Farmacia*, 65(2), 189–199.

Santana, N. A. De, Silva, R. C. S. da, Fourmentin, S., Anjos, K. F. L. dos, Ootan, M. A., Silva, A. G. da, Araújo, B. G. P., Correia, M. T. dos S., Silva, M. V. da, & Machado, G. (2020). Synthesis, characterization and cytotoxicity of the *Eugenia brejoensis* essential oil inclusion complex with β-cyclodextrin. *Journal of Drug Delivery Science and Technology*, 60, 101876.

Satpathy, B., Sa, N., Behera, A., & Sahu, P. K. (2024). Dose-dependent attenuation of the efficacy of Clitoria ternatea by cobalt oxide nanoparticles against diabetes-induced cognitive impairment. *Molecular Neurobiology*, 62, 2601–2616.

Sebaaly, C., Jraij, A., Fessi, H., Charcosset, C., & Greige-Gerges, H. (2015). Preparation and characterization of clove essential oil-loaded liposomes. *Food Chemistry*, 178, 52–62.

Sharma, M., Grewal, K., Jandrotia, R., Batish, D. R., Singh, H. P., & Kohli, R. K. (2022). Essential oils as anticancer agents: Potential role in malignancies, drug delivery mechanisms, and immune system enhancement. *Biomedicine & Pharmacotherapy*, 146, 112514.

Shetty, V., Jakhade, A., Shinde, K., Chikate, R., & Kaul-Ghanekar, R. (2021). Folate mediated targeted delivery of cinnamaldehyde loaded and FITC functionalized magnetic nanoparticles in breast cancer: In vitro, in vivo and pharmacokinetic studies. *New Journal of Chemistry*, 45, 1500–1515.

Shi, X., Wang, Z., Liu, S., Xia, Q., Liu, Y., Chen, W., Yu, H., & Zhang, K. (2024). Scalable production of carboxylated cellulose nanofibres using a green and recyclable solvent. *Nature Sustainability*, 7(3), 315–325.

Shreya, A. B., Raut, S. Y., Managuli, R. S., Udupa, N., & Mutalik, S. (2019). Active targeting of drugs and bioactive molecules via oral administration by ligand-conjugated lipidic nanocarriers: Recent advances. *AAPS PharmSciTech*, 20(1), 30564942.

Silva-Filho, C. J. A., Freitas, P. G. C., Oliveira, F. C. E., Barbosa, F. G., Oliveira, M. C. F., Eloy, J. O., Pessoa, C., & Mafezoli, J. (2020). Nanoencapsulation of triterpene 3β,6β,16β-trihydro xylup-20(29)-ene from Combretum leprosum as strategy to improve its cytotoxicity against cancer cell lines. *Bioorganic and Medicinal Chemistry Letters*, 30(20), 127469.

Spyridopoulou, K., Tiptiri-Kourpeti, A., Lampri, E., Fitsiou, E., Vasileiadis, S., Vamvakias, M., Bardouki, H., Goussia, A., Malamou-Mitsi, V., Panayiotidis, M. I., Galanis, A., Pappa, A., & Chlichlia, K. (2017). Dietary mastic oil extracted from Pistacia lentiscus var. chia suppresses tumour growth in experimental colon cancer models. *Scientific Reports*, 7(1), 1–14.

Sun, J., Bi, C., Chan, H. M., Sun, S., Zhang, Q., & Zheng, Y. (2013). Curcumin-loaded solid lipid nanoparticles have prolonged in vitro antitumour activity, cellular uptake and improved in vivo bioavailability. *Colloids and Surfaces B: Biointerfaces*, 111, 367–375.

Sun, X. B., Wang, S. M., Li, T., & Yang, Y. Q. (2015). Anticancer activity of linalool terpenoid: Apoptosis induction and cell cycle arrest in prostate cancer cells. *Tropical Journal of Pharmaceutical Research*, 14(4), 619–625.

Tamil Selvi, M., Thirugnanasampandan, R., & Sundarammal, S. (2015). Antioxidant and cytotoxic activities of essential oil of Ocimum canum Sims. from India. *Journal of Saudi Chemical Society*, 19(1), 97–100.

Thapa, D., Richardson, A. J., Zweifel, B., Wallace, R. J., & Gratz, S. W. (2019). Genoprotective effects of essential oil compounds against oxidative and methylated DNA damage in human colon cancer cells. *Journal of Food Science*, 84(7), 1979–1985.

Torres-Alvarez, C., Castillo, S., Sánchez-García, E., Aguilera González, C., Galindo-Rodríguez, S. A., Gabaldón-Hernández, J. A., & Báez-González, J. G. (2020). Inclusion complexes of concentrated orange oils and β-cyclodextrin: Physicochemical and biological characterizations. *Molecules*, 25(21), 1–17.

Trindadea, G. G. G., Thrivikraman, G., Menezesa, P. P., Françab, M. C., Limaa, S. B., Carvalhoa, M. B. G. Y., Souzaa, P. B. S. S. E., Duartea, C. M., Shanmugama, S., Quintans-Júniorc, J. L., Bezerrad, P. D., Bertassoni, E. L., & Araújo, A. A. S. (2019). Carvacrol/β-cyclodextrin inclusion complex inhibits cell proliferation and migration of prostate cancer cells. *Food and Chemical Toxicology*, 125, 198–209.

Turan, I., Demir, S., Kilinc, K., Yaman, S. O., Misir, S., Kara, H., Genc, B., Mentese, A., Aliyazicioglu, Y., & Deger, O. (2018). Cytotoxic effect of Rosa canina extract on human colon cancer cells through repression of telomerase expression. *Journal of Pharmaceutical Analysis*, 8(6), 394–399.

Uma Maheswari, P., Muthappa, R., Bindhya, K. P., & Meera Sheriffa Begum, K. M. (2020). Evaluation of folic acid functionalized BSA-$CaFe_2O_4$ nanohybrid carrier for the controlled delivery of natural cytotoxic drugs hesperidin and eugenol. *Journal of Drug Delivery Science and Technology*, 61, 1–13.

Wang, W., Chen, T., Xu, H., Ren, B., Cheng, X., Qi, R., Liu, H., Wang, Y., Yan, L., Chen, S., Yang, Q., & Chen, C. (2018). Curcumin-loaded solid lipid nanoparticles enhanced anticancer efficiency in breast cancer. *Molecules Article*, 23, 1–13.

Wang, W., Li, N., Luo, M., Zu, Y., & Efferth, T. (2012). Antibacterial activity and anticancer activity of Rosmarinus officinalis L. essential oil compared to that of its main components. *Molecules*, 17(3), 2704–2713.

Weisany, W., Yousefi, S., Tahir, N. A. R., Golestanehzadeh, N., McClements, D. J., Adhikari, B., & Ghasemlou, M. (2022). Targeted delivery and controlled released of essential oils using nanoencapsulation: A review. *Advances in Colloid and Interface Science*, 303, 102655.

Yang, B., Jiang, J., Jiang, L., Zheng, P., Wang, F., Zhou, Y., Chen, Z., ILi, M., Lian, M., Tang, S., Liu, X., Peng, H., & Wang, Q. (2020). Chitosan mediated solid lipid nanoparticles for enhanced liver delivery ofzedoary turmeric oilin vivo. *International Journal of Biological Macromolecules*, 149, 108–115.

Yessential Oils, S. K., Ali, A. Y., Hayward, O. A., Turnham, D., Jackson, T., Bowen, I. D., & Clarkson, R. (2016). β-Bisabolene, a sesquiterpene from the essential oil extract of Opoponax (Commiphora guidottii) exhibits cytotoxicity in breast cancer cell lines. *Phytotherapy Research*, 30(3), 418–425.

Yousefian Rad, E., Homayouni Tabrizi, M., Ardalan, P., Seyedi, S. M. R., Yadamani, S., Zamani-Esmati, P., & Haghani Sereshkeh, N. (2020). Citrus lemon essential oil nanoemulsion (CLESSENTIAL OILS-NE), a safe cell-depended apoptosis inducer in human A549 lung cancer cells with anti-angiogenic activity. *Journal of Microencapsulation*, 37(5), 394–402.

Zhao, Y., Chen, R., Wang, Y., & Yang, Y. (2018). A-pinene inhibits human prostate cancer growth in a mouse xenograft model. *Chemotherapy*, 63(1), 1–7.

Zhao, Y., Chen, R., Wang, Y., Qing, C., Wang, W., & Yang, Y. (2017). In vitro and in vivo efficacy studies of Lavender angustifolia essential oil and its active constituents on the proliferation of human prostate cancer. *Integrative Cancer Therapies*, 16(2), 215–226.

19 Nanoenabled Systems for the Delivery of Carotenoids as Anticancer Agents

Sumit Sharma[1], *Sonali Batra*[2], *Lalit Kumar*[3], *Valamla Bhavana*[4], *and Neelesh Kumar Mehra*[4]

[1]School of Pharmaceutical Sciences, Delhi Pharmaceutical Sciences and Research University, New Delhi, India

[2]Department of Pharmaceutical Sciences, Indira Gandhi University, Meerpur, Rewari, Haryana, India

[3]Department of Pharmaceutics, Shiva Institute of Pharmacy, Near Chandpur, Bilaspur, Himachal Pradesh, India

[4]Pharmaceutical Nanotechnology Research Laboratory, Department of Pharmaceutics, National Institute of Pharmaceutical Education and Research, Hyderabad, India

19.1 INTRODUCTION

Carotenoids are natural compounds found in plants as coloured pigments and are part of the human diet as a micronutrient. Scientific reports have shown that carotenoids are biologically active compounds with profound anticancer activity. This makes carotenoids a new frontier in anticancer research. Chemically, carotenoids are tetraterpenoids commonly synthesized in plants, photosynthetic organisms, and non-photosynthetic bacteria and fungi (Niranjana et al., 2015). Epidemiological studies, supported by *in vitro* and *in vivo* studies, support the linkage of carotenoids with apoptosis and reducing metastasis in cancer cells. For instance, β-carotene showed an antigenotoxic effect (Aissa et al., 2012), lycopene in combination with other carotenoids has shown an inhibitory effect on androgen receptor activity in prostate cancer (Zu et al., 2014), curcumin inhibited progression of glioblastoma (Chen et al., 2020), lutein showed anticancer activity against breast cancer (Gong et al., 2018), the antiproliferative effect of zeaxanthin against uveal melanoma cells (Juin et al., 2018) and β-cryptoxanthin showed antitumour effect against 4-[methyl nitrosamine]-1-[3-pyridyl]-1-butanone induced lung tumorigenesis (Iskandar et al., 2016), etc.

Among the challenges associated with carotenoids, which may influence the therapeutic activity and pharmaceutical applications of these bioactive compounds, is their aqueous solubility, which is compromised for most carotenoids. Secondly, the size and shape of these compounds may interfere with cellular membrane internalization; for instance, a - trans isomer of carotenoids can aggregate as compared to the - cis isomer. In general, *cis* isomers of carotenoids are more polar and possess better bioavailability and biological function (Honda et al., 2019). Moreover, polar groups at the end of the chain, as observed in lutein and zeaxanthin, also facilitate the aqueous interaction of carotenoids. Chirality is another characteristic attribute followed in carotenoids with C-3 of the 3-hydroxy-β-ring (Black et al., 2020). A conjugated polyene chain in the chemical structure of carotenoids makes them susceptible to oxidation during storage, another matter of concern with carotenoids that reduces their bioactivity. A characteristic instability of carotenoids is observed towards light, oxygen, and temperature. The shortcomings of these bioactive compounds lead to low oral bioavailability and compromised therapeutic efficacy. Considering the shortfalls mentioned above, nanotechnology confers improved stability and aqueous solubility, targeting cancerous tissues, enhancing bio-accessibility to carotenoids, and thus increasing their pharmaceutical applications (Padhi et al., 2016; Kundu et al., 2022). In context to the application in the treatment of cancer, nanostructured carriers like micelles, nanoliposomes, nanoemulsions, biopolymeric nanoparticles, and solid lipid nanoparticles must be evaluated for polydispersity in nanosize range, stability, morphology, drug loading, and tethering capability to facilitate targeting (Dos Santos et al., 2018; Rostamabadi et al., 2019; Pradhan et al., 2021; Behera and Padhi, 2022; Padhi et al., 2018).

19.2 MECHANISMS OF CAROTENOIDS ACTING AS AN ANTICANCER AGENT

Certain pathologic variables and pathways are involved in the development of cancer. The unique mechanism of action of various anticancer drugs can be better comprehended with the concourse of an in-depth investigation of these pathways. The following processes are listed in the literature as ways that carotenoids are said to have an anticancer effect.

19.2.1 Matrix Metalloproteinase-9 Mediated Apoptosis

Matrix metalloproteinase-9 is classified as a gelatinase subgroup of the matrix metalloproteinase family,

DOI: 10.1201/9781003611516-24

zinc-dependent endopeptidases. Matrix metalloproteinase-9 is one of the biomarkers that play a vital role in the pathological processes of cancer, such as cancer cell invasion, tumour metastasis, and angiogenesis. Matrix metalloproteinase-9 has a prominent effect on extracellular matrix remodelling and membrane protein cleavage, mediating the tumour microenvironment. Moreover, matrix metalloproteinase-9 is most frequently known to promote cancer progression by multiple mechanisms, like growth factor, cytokine activation, and extracellular matrix remodelling (Juric et al., 2018). For this reason, matrix metalloproteinase-9 is one of the target biomarkers for sensing cancer. It is observed that matrix metalloproteinase-9 has both promoting and suppressive roles (in special cases like colon cancer) in cancer progression (Garg et al., 2010; Lewins Walter et al., 2017). This dual role of matrix metalloproteinase-9 depends on the factors of the tumour stage, enzyme localization (tumour cells, stroma), tumour site (primary or metastasis), and substrate profile (Deryugina & Quigley, 2006; Gialeli et al., 2011; Hofmann et al., 2000; Huang, 2018). This concluded that more aggressive cancer progression was observed in matrix metalloproteinase-9 deficient mice, and a comparatively low cancer incidence was observed in matrix metalloproteinase-9 knockdown mice (Sharma, Kaur et al., 2025; Deryugina & Quigley, 2006).

A study observed that with an IC_{50} value of approximately 0.484 µg/mL, the IC_{50} carotenoid of *Natrialba* sp was observed. M6 showed potent anticancer activity compared to 5-fluorouracil with an IC_{50} of about 175 µg/mL. The cytotoxicity of the extracted pigment was confirmed by a significant inhibitory effect on matrix metalloproteinase-9, responsible for the progression of cancer pathological processes. Moreover, the extracted carotenoids also showed caspase 3/7 mediated apoptosis anticancer effect (Hegazy et al., 2020).

19.2.2 Pro-apoptotic and Anti-apoptotic Proteins Mediated Apoptosis

Bcl-2-associated X protein and Bcl-2 antagonist killer 1 are pro-apoptotic factors that promote the release of mitochondrial cytochrome C in the cytosol and increase its outer membrane permeability. Consequently, cytochrome C and other factors like pro-caspase-9 and apoptosis protease activating factor-1 form a macromolecular complex that further activates caspase-9. Further, caspase-9, an intrinsic pathway, leads to the cleavage of cellular substrates and causes apoptosis through the intrinsic pathway. On the other hand, the anti-apoptotic factors like B-cell lymphoma 2, B-cell lymphoma-extra large, and myeloid cell leukaemia sequence 1 bind to the outer mitochondrial membrane, inhibit the cytochrome C release into the cytosol and prevent apoptosis. In conclusion, the pro-apoptotic proteins induce caspase activation by facilitating apoptosome complexation, whereas anti-apoptotic proteins promote the process. Hence, the ratio of anti-apoptotic (B-cell lymphoma 2) and pro-apoptotic (Bcl-2-associated X protein) proteins is a critical determinant in determining the survival of cells (Khodapasand et al., 2015; Papaliagkas et al., 2007).

Carotenoids like β-carotene are reported in the literature to reduce the expression of B-cell lymphoma 2. In contrast, no significant activity is reported against the levels of Bcl-2-associated X protein (anti-apoptotic protein). Therefore, the ratio of Bcl-2-associated X protein/B-cell lymphoma 2 is significantly increased after β-carotene treatment, which further increases the expression of caspase-9 and, subsequently, caspase-3 mediated apoptosis (Koklesova et al., 2020; Sowmya Shree et al., 2017). Similarly, lycopene, A nonoxygenated red-coloured carotenoid, and fucoxanthin are also reported to induce apoptosis through the same mechanism, *i.e.* expressing caspase-3, -9 and increasing Bcl-2-associated X protein/B-cell lymphoma 2 ratios in pancreatic and lung cancer cells (Chen et al., 2019; Jeong et al., 2019).

19.2.3 Reactive Oxygen Species (ROS) Mediated Apoptosis

Carotenoids like β-cryptoxanthin (≤10 µM) (Gansukh et al., 2020), Bixin (50–100 µM) (de Oliveira Júnior et al., 2019), lycopene (2 µM) (Yeh & Hu, 2000), lutein (~5 µM), astaxanthin (20 µM) (Sowmya et al., 2017), and fucoxanthin (≥5 µM) (Liu et al., 2011) are suggested to show pro-oxidant action, particularly in cancer cells. In normal physiological conditions, carotenoids give electrons with lipid peroxyl radical and are regenerated into neutral carotenoids upon interaction with endogenous antioxidants like ascorbate and tocopherol. Hence, in normal conditions, carotenoids act as an antioxidant. In contrast, in cancer cells, there are reduced levels of their redox partners, i.e., endogenous antioxidant enzymes, and hence, they cannot regenerate into neutral carotenoids. Consequently, instead of antioxidant activity, these carotenoids increase the level of ROS in cancer cells and catalyse the radical chain reactions. The elevated levels of carotenoid radicals in cancer cells result in significant cellular damage and act as pro-oxidants by triggering oxidative stress (Arathi et al., 2018; Shin et al., 2020).

19.3 NANO-ENABLED TARGETING STRATEGIES FOR THE CAROTENOIDS

Targeted strategies for cancer provide a convenient approach that takes the edge off by reducing the adverse effects of the conventional approach. After drug administration through any route (orally or parenterally), the drug reaches the site of cancer cells either by passive targeting or by active targeting (Behera & Padhi, 2020; Behera & Padhi, 2022; Padhi et al., 2018). Figure 19.1 demonstrates the mechanism of drug targeting. In passive targeting, the drug utilizes the leaky vasculature of the cancer site where the drug accumulates in high concentration compared to normal cells. This process is popularly known as the enhanced permeation

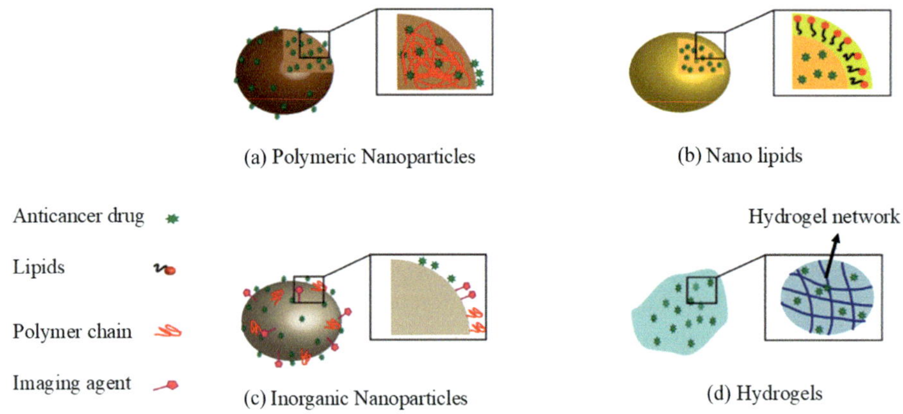

FIGURE 19.1 Nano-enabled targeting approaches for cancer. (Source: Open access from Ketabat et al. (2019).)

and retention effect. The drug reaches through diffusion into the extracellular matrix and then into the cytoplasm through endocytosis. On the other hand, in active targeting, the nanocarriers, through functionalization with specific ligands, utilize the specific receptors of cancer cells and render the normal cells unaffected. The additional advantage of active targeting over passive targeting is that it increases the specificity of cancer cells and offers increased efficacy and reduced adverse effects (Hirsjärvi et al., 2011; Padhi & Behera, 2020).

19.3.1 Passive Targeting Strategy

Among the various strategies to target chemotherapeutic agents to cancer cells, passive targeting has been the most explored strategy. As discussed, passive targeting utilizes cancer cells' leaky vasculature for internalization and circumvents the normal cells (Sharma, Batra et al., 2023; Padhi & Behera, 2022; Behera et al., 2023). In this strategy, the major challenge is recognizing the cancer site and enabling precise drug targeting. This is an important concern with drug molecules having limited solubility, poor pharmacokinetic profile, and high toxicity. Nanocarriers hold a promising approach to circumventing this challenge, where such delivery systems not only improve the solubility but also facilitate precise drug targeting of the compromised vasculature of cancer cells.

Moreover, defective lymphatic drainage systems, particularly in solid tumours and some metastatic tumours, offer enhanced nanocarrier permeability and retention (Figure 19.1) (Patel & Patel, 2019). The enhanced permeability and retention phenomenon has been shown to increase drug accumulation in a tumour by approximately 70-folds in some cases (Duncan, 2003). Therefore, such tumours permit the accumulation of high molecular weight drug molecules and nanocarriers, particularly in the 20–500 nm range. The circulation of these nanocarriers holds another crucial factor that influences passive targeting. Nanoparticles with a size of more than 10 nm are thought to bypass the clearance through filtration by the kidney. However, it was also observed that nanoparticles with a size range of approximately 100–200 nm undergo faster clearance than nanoparticles with a size less than 100 nm. The probable reason for such observation was rapid clearance by opsonization and then the reticuloendothelial system. Hence, an appropriate size range of nanocarriers increases residence time in circulation and facilitates passive targeting (Liechty & Peppas, 2012; Patnaik et al., 2021; Behera & Padhi, 2022).

19.3.2 Active Targeting Strategy

Passive targeting ensures drug accumulation in the cancer cells. It enhances the therapeutic efficiency of nanocarriers loaded with cytotoxic molecules and reduces adverse effects (Sharma, Batra et al., 2023; Padhi et al., 2022). Furthermore, after drug accumulation at the tumour site, the therapeutic efficiency can be increased by the phenomenon known as active targeting. In active targeting, nanocarriers generally conjugate with specific potential ligands. These ligands are usually referred to as stimuli-responsive, which means ligands are supposed to sense the stimulus, which can be either any overexpressed receptor or a specific microenvironment at the targeting site. Hence, these ligands explore the type of stimulus, guide the nanocarriers to the targeting site, facilitate targeting leading to drug accumulation, and prevent toxicity to normal cells. Some of the commonly used ligands for active targeting in different cancer cells are anti-human epidermal growth factor receptor 2 monoclonal antibodies in breast cancer cells against human epidermal growth factor receptor 2 receptor, transferrin against transferrin receptor, and folic acid against over-expressed folate receptors (Sharma, Batra et al., 2023; Salahpour, 2019).

The tumour microenvironment is another distinct feature of cancer cells that assists in active targeting. Due to the high metabolic rate, the cancer cells develop an acidic environment at the site. Therefore, after passive absorption in the cancer tissues, the nanocarriers are exposed to the acidic microenvironment of cancer cells. Here, pH-sensitive polymeric nanocarriers loaded with the drug ensure the drug targets and releases the maximum amount of drug at the cancerous site, leaving normal cells unaffected. Moreover, cancer cells are reported to release pro-inflammatory and inflammatory cytokines, leading to a cascade of inflammation resulting in hyperthermia at the local site. Hence, researchers have explored this pathological condition for active targeting of carotenoids through a local application. For instance, PEGylated thermos-sensitive poly(*N*-isopropyl acrylamide) based nanoparticles were developed to improve the therapeutic activity of poorly water-soluble lycopene. A hyperthermia condition at the inflamed tumour site was the motivational factor for developing such a system. PEGylated-poly(*N*-isopropyl acrylamide) has an edge over the poly(*N*-isopropyl acrylamide) nanoparticles, as PEGylated-poly(*N*-isopropyl acrylamide) has a higher critical solution temperature as compared to poly(*N*-isopropyl acrylamide). This ensures the selective drug targeting in the tumour comprising hyperthermia and an inflamed condition. Thereby results in increased drug accumulation in the tumour by thermosensitive drug release mechanism. The study reported a better therapeutic chemopreventive activity of lycopene by PEGylated thermosensitive nanoparticles than free lycopene in the melanoma B16 cell line (Bano et al., 2020).

19.3.3 Nanocarriers for Passive and Active Targeting of Carotenoids

Despite several health benefits of carotenoids, including promising anticancer activity, these bioactives are associated with numerous challenges like instability during storage and processing, variable pharmacokinetics, poor water solubility, and biopharmaceutical properties (Kumar, Verma et al., 2023; Jain et al., 2015). All these challenges limit their pharmaceutical application. Nanotechnology offers remarkable carrier systems like nanoemulsions, liposomes, and lipid or polymer-based nanoparticles, which are explored experimentally to improve the aqueous solubility and avoid physicochemical instability during the processing and storage of carotenoids. For instance, nanosized amphiphilic colloidal dispersions like nanoemulsions and liposomes are promising frontiers of research among scientists and pharmaceutical product developers. This is because such systems have unique core-shell amphiphilic nanostructures presenting an ideal carrier for poorly water-soluble carotenoids and the ability to facilitate cell membrane or tissue internalization. Thus, offering a targeted approach, particularly in tumours with leaky vasculature and a compromised clearance system, enhances cytotoxic carotenoid accumulation in cancer cells. Moreover, along with possibilities to improve bioaccessibility, pharmacodynamic and pharmacokinetic attributes, and prevent susceptible degradation of these poorly water-soluble compounds, this technology unlocks the possibility of combination therapy with standard cytotoxic drugs. Hence, as an adjunct, the carotenoids also reduce the dose of standard cytotoxic drugs, thus reducing side effects caused to normal cells due to higher doses of these drugs. Various nanostructure assemblies are being studied to deliver carotenoids for the treatment of cancer, as highlighted below:

19.3.3.1 Liposomes

Liposomes are colloidal nanodispersions with amphiphilic attributes best suited for encapsulating hydrophobic and hydrophilic molecules. These nanostructures impart chemical stability to the encapsulated molecules enclosed in their unique bilayer structure and also facilitate targeting the cancer cells. One promising approach is using pH-sensitive liposomes, which usually remain stable in normal cells' alkaline or neutral pH (~7.0). In contrast, these liposomes tend to dissociate in the acidic environment of tumour tissues. The acidic pH of the extracellular matrix in the proximity of tumour cells ranges from 5.5 to 7.6 (Alshahrani, Thotakura et al., 2023; Tannock & Rotin, 1989). In addition to this, liposomes also allow ligand attachment to their basic structure, such as folate group coating. Higher cellular uptake of folate-coated liposomes was observed in breast cancer cells of BALB/c mice, which ensures the targeting of cancer cells. Moreover, pH sensitivity imparted by di oleoyl phosphatidyl ethanolamine in liposomes provides the maximum drug release into the acidic microenvironment of tumour cells (de Oliveira Silva et al., 2019).

Another category is cationic liposomes, which have an affinity to interact with negatively charged genetic components like DNA. Also, these nanocarriers potentiate the permeation through negatively charged cellular membranes through the transcellular adsorptive-mediated endocytosis pathway (Joshi et al., 2015; Shikha et al., 2012). Thus, cationic liposomes doped with carotenoids enhance their anticancer activity. For instance, lutein and β-carotene were successfully loaded into the cationic liposomes with approximately >90% entrapment efficiency where the cationic charge was incorporated by stearyl amine. Carotenoids are classified into two groups depending on the presence of oxygen in their chemical structure: oxygen deficit, i.e. carotenes and oxygen-containing, i.e. xanthophylls. This was explained by Elkholy et al. (2020) as the xanthophylls impart remarkable rigidity to the lipidic-bilayer membrane fluidity compared to carotenes. The possible reason suggested was polar xanthophylls like lutein tend to form strong hydrogen bonds with the polar head of liposomes at the interior side and are usually oriented perpendicularly to the membrane. As a result, it immobilizes the compound inside the core, reduces membrane fluidity, and imparts structural stability to liposomes. On the other hand, carotene-like β-carotene does not present such a rigidifying effect on the membrane. Moreover, due to this structural specificity, lutein showed comparatively higher encapsulation efficiency, slow drug release, and ultimately reduced rate of drug leakage than β-carotene (Elkholy et al., 2020).

19.3.3.2 Nanoemulsion

Nanoemulsions are classified as colloidal dispersions commonly preferred for poorly water-soluble molecules using generally safe excipients. Due to nano size, amphiphilic nature, and encapsulation of the molecules, these carriers offer advantages like enhanced solubility and payload delivery of hydrophobic compounds, reduced degradation, enhanced plasma half-life, and better absorption into the tissues (Sharma & Sinha, 2018; Sharma, Batra et al., 2022; Behera et al., 2020). Research focusing on targeting cancer using nanoemulsion holds great importance because of promising results in therapeutic efficacy. This can be explained as cancer cells are highly vascularized tissues and compromised lymphatic drainage systems, leading to enhanced permeability and retention. Hence, due to the nano size and amphiphilic characteristics, the nanoemulsion can easily invade, bypass the barriers, and accumulate.

Considering the specificity of cancer cells, nanoemulsions also offer ligand attachment to the surface for targeting therapeutic molecules. The passive targeting of the molecules cannot differentiate between normal healthy cells and cancer cells. On the other hand, active targeting holds the advantage of specifically recognizing the characteristic cancerous sites (overexpressed receptors such as folate, epidermal growth factor, transferrin, or prostate-specific membrane antigen) and the environment surrounding the cancerous tissue. Hence, the active targeting approach holds an edge over passive targeting, particularly for cancer cells (Sánchez-López et al., 2019). Nanoemulsion can be attached to antibodies like AMB8LK, which has an affinity for overexpressed H-ferritin receptors, thus increasing cellular uptake (Goldstein et al., 2005). Also, due to the well-described distribution of folate receptors in cancer cells, folic acid (low molecular weight) is the most preferred biomarker used in targeting cancer cells in conjugation with nanoemulsion.

Moreover, folate receptors are overexpressed in epithelial, cervical, ovarian, breast, kidney, lung, colorectal, and brain tumours. However, it is less expressed in cancers of the pancreas, prostate, bladder, lymphomas, and sarcomas (Sharma, Batra et al., 2022; Kumar & Divya, 2015). The efficacy of nanoemulsion loaded with a zeaxanthin-dominated carotenoid extract from *Lycium barbarum* was studied against colon cancer cells (HT-29). The formulation showed a rapid pH-dependent release at an acidic environment (pH 5.2) compared to normal physiological pH, i.e., pH 7.4. Thus, pH-dependent targeting in cancer cells improves efficacy with nanoemulsion at IC_{50} of 4.5 μg/mL, whereas the extract alone shows an inhibitory effect with IC_{50} 4.9 μg/mL in colon cancer cells (Hsu et al., 2017).

19.3.3.3 Lipid-Based Nanoparticles

Lipid-based nanoparticles offer remarkable pharmaceutical advantages such as targeted delivery, controlled release, high payload, reduced toxicity, omitting organic solvents, and ease of scale-up. The therapeutic efficacy of lipid nanoparticles is due to bypassing P-glycoprotein efflux, particularly in tumours with multi-drug resistance, facilitating enhanced circulation and allowing ligand conjugation for targeting. Solid lipid nanoparticles are generally prepared from glyceryl monostearate, stearic acid, tristearin, and cholesterol dispersed in surfactant-containing aqueous solutions. On the other hand, nanostructured lipid carriers comprised liquid oil within the solid lipid network, thereby allowing carotenoid loading and preventing surface diffusion from nanoparticles (Zare et al., 2021).

19.3.3.4 Polymeric Nanoparticles

Lycopene has been reported in the literature for its chemopreventive efficacy in cancers. Simultaneously, it presents exceptionally high lipophilicity and poor water solubility, compromising bioavailability and therapeutic efficacy against cancer. Therefore, to address the poor bioavailability of lycopene, a novel approach for developing thermoresponsive polymeric nanoparticles loaded with lycopene has been developed for skin tumours. PEGylated *N*-isopropyl acrylamide-based thermosensitive polymeric nanoparticles loaded with lycopene and subjected to application on inflamed skin. The study revealed better antitumour activity of lycopene with nanocarrier as it showed improved expression of pro-apoptotic Bax protein and reduced expression of anti-apoptotic B-cell leukaemia 2 protein in 12-*O*-tetra decanoyl phorbol-13-acetate induced tumour in mouse skin model at low doses as compared to free lycopene. This is attributed to the targeted and tissue internalization properties of nanocarriers, which have increased the antitumour efficacy of lycopene (Bano et al., 2020).

Amphiphilic material has an additional advantage for developing nanoparticles to enhance cellular uptake, which is particularly required in cancer. Such material interacts better with the cell membrane and promotes permeation into the cells in combination with nanotechnology. For instance, in a study, zein, a natural corn protein, has been utilized for developing nanoparticles with an average size of ~130 nm for β-carotene. The *in vitro* cellular uptake studies showed enhanced cellular uptake in MCF-7 breast cancer cell lines compared to free β-carotene. Apart from the nanosize range, the enhanced cytotoxicity of β-carotene was accredited to an amphiphilic attribute of zein and suggested improved oral bioavailability of β-carotene (Jain et al., 2018).

Crocetin is a bioactive carotenoid that is one of the chief components of saffron. Crocetin is of low molecular weight with multiple bonds and dicarboxylic acid groups representing an amphiphilic nature (Gutheil et al., 2012). In literature, it is reported as an anticancer agent where it was observed that it prevents the synthesis of protein, DNA, and RNA. Also, it prevents signalling pathways and cell proliferation. However, due to its limited aqueous solubility and predominant hydrophobic attribute, it has limited pharmacokinetic and biological activity. Poly(lactic-co-glycolic acid), an FDA-approved biocompatible and biodegradable polymer, has been successfully utilized for the targeted delivery of a wide variety of poorly aqueous soluble phytochemicals, including carotenoids. Crocetin loaded with

PLGA nanoparticles was studied for its cytotoxicity and apoptotic activity on MCF-7 cell lines. The results showed superior antitumour activity with PLGA nanoparticles compared to free crocetin, which may be attributed to better cellular uptake and increased crocetin solubility (Hafezi Ghahestani et al., 2017).

Lutein-loaded folate conjugated polymeric nanoparticles (poly-D,L-lactide-co-glycolide)-(polyethylene glycol)-folate nanoparticles were fabricated and evaluated for its potential treatment in hypoxic ischaemic encephalopathy by Bolla and co-workers. The fabricated nanoparticles were internalized into the cells by folate receptor-mediated endocytosis. In the cytosol, lutein was released due to the degradation of nanoparticles, which further upregulated/downregulated several cellular mechanisms and enzymes, ultimately causing therapeutic benefit. The *in vitro* uptake of lutein nanoparticles and fluorescein isothiocyanate-labeled lutein was noted at 6 and 12 hours, respectively. It was also reported that the high fluorescence uptake at all time points with fluorescein isothiocyanate-labeled lutein-loaded polymeric nanoparticles than fluorescein isothiocyanate-labeled lutein (Bolla et al., 2020).

Bano et al. synthesized co-polymeric nanoparticles' formulation of commercial nano lycopene and nano-extracted lycopene. They further evaluated *in vitro* anticancer activity, inhibitory effect on 12-*O*-tetradecanoylphorbo l-13-acetate-promoted skin inflammation, and tumorigenesis in Swiss albino mice. The particle size was less than 100 nm with a spherical morphology, and an average polydispersity index of 0.594 was demonstrated by the dynamic light scattering and Transmission electron microscopic images (Figure 19.2). Skin tumour promotion was directly related to inflammation and the effect of the pre-treatment of extracted lycopene and nano-extracted lycopene on 12-*O*-tetradecanoylphorbol-13-acetate-induced skin inflammation (Figure 19.3). The results obtained confirmed that the pre-treatment of nano-extracted lycopene at low doses more significantly inhibited the 12-*O*-tetradecanoylphorbol-13-acetate-induced induction of cyclooxygenase-2 in epidermal cells as well as 12-*O*-tetradecanoylphorbol-13-acetate-induced skin oedema and the infiltration of inflammatory cells than extracted lycopene (Bano et al., 2020).

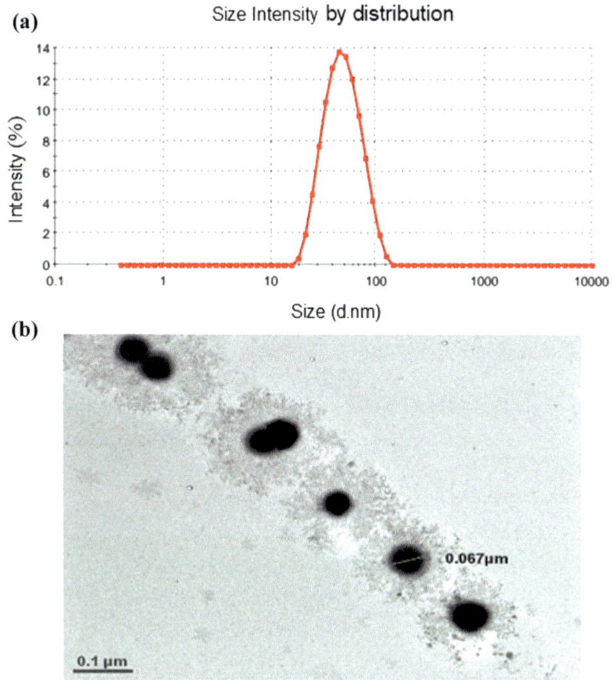

FIGURE 19.2 Size characterization of the polymeric nanoparticles. (a) Dynamic light scattering of the polymeric nanoparticles confirmed a narrow size distribution in the 65 nm range with an average polydispersity index of 0.594. (b) Transmission electron microscopy image demonstrating spherical morphology of the nanoparticle as compared to the dynamic light scattering studies. (Reproduced with open access source. Bano et al. (2020).)

19.3.3.5 Hybrid Nanoparticles

Hybrid nanoparticles are a nanotechnological approach with the characteristic properties of both polymer and lipid. Lipid coating on the outer surface of polymeric nanoparticles restricts the water diffusion into the core. Thus, the lipid layer prevents the immediate drug release from the nanoparticle and provides a sustained release. Also, lipid-polymer hybrid nanoparticles are well recognized for their high drug-loading capacity. The lipid layer on the polymer surface increases the utility by providing biocompatibility and retards the rate of biodegradation.

On the other hand, polymers facilitate controlled release by providing stability to nanoparticles and increasing their shelf life. Lipid-polymer hybrid nanoparticles allow surface modification with ligands that add value to the delivery system by imparting targeting properties. For instance, β-carotene, along with methotrexate (anticancer drug), was incorporated into the polylactic acid (polymer) and 1,2-Distearoyl-sn-glycero-3-phosphoethanolamine-polyethylene glycol$_{3000}$ (lipid) along with stearic acid and gelucire (lipid surfactant) based hybrid nanoparticle. Additionally, a fructose ligand was conjugated with nanoparticles to facilitate targeting as fructose is selective to the cells that over-express GLUT 5 receptors (Zhao et al., 2014). Moreover, the hybrid nanoparticle has shown approximately 60% entrapment efficiency for the β-carotene. The hybrid nanoparticle co-loaded with β-carotene and methotrexate showed controlled release and superior *in-vitro* cell toxicity against MCF-7 breast cancer cell lines (Jain et al., 2017).

19.3.3.6 Dendrimers

The amalgamation of nanotechnology and the unique properties of dendrimers makes it a potential choice for nano drug delivery of sensitive hydrophobic carotenoids, targeting, improving solubility, controlled drug delivery, and enhancing permeation. Dendrimers also provide certain possibilities of ligand attachment, like polyethylene glycol, acetylation, hydroxylation, methylation, and oligosaccharides, thus adding biocompatibility and

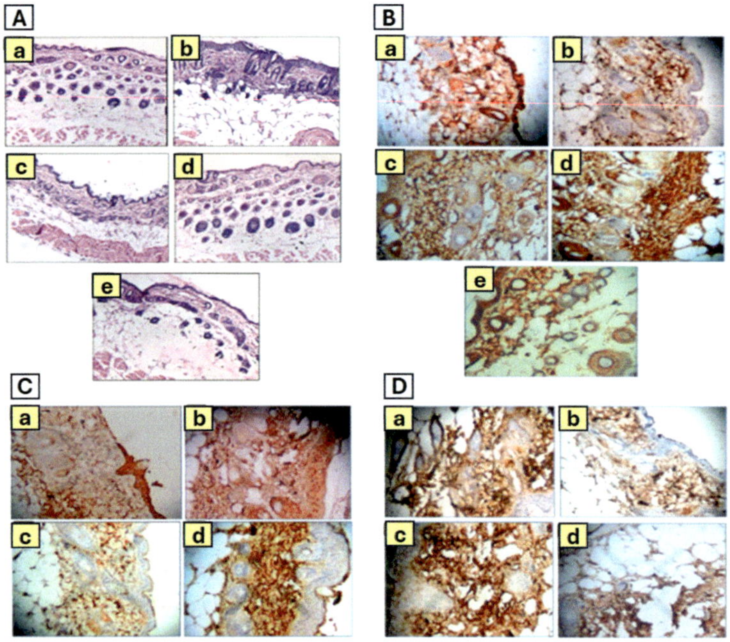

FIGURE 19.3 Histopathological sections of skin showing the effect of the pre-treatment of extracted lycopene and nano-extracted lycopene on 12-*O*-tetradecanoylphorbol-13-acetate-induced histological alteration in the skin (A)(a–e), and on the immunohistochemical expression of COX-2 in the skin (B) (a–e). Effect of the pre-treatment of nano-extracted lycopene on the 7,12-Dimethylbenz[*a*]anthracene/12-*O*-tetradecanoylphorbol-13-acetate-induced immuno-histochemical expression of Bax and Bcl-2 in skin/tumours (C) (a–d) and (D) (a–d), respectively. (Reproduced with open access source. Bano et al. (2020).)

reducing toxicity (Batra & Sharma, 2021). Dendrimers are nanosized 3-dimensional architectural structures possessing extremely branched polyvalent organic surfaces. This unique complex structure allows conjugation with various drug molecules (peptides, antibodies, carotenoids, and nucleic acids, etc.) to overcome associated issues like bio-distribution and its solubility, which influence the pharmacodynamic effect of drug molecules (Batra et al., 2021). Dendrimers are known to improve water solubility either by making hydrogen bonds or electrostatic interactions. Crocetin is one of the potential antitumour amphiphilic carotenoids with poor water solubility. It is naturally synthesized by enzymatic cleavage of zeaxanthin. Poly(amidoamine) dendrimers have been widely studied for drug delivery, targeting, and solubility enhancer. A study demonstrated that the 4th generation poly(amidoamine) dendrimer alkylated with 30% of 10-bromodecanoate has successfully increased the water solubility and cytotoxic property of crocetin at a dose of 6.25 µM on MCF-7 cell lines (Soltani et al., 2017).

19.4 REPORT ON CLINICAL TRIALS AND PATENTS OF THE CAROTENOIDS

Plant-based phytochemicals and derivatives can potentially increase treatment efficacy and reduce side effects in cancer patients. Several of these phytochemicals are biologically active compounds with major antitumour potential. The evaluation of natural extracts (from dry/wet plant material) for potential anticancer biological activity, followed by purification of active phytochemicals based on bioassay-guided fractionation and testing for *in vitro* and *in vivo* effects, is the first step in the production of successful and side-effect-free phytochemical-based anticancer therapy. Clinical trials using phytochemicals to fight cancer are still in their early stages, even though many anticancer compounds are widely deployed (Choudhari et al., 2020).

Nancy Ordaz-Trinidad et al. provided a detailed overview of patents issued in the last 5 years related to phytochemical extraction and their use in the food and pharmaceutical industries. Thirty percent (30%) methodologies for phytochemical extraction, 16% application of phytochemicals in food matrixes to obtain usable foods, 18% biological activity of extracts or compounds, and 36% application in the prevention and treatment of an illness are among the patents that have been discussed in the review (Ordaz-Trinidad et al., 2015).

The patent entitled "Production of carotenoids in oleaginous yeast and fungi" (US8288149B2) provides information about the systems used for producing engineered oleaginous yeast or fungi that express carotenoids (Bailey et al., 2012; US8288149B2). The patent entitled "Multi-carotenoids compositions and uses, therefore"

(US9044407B2) provides information about the methods for treating lower urinary tract in men caused by benign prostate hyperplasia, including orally delivering an appropriate amount of multi-carotenoid compositions (Kuo, 2015, US9044407B2). Clinical trials involving phytochemicals are based on three main aspects of cancer research: (1) enhancing cancer cell response to normal radiotherapy and chemotherapy, and (2) reducing the serious side effects of conventional cancer treatment. The phytochemicals currently under clinical trials against various cancers are summarized in Table 19.1.

19.5 FUTURE PERSPECTIVES

Nano-enabled delivery of hydrophobic carotenoids possesses several advantages over free form and has become a new vogue emerging in nanomedicine for cancer treatment. Nanocarriers not only resolve the solubility and stability issues of carotenoids but also target cancer tissues, resulting in better efficacy of potential carotenoids. Although there is no universally accepted treatment for cancer due to its heterogeneity, different approaches are applied simultaneously. Administration of cytotoxic compounds in systemic

TABLE 19.1
Clinical Trials to Determine the Effect of Carotenoids on Different Types of Cancer

Name of Carotenoid	Population Enrolled/Phase of Clinical Trial	Type of Cancer	Intervention	Objectives	References
β-carotene and retinol	18,314 participants/ Phase 4	Lung cancer	• *Asbestos-exposed participants:* 15 mg/day beta-carotene+25,000 IU/ day retinol • *Heavy smokers:* 30 mg/ day β-carotene+25,000 IU/day retinol	To investigate the maximum cancer prevention efficacy and safety of β-carotene and retinyl palmitate in people with high risk for lung cancer.	https://clinicaltrials.gov/ct2/show/ NCT00712647l
Serum carotenoids	40,641 participants/ Meta-analysis of ten studies	Gastrointestinal tract cancer	• Serum carotenoids	To investigate the risk associated with serum carotenoids for gastrointestinal tract cancer	https://ClinicalTrials.gov/show/ NCT03433092
α-Tocopherol and β-carotene	12 participants/ Phase 2	Lung cancer	α-tocopherol (50 mg/day) daily and β-carotene (20 mg/day) to male smokers in Southwestern Finland	To investigate the preventive effects of α-tocopherol and β-carotene on cancer incidence and mortality.	https://ClinicalTrials.gov/show/ NCT0034299
Multi carotenoids rich	702 participants/ Phase 2	Prostate cancer	30, 15, and 0 mg (placebo) of multi-carotenoid rich (1:1:1) taken daily orally.	To investigate the preventive effect on the risk of prostate cancer.	https://ClinicalTrials.gov/show/ NCT02426216
α-carotene, β-carotene, β-cryptoxanthin and lutein/ zeaxanthin	561 cases and 561 control/ Case-control study	Breast cancer in Chinese women	The information on the dietary intake of carotenoids was collected from the hospital through a face-to-face interview.	To investigate the preventive effect of carotenoid intake on the risk of breast cancer among Chinese women.	Wang et al. (2014)
α-carotene, β-carotene and lutein/zeaxanthin	31 articles/ Meta-analysis of 31 studies	Non-Hodgkin lymphoma	Case-control and cohort studies representing the dose-response between the use of carotenoids and non-Hodgkin lymphoma	Analysis report suggested a correlation of dietary intake of α-carotene, β-carotene and lutein/zeaxanthin in reducing the risk of non-Hodgkin lymphoma.	Chen et al. (2017)
Tangerine rich in β-carotene and cryptoxanthin and red tomato juice rich in lycopene content	45 male patients	Prostate cancer	Two 5.5 oz. cans of tangerine and red tomato juice were consumed for approximately 4 weeks before surgery.	To determine whether tangerine and red tomato juice rich in lycopene content can prevent normal cells from transforming into cancer cells after surgical removal of prostate cancer.	https://ClinicalTrials.gov/show/ NCT02144649
Lycopene	Phase 2	Prostate cancer	Patients were given low and high doses of lycopene orally for 4–7 weeks before radical prostatectomy.	To determine the dose-dependent response of lycopene to prevent prostate cancer growth.	https://ClinicalTrials.gov/show/ NCT00450749

circulation, surgery, and radiotherapy is a general approach for treating cancer. Several synthetic and natural cytotoxic compounds are studied for their anticancer potential, and nanotechnology further widens their utility.

19.6 CONCLUSION

Despite numerous studies and meta-analyses suggesting the potential of carotenoids and improved therapeutic efficacy when coupled with nanotechnology in cancer, industrial production is still in the preliminary phases. Therefore, future strategies should focus on applying carotenoids using nanocarriers for cancer and their industrial scale-up. Nanotechnology offers a wide range of nanocarriers like liposomes, nanoemulsions, dendrimers, and particulates, resolving almost all carotenoid issues. Thereby, it opens opportunities in product development for carotenoids to increase their pharmaceutical applications and allows their applicability in preventing relapse and further growth of cancer.

REFERENCES

Aissa, A. F., Bianchi, M. L., Ribeiro, J. C., Hernandes, L. C., de Faria, A. F., Mercadante, A. Z., & Antunes, L. M. (2012). Comparative study of β-carotene and microencapsulated β-carotene: evaluation of their genotoxic and antigenotoxic effects. *Food Chem Toxicol*, *50*(5), 1418–1424.

Alpha-Tocopherol, Beta-Carotene Cancer Prevention (ATBC) Study. https://ClinicalTrials.gov/show/NCT00342992.

Alshahrani, S. M., Thotakura, N., Sharma, S., Quadir, S. S., Chaurawal, N., Sharma, S., Chitkara, D., & Raza, K. (2023). Influence of nanocarrier type on the drug delivery aspects of docetaxel: empirical evidences. *J Pharm Innov*, *18*(2), 641–652.

Arathi, B. P., Raghavendra-Rao Sowmya, P., Kuriakose, G. C., Shilpa, S., Shwetha, H. J., Kumar, S., … & Lakshminarayana, R. (2018). Fractionation and characterization of lycopene-oxidation products by LC-MS/MS (ESI)(+): elucidation of the chemopreventative potency of oxidized lycopene in breast-cancer cell lines. *J Agric Food Chem*, *66*(43), 11362–11371.

Bailey, R. B., Madden, K. T., & Trueheart, J. (2012). Production of carotenoids in oleaginous yeast and fungi. US8,288,149 B2, 16, Oct 2012.

Bano, S., Ahmed, F., Khan, F., Chaudhary, S. C., & Samim, M. (2020). Targeted delivery of thermoresponsive polymeric nanoparticle-encapsulated lycopene: in vitro anticancer activity and chemopreventive effect on murine skin inflammation and tumorigenesis. *RSC Adv*, *10*(28), 16637–16649.

Batra, S., & Sharma, S. (2021). Biomedical applications of dendrimers. In: Neelesh Kumar Mehra, K. J. (Ed.), *Dendrimers in Nanomedicine: Concept, Theory and Regulatory* Perspectives. CRC Press, pp. 277–285.

Batra, S., Thakral, S., Singh, A., & Sharma, S. (2021). Dendrimer–nanomaterial conjugation: concept, chemistry and applications. In: Neelesh Kumar Mehra, K. J. (Ed.), *Dendrimers in Nanomedicine: Concept, Theory and Regulatory* Perspectives. CRC Press, pp. 217–232.

Behera, A., & Padhi, S. (2020). Passive and active targeting strategies for the delivery of the camptothecin anticancer drug: a review. *Environ Chem Lett*, 18(5), 1557–1567.

Behera, A., & Padhi, S. (2022). pH-Sensitive polymeric nanoparticles for cancer treatment. In: Padhi, S., Behera, A., Lichtfouse, E. (Eds.), *Polymeric Nanoparticles for the Treatment of Solid Tumours*. Environmental Chemistry for a Sustainable World, vol 71. Springer, Cham, pp. 401–425.

Behera, A., Mittu, B., Padhi, S., & Singh, A. (2020). Antimicrobial efficacy of essential oil nanoemulsions. In: Dhull, S. B., Chawla, P., Kaushik, R. (Eds.), *Nanotechnological Approaches in Food Microbiology*, CRC Press, pp. 293–309.

Behera, A., Padhi, S., Nayak, A. K. (2023). Engineered liposomes as drug delivery and imaging agents. In: Ray, S., Nayak, A. K. (Eds.), *Design and Applications of Theranostic Nanomedicines*, Woodhead Publishing, pp. 75–108.

Black, H. S., Boehm, F., Edge, R., & Truscott, T. G. (2020). The benefits and risks of certain dietary carotenoids that exhibit both anti- and pro-oxidative mechanisms: a comprehensive review. *Antioxidants*, *9*(3), 264.

Bolla, P. K., Gote, V., Singh, M., Yellepeddi, V. K., Patel, M., Pal, D., Gong, X., Sambalingam, D., & Renukuntla, J. (2020). Preparation and characterization of lutein loaded folate conjugated polymeric nanoparticles. *J Microencapsul*, *37*(7), 502–516.

Carotene and Retinol Efficacy Trial. https://ClinicalTrials.gov/show/NCT00712647.

Chen, F., Hu, J., Liu, P., Li, J., Wei, Z., & Liu, P. (2017). Carotenoid intake and risk of non-Hodgkin lymphoma: a systematic review and dose-response meta-analysis of observational studies. *Ann Hematol*, *96*(6), 957–965.

Chen, T.-C., Chuang, J.-Y., Ko, C.-Y., Kao, T.-J., Yang, P.-Y., Yu, C.-H., … & Hsu, T.-I. (2020). AR ubiquitination induced by the curcumin analog suppresses growth of temozolomide-resistant glioblastoma through disrupting GPX4-Mediated redox homeostasis. *Redox Biol*, *30*, 101413.

Chen, W., Zhang, H., & Liu, Y. (2019). Anti-inflammatory and apoptotic signaling effect of fucoxanthin on benzo(a)pyrene-induced lung cancer in mice. *J Environ Pathol Toxicol Oncol*, *38*(3), 239–251.

Choudhari, A. S., Mandave, P. C., Deshpande, M., Ranjekar, P., & Prakash, O. (2020). Phytochemicals in cancer treatment: from preclinical studies to clinical practice. *Front Pharmacol*, *10*, 1614.

de Oliveira Júnior, R. G., Bonnet, A., Braconnier, E., Groult, H., Prunier, G., Beaugeard, L., … & Picot, L. (2019). Bixin, an apocarotenoid isolated from *Bixa orellana* L., sensitizes human melanoma cells to dacarbazine-induced apoptosis through ROS-mediated cytotoxicity. *Food Chem Toxicol*, *125*, 549–561.

de Oliveira Silva, J., Fernandes, R. S., Ramos Oda, C. M., Ferreira, T. H., Machado Botelho, A. F., Martins Melo, M., … & de Barros, A. L. B. (2019). Folate-coated, long-circulating and pH-sensitive liposomes enhance doxorubicin antitumor effect in a breast cancer animal model. *Biomed Pharmacother*, *118*, 109323.

Deryugina, E. I., & Quigley, J. P. (2006). Matrix metalloproteinases and tumor metastasis. *Cancer Metastasis Rev*, *25*(1), 9–34.

Dos Santos, P. P., Andrade, L. d. A., Flôres, S. H., & Rios, A. d. O. (2018). Nanoencapsulation of carotenoids: a focus on different delivery systems and evaluation parameters. *J Food Sci Technol*, *55*(10), 3851–3860.

Duncan, R. (2003). The dawning era of polymer therapeutics. *Nat Rev Drug Discov*, *2*(5), 347–360.

Elkholy, N. S., Shafaa, M. W., & Mohammed, H. S. (2020). Biophysical characterization of lutein or beta carotene-loaded cationic liposomes. *RSC Adv*, *10*(54), 32409–32422.

Gansukh, E., Nile, A., Sivanesan, I., Rengasamy, K. R., Kim, D.-H., Keum, Y.-S., & Saini, R. K. (2020). Chemopreventive effect of β-cryptoxanthin on human cervical carcinoma (HeLa) cells is modulated through oxidative stress-induced apoptosis. *Antioxidants*, *9*(1), 28.

Garg, P., Sarma, D., Jeppsson, S., Patel, N. R., Gewirtz, A. T., Merlin, D., & Sitaraman, S. V. (2010). Matrix metalloproteinase-9 functions as a tumor suppressor in colitis-associated cancer. *Cancer Res*, *70*(2), 792–801.

Gialeli, C., Theocharis, A. D., & Karamanos, N. K. (2011). Roles of matrix metalloproteinases in cancer progression and their pharmacological targeting. *FEBS J*, *278*(1), 16–27.

Goldstein, D., Nassar, T., Lambert, G., Kadouche, J., & Benita, S. (2005). The design and evaluation of a novel targeted drug delivery system using cationic emulsion-antibody conjugates. *J Control Release*, *108*(2–3), 418–432.

Gong, X., Smith, J. R., Swanson, H. M., & Rubin, L. P. (2018). Carotenoid lutein selectively inhibits breast cancer cell growth and potentiates the effect of chemotherapeutic agents through ROS-mediated mechanisms. *Molecules*, *23*(4), 905.

Gutheil, W. G., Reed, G., Ray, A., Anant, S., & Dhar, A. (2012). Crocetin: an agent derived from saffron for prevention and therapy for cancer. *Current Pharm Biotechnol*, 13(1), 173–179.

Hafezi Ghahestani, Z., Alebooye Langroodi, F., Mokhtarzadeh, A., Ramezani, M., & Hashemi, M. (2017). Evaluation of anticancer activity of PLGA nanoparticles containing crocetin. *Artif Cells Nanomed Biotechnol*, 45(5), 955–960.

Hegazy, G. E., Abu-Serie, M. M., Abo-Elela, G. M., Ghozlan, H., Sabry, S. A., Soliman, N. A., & Abdel-Fattah, Y. R. (2020). In vitro dual (anticancer and antiviral) activity of the carotenoids produced by haloalkaliphilic archaeon Natrialba sp. M6. *Sci Rep*, *10*(1), 5986.

Hirsjärvi, S., Passirani, C., & Benoit, J. P. (2011). Passive and active tumour targeting with nanocarriers. *Curr Drug Discov Technol*, *8*(3), 188–196.

Hofmann, U. B., Westphal, J. R., Zendman, A. J., Becker, J. C., Ruiter, D. J., & van Muijen, G. N. (2000). expression and activation of matrix metalloproteinase-2 (MMP-2) and its co-localization with membrane-type 1 matrix metalloproteinase (MT1-MMP) correlate with melanoma progression. *J Pathol*, *191*(3), 245–256.

Honda, M., Kageyama, H., Hibino, T., Zhang, Y., Diono, W., Kanda, H., ... & Goto, M. (2019). Improved carotenoid processing with sustainable solvents utilizing Z-isomerization-induced alteration in physicochemical properties: a review and future directions. *Molecules*, *24*(11), 2149.

Hsu, H. J., Huang, R. F., Kao, T. H., Inbaraj, B. S., & Chen, B. H. (2017). Preparation of carotenoid extracts and nanoemulsions from *Lycium barbarum* L. and their effects on growth of HT-29 colon cancer cells. *Nanotechnology*, *28*(13), 135103.

Huang, H. (2018). Matrix metalloproteinase-9 (MMP-9) as a cancer biomarker and MMP 9 biosensors: recent advances. *Sensors, 18*(10), 3249.

Iskandar, A. R., Miao, B., Li, X., Hu, K.-Q., Liu, C., & Wang, X.-D. (2016). β-Cryptoxanthin reduced lung tumor multiplicity and inhibited lung cancer cell motility by downregulating nicotinic acetylcholine receptor α7 signaling. *Cancer Prev Res*, *9*(11), 875–886.

Jain, A. K., Mehra, N. K., & Swarnakar, N. K. (2015). Role of antioxidants for the treatment of cardiovascular diseases: challenges and opportunities. *Curr Pharm Des*, *21*(30), 4441–4455.

Jain, A., Sharma, G., Kushwah, V., Garg, N. K., Kesharwani, P., Ghoshal, G., ... & Katare, O. P. (2017). Methotrexate and beta-carotene loaded-lipid polymer hybrid nanoparticles: a preclinical study for breast cancer. *Nanomedicine*, *12*(15), 1851–1872.

Jain, A., Sharma, G., Kushwah, V., Ghoshal, G., Jain, A., Singh, B., ... & Katare, O. P. (2018). Beta carotene-loaded zein nanoparticles to improve the biopharmaceutical attributes and to abolish the toxicity of methotrexate: a preclinical study for breast cancer. *Artif Cells Nanomed Biotechnol*, *46*(sup1), 402–412.

Jeong, Y., Lim, J. W., & Kim, H. (2019). Lycopene inhibits reactive oxygen species-mediated NF-κB signaling and induces apoptosis in pancreatic cancer cells. *Nutrients*, *11*(4), 762.

Joshi, S., Singh-Moon, R. P., Ellis, J. A., Chaudhuri, D. B., Wang, M., Reif, R., ... & Straubinger, R. M. (2015). Cerebral hypoperfusion-assisted intra-arterial deposition of liposomes in normal and glioma-bearing rats. *Neurosurgery*, *76*(1), 92–100.

Juin, C., Oliveira Junior, R. G. d., Fleury, A., Oudinet, C., Pytowski, L., Bérard, J.-B., ... & Picot, L. (2018). Zeaxanthin from *Porphyridium purpureum* induces apoptosis in human melanoma cells expressing the oncogenic BRAF V600E mutation and sensitizes them to the BRAF inhibitor vemurafenib. *Rev Bras Farmacogn*, *28*(4), 457–467.

Juric, V., O'Sullivan, C., Stefanutti, E., Kovalenko, M., Greenstein, A., Barry-Hamilton, V., ... & Mikels-Vigdal, A. (2018). MMP-9 inhibition promotes antitumor immunity through disruption of biochemical and physical barriers to T-cell trafficking to tumors. *PLoS One*, *13*(11), e0207255.

Ketabat, F., Pundir, M., Mohabatpour, F., Lobanova, L., Koutsopoulos, S., Hadjiiski, L., ... & Papagerakis, S. (2019). Controlled drug delivery systems for oral cancer treatment-current status and future perspectives. *Pharmaceutics*, *11*(7), 302.

Khodapasand, E., Jafarzadeh, N., Farrokhi, F., Kamalidehghan, B., & Houshmand, M. (2015). Is Bax/Bcl-2 ratio considered as a prognostic marker with age and tumor location in colorectal cancer? *Iran Biomed J*, *19*(2), 69–75.

Koklesova, L., Liskova, A., Samec, M., Buhrmann, C., & Samuel, S. M. (2020). Carotenoids in cancer apoptosis: the road from bench to bedside and back. *Cancers*, *12*(9), 2425.

Kumar, G. P., & Divya, A. (2015). Nanoemulsion based targeting in cancer therapeutics. *Med Chem*, *5*(6), 272–284.

Kumar, L., Verma, S., Joshi, K., & Sharma, S. (2023). Exploring the therapeutic potential of anticancer heterocyclic compounds: role in nanoscale pharmacotherapy. *Crit Rev Ther Drug Carrier Syst*, *40*(3), 47–94.

Kundu, A., Padhi, S., Behera, A., Hasnain, M. S., & Nayak, A. K. (2022). Tumour targeting strategies by chitosan-based nanocarriers. In: Hasnain, M. S., Beg, S., & Nayak, A. K. (Eds.), *Chitosan in Biomedical Applications*, Academic Press, London, pp. 163–188.

Kuo, F. F. (2015). Multi-Carotenoids Compositions and Uses Therefor. US Patents. US 9,044,407 B2. June 2, 2015.

Liechty, W. B., & Peppas, N. A. (2012). Expert opinion: responsive polymer nanoparticles in cancer therapy. *Eur J Pharm Biopharm*, *80*(2), 241–246.

Liu, C. L., Chiu, Y. T., & Hu, M. L. (2011). Fucoxanthin enhances HO-1 and NQO1 expression in murine hepatic BNL CL.2 cells through activation of the Nrf2/ARE system partially by its pro-oxidant activity. *J Agric Food Chem*, *59*(20), 11344–11351.

Lycopene in Treating Patients Undergoing Radical Prostatectomy for Prostate Cancer. https://ClinicalTrials.gov/show/NCT00450749.

Molecular Effects of a Multi-carotenoids (MCS) New Agent on Prostate Cancer Chemoprevention. https://ClinicalTrials.gov/show/NCT02426216.

Niranjana, R., Gayathri, R., Nimish Mol, S., Sugawara, T., Hirata, T., Miyashita, K., & Ganesan, P. (2015). Carotenoids modulate the hallmarks of cancer cells. *J Funct Foods*, *18*, 968–985.

Ordaz-Trinidad, N., Dorantes-Alvarez, L., & Salas-Benito, J. (2015). Patents on phytochemicals: methodologies of extraction, application in food and pharmaceutical industry. *Recent Patents Biotechnol*, *9*(3), 158–167.

Padhi, S., & Behera, A. (2020). Nanotechnology based targeting strategies for the delivery of camptothecin. In: Saneja, A., Panda, A., Lichtfouse, E. (Eds.), *Sustainable Agriculture Reviews*, vol 44. Springer, Cham, pp. 243–272.

Padhi, S., & Behera, A. (2022). Cellular internalization and toxicity of polymeric nanoparticles. In: Padhi, S., Behera, A., Lichtfouse, E. (Eds.), *Polymeric Nanoparticles for Treating Solid Tumours*. Environmental Chemistry for a Sustainable World, vol 71. Springer, Cham, pp. 473–488.

Padhi, S., Behera, A., Hasnain, M. S., & Nayak, A. K. (2022). Chitosan-based drug delivery systems in cancer therapeutics. In: Hasnain, M. S., Beg, S., Nayak, A. K. (Eds.), *Chitosan in Drug Delivery*, Academic Press, London, pp. 159–193.

Padhi, S., Kapoor, R., Verma, D., Panda, A. K., & Iqbal, Z. (2018). Formulation and optimization of topotecan nanoparticles: in vitro characterization, cytotoxicity, cellular uptake and pharmacokinetic outcomes. *J Photochem Photobiol B Biol*, 183, 222–232.

Padhi, S., Mirza, M. A., Verma, D., Khuroo, T., Panda, A. K., Talegaonkar, S., Khar, R. K., & Iqbal, Z. (2016). Revisiting the nanoformulation design approach for effective delivery of Topotecan in its stable form: an appraisal of its in vitro behavior and tumour amelioration potential. *Drug Deliv*, *23*(8), 2827–2837.

Papaliagkas, V., Anogianaki, A., Anogianakis, G., & Ilonidis, G. (2007). The proteins and the mechanisms of apoptosis: a mini-review of the fundamentals. *Hippokratia*, *11*(3), 108–113.

Patel, J. K., & Patel, A. P. (2019). Passive targeting of nanoparticles to cancer. In: Pathak, Y. (Ed.), *Surface Modification of Nanoparticles for Targeted Drug Delivery*, Springer, Cham, pp. 125–143.

Patnaik, S., Gorain, B., Padhi, S., Choudhury, H., Gabr, G. A., Md, S., ... & Kesharwani, P. (2021). Recent update of toxicity aspects of nanoparticulate systems for drug delivery. *European Journal of Pharmaceutics and Biopharmaceutics*, *161*, 100–119.

Pradhan, S. P., Padhi, S., Dash, M., Heena, Mittu, M., & Behera, A. (2021). Carotenoids. In: Kaur, J., Nayik, G. A. (Eds.), *Nutraceuticals and Health Care*, Elsevier, Amsterdam, pp.135–157.

Rostamabadi, H., Falsafi, S. R., & Jafari, S. (2019). Nanoencapsulation of carotenoids within lipid-based nanocarriers. *J Control Release*, *298*, 38–67.

Salahpour Anarjan, F. (2019). Active targeting drug delivery nanocarriers: ligands. *Nanostruct Nano-Objects*, *19*, 100370.

Sánchez-López, E., Guerra, M., Dias-Ferreira, J., Lopez-Machado, A., Ettcheto, M., Cano, A., … & Souto, E. B. (2019). Current applications of nanoemulsions in cancer therapeutics. *Nanomaterials*, *9*(6), 821.

Serum Carotenoids and Risk of Gastrointestinal Tract Cancer: Meta-analysis of Ten Studies. https://ClinicalTrials.gov/show/NCT03433092.

Sharma, S., & Sinha, V. R. (2018). Liquid nanosize emulsion-filled enteric-coated capsules for colon delivery of immunosuppressant peptide. *AAPS PharmSciTech*, *19*(2), 881–885.

Sharma, S., Batra, S., & Sinha, V. R. (2022). Nanoemulsions: a potential advanced nanocarrier platform for herbal drug delivery. In: Sobti, R. C., Dhalla, N. S. (Eds.), *Biomedical Translational Research: Drug Design and Discovery*, Springer Nature, Singapore, pp. 351–368.

Sharma, S., Batra, S., Chauhan, M. K., & Kumar, V. (2023). Photothermal therapy for cancer treatment. In: Malviya, R., Sundram, S. (Eds.), *Targeted Cancer Therapy in Biomedical Engineering*, Springer Nature, Singapore, pp. 755–780.

Sharma, S., Kaur, A., Batra, S., Marisetti, A. L., Kumar, L., & Tewari (2025). Unveiling the potential of combining photodynamic with anti-vascular therapy for cancer treatment. In: Manivasagan, P., Venkatesan, J., Jang, E.-S. (Eds.), *Nanophototherapy*, Elsevier, Amsterdam, pp. 595–610.

Shikha, L., Sumit, S., & Murthy, R. S. R. (2012). Ligands as tools for brain targeting. *Int J Res Pharm Sci*, *3*(4), 559–571.

Shin, J., Song, M.-H., Oh, J.-W., Keum, Y.-S., & Saini, R. K. (2020). Pro-oxidant actions of carotenoids in triggering apoptosis of cancer cells: a review of emerging evidence. *Antioxidants*, *9*(6), 532.

Soltani, F., Ramezani, M., Amel Farzad, S., Mokhtarzadeh, A., & Hashemi, M. (2017). Comparison study of the effect of alkyl-modified and unmodified PAMAM and PPI dendrimers on solubility and antitumor activity of crocetin. *Artif Cells Nanomed Biotechnol*, *45*(7), 1356–1362.

Sowmya Shree, G., Yogendra Prasad, K., Arpitha, H. S., Deepika, U. R., Nawneet Kumar, K., Mondal, P., & Ganesan, P. (2017). β-Carotene at physiologically attainable concentration induces apoptosis and down-regulates cell survival and antioxidant markers in human breast cancer (MCF-7) cells. *Mol Cell Biochem*, *436*(1–2), 1–12.

Sowmya, P. R., Arathi, B. P., Vijay, K., Baskaran, V., & Lakshminarayana, R. (2017). Astaxanthin from shrimp efficiently modulates oxidative stress and allied cell death progression in MCF-7 cells treated synergistically with β-carotene and lutein from greens. *Food Chem Toxicol*, *106*(Pt A), 58–69.

Tangerine or Red Tomato Juice in Treating Patients With Prostate Cancer Undergoing Surgery. https://ClinicalTrials.gov/show/NCT02144649.

Tannock, I. F., & Rotin, D. (1989). Acid pH in tumors and its potential for therapeutic exploitation. *Cancer Res*, *49*(16), 4373–4384.

Walter, L., Pujada, A., Bhatnagar, N., Bialkowska, A. B., Yang, V. W., Laroui, H., & Garg, P. (2017). Epithelial derived-matrix metalloproteinase (MMP9) exhibits a novel defensive role of tumor suppressor in colitis associated cancer by activating MMP9-Notch1-ARF-p53 axis. *Oncotarget*, *8*, 364–378.

Wang, L., Li, B., Pan, M.-X., Mo, X.-F., Chen, Y.-M., & Zhang, C.-X. (2014). Specific carotenoid intake is inversely associated with the risk of breast cancer among Chinese women. *Br J Nutr*, *111*(9), 1686–1695.

Yeh, S., & Hu, M. (2000). Antioxidant and pro-oxidant effects of lycopene in comparison with beta-carotene on oxidant-induced damage in Hs68 cells. *J Nutr Biochem*, *11*(11–12), 548–554.

Zare, M., Norouzi Roshan, Z., Assadpour, E., & Jafari, S. M. (2021). Improving the cancer prevention/treatment role of carotenoids through various nano-delivery systems. *Crit Rev Food Sci Nutr*, *61*(3), 522–534.

Zhao, J., Babiuch, K., Lu, H., Dag, A., Gottschaldt, M., & Stenzel, M. H. (2014). Fructose-coated nanoparticles: a promising drug nanocarrier for triple-negative breast cancer therapy. *Chem Commun*, *50*(100), 15928–15931.

Zu, K., Mucci, L., Rosner, B. A., Clinton, S. K., Loda, M., Stampfer, M. J., & Giovannucci, E. (2014). Dietary lycopene, angiogenesis, and prostate cancer: a prospective study in the prostate-specific antigen era. *J Natl Cancer Inst*, *106*(2), djt430.

20 Nanoenabled Systems for the Delivery of Docetaxel as an Anticancer Agent

Dipak N. Raut[1], *Sadikali F. Sayyad*[2], *and Anita S. Wagh*[3]
[1]Department of Pharmacognosy, Amrutvahini College of Pharmacy, Sangamner, India
[2]Department of Pharmaceutics, Amrutvahini College of Pharmacy, Sangamner, India
[3]Department of Pharmacognosy, Srinath College of Pharmacy, Aurangabad, India

20.1 INTRODUCTION

National Cancer Institute discovered the first taxoid in the 1960s (Cortes & Pazdur, 1995; Feng & Mumper, 2013). Taxanes are originally derived from the plant *Taxus brevifolia* and then synthetically derivatized to get paclitaxel (Taxol) and docetaxel (taxotere®) (Safavy et al., 1999). Docetaxel was first developed by Pierre Potier (Chast, 2006). It is a semisynthetic compound obtained from the non-cytotoxic precursor compound 10-deacetyl-baccatin III extracted from the needles of *Taxus baccata* L., the European yew tree. It is commonly used as an herbal anticancer agent for breast cancer (Chang, 2002; Wani et al., 2015). Docetaxel is a second-generation anticancer taxane derived from the needles of the European yew tree (Bissery, 1995). The paclitaxel, docetaxel, and analogue compounds possess a taxane skeleton in the family (Figure 20.1).

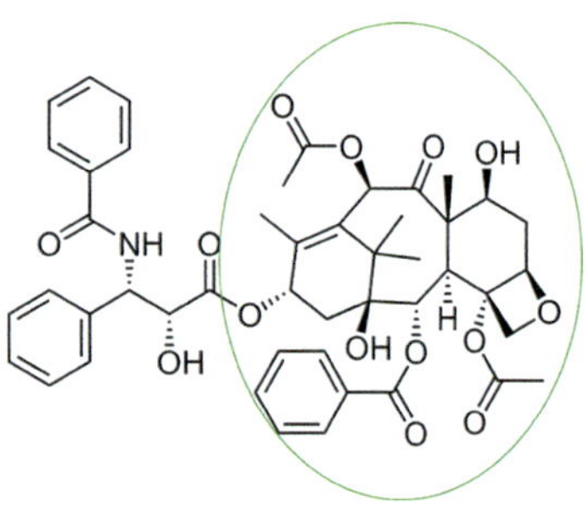

FIGURE 20.1 Structure of taxol skeleton of paclitaxel molecule shown in green circle.

Paclitaxel and docetaxel (Figure 20.2) are chemically bulky and fused ring structures with several lipophilic substituents (Figure 20.3). Docetaxel possesses high lipophilicity and very limited (0.3–25 µg/mL) aqueous solubility (Du et al., 2007; Ma et al., 2018). Also, paclitaxel has reported water solubility of 0.35–0.7 µg/mL. Due to the structural differences between docetaxel and paclitaxel, Docetaxel showed about tenfold more water solubility than paclitaxel (Yuan et al., 2025).

Paclitaxel, administered under the name 'Taxol' (Figure 20.4A4a) through the intravenous route, is rapidly eliminated (with an average elimination half-life of 5.8 hours) from circulation, exhibiting an average distribution half-life of 0.34 hours (Schiff et al., 1979). The non-linear disposition of paclitaxel is due to polyoxyethylated castor oil as the formulation vehicle. The mechanisms are postulated as alteration of hepatic transport function (Ellis et al., 1996) and change of erythrocyte accumulation (Sparreboom et al., 1999). The polyoxyethylated castor oil in the formulation is more likely to change hepatic transporter activity, thereby intensely influencing hepatic uptake and biliary excretion rates of co-administered compounds, including paclitaxel. Polyoxyethylated castor oil entraps paclitaxel in micelles so that less free drug fraction is available for cells, altering paclitaxel accumulation in erythrocytes and showing dose-dependent disposition (Weiss et al., 1990). Taxotere® parenteral formulation (Figure 20.4b) of docetaxel concentration 40 mg/mL in polysorbate 80 needs two dilutions before administration (Ibrahim et al., 2002; Loos et al., 2003). Polysorbate 80 has a high binding affinity

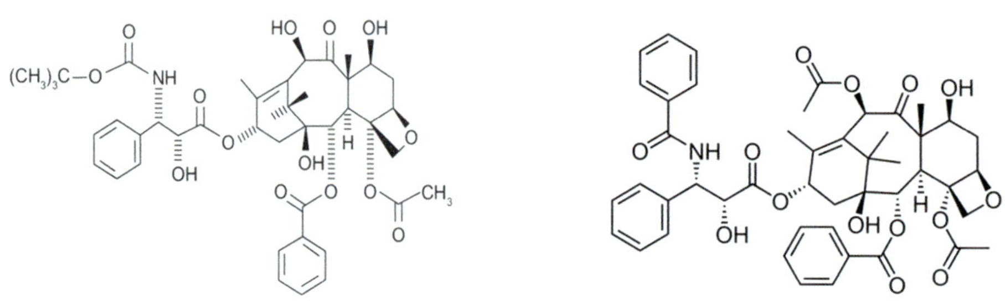

a) Docetaxel   b) Paclitaxel

FIGURE 20.2 Structure of docetaxel and paclitaxel.

DOI: 10.1201/9781003611516-25

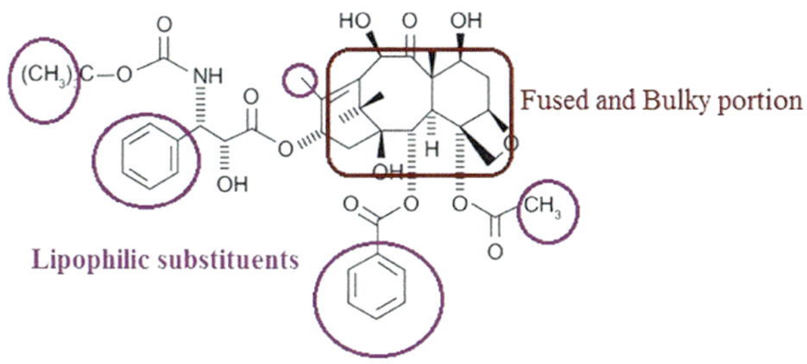

FIGURE 20.3 Lipophilic substituents and fused ring of docetaxel.

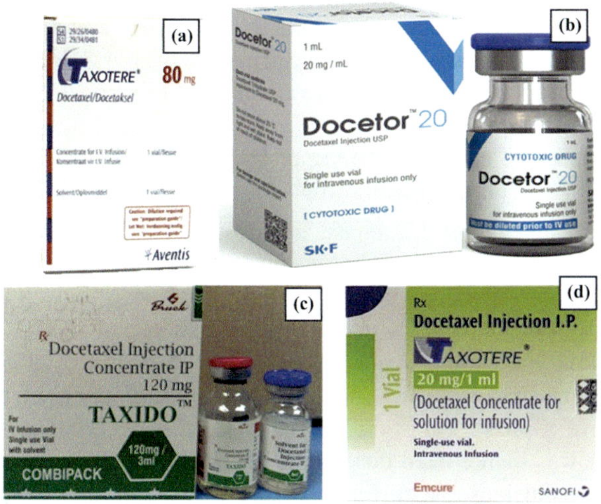

FIGURE 20.4 Different marketed formulations containing docetaxel (a) Taxol 300mg (6 mg/mL) injection, (b) Taxotere Docetaxel 80 mg Injection, (c) Taxotere (20 mg/mL) (Docetaxel concentrate for solution for infusion, (d) Taxotere (Docetaxel) Injection concentrate – ready to add to infusion solution..

for docetaxel and solubilizes docetaxel into water through micelle formation, entrapping docetaxel inside and increasing the protein-unbound docetaxel fraction. The supersaturated micellar solution facilitates the drug to crystallize over time. The stability of the formulation is increased by providing additional vials containing 13% ethanol in water as diluents separately. The infusion must be completed within 4 hours after the dosage preparation (Loos et al., 2003). Sanofi-Aventis launched a one-vial formulation of Taxotere (Figure 20.4c and d) in 2010 that removes initial dilution and is ready to be placed straight into the infusion solution. It contains 20 mg/mL docetaxel in 50/50 (v/v) polysorbate 80/dehydrated ethanol. The stability was achieved using % dehydrated ethanol with reduced viscosity, making it easier to extract and add to the infusion solution. However, there is still a problem with the excipients (Pinguet et al., 2010).

To overcome the short half-life and low therapeutic index of taxanes, nanotechnology-based targeted drug delivery systems have received much attention in the past two decades because of their wide applications and attractive and unique properties (Khuroo et al., 2014). From a pharmaceutical point of view, nanoformulations provide physical and chemical stability for water-insoluble and labile drugs, including taxanes (Pradhan et al., 2022, 2023; Patra et al., 2018). Parenteral administration of the drug essentially requires excellent solubility in water.

Nanotechnology has some significant advantages in the delivery of taxane-like compounds. Poor water-soluble taxanes could be incorporated into nanoparticles to attain large concentrations in parenteral aqueous vehicles (Simeonova et al., 1991; Souza et al., 2011; Behera et al., 2023). Chemically unstable medications, such as camptothecin, SN-38, all-trans retinoic acid, peptides, proteins, and nucleotides, are protected by minimizing their exposure to water or the biological environment (Williams et al., 2003; Yang & Zhu, 2002). Through diverse polyethylene glycol-based coatings, the technique accomplishes long circulation of delivery vehicles and long retention of anticancer agents in the delivery vehicles, which can improve the pharmacokinetics of anticancer medicines such as docetaxel (Cole et al., 2011). In the case of passive targeting of nanoparticles, nano-based formulations improve the permeability, retention, and biodistribution of anticancer drugs. The surface chemistry of nanoparticles allows for covalent or non-covalent insertion of targeted ligands. The overexpression of a certain receptor in tumour cells or at a specific place is thought to "attract" additional nanoparticles. Once the drug-loaded nanoparticles reach the tumour interstitial region, internalization is boosted in tumour cells (Benhabbour et al., 2012; Padhi et al., 2016).

The passive and active targeting properties of nanoparticles decrease penetration into normal tissue and increase the accumulation of drugs at the tumour site (Verma et al., 2017). This leads to increased efficacy and reduced systemic toxicity. The nanotechnology system is versatile and multifunctional, enabling the co-administration of multiple agents through the same nanocarrier to gain synergistic anticancer effects or multi-functions. Several alterations can be made to the surface of nanoparticles to broaden their application in drug delivery systems, which can be engineered to be positively or negatively charged (Padhi et al., 2018). Active targeting ligands covalently or physically coupled to one end of the polyethylene glycol chain or lipids or polymers improve the therapeutic efficacy of these nano-enabled systems (Kundu et al., 2022). Chelating of nanoparticle surface with Ni and incorporation of His-tagged antibody could be employed in this aspect (Cui et al., 2004; Patel et al., 2007).

20.2 MECHANISM OF ACTION OF DOCETAXEL

20.2.1 Mechanism of Action of Taxanes

Docetaxel, which was first available in the taxotere® formulation, is a product of Aventis Pharmaceuticals Inc., Bridgewater, NJ, and is primarily used as a monotherapy or in combination for the treatment of androgen-independent prostate cancer (Pienta, 2001). Docetaxel exerts its anticancer effect mainly by two modes of action (Haldar et al., 1997; Stein, 1999). The first is reducing the impact of B-cell lymphoma 2 and B-cell lymphoma extra-large gene expression, and the second is the suppression of microtubular depolymerization. Microtubules normally undergo polymerization in the presence of microtubule-associated proteins and guanosine triphosphate, which interacts with β-tubulin and promotes cell division through microtubule disassembly (Rowinsky & Eric, 1997). Docetaxel binds preferentially to β-tubulin and promotes microtubule assembly without guanosine-5′-triphosphate and other cofactor proteins. Due to these reasons, microtubules cannot be disassembled in vitro even at 4°C or in the presence of calcium ions that arrest the cell in the G2/M phase cycle, thereby disrupting mitosis and leading to apoptosis (Manfredi et al., 1982). According to Schiff et al. (1979), taxane binding to microtubules stabilized them, thereby preventing their depolymerization. This is a reason for cell growth inhibition (Ringel & Horwitz, 1991). Docetaxel has a 1.9-fold stronger binding affinity to microtubules than paclitaxel and has superior cellular accumulation, resulting in a potency that is twice that of paclitaxel.

The intrinsic pathway is the second method by which docetaxel induces apoptosis. Naturally, the combination of proapoptotic and antiapoptotic molecules facilitates programmed cell death, also known as apoptosis. The proapoptotic Bax proteins activate caspase to expedite the caspase cascade, leading to cell death. The active B-cell lymphoma 2 protein family attaches to Bcl-2-associated X protein and prevents it from triggering the cell death mechanism. As a result, anticancer medications such as docetaxel target the B-cell lymphoma 2 protein. B-cell lymphoma 2 phosphorylation at serine residues reduces binding to the Bcl-2-associated X protein, resulting in the loss of B-cell lymphoma 2's antiapoptotic activity. Overexpression of B-cell lymphoma 2 in cells has anticancer properties.

Tumours develop new capillary blood vessels by the process of angiogenesis. Vacca et al. (2002) studied that docetaxel inhibits angiogenesis *in vitro* and *in vivo*. In conclusion, the overall effect of docetaxel used as an anticancer agent is apoptosis, angiogenesis, and gene expression, which are complex processes involving numerous components. The drug can induce signalling aberrations likely to trigger multiple messages within tumour cells.

20.2.2 Structure-Activity Relationship Study of Docetaxel

In the structure-activity relationship study, the 2′-OH docetaxel group is more reactive than other hydroxyl groups (Parness et al., 1982). The 2′-OH group is critical for microtubule binding and cytotoxic activity. It is also clear that 7-OH is not required for cytotoxicity, but its derivatives at the 7-OH position are more stable under physiological conditions. Carrying out derivatives at 2′-OH is more beneficial to produce less toxic taxane prodrugs.

20.2.3 Targeted Delivery of Docetaxel

Passive and active targeting approaches have been established to target a drug to the solid tumour (Behera & Padhi, 2020). In passive targeting approaches, the effective concentration of the drug or formulation reaches the tumour cells through leaky vasculature between the tissues or direct intratumoral administration by enhanced permeation and retention effect (Padhi et al., 2018). Active targeting involves the chemical conjugation of the anticancer drug, carrier material, or linkers with receptor-specific ligands specific to overexpressed receptors for the particular type of tumour or conjugation by adsorption (Behera and Padhi, 2022). As the drug-ligand combination penetrates the tumour cell via endocytosis, the drug is delivered inside the cell, away from healthy tissues (Padhi and Behera, 2022).

20.2.3.1 Passive Targeting of Docetaxel Nanoformulations

Khalid et al. used a phase inversion-based approach to incorporate the hydrophobic anticancer drug docetaxel into poly(ethylene glycol)-coated lipid nanocapsules. Under magnetic stirring, polyethylene glycol 660–12 hydroxyl stearate, hydrogenated soya phosphatidylcholine, tricapryline, and docetaxel were mixed and heated above a phase transition temperature of 80°C. The mixture was then cooled down to 60°C to get an o/w type of nanoemulsions and further heated above the phase transition temperature to produce a w/o system. The PEGylated lipid nanocapsules were formed by incubating the above mixture with 1,2-distearoyl-sn-glycerol-3-phosphatidylethanolamine-*N*-monomethoxy-poly(ethylene glycol)$_{2000}$ at different concentrations. As a result, the lipid colloidal carrier containing docetaxel would be effectively targeted to malignant cells (Khalid et al., 2006).

Gao et al. (2008a) formulated docetaxel using a high-pressure homogenization method in a nanoemulsion system. The study utilized submicron-sized lipid emulsions as carriers for docetaxel. The entrapment efficiency determined by ultrafiltration and ultracentrifugation was greater than 90%. Animal studies showed that docetaxel emulsion has a significantly threefold higher area under the curve and Cmax in rats than in its micellar solution. The results suggested that the submicron lipid emulsion may be a promising intravenous carrier for docetaxel compared to the commercially available docetaxel micellar solution.

To increase docetaxel solubilization capacity and oral bioavailability, an o/w microemulsion system including the drug has been developed (Yin et al., 2009). To construct a microemulsion, a lipid (Capryol 90 oil), a surfactant (Cremophor EL), and a co-surfactant (Transcutol) were combined in a ratio of 29.4:24.9:12.4 w/w. In comparison to taxotere®, the formulation had a higher solubilization capacity. Compared to taxotere®, the apical to basolateral transport of docetaxel across the colon cancer cells (CACO-2) monolayer in rats was observed to be improved (34.4%). This may be due to increased docetaxel transfer.

The increased oral bioavailability may be attributed to enhanced solubility, suppression of the P-gp efflux mechanism, and increased permeability. Docetaxel's limited aqueous solubility necessitates the addition of Tween 80 and ethanol to its composition. This method may result in harmful repercussions. Submicron lipid emulsions may be used as a docetaxel carrier to circumvent these potentially dangerous effects.

Huynh et al. (2009) investigated the delivery of docetaxel using a prodrug approach. The theoretical computer-based investigation was used to prepare appropriate docetaxel conjugate synthesis to improve solubility in the internal phase of a nanoemulsion formulation. This was accomplished by attaching lauroyl moieties to docetaxel via an ester bond. For the docetaxel conjugates, the results revealed high entrapment efficiency of up to 97% and a loading capacity of up to 5.7% w/w. In human cancer cells, mono-substituted docetaxel had a toxicity reduction of 37–46 times that of the neat drug. Adding a lauroyl group to docetaxel via an ester bond produced a prodrug of docetaxel to improve solubility in propylene glycol dicaprylate/dicaprate (Labrafac). When compared to neat docetaxel, 2′-lauroyl docetaxel had an oil solubility of more than 8 times, and the nanoemulsion has a high loading capacity (5.7% w/w) and entrapment efficiency (97%). Under biological conditions, the ester group at the C-2′ may be hydrolysed more quickly than other conjugation locations, implying better bioactivity. With various nanocarriers, this technique offers potential applications in the said domain.

Zhigaltsev et al. (2010) prepared a weak-base docetaxel derivative and further incorporated it into lipid nanoparticles as a prodrug prepared by the pH gradient loading technique. The prodrug was loaded into liposomes to generate a stable encapsulated product with activity comparable to neat docetaxel. Compared to taxotere®, the prodrug 2′-*O*-(*N*-methyl-piperazinylbutanoyl) docetaxel allowed a longer circulation half-life and higher plasma exposure. Compared to neat drug and taxotere®, the docetaxel encapsulated lipid nanoparticle prodrug possessed the extended drug circulation half-life required for effective tumour accumulation, with 50–100 times greater drug plasma levels (Zhigaltsev et al., 2010).

Liu et al. (2011) synthesized docetaxel-loaded nanostructured lipid carriers to minimize the associated toxicity and improve therapeutic efficacy. Desired amounts of excipients and docetaxel were dissolved in ethanol. Then, the ethanol was evaporated under decreased pressure in a water bath for half an hour at 50°C to form the oil phase. Water was mixed with soya lecithin and F68 to make up the aqueous phase. The aqueous phase was introduced to the oil phase by ultrasonication for five minutes at 80°C with continuous stirring. The nanostructured lipid carriers were produced by solidifying the mixture in an ice bath and then lyophilizing it. The formulation outperformed the commercially available 'Duopafei' in terms of cytotoxicity against lung cancer cell lines (A549). Superior apoptosis and G2/M arrest were noted. Because of the enhanced permeation and retention effects, docetaxel-encapsulated nanostructured lipid carriers accumulated and penetrated tumour tissue, resulting in a larger concentration of the formulation in malignant tissue than in normal tissue.

Ernsting et al. (2012) developed a polymer conjugate composed of acetylated carboxymethylcellulose, docetaxel, and polythylene glycol to enhance the pharmacokinetics and antitumour efficacy of docetaxel. PEGylation protected docetaxel from adsorption onto serum. The nanoparticles accumulated in tumour tissue due to enhanced permeation and retention effects that enhanced pharmacokinetic characteristics and inhibited tumour growth. By limiting drug-polymer interactions with opsonins and reticuloendothelial system clearance, the formulation was designed and observed to improve solubility and lengthen pharmacokinetic profiles. It increased tumour uptake through the enhanced permeation and retention effects-based targeting mechanism while minimizing exposure to healthy tissues.

Wang et al. investigated a docetaxel-loaded pH-responsive hydrogel for oral drug administration to improve anticancer efficacy. The solubility and permeability of docetaxel were improved by encapsulating it in micelles based on the amphiphilic copolymer poly(-caprolactone)-poly(ethylene glycol)-poly(-caprolactone) copolymer using co-monomers methoxyl poly(ethylene glycol)-poly(-caprolactone)-acryloyl chloride, poly(ethylene glycol)methyl ether methacrylate. The formulation was passively targeted to tumour tissue (Wang et al., 2014). The heightened enhanced permeation and retention effect was due to increased vascular permeability, dense vascular density, and fast angiogenesis (Maeda et al., 2009). The polymer provided targeted delivery and improved tumour suppression against breast cancer cells (4T1). A tenfold increase in oral bioavailability of docetaxel was noted, possibly due to the hydrogel's controlled release pattern.

Docetaxel-loaded mixed micelle of monomethylol poly(ethylene glycol)-poly(D,L-lactic acid), tocopheryl polyethylene glycol 1000 succinate, and chitosan oligosaccharide grafted with stearic acid copolymers were prepared to enhance oral absorption of the drug using the thin film hydration method (Dou et al., 2014). The polymeric micelles increased drug solubility through passive diffusion into the GIT membrane while suppressing the P-gp efflux pump, which protected the drug from the P-gp efflux pump. Compared to neat docetaxel, the oral bioavailability revealed delayed release characteristics and a greater than two-fold rise in oral bioavailability.

Xu et al. (2016) loaded gambogenic acid and docetaxel simultaneously in poly(lactic-co-glycolic acid) nanoparticles in an optimal ratio to achieve the maximal *In-vitro* anticancer efficiency. The formulation conferred therapeutic efficacy in drug-resistant breast cancer cells. P-glycoprotein suppression of gambogic acid, synchronized release, and lengthy retention in vivo have all been linked to the reversal of multidrug resistance. Gambogic acid straightaway enhanced the post-translational proteasome pathway, which suppressed P-glycoprotein expression. Hence, the poly(lactic-co-glycolic acid) nanoparticulate delivery system may be a platform for simultaneously delivering drugs.

As reported by Dawoud and coworkers, the high hydrophobicity and systemic toxicity issues due to the nonspecific distribution of docetaxel were solved by the preparation of cubic nanoparticles called as cubosomes. Due to their particle size and release characteristics, monoolein cubic nanoparticles appeared to be an effective carrier for docetaxel, and these cubic nanoparticles demonstrated superior tumour regression. As amphiphiles, monoolein and poloxamer in various concentrations were utilized, and the mixture was kept constant at 12% poloxamer. A homogenization process was used to make docetaxel-loaded cubic nanoparticles. Amphiphiles were introduced dropwise into water and stirred at room temperature to make coarse dispersions. The dispersion was stored at room temperature for one day under magnetic stirring, protected from sunlight, before being homogenized in a microfluidizer at 350 bar for 15 minutes at 40°C and then sterilized in an autoclave. To make monoolein cubic nanoparticles, 10 mL of monoolein dispersion was mixed with docetaxel in methanol and agitated for three days at 25°C. In chloroform, egg phosphatidylcholine and cholesterol (4:1 mol/mol) are also used to make multi-lamellar vesicles. The strong inhibitory efficacy of the formulation was a consequence of passive targeting, also known as the enhanced permeation and retention effect, which occurs as a result of the nanoparticles' small particle sizes in the nanometer size range accumulating in the tumour. For preferential accumulation in tumours, circulating nanocarriers frequently took advantage of leaky vasculature and/or inadequate lymphatic drainage in tumour tissues (Dawoud et al., 2020).

20.2.3.2 Active Targeting of Docetaxel Nanoformulations

Gao et al. (2008b) reported that integrating docetaxel in nanoparticles modified with recombinant human epidermal growth factor improved cellular internalization and cytotoxic activity against breast cancer cells (MDA-MB-468) by inducing apoptosis and inducing more G2/M arrest at high drug concentrations, as compared to the marketed formulation 'taxotere®.' Polyethylene glycol-distearoylphosphatidyl ethanolamine and epidermal growth factor were used in the targeted nanoassembly encapsulating docetaxel. Compared to taxotere®, the formulation showed extended circulation time *in-vivo* and boosted drug concentration in tumour cells. Epidermal growth factor as a ligand was noted to bind specifically to epidermal growth factor receptors overexpressed in breast cancer cells (MDA-MB-468) and stimulated cellular uptake and internalization via epidermal growth factor receptor-driven endocytosis, demonstrating active targeting of the formulation.

Dhar et al. (2008) used the nanoprecipitation process to develop nanoparticles employing cisplatin and docetaxel incorporated efficiently in poly(lactic-co-glycolic acid)-poly(ethylene glycol) polymer. The dual loading of drugs with continuous feed ratios resulted in increased encapsulation, which varied depending on the drug-polymer ratio. Prostate-specific membrane antigen is overexpressed in prostate cancer. Using an antibody specific for endosome formation, the A10 aptamer was used to decorate the surface of the polymeric nanoparticles, which bound to the extracellular domain of prostate-specific membrane antigen, facilitating cellular reception of the platinum(IV)-encapsulated nanoparticles via endocytosis.

To target hepatocellular cancer, docetaxel-loaded novel hepatoma-targeted solid lipid nanoparticles were created by Xu et al. (2009). The solid lipid nanoparticles produced with galactosylated dioleoylphosphatidyl ethanolamine had a 90% encapsulation efficiency with a 29-day sustained release characteristic. The docetaxel-encapsulated solid lipid nanoparticles were noted to accumulate in the liver, causing hepatocyte destruction. Galactosylation of the docetaxel-loaded nanoparticles exhibited enhanced cellular uptake through overexpressed asialoglycoprotein receptors on hepatocellular carcinoma cells.

Mei et al. (2009) studied docetaxel-loaded poly(ε-caprolactone)/pluronic F68 nanoparticles to overcome multidrug resistance in docetaxel-resistant human breast cancer cell lines. In human breast cancer cells, the nanoparticles had a substantially higher level of cytotoxic effect than taxotere® and could overcome multidrug resistance. The nanoparticles were internalized faster in a docetaxel-resistant human breast cancer cell line. The pluronic block copolymers used in the formulation have functioned in changing the micro-viscosity of membranes, inducing a reduction in ATP levels in cancer cells, inhibiting P-glycoprotein, breast cancer resistance protein, and multidrug resistance proteins, releasing cytochrome C and increasing reactive oxygen species levels in the cytoplasm, and enhancing proapoptotic signalling.

Core-shell nanoparticles with a lipid monolayer shell and a biodegradable polymer core containing docetaxel were reported by Liu et al. (2010). The core-shell polymeric nanoparticles were assessed for their anticancer efficacy on two cell lines. The breast cancer cells (MCF-7) with overexpressed folate receptors and fibroblast cells (NIH/3T3) lacking overexpressed folate receptors were assessed in vitro for cellular uptake of docetaxel-loaded core-shell nanoparticles. The evaluation showed a controlled and targeted delivery of docetaxel to folate overexpressed cancer cells.

The targeted codelivery of docetaxel and cisplatin decorated with Herceptin was studied to target the over-expressed high human epidermal growth factor receptor 2 in breast cancer cells (Mi et al., 2013). The nanoprecipitation process was used to design the formulation with a variable docetaxel-cisplatin ratio, which resulted in a pH-sensitive drug release. The higher the docetaxel to cisplatin ratio, the better the efficacy was observed. The prodrug system concept for nanoformulation can deliver various hydrophilic and hydrophobic drugs to intended sites. Conjugated nanoparticles are the most effective for human epidermal growth factor receptor 2 - positive cancer cells. The nanoparticles' targeting effects were evidenced by the IC_{50} values being substantially lower for breast cancer cells (SK-BR-3) with high human epidermal growth factor receptor 2 overexpression than for fibroblast cells (NIH/3T3 cells and MCF7 cells) with little or low human epidermal growth factor receptor 2 overexpression.

Song et al. (2014) studied the cyclic tetrapeptide-decorated co-entrapped docetaxel and cisplatin-based micelles. Docetaxel and cisplatin were loaded into an amphiphilic graft copolymer of a-tocopherol and polyethylene glycol to poly(L-glutamic acid) via hydrophobic and chelating effects. The cytotoxicity of the dual drug-loaded micelle decorated with cyclic tetrapeptide for $\alpha v\beta 3$ integrin targeting peptide was synergistic. The micelles' endocytosis rate and retention effect in tumour tissue were improved by cyclic tetrapeptide conjugation. Longer circulation times and superior antitumour efficacy with fewer side effects were noted.

Ren et al. (2016) developed a redox-responsive docetaxel - vitamin E prodrug formulated into liposomes by the thin film hydration method, which has high drug loading and good stability. The disulphide linkage was formed between docetaxel and Vitamin E to encapsulate the drug in the liposomes. The docetaxel liposome showed nearly complete tumour suppression. This redox-responsive prodrug was hydrolysed in a high glutathione concentration and the released docetaxel induced cell apoptosis.

Yan et al. (2017) investigated epidermal growth factor receptor peptide decorated nanoparticles loaded with docetaxel and pH-sensitive curcumin prodrug to ameliorate prostate cancer. The formulated nanoparticles improved cancer cell inhibition and tumour tissue growth. The encapsulated drugs were effectively targeted into tumour cells. RIPL peptide (IPLVVPLRRRRRRRRC, 16-mer) conjugated docetaxel liposomes were synthesized to enable selective delivery into hepsin-overexpressed cancer cells. Docetaxel-conjugated RIPL – malemine derivatized liposomes were noted to bind to hepsin-positive cancer cells selectively, efficiently transporting the drug into the cells. Due to components in the interstitium, such as proteoglycans and the glycocalyx, positively charged docetaxel-RIPL liposomes effectively adhered to negatively charged tumour surfaces, making it a strong candidate for efficient drug targeting to hepsin-expressing cancer cells (Yoon et al., 2017).

Singh et al. (2018) formulated folic acid-conjugated docetaxel co-encapsulated with resveratrol nanoparticles to treat prostate cancer. In prostate cancer cells (PaCa cells), it exerted cytotoxic effects by downregulating survivin and increasing the expression of cleaved Caspase-3. The folic acid-attached nanoparticles operated as effective inhibitors of drug efflux, increasing intracellular drug concentration and thereby increasing cytotoxicity in the tumour cells (Singh et al., 2018).

Zou et al. (2020) developed cyclic NGQ peptide-decorated polymersome entrapping docetaxel as a targeted delivery system for non-small cell lung cancer (A549). The formulation presented good drug loading. This $\alpha 3\beta 1$ integrin-targeted polymersomal docetaxel promoted selective uptake by A549 cells, which potentiated antitumour activity.

20.3 HIGHLIGHT ON CLINICAL TRIALS AND PATENTS ON DOCETAXEL AND ITS FORMULATIONS

Recently, numerous studies on new docetaxel formulations show significantly enhanced antitumour and targeting activity compared with neat docetaxel in preclinical studies (Bissery, 1992). However, only a few have entered clinical trials, which are still not marketed. The clinical translation of experimental drugs faces many hurdles, including safety issues, uncontrollable and reproducible production processes (Zhang et al., 2019). Few docetaxel-based formulations are under either phase I or phase II clinical trials, as mentioned in Table 20.1 (Zhang et al., 2019). These trials on different formulations are going on for indications such as ovarian cancer, advanced solid tumour, metastatic

TABLE 20.1
Docetaxel-Based Formulations under Clinical Trials

Name of Company	Types of Formulation	Clinical Trial Phase	References
Neo Pharm	Liposome	Phase II	Louage et al. (2017)
Azaya Therapeutics	Liposome	Phase I	Mahalingam et al. (2014)
BIND Therapeutics	Polymeric nanoparticle	Phase I/II	Ledford (2016)
Celgene	Polymeric nanoparticle	Phase II	Ren et al. (2013)
Samyang Pharma.	Polymeric nanoparticle	Phase I	Conte et al. (2014)
Samyang Pharma	Polymeric micelle	Phase I	Conte et al. (2014)
Cristal Therapeutics	Polymeric micelle	Phase I	Hu et al. (2016)
Nektar	Polymer conjugate	Phase I/II	Louage et al. (2017)
Cerulean Pharma	Polymer conjugate	Phase I/II	Markman et al. (2016)
Starpharma	Dendrimer	Phase II	Webpage, Cited (n.d.)
Modra Pharmaceuticals	Solid dispersion	Phase II	Webpage (n.d.)

TABLE 20.2
Some of the Patents on Docetaxel and Its Formulation

Patent Application	Title of Patent	Inventor	References
US 8.476,310 B2; July 2, 2013, Appl.-12/721,564	Docetaxel formulations with lipoic acid	Nagesh Palepu (US)	Palepu (2013)
WO 2012/156999 Al; Nov 2012, Appl.-PCT/IN2012/000358	Ready to use docetaxel formulation	Chaudhary, Manu, (IN)	Chaudhary (2012)
WO 2009/002425A3; Dec 2008, Appl.-PCT/US2008/007619	Solubilized formulation of docetaxel without Tween 80	Palepu Nageshwara R. (US) BulusuBhamuTeja (IN)	Palepu Nageshwara R. (2008)
US 7,332,617 B2 Feb 2008, Appl.-10/555,889	Process for the preparation of docetaxel, trihydrate	Jinliang Li, Shanghai (CN); Zuwang Wu, Shanghai (CN)	Li and Wu (2008)
WO 2000/061185 (19.10.2000 Gazette 2000/42); Oct 2005, Appl.-PCT/US2000/009247	Docetaxel in combination with rhuMAb HER2 for the treatment of cancers	Vogel, Charles LE (US) Bellet, Robert E (US)	Vogel and Bellet (2005)

castration-resistant prostate cancer, metastatic breast cancer, hormone-refractory prostate cancer, various solid malignancies, breast cancer, refractory tumour, advanced cancer, and metastatic prostate cancer.

Some of the patents are purely on docetaxel and its formulation as an anticancer agent, as listed in Table 20.2.

20.4 FUTURE PERSPECTIVES

Cancer cells are a diverse group of cells that develop uncontrollably. Clinical oncologists hope to get the best anticancer impact with the least side effects and no unintended effects on normal cells. It is preferable to have therapies that address as many potential targets or pathways as possible. Drug developers and formulators must look for a delivery system that can administer a mix of anticancer medications, such as cytotoxic pharmaceuticals, drugs targeting specific biomarkers, radio-therapeutics, and even small interfering RNAs. Highly sensitive diagnostic probes allow for early drug response detection, quick regimen adjustments, and customized medicine. Taxanes' toxicity to rapidly developing intestinal epithelium has hindered the development of oral dosage forms. Before the development of oral dosage for the distribution of taxane, further work needs to be done to reduce drug-associated gastrointestinal toxicity and carrier-associated toxicity.

The stimuli-responsive system, achievable and established in nano drug delivery systems for other anticancer medications, is also a challenge in taxane delivery. Codelivery with other potent and particular herbal anticancer drugs or groups of chemicals should also be investigated. There is a need to find more specific new targets and ligands that will aid in delivering active taxane targeting. For example, new generation binding ligands have been generated, such as monobodies, affibodies, heptameric binding domains, receptors, integrins, and enzymes, which could pique interest in active pharmaceutical delivery. Before the potential advantage of these formulations can be fully realized, there are still several noticeable gaps to fill.

20.5 CONCLUSION

The discovery of taxoid in the 1960s led to the discovery of its family of bioactive chemicals, which include paclitaxel, docetaxel, and its analogues, all of which have a taxane skeleton. Taxanes have a one-of-a-kind pharmacological mechanism for inhibiting cell proliferation. Docetaxel contains structural changes that give it superior water solubility, stronger binding affinity, and increased potency than paclitaxel despite having few chemical, physical, and pharmacological similarities. It is tenfold more water soluble than paclitaxel, has a 1.9-fold higher microtubule binding affinity, which leads to better cellular accumulation, and is twice as effective.

Docetaxel vectorization employing a carrier coupled to a targeted agent has been the subject of extensive research. Different investigations in vitro and in vivo have shown benefits and a significant increase in docetaxel delivery utilizing various techniques. Biocompatible polymers were used as transporters, while antibodies were used as targeting agents, with positive results.

Due to their high lipophilicity and lattice energy, paclitaxel and docetaxel have low water solubility. Taxotere is a commercially available docetaxel dosage form based on an emulsion and requires special mixing before administration. Furthermore, due to quick clearance, short half-lives, and vast volumes of distribution, the existing dosage forms of docetaxel have unfavourable pharmacokinetic characteristics, resulting in limited drug accumulation in tumour sites and relatively substantial drug exposure in normal organs. Docetaxel formulations based on nanotechnology benefits are being tested to address these concerns. Docetaxel formulations include micelles, nanoemulsions, microemulsions, solid lipid nanoparticles, nanocapsules, and others. A prodrug with docetaxel codelivery is a must for improved docetaxel encapsulation in nanoparticles.

The complication could occur in taxane delivery in terms of the drug release pattern. Premature cleavage or release, or too slow cleavage and ineffective release of active drugs, could cause this. As a result, while creating taxane prodrugs, the appropriate drug cleavage rate is extremely important. The prodrugs strategy is a response to such issues or limitations with current drug carriers. It reduces systemic toxicity, has the potential for site-specific drug release based on conjugation chemistry, and the prodrug formulation has additional targeting properties and a higher therapeutic index when the conjugation linkage is cleaved by some enzymes expressed specifically in the tumour site.

REFERENCES

Behera, A., & Padhi, S. (2020). Passive and active targeting strategies for the delivery of the camptothecin anticancer drug: A review. *Environmental Chemistry Letters*, 18(5), 1557–1567.

Behera, A., & Padhi, S. (2022). pH-Sensitive polymeric nanoparticles for cancer treatment. In: Padhi, S., Behera, A., Lichtfouse, E. (Eds.) *Polymeric Nanoparticles for the Treatment of Solid Tumours*. Environmental Chemistry for a Sustainable World, vol 71. Springer, Cham, pp. 401–425.

Behera, A., Pradhan, S. P., Tejaswani, P., Sa, N., Pattnaik, S., & Sahu, P. K. (2023). Ameliorative and neuroprotective effect of core-shell type Se@ Au conjugated hesperidin nanoparticles in diabetes-induced cognitive impairment. *Molecular Neurobiology*, 60(12), 7329–7345.

Benhabbour, S. R., Luft, J. C., Kim, D., Jain, A., Wadhwa, S., Parrott, M. C., Liu, R., DeSimone, J. M., & Mumper, R. J. (2012). In vitro and in vivo assessment of targeting lipid-based nanoparticles to the epidermal growth factor-receptor (EGFR) using a novel Heptameric ZEGFR domain. *Journal of Controlled Release*, 158(1), 63–71.

Bissery, M. (1992). Preclinical pharmacology and toxicology of Taxotere (RP56976, NSC628503). *Annals of Oncology*, 3, 121.

Bissery, M.-C. (1995). Preclinical pharmacology of docetaxel. *European Journal of Cancer*, 31, S1–S6.

Chang, R. (2002). Bioactive polysaccharides from traditional Chinese medicine herbs as anticancer adjuvants. *The Journal of Alternative & Complementary Medicine*, 8(5), 559–565.

Chast, F. (2006). Pierre Potier (1934–2006). *Nature*, 440(7082), 291–291.

Chaudhary, M. (2012). *Patent on 'Ready to use docetaxel formulation'* (IN Patent No. WO 2012/156999 Al).

Cole, A. J., David, A. E., Wang, J., Galbán, C. J., Hill, H. L., & Yang, V. C. (2011). Polyethylene glycol modified, cross-linked starch-coated iron oxide nanoparticles for enhanced magnetic tumour targeting. *Biomaterials*, 32(8), 2183–2193.

Conte, C., d'Angelo, I., Miro, A., Ungaro, F., & Quaglia, F. (2014). PEGylated polyester-based nanoncologicals. *Current Topics in Medicinal Chemistry*, 14(9), 1097–1114.

Cortes, J. E., & Pazdur, R. (1995). Docetaxel. *Journal of Clinical Oncology*, 13(10), 2643–2655.

Cui, Z., Patel, J., Tuzova, M., Ray, P., Phillips, R., Woodward, J. G., Nath, A., & Mumper, R. J. (2004). Strong T cell type-1 immune responses to HIV-1 Tat (1–72) protein-coated nanoparticles. *Vaccine*, 22(20), 2631–2640.

Dawoud, M., Abourehab, M. A., & Abdou, R. (2020). Monoolein cubic nanoparticles as novel carriers for docetaxel. *Journal of Drug Delivery Science and Technology*, 56, 101501.

Dhar, S., Gu, F. X., Langer, R., Farokhzad, O. C., & Lippard, S. J. (2008). Targeted delivery of cisplatin to prostate cancer cells by aptamer functionalized Pt (IV) prodrug-PLGA–PEG nanoparticles. *Proceedings of the National Academy of Sciences of the United States of America*, 105(45), 17356–17361.

Dou, J., Zhang, H., Liu, X., Zhang, M., & Zhai, G. (2014). Preparation and evaluation in vitro and in vivo of docetaxel loaded mixed micelles for oral administration. *Colloids and Surfaces B: Biointerfaces*, 114, 20–27.

Du, W., Hong, L., Yao, T., Yang, X., He, Q., Yang, B., & Hu, Y. (2007). Synthesis and evaluation of water-soluble docetaxel prodrugs-docetaxel esters of malic acid. *Bioorganic & Medicinal Chemistry*, 15(18), 6323–6330.

Ellis, A. G., Crinis, N. A., & Webster, L. K. (1996). Inhibition of etoposide elimination in the isolated perfused rat liver by Cremophor EL and Tween 80. *Cancer Chemotherapy and Pharmacology*, 38(1), 81–87.

Ernsting, M. J., Tang, W.-L., MacCallum, N. W., & Li, S.-D. (2012). Preclinical pharmacokinetic, biodistribution, and anticancer efficacy studies of a docetaxel-carboxymethylcellulose nanoparticle in mouse models. *Biomaterials*, 33(5), 1445–1454.

Feng, L., & Mumper, R. J. (2013). A critical review of lipid-based nanoparticles for taxane delivery. *Cancer Letters*, 334(2), 157–175.

Gao, K., Sun, J., Liu, K., Liu, X., & He, Z. (2008a). Preparation and characterization of a submicron lipid emulsion of docetaxel: Submicron lipid emulsion of docetaxel. *Drug Development and Industrial Pharmacy*, 34(11), 1227–1237.

Gao, Y., Chen, L., Gu, W., Xi, Y., Lin, L., & Li, Y. (2008b). Targeted nanoassembly loaded with docetaxel improves intracellular drug delivery and efficacy in murine breast cancer model. *Molecular Pharmaceutics*, 5(6), 1044–1054.

Haldar, S., Basu, A., & Croce, C. M. (1997). Bcl2 is the guardian of microtubule integrity. *Cancer Research*, 57(2), 229–233.

Hu, Q., Rijcken, C. J., van Gaal, E., Brundel, P., Kostkova, H., Etrych, T., Weber, B., Barz, M., Kiessling, F., & Prakash, J. (2016). Tailoring the physicochemical properties of core-crosslinked polymeric micelles for pharmaceutical applications. *Journal of Controlled Release*, 244, 314–325.

Huynh, L., Leroux, J.-C., & Allen, C. (2009). Enhancement of docetaxel solubility via conjugation of formulation-compatible moieties. *Organic & Biomolecular Chemistry*, 7(17), 3437–3446.

Ibrahim, N. K., Desai, N., Legha, S., Soon-Shiong, P., Theriault, R. L., Rivera, E., Esmaeli, B., Ring, S. E., Bedikian, A., & Hortobagyi, G. N. (2002). Phase I and pharmacokinetic study of ABI-007, a Cremophor-free, protein-stabilized, nanoparticle formulation of paclitaxel. *Clinical Cancer Research*, 8(5), 1038–1044.

Khalid, M. N., Simard, P., Hoarau, D., Dragomir, A., & Leroux, J.-C. (2006). Long circulating poly (ethylene glycol)-decorated lipid nanocapsules deliver docetaxel to solid tumours. *Pharmaceutical Research*, 23(4), 752–758.

Khuroo, T., Verma, D., Talegaonkar, S., Padhi, S., Panda, A. K., & Iqbal, Z. (2014). Topotecan–tamoxifen duple PLGA polymeric nanoparticles: Investigation of in vitro, in vivo and cellular uptake potential. *International Journal of Pharmaceutics*, 473(1–2), 384–394.

Kundu, A., Padhi, S., Behera, A., Hasnain, M. S., & Nayak, A. K. (2022). Tumour targeting strategies by chitosan-based nanocarriers. In: Hasnain, M. S., Beg, S., Nayak, A. K. (Eds.) *Chitosan in Biomedical Applications*. Academic Press, London, pp. 163–188.

Ledford, H. (2016). Bankruptcy filing worries developers of nanoparticle cancer drugs. *Nature News*, 533(7603), 304.

Li, J., & Wu, Z. (2008). U.S. Patent No. 7,332,617. Washington, DC: US Patent and Trademark Office.

Liu, D., Liu, Z., Wang, L., Zhang, C., & Zhang, N. (2011). Nanostructured lipid carriers as novel carrier for parenteral delivery of docetaxel. *Colloids and Surfaces B: Biointerfaces*, 85(2), 262–269.

Liu, Y., Li, K., Pan, J., Liu, B., & Feng, S.-S. (2010). Folic acid conjugated nanoparticles of mixed lipid monolayer shell and biodegradable polymer core for targeted delivery of Docetaxel. *Biomaterials*, 31(2), 330–338.

Loos, W. J., Baker, S. D., Verweij, J., Boonstra, J. G., & Sparreboom, A. (2003). Clinical pharmacokinetics of unbound docetaxel: Role of polysorbate 80 and serum proteins. *Clinical Pharmacology & Therapeutics*, 74(4), 364–371.

Louage, B., De Wever, O., Hennink, W. E., & De Geest, B. G. (2017). Developments and future clinical outlook of taxane nanomedicines. *Journal of Controlled Release*, 253, 137–152.

Ma, Y.-T., Yang, Y., Cai, P., Sun, D.-Y., Sánchez-Murcia, P. A., Zhang, X.-Y., Jia, W.-Q., Lei, L., Guo, M., & Gago, F. (2018). A series of enthalpically optimized docetaxel analogues exhibiting enhanced antitumour activity and water solubility. *Journal of Natural Products*, 81(3), 524–533.

Maeda, H., Bharate, G., & Daruwalla, J. (2009). Polymeric drugs for efficient tumour-targeted drug delivery based on EPR-effect. *European Journal of Pharmaceutics and Biopharmaceutics*, 71(3), 409–419.

Mahalingam, D., Nemunaitis, J. J., Malik, L., Sarantopoulos, J., Weitman, S., Sankhala, K., Hart, J., Kousba, A., Gallegos, N. S., & Anderson, G. (2014). Phase I study of intravenously administered ATI-1123, a liposomal docetaxel formulation in patients with advanced solid tumours. *Cancer Chemotherapy and Pharmacology*, 74(6), 1241–1250.

Manfredi, J. J., Parness, J., & Horwitz, S. B. (1982). Taxol binds to cellular microtubules. *The Journal of Cell Biology*, 94(3), 688–696.

Markman, B., De Souza, P. L., Dees, E. C., Gangadhar, T. C., Cooper, A., Roohullah, A., Boolell, V., Zamboni, W., Murphy, C., & Senderowicz, A. M. (2016). A phase 1 study of CRLX301, a novel nanoparticle-drug conjugate (NDC) containing docetaxel (DOC), in patients with refractory solid tumours. *Journal of Clinical Oncology*, 34, 2526.

Mei, L., Zhang, Y., Zheng, Y., Tian, G., Song, C., Yang, D., Chen, H., Sun, H., Tian, Y., & Liu, K. (2009). A novel docetaxel-loaded poly(ε-caprolactone)/pluronic F68 nanoparticle overcoming multidrug resistance for breast cancer treatment. *Nanoscale Research Letters*, 4(12), 1530–1539.

Mi, Y., Zhao, J., & Feng, S.-S. (2013). Targeted codelivery of docetaxel, cisplatin and herceptin by vitamin E TPGS-cisplatin prodrug nanoparticles for multimodality treatment of cancer. *Journal of Controlled Release*, 169(3), 185–192.

Padhi, S., & Behera, A. (2022). Cellular internalization and toxicity of polymeric nanoparticles. In: Padhi, S., Behera, A., Lichtfouse, E. (Eds.) *Polymeric Nanoparticles for Treating Solid Tumours*. Environmental Chemistry for a Sustainable World, vol 71. Springer, Cham, pp. 473–488.

Padhi, S., Kapoor, R., Verma, D., Panda, A. K., & Iqbal, Z. (2018). Formulation and optimisation of topotecan nanoparticles: In vitro characterisation, cytotoxicity, cellular uptake and pharmacokinetic outcomes. *Journal of Photochemistry and Photobiology B: Biology*, 183, 222–232.

Padhi, S., Mirza, M. A., Verma, D., Khuroo, T., Panda, A. K., Talegaonkar, S., Khar, R. K., & Iqbal, Z. (2016). Revisiting the nanoformulation design approach for effective delivery of Topotecan in its stable form: An appraisal of its in vitro behavior and tumour amelioration potential. *Drug Delivery*, 23(8), 2827–2837.

Palepu Nageshwara R., B. B. T. (2008). *Patent on 'Solubilized formulation of docetaxel without tween 80'*. W. PCT.

Palepu, N. (2013). *Patent on 'Docetaxel formulations with lipoic acid'* (US Patent No. US 8.476,310 B2). U. Patent.

Parness, J., Kingston, D. G., Powell, R. G., Harracksingh, C., & Horwitz, S. B. (1982). Structure-activity study of cytotoxicity and microtubule assembly in vitro by taxol and related taxanes. *Biochemical and Biophysical Research Communications*, 105(3), 1082–1089.

Patel, J. D., O'Carra, R., Jones, J., Woodward, J. G., & Mumper, R. J. (2007). Preparation and characterization of nickel nanoparticles for binding to his-tag proteins and antigens. *Pharmaceutical Research*, 24(2), 343–352.

Patra, N., Dehury, N., Pal, A., Behera, A., & Patra, S. (2018). Preparation and mechanistic aspect of natural xanthone functionalized gold nanoparticle. *Materials Science and Engineering: C*, 90, 439–445.

Pienta, K. J. (2001). Preclinical mechanisms of action of docetaxel and docetaxel combinations in prostate cancer. *Seminars in Oncology*, 28, 3–7.

Pinguet, F., Agustin, M., Brown, P., Soefje, S., Needleman, R., Sanchez, M., Sommer, U., & Tranter, B. (2010). New formulation of Taxotere® reduces preparation time and supports the physical stability of the product in infusion bags: Data from the RELIUS international observational study. *European Journal of Hospital Pharmacy*, 16(Suppl), 1–4.

Pradhan, S. P., Sahoo, S., Behera, A., Sahoo, R., & Sahu, P. K. (2022). Memory amelioration by hesperidin conjugated gold nanoparticles in diabetes induced cognitive impaired rats. *Journal of Drug Delivery Science and Technology*, 69, 103145.

Pradhan, S. P., Tejaswani, P., Sa, N., Behera, A., Sahoo, R. K., & Sahu, P. K. (2023). Mechanistic study of gold nanoparticles of vildagliptin and vitamin E in diabetic cognitive impairment. *Journal of Drug Delivery Science and Technology*, 84, 104508.

Ren, G., Liu, D., Guo, W., Wang, M., Wu, C., Guo, M., Ai, X., Wang, Y., & He, Z. (2016). Docetaxel prodrug liposomes for tumour therapy: Characterization, in vitro and in vivo evaluation. *Drug Delivery*, 23(4), 1272–1281.

Ren, K., Dusad, A., Dong, R., & Quan, L. (2013). Albumin as a delivery carrier for rheumatoid arthritis. *Journal of Nanomedicine & Nanotechnology*, 4(4), 176.

Ringel, I., & Horwitz, S. B. (1991). Studies with RP 56976 (taxotere): A semisynthetic analogue of taxol. *JNCI: Journal of the National Cancer Institute*, 83(4), 288–291.

Rowinsky, M., & Eric, K. (1997). The development and clinical utility of the taxane class of antimicrotubule chemotherapy agents. *Annual Review of Medicine*, 48(1), 353–374.

Safavy, A., Raisch, K. P., Khazaeli, M., Buchsbaum, D. J., & Bonner, J. A. (1999). Paclitaxel derivatives for targeted therapy of cancer: Toward the development of smart taxanes. *Journal of Medicinal Chemistry*, 42(23), 4919–4924.

Schiff, P. B., Fant, J., & Horwitz, S. B. (1979). Promotion of microtubule assembly in vitro by taxol. *Nature*, 277(5698), 665–667.

Simeonova, M., Ilarionova, M., Ivanova, T., Konstantinov, C., & Todorov, D. (1991). Nanoparticles as drug carriers for vinblastine. Acute toxicity of vinblastine in a free form and associated to polybutylcyanoacrylate nanoparticles. *Acta Physiologica et Pharmacologica Bulgarica*, 17(4), 43–49.

Singh, S. K., Lillard Jr, J. W., & Singh, R. (2018). Reversal of drug resistance by planetary ball milled (PBM) nanoparticle loaded with resveratrol and docetaxel in prostate cancer. *Cancer Letters*, 427, 49–62.

Song, W., Tang, Z., Zhang, D., Zhang, Y., Yu, H., Li, M., Lv, S., Sun, H., Deng, M., & Chen, X. (2014). Antitumour efficacy of c(RGDfK)-decorated polypeptide-based micelles co-loaded with docetaxel and cisplatin. *Biomaterials*, 35(9), 3005–3014.

Souza, L., Silva, E., Martins, A., Mota, M., Braga, R., Lima, E., Valadares, M., Taveira, S., & Marreto, R. (2011). Development of topotecan loaded lipid nanoparticles for chemical stabilization and prolonged release. *European Journal of Pharmaceutics and Biopharmaceutics*, 79(1), 189–196.

Sparreboom, A., Van Zuylen, L., Brouwer, E., Loos, W. J., De Bruijn, P., Gelderblom, H., Pillay, M., Nooter, K., Stoter, G., & Verweij, J. (1999). Cremophor EL-mediated alteration of paclitaxel distribution in human blood: Clinical pharmacokinetic implications. *Cancer Research*, 59(7), 1454–1457.

Stein, C. (1999). Mechanisms of action of taxanes in prostate cancer. *Seminars in Oncology*, 26, 3–7.

Vacca, A., Ribatti, D., Iurlaro, M., Merchionne, F., Nico, B., Ria, R., & Dammacco, F. (2002). Docetaxel versus paclitaxel for antiangiogenesis. *Journal of Hematotherapy & Stem Cell Research*, 11(1), 103–118.

Verma, D., Thakur, P. S., Padhi, S., Khuroo, T., Talegaonkar, S., & Iqbal, Z. (2017). Design expert assisted nanoformulation design for codelivery of Topotecan and thymoquinone: Optimization, in vitro characterization, and stability assessment. *Journal of Molecular Liquids*, 242, 382–394.

Vogel, C. L., & Bellet R. E. (2005). *Patent on 'Docetaxel in combination with rhumab her2 for the treatment of cancers'* WO 2000/061185). E. P. Specification.

Wang, Y., Chen, L., Tan, L., Zhao, Q., Luo, F., Wei, Y., & Qian, Z. (2014). PEG–PCL based micelle hydrogels as oral docetaxel delivery systems for breast cancer therapy. *Biomaterials*, 35(25), 6972–6985.

Wani, K., Tarawadi, K., & Kaul-Ghanekar, R. (2015). Nanocarriers for delivery of herbal based drugs in breast cancer-an overview. *Journal of Nano Research*, 34, 29–40.

Webpage. (Cited). (n.d.). *DEP® docetaxel positive phase 1 results; phase 2 commences*. https://www.starpharma.com/news/339

Weiss, R. B., Donehower, R., Wiernik, P., Ohnuma, T., Gralla, R., Trump, D., Baker Jr, J., Van Echo, D., Von Hoff, D., & Leyland-Jones, B. (1990). Hypersensitivity reactions from taxol. *Journal of Clinical Oncology*, 8(7), 1263–1268.

Williams, J., Lansdown, R., Sweitzer, R., Romanowski, M., LaBell, R., Ramaswami, R., & Unger, E. (2003). Nanoparticle drug delivery system for intravenous delivery of topoisomerase inhibitors. *Journal of Controlled Release*, 91(1–2), 167–172.

Xu, Y., Wang, C., Ding, Y., Wang, Y., Liu, K., Tian, Y., Gao, M., Li, Z., Zhang, J., & Li, L. (2016). Nanoparticles with optimal ratiometric codelivery of docetaxel with gambogic acid for treatment of multidrug-resistant breast cancer. *Journal of Biomedical Nanotechnology*, 12(9), 1774–1781.

Xu, Z., Chen, L., Gu, W., Gao, Y., Lin, L., Zhang, Z., Xi, Y., & Li, Y. (2009). The performance of docetaxel-loaded solid lipid nanoparticles targeted to hepatocellular carcinoma. *Biomaterials*, 30(2), 226–232.

Yan, J., Wang, Y., Jia, Y., Liu, S., Tian, C., Pan, W., Liu, X., & Wang, H. (2017). Codelivery of docetaxel and curcumin prodrug via dual-targeted nanoparticles with synergistic antitumour activity against prostate cancer. *Biomedicine & Pharmacotherapy*, 88, 374–383.

Yang, S. C., & Zhu, J. B. (2002). Preparation and characterization of camptothecin solid lipid nanoparticles. *Drug Development and Industrial Pharmacy*, 28(3), 265–274.

Yin, Y.-M., Cui, F.-D., Mu, C.-F., Choi, M.-K., Kim, J. S., Chung, S.-J., Shim, C.-K., & Kim, D.-D. (2009). Docetaxel microemulsion for enhanced oral bioavailability: Preparation and in vitro and in vivo evaluation. *Journal of Controlled Release*, 140(2), 86–94.

Yuan, Y., Long, X., Wei, M., Chen, L., Zhang, J., & Liu, X. (2025). Long and short-term efficacy and safety comparison of nab-paclitaxel versus paclitaxel combined with trastuzumab and pertuzumab for neoadjuvant treatment of HER2-positive breast cancer: A systematic review and meta-analysis. *Cancer Treatment Reviews*, 102975.

Yoon, H. Y., Kwak, S. S., Jang, M. H., Kang, M. H., Sung, S. W., Kim, C. H., Kim, S. R., Yeom, D. W., Kang, M. J., & Choi, Y. W. (2017). Docetaxel-loaded RIPL peptide (IPLVVPLRRRRRRRRC)-conjugated liposomes: Drug release, cytotoxicity, and antitumour efficacy. *International Journal of Pharmaceutics*, 523(1), 229–237.

Zhang, E., Xing, R., Liu, S., & Li, P. (2019). Current advances in development of new docetaxel formulations. *Expert Opinion on Drug Delivery*, 16(3), 301–312.

Zhigaltsev, I. V., Winters, G., Srinivasulu, M., Crawford, J., Wong, M., Amankwa, L., Waterhouse, D., Masin, D., Webb, M., & Harasym, N. (2010). Development of a weak-base docetaxel derivative that can be loaded into lipid nanoparticles. *Journal of Controlled Release*, 144(3), 332–340.

Zou, Y., Sun, Y., Guo, B., Wei, Y., Xia, Y., Huangfu, Z., Meng, F., van Hest, J. C., Yuan, J., & Zhong, Z. (2020). $\alpha3\beta1$ Integrin-targeting polymersomal docetaxel as an advanced nanotherapeutic for non-small cell lung cancer treatment. *ACS Applied Materials & Interfaces*, 12(13), 14905–14913.

Section VI

Co-Delivery of Phytomolecules in Cancer Chemotherapy

21 Codelivery of Natural Drugs with Other Anticancer Drugs Entrapping Nanodelivery Systems for Cancer

Pooja Mahale[1] *and Hrushikesh Velis*[2]
[1]Department of Pharmacology, Dr. Bhanudas Dere College of Pharmacy, Sangamner, India
[2]Department of Pharmacology, Amrutvahini College of Pharmacy, Sangamner, India

21.1 INTRODUCTION

Chemotherapy is the first line of treatment for most malignancies due to its efficacy and safety. Although cancer chemotherapy is very effective in treating cancer, it has various drawbacks, including lack of specificity, organ toxicity, cancer cell resistance, bone marrow suppression, teratogenicity, and so on (Padhi and Behera, 2022; Behera et al., 2023; Patnaik et al., 2021). Chemotherapeutic entrapped drug delivery systems are presently being developed to address these constraints by targeting cancer cells specifically. Nanotechnology has been exploited to generate novel dosage forms such as polymeric nanoparticles, liposomes, niosomes, dendrimers, and so on (Padhi et al., 2016; Kundu et al., 2022). The problems of standard chemotherapy medication have been overcome by combining natural drugs with conventional chemotherapeutic agents. Anticancer medications integrated with a natural drug have improved tumour cell uptake, accumulation, and cytotoxicity when co-delivered (Zou et al., 2017). It not only enhances cellular internalization (Ma et al., 2018), but it also has a synergistic impact by targeting cancer cells in multiple ways (Cheng et al., 2018) while preventing multi-drug resistance (Zhao et al., 2016) by reducing toxicity associated with anticancer drugs (Zhang et al., 2018).

The present chapter highlights the current developments in the codelivery nanosystems comprising natural chemotherapeutics and other anticancer drugs for cancer therapy in the area of anticancer drug delivery systems.

21.2 DIFFERENT TYPES OF ANTICANCER AGENTS, ALONG WITH THEIR LIMITATIONS

21.2.1 Different Types of Anticancer Agents

Anticancer drugs have typically been categorized according to their mode of action. Alkylating agents, natural compounds, antibiotics, platinum-based compounds, and other cytotoxic drugs are among them. Anticancer action is mediated by folate, purine, and pyrimidine antagonists, which interfere with metabolic pathways involved in the biosynthesis of RNA and DNA. Cancer chemotherapy is supported by adjuvant treatment with drugs, which influence hormonal milieus like estrogen, progesterone, and androgen agonists and antagonists. A list of anticancer drugs and their mode of action is presented in Table 21.1.

21.2.2 Limitations Associated with the Use of Anticancer Drugs

Chemotherapeutic agents are commonly associated with two major limitations: cancer cell resistance and adverse drug reactions to normal cells.

21.2.2.1 Cancer Cell Resistance

Multi-drug resistance, caused by cancer cells' inherent or acquired mutations, is a common challenge with anticancer drugs (Kelderman et al., 2014; Padhi et al., 2018). Cancer stem cells are a small group of cells in a tumour that have self-renewal and differentiation properties and are involved in tumour initiation and progression. The cancer stem cells are highly resistant to the chemotherapeutic agents that are commonly used for the treatment of different types of cancers, like leukaemia (Viale et al., 2009), glioblastoma (Eramo et al., 2006), and pancreatic cancer (Hermann et al., 2007). Overexpression of human epidermal growth factor receptor 2 was discovered to be linked to the failure of cisplatin chemotherapy in the treatment of gastric cancer. The human epidermal growth factor receptor-2 gene is responsible for the up-regulation of epithelial-mesenchymal transition-related transcription factor called Snail (Zinc finger protein SNAI1), which is a morphological change similar to epithelial-mesenchymal transition, which makes cancer cells more resistant.

Moreover, in human epidermal growth factor receptor 2/epithelial-mesenchymal transition-related transcription factor, double-positive patients have a lower survival rate than patients with only single-positive or double-negative genetic make-up. Another feature of the transcriptional repressor Snail and Slug (also an epithelial-mesenchymal transition-related transcription factor) is that it not only mediates epithelial-mesenchymal transition but also produces

DOI: 10.1201/9781003611516-27

TABLE 21.1
Different Types of Anticancer Drugs

Type of Anticancer Drug	Examples	Mechanism of Action	Indications	References
Alkylating agents	*Nitrogen mustards:* Mechlorethamine, Cyclophosphamide, Ifosfamide, Chlorambucil, Melphalan *Ethylenimine:* Thio-TEPA *Alkyl sulfonate:* Busulfan *Nitrosoureas:* Carmustine, Lomustine *Triazine:* Dacarbazine	They form reactive molecular species that alkylate nucleophilic groups on DNA bases, particularly the N-7 position of guanine.	Leukaemia, non-Hodgkin's lymphoma, breast and ovarian cancers, and neuroblastoma.	Meegan and O'Boyle (2019) and More et al. (2019)
Antimetabolites	Methotrexate	Folate antagonist	Choriocarcinoma, acute leukaemia, non-Hodgkin's lymphomas, breast cancer, head, neck, and bladder cancer	Espinosa et al. (2003) and Oliveira et al. (2018)
	6-Mercaptopurine	Purine antagonist	Acute lymphocytic leukaemia	Nadhum et al. (2020)
	6-Thioguanine	Purine antagonist	Chronic myelocytic leukaemia.	Karimi-Maleh et al. (2018)
	Fludarabine	Pyrimidine antagonist	Non-Hodgkin Lymphoma	Zhang et al. (2017)
	5-Fluorouracil	Pyrimidine antagonist	Colorectal cancer	Malesci et al. (2017)
Natural compound antibiotics	*Vinca alkaloids* Vincristine (Oncovin) Vinblastine Vindesine Vinflunine Vinorelbine	Inhibit microtubule polymerization	Acute leukaemia Breast cancer Non-small-cell lung carcinoma	Tok and Gowder (2013), Taher et al. (2017), and Choudhari et al. (2020)
	Taxols Paclitaxel Docetaxel	Inhibit microtubule function	Early breast cancer, prostate cancer Non-small cell lung cancer	Willson et al. (2019) and Oudard et al. (2017)
	Camptothecin analogues Topotecan Irinotecan	Topoisomerase-1 inhibitors	Second-line therapy for advanced ovarian cancer & non-small cell lung cancer	Moon et al. (2018), Prasad and Banerjee (2019), and Rahman et al. (2017)
	Epipodophyllo toxin Etoposide	Topoisomerase-2 inhibitors	Combination drug regimens for lymphoma, & lung, germ cell, & gastric cancers Metastatic breast cancer	Sun et al. (2018) and Voutsadakis (2018)
	Actinomycin D Daunorubicin Doxorubicin Epirubicin Idarubicin	Intercalate between base pairs, inhibit topoisomerase II, & generate free radicals. Block the synthesis of RNA and DNA & membrane disruption.	Rhabdomyosarcoma and Wilms tumour, Ewing, Kaposi, and soft-tissue sarcomas, breast cancer	Tok and Gowder (2013), Ramalingam et al. (2018), Tarpgaard et al. (2021) and de Souza Braga et al. (2018)
Other cytotoxic drugs	Hydroxyurea	Block biosynthesis of DNA. Radiation sensitizer	Chronic myelomonocytic leukaemia	Tok and Gowder (2013) and Solary and Itzykson (2017)
	L-Asparaginase	Hydrolysis of amino acid asparagine	Acute myeloid leukaemia Use in combination with other anticancer drugs	Chen et al. (2020) and Kaspers (2019)
	Platinum-based compounds Carboplatin Cisplatin	DNA strand intercalation	Advanced epithelial ovarian cancer	Marth et al. (2017) and Zhu et al. (2019)

(*Continued*)

TABLE 21.1 (*Continued*)
Different Types of Anticancer Drugs

Type of Anticancer Drug	Examples	Mechanism of Action	Indications	References
Hormone-based drugs	Prednisolone	Glucocorticoid	Immunosuppressant Breast cancer	Tripathi (2013) and Ricciuti et al. (2019)
	Fosfestrol Ethinylestradiol	Estrogenic	Adjuvant treatment of breast and prostate cancer	Roshan and Ravindranath (2019)
	Tamoxifen Toremifene	Selective estrogen receptor modulators	Premenopausal breast cancer & prevention of breast cancer	Hong et al. (2020)
	Fulvestrant	Selective estrogen receptor downregulator	Breast cancer	Nathan and Schmid (2017)
	Letrozole, Anastrozole, Exemestane	Aromatase inhibitors	Luminal breast cancer	Cottu et al. (2018)
	Flutamide Bicalutamide	Antiandrogen	Prostate cancer	Nakai et al. (2015)
	Finasteride Dutasteride	5-α reductase inhibitor	Benign prostatic hyperplasia Androgenic alopecia Hirsutism Castration-resistant prostate cancer	Hirshburg et al. (2016) and Azuma et al. (2017)
	Nafarelin, Triptorelin	Gonadotropin hormone-releasing hormone analogues	Prostate cancer Endometriosis	Woo et al. (2017)
Targeted drugs	Imatinib, Nilotinib, Sorafenib, Dasatinib	Protein tyrosine-protein kinase inhibitors	Chronic myeloid leukaemia	Bhanumathy et al. (2021), Ling et al. (2018), and Tripathi (2013)
	Gefitinib, Erlotinib, Cetuximab, Panitumumab	Epidermal growth factor receptor inhibitors	Colorectal cancer, pancreatic cancer, and squamous cell carcinoma of the head and neck	Nagano et al. (2018), and Wee and Wang (2017)
	Bevacizumab Sunitinib	Angiogenesis inhibitors	Gastric & gastroesophageal junction cancer Colorectal cancer	Roviello et al. (2016) and Mousa et al. (2015)
	Bortezomib, Ixazomib Carfilzomib	Proteasome inhibitors	Multiple myeloma or mantle-cell lymphoma, solid malignancies	Wu et al. (2017) and Roeten et al. (2018)
	Rituximab Trastuzumab	Unarmed monoclonal antibodies	Gastric cancer Breast cancer Chronic lymphocytic leukaemia	Jain et al. (2017), Modi et al. (2020), and Eto et al. (2015)
	Ozogamicin, Gemtuzumab	Toxin-linked monoclonal antibodies (immunotoxines)	Breast cancer with other anticancer drugs, Acute Lymphoblastic Leukaemia	Jen et al. (2018) and Jabbour et al. (2018)
	Ibritumomab (^{90}Y), Tositumomab (^{135}I)	Radioisotope carrying monoclonal antibodies	Leukaemias and lymphomas	Hosono (2019) and Iino et al. (2019)
	Alemtuzumab	Depleting leucocytes	Chronic lymphocytic leukaemia, B-cell chronic, lymphocytic leukaemia Non-Hodgkin lymphoma	Zent et al. (2016)

resistance to p53-induced apoptosis and self-renewal (Huang et al., 2016). After repeated phosphorylation processes, cytarabine (a nucleotide medication) is activated to generate cytarabine triphosphate, which treats acute myelogenous leukaemia. A decrease in the activation of cytarabine may cause downregulation or mutation in this pathway, which can lead to drug resistance (Bukowski et al., 2020). P-glycoproteins are overexpressed in lung, liver, kidney, rectum, and colon tumours, resulting in intrinsic resistance to medications such as epipodophyllotoxins, anthracyclines, vinca alkaloids, bisantrene, and colchicine (Wang et al., 2019). Drug resistance is acquired as a result of long-term or unreasonable chemotherapeutic usage.

Chemotherapeutic agents cause acquired resistance by activating a second proto-oncogene, which acts as a new driver gene for drug target mutations or changes in expression levels, as well as changes in the tumour microenvironment following treatment. When the genes encoding target proteins generate new mutations or their expression levels change, cancer cells can develop resistance to targeted therapies (Challagundla et al., 2015). Imatinib is a tyrosine kinase inhibitor used to treat chronic myelogenous leukaemia. Imatinib suffers from resistance as a result of a point mutation in the fusion tyrosine kinase protein T315I (Wang et al., 2019). The tyrosine kinase inhibitors for epidermal growth factor receptors, such as erlotinib and gefitinib, have a high response rate in the early phases of treatment, but 20%–30% of patients develop resistance and relapse following treatment (Cuellar et al., 2018).

Furthermore, the tumour suppressor protein (Tp53), implicated in chemotherapy-induced apoptosis, is mutated in 50% of cancer patients. When Tp53 is mutated, it is possible to exhibit non-functional drug resistance. Inactivation of p53 regulators such as caspase-9 and its cofactor, apoptotic protease-activating factor, on the other hand, may lead to treatment resistance (Bukowski et al.,2020). The efficacy of a drug is determined by its molecular target and any changes to that target, such as mutations or changes in expression levels. Certain anticancer medications, for example, target topoisomerase-II, an enzyme that stops DNA from becoming supercoiled or undercoiled, resulting in treatment resistance in cancer. The complex of DNA and topoisomerase-II usually stays for a very short time, but anticancer drugs stabilize it, causing DNA damage, inhibiting DNA synthesis, and making the mitotic process hesitant. Mutations in the topoisomerase II gene have made some cell lines resistant to inhibitors of topoisomerase II (Stavrovskaya, 2000). Drug target modifications have also been found with the androgen receptor. The androgen receptor is genomically amplified in around 30% of prostate tumours, allowing the cancer to become resistant to androgen deprivation therapy with the drugs, namely leuprolide and bicalutamide. While changes in the signal transduction pathway cause drug resistance, trastuzumab (Herceptin), which is used to treat human epidermal growth factor receptor-2 positive breast cancer, has been shown to be highly effective when combined with chemotherapy.

When comparing the trastuzumab-resistant cell lines to the nonresistant parental cell lines, insulin-like growth factor 1 receptor levels were considerably higher in the trastuzumab-resistant cell lines. This finding indicates that inhibiting the insulin-like growth factor 1 receptor improves trastuzumab responsiveness in human epidermal growth factor receptor 2-positive breast cancer cells (Razis et al., 2011). Multi-drug resistance protein 1 and P-glycoprotein are two of the most well-studied ATP-binding cassette subfamily B transporters. It possesses several drug-binding sites that can bind and pump out drugs from the cell, like etoposide, doxorubicin, paclitaxel, and vinblastine (Lagas et al., 2010). It can transport both positively and negatively charged drugs, such as mitoxantrone, bisantrene, epipodophyllotoxin, camptothecin, flavopiridol, anthracyclines, imatinib, and gefitinib, among others. Overexpression of the ATP-binding cassette transporter G2 has also been seen in lung cancer and leukaemia (Yin & Zhang, 2011; Munoz et al., 2007; Cho et al., 2011). The epithelial-mesenchymal transition and metastasis pathway transforms solid tumours into metastatic tumours. In an epithelial-mesenchymal transition lineage-tracing animal system, drug resistance to cyclophosphamide was observed in the epithelial-mesenchymal transition (Shang et al. 2014).

21.2.2.2 Toxicities and Adverse Drug Reactions of Anticancer Drugs

Damage to growing stem cells causes bone marrow depression and a drop in white blood cell, platelet, and red blood cell levels, presenting a risk for infections, bleeding, and anaemia. Bone deterioration, such as jaw osteonecrosis, has been documented in rare cases (Sarin et al., 2008). Anthracyclines cause cardiotoxicity, while paclitaxel and vincristine cause peripheral neuropathy (Bajelan et al., 2012). Cinnamaldehyde (an anticancer drug) at a daily dose of 73.5 mg/kg body weight exhibited histopathological alterations in rats' kidneys, including elevated marker enzyme activity and an antioxidant imbalance (Gowder & Devaraj, 2006). Anticancer drugs like methotrexate, cytarabine, 5-fluorouracil, and ifosfamide provoke central nervous system toxicities such as encephalopathy, meningitis, and myelitis (Peddi et al., 2014). Cisplatin, carboplatin, and oxaliplatin are platinum-based first-line chemotherapy drugs. Severe toxicities such as neurotoxicity, cardiotoxicity, nephrotoxicity, and severe gastrointestinal side effects such as nausea, vomiting, constipation, and diarrhoea are major barriers in platinum-based chemotherapy (Stojanovska et al., 2015).

Dermatological toxicities such as alopecia, erythema, urticaria, photosensitivity, hyperpigmentation, and alterations to the nails were discovered in clinical research involving women receiving chemotherapy for breast cancer with drugs such as epirubicin, docetaxel, and cyclophosphamide (De Oliveira et al., 2010). When administered during the first trimester of pregnancy, the antimetabolite medication methotrexate was discovered to cause significant teratogenicity. Dysostosis with ossification, wide nasal bridge, hypertelorism, micrognathia, and ear abnormalities are all deformities that are related to aminopterin syndrome. Anticancer drugs such as dacarbazine, bleomycin, epirubicin, rituximab, and others may have teratogenic effects (Koren et al., 2013). Chemotherapy increases the risk of leukaemia genesis more than radiation therapy. Alkylating drugs, which are used to treat diseases such as Hodgkin's disease, non-lymphoma, and Hodgkin's ovarian, lung, and breast cancer, have been proven to increase the chance of acquiring a secondary malignancy termed acute myelogenous leukaemia.

21.3 REQUIREMENT OF CODELIVERY OF PHYTOMOLECULES WITH OTHER ANTICANCER DRUGS

Phytomolecules have played an important role in the treatment of human diseases since ancient times. Penicillin, salicylates, morphine, willow, and quinine, for example, have all impacted people's lives globally. The high antitumour potential of phytomolecules was justified by scientific findings (Cragg & Pezzuto, 2016; Behera and Padhi, 2022). From 1940 to 2014, natural-origin medications accounted for nearly half of all approved anticancer treatments. All of these anticancer medications are effective in both *in-vivo* and *in-vitro* studies. They demonstrated complementary and overlapping antitumour mechanisms. All of these medications act on a variety of molecular targets and signalling pathways, including membrane receptors, kinases, downstream tumour activator or suppressor proteins, transcriptional factors, microRNAs, cyclins, and caspases, among others (Choudhari et al., 2020).

The most effective chemotherapeutics currently accessible include plant-derived anticancer drugs such as vinblastine and vincristine, etoposide, paclitaxel, docetaxel, topotecan, and irinotecan. Plant-derived compounds have a significant therapeutic impact, either alone or in combination, and may have anticancer activities that are synergistic with standard chemotherapeutic treatments. When correctly combined with conventional medications, phytomolecules have a beneficial impact. They improve the quality of life. The synergistic action of plant-derived bioactives and traditional chemotherapeutic agents is primarily responsible for combating drug resistance, improving pharmacological characteristics, and reducing side effects (Cragg & Pezzuto, 2016).

Multi-drug therapy or combination therapy is a helpful strategy focused on the direct blocking and killing of a cancer cell or pathogens, as well as on the human body's defence or repair mechanism (Doos et al., 2014). Over the last two decades, the world has seen an increase in the use of conventional medicine in conjunction with complementary and alternative medicine, including homeopathy, naturopathy, chiropractic, ethnopharmacology, and phytotherapy. Many diseases have complicated aetiology that can be treated more effectively with a pharmacological combination strategy than a single treatment. When correctly paired with conventional medications, phytomedicine positively impacts the patient's survival and improves the patient's quality of life. The capacity of natural derivatives to antagonize drug resistance, increase pharmacological characteristics, or mitigate side effects is primarily responsible for the synergistic impact of plant-derived bioactive and conventional chemotherapeutic agents. Geraniol, an essential oil, reduced tumour mass volume by 70% (Pezzani et al., 2019).

Cisplatin, when combined with methyl eugenol, significantly enhanced the anticancer activity against cervical cancer cells (HeLa) by inducing cell apoptosis, caspase-3 activity, cell cycle arrest, and mitochondrial membrane potential loss (Gautam et al., 2014). β-caryophyllene, a sesquiterpene present in essential oils of various plants, and paclitaxel were tested against human breast cancer cells (MCF-7), human colon cancer cells (DLD-1), and mouse fibroblasts (L-929) tumour cell lines. β-caryophyllene potentiated the anticancer activity of paclitaxel on those cell lines and facilitated the passage of paclitaxel through the plasma membrane. The combination of β-caryophyllene oxide and doxorubicin showed synergistic anticancer activity when evaluated on ovarian cancer cell line (SKOV-3) and lymphoblastic cell line (CCRF/CEM-T) (Legault & Pichette, 2007).

Moreover, D-limonene, a cyclic terpene combined with docetaxel, was effective against human prostate carcinoma (DU-145) and normal prostate epithelial cells (PZ-HPV-7) (Rabi & Bishayee, 2009). Thymoquinone, a constituent of the *Nigella sativa* essential oil, has been studied in combination with doxorubicin on HL-60 (leukaemia), 518A2 (melanoma), HT-29 (colon), KB-vl(cervix), and MCF-7 (breast cancer) cell lines. Thymoquinone potentiates the anticancer activity of doxorubicin by inhibiting cancer cell growth. Similarly, thymoquinone also synergistically enhanced the cisplatin and docetaxel-induced cytotoxicity in triple-negative breast cancer cell lines (with mutant p53). Topotecan was used in combination with thymoquinone in acute myelogenous leukaemia cells. The combination regimen (better with pre-exposure) reduced cell proliferation with increased B-cell lymphoma 2-associated X protein/B-cell lymphoma 2, p53, and caspase-9 expression levels (Rahmani et al., 2014).

Polyphenol derivatives are natural substances containing multiple phenol structural units that can modulate oxidative stress in cancer cells. In conjunction with conventional chemotherapy medications, polyphenols such as resveratrol, genistein, curcumin, and quercetin have been demonstrated to be effective. Resveratrol suppresses the expression of multi-drug resistance-associated protein 1 in doxorubicin-resistant acute myeloid leukaemia cells by causing cell growth arrest and apoptosis. Resveratrol also reduces the proliferation of several human myeloma cell lines and boosts the apoptotic effect of bortezomib and thalidomide. Curcumin, a derivative of *Curcuma longa* (turmeric) rhizomes, has been employed in human (T98G, U87MG, and T67) and rat (C6) glioma cell lines due to its anticancer potential. Curcumin, cisplatin, and oxaliplatin improve drug sensitivity and apoptosis in ovarian cancer cells. The combination of curcumin and 5-fluorouracil inhibited cell proliferation in the human colon cancer cell line (HT-29). When curcumin was combined with ixabepi lone, cisplatin, vinorelbine, or everolimus, cell cycle arrest and cell death were seen (Mileo & Miccadei, 2016). In the human malignant mesothelioma cell line, quercetin had a synergistic impact with cisplatin, and in neuroblastoma and Ewing's sarcoma cell lines, it demonstrated a synergistic effect with doxorubicin. Sulforaphane has been utilized as an adjuvant in the chemotherapeutic regimen along with paclitaxel or telomerase inhibitors. Sulforaphane inhibits multi-drug resistance protein expression, reduces drug efflux, and boosts the anticancer potential of other antiproliferative drugs like paclitaxel (Demiroglu-Zergeroglu et al., 2010). Pterostilbene, a phenolic compound, potentiates the effect of megestrol acetate in endometrial cancer (Wen et al., 2017).

21.3.1 Current Cancer Therapy and Phytochemicals

1. *Vinca alkaloids:* They are obtained from *Catharanthus* (Apocynaceae) and are employed in treating various types of cancer, like breast, liver, leukaemia, and lung cancer. Vinca alkaloids disturb the microtubule function and arrest the metaphase of the cell cycle. Currently, semi-synthetic derivatives of vinca alkaloids are vinorelbine, vindesine, and vinfosiltine vinorelbine, which have been introduced in the market. These derivatives are used alone or with another phytochemical agent against several cancers (Moudi et al., 2013; Banyal et al., 2023).
2. *Taxanes:* Most widely used drugs of this class are paclitaxel (Taxol), obtained from the bark of the *Taxus brevifolia* (Taxaceae), and docetaxel (Taxotere), a semi-synthetic analogue synthesized from 10-diacetyl baccatin III, isolated from the leaves of the European yew, *Taxus baccata*. Taxanes act by binding to microtubules and have a role in cell division. They are responsible for the polymerization of tubulin heterodimers to microtubules and suppress the dynamic changes in microtubules, resulting in mitotic arrest. Paclitaxel is used in the treatment of breast, ovarian, and non-small cell lung cancer. It has shown efficacy against Kaposi's sarcoma, while docetaxel is used mainly in the treatment of breast cancer and non-small cell lung cancer. The paclitaxel analogs undergoing clinical trials are larotaxel, milataxel, ortataxel, and tasetaxel. Lorataxel is used alone or in combination with other therapies for the treatment of urethral, bladder, pancreatic, lung, and breast cancer (Kaur et al., 2011; Wang et al., 2022).
3. *Camptothecin derivatives:* It binds with topoisomerase-I DNA binary complex, thereby preventing DNA re-ligation and causing DNA damage, which results in apoptosis. Researchers found that camptothecin derivatives like topotecan and irinotecan are used to treat ovarian and lung cancer (Kim et al., 2015). In addition, modifications in the A, B, and E rings of camptothecin generated derivatives that have shown efficient action in various cancers (Wahid & Bano, 2014; Behera and Padhi, 2020). A new camptothecin derivative, CPT 417, has proven effective in glioblastoma and colon cancer cell lines (Yount et al., 2007; Kamle et al., 2024).
4. *Cephalotaxus:* It is an alkaloid used in various cancers, including lung and gastric cancer. They act by inhibiting protein synthesis. Harringtonine and isoharringtonine are anticancer agents obtained from *Cephalotaxux harringtonii*. Homoharingtonine is very effective in chronic and acute myelogenous leukaemia. A combination of homoharringtonine and harringtonine is successfully used in China for the treatment of chronic myelogenous leukaemia and acute myelogenous leukaemia (Kaur et al., 2011; Wu et al., 2025).
5. *Colchicine:* Colchicine is a natural bioactive compound isolated from *Colchicum autumnale* (Colchicaceae). It arrests the cell cycle and induces apoptosis by binding with tubulin and stabilizing the microtubule formation. Colchicine's action is not specific and can target rapidly dividing normal cells. Hence, semi-synthetic derivatives of colchicines, which are more specific towards cancer cells and less toxic to normal cells, were developed (Iqbal et al., 2017). Colchicine also promotes apoptosis of adenocarcinoma gastric carcinoma cells and NCI-N87-derived epithelial cell lines in a dose-dependent manner (Zhang et al., 2019). A study has shown that colchicine induces mutation and reactive oxygen species-mediated autophagy in lung cancer cells, has no cytotoxic effect, and does not halt the G2/M phase in normal cells (Kumar et al., 2016; Doan et al., 2024).
6. *Ellipticine:* Ellipticine (topoisomerase II inhibitors) is a naturally occurring anticancer compound obtained from the stem, bark, leaf, and root of *Bleekeria vitensis* and *Ochrosia elliptical*. It exhibits a DNA-damaging effect due to topoisomerase-II inhibition and intercalation with DNA, and thus inhibits the proliferation of tumour cells. Ellipticine and its derivatives, *N*-2-(diethylaminoethyl)-9-hydroxyellipticinium chloride, 2-*N*-methyl-9-hydroxy ellipticine, are efficient anticancer compounds and are used to cure ependymoblastoma, leukaemia, myeloma, melanoma, breast, and colon cancer (Iqbal et al., 2017; Costa de Oliveira et al., 2020).
7. *Berberine:* Berberine is obtained from roots and ribosomes of *Tinospora Cordifolia, Berberis vulgaris, Berberis aquifolium, and Rhizoma coptidis* (Xu et al., 2022). It is used in various cancers like breast, prostate, and colorectal cancer. It induces cell cycle arrest and apoptosis at the G2/M phase. It also inhibits anti-apoptotic proteins (p21, p53, caspase-3, and caspase-9) (Iqbal et al., 2017). The cellular targets for berberine are protein kinase B, mitogen-activated protein kinase, and cell cycle checkpoint kinases. These regulate transcription factors for gene and protein regulators required for cell survival, motility, and death. Due to poor absorption of berberine, it is used with other

anticancer agents like doxorubicin as the combination is effective in treatment strategies for bladder, breast, lymphoma, and leukaemia (Habtemariam, 2020; Hsu et al., 2024).

8. *Capsaicin:* It is a natural phytochemical obtained from pepper and has chemopreventive properties in pancreatic, prosthetic, liver, skin, leukaemia, lung, bladder, colon, and endothelial cancer cells. It regulates different molecular targets in breast cancer, like caspase-3, reactive oxygen species, Ras-related C3 botulinum toxin substrate-1, human epidermal growth factor receptor-2, etc. (Prakash et al., 2013). Capsaicin possesses anti-angiogenic properties and induces G0/G1 phase arrest in human oesophagal carcinoma cells with the increased expression of p21gene and decreased CDK4, CDK6, and cyclin expression (Thoennissen et al., 2010; Lin et al., 2013). Some studies have revealed cell cycle arrest by capsaicin in human epithelial carcinoma cells (KB cells) and inhibition of the migration of melanoma cells without any cellular toxicity to normal cells (Al-Samydai et al., 2019; Iraci et al., 2022).
9. *Flavones/Flavonoids:* They are secondary metabolites with 800 diverse compounds largely distributed in fruits, grains, tea, and vegetables. Flavopiridol is a plant-derived semi-synthetic flavone that inhibits cyclin-dependent kinase with anticancer activity against oesophageal and gastric cancer (Ververidis et al., 2007). Flavonoids such as quercetin and resveratrol impede the proliferation of hormone-sensitive breast cancer cell lines (ER+ human BC T47D and MCF-7) (Damianaki et al., 2000). The flavanone naringenin induces reactive oxygen species-dependent apoptosis in human triple-negative breast cancer cell lines (MDA-MB-468) (Sudhakaran et al., 2019). Hesperidin or 5,7,3-trihydroxy-4-methoxy flavone possesses neuroprotective, vascular, anti-inflammatory, anti-allergic, antioxidant, and anticarcinogenic effects by increasing reactive oxygen species and DNA fragmentation or by overexpressing Fas cell surface death receptor and Fas-associated protein (Zhang et al., 2015; Çetinkaya et al., 2022).
10. *Gingerol:* Gingerol is isolated from fresh rhizomes of *Zingiber officinale* and has strong anticancer activity in the colon, pancreas, ovarian, and breast cancers. It downregulates the expression of inducible nitric oxide synthase and tumour necrosis factor-alpha through suppressing nuclear factor kappa-light-chain-enhancer of activated B cells nuclear translocation and IγBα phosphorylation (Iqbal et al., 2017). 6-Gingerol and 6-shogaol were reported to have antitumour activity in cancer by inhibiting cell proliferation, migration, and invasion or induction of apoptosis (Joo et al., 2016). 10-Gingerol, a phenolic compound from ginger, possesses antitumour activity against ovarian, colon, lung, and prostate cancer cells by inhibiting cell proliferation or inducing apoptosis (Joo et al., 2016).
11. *Lycopene:* It is a bright red pigment with anticancer potential obtained from tomatoes, watermelons, red papayas, and red carrots, and plays an important role in targeting phosphoinositide-3-kinase/Protein kinase B signalling pathway in pancreatic and stomach cancer by downregulating extracellular signalling-regulated kinase and B-cell lymphoma-2 proteins. It upregulates antioxidant enzymes and removes oxidative damage caused by carcinogens in breast, endometrial, prostate, and colon cancer (Kaur et al., 2011; Wang et al., 2023). Lycopene suppresses prostate cancer by downregulating the expression of protein kinase B and upregulating in the prostate cancer-3 cell line (microRNA-let-7f1) (Mirahmadi et al., 2020). Lycopene inhibits the phosphorylation of extracellular signal-regulated kinase (cell proliferator and regulator of apoptosis) in gastric cells and hepatocarcinoma cells (Kim & Kim, 2015). Apo-10-lycopenoic acid, an active metabolite of lycopene, has been reported for its inhibition of lung carcinogenesis activity (Ono et al., 2015).
12. *Saffron (Crocetin):* It is a potent anticancer compound and regulates different nuclear and cellular factors, which downregulates inducible nitric oxide synthase, cyclooxygenase-2 reduces serum–interleukin-β, Tissue necrosis factor-α, cyclin B, cyclin A, and cyclin-dependent kinase 2. Saffron is beneficial in 1-methyl-3-nitro-1-nitrosoguanidine-induced gastric cancer in rats (Bhandari, 2015). Lung cancer research in cancer-bearing mice revealed a significant reduction in elevated levels of lipid peroxidation and a marker enzyme in carcinogen-treated mice (Chun et al., 2014). Nair et al. (1991) reported that oral administration of saffron showed antitumour activity in mice.
13. *Resveratrol:* It is a polyphenol compound that is employed in a variety of cancers, like breast, colorectal, liver, pancreatic, prostate, etc. Topical application of resveratrol protects the mice from tumourigenesis in skin cancer models (Jang et al., 1997; Yi et al., 2022). Resveratrol, combined with ionizing radiation for pancreatic cancer treatment, was more efficient by inducing apoptosis via the mitochondrial pathway (Sun et al., 2008).
14. *Curcumin:* Curcumin is obtained from *Curcuma longa*. It inhibits the growth of human glioblastoma cells by modulating different nuclear and cellular factors and upregulates the expression of various genes (Kaur et al., 2011). Curcumin decreases the IC_{50} value of other anticancer drugs (paclitaxel, cisplatin, or doxorubicin) and the number of stem cells. Some enzymes responsible for inflammation

and cancer, like cyclooxygenase-2 inducible nitric oxide, 5-lipoxygenase, phospholipases A2, and cyclooxygenase-2, are modified by curcumin administration in inflamed and neoplastic tissue. Curcumin markedly inhibits 7,12 dimethylbenz(*a*) anthracine and 12-*O*-tetradecanoyl phorbol 13-acetate-induced skin tumour formation in mice (Azuine et al., 1992). Curcumin inhibits the development of cancerous and precancerous lesions induced by *N*-methyl-*N*-nitro-*N*-nitrosoguanosine, which causes DNA methylation in animal models of gastric and colon cancer (Pongrakhananon & Rojanasakul, 2011).

21.4 REQUIREMENT OF NANO-DELIVERY SYSTEMS FOR CODELIVERY OF PHYTOMOLECULES WITH ANTICANCER AGENTS

The treatment effect is still unsatisfactory despite the benefits of co-delivering phytomolecules with other anticancer medications (Verma et al., 2017). Due to varying pharmacokinetic features, the amount of drugs delivered to the tumour site is unequal, resulting in uneven drug distribution. To address these issues, researchers discovered that a stimulus-sensitive polymeric nanoparticle is useful in co-delivering anticancer drugs in an optimal ratio, allowing for high drug loading, an improved release profile, and increased dosage form stability (Pan et al., 2019). Furthermore, nano-drug delivery systems can efficiently penetrate biological barriers, increasing drug penetration and targeting tumour cells selectively (Padhi et al., 2022; Padhi and Behera, 2021).

Chemosensitizers can be incorporated into a nano-drug delivery system with an anticancer agent, delaying cancer cell adaptation and lowering the risk of mutation (Qi et al., 2017). A nanoparticle codelivery method based on polylactic-co-glycolic acid was designed to deliver the optimal amount of docetaxel and gambogic acid anticancer medications. P-glycoprotein expression was downregulated by this nano-co-delivery method, which promoted cell death and prevented multi-drug resistance (Xu et al., 2016). The codelivery of anticancer medicines with differing polarities can be difficult using conventional dose forms. Drugs with opposite polarities have different pharmacokinetic characteristics; therefore, combining them can have a major impact on their bioavailability. Developing a nano-based codelivery system, such as the folate-based multi-functionalized nanocarrier delivery system, can alleviate such challenges. Using a folic acid conjugated polymer, the multi-functionalized nanocarrier delivery systems were developed by entrapping hydrophilic gemcitabine and hydrophobic paclitaxel. These smart polymer-based nanoparticles were discovered to be highly successful in the treatment of breast cancer, with excellent stability and biocompatibility (Lei et al., 2019).

Doxorubicin is a highly potent anticancer antibiotic that is used in the treatment of a variety of malignancies. The use of doxorubicin for cancer treatment, on the other hand, has the distinct disadvantage of cardiotoxicity, which restricts its therapeutic efficacy. Although doxorubicin's cardiotoxicity can be minimized by combining it with the antioxidant curcumin, the traditional combination has low bioavailability due to rapid drug clearance. Self-assembled polymeric micelles were generated, which increased steric stability by lowering opsonization in the bloodstream. The polymeric micelles selectively and effectively aggregated at the tumour's location, boosting doxorubicin's therapeutic efficiency. Through the antioxidant action of curcumin, the polymeric micelles also reduced doxorubicin-induced cardiotoxicity (Zhang et al., 2018). Chemotherapeutic agents such as cisplatin, paclitaxel, bortezomib, resveratrol, piperine, and others have been successfully entrapped using similar curcumin-based polymeric nano-co-delivery systems (Alven & Aderibigbe, 2020). Furthermore, nano-codelivery systems have been discovered to reverse anticancer drug resistance. For example, a composite liposomal system entrapping paclitaxel with chloroquine reversed paclitaxel-resistant malignancy. The chloroquine entrapped in the composite liposomal system was found to re-sensitize paclitaxel-resistant cancer cells and was more effective than a combination of paclitaxel and chloroquine phosphate liposomes (Gao et al., 2015).

21.5 TARGETED CODELIVERY OF NATURAL DRUGS WITH ANTICANCER DRUGS

Drug delivery systems are intended to increase the drug's safety and therapeutic efficacy. Targeted drug delivery systems, instead of conventional drug delivery systems, deliver a higher drug concentration to the targeted cells. Because cancer is a leading cause of death worldwide, novel methods such as tailored medication delivery systems have been shown to be a safe and effective way of administering the desired drug (Li et al., 2019). Reduced dose frequency, homogeneity in drug action, maintenance of maximum effective concentration, and minimum unwanted effects are the major advantages of targeted drug delivery systems (Behera et al., 2020; Sahoo et al., 2022; Pradhan et al., 2023, 2024).

The two main ways to achieve targeted drug delivery are active and passive drug targeting strategies (Moorthi et al., 2011; Leyland-Jones, 1993) (Figure 21.1).

[The passive targeting is based on the enhanced permeability and retention (EPR) effect. The capillary endothelium of the newly grown blood vessels in the tumour is highly permeable. The nanoparticles embedded with the drug are injected intravenously, which reach to the tumour site, and cross the capillary endothelium to target the cancer cells. In the active targeting system, the nanoparticles embedded with the drug are attached to the ligands, which specifically target the receptors overexpressed on the tumour cells like epidermal growth factor receptor (EGFR)]

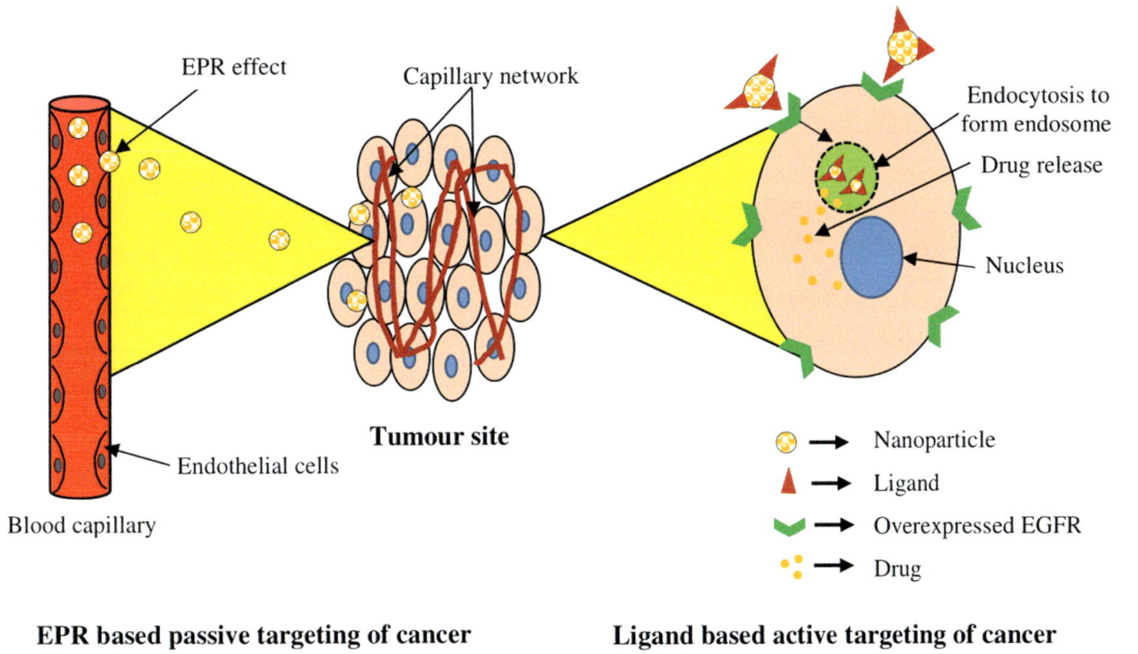

FIGURE 21.1 Mechanism of EPR-based passive and ligand-based active targeting of the cancer cells.

21.5.1 Active Targeting Strategy for Codelivery of Phytomolecules with Anticancer Drugs

Paul Ehrlich coined the term "magic bullet" for the homing device, a compact packaging system attached to a carrier system to deliver the drug to the target site. This has become possible through an active targeting strategy, which involves the interaction of targeting molecules (like antibiotics, ligands, peptides, nucleic acid, etc.) ligated on nanoparticle surfaces with overexpressed tumour cell receptors (Figure 21.1). Tumour-targeting molecules bind with a cell using an endosome-dependent mechanism, allowing drug influx (Moorthi et al., 2011; Padhi and Behera, 2020). Taking advantage of the active drug targeting strategy, several phyto-molecules have been co-delivered with anticancer drugs to enhance their therapeutic effectiveness. Some relevant studies have been cited below to highlight the current research in the codelivery of phytomolecules with anticancer drugs.

Active targeting of paclitaxel and trastuzumab combination showed greater efficacy. Based on the large pivotal study on advanced breast cancer patients, three combinations of trastuzumab and paclitaxel have been approved for the treatment. However, mono and combination therapy suggested that weakly paclitaxel has a better therapeutic index, especially in palliative settings (Tolaney et al., 2015). In Phase II research, female subjects with normal and over-expressed human epidermal growth factor receptor-2 (metastatic breast cancer) were given weekly trastuzumab and paclitaxel therapy. The findings of immune histochemical and fluorescence in situ hybridization assays were compared to determine effectiveness (Han et al., 2009). Doxorubicin and cetuximab combination in nanoparticles, which have been employed for cancer therapy in recent years, efficiently target and improve cytotoxicity on tumour cells. The effect of doxorubicin and cetuximab conjugated to dextran-coated iron oxide magnetic nanoparticles on lung cancer cells (A549) that overexpress human epidermal growth factor receptors was evaluated. After the treatment, the targeting of conjugates resulted in cytotoxicity, cell growth inhibition, and enhanced intracellular uptake. When compared to A549 cells treated with nanoparticles conjugated with doxorubicin, the results showed that cetuximab plus doxorubicin-loaded dextran nanoparticles greatly inhibited cell proliferation (Zhang et al., 2019).

To treat paclitaxel-resistant cancer, a novel liposomal system was developed using an active targeting technique to encapsulate paclitaxel with chloroquine phosphate. The study found that liposomal chloroquine phosphate can sensitize paclitaxel by inhibiting autophagy and competitively binding to multi-drug resistance transporters (Gao et al., 2015). Curcumin can prevent the activation of chemo-resistance-related transcription factors and drug transporters. Curcumin has limited water solubility and bioavailability despite its anticarcinogenic capabilities. As a result, polymeric nanocapsules have been used to solve these issues. Curcumin encapsulated in polymeric nanoparticles with piperine functions as a glucuronidation inhibitor and enhances curcumin bioavailability. The drug release profile of nanoparticles was studied at various pH levels. In vitro tests against colon carcinoma cells (CACO-2) and in vivo tests on colon cancer in mice were used to determine their cytotoxic effect. Curcumin nanoparticles demonstrated a selective and potentially lethal action against colon cancer cells in both *in vivo* and *in vitro* investigations. The addition of piperine aided curcumin encapsulation and drug loading. Curcumin encapsulation improved curcumin solubility and bioavailability in the battle against colon cancer (Slika et al., 2020). For colon cancer, camptothecin and curcumin were co-delivered utilizing an active method involving cationic polymeric nanoparticles. Both camptothecin and curcumin demonstrated a simultaneous sustained release profile in these nanoparticles.

In addition, cellular absorption was increased, indicating a synergistic impact compared to other drug formulations (Xiao et al., 2015). Yan et al. (2016) co-encapsulated doxorubicin and curcumin, which was easily achieved by including poly-caprolactone groups. The greatest drug loading content was 19.8% (w/w) for doxorubicin and 8.9% (w/w) for curcumin, respectively. At pH 6.8, cell absorption of micelles was also maximum. The findings revealed that these two drugs had a considerable synergistic impact. A pro-apoptotic drug combined with an antiangiogenic agent is the most effective drug combination against human hepatocellular cancer. The co-delivery of pro-apoptotic drugs like doxorubicin and antiangiogenic drugs like curcumin in pH-sensitive nanoparticles with an amphiphilic poly-amino ester copolymer showed minimal polydispersity, high encapsulation efficacy, and improved release in the cancer cell's acidic environment. Curcumin and doxorubicin nanoparticles were synthesized, which were evaluated on breast cancer-resistant cells (MCF-7) to determine if they might limit adaptive treatment tolerance and kill cancer cells (Motevalli et al., 2019). Curcumin and paclitaxel were delivered together in a novel structure of dynamically targeted nanoparticles, which increased anticancer activity without raising side effects. Polyethylenimine, stearic acid, and hyaluronic acid were utilized to formulate nanoparticles. Endocytosis was enabled with these modified nanoparticles, which targeted the CD22 receptor on the ovarian cancer cell membrane. Curcumin and paclitaxel delivered together revealed synergistic anti-ovarian cancer effects in chemosensitive human ovarian cancer cells and multi-drug resistant variants in vitro, as well as a good antitumour impact in nude mice with ovarian tumours (Zhao et al., 2016).

In another study, resveratrol was combined with paclitaxel in PEGylated liposomes to provide a carrier-delivered type of drug-resistant tumour combination therapy. The combinatorial drug therapy had a strong cytotoxic effect on drug-resistant breast tumour cells (MCF-7/ADR) and improved bioavailability. Systemic treatment with this combination of liposomes effectively inhibited drug-resistant tumours in treated mice (Meng et al., 2016). Curcumin and 5-fluorouracil were encapsulated in poly(D,L-lactic-co-glycolic acid) nanoparticles and further tested for anticancer effectiveness. The uptake of nanoparticles by cancer cells was enhanced by magnetic hypothermia. Both drugs acted synergistically to destroy cancer cells quickly (Balasubramanian et al., 2014).

21.5.1.1 Mechanism of Active Targeting of Anticancer Drugs

In active targeting, specific ligands attached to nanostructures are recognized and identified by the overexpressed receptors present at the disease site. Ternary structure plays a very important role in active targeting. The ternary structure is made up of targeting moiety (ligand or antibody), carrier (polymers or lipids), and active pharmaceutical ingredient (anticancer drug). During the preparation of the ternary structure, the following factors must be considered for the effective delivery of active pharmaceuticals (Yingchoncharoen et al., 2016).

- Cellular receptor and ligand properties should be suitable as tumour-specific targets. For example, receptors should be overexpressed on cancer cells and not expressed on a normal cell.
- The selected receptor should be expressed homogeneously on the surface of all target cells.
- After ligand-receptor binding, the complex should not be discharged into the blood circulation.
- The ligand-receptor complex must be internalized into the targeted cell after binding.

The internalization of the ligand-receptor complex takes place via receptor-mediated endocytosis. The endocytosis process speeds up with proteins like caveolin-1 or clathrin, resulting in endocytic vesicles. The resultant vesicle is coated with caveolin or clathrin. Then, this coated vesicle gets dissociated from the membrane, travels through the cytoplasm, and fuses with an early endosome, also known as a sorting endosome. As the early endosome matures, vascular ATPase pumps are recruited to pump H^+ ions into its lumen and decrease the pH to 5–6. A pH decrease facilitates the ligand dissociation from its receptor by inducing conformational changes to the receptor. The reduced pH also promotes the release of the encapsulated drug from the nanoformulation by partial degradation. At this point, the receptor and its ligand get recycled back to the plasma membrane or continue along the endo-lysosomal pathway for lysosomal degradation in a late endosome stage. Recycled endosomes accumulate for therapeutic action, and if continued, they will enter the multivesicular body, which matures into the late endosome. In the maturation process, from an early endosome into the late endosome, the pH decreases further to 5, allowing the encapsulated drug to be released from the nanoformulations. Finally, the late endosome fuses with the lysosome, where its content is degraded by the lysosomal enzymes in the acidic pH. This allows the more encapsulated drug to be released from nanoparticles.

21.5.1.2 Ligands Used for Active Targeting Strategy

Antibodies: Antibodies are the most commonly used targeting molecules for epithelial and anti-epidermal growth factor receptors. As the epidermal growth factor receptor is expressed in all epithelial cells, it becomes a special target when delivering a drug to human epidermal growth factor receptor 2. They are overexpressed in several cancer cells and, therefore, commonly targeted in nano-enabled targeted drug delivery. The human epidermal growth factor receptor 2/neu human epidermal growth factor receptor 2-oncogene is another epidermal growth factor receptor family member and gets upregulated in various cancers like breast, ovarian, lung, gastric, and oral cancers. Most breast cancer patients show increased malignancy and poor survival rate, which is associated with the up-regulation of human epithelial receptor-2, which is responsible for resistance development in the cancer cell (Tang et al., 2006).

Peptides: Though antibodies have affinity and selectivity for binding, they are potentially immunogenic. To overcome this drawback, antibodies may be engineered into acquired humanized or chimeric properties to avoid immune detection. Also, the production and isolation of antibodies are difficult and tedious, so alternative targeting molecules like peptides, aptamers, and some small molecules are used. Peptides are proving to be more adventitious than antibodies, such as lacking immunogenicity and having a lower cost of production. However, disadvantages include lower targeting affinity, increased chance of non-specific binding, and increased probability of proteolytic cleavage. This issue can be solved by displaying the peptides as multivalent to increase the binding affinity and using D-amino acid, evading proteolysis. Peptide-based targeting ligands are identified from the binding region of proteins for the target of interest. They can also be produced using the phage display technique, which is used for efficient and rapid screening of peptide processing specific binding to a target of interest. This method uses bacteriophages and viruses that only infect bacteria. They are modified to express multiple copies of a single peptide on their surface, forming a library containing log sequences. The tripeptide arginine-glycine-aspartate is the most popular peptide sequence to target diagnostic therapeutic moieties. The tripeptide arginine-glycine-aspartate binds with ANB3 integrin, which is upregulated in activated endothelial cells during inflammation or angiogenesis, both observed in cancer cases (McCarthy et al., 2007).

Aptamers: They are short oligonucleotides (15–40 bases) and are identified via the selection process called systematic evaluation of ligands by exponential enrichment that can be used as both therapeutic and targeting entities. Just like antibodies, aptamers are highly specific for their targets and have high binding affinity due to the ability of the molecule's complex structure (Levy-Nissenbaum et al., 2008).

21.5.2 Passive Targeting Strategy for Codelivery of Phytomolecules with Anticancer Drugs

The nanocarrier's size and the tumour tissue's vasculature both play passive targeting roles. The parenchyma and stroma of a solid tumour contain vascular and other supportive cells. As a cancer cell grows, its metabolic requirements increase, putting angiogenic pressure on pre-existing blood vessels. This creates new capillaries to the tumour, a process known as angiogenesis. Scanning electron microscopic examinations of produced tumour capillaries reveal that the tumour cell's general architecture changes. Normal tissue vasculature has tight endothelial cell junctions; thus, entering nanoformulations in normal cells is impossible. Tumour cells, on the other hand, have intercellular gaps of 200–2,000 nm. There is a buildup of nanoparticles (carriers) in the tumour intestitial region due to poor vasculature, increased permeability, and retention effect (Figure 21.1). The accumulation of nanoparticles in tumours is caused by increased interstitial pressure in tumour tissue compared to normal and benign tissue (Behera and Padhi, 2020). The size of the nanoparticle, its surface characteristics, its circulatory half-life, and the degree of angiogenesis at the tumour location all influence the accumulation of nanoparticles in tumour tissue. The drug delivery systems employed for passive targeting are polymeric nanoparticles, micelles, liposomes, nanocrystals, inorganic nanoparticles, and dendrimers (Moorthi et al., 2011).

Barui et al. reported on the delivery of curcumin and doxorubicin together at the tumour site. They developed a pegylated RGDK-lipopeptide nano system targeting tumour vasculature. Curcumin and doxorubicin were co-encapsulated in liposomes targeted with PEGylated RGDK-tetrapeptide, and the nanoformulation exhibited synergism in inhibiting the proliferation of tumour cells. In mice treated with a formulation comprising only curcumin or doxorubicin, tumour growth inhibition was shown to be 2–3 times lower (Barui et al., 2014).

A passive targeting strategy was employed for enhanced cellular internalization of doxorubicin and curcumin nanoparticles into liver cancer cells (SMMC 7721) and human umbilical vein endothelial cells (HUVEC) (Zhang et al., 2018). The doxorubicin and curcumin codelivery in nanoparticles resulted in a high apoptosis rate in SMMC 7721 cells through decreased mitochondrial membrane potential. The combination therapy also exerted stronger antiangiogenic effects, including inhibition of human umbilical vein endothelial cell proliferation, migration, invasion, and tube formation *in vitro* and *in vivo*.

Long circulating liposomes containing curcumin and doxorubicin were synthesized by Tefas and colleagues to evaluate the cytotoxic potential using C26 murine colon carcinoma cell lines. On colon cancer cell lines, dual drug therapy had a stronger cytotoxic effect than doxorubicin-loaded liposomes alone. By incorporating doxorubicin into PEGylated liposomes, researchers minimized the drug's toxicity while lengthening its circulation time in the bloodstream. This resulted in better tumour accumulation due to increased permeability and retention (Tefas et al., 2017). Moradi et al. investigated the use of a combination of curcumin and 5-fluorouracil in treating colitis-related colon cancer. Combination therapy targeted colon cancer cells by a passive targeting mechanism, resulting in a decrease in tumour size in both the distal and middle regions of the colon in colitis-associated colon cancer, as well as a reduction of disease activity index (Moradi et al., 2018).

Another study employed a passive targeting technique to deliver curcumin and docetaxel for the treatment of prostate cancer. A glycyrrhetinic acid-modified chitosan-cystamine -poly(ε-caprolactone) copolymer micelle was developed by Yan and his co-workers for the codelivery of doxorubicin and curcumin. Both drugs were combined into lipid polymer hybrid nanoparticles to improve anticancer activity in vitro and in vivo. The use of both drugs together had a synergistic anticancer impact, maximizing the therapy benefit and overcoming multi-drug resistance. In-vitro cytotoxicity of docetaxel and curcumin lipid polymer hybrid nanoparticles was highest in tumour cells. These nanoparticles reduced tumour growth more effectively without causing adverse side effects (Yan et al., 2016).

In another study, lipid polymer hybrid nanoparticles and polymeric nanoparticles were constructed for the codelivery of cisplatin and curcumin to ameliorate cervical cancer. The cytotoxicity and in-vivo antitumour efficacy were studied in human cervix adenocarcinoma cell lines and mice-bearing cervical cancer models. The lipid polymer hybrid nanoparticles entrapping cisplatin and curcumin showed superior antitumour activity than polymeric nanoparticles (Li et al., 2017). The antiangiogenic and anti-inflammatory effect of long-circulating liposomes containing curcumin and doxorubicin was studied on murine colon cancer cells (C26) by Sesarman et al. (2017). PEGylated long-circulating liposomes of curcumin and doxorubicin could reach the target site passively. They exerted a strong antiproliferative effect compared to drugs administered alone in their free forms on C26 colon cancer cells. Liposomal codelivery of curcumin and albumin/paclitaxel nanoparticles for increased synergistic antitumour efficacy was studied by Ruttala and co-workers. This combination of drugs was significantly more cytotoxic than individual drug administration (Ruttala & Ko, 2015).

21.6 FUTURE PROSPECTIVE

Codelivery of phytochemicals with conventional chemotherapeutic agents using nano-drug delivery systems is a new and interesting area of formulation and development. The understanding of the physiology and genetics of cancer has increased manyfold, which has greatly helped in designing nanoscale dosage forms. These nano-drug delivery systems have the unique ability to deliver the desired drugs in the macro and microenvironment of the cancer cells. In the future, successful cancer treatment will require a more holistic and individualized approach. The codelivery of natural compounds with anticancer drugs assisted with nanotechnology has successfully targeted multiple cancer pathways, reducing cancer cell resistance and toxicity. Advanced technology like folate-based multi-functionalized nanoparticles has shown great potential for delivering numerous drugs at the tumour site. However, systems like this will require efficacy and safety assessments to be accepted by the regulatory authorities and approved for their clinical use. The nano delivery of phytomolecules like curcumin, resveratrol, gingerol, etc., has shown great anticancer potential. This has certainly attracted the attention of the researchers to co-deliver these drugs with conventional anticancer drugs. Nano-codelivery systems are certainly the future of drug delivery systems, which can smartly and safely deliver a cocktail of natural and other anticancer drugs at the desired dose ratio to get better therapeutic outcomes.

21.7 CONCLUSION

The pathophysiology of cancer is complex and hence needs a more holistic and comprehensive approach for the success of any treatment. Combining phytomolecules with other anticancer drugs is an attractive strategy with several advantages. However, combining effective anticancer drugs in a conventional dosage form has not yielded the desired therapeutic results. The current chapter has discussed various strategies to co-deliver anticancer drugs through nano-drug delivery systems. The active approach is used to codelivery ligands such as peptides, nucleic acid, antibiotics, etc. The passive strategy is more commonly used to design nano-drug delivery systems employing polymeric nanoparticles, micelles, liposomes, and nanocrystals by targeting the unique tumour microenvironment. Both approaches have been successfully utilized to formulate nano-drug delivery systems. These nano-drug delivery systems can successfully co-deliver the phytomolecules with other anticancer drugs and have shown encouraging results in preclinical and clinical studies. However, further research is needed to design nano-delivery systems that deliver the drugs in the desired dose ratio at the tumour site to ensure effective treatment of various types of cancers without compromising safety. Nano-delivery systems are still in their budding phases, and very few clinical studies and patents have been reported. Hence, more work in this area is needed to formulate safe and effective nano-delivery systems, which will be a significant leap in cancer chemotherapy.

REFERENCES

Al-Samydai, A., Al-Mamoori, F., Abdelnabi, H., & Aburjai, T. (2019). An updated review on anticancer activity of capsaicin. *International Journal of Scientific and Technology Research*, *8*(12), 2625–2630.

Alven, S., & Aderibigbe, B. A. (2020). Efficacy of polymer-based nanocarriers for codelivery of curcumin and selected anticancer drugs. *Nanomaterials*, *10*(8), 1–28.

Azuine, M. A., Kayal, J. J., & Bhide, S. V. (1992). Protective role of aqueous turmeric extract against mutagenicity of direct-acting carcinogens as well as benzo[a]pyrene-induced genotoxicity and carcinogenicity. *Journal of Cancer Research and Clinical Oncology*, *118*(6), 447–452.

Azuma, T., Matayoshi, Y., Sato, Y., & Nagase, Y. (2017). Effect of dutasteride on castration-resistant prostate cancer. *Molecular and Clinical Oncology*, *8*, 133–136.

Bajelan, E., Haeri, A., Vali, A. M., Ostad, S. N., & Dadashzadeh, S. (2012). Codelivery of doxorubicin and PSC 833 (Valspodar) by stealth nanoliposomes for efficient overcoming of multi-drug resistance. *Journal of Pharmacy and Pharmaceutical Sciences*, *15*(4), 568–582.

Balasubramanian, S., Ravindran Girija, A., Nagaoka, Y., Iwai, S., Suzuki, M., Kizhikkilot, V., … & Dasappan Nair, S. (2014). Curcumin and 5-fluorouracil-loaded, folate- and transferrin-decorated polymeric magnetic nanoformulation: A synergistic cancer therapeutic approach, accelerated by magnetic hyperthermia. *International Journal of Nanomedicine*, *9*(1), 437–459.

Banyal, A., Tiwari, S., Sharma, A., Chanana, I., Patel, S. K. S., Kulshrestha, S., & Kumar, P. (2023). Vinca alkaloids as a potential cancer therapeutics: Recent update and future challenges. *3 Biotech*, *13*(6), 211.

Barui, S., Saha, S., Mondal, G., Haseena, S., & Chaudhuri, A. (2014). Simultaneous delivery of doxorubicin and curcumin encapsulated in liposomes of pegylated RGDK-lipopeptide to tumor vasculature. *Biomaterials*, *35*(5), 1643–1656.

Behera, A., & Padhi, S. (2020). Passive and active targeting strategies for the delivery of the camptothecin anticancer drug: A review. *Environmental Chemistry Letters*, *18*(5), 1557–1567.

Behera, A., & Padhi, S. (2022). pH-Sensitive polymeric nanoparticles for cancer treatment. In: Padhi, S., Behera, A., Lichtfouse, E. (Eds.) *Polymeric Nanoparticles for the Treatment of Solid Tumours*. Environmental Chemistry for a Sustainable World, vol 71. Springer, Cham, pp. 401–425.

Behera, A., Mittu, B., Padhi, S., & Singh, A. (2020). Antimicrobial efficacy of essential oil nanoemulsions. In: Dhull, S. B., Chawla, P., Kaushik, R. (Eds.) *Nanotechnological Approaches in Food Microbiology*, CRC Press, Boca Raton, FL, pp. 293–309.

Behera, A., Padhi, S., & Nayak, A. K. (2023). Engineered liposomes as drug delivery and imaging agents. In: Ray, S., Nayak, A. K. (Eds.) *Design and Applications of Theranostic Nanomedicines*, Woodhead Publishing, Cambridge, MA, pp. 75–108.

Bhandari, P. R. (2015). Crocus sativus L. (saffron) for cancer chemoprevention: A mini review. *Journal of Traditional and Complementary Medicine*, *5*(2), 81–87.

Bhanumathy, K. K., Balagopal, A., Vizeacoumar, F. S., Vizeacoumar, F. J., Freywald, A., & Giambra, V. (2021). Review protein tyrosine kinases: Their roles and their targeting in leukaemia. *Cancers*, *13*(2), 1–21.

Bukowski, K., Kciuk, M., & Kontek, R. (2020). Mechanisms of multi-drug resistance in cancer chemotherapy. *International Journal of Molecular Sciences*, *21*(9), 3233.

Çetinkaya, S., Akça, K. T., & Süntar, I. (2022). Flavonoids and anticancer activity: Structure–activity relationship. *Studies in Natural Products Chemistry*, *74*, 81–115.

Challagundla, K. B., Wise, P. M., Neviani, P., Chava, H., Murtadha, M., Xu, T., … & Fabbri, M. (2015). Exosome-mediated transfer of microRNAs within the tumour microenvironment and neuroblastoma resistance to chemotherapy. *Journal of the National Cancer Institute*, *107*(7), 1–13.

Chen, T., Zhang, J., Zeng, H., Zhang, Y., Zhang, Y., Zhou, X., & Zhou, H. (2020). Antiproliferative effects of L-asparaginase in acute myeloid leukaemia. *Experimental and Therapeutic Medicine*, *8*, 2070–2078.

Cheng, Y., Zhao, P., Wu, S., Yang, T., Chen, Y., Zhang, X., … & Xiang, G. (2018). Cisplatin and curcumin co-loaded nano-liposomes for the treatment of hepatocellular carcinoma. *International Journal of Pharmaceutics*, *545*, 261–273.

Cho, S., Lu, M., He, X., Ee, P. L. R., Bhat, U., Schneider, E., … & Beck, W. T. (2011). Notch1 regulates the expression of the multi-drug resistance gene ABCC1/MRP1 in cultured cancer cells. *Proceedings of the National Academy of Sciences of the United States of America*, *108*(51), 20778–20783.

Choudhari, A. S., Mandave, P. C., Deshpande, M., Ranjekar, P., & Prakash, O. (2020). Phytochemicals in cancer treatment: From preclinical studies to clinical practice. *Frontiers in Pharmacology*, *10*, 1614.

Chun, K. S., Kundu, J., Kundu, J. K., & Surh, Y. J. (2014). Targeting Nrf2-Keap1 signaling for chemoprevention of skin carcinogenesis with bioactive phytochemicals. *Toxicology Letters*, *229*(1), 73–84.

Costa de Oliveira, R., Soares Pontes, G., Kostyuk, A., Coutinho Camargo, G. B., Dhyani, A., Shvydenko, T., … & Grafov, A. (2020). Anticancer and immunomodulatory activities of a novel water-soluble derivative of ellipticine. *Molecules*, *25*(9), 2130.

Cottu, P., D'Hondt, V., Dureau, S., Lerebours, F., Desmoulins, I., Heudel, P. E., … Delaloge, S. (2018). Letrozole and palbociclib versus chemotherapy as neoadjuvant therapy of high-risk luminal breast cancer. *Annals of Oncology*, *29*(12), 2334–2340.

Cragg, G. M., & Pezzuto, J. M. (2016). Natural products as a vital source for the discovery of cancer chemotherapeutic and chemopreventive agents. *Medical Principles and Practice*, *25*(2), 41–59.

Cuellar, S., Vozniak, M., Rhodes, J., Forcello, N., & Olszta, D. (2018). BCR-ABL1 tyrosine kinase inhibitors for the treatment of chronic myeloid leukaemia. *Journal of Oncology Pharmacy Practice*, *24*(6), 433–452.

Damianaki, A., Bakogeorgou, E., Kampa, M., Notas, G., Hatzoglou, A., Panagiotou, S., … & Castanas, E. (2000). Potent inhibitory action of red wine polyphenols on human breast cancer cells. *Journal of Cellular Biochemistry*, *78*(3), 429–441.

de Oliveira Gozzo, T., Panobianco, M. S., Clapis, M. J., & de Almeida, A. M. (2010). Dermatological toxicity in women with breast cancer undergoing chemotherapy treatment. *Revista Latino-Americana de Enfermagem*, *18*(4), 681–687.

de Souza Braga, M., Carvalho, G. C., Dua, K., Gupta, G., Oliveira, A., Cerize, N., & Pinto1, T. J. A. (2018). Idarubicin loaded nanoparticles for breast cancer. *World Congress on Recent Advances in Nanotechnology*, (May), 18–20.

Demiroglu-Zergeroglu, A., Basara-Cigerim, B., Kilic, E., & Yanikkaya-Demirel, G. (2010). The investigation of effects of quercetin and its combination with cisplatin on malignant mesothelioma cells in vitro. *Journal of Biomedicine and Biotechnology*, *2010*, 851589.

Doan, N. Q., Tran, H. N., Nguyen, N. T., Nguyen, K. D., Tao, V. M., Lai, N. N., … & Luu, P. H. (2024). Design, synthesis, and evaluation of anti breast cancer activity of colchicine-combretastatin A-4 analogues containing quinoline as microtubule-targeting agents. *Journal of Molecular Structure*, *1312*, 138465.

Doos, L., Roberts, E. O., Corp, N., & Kadam, U. T. (2014). Multi-drug therapy in chronic condition multimorbidity: A systematic review. *Family Practice*, *31*(6), 654–663.

Eramo, A., Ricci-Vitiani, L., Zeuner, A., Pallini, R., Lotti, F., Sette, G., … & De Maria, R. (2006). Chemotherapy resistance of glioblastoma stem cells [2]. *Cell Death and Differentiation*, *13*(7), 1238–1241.

Espinosa, E., Zamora, P., Feliu, J., & González Barón, M. (2003). Classification of anticancer drugs: A new system based on therapeutic targets. *Cancer Treatment Reviews*, *29*(6), 515–523.

Eto, K., Iwatsuki, M., Watanabe, M., Ishimoto, T., Ida, S., Imamura, Y., … & Baba, H. (2015). The sensitivity of gastric cancer to trastuzumab is regulated by the miR-223/FBXW7 pathway. *International Journal of Cancer*, *136*(7), 1537–1545.

Gao, M., Xu, Y., & Qiu, L. (2015). Enhanced combination therapy effect on paclitaxel-resistant carcinoma by chloroquine codelivery via liposomes. *International Journal of Nanomedicine*, *10*, 6615–6632.

Gautam, N., Mantha, A. K., & Mittal, S. (2014). Essential oils and their constituents as anticancer agents: A mechanistic view. *BioMed Research International*, *2014*, 154106.

Gowder, S. J. T., & Devaraj, H. (2006). Effect of the food flavour cinnamaldehyde on the antioxidant status of rat kidney. *Basic and Clinical Pharmacology and Toxicology*, *99*(5), 379–382.

Habtemariam, S. (2020). Recent advances in berberine inspired anticancer approaches: From drug combination to novel formulation technology and derivatization. *Molecules*, *25*(6), 1426.

Han, H. S., Kim, J. S., Park, J. H., Jeon, Y. K., Lee, K. W., Oh, D. Y., ... & Bang, Y. J. (2009). Weekly paclitaxel and trastuzumab as a first-line therapy in patients with HER2-overexpressing metastatic breast cancer: magnitude of HER2/neu amplification as a predictive factor for efficacy. *Journal of Korean Medical Science*, *24*(5), 910–917.

Hermann, P. C., Huber, S. L., Herrler, T., Aicher, A., Ellwart, J. W., Guba, M., … & Heeschen, C. (2007). Distinct populations of cancer stem cells determine tumour growth and metastatic activity in human pancreatic cancer. *Cell Stem Cell*, *1*(3), 313–323.

Hirshburg, J. M., Kelsey, P. A., Therrien, C. A., Gavino, A. C., & Reichenberg, J. S. (2016). Adverse effects and safety of 5-alpha reductase inhibitors (finasteride, dutasteride): A systematic review. *Journal of Clinical and Aesthetic Dermatology*, *9*(7), 56–62.

Hong, J., Huang, J., Shen, L., Zhu, S., Gao, W., Wu, J., … & Shen, K. (2020). A prospective, randomized study of Toremifene vs. tamoxifen for the treatment of premenopausal breast cancer: Safety and genital symptom analysis. *BMC Cancer*, *20*(1), 1–10.

Hosono, M. (2019). Radiation protection in therapy with radiopharmaceuticals. *International Journal of Radiation Biology*, *95*(10), 1427–1430.

Hsu, C. Y., Pallathadka, H., Gupta, J., Ma, H., Al-Shukri, H. H., Kareem, A. K., … & Mustafa, Y. F. (2024). Berberine and berberine nanoformulations in cancer therapy: Focusing on lung cancer. *Phytotherapy Research*, *38*(8), 4336–4350.

Huang, D., Duan, H., Huang, H., Tong, X., Han, Y., Ru, G., … & Zhao, Z. (2016). Cisplatin resistance in gastric cancer cells is associated with HER2 upregulation-induced epithelial-mesenchymal transition. *Scientific Reports*, *6*(September 2015), 1–12.

Iino, M., Sakamoto, Y., & Sato, T. (2019). Yttrium-90 ibritumomab tiuxetan consolidation versus rituximab maintenance therapy after induction chemotherapy in patients with indolent non-Hodgkin lymphoma: A single-institution experience. *Hematology (United Kingdom)*, *24*(1), 623–630.

Iqbal, J., Abbasi, B. A., Mahmood, T., Kanwal, S., Ali, B., Shah, S. A., & Khalil, A. T. (2017). Plant-derived anticancer agents: A green anticancer approach. *Asian Pacific Journal of Tropical Biomedicine*, *7*(12), 1129–1150.

Iraci, N., Ostacolo, C., Medina-Peris, A., Ciaglia, T., Novoselov, A. M., Altieri, A., … & Kurkin, A. V. (2022). In vitro and in vivo pharmacological characterization of a novel trpm8 inhibitor chemotype identified by small-scale preclinical screening. *International Journal of Molecular Sciences*, 23(4), 2070.

Jabbour, E. J., DeAngelo, D. J., Stelljes, M., Stock, W., Liedtke, M., Gökbuget, N., … & Kantarjian, H. M. (2018). Efficacy and safety analysis by age cohort of inotuzumab ozogamicin in patients with relapsed or refractory acute lymphoblastic leukaemia enrolled in INO-VATE. *Cancer*, *124*(8), 1722–1732.

Jain, P., Keating, M. J., Wierda, W. G., Sivina, M., Thompson, P. A., Ferrajoli, A., … & Burger, J. A. (2017). Long-term follow-up of treatment with ibrutinib and rituximab in patients with high-risk chronic lymphocytic leukaemia. *Clinical Cancer Research*, *23*(9), 2154–2158.

Jang, M., Cai, L., Udeani, G. O., Slowing, K. V., Thomas, C. F., Beecher, C. W. W., … & Pezzuto, J. M. (1997). Cancer chemopreventive activity of resveratrol, a natural product derived from grapes. *Science*, *275*(5297), 218–220.

Jen, E. Y., Ko, C. W., Eun Lee, J., Del Valle, P. L., Aydanian, A., Jewell, C., … & Pazdur, R. (2018). Fda approval: Gemtuzumab ozogamicin for the treatment of adults with newly diagnosed cd33-positive acute myeloid leukaemia. *Clinical Cancer Research*, *24*(14), 3242–3246.

Joo, J. H., Hong, S. S., Cho, Y. R., & Seo, D. W. (2016). 10-Gingerol inhibits proliferation and invasion of MDA-MB-231 breast cancer cells through suppression of Akt and p38MAPK activity. *Oncology Reports*, *35*(2), 779–784.

Kamle, M., Pandhi, S., Mishra, S., Barua, S., Kurian, A., Mahato, D. K., … & Sharifi-Rad, J. (2024). Camptothecin and its derivatives: Advancements, mechanisms and clinical potential in cancer therapy. *Medical Oncology*, 41(11), 263.

Karimi-Maleh, H., Shojaei, A. F., Karimi, F., Tabatabaeia, K., & Shakeri, S. (2018). Au nanoparticle loaded with 6-thioguanine anticancer drug as a new strategy for drug delivery. *Journal of Nanostructures*, *8*(4), 417–424.

Kaspers, G. J. L. (2019). Acute myeloid leukaemia niche regulates response to L-asparaginase. *British Journal of Haematology*, *186*(3), 397–399.

Kaur, R., Kapoor, K., & Kaur, H. (2011). Plants as a source of anticancer agents. *Journal of Ethnopharmacology*, *1*(1), 119–124.

Kelderman, S., Schumacher, T. N. M., & Haanen, J. B. A. G. (2014). Acquired and intrinsic resistance in cancer immunotherapy. *Molecular Oncology*, *8*(6), 1132–1139.

Kim, M. J., & Kim, H. (2015). Anticancer effect of lycopene in gastric carcinogenesis. *Journal of Cancer Prevention*, *20*(2), 92–96.

Kim, S. H., Kaplan, J. A., Sun, Y., Shieh, A., Sun, H. L., Croce, C. M., … & Parquette, J. R. (2015). The self-assembly of anticancer camptothecin-dipeptide nanotubes: A minimalistic and high drug loading approach to increased efficacy. *Chemistry - A European Journal*, *21*(1), 101–105.

Koren, G., Carey, N., Maxwell, C., Nulman, I., Gagnon, R., & Senikas, V. (2013). Cancer chemotherapy and pregnancy. *Journal of Obstetrics and Gynaecology Canada*, *35*(3), 263–278.

Kumar, A., Sharma, P. R., & Mondhe, D. M. (2016). Potential anticancer role of colchicine-based derivatives: An overview. *Anticancer Drugs*, *28*(3), 250–262.

Kundu, A., Padhi, S., Behera, A., Hasnain, M. S., & Nayak, A. K. (2022). Tumour targeting strategies by chitosan-based nanocarriers. In: Hasnain, M. S., Beg, S., Nayak, A. K. (Eds.) *Chitosan in Biomedical Applications*, Academic Press, London, pp. 163–188.

Lagas, J. S., Fan, L., Wagenaar, E., Vlaming, M. L. H., Van Tellingen, O., Beijnen, J. H., & Schinkel, A. H. (2010). P-Glycoprotein (P-gp/Abcb1), Abcc2, and Abcc3 determine the pharmacokinetics of etoposide. *Clinical Cancer Research*, *16*(1), 130–140.

Legault, J., & Pichette, A. (2007). Potentiating effect of β-caryophyllene on anticancer activity of α-humulene, isocaryophyllene and paclitaxel. *Journal of Pharmacy and Pharmacology*, *59*(12), 1643–1647.

Lei, M., Sha, S., Wang, X., Wang, J., Du, X., Miao, H., … & Zhu, Y. (2019). Codelivery of paclitaxel and gemcitabine via a self-assembling nanoparticle for targeted treatment of breast cancer. *RSC Advances*, *9*(10), 5512–5520.

Levy-Nissenbaum, E., Radovic-Moreno, A. F., Wang, A. Z., Langer, R., & Farokhzad, O. C. (2008). Nanotechnology and aptamers: Applications in drug delivery. *Trends in Biotechnology*, *26*(8), 442–449.

Leyland-Jones, B. (1993). Targeted drug delivery. *Seminars in Oncology*, *20*, 12–17.

Li, C., Ge, X., & Wang, L. (2017). Construction and comparison of different nanocarriers for codelivery of cisplatin and curcumin: A synergistic combination nanotherapy for cervical cancer. *Biomedicine et Pharmacotherapy*, *86*(27), 628–636.

Li, Q., Xiong, Y., Ji, C., & Yan, Z. (2019). The application of nanotechnology in the codelivery of active constituents of plants and chemotherapeutics for overcoming physiological barriers during antitumour treatment. *BioMed Research International*, *2019*, 9083068.

Lin, C. H., Lu, W. C., Wang, C. W., Chan, Y. C., & Chen, M. K. (2013). Capsaicin induces cell cycle arrest and apoptosis in human KB cancer cells. *BMC Complementary and Alternative Medicine*, *13*, 46.

Ling, Y., Xie, Q., Zhang, Z., & Zhang, H. (2018). Protein kinase inhibitors for acute leukaemia. *Biomarker Research*, *6*(1), 1–7.

Ma, X., Hu, M., Wang, H., & Li, J. (2018). Discovery of traditional Chinese medicine monomers and their synthetic intermediates, analogs or derivatives for battling P-gp-mediated multi-drug resistance. *European Journal of Medicinal Chemistry*, *159*, 381–392.

Malesci, A., Bianchi, P., Celesti, G., Basso, G., Marchesi, F., Grizzi, F., … & Laghi, L. (2017). Tumour-associated macrophages and response to 5-fluorouracil adjuvant therapy in stage III colorectal cancer. *OncoImmunology*, *6*(12), e1342918.

Marth, C., Reimer, D., & Zeimet, A. G. (2017). Front-line therapy of advanced epithelial ovarian cancer: Standard treatment. *Annals of Oncology*, *28*(Supplement 8), viii36–viii39.

McCarthy, J. R., Kelly, K. A., Sun, E. Y., & Weissleder, R. (2007). Targeted delivery of multifunctional magnetic nanoparticles. *Nanomedicine*, *2*(2), 153–167.

Meegan, M. J., & O'Boyle, N. M. (2019). Special issue "anticancer drugs". *Pharmaceuticals*, *12*(3), 134.

Meng, J., Guo, F., Xu, H., Liang, W., Wang, C., & Yang, X. D. (2016). Combination therapy using co-encapsulated resveratrol and paclitaxel in liposomes for drug resistance reversal in breast cancer cells in vivo. *Scientific Reports*, *6*(September 2015), 1–11.

Mileo, A. M., & Miccadei, S. (2016). Polyphenols as modulator of oxidative stress in cancer disease: New therapeutic strategies. *Oxidative Medicine and Cellular Longevity*, *2016*, 6475624.

Mirahmadi, M., Azimi-Hashemi, S., Saburi, E., Kamali, H., Pishbin, M., & Hadizadeh, F. (2020). Potential inhibitory effect of lycopene on prostate cancer. *Biomedicine & Pharmacotherapy*, 129, 110459.

Modi, S., Saura, C., Yamashita, T., Park, Y. H., Kim, S.-B., Tamura, K., … & Krop, I. (2020). Trastuzumab deruxtecan in previously treated HER2-positive breast cancer. *New England Journal of Medicine*, *382*(7), 610–621.

Moon, J. Y., Song, I. C., Ko, Y. B., & Lee, H. J. (2018). The combination of cisplatin and topotecan as a second-line treatment for patients with advanced/recurrent uterine cervix cancer. *Medicine (United States)*, *97*(14), 4–9.

Moorthi, C., Manavalan, R., & Kathiresan, K. (2011). Nanotherapeutics to overcome conventional cancer chemotherapy limitations. *Journal of Pharmacy and Pharmaceutical Sciences*, *14*(1), 67–77.

Moradi, R., Farzad, M., Seyed, R., Hassanian, M., Rezaei, N., Hashemzehi, M., … & Khazaei, M. (2018). Phytosomal curcumin inhibits tumour growth in colitis-associated colorectal cancer. *Journal of Cellular Physiology*, *233*, 1–14.

More, G. S., Thomas, A. B., Chitlange, S. S., Nanda, R. K., & Gajbhiye, R. L. (2019). Nitrogen mustards as alkylating agents: A review on chemistry, mechanism of action and current USFDA status of drugs. *Anticancer Agents in Medicinal Chemistry*, *19*(9), 1080–1102.

Motevalli, S. M., Eltahan, A. S., Liu, L., Magrini, A., Rosato, N., Guo, W., … Liang, X.-J. (2019). Co-encapsulation of curcumin and doxorubicin in albumin nanoparticles blocks the adaptive treatment tolerance of cancer cells. *Biophysics Reports*, *5*(1), 19–30.

Moudi, M., Go, R., Yien, C. Y. S., & Nazre, M. (2013). Vinca alkaloids. *International Journal of Preventive Medicine*, *4*(11), 1131–1135.

Mousa, L., Salem, M. E., & Mikhail, S. (2015). Biomarkers of angiogenesis in colorectal cancer. *Biomarkers in Cancer*, *7s1*, BIC.S25250.

Munoz, M., Henderson, M., Haber, M., & Norris, M. (2007). Role of the MRP1/ABCC1 multi-drug transporter protein in cancer. *IUBMB Life*, *59*(12), 752–757.

Nadhum, S. A., Kamoon, R. A., & Mohammed, M. H. (2020). 6-Mercaptopurine derivatives: Maintenance therapy of acute lymphoblastic leukaemia: A review. *Biochemical and Cellular Archives*, *20*(1), 2081–2090.

Nagano, T., Tachihara, M., & Nishimura, Y. (2018). Mechanism of resistance to epidermal growth factor receptor-tyrosine kinase inhibitors and a potential treatment strategy. *Cells*, *7*(11), 212.

Nair, S. C., Pannikar, B., & Panikkar, K. R. (1991). Antitumour activity of saffron (Crocus sativus). *Cancer Letters*, *57*(2), 109–114.

Nakai, Y., Tanaka, N., Anai, S., Miyake, M., Tatsumi, Y., & Fujimoto, K. (2015). A randomized control trial comparing the efficacy of antiandrogen monotherapy: Flutamide vs. Bicalutamide. *Hormones and Cancer*, *6*(4), 161–167.

Nathan, M. R., & Schmid, P. (2017). A review of fulvestrant in breast cancer. *Oncology and Therapy*, *5*(1), 17–29.

Oliveira, P. D., Bütten Bender, S. L., Prado, W. A., Beckenkamp, A., Id, A. C. A., Guterres, S. S., & Pohlmann, A. R. (2018). Enhanced and selective antiproliferative activity of methotrexate-functionalized-nanocapsules to human breast cancer cells (MCF-7). *Nanomaterials (Basel)*, *8*, 1–19.

Ono, M., Takeshima, M., & Nakano, S. (2015). Mechanism of the anticancer effect of lycopene (tetraterpenoids). *Enzymes*, *37*, 139–166.

Oudard, S., Fizazi, K., Sengeløv, L., Daugaard, G., Saad, F., Hansen, S., … & Sartor, O. (2017). Cabazitaxel versus docetaxel as first-line therapy for patients with metastatic castration-resistant prostate cancer: A randomized phase III trial-FIRSTANA. *Journal of Clinical Oncology*, *35*(28), 3189–3197.

Padhi, S., & Behera, A. (2020). Nanotechnology based targeting strategies for the delivery of camptothecin. In: Saneja, A., Panda, A., Lichtfouse, E. (Eds.) *Sustainable Agriculture Reviews*, vol 44. Springer, Cham, pp. 243–272.

Padhi, S., & Behera, A. (2021). Advanced drug delivery systems in the treatment of ovarian cancer. In: Dua, K., Mehta, M., de Jesus Andreoli Pinto, T., Pont, L. G., Williams, K. A., Rathbone, M. J. (Eds.) *Advanced Drug Delivery Systems in the Management of Cancer*, Academic Press, London, pp. 127–139.

Padhi, S., & Behera, A. (2022). Cellular internalization and toxicity of polymeric nanoparticles. In: Padhi, S., Behera, A., Lichtfouse, E. (Eds.) *Polymeric Nanoparticles for Treating Solid Tumours*. Environmental Chemistry for a Sustainable World, vol 71. Springer, Cham, pp. 473–488.

Padhi, S., Behera, A., Hasnain, M. S., & Nayak, A. K. (2022). Chitosan-based drug delivery systems in cancer therapeutics. In: Hasnain, M. S., Beg, S., Nayak, A. K. (Eds.) *Chitosan in Drug Delivery*, Academic Press, London, pp. 159–193.

Padhi, S., Kapoor, R., Verma, D., Panda, A. K., Iqbal, Z. (2018). Formulation and optimization of topotecan nanoparticles: In vitro characterization, cytotoxicity, cellular uptake and pharmacokinetic outcomes. *Journal of Photochemistry and Photobiology B: Biology*, *183*, 222–232.

Padhi, S., Mirza, M. A., Verma, D., Khuroo, T., Panda, A. K., Talegaonkar, S., Khar, R. K., & Iqbal, Z. (2016). Revisiting the nanoformulation design approach for effective delivery of topotecan in its stable form: An appraisal of its in vitro behavior and tumour amelioration potential. *Drug Delivery*, *23*(8), 2827–2837.

Pan, J., Rostamizadeh, K., Filipczak, N., & Torchilin, V. P. (2019). Polymeric codelivery systems in cancer treatment: An overview on component drugs' dosage ratio effect. *Molecules*, *24*(6), 1035.

Patnaik, S., Gorain, B., Padhi, S., Choudhury, H., Gabr, G. A., Md, S., Kumar Mishra, D., & Kesharwani, P. (2021). Recent update of toxicity aspects of nanoparticulate systems for drug delivery. *European Journal of Pharmaceutics and Biopharmaceutics*, *161*, 100–119.

Peddi, P. F., Peddi, S., Santos, E. S., & Morgensztern, D. (2014). Central nervous system toxicities of chemotherapeutic agents. *Expert Review of Anticancer Therapy*, 14(7), 857–863. https://doi.org/10.1586/14737140.2014.911089

Pezzani, R., Salehi, B., Vitalini, S., Iriti, M., Zuñiga, F. A., Sharifi-Rad, J., … & Martins, N. (2019). Synergistic effects of plant derivatives and conventional chemotherapeutic agents: An update on the cancer perspective. *Medicina (Lithuania)*, *55*(4), 1–16.

Pongrakhananon, V., & Rojanasakul, Y. (2011). Anticancer properties of curcumin. In: Gali-Muhtasib, H. (Ed.) *Advances in Cancer Therapy*, InTech Open, pp. 345–368.

Pradhan, S. P., Sahu, P. K., & Behera, A. (2023). New insights toward molecular and nanotechnological approaches to antidiabetic agents for Alzheimer's disease. *Molecular and Cellular Biochemistry*, *478*, 1–24.

Pradhan, S. P., Tejaswani, P., Behera, A., & Sahu, P. K. (2024). Phytomolecules from conventional to nano form: Next-generation approach for Parkinson's disease. *Ageing Research Reviews*, *93*, 102136.

Prakash, O., Kumar, A., Kumar, P., & Ajeet, A. (2013). Anticancer potential of plants and natural products: A review. *American Journal of Pharmacological Sciences*, *1*(6), 104–115.

Prasad, C., & Banerjee, R. (2019). Ultrasound-triggered spatiotemporal delivery of topotecan and curcumin as combination therapy for cancer. *Journal of Pharmacology and Experimental Therapeutics*, *370*(3), 876–893.

Qi, S. S., Sun, J. H., Yu, H. H., & Yu, S. Q. (2017). Codelivery nanoparticles of anticancer drugs for improving chemotherapy efficacy. *Drug Delivery*, *24*(1), 1909–1926.

Rabi, T., & Bishayee, A. (2009). d-Limonene sensitizes docetaxel-induced cytotoxicity in human prostate cancer cells: Generation of reactive oxygen species and induction of apoptosis. *Journal of Carcinogenesis*, *8*.

Rahman, F. N. U. A. U., Ali, S., & Saif, M. W. (2017). Update on the role of nanoliposomal irinotecan in the treatment of metastatic pancreatic cancer. *Therapeutic Advances in Gastroenterology*, *10*(7), 563–572.

Rahmani, A. H., Alzohairy, M. A., Khan, M. A., & Aly, S. M. (2014). Therapeutic implications of black seed and its constituent thymoquinone in the prevention of cancer through inactivation and activation of molecular pathways. *Evidence-Based Complementary and Alternative Medicine*, *2014*, 724658.

Ramalingam, V., Varunkumar, K., Ravikumar, V., & Rajaram, R. (2018). Target delivery of doxorubicin tethered with PVP stabilized gold nanoparticles for effective treatment of lung cancer. *Scientific Reports*, *8*(1), 1–12.

Razis, E., Bobos, M., Kotoula, V., Eleftheraki, A. G., Kalofonos, H. P., Pavlakis, K., … & Fountzilas, G. (2011). Evaluation of the association of PIK3CA mutations and PTEN loss with efficacy of trastuzumab therapy in metastatic breast cancer. *Breast Cancer Research and Treatment*, *128*(2), 447–456.

Ricciuti, B., Dahlberg, S. E., Adeni, A., Sholl, L. M., Nishino, M., & Awad, M. M. (2019). Immune checkpoint inhibitor outcomes for patients with non-small-cell lung cancer receiving baseline corticosteroids for palliative versus non-palliative indications. *Journal of Clinical Oncology*, *37*(22), 1927–1934.

Roeten, M. S. F., Cloos, J., & Jansen, G. (2018). Positioning of proteasome inhibitors in therapy of solid malignancies. *Cancer Chemotherapy and Pharmacology*, *81*(2), 227–243.

Roshan, B., & Ravindranath, S. (2019). A review on recent anticancer agents and drugs containing pyrazole. *Journal of Drug Delivery and Therapeutics*, *9*(3), 753–759.

Roviello, G., Petrioli, R., Marano, L., Polom, K., Marrelli, D., Perrella, A., & Roviello, F. (2016). Angiogenesis inhibitors in gastric and gastroesophageal junction cancer. *Gastric Cancer*, *19*(1), 31–41.

Ruttala, H. B., & Ko, Y. T. (2015). Liposomal co-delivery of curcumin and albumin/paclitaxel nanoparticle for enhanced synergistic antitumor efficacy. *Colloids and Surfaces B: Biointerfaces*, *128*, 419–426.

Sahoo, P., Dey, J., Mahapatra, S. R., Ghosh, A., Jaiswal, A., Padhi, S., … & Suar, M. (2022). Nanotechnology and COVID-19 convergence: Toward new planetary health interventions against the pandemic. *OMICS: A Journal of Integrative Biology*, 26(9), 473–488.

Sarin, J., DeRossi, S. S., & Akintoye, S. O. (2008). Updates on bisphosphonates and potential pathobiology of bisphosphonate-induced jaw osteonecrosis. *Oral Diseases*, *14*(3), 277–285.

Sesarman, A. A., Tefas, L., Sylvester, B., Licarete, E., Rauca, V., Luput, L., … & Porfire, A. (2017). Antiangiogenic and anti-inflammatory effects of long-circulating liposomes co-encapsulating curcumin and doxorubicin on C26 murine colon cancer cells. *Pharmacological Reports*, *70*, 331–339.

Shang, Y., Cai, X., & Fan, D. (2014). Roles of epithelial-mesenchymal transition in cancer drug resistance. *Current Cancer Drug Targets*, *13*(9), 915–929.

Slika, L., Moubarak, A., Borjac, J., Baydoun, E., & Patra, D. (2020). Preparation of curcumin-poly (allyl amine) hydrochloride based nanocapsules: Piperine in nanocapsules accelerates encapsulation and release of curcumin and effectiveness against colon cancer cells. *Materials Science and Engineering: C*, *109*, 110550.

Solary, E., & Itzykson, R. (2017). How I treat chronic myelomonocytic leukaemia. *Blood*, *130*(2), 126–133.

Stavrovskaya, A. A. (2000). Cellular mechanisms of multi-drug resistance of tumour cells. *Biochemistry (Moscow)*, *65*(1), 95–106.

Stojanovska, V., Sakkal, S., & Nurgali, K. (2015). Platinum-based chemotherapy: Gastrointestinal immunomodulation and enteric nervous system toxicity. *American Journal of Physiology - Gastrointestinal and Liver Physiology*, *308*(4), G223–G232.

Sudhakaran, M., Sardesai, S., & Doseff, A. I. (2019). Flavonoids: New frontier for immuno-regulation and breast cancer control. *Antioxidants*, *8*(4), 1–27.

Sun, J. M., Lee, K. H., Kim, B. S., Kim, H. G., Min, Y. J., Yi, S. Y., … & Ahn, M. J. (2018). Pazopanib maintenance after first-line etoposide and platinum chemotherapy in patients with extensive disease small-cell lung cancer: A multicentre, randomized, placebo-controlled Phase II study (KCSG-LU12–07). *British Journal of Cancer*, *118*(5), 648–653.

Sun, W., Wang, W., Kim, J., Keng, P., Yang, S., Zhang, H., … & Zhang, L. (2008). Anticancer effect of resveratrol is associated with induction of apoptosis via a mitochondrial pathway alignment. *Advances in Experimental Medicine and Biology*, *614*, 179–186.

Taher, M. A., Abu, M., Nyeem, B., & Billah, M. (2017). Vinca alkaloid- the second most used alkaloid for cancer treatment: A review. *International Journal of Physiology, Nutrition and Physical Education*, *2*(2), 723–727.

Tang, P., Wang, X., Schiffhauer, L., Wang, J., Bourne, P., Yang, Q., … & Hajdu, S. (2006). Expression patterns of ER-α, PR, HER-2/neu, and EGFR in different cell origin subtypes of high grade and non-high grade ductal carcinoma in situ. *Annals of Clinical & Laboratory Science*, *36*(2), 137–143.

Tarpgaard, L. S., Qvortrup, C., Nielsen, S. L., Stenvang, J., Detlefsen, S., Brünner, N., & Pfeiffer, P. (2021). New use for old drugs: Epirubicin in colorectal cancer. *Acta Oncologica*, *60*(7), 954–956.

Tefas, L. R., Sylvester, B., Sesarman, A., Licarete, E., & Banciu, M. (2017). Development of antiproliferative long-circulating liposomes co-encapsulating doxorubicin and curcumin, through the use of a quality-by-design approach. *Drug Design, Development and Theraphy*, 11, 1605–1621.

Thoennissen, N. H., O'Kelly, J., Lu, D., Iwanski, G. B., La, D. T., Abbassi, S., … & Koeffler, H. P. (2010). Capsaicin causes cell-cycle arrest and apoptosis in ER-positive and-negative breast cancer cells by modulating the EGFR/HER-2 pathway. *Oncogene*, *29*(2), 285–296.

Tolaney, S. M., Barry, W. T., Dang, C. T., Yardley, D. A., Moy, B., Marcom, P. K., ... & Winer, E. P. (2015). Adjuvant paclitaxel and trastuzumab for node-negative, HER2-positive breast cancer. *New England Journal of Medicine*, *372*(2), 134–141.

Verma, D., Thakur, P. S., Padhi, S., Khuroo, T., Talegaonkar, S., & Iqbal, Z. (2017). Design expert assisted nanoformulation design for codelivery of Topotecan and thymoquinone: Optimization, in vitro characterization, and stability assessment. *Journal of Molecular Liquids*, *242*, 382–394.

Ververidis, F., Trantas, E., Douglas, C., Vollmer, G., Kretzschmar, G., & Panopoulos, N. (2007). Biotechnology of flavonoids and other phenylpropanoid-derived natural products. Part I: Chemical diversity, impacts on plant biology and human health. *Biotechnology Journal*, *2*(10), 1214–1234.

Viale, A., De Franco, F., Orleth, A., Cambiaghi, V., Giuliani, V., Bossi, D., … & Pelicci, P. G. (2009). Cell-cycle restriction limits DNA damage and maintains self-renewal of leukaemia stem cells. *Nature*, *457*(7225), 51–56.

Voutsadakis, I. A. (2018). A systematic review and pooled analysis of studies of oral etoposide in metastatic breast cancer. *European Journal of Breast Health*, *14*(1), 10–16.

Wahid, M., & Bano, Q. (2014). Structure activity relationship and clinical development perspective of analogs. *Journal of Applied Pharmacology*, *6*, 286–295.

Wang, C., Aguilar, A., & Ojima, I. (2022). Strategies for the drug discovery and development of taxane anticancer therapeutics. *Expert Opinion on Drug Discovery*, *17*(11), 1193–1207.

Wang, H., Lin, Y., Liu, Q., Zhou, A., Bian, H., Zhang, W., … & Wu, Z. (2023). Antioxidant, anticancer activity and molecular docking study of lycopene with different ratios of Z-isomers. *Current Research in Food Science*, *6*, 100455.

Wang, X., Zhang, H., & Chen, X. (2019). Drug resistance and combating drug resistance in cancer. *Cancer Drug Resistance*, *2*(2), 141.

Wee, P., & Wang, Z. (2017). Epidermal growth factor receptor cell proliferation signaling pathways. *Cancers*, *9*(5), 1–45.

Wen, W., Lowe, G., Roberts, C. M., Finlay, J., Han, E. S., Glackin, C. A., & Dellinger, T. H. (2017). Pterostilbene, a natural phenolic compound, synergizes the antineoplastic effects of megestrol acetate in endometrial cancer. *Scientific Reports*, *7*(1), 1–10.

Willson, M. L., Burke, L., Ferguson, T., Ghersi, D., Nowak, A. K., & Wilcken, N. (2019). Taxanes for adjuvant treatment of early breast cancer. *Cochrane Database of Systematic Reviews*, *2019*(9), CD004421.

Woo, H. H., Murphy, D. G., Testa, G. M., Grummet, J. P., Chong, M., & Stork, A. P. (2017). Effect of triptorelin on lower urinary tract symptoms in Australian prostate cancer patients. *Research and Reports in Urology*, *9*, 27–35.

Wu, J., Liu, X. Y., Bao, M. F., Shi, B. B., Li, M. X., & Cai, X. H. (2025). Rearranged and enantiomeric alkaloids from Cephalotaxus sinensis. *Tetrahedron Letters*, *155*, 155437.

Wu, T., Chen, W., Zhong, Y., Hou, X., Fang, S., Liu, C. Y., … & Yang, Y. (2017). Nuclear export of ubiquitinated proteins determines the sensitivity of colorectal cancer to proteasome inhibitor. *Molecular Cancer Therapeutics*, *16*(4), 717–728.

Xiao, B., Si, X., Han, M. K., Viennois, E., Zhang, M., & Merlin, D. (2015). Co-delivery of camptothecin and curcumin by cationic polymeric nanoparticles for synergistic colon cancer combination chemotherapy. *Journal of Materials Chemistry B*, *3*(39), 7724–7733.

Xu, F., Liu, M., Liao, Y., Zhou, Y., Zhang, P., Zeng, Y., & Liu, Z. (2022). Improvement of anticancer effect of berberine by salt formation modifications. *Phytomedicine*, *104*, 154314.

Xu, Y., Wang, C., Ding, Y., Wang, Y., Liu, K., Tian, Y., … & Li, L. (2016). Nanoparticles with optimal ratiometric codelivery of docetaxel with gambogic acid for treatment of multidrug-resistant breast cancer. *Journal of Biomedical Nanotechnology*, *12*(9), 1774–1781.

Yan, J., Wang, Y., Zhang, X., Liu, S., & Tian, C. (2016). Targeted nanomedicine for prostate cancer therapy: Docetaxel and curcumin co-encapsulated lipid–polymer hybrid nanoparticles for the enhanced antitumour activity in vitro and in vivo. *Drug Delivery*, *7544*, 1757–1762.

Yi, J., He, Q., Peng, G., & Fan, Y. (2022). Improved water solubility, chemical stability, antioxidant and anticancer activity of resveratrol via nanoencapsulation with pea protein nanofibrils. *Food Chemistry*, *377*, 131942.

Yin, J., & Zhang, J. (2011). Multi-drug resistance-associated protein 1 (MRP1/ABCC1) polymorphism: From discovery to clinical application. *Journal of Central South University (Medical Sciences)*, *36*(10), 927–938.

Yingchoncharoen, P., Kalinowski, D. S., & Richardson, D. R. (2016). Lipid-based drug delivery systems in cancer therapy: What is available and what is yet to come. *Pharmacological Reviews*, *68*(3), 701–787.

Yount, G., Yang, Y., Wong, B., Wang, H. J., & Yang, L. X. (2007). A novel camptothecin analog with enhanced antitumour activity. *Anticancer Research*, *27*(5A), 3173–3178.

Zent, C. S., Victoria Wang, X., Ketterling, R. P., Hanson, C. A., Libby, E. N., Barrientos, J. C., … & Tallman, M. S. (2016). A phase II randomized trial comparing standard and low dose rituximab combined with alemtuzumab as initial treatment of progressive chronic lymphocytic leukaemia in older patients: A trial of the ECOG-ACRIN cancer research group (E1908). *American Journal of Hematology*, *91*(3), 308–312.

Zhang, Q., Liu, Q., Du, M., Vermorken, A., Cui, Y., Zhang, L., ... & Chen, M. (2019). Cetuximab and Doxorubicin loaded dextran-coated Fe3O4 magnetic nanoparticles as novel targeted nanocarriers for non-small cell lung cancer. *Journal of Magnetism and Magnetic Materials*, 481, 122–128.

Zhang, D., Xu, Q., Wang, N., Yang, Y., Liu, J., Yu, G., … & Wang, H. (2018). A complex micellar system co-delivering curcumin with doxorubicin against cardiotoxicity and tumour growth. *International Journal of Nanomedicine*, *13*, 4549–4561.

Zhang, J., Wu, D., Vikash, Song, J., Wang, J., Yi, J., & Dong, W. (2015). Hesperetin induces the apoptosis of gastric cancer cells via activating mitochondrial pathway by increasing reactive oxygen species. *Digestive Diseases and Sciences*, *60*(10), 2985–2995.

Zhang, T., Chen, W., Jiang, X., Liu, L., Wei, K., Du, H., … & Li, J. (2019). Anticancer effects and underlying mechanism of colchicine on human gastric cancer cell lines in vitro and in vivo. *Bioscience Reports*, *39*(1), 1–10.

Zhao, Y., Huan, M. L., Liu, M., Cheng, Y., Sun, Y., Cui, H., … & Zhou, S. Y. (2016). Doxorubicin and resveratrol codelivery nanoparticle to overcome doxorubicin resistance. *Scientific Reports*, *6*, 1–15.

Zhu, Z., Wang, Z., Zhang, C., Wang, Y., Zhang, H., Gan, Z., … & Wang, X. (2019). Mitochondrion-targeted platinum complexes suppressing lung cancer through multiple pathways involving energy metabolism. *Chemical Science*, *10*(10), 3089–3095.

Zou, L., Di Wang, Hu, Y., Fu, C., Li, W., Dai, L., … & Zhang, J. (2017). Drug resistance reversal in ovarian cancer cells of paclitaxel and borneol combination therapy mediated by PEG-PAMAM nanoparticles. *Oncotarget*, *8*(36), 60453–60468.

22 Drug Delivery Systems Entrapping Natural Drugs and Derivatives

Recent Advancements and Future Challenges

Vidhi Shah, Radhika Pandya, and Ketan M. Ranch
Department of Pharmaceutics, L. M. College of Pharmacy, Ahmedabad, India

22.1 INTRODUCTION

Natural products have been used as herbal medicines throughout human history. Ancient literature works, like Ebers Papyrus, Chinese Materia Medica, Shennong, Tang Herbal, and Charaka Samhita, specify the use of many natural molecules from different sources, ranging from plant extracts and fungi to inorganic chemicals, as potential medicines to treat various ailments before the initiation of Western/synthetic medicine. Natural products or their derivatives comprise around one-third of the best-selling medications today. Many of the medications used today are derived from natural components, although for several reasons, big pharmaceutical corporations have failed to give those substances the attention they deserve (Lu et al., 2020).

Although the benefits of natural compounds attracted much attention, their low permeability, poor solubility, and poor chemical stability became major obstacles to their development as clinical candidates. Developing a suitable delivery system for natural molecules may subdue these associated challenges and help to achieve clinical success (Majolo et al., 2019). The problem with using natural products in disease treatment is their poor bioavailability, which causes issues in clinical trials. The use of nanotechnology has shown immense success in the field of drug delivery. The array of characteristics and interactions peculiar to the nanoscale structure accounts for the advantages of nanotechnology (Enrico, 2019). Polymeric nanoparticles have been the most tested ones in combination with natural products. Poly(lactic-co-glycolic acid), polyethylene glycol, polyvinyl alcohol, poly-L-lactic acid, polycaprolactone, and chitosan are the most common polymers used due to their biocompatibility, biodegradability, and the fact that they are easy to functionalize (Wong et al., 2020). The benefits of drug delivery methods and how they can increase the therapeutic efficacy of different natural products are discussed in this chapter, along with pertinent examples.

22.2 DRUG DELIVERY SYSTEMS

22.2.1 Nanoparticles

Nanotechnology provides promising avenues for the treatment and diagnosis of cancer. Due to their smaller size, nanoparticles possess several advantages, like higher drug payloads, prolonged plasma circulation time, and enhanced permeation and retention effect by passive targeting (Sharma et al., 2015). A group of researchers developed curcumin-entrapped poly lactic-co-glycolic acid nanoparticles for breast and lung carcinoma. This resulted in a tenfold increase in the solubility of curcumin in nanoparticles and a threefold improvement in anticancer activity compared to free curcumin formulation (Khan et al., 2018).

The ease of surface modification and functionalization with different ligands can offer active targeting, thereby minimizing the off-target toxicities. Various ligands like epidermal growth factor, human epidermal growth factor receptor 2, folic acid, transferrin receptors, antibodies, peptides, and aptamers are used in nanoparticle conjugation. Vandghanooni et al. (2020) have reported using aptamer-conjugated mesoporous silica nanoparticles for cancer therapy.

Compared to healthy cells, cancer cells have a distinct microenvironment. Stimuli-sensitive nanoparticles can be used to take advantage of the said condition. Lai et al. (2021) developed pH-responsive hyaluronic acid-based nanoparticles entrapping curcumin. These nanoparticles entered the lysosome, releasing the curcumin at low pH, resulting in the necrosis of tumour cells. These nanoparticles exhibited notable efficiency in effectively delivering curcumin to the tumour cells, thereby inhibiting the growth of breast cancer cells. The cancer cells are known to have less blood supply than normal cells. Considering this, hypoxia-responsive formulations were reported where a near-infrared dye (Cy7) moiety was conjugated to an antitumour agent, gambogic acid. This conjugate could self-assemble into nanoparticles in the presence of surfactin, and the cationic Cy7 group

DOI: 10.1201/9781003611516-28

was embedded on the surface of anionic surfactin. The Cy7 groups trigger the release of gambogic acid-Cy7 conjugates from nanoparticles by interacting with the organic anion-transporting polypeptides on the cytomembrane, consequently enhancing the cellular uptake of the conjugated drugs. The *in vivo* study indicated that GA-Cy7-NP were localized in the tumours, providing improved tumour efficacy with decreased off-target accumulation in livers compared to the original gambogic acid and free gambogic acid-Cy7 conjugate (Wang et al., 2019).

Photodynamic therapy is a novel technique where a specific wavelength of light activates photosensitive agents, transferring the energy to free oxygen, thereby producing reactive oxygen species that result in cell apoptosis. Combining this therapy with nanoparticles for cancer treatment is expected to provide a synergistic antitumour effect. Huang et al. (2018) formulated cetuximab encapsulated in FKR560 dye-containing poly lactic-co-glycolic acid nanoparticles for managing ovarian cancer. These photo-immunoconjugates of nanoparticles showed double the accumulation in the tumour and resulted in rapid shrinking of the tumour, compared with the unconjugated immunoconjugates.

The biggest obstacle in cancer treatment is multidrug resistance. Redox-responsive nanoparticles of chondroitin sulphate encapsulating quercetin, chlorin e6, and paclitaxel as chemosensitizers, photosensitizers, and chemotherapeutic agents, respectively, were developed by Shi et al. (2021). This allowed for greater chemo-photodynamic therapy to overcome multidrug resistance. According to Shi et al. (2021), quercetin possesses P-gp inhibitory activity, which, combined with paclitaxel, can overcome its suppression of anticancer potency against multidrug-resistant cells. Furthermore, exposure to a near-infrared laser can raise the quantity of intracellular reactive oxygen species produced by the nanoparticles, aiding in the loss of mitochondrial membrane potential and ultimately triggering cell apoptosis.

The inorganic nanoparticles, like gold and quantum dots, exhibit size-dependent light scattering, leading to different colours that enable the diagnosis based on the localization of the nanoparticles in a particular site. The gold nanoparticles coated with curcumin and isonicotinic acid hydrazide were used to treat lung cancer. This study illustrated the exceptional stabilization of isonicotinic acid hydrazide and curcumin molecules by gold nanoparticles. Studies on human lung cancer and fibroblasts confirmed that gold nanoparticles coated with curcumin and isonicotinic acid hydrazide have selective toxicity due to the stable surface corona (Umapathi et al., 2020). Jin et al. (2017) prepared polyethylene glycol-coated poly lactic-co-glycolic acid nanoparticles to target peptides like epidermal growth factor receptors for specific cancer cell targeting. This resulted in an approximately 40-fold increase in the curcumin's half-life and more than 50% reduction in tumour burden in the mouse xenograft tumour model.

22.2.2 Microemulsion

Microemulsions are transparent, thermodynamically stable, isotropic colloidal dispersions of oil and water stabilized by surfactant and co-surfactant molecules. Microemulsions require a relatively large amount of surfactants to stabilize the oil-water interface of nanodroplets (usually 10–100 nm), where co-surfactants impart fluidity to this interfacial film (Brown et al., 2012). Microemulsions are most widely used due to increased drug solubilization, enhanced permeation, prolonged circulation, enhanced dissolution offered by larger surface area, and enhanced drug bioavailability (Koli et al., 2021; Shah et al., 2017).

Unlike emulsions, microemulsions offer greater stability, which can be further improved by the gel-loaded microemulsion, where aggregation of the globules is restricted. Wang et al. (2016) developed the thermoresponsive gel-based microemulsion of 5-fluorouracil for colorectal cancer. This resulted in the sustained release of 5-fluorouracil for 10 hours and improved stability. No morphological change of microemulsions or 5-fluorouracil content change was observed during the 4-month storage period.

The targeted delivery of the microemulsions is also possible by employing the component possessing the targeting potential. Zhao et al. (2020a) developed the targeted tumour therapy using tripterine, brucea oil, glycyrrhizin microemulsion, and targeting ligands, transferrin and cell-penetrating peptide SA-R6H4. The ligand-based microemulsion showed 1.56 times higher cellular uptake of tripterine compared to the non-ligand-based microemulsion. Additionally, tripterine against the kidneys and liver was reduced significantly.

The combined formulation of nanoparticle-loaded microemulsions was further explored to provide greater benefit. Natesan et al. (2014) formulated a camptothecin-loaded magnetic microemulsion for active targeting of breast cancer. The magnetite particles of ~70 nm were added to the oil phase containing camptothecin. The active targeting of camptothecin-loaded magnetic microemulsion was achieved with the help of an external magnetic field (1,000 gauss) to determine the localization of drugs in the cancer cells. This provided controlled release of the drug using a magnetic field. Thapliyal et al. (2020) reported the development of gallic acid containing silver nanoparticles loaded microemulsion, which inhibited breast cancer cells (MCF-7) significantly.

Recent advances include the development of stimuli-responsive microemulsions (Zhang & Feng, 2020). The modified silica nanoparticles with selenium-containing tertiary amine exhibit stimuli-responsive groups. The tertiary amine group acts as a CO_2-responsive group, and the selenium atom acts as a redox-responsive system (Zhang et al., 2017). Microemulsions require a high concentration of surfactant to stabilize the system. This high surfactant is a major concern while considering its application in drug delivery due to the toxicity of the surfactant. Recent advancements in microemulsion include surfactant-free microemulsions that contain oil, water, and an "amphi-solvent". It consists of solvents like ethanol, propanol or acetone having complete or partial miscibility with oil and water (Hou & Xu, 2016).22.2.3 Micelles

Amphiphilic block copolymers have gained considerable attention in recent decades as they self-assemble into spherical micelles above the critical micelle concentration. The micelles provide various advantages, such as solubilizing poorly water-soluble drugs in the micellar core (Shah et al., 2019). Micelles prolong the circulation time in the blood due to their hydrophilic shell, and their smaller size assists in enhanced permeability and retention effect (Chen et al., 2020c). The solubility of curcumin in micelle containing hydroxyethyl starch was noted to be 1,000 times more than that of free curcumin (Chen et al., 2020a). Raval et al. (2017) reported an increased solubility of genistein, paclitaxel, and quercetin using six different pluronics.

Additionally, pluronic block copolymers have been reported to exhibit an inhibitory effect on transmembrane drug efflux transporter P-glycoprotein, which is overexpressed in multidrug-resistant cells, resulting in increased drug accumulation and cytotoxicity. Inhibition of ATP-binding P-glycoprotein by pluronics correlates with the observed ATP depletion in multidrug-resistant tumour cells observed upon exposure to pluronics. Amphiphilic block copolymers like pluronics P85 and P123 inhibited pulmonary metastasis of 4T1 mammary tumours in vitro and in vivo. Pluronic L61 also influenced doxorubicin distribution within the cells. While doxorubicin remained trapped in cytoplasmic vesicles, incubation with doxorubicin-L61 resulted in drug release and a shift toward the nucleus. Polymers with lower hydrophilic-lipophilic balance act as membrane permeabilizers, and more hydrophilic polymers restore damaged cell membranes (Jarak et al., 2020). Thus, selecting appropriate surfactants for micelle may also help provide synergistic anticancer activity (Patel et al., 2023).

The controlled release can be achieved using stimuli-responsive micelles like thermo-responsive, pH-responsive, redox-responsive, or enzyme-responsive systems (Qiu et al., 2021). The redox-responsive micelles are designed based on the high concentration gradient of reduced glutathione between extracellular (~0.002 mM) and intracellular (~10 mM) environments. Furthermore, there are few reports on higher reduced glutathione concentrations in tumour tissues (~4 folds) than normal tissue when evaluated in mice. As a result, the stability of micelle is increased in systemic circulation with nominal drug release while intracellularly, they disintegrate rapidly, providing burst release of the entrapped drug (Wang et al., 2013). Yin et al. (2015) developed conjugated micelles of hyaluronic acid-disulphide paclitaxel, resulting in higher drug loading of ~ 40%. Compared to non-redox-sensitive micelles, redox-sensitive micelles demonstrated higher drug release, intracellular uptake, and reduced systemic toxicity.

However, the major concern limiting the application of polymeric micelles is the instability of micelles, as they might disassemble into monomers owing to dilution. Thus, various approaches to improve micelle stability involve reducing the critical micelle concentration, covalent attachment of hydrophobic drugs or cross-linking of the core or shell part. Gao et al. (2016) reported the injectable curcumin-loaded shell-crosslinked F127 micelle dispersed in 5-fluorouracil-containing hydrogel. The shell of the F127 micelle was cross-linked with disulfide bonds, allowing for redox-responsive release. Schiff base bonds conjugated the shell-crosslinked F127 micelle to the hydrogel polymer, resulting in pH-sensitive release. 3-[4,5-dimethylthiazol-2-yl]-2,5-diphenyl tetrazolium bromide assay illustrated excellent compatibility of hydrogels with cells.

The popular approach of ligand-based targeting of anticancer drugs is also possible with micelle-based delivery. Jelonek et al. (2019) developed dual-targeted micelles of paclitaxel using folic acid and biotin. The mixed micelle of poly(L-lactide)-co-poly (ethylene glycol)-folic acid and poly(L-lactide)-co-poly (ethylene glycol)-biotin exhibited a very low critical micelle concentration value (0.001 mg/mL), thereby increasing the micellar stability by minimizing the dilution in the bloodstream. The biphasic release pattern was observed as the initial 96 hours showed faster release followed by slow release of paclitaxel for more than 500 hours. Gothwal et al. (2016) developed a hydrophilic hyaluronic acid derivative grafted with hydrophobic moiety like octadecyl, conjugated to folic acid as a ligand. The higher cytotoxicity of micelle loaded with paclitaxel was observed when evaluated in breast cancer cells (MCF-7) and lung cancer (A549) cell lines due to higher uptake through CD44 receptors.

Considering the benefits of the micellar system, the Food and Drug Administration has approved the polymeric micelles encapsulating paclitaxel. Paclitaxel possesses a lower water solubility of 0.3 μg/mL. Albumin-bound paclitaxel nanoparticles (Abraxane®) and polyethylene glycol-polylactic acid polymeric micelles (Genexol® PM) are marketed. These formulations result in greater efficacy, bioavailability, and reduced severe side effects during chemotherapy. *In vivo,* studies on human ovarian and breast cancer cells carried out in nude mice by administration of Genexol® PM at its maximum tolerated dose (three times higher than Taxol) showed a reduction in tumour growth after 48 hours and 14 days, respectively (Kim et al., 2001).

22.2.3 Vesicular Drug Delivery

Vesicular drug delivery systems have recently gained wide usage for its versatility. It can be used for drug-targeting controlled and sustained-release formulations. They are high-order self-assemblies of amphiphilic molecules resulting in one or more concentric lipid bilayers. Several types of vesicular systems are liposomes, niosomes, transferosomes, and pharmacosomes (Biju et al., 2006).

The application of natural anticancer agents has gained much appreciation in the past few years. Many reports have reported about the use of vesicular delivery of several anticancer phytochemicals like polyphenols (e.g., curcumin, quercetin, ellagic acid, and resveratrol), terpenoids (e.g., lycopene and squalene), and essential oils (e.g., carvacrol, linalool, cymene, and thymol) (Le et al., 2019 Van Kiem et al., 2017; Sarker and Nahar, 2017).

Vesicular drug delivery systems have the potential to enhance the effectiveness of phytoconstituents and overcome the challenges associated with them. They are useful in increasing solubility, stability, absorption, bioavailability, and controlled and targeted release. The targeted delivery aids in high differential uptake in the cancer cells compared to the normal tissues. They are biocompatible and biodegradable and have high cell affinity. They are especially favourable for the delivery of anticancer drugs due to their enhanced permeation and retention effect. Along with its appended advantages, it also suffers from the disadvantage of short blood circulation time and lack of physical and biological stability (Kumar et al., 2022, 2023).

22.2.4 Liposomes

Liposomes possess the unique feature of encapsulating both water-soluble and lipid-soluble drugs. They are well known for their high bioavailability and intracellular uptake (Banerjee & Sen, 2024). Zaremba-Czogalla et al. (2020) formulated liposomes composed of soybean phosphatidylcholine and 1,2-distearoyl-sn-glycero-3-phosphoethanolamine-*N*-[amino (polyethylene glycol)-2000] entrapping caffeic acid derivatives for the treatment of pancreatic cancer. Caffeic acid is a phenolic compound primarily found in coffee. Molecular mechanisms underlying caffeic acid's anticancer effect are connected with decreasing cell proliferation (antiproliferative effect), increasing intracellular reactive oxygen species levels, changing mitochondrial membrane potential, lipid peroxidation, and induction of apoptosis. Caffeic acid derivatives have also been considered selective matrix metalloproteinases-9 and matrix metalloproteinases-2 inhibitors, antiangiogenic activators, and DNA methylation suppressors. The formulated liposomes were evaluated for physical parameters, hemolytic and cytotoxic activity, and cellular uptake. Favourable results indicated a promising application of liposomes as drug delivery vehicles for caffeic acid derivatives in treating pancreatic cancer.

Thangapazham et al. (2008) formulated liposomes containing curcumin and coated them with prostate membrane-specific antigen-specific antibodies to treat prostate cancer. The formulations were evaluated in the human prostate cancer cell line (LNCaP). The PEGylation of curcumin resulted in significant anticancer activity compared to free curcumin.

Watanabe et al. (2008) formulated PEGylated liposomes entrapping camptothecin. PEGylated form increased the solubility and stability of camptothecin compared to conventional liposomes. The tumour growth in mice with colon adenocarcinoma was inhibited significantly upon treatment with the liposomal formulation compared to the control camptothecin solution.

Talking about innovations in liposomes, light-induced liposome technology has been explored as a novel tool for spatiotemporal control of cargo release. They enhanced the therapeutic efficacy and minimized the side effects of the therapeutic substance in comparison to conventional liposomes (Agiba et al., 2024). For a rational design of light-sensitive liposomes, some parameters have to be considered, like membrane permeabilization, which is influenced by the molar ratio of unsaturated lipids in liposomes, and lipid polymerization is another property to be considered. The hydrocarbon chain length and degree of unsaturation are important factors that govern the photopolymerization properties of liposomes. All these parameters must be properly assessed and used for the formulation of liposomes as they could affect the stability, payload capacity, light-triggered release profile, and eventually the therapeutic performance (Chen et al., 2021).

Reports have shown liposomes loaded with indocyanin green and immune stimulatory molecule polyinosinic: polycytidylic acid for cancer therapy. The light-induced photothermal effect from indocyanine green destabilized the liposome structure, triggering the fast burst of polyinosinic: polycytidylic acid and promoting cancer cell apoptosis in the colorectal carcinoma (CT-26) and melanoma tumour (B16) models when irradiated with 808 nm laser. Thus, this approach offers controlled payload release at the target site, reduced toxicity to other healthy tissues, and increased efficacy due to precise light illumination control. Therefore, this approach can deliver natural drugs and their derivatives for any treatment, especially cancer (Chen et al., 2020b).

The antitumour activity of artemisinin derivatives is increasingly gaining attention, but the lack of tumour-targeting properties is becoming a hurdle in successful formulation of artemisinin derivatives. Liposomes were explored in the context where alkyl glycoside-modified dihydroartemisinin liposomes were prepared. The glucose group in alkyl glycoside acts as the GLUT1-targeting moiety on liposomes. These targeted liposomes resulted in better targeting ability to human liver carcinoma cells (HepG2). Such findings provide fundamentals for developing artemisinin derivatives as antitumour nano-agents (Shen et al., 2020).

Zhao et al. (2020b) developed curcumin liposomes to treat small-cell lung cancer. The solubility limitation of curcumin was surpassed by its dissolution in the ethanoic core, which resulted in an 11.84-fold increase in bioavailability, and the tumour growth inhibition rate was noted to be ~64%.

22.2.5 Niosomes or Nonionic Surfactant Vesicles

Unlike phospholipids used in liposomes, niosomes are comprised of nonionic surfactants. The surfactants with high critical packing parameters (0.5–1.0) result in the bi-layered vesicular structure where the assembly of hydrophobic tails of surfactant monomer forms the bilayer, and the hydrophilic head group is in contact with the aqueous space located in the centre and outer media. Adding cholesterol helps develop an ordered liquid phase, providing rigidity to the bilayer and minimizing leakage. Niosomes have unique advantages over liposomes in terms of stability. Niosomes are quite stable structures, even in the emulsified form (Bhardwaj et al., 2020; Kumar & Rajeshwarrao, 2011). They require no special conditions, such as low temperature or inert atmosphere, for protection or storage, and are chemically stable. Moreover, the low cost of surfactants makes this formulation a choice of material for industrial manufacturers. Various non-ionic surfactants reported to prepare vesicles are glucosyl dialkyl ethers, polyglycerol alkyl ethers, ester-linked surfactants, crown ethers, polyoxyethylene alkyl ethers, Brij, and a series of spans and tweens (Baillie et al., 1986; Hunter et al., 1988).

Drug delivery of anticancer drugs can be targeted using niosomes either through passive, physical or active methods. Various research works have been published for anticancer delivery of natural drugs using niosomes like oral delivery of paclitaxel using surfactants like tweens and spans (Psimadas et al., 2012), PEGylated niosomes of gambogenic acid formulated with span 60, cholesterol, and dicetyl phosphate by ethanol injection method (Lin et al., 2013). A group of researchers developed paclitaxel and curcumin-loaded cationic niosomes to treat breast carcinoma (Mehta, 2020).

22.2.6 Dendrimers

Dendrimers are monodispersed, nano-sized, and radially symmetric hyperbranched molecules with host-guest and molecular weight traits. The three-dimensional structure of a dendrimer consists of a central core, branching units, terminal functionalities and surface or end groups that interact with the drug. Depending on the layer of branching units, dendrimers are classified into 'generation' (G*n*) (where *n* is generally 1, 2, 3, 4) composed of a regular repeating of branching units attached to the core. Various approaches are reported for drug delivery using dendrimers, such as prodrug dendrimer formation using the covalent drug attachment to the terminal groups of dendrimers. Ionic interactions of outer functional groups with the drug is another host-guest interaction approach, encapsulating the drug in unimolecular micelle dendrimer (Abbasi et al., 2014).

The advantages of dendrimers for drug delivery include increased solubility, improved delivery profiles, reduced non-specific toxicities, enhanced stability and bioavailability and targeted delivery of drugs (Mignani et al., 2018). The 10-hydroxy camptothecin was encapsulated in carboxylated poly (glycerol succinic acid) dendrimer and a PEG_{3400}-based tri-block of two poly(glycerol succinic acid) where tri-block polymer was found to increase the water solubility of 10-hydroxy camptothecin by 20 folds (Santos Miranda et al., 2006). Rompicharla et al. (2019) developed biotin-tagged PEGylated poly(amidoamine) dendrimers of paclitaxel for lung cancer. In-vitro cytotoxicity study indicated greater cytotoxicity with poly(amidoamine)-paclitaxel-polyethylene glycol-biotin with a cell viability of 39.35% ± 2.6% whereas the cell viability with poly(amidoamine)-paclitaxel-polyethylene glycol was ~50.5% and free PTX displayed a cellular viability of ~68.9%. The functionalized biotin ligand increased the targeting of the cancer cells, thereby providing superior penetration, cytotoxicity, and tumour growth inhibition.

The pH-sensitive dendrimers are the next-generation dendrimers that can release the drug at the acidic pH of the tumour (pH 5–6). Various pH-responsive dendrimers functionalized with β-carboxylic amides like succinic anhydride, 1,2-dicarboxylic-cyclohexene anhydride, or 2,3-dimethyl-maleic anhydride containing camptothecin were prepared for the treatment of ovarian cancer (Behera & Padhi, 2022; Ma et al., 2024).

Another study reported the use of poly(3-caprolactone) dendrimer containing quercetin. The results showed increased drug release with decreased pH, which is responsible for controlled release at tumour pH. An in-vitro drug release study for 8 hours demonstrated 38% release by bis-hydroxylated core-based poly(3-caprolactone) nanocarriers with 12 arms, and a similar dendrimer with 8 arms had 28% drug release. Similar results were observed with tris-hydroxylated core-based poly(3-caprolactone) dendrimer with 12 arms (50% in 8 hours), which showed faster release than other analogues (Sedghi et al., 2013).

Alibolandi et al. reported the poly(amidoamine) dendrimer encapsulating curcumin, mucin-1 aptamer (targeting moiety), and gold nanoparticles (diagnostic agent) for colon adenocarcinoma (Alibolandi et al., 2018). The formulated dendrimers demonstrated good X-ray attenuation and a high therapeutic index against colorectal adenocarcinoma.

22.3 CONCLUSIONS

Nano-drug delivery systems are designed using cutting-edge technology to maximize treatment efficacy and reduce off-target accumulation in the body by speeding systemic drug delivery to the precise target site. They thus have a significant impact on the management and treatment of malignancies. These are more advantageous than traditional ones because of their improved efficacy, automation, performance, and precision. They comprise multifunctional nanomaterials or tiny components with long circulating half-lives that are biocompatible, biodegradable, and have high viscoelasticity. A nano-drug delivery technology like this could increase the use of herbal medicines in the future. More studies are necessary to address the safety concerns associated with these technologies.

REFERENCES

Abbasi, E., Aval, S. F., Akbarzadeh, A., Milani, M., Nasrabadi, H. T., Joo, S. W., Hanifehpour, Y., Nejati-Koshki, K., & Pashaei-Asl, R. (2014). Dendrimers: synthesis, applications, and properties. *Nanoscale Research Letters*, *9*(1), 247.

Agiba, A. M., Arreola-Ramírez, J. L., Carbajal, V., & Segura-Medina, P. (2024). Light-responsive and dual-targeting liposomes: From mechanisms to targeting strategies. *Molecules*, *29*(3), 636.

Alibolandi, M., Hoseini, F., Mohammadi, M., Ramezani, P., Einafshar, E., Taghdisi, S. M., Ramezani, M., & Abnous, K. (2018). Curcumin-entrapped MUC-1 aptamer targeted dendrimer-gold hybrid nanostructure as a theranostic system for colon adenocarcinoma. *International Journal of Pharmaceutics*, *549*(1–2), 67–75.

Baillie, A. J., Coombs, G. H., Dolan, T. F., & Laurie, J. (1986). Nonionic surfactant vesicles, niosomes, as a delivery system for the anti-leishmanial drug, sodium stibogluconate. *Journal of Pharmacy and Pharmacology*, *38*(7), 502–505.

Banerjee, T., & Sen, K. (2024). Liposomal encapsulation of different anticancer drugs: an effective drug delivery technique. *BioNanoScience*, *14*(3), 3476–3495.

Behera, A., & Padhi, S. (2022). pH-Sensitive polymeric nanoparticles for cancer treatment. In: Padhi, S., Behera, A., Lichtfouse, A. (Eds.) *Polymeric Nanoparticles for the Treatment of Solid Tumours*, Springer Nature, pp. 401–425.

Bhardwaj, P., Tripathi, P., Gupta, R., & Pandey, S. (2020). Niosomes: a review on niosomal research in the last decade. *Journal of Drug Delivery Science and Technology*, *56*, 101581.

Biju, S., Talegaonkar, S., Mishra, P., & Khar, R. (2006). Vesicular systems: An overview. *Indian Journal of Pharmaceutical Sciences*, *68*(2), 141–153.

Brown, P., Butts, C. P., Eastoe, J., Glatzel, S., Grillo, I., Hall, S. H., Rogers, S., & Trickett, K. (2012). Microemulsions as tunable nanomagnets. *Soft Matter*, *8*(46), 11609–11612.

Chen, S., Wu, J., Tang, Q., Xu, C., Huang, Y., Huang, D., Luo, F., Wu, Y., Yan, F., Weng, Z., & Wang, S. (2020a). Nano-micelles based on hydroxyethyl starch-curcumin conjugates for improved stability, antioxidant and anticancer activity of curcumin. *Carbohydrate Polymers*, *228*, 115398.

Chen, T., Tu, L., Wang, G., Qi, N., Wu, W., Zhang, W., & Feng, J. (2020b). Multifunctional chitosan polymeric micelles as oral paclitaxel delivery systems for enhanced bioavailability and antitumour efficacy. *International Journal of Pharmaceutics*, *578*, 119105.

Chen, W., Deng, W., & Goldys, E. M. (2021). Recent developments of light: triggered liposome nanosystems for cancer treatments. In: Lúcio, M., Lopes, C. M., Maria Elisabete, C. D., Oliveira, R. (Eds.) *Functional Lipid Nanosystems in Cancer*, Jenny Stanford Publishing, pp. 485–511.

Chen, W., Goldys, E. M., & Deng, W. (2020c). Light-induced liposomes for cancer therapeutics. *Progress in Lipid Research*, *79*(July), 101052.

Enrico, C. (2019). Nanotechnology-based drug delivery of natural compounds and phytochemicals for the treatment of cancer and other diseases. *Studies in Natural Products Chemistry*, *62*, 91–123.

Gao, N., Lü, S., Gao, C., Wang, X., Xu, X., Bai, X., Feng, C., & Liu, M. (2016). Injectable shell-crosslinked F127 micelle/hydrogel composites with pH and redox sensitivity for combined release of anticancer drugs. *Chemical Engineering Journal*, *287*, 20–29.

Gothwal, A., Khan, I., & Gupta, U. (2016). Polymeric micelles: recent advancements in the delivery of anticancer drugs. *Pharmaceutical Research*, *33*(1), 18–39.

Hou, W., & Xu, J. (2016). Surfactant-free microemulsions. *Current Opinion in Colloid and Interface Science*, *25*, 67–74.

Huang, H. C., Pigula, M., Fang, Y., & Hasan, T. (2018). Immobilization of photo-immunoconjugates on nanoparticles leads to enhanced light-activated biological effects. *Small*, *14*(31), 1–11.

Hunter, C. A., Dolan, T. F., Coombs, G. H., & Baillie, A. J. (1988). Vesicular systems (niosomes and liposomes) for delivery of sodium stibogluconate in experimental murine visceral leishmaniasis. *The Journal of Pharmacy and Pharmacology*, *40*(3), 161–165.

Jarak, I., Varela, C. L., Tavares da Silva, E., Roleira, F. F. M., Veiga, F., & Figueiras, A. (2020). Pluronic-based nanovehicles: recent advances in anticancer therapeutic applications. *European Journal of Medicinal Chemistry*, *206*, 112526.

Jelonek, K., Zajdel, A., Wilczok, A., Latocha, M., Musiał-Kulik, M., Foryś, A., & Kasperczyk, J. (2019). Dual-targeted biodegradable micelles for anticancer drug delivery. *Materials Letters*, *241*, 187–189.

Jin, H., Pi, J., Zhao, Y., Jiang, J., Li, T., Zeng, X., Yang, P., Evans, C. E., & Cai, J. (2017). EGFR-targeting PLGA-PEG nanoparticles as a curcumin delivery system for breast cancer therapy. *Nanoscale*, *9*(42), 16365–16374.

Khan, M. N., Haggag, Y. A., Lane, M. E., McCarron, P. A., & Tambuwala, M. M. (2018). Polymeric nano-encapsulation of curcumin enhances its anticancer activity in breast (MDA-MB231) and lung (A549) cancer cells through reduction in expression of HIF-1α and nuclear p65 (Rel A). *Current Drug Delivery*, *15*(2), 286–295.

Kim, S. C., Kim, D. W., Shim, Y. H., Bang, J. S., Oh, H. S., Wan Kim, S., & Seo, M. H. (2001). In vivo evaluation of polymeric micellar paclitaxel formulation: toxicity and efficacy. *Journal of Controlled Release: Official Journal of the Controlled Release Society*, *72*(1–3), 191–202.

Koli, A. R., Ranch, K. M., Patel, H. P., Parikh, R. K., Shah, D. O., & Maulvi, F. A. (2021). Oral bioavailability improvement of felodipine using tailored microemulsion: surface science, ex vivo and in vivo studies. *International Journal of Pharmaceutics*, *596*, 120202.

Kumar, G. P., & Rajeshwarrao, P. (2011). Nonionic surfactant vesicular systems for effective drug delivery: an overview. *Acta Pharmaceutica Sinica B*, *1*(4), 208–219.

Kumar, G., Virmani, T., Sharma, A., & Pathak, K. (2023). Codelivery of phytochemicals with conventional anticancer drugs in form of nanocarriers. *Pharmaceutics*, *15*(3), 889.

Kumar, P., Yadav, N., Chaudhary, B., Jain, V., Balaramnavar, V. M., Alharbi, K. S., … & Gupta, M. M. (2022). Promises of phytochemical-based nano drug delivery systems in the management of cancer. *Chemico-Biological Interactions*, *351*, 109745.

Lai, H., Ding, X., Ye, J., Deng, J., & Cui, S. (2021). pH-responsive hyaluronic acid-based nanoparticles for targeted curcumin delivery and enhanced cancer therapy. *Colloids and Surfaces B: Biointerfaces*, *198*(June), 111455.

Le, N. T. T., Pham, L. P. T., Nguyen, D. H. T., Le, N. H., Tran, T. V., Nguyen, C. K., & Nguyen, D. H. (2019). Liposome-based nanocarrier system for phytoconstituents. In: *Novel Drug Delivery Systems for Phytoconstituents*. Edited By Madhu Gupta, Durgesh Nandini Chauhan, Vikas Sharma, Nagendra Singh Chauhan, CRC Press, pp. 45–68.

Lin, T., Fang, Q., Peng, D., Huang, X., Zhu, T., Luo, Q., Zhou, K., & Chen, W. (2013). PEGylated nonionic surfactant vesicles as drug delivery systems for Gambogenic acid. *Drug Delivery*, *20*(7), 277–284.

Lu, D. Y., Lu, T. R., Yarla, N. S., Lu, Y., Che, J. Y., Ding, J., … & Wu, H. Y. (2020). Natural drug cancer treatments, strategies from herbal medicine to chemical or biological drugs. *Studies in Natural Products Chemistry*, *66*, 91–115.

Ma, B., Li, Q., Mi, Y., Zhang, J., Tan, W., & Guo, Z. (2024). pH-responsive nanogels with enhanced antioxidant and antitumor activities on drug delivery and smart drug release. *International Journal of Biological Macromolecules*, *257*, 128590.

Majolo, F., Delwing, L. K. D. O. B., Marmitt, D. J., Bustamante-Filho, I. C., & Goettert, M. I. (2019). Medicinal plants and bioactive natural compounds for cancer treatment: important advances for drug discovery. *Phytochemistry Letters*, *31*, 196–207.

Mehta, S. (2020). Anticancer drugs targeting using nanocarriers; niosomes: a review. *Traditional Medicine Research*, *3*, 169–174.

Mignani, S., Rodrigues, J., Tomas, H., Zablocka, M., Shi, X., Caminade, A. M., & Majoral, J. P. (2018). Dendrimers in combination with natural products and analogues as anti-cancer agents. *Chemical Society Reviews*, *47*(2), 514–532.

Natesan, S., Sugumaran, A., Ponnusamy, C., Jeevanesan, V., Girija, G., & Palanichamy, R. (2014). Development and evaluation of magnetic microemulsion: tool for targeted delivery of camptothecin to BALB/c mice-bearing breast cancer. *Journal of Drug Targeting*, *22*(10), 913–926.

Patel, V., Parekh, P., Khimani, M., Yusa, S. I., & Bahadur, P. (2023). Pluronics® based Penta block copolymer micelles as a precursor of smart aggregates for various applications: a review. *Journal of Molecular Liquids*, *372*, 121140.

Psimadas, D., Georgoulias, P., Valotassiou, V., & Loudos, G. (2012). Molecular nanomedicine towards cancer. *Journal of Pharmaceutical Sciences*, *101*(7), 2271–2280.

Qiu, N., Du, X., Ji, J., & Zhai, G. (2021). A review of stimuli-responsive polymeric micelles for tumor-targeted delivery of curcumin. *Drug Development and Industrial Pharmacy*, *47*(6), 839–856.

Raval, A., Pillai, S. A., Bahadur, A., & Bahadur, P. (2017). Systematic characterization of Pluronic® micelles and their application for solubilization and in vitro release of some hydrophobic anticancer drugs. *Journal of Molecular Liquids*, *230*, 473–481.

Rompicharla, S. V. K., Kumari, P., Bhatt, H., Ghosh, B., & Biswas, S. (2019). Biotin functionalized PEGylated poly(amidoamine) dendrimer conjugate for active targeting of paclitaxel in cancer. *International Journal of Pharmaceutics*, *557*(December 2018), 329–341.

Santos Miranda, M. E., Marcolla, C., Rodriguez, C. A., Wilhelm, H. M., Sierakowski, M. R., BelleBresolin, T. M., & Alves de Freitas, R. (2006). I. The role of N-carboxymethylation of chitosan in the thermal stability and dynamic. *Polymer International*, *55*(September 2006), 961–969.

Sarker, S. D., & Nahar, L. (2017). Progress in the chemistry of naturally occurring coumarins. *Progress in the Chemistry of Organic Natural Products*, *106*, 241–304.

Sedghi, R., Oskooie, H. A., Heravi, M. M., Nabid, M. R., & Zarnani, A. H. (2013). Divergent synthesis of dendrimer-like pH-responsive macromolecules through a combination of ATRP and ROP for controlled release of anticancer drug. *Journal of Materials Chemistry B*, *1*(6), 773–786.

Shah, V., Bharatiya, B., Patel, V., Mishra, M. K., Shukla, A. D., & Shah, D. O. (2019). Interaction of salicylic acid analogues with Pluronic® micelles: investigations on micellar growth and morphological transition. *Journal of Molecular Liquids*, *277*, 563–570.

Shah, Vidhi, Sharma, M., Pandya, R., Parikh, R. K., Bharatiya, B., Shukla, A., & Tsai, H. C. (2017). Quality by design approach for an in situ gelling microemulsion of Lorazepam via intranasal route. *Materials Science and Engineering C*, *75*, 1231–1241.

Sharma, M., Shah, V., Pandya, R., Parikh, R., Bharatiya, B., Suhagia, B., & Shukla, A. (2015). Solid lipid nanospheres for Atrovastatin calcium: formulation design, optimization and drug release dynamics. *Pharmaceutical Nanotechnology*, *3*(3), 188–204.

Shen, S., Du, M., Liu, Q., Gao, P., Wang, J., Liu, S., & Gu, L. (2020). Development of GLUT1-targeting alkyl glucoside-modified dihydroartemisinin liposomes for cancer therapy. *Nanoscale*, *12*(42), 21901–21912.

Shi, X., Yang, X., Liu, M., Wang, R., Qiu, N., Liu, Y., Yang, H., Ji, J., & Zhai, G. (2021). Chondroitin sulfate-based nanoparticles for enhanced chemo-photodynamic therapy overcoming multidrug resistance and lung metastasis of breast cancer. *Carbohydrate Polymers*, *254*(May 2020), 117459.

Thangapazham, R. L., Puri, A., Tele, S., Blumenthal, R., & Maheshwari, R. K. (2008). Evaluation of a nanotechnology-based carrier for delivery of curcumin in prostate cancer cells. *International Journal of Oncology*, *32*(5), 1119–1123.

Thapliyal, A., Khar, R. K., & Chandra, A. (2020). AgNanoparticles loaded microemulsion using gallic acid inhibits MCF-7 breast cancer cell line and solid ehrlich carcinoma. *International Journal of Polymeric Materials and Polymeric Biomaterials*, *69*(5), 292–316.

Umapathi, A., PN, N., Madhyastha, H., Singh, M., Madhyastha, R., Maruyama, M., & Daima, H. K. (2020). Curcumin and isonicotinic acid hydrazide functionalized gold nanoparticles for selective anticancer action. *Colloids and Surfaces A: Physicochemical and Engineering Aspects*, *607*(June), 125484.

Vandghanooni, S., Barar, J., Eskandani, M., & Omidi, Y. (2020). Aptamer-conjugated mesoporous silica nanoparticles for simultaneous imaging and therapy of cancer. *TrAC - Trends in Analytical Chemistry*, *123*, 115759.

Van Kiem, P., Thi Dung, D., Thi, T., Hong Quang, T., Thi Thanh Ngan, N., Minh Ha, T., ... & Van Minh, C. (2017). Constituents from Ircinia echinata and their antiproliferative effect on six human cancer cell strains. *Letters in Organic Chemistry*, *14*(4), 248–253.

Wang, H., Tang, L., Tu, C., Song, Z., Yin, Q., Yin, L., Zhang, Z., & Cheng, J. (2013). Redox-responsive, core-cross-linked micelles capable of on-demand, concurrent drug release and structure disassembly. *Biomacromolecules*, *14*(10), 3706–3712.

Wang, L. L., Huang, S., Guo, H. H., Han, Y. X., Zheng, W. S., & Jiang, J. D. (2016). In situ delivery of thermosensitive gel-mediated 5-fluorouracil microemulsion for the treatment of colorectal cancer. *Drug Design, Development and Therapy*, *10*, 2855–2867.

Wang, W., Li, X., Wang, Z., Zhang, J., Dong, X., Wu, Y., Fang, C., Zhou, A., & Wu, Y. (2019). A novel "mosaic-type" nanoparticle for selective drug release targeting hypoxic cancer cells. *Nanoscale*, *11*(5), 2211–2222.

Watanabe, M., Kawano, K., Toma, K., Hattori, Y., & Maitani, Y. (2008). In vivo antitumour activity of camptothecin incorporated in liposomes formulated with an artificial lipid and human serum albumin. *Journal of Controlled Release*, *127*(3), 231–238.

Wong, K. H., Lu, A., Chen, X., & Yang, Z. (2020). Natural ingredient-based polymeric nanoparticles for cancer treatment. *Molecules*, *25*(16), 3620.

Yin, T., Wang, J., Yin, L., Shen, L., Zhou, J., & Huo, M. (2015). Redox-sensitive hyaluronic acid-paclitaxel conjugate micelles with high physical drug loading for efficient tumour therapy. *Polymer Chemistry*, *6*(46), 8047–8059.

Zaremba-Czogalla, M., Jaromin, A., Sidoryk, K., Zagórska, A., Cybulski, M., &Gubernator, J. (2020). Evaluation of the in vitro cytotoxic activity of caffeic acid derivatives and liposomal formulation against pancreatic cancer cell lines. *Materials*, *13*(24), 1–18.

Zhang, Y., & Feng, Y. (2020). Stimuli-responsive microemulsions: state-of-the-art and future prospects. *Current Opinion in Colloid and Interface Science*, *49*, 27–41.

Zhang, Y., Guo, S., Ren, X., Liu, X., & Fang, Y. (2017). CO_2 and redox dual responsive pickering emulsion. *Langmuir*, *33*(45), 12973–12981.

Zhao, H., Chen, M., Zhao, Z., Zhu, L., & Yuan, S. (2020a). A multicomponent-based microemulsion for boosting ovarian cancer therapy through dual modification with transferrin and SA-R6H4. *Drug Delivery and Translational Research*, *11*, 1969–1982.

Zhao, J., Li, Y., He, D., Hu, X., Li, K., Yang, Q., Fang, C., Zhong, C., Yang, J., Tan, Q., & Zhang, J. (2020b). Natural oral anticancer medication in small ethanol nanosomes coated with a natural alkaline polysaccharide. *ACS Applied Materials and Interfaces*, *12*(14), 16159–16167.

Zhou, Z., Ma, X., Murphy, C. J., Jin, E., Sun, Q., Shen, Y., Vankirk, E. A., & Murdoch, W. J. (2014). Molecularly precise dendrimer-drug conjugates with tunable drug release for cancer therapy. *Angewandte Chemie-International Edition*, *53*(41), 10949–10955.

23 Phytomolecules Targeting Resistant Cancer Cells through Nano-Delivery Systems

Sopan N. Nangare[1], *Sameer Nadaf*[2], *Preeti Karade*[3], *Pranali Takale*[4], *Prajwala Khapale*[5], *and Namdeo R. Jadhav*[1]

[1]Department of Pharmaceutics, Krishna Institute of Pharmacy, Krishna Vishwa Vidyapeeth (Deemed to be University) Karad, India

[2]Department of Pharmaceutics, Bharati Vidyapeeth College of Pharmacy, Palus, India

[3]Department of Pharmaceutics, Appasaheb Birnale College of Pharmacy, Sangli, India

[4]Department of Pharmaceutics, Bharati Vidyapeeth College of Pharmacy, Kolhapur, India

[5]Department of Pharmacology, R. C. Patel Institute of Pharmaceutical Education and Research, Shirpur, India

23.1 INTRODUCTION

Cancer is marked by the uncontrolled growth of cells, which eventually results in tumour formation. This process begins with genetic mutations in genes that regulate the cell cycle, along with somatic changes in upstream signalling pathways (Yahya and Alqadhi 2021; Brown, Amend et al. 2023). Cancer can be broadly categorized into five major types: leukemia, sarcoma, carcinoma, lymphoma, and central nervous system cancer. Its occurrence is linked to various risk factors, including smoking, obesity, poor nutrition, alcohol consumption, and viral infections. According to GLOBOCAN (Global Cancer Observatory) 2018, there were an estimated 18.1 million new cancer cases globally (17.0 million excluding non-melanoma skin cancers) and 9.6 million cancer-related deaths (9.5 million excluding non-melanoma skin cancers). Lung cancer is the most frequently diagnosed cancer in both men and women, accounting for 11.6% of all cases and 18.4% of all cancer-related deaths. Following lung cancer, breast cancer in women (11.6%), prostate cancer (7.1%), and colorectal cancer (6.1%) are among the most prevalent. In terms of mortality, colorectal cancer (9.2%), stomach cancer (8.2%), and liver cancer (8.2%) are significant contributors. Among younger individuals, leukaemia, brain tumours, and lymphomas are the most common cancers (Hassanpour and Dehghani 2017; Bray, Ferlay et al. 2018; WHO 2020). Recent estimates indicate that by 2022, there will be 20 million new cancer cases and 9.7 million cancer-related deaths globally. Approximately 53.5 million people are projected to be living 5 years after a cancer diagnosis. It is anticipated that one in five people will develop cancer at some point in their lives, with one in nine men and one in 12 women expected to die from the disease (Zubair, Wang et al. 2021; Bray, Laversanne et al. 2024). Fowler and colleagues have emphasized that many cancer patients suffer from multiple comorbidities (Fowler, Belot et al. 2020). Additionally, a study involving 276 patients with superficial vein thrombosis (SVT) found a cancer prevalence rate of 8.7% (Hirmerová, Seidlerová et al. 2022). Ma et al. reported that 52% of individuals diagnosed with cancer experience fatigue. Notably, low-performance status, concurrent radiation and chemotherapy, female sex, sleep deprivation, neuroticism, pain, and depressive disorders have been identified as potential risk factors for cancer-related fatigue (Urruticoechea, Alemany et al. 2010; Ma, He et al. 2020). Depending on the type and stage of cancer, treatment options may include chemotherapy, hormone therapy, surgery, biomarker testing, hyperthermia, immunotherapy, photodynamic therapy (PDT), radiation therapy, stem cell transplantation, and targeted therapy (Figure 23.1) (Urruticoechea, Alemany et al. 2010; Chu, Nguyen et al. 2020).

Chemotherapy is a widely used cancer treatment that employs chemicals to destroy cancer cells by primarily inhibiting their growth and division. Because cancer cells reproduce and grow at a much faster rate than normal cells, they experience higher levels of stress, which makes them more vulnerable to the effects of chemotherapy. Additionally, combination therapy is emerging as a promising approach for treating various tumours. This strategy involves combining chemotherapy with targeted drug delivery, immunotherapy, and personalized medicine to enhance therapeutic outcomes (Chu, Nguyen et al. 2020). Despite the advancements, several promising inhibitor therapies are being explored for the treatment of solid tumours. These include poly(adenosine diphosphate-ribose) polymerase (PARP) inhibitors, p53/mouse double minute 2 homolog (MDM2) inhibitors, hedgehog pathway blockers, angiogenesis inhibitors, tyrosine kinase inhibitors, mechanistic target of rapamycin (mTOR) inhibitors, histone deacetylase (HDAC) inhibitors, and proteasome inhibitors (Anand, Dey et al. 2023).

DOI: 10.1201/9781003611516-29

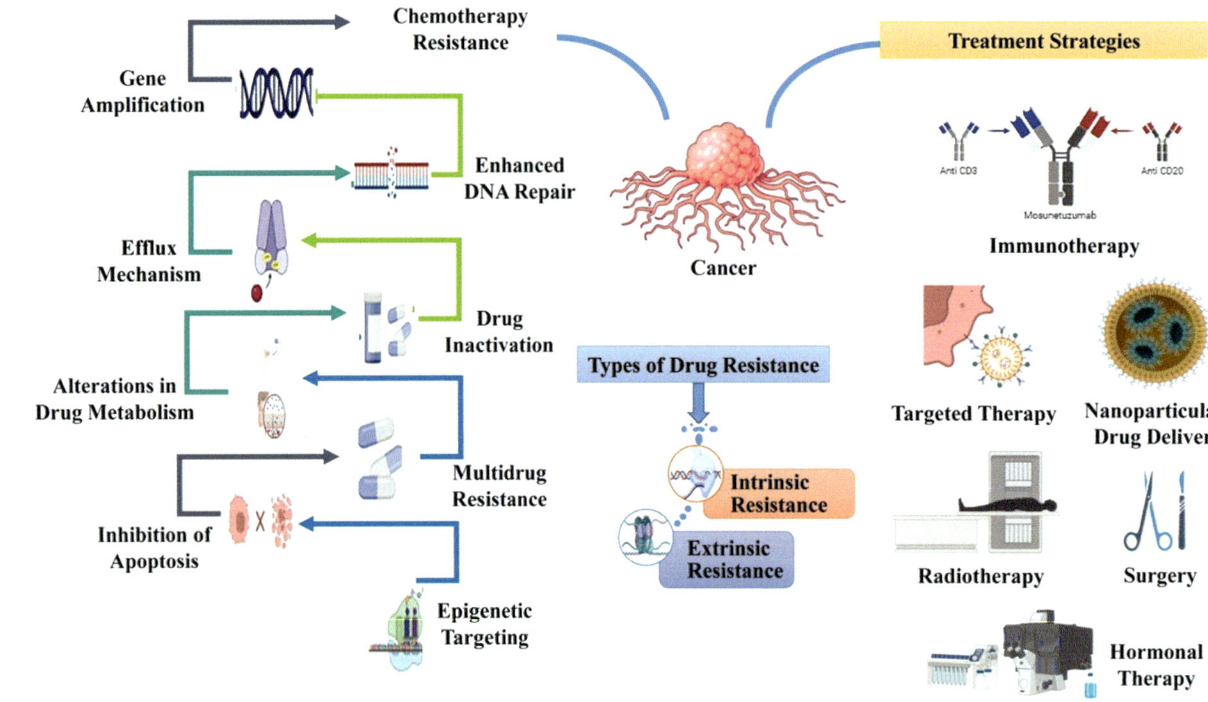

FIGURE 23.1 Types of drug resistance and treatment strategies.

Surgery is a standard cancer treatment involving the removal of tumours from the body. Additionally, biomarker testing plays a crucial role in identifying genes, proteins, and other substances (known as biomarkers or tumour markers) that offer valuable insights into the cancer's characteristics and progression. Hormone therapy, another treatment option, works by slowing or stopping the progression of certain cancers. Research indicates that appropriately selected hormone replacement therapy can be beneficial for some cancers, such as endometrial cancer, hematologic malignancies, and colorectal carcinoma in females (Deli, Orosz et al. 2020). Hyperthermia is another treatment modality, where body tissues are heated to temperatures as high as 113°F to disrupt and kill cancer cells, with minimal damage to normal tissues. Immunotherapy is also a crucial aspect of cancer treatment, which involves stimulating the immune system to combat cancer.

Alongside surgery, radiation, and chemotherapy, monoclonal antibody-based immunotherapy has become a vital component of cancer therapy. These antibodies can specifically target tumour cells and elicit long-lasting immune responses. The distinct properties of antibodies as therapeutic agents have paved the way for innovative cancer treatment strategies, which are anticipated to make a significant impact on cancer care (Zugazagoitia, Guedes et al. 2016; Zahavi and Weiner 2020). Data from 300 studies involving over 70,000 patients were analysed to determine the relevance of 15 immune cell types: CD8+ (cytotoxic) T lymphocytes, M1 macrophages, macrophages, M2 macrophages, tertiary lymphoid structures (TLSs), T helper 1 (TH1) cells, regulatory T cells (Treg), T helper 17 (TH17) cells, T helper 2 (TH2) cells, T follicular helper (TFH) cells, B cells, and natural killer (NK) cells (Bruni, Angell et al. 2020). Here, the PDT is a technique that destroys cancer and other abnormal cells by activating a photosensitizing drug with light. In this, PDT has been recognized for over a century and is considered an alternative to traditional cancer treatments (Li, Lovell et al. 2020; Ostańska, Aebisher et al. 2021). Radiation therapy is utilized both as a curative and palliative treatment. It involves the use of high doses of radiation to destroy cancer cells and reduce the size of solid tumours (Wang and Tepper 2021). Another treatment approach involves stem cell transplants, which replace damaged embryonic stem cells that develop into blood cells in patients who have undergone high-dose chemotherapy or radiation therapy.

Many chemotherapeutic agents have low water solubility, and a significant number of these drugs fall into BCS (Biopharmaceutics Classification System) classes III or IV. Their poor water solubility can lead to reduced bioavailability and increased systemic toxicity (Ghosh and Biswas 2021). Chemotherapy drugs often have limited selectivity, leading to significant damage to healthy tissues and causing severe side effects. In addition to these challenges, the early detection of cancer remains a significant hurdle. Despite advancements in treatment methods, patients still face poor long-term survival rates and a high risk of adverse reactions from chemotherapy. As a result, targeted therapy is increasingly utilized, tailored to the disease's pathological stages and clinical indicators. This includes novel anticancer agents that work by inhibiting angiogenesis, disrupting signal transduction pathways, regulating gene transcription, and inducing apoptotic dysfunction in the cell cycle (Pistritto, Trisciuoglio et al. 2016; Gomes, Martins et al. 2021). In conclusion, an increasing number of studies suggest that new technologies, particularly those based on nanotechnology, may help reduce many of these adverse effects (Piktel, Niemirowicz et al. 2016; Klochkov, Neganova et al. 2021).

23.2 RESISTANCE TO CANCER

In recent decades, significant progress has been made in cancer treatment. Common therapies to combat cancer include surgery, chemotherapy, immunotherapy, and radiotherapy, among others (Zhang and Chen 2018). Unfortunately, despite these advancements, drug resistance remains a major obstacle to the successful eradication of cancer. Chemoresistance is a major cause of treatment failure and contributes significantly to cancer-related deaths (Elgendy, Alyammahi et al. 2020). Resistance to chemotherapy results in an inadequate response to chemotherapeutic agents, diminishing the effectiveness of the treatment. This resistance is associated with the evolution of tumour cells into more aggressive forms, which is a primary reason for cancer-related mortality. Nearly nine out of ten cancer deaths are attributed to this phenomenon (Rezayatmand, Razmkhah et al. 2022).

Despite extensive efforts, the complete eradication of cancer remains challenging due to the unique characteristics of tumours that enable them to evade therapy. Drug resistance can be categorized into two main types: intrinsic resistance and extrinsic (or acquired) resistance (Figure 23.1). Intrinsic resistance refers to the pre-existing factors within cancer cells and tissues that reduce the effectiveness of chemotherapeutic agents even before treatment starts. In contrast, extrinsic or acquired resistance develops during treatment in tumours that were initially responsive to antineoplastic agents. Acquired resistance often results from various adaptive responses, such as the activation of alternative signalling pathways and increased expression of the therapeutic target. Tumours exhibit high adaptability, deactivation of defunct signalling pathways, and the ability to escape growth inhibition, all of which contribute to resistance against anticancer drugs (Haider, Pandey et al. 2020). To overcome chemotherapy resistance, it is crucial to develop strategies that effectively target these proliferative cells and address the challenges posed by recurring, aggressive tumours (Elgendy, Alyammahi et al. 2020). Alterations in signalling pathways, driven by genetic changes, contribute significantly to the increase of resistance in cancer cells. These cells often exhibit concurrent resistance to multiple drugs, leading to the failure of combination treatments (Safhi, Sivakumar et al. 2017). When exposed to chemotherapeutic agents, cancer cells can develop resistance as well as cross-resistance to a wide array of functionally and structurally unrelated drugs (Wang, Seebacher et al. 2017).

Resistance can develop through several mechanisms, including drug inactivation, multidrug resistance (MDR), inhibition of apoptosis, alterations in drug metabolism, epigenetic modifications, enhanced DNA repair, efflux mechanisms, and gene amplification (Marques et al. 2024). Drug activation *in vivo* involves complex interactions with multiple proteins. These interactions can alter, degrade, or activate the drug. For instance, Cytarabine (AraC) is a nucleoside drug used to treat acute myelogenous leukemia. Here, it undergoes multiple phosphorylation steps to form AraC-triphosphate, its active form. Similarly, the cytochrome (CYP) system, particularly enzymes such as CYP1A1, CYP1A2, CYP2E1, and CYP3A4, plays a critical role in drug and procarcinogen metabolism. CYP1A1 and CYP1A2, for instance, convert procarcinogens into carcinogenic forms in the liver, a process essential for the metabolism of many anticancer drugs. Additionally, the UDP-glucuronosyltransferases (UGT) superfamily consists of enzymes responsible for catalysing glucuronidation, a mechanism that converts bile acids, steroids, and xenobiotics, including ecological carcinogens and cytotoxins, into inactive hydrophilic glucuronides. Modifications to a drug's molecular target, such as mutations or changes in expression levels, can also lead to treatment resistance. One of the most extensively studied mechanisms of cancer drug resistance is the reduction of drug accumulation through increased efflux, a process managed by the ATP-binding cassette (ABC) transporter family at the plasma membrane. Additionally, DNA repair mechanisms are crucial in anticancer drug resistance.

The DNA damage response (DDR) pathways can repair damage inflicted by chemotherapeutic agents, either directly or indirectly. Moreover, cell death processes, such as apoptosis and autophagy, play key regulatory roles in determining cell fate, despite their opposing functions (Housman, Byler et al. 2014; Labrie, Brugge et al. 2022; Garg, Malhotra et al. 2024). Among the various factors involved in cancer treatment, cancer stem cells (CSCs) stand out for their significant impact. CSCs are believed to play a crucial role in chemotherapy resistance, tumour recurrence, and the aggressiveness of cancer (Elgendy, Alyammahi et al. 2020). A recent study on breast cancer treatment highlighted that metastasis and drug resistance are responsible for 90% of cancer-related deaths. Consequently, targeting breast cancer stem cells (BCSCs) has become a top priority in developing more effective treatments (Muley, Fado et al. 2020). Drug resistance in breast cancer stem cells (BCSCs) is multifaceted, with mechanisms including drug efflux, avoidance of apoptosis, enhanced DNA damage repair, and alterations in drug metabolism. However, the role of impaired drug uptake in resistance remains largely unexplored. Notably, the malfunctioning PI3K/AKT pathway is crucial in regulating various cellular processes involved in Class III multidrug resistance (MDR3). Targeting this pathway and its associated proteins could potentially overcome MDR in various clinical contexts. For instance, continuous activation of the PI3K/AKT/mTORC1 pathway may be linked to changes in microRNA (miRNA) expression, which can induce epigenetic modifications contributing to sustained MDR. Although no miRNA-targeting therapies have yet reached clinical use, miRNA inhibitors may offer a way to address MDR by indirectly inhibiting the PI3K/AKT pathway.

Combination therapies targeting cancer cells have also been suggested as a strategy to improve treatment outcomes. Combining different anti-angiogenic agents could enhance efficacy, although resistance mechanisms may still develop, leading to excessive hypoxia and further resistance. Therefore, initiating clinical trials to evaluate the effectiveness and safety of these novel combination approaches is crucial. Given the complexity of these mechanisms, direct targeting strategies must be thoroughly researched (Haibe, Kreidieh et al. 2020). To effectively combat cancer resistance, it is essential to assess the tumour's physical characteristics, such as tumour burden, growth rate, and localization, as well as to analyse its drivers, vulnerabilities, dependencies, druggability, and prior detection. Proper monitoring and the use of powerful analytics based on patient databases are also necessary. Although challenging, this approach is promising and developmental (Vasan, Baselga et al. 2019).

Advances in technology, such as ultra-modern software programming and robotic platforms, have significantly contributed to the identification and treatment of cancer. In this context, immunotherapy, targeted drug delivery, and personalized medicine have become increasingly important. However, combining immunotherapy with molecularly targeted therapies, whether concurrently or sequentially, has raised concerns about potential hepatotoxicity, which must be carefully evaluated. For example, combining anti-PD-1, dabrafenib, and trametinib treatments in melanoma has been connected with a higher risk of adverse events in contrast to targeted therapy alone (Anand, Dey et al. 2023). Currently, research is expanding to explore molecular targets such as oncogenes, tumour suppressor genes, and RNA interference (RNAi) to develop new cancer treatments. These therapies focus on inhibiting kinases involved in cell growth, boosting rapid immune responses to cancer, creating specialized pharmaceuticals, delivering therapeutics directly into cancer cells, and minimizing the adverse effects of anticancer agents (Mansoori, Mohammadi et al. 2017).

This research has led to the development of nanomedicine, the medical application of nanotechnology, to provide effective and safe cancer treatments. Nanostructured medicines offer several advantages, such as overcoming resistance, improving solubility, enhancing pharmacological profiles, and minimizing the side effects of chemotherapeutic drugs. Their application in therapeutic settings has grown significantly over the past decade. Among the various nanoparticulate systems, metallic nanoparticles, polymeric nanoparticles, and hybrid nanoparticles have shown great promise as platforms for targeted therapies. These systems have the potential to revolutionize conventional medicine by minimizing side effects and enabling more precise delivery of therapeutic substances, with their small size aiding intracellular absorption (Gavas, Quazi et al. 2021; Ulldemolins, Seras-Franzoso et al. 2021). In addition, Bagheri et al. demonstrated in their research that exosomes, due to their inherently low immunogenicity, have tremendous potential as therapeutic delivery vehicles for cancer treatment. Exosomes, secreted by most cell types, are biocompatible and hold promise for use in biomanufacturing, particularly for drug delivery and cancer therapy. In a study using an ectopic C26 (mouse colon cancer) model in BALB/c mice, it was found that a single intravenous injection of doxorubicin (DOX)-loaded exosomes (DOX@exosome) significantly reduced tumour growth compared to free DOX (Bagheri, Abnous et al. 2020).

To avoid hepatotoxicity and effectively treat liver cancer, various natural compounds such as piperine (PIP), curcumin (CUR), oleocanthal, and allium extracts have been employed. Moreover, Maleki Dana et al. (2022) reported that a natural compound known as polyphenol shows promise as a potent agent for treating resistant tumours. Polyphenols help inhibit cell growth, reduce inflammation, and induce apoptosis. Additionally, they can mitigate oxidative stress, inhibit epithelial-mesenchymal transition (EMT), reduce drug metabolism and efflux, and enhance drug uptake by cancer cells. As a result, natural products combined with nanoparticulate systems are emerging as prime candidates for treating resistant tumours (Anwanwan, Singh et al. 2020).

23.3 PRECLINICAL AND CELL LINE STUDIES OF PHYTOMOLECULES AGAINST RESISTANT CANCER CELLS

Pre-clinical *in vitro* and *in vivo* models are indispensable in cancer research. They play a vital role in uncovering the molecular mechanisms behind critical processes like tumour growth, metastasis, drug resistance, and immune evasion (Katt, Placone et al. 2016). They are also crucial for screening and developing anti-cancer drugs, as only about 10% of potential anti-cancer drugs succeed in clinical trials. This high failure rate is often due to insufficient efficacy or intolerable toxicity that is identified during preclinical studies (Kola 2008). To advance precision medicine for cancer, it is essential to use pre-clinical models that faithfully replicate *in vivo* biology and microenvironmental factors, while also upholding ethical standards concerning animal welfare (Katt, Placone et al. 2016). Preclinical studies enable researchers to uncover mechanisms of drug resistance. For example, these studies have identified key resistance mechanisms such as MET gene amplification in lung cancer and mutations in the androgen receptor in prostate cancer (Michaelis, Wass et al. 2019).

Moreover, preclinical models accelerate the testing of novel therapeutic approaches. Patient-derived xenografts and organoids are particularly valuable for evaluating the efficacy of targeted therapies and immunomodulators, as they preserve tumour cell heterogeneity and are useful for assessing treatments specific to genetic profiles and resistance mechanisms (Sajjad, Imtiaz et al. 2021). These models can also predict tumour cell responses to drug treatments. For instance, patient-derived xenografts from colorectal cancer have been employed to classify patients into different molecular subtypes, aiding in the selection of targeted therapies (Durinikova, Buzo et al. 2021). Organoids, in particular, retain the interactions between cancer cells and their microenvironment. It is crucial for understanding how the tumour microenvironment contributes to drug resistance and for developing effective therapies (Sharma, Lee et al. 2010). Preclinical studies facilitate high-throughput screening of potential therapeutic molecules, allowing for the rapid identification of promising compounds for cancer treatment and the discovery of drugs that can overcome resistance mechanisms (Garg, Malhotra et al. 2024). By simulating clinical resistance mechanisms, preclinical models help validate findings from patient studies and assess the potential for similar effects in patients. This validation is critical for the development of effective treatments (Michaelis, Wass et al. 2019).

Although models such as patient-derived xenografts (PDXs) and organoids mimic human tumour cells, they cannot fully replicate the complex microenvironment of the human body. Variability between animal models and humans, including differences in genetics and tumour behavior, can lead to discrepancies in how tumours respond to treatments. For instance, mechanisms of mutations, cycles, or pathways that drive drug resistance in humans may differ from those observed in animal models. This variability can result in preclinical models that are not always representative of human tumours. In patient-derived xenograft models, genetic variations can affect study outcomes and their relevance to clinical settings. Proper clonal selection is essential to avoid the dominance of subclones, which may not accurately reflect the original tumour's heterogeneity and complexity. Additionally, developing and maintaining preclinical models can be time-consuming and costly, which may influence the speed and economic feasibility of research. Many preclinical models do not accurately predict clinical outcomes, leading to a high rate of failure in clinical trials and significant setbacks in drug development. The use of animals in research also raises ethical issues and controversies, which can lead to regulatory and public scrutiny, further affecting the pace of research (Michaelis, Wass et al. 2019; Durinikova, Buzo et al. 2021; Sajjad, Imtiaz et al. 2021).

Cancer cell line models are crucial for studying drug resistance mechanisms in humans. These models replicate the drug-resistant phenotypes observed in patients and provide valuable insights into how cancer cells tolerate chemotherapeutic agents. Key mechanisms identified through cell line studies include overexpression of drug efflux pumps, mutations in drug targets, and activation of alternative pathways (Gillet and Gottesman 2010). Cell line studies are an essential first step in the screening of new therapeutic agents. By evaluating efficacy in the early phases, these studies help identify effective compounds against resistant cancer cells (Yager 2012). They also elucidate the molecular and cellular pathways involved in drug resistance, including protein binding, genetic makeup, and signalling pathways, which are critical for developing new drugs to treat resistant cancer cells (Holohan, Van Schaeybroeck et al. 2013). Cell line studies are foundational for applicable research, bridging the gap between laboratory research and clinical applications. These studies play a crucial role in developing biomarkers for resistance, establishing the basis for clinical trials, and enhancing therapeutic strategies (Sharma, Lee et al. 2010). They help identify synergistic effects of combined drug treatments in resistant cancer cells and are essential for optimization before progressing to preclinical and clinical studies (Vila-Leahey, Oldford et al. 2016).

Cancer cell line studies are conducted under controlled conditions to ensure reproducibility across different studies and laboratories. They facilitate high-throughput screening of therapeutic agents, allowing for the screening of numerous molecules. These studies also provide detailed insights into the molecular mechanisms behind drug resistance. For example, drug-adapted cell lines have been used to uncover mechanisms such as ABC transporter-mediated drug efflux and mutations in cancer cell resistance or tumour suppressors.

Compared to animal models, cell lines are more cost-effective and easier to maintain, with minimal ethical concerns. They can be modified with specific genetic alterations to study how these changes affect drug resistance in cancer. Understanding the role of specific genes and pathways in cancer biology is crucial for developing treatments for drug-resistant cancer. Cell line experiments produce results more quickly than *in vivo* studies, accelerating research and the development of effective treatments. Patient-derived cell lines are valuable for personalized drug therapy, and resistant cell lines can serve as models to study acquired resistance in other research contexts. These models are instrumental in identifying resistance mechanisms and developing strategies to overcome resistance in various cancers (Michaelis, Wass et al. 2019; Sajjad, Imtiaz et al. 2021).

However, cell lines do not fully replicate the tumour microenvironment found in living organisms, which significantly influences drug response and resistance. Genetic changes and clonal selection in cell lines can lead to discrepancies in results. Additionally, the response of cancer cell lines to drugs often differs from clinical outcomes due to pharmacokinetic variations in the human body. Cell lines lack an immune system, making it difficult to study drugs that rely on immunotherapy (Michaelis, Wass et al. 2019; Durinikova, Buzo et al. 2021; Sajjad, Imtiaz et al. 2021). For example, colorectal cell lines have been used to investigate chemoresistance, poor prognosis, and the effects of ursolic acid on colorectal cancer growth, metastasis, and the suppression of inflammatory, metastatic, and invasive biomarkers (Prasad, Yadav et al. 2012). Taxanes, a class of promising anticancer drugs initially derived from the bark of the Yew tree, include cabazitaxel, a second-generation derivative of docetaxel. Cabazitaxel demonstrates cytotoxic activity against a range of docetaxel-resistant tumours while showing reduced overall toxicity in cell line studies (Kotsakis, Matikas et al. 2016). Podophyllotoxin is a natural compound that binds reversibly to tubulin, and its derivatives, etoposide and teniposide, inhibit topoisomerase II-mediated DNA cleavage. These compounds have shown promise against multidrug-resistant (MDR) tumour cells. For example, CIP-36, a derivative of podophyllotoxin, effectively overcomes MDR in the adriamycin-resistant human leukemic cell line K562/ADR by modulating topoisomerase IIα activity (Cao, Chen et al. 2015).

Epigenetic deregulation of gene expression contributes to acquired drug resistance and metastasis in breast cancer. Epigenetic drugs used to address these aberrant transcriptional patterns have revealed promise in sensitizing cancer cells and providing new treatment strategies. Cell line studies assessing cytotoxicity, genotoxicity, apoptosis, and migration capacity have demonstrated that combination therapies can be as effective as pegylated liposomal DOX in treating metastatic breast cancer (Talib, Alsayed et al. 2021). The CUR is a natural derivative of turmeric that exhibits significant anticancer properties. It has shown chemopreventive effects in preclinical models of carcinogenesis in the mammary gland, lung, colorectal, and hepatic tissues. It induces cell death through apoptosis and autophagy. Hence, proving its potential as a potent drug in overcoming cancer drug resistance and improving treatment outcomes (Sharma, Gescher et al. 2005; BB 2007).

Resveratrol (RES), a natural polyphenolic compound, has demonstrated cardioprotective properties, as well as benefits in combating drug-resistant cancer cells and improving metabolism. Studies have highlighted its effects in the treatment of prostate, liver, mammary, and hepatic cancers (Fulda 2010; Carter, D'Orazio et al. 2014). Quercetin (QUE) is a flavonoid found in many fruits and vegetables. It has shown promise in overcoming resistance in cancer cells. Research indicates that QUE can reverse resistance by inhibiting the function of *P*-glycoprotein, inducing cell cycle arrest, and promoting apoptosis in resistant cancer cell lines (Almatroodi, Alsahli et al. 2021; Asgharian, Tazekand et al. 2022). Berberine, a naturally occurring alkaloid, has demonstrated potential in treating drug-resistant cancers, particularly leukemia and breast cancer. It inhibits cell proliferation by targeting key signaling pathways and inducing apoptosis, suggesting that berberine could be an effective agent for managing resistant cancers (Liu, Liu et al. 2009; Liu, Nwafor et al. 2022). Epigallocatechin gallate (EGCG), a catechin present in green tea, has shown potential in modulating drug resistance in cancer cells. Here, EGCG affects resistant cancer cell lines by downregulating efflux transporters, inhibiting survival pathways, and inducing apoptosis (Yang, Chung et al. 2000; Khan, Afaq et al. 2006). Lycopene, a carotenoid found in tomatoes, also plays a role in addressing drug resistance in cancer cells. The results indicate that lycopene could be a valuable addition to cancer therapy, particularly in resistant cases, by inhibiting cell proliferation and modulating signalling pathways (Giovannucci 1999; Story, Kopec et al. 2010; Talib, Alsayed et al. 2021).

23.4 OVERVIEW OF PHYTOMOLECULES IN RESISTANT CANCER TREATMENT

Cancer treatment typically involves surgery, radiotherapy, and chemotherapy, depending on the location and stage of the tumour. Chemotherapy, which includes cytotoxic drugs, has proven effective when used in combination with other treatments. However, most anticancer drugs approved by the Food and Drug Administration (FDA) (over 300) are toxic and can induce chemoresistance in cancer cells. This resistance leads to tumour recurrence and metastasis, limiting their anticancer efficacy (Banik, Ranaware et al. 2019; Khatoon, Banik et al. 2022). Existing anticancer drugs target rapidly dividing cells, but this also results in toxic effects on non-targeted tissues. Common side effects include severe damage to normal cells, myelosuppression, anaemia due to bone marrow suppression, hair loss, gastrointestinal issues (such as constipation and diarrhoea), digestive tract infections, mucositis, immunosuppression, and neurotoxicity. Here, all of which impair the patient's quality of life. Additionally, the frequent mutation of cancer cells often results in resistance and reduced responsiveness to the drugs administered. Drug resistance can be classified as either acquired or intrinsic.

Acquired resistance develops after exposure to chemotherapy and is induced by the drug itself, while intrinsic resistance exists before drug administration, being an inherent characteristic of the cancer cells (Wang, Zhang et al. 2019). Evidence suggests that multiple mechanisms contribute to multidrug resistance (MDR) in tumour cells. These include reduced drug uptake, increased drug efflux, enhanced DNA repair capacity, altered drug metabolism, mutations or changes in tubulin, modifications in microtubule-associated proteins, suppression of apoptosis, mutated p53 protein, and genetic factors such as gene mutations, amplifications, and epigenetic modifications (Zhang, Lu et al. 2020). These mechanisms can operate independently or in concert. While numerous reversal agents have been tested for their ability to combat MDR cancers, many have demonstrated limited therapeutic potential due to their high toxicity at the doses needed to reverse MDR, as seen with verapamil (Cao, Chen et al. 2015). Consequently, there is a growing focus on developing new anticancer drugs that are multitargeted, more effective, and have fewer side effects to address the limitations of current treatments (Dehelean, Marcovici et al. 2021).

Phytochemicals and their derivatives found in plants offer promising prospects for improving treatment and reducing side effects in cancer patients. Plant-based bioactive molecules, including phenolic compounds, steroids, organosulfides, flavonoids, alkaloids, glucosinolates, and terpenes, exhibit anticancer activities (Situmourang,

Ilyas et al. 2024). Naturally occurring chemopreventive agents can inhibit or terminate carcinogenesis at various stages. The health benefits of plant-based compounds are often linked to a reduced risk of many chronic diseases. Phytochemicals are considered safe, easily accessible, less toxic, and inexpensive, targeting multiple pathways involved in malignancies. They possess various pharmacological effects, including antimicrobial, antitumour, antimutagenic, and antioxidant or pro-oxidant activities (Moosavi, Haghi et al. 2018).

In the context of cancer, phytochemicals may strengthen the immune system, slow cancer cell growth, reduce inflammation, and prevent the replication of damaged cells. Plant-derived phytochemicals are valuable sources for new drug development and cancer treatment. These compounds are bioactive non-nutrient substances that have been linked with a lower risk of major chronic diseases. Traditional medicine practitioners use a range of herbal preparations, each with its own unique cultural and philosophical background, to treat various ailments. Plant-based natural products are non-toxic to normal cells and generally better tolerated, making them preferable. Ideally, an anticancer drug should selectively target cancer cells, and research indicates that natural compounds found in plants hold significant potential for developing safer and more effective cancer treatments.

Over 5,000 phytomolecules from approximately 3,000 plant species have been identified as playing an important role in cancer treatment (Graham, Quinn et al. 2000; Prakash, Kumar et al. 2013). Phytochemicals are classified based on their structure and function into various categories, including polyphenols, phytoestrogens, salicylates, phytosterols, glucosinolates, saponins, sulfides, protease inhibitors, terpenes, monoterpenes, indoles, lectins, and isothiocyanates. The development of phytochemicals with anticancer activity involves several steps, such as the extraction, separation, and purification of these compounds. Once isolated, these substances are tested both *in vitro* and *in vivo* on various cell lines to assess their efficacy. Evidence has confirmed that a wide range of phytochemicals play crucial roles in blocking the cancer cell cycle, causing DNA damage, inhibiting proliferation, activating pro-apoptotic proteins, suppressing anti-apoptotic proteins, and inducing apoptosis. Extensive research has highlighted their potential in both chemoprevention and chemosensitization.

Additionally, these bioactive compounds can halt or delay the malignant transformation of precancerous cells (Aggarwal, Kunnumakkara et al. 2008). According to the literature, phenolics have demonstrated significant efficacy in inhibiting cancer cell proliferation, migration, and invasion (Tsakiroglou, VandenAkker et al. 2019). Phenolics can bolster the body's immune system to recognize and eliminate cancer cells, reduce the adhesiveness and invasiveness of malignant cells, and inhibit angiogenesis, a critical process for tumour growth. For example, the anticancer mechanism of curcumin involves the ruling of multiple cellular signalling pathways. Here, it includes NF-κB, CDKs, JAK/STAT, cyclooxygenase-2 (COX-2), tumour necrosis factor-alpha (TNF-α), PI3K/Akt, p53, and MAPK. Various studies have reported that RES modulates different signalling pathways involved in cancer onset, inhibits proliferation, induces programmed cell death, and arrests the cancer cell cycle. Additionally, resveratrol has been shown to enhance the chemotherapeutic potential of doxorubicin by upregulating the tumour suppressor gene p53 in chemoresistant B16 melanoma cells (Gupta, Kannappan et al. 2011). Numerous studies recommend that using resveratrol in amalgamation with other treatments can reduce the side effects of cancer therapies (Berman, Motechin et al. 2017).

The major bioactive compound in green tea, epigallocatechin gallate (EGCG), inhibits cancer cell growth by blocking Bcl-2 and COX-2 expression and NF-κB activation in prostate carcinoma cells, thereby inducing apoptosis. EGCG also suppresses the production of vascular endothelial growth factor in head and neck carcinoma cells and targets JAK/STAT, Janus kinases (JAK), while downregulating the protein levels of Cyclin D1, PCNA, and Bcl-2 (Chen, Wang et al. 2017). Several studies have verified the efficacy of plant-based alkaloids in inhibiting cancer. Alkaloids have been shown to modulate pathways involved in proliferation, metastasis, and cell cycle regulation. Different indole alkaloids exhibit antiproliferative effects through mechanisms such as inducing apoptosis or arresting the cell cycle in cancer cell lines as well as in MDR phenotypes. Among the numerous anticancer alkaloids, vinca alkaloids, colchicine, and paclitaxel (PTX) are particularly important. Min et al. investigated the MDR-reversing activity of quinolines and isoquinolines in MES-SA/DX5 and HCT15 cell lines. Colchicine, an anti-mitotic agent, inhibits cancer cell movement and angiogenesis, possesses metastatic potential, forms a complex with tubulin, causes cell blebbing, and releases caspase protease and cytochrome-c, leading to apoptosis.

Vincristine is used to treat leukaemia, myeloma, lymphoma, and cancers of the neck, breast, and head. Vinblastine's anticancer potential arises from its interaction with tubulin, which inhibits mitosis at metaphase. By binding to the microtubular proteins of the mitotic spindle, vinblastine causes microtubule crystallization, leading to cell cycle arrest and subsequent apoptosis (Dhyani, Quispe et al. 2022). Vinorelbine and vindesine are two effective semisynthetic analogues that have received clinical approval. Additionally, irinotecan and topotecan are FDA-approved semisynthetic derivatives of camptothecin that are clinically active and less toxic than the parent compound. Irinotecan is used for treating advanced colorectal cancer, while topotecan is approved for recurrent ovarian cancer, small-cell lung cancer, and cervical cancer.

Organosulfur compounds are reported to possess strong antioxidant properties. These compounds enhance cell death, inhibit breast cancer formation, and retard cancer cell growth. Among the various organosulfur compounds, garlic-derived compounds are the most prominent. Oil-soluble garlic compounds exhibit stronger anticancer activity than their water-soluble counterparts. Wattenberg et al. (1989) evaluated various organosulfur compounds for their ability to inhibit carcinogenesis induced by *N*-nitrosodiethylamine, identifying diallyl disulfide as the most potent.

Carotenoids, the most prevalent natural pigments, have garnered significant attention for their provitamin and antioxidant properties (Tsakiroglou, VandenAkker et al. 2019). Over 600 different carotenoids have been identified from various sources, including plants, animals, fungi, fish flesh, algae, and bacteria. Carotenoids are a class of phytochemicals that include carotenes, xanthophylls, and their derivatives. The cellular mechanisms by which carotenoids exert anticancer activity are multifaceted and include cell cycle arrest, apoptosis, anti-invasive, anti-angiogenic, anti-proliferative, and anti-metastatic effects, as well as the inhibition of tumour growth. For instance, foods containing β-carotene or vitamin C appear to offer protection against cancer. β-Carotene decreases the expression of Bcl-2, NF-κB, and p-Akt while increasing the expression of caspase-3. Moreover, β-carotene, in combination with 5-fluorouracil (5-FU), has been shown to induce apoptosis (Koklesova, Liskova et al. 2020).

Gingerol, a compound derived from ginger, attenuates the invasiveness of lung and breast cancer cells by suppressing the AKT and p38 MAPK pathways. A published study demonstrated that mice treated with gingerol exhibited inhibition of tumour growth and metastasis by suppressing caspase-3 expression (Joo, Hong et al. 2016). NF-κB, a master transcription factor, plays a critical role in the induction and proliferation of cancer cells by modulating growth signalling and apoptosis pathways. Tumour cell proliferation and the suppression of apoptosis are promoted by NF-κB activity. The direct association between NF-κB activation and chemoresistance in bladder cancer cells has been reported. Laurindo et al. included 56 phytochemicals in their study and observed that the effects related to NF-κB pathways included reductions in interferon-gamma (IFN-γ), interleukin (IL)-1β, tumour necrosis factor-alpha (TNF-α), IL-6, and cyclooxygenase-2 (COX-2). Several phytochemicals, such as gingerol, apigenin, RES, epigallocatechin gallate, CUR, terostilbene, QUE, anatabine, berberine, and galantamine, are known to modulate NF-κB pathways (Chauhan, Islam et al. 2022). Cho et al. reported increased cytotoxicity and additive efficacy in T24-GCB cells using a combination treatment of gemcitabine with potential phytochemicals (capsaicin, QUE, and RES) in both *in vitro* and xenografted animal tumour models. They found that this combination therapy had enhanced cytotoxic effects on T24-GCB cells compared to gemcitabine treatment alone, indicating a synergistic effect on gemcitabine cytotoxicity (Cho, Yu et al. 2021).

Efferth et al. studied natural products with activity against MDR cells and investigated the efficacy of 102 cytotoxic medicinal plants and 228 isolated cytotoxic phytochemicals from 16 different countries in killing MDR cells (Efferth, Kadioglu et al. 2021). Several studies have reported that various compounds effectively suppress both sensitive and MDR cancer cells with similar efficacy. Interestingly, a few phytochemicals (3%–8%) and medicinal plants have been shown to inhibit MDR cancer cells more efficiently than conventional drugs. Tuorkey investigated the efficacy of phytomolecules against molecular targets in cancer treatment, particularly from non-edible plants such as rottlerin, berberine, sparstolonin B, sulforaphane, plumbagin, and 6-shogaol. The findings suggest that these phytomolecules could be used in combination with current cancer treatments to enhance efficacy and prevent tumour recurrence. For example, berbamine, isolated from the *Berberis amurensis* plant, reduces the Bcl-2/Bax protein ratio, downregulates bcr/abl gene expression, and increases the expression of Fas and P53, leading to the activation of the apoptotic pathway. Notably, berbamine induces apoptosis in both estrogen receptor-negative MDA-MB-231 cells and estrogen receptor-alpha-positive MCF-7 breast cancer cells, while sparing normal human mammary epithelial cells.

Sparstolonin B (SsnB) is a natural compound derived from *Sparganium stoloniferum*. It exhibits apoptotic activity in cancer cells by down-regulating mRNA levels associated with the cell cycle, inhibiting the expression of inflammatory cytokines, and reducing the phosphorylation of Erk1/2, p38a, IκBα, and JNK in macrophage cell lines (Tuorkey 2015). In this context, the therapeutic potential of phytochemicals as cytotoxic agents cannot be overlooked. The examples mentioned above are just a few of the many plant-based anticancer agents. There is an abundance of literature indicating that phytomolecules are also valuable for targeted cancer therapy and overcoming MDR in cancer chemotherapy.

23.5 NANO-DELIVERY SYSTEMS IN RESISTANT CANCER

Nanomedicine involves the use of nano-sized drugs or biologics (1–1,000 nm) that possess elevated surface-to-volume ratios and unique chemical, physical, and biological properties. Here, it allows for the specific delivery of drug molecules to pathological sites, enabling them to accumulate precisely at the target location (Gwinn and Vallyathan 2006; Tabassum, Verma et al. 2018). Nanoparticles can protect drugs from enzymatic degradation, prolong drug release, and enhance the half-life and bioavailability of therapeutics. Studies have demonstrated that nanoparticle-mediated drug delivery offers several advantages over conventional therapies, including enhanced drug solubility and permeability, targeting capabilities, retention effects, stability, improved biocompatibility, and controlled release.

Nanoparticles can take on various shapes, such as spheres, cubes, needles, octahedrons, flowers, rods, ellipses, stars, cylinders, hexagons, pentagons, clusters, triangles, branched chains, and sheets, depending on the composition of their inner and outer cores. Additionally, nanoparticles are often coated with suitable excipients to improve their biocompatibility and biodegradability. Nanomaterials can be broadly classified into organic and inorganic categories.

Organic nanomaterials include micelles, nanofibers, dendrimers, nanotubes, liposomes, nanogels, polymeric nanoparticles, and layered biopolymers. In contrast, inorganic nanomaterials comprise iron oxide nanoparticles, mesoporous silica-based agents, gold nanoparticulate systems, and alloys, with inorganic nanoparticles generally exhibiting superior stability compared to their organic counterparts (Khalid, Tan et al. 2020; Lan 2022). Two commonly employed methods for synthesizing nanoparticles are the top-down as well as bottom-up approaches, which are further categorized into physical, chemical, and biological methods (Kumbhar, Patil et al. 2024).

In cancer treatment, it is crucial to develop drugs that can effectively target cancer cells while being safe for healthy cells. This goal can be accomplished through the systematic delivery of nanoparticles into the tumour microenvironment, where they encapsulate chemotherapeutic molecules. In this, it enhances treatment efficacy and minimizes side effects. Moreover, nanoparticles can be employed to heat and destroy cancer cells via hyperthermia, advanced medical imaging techniques, and enable better tumour monitoring (Elumalai, Srinivasan et al. 2024). Nanoparticle-based drug delivery systems have demonstrated numerous benefits in cancer treatment, such as targeting tumour cells, reducing side effects, improving pharmacokinetics, and overcoming drug resistance (Dadwal, Baldi et al. 2018; Palazzolo, Bayda et al. 2018). Ideal characteristics of nanoscale materials include their capability to evade clearance by the reticuloendothelial system and avoid capture by the mononuclear phagocyte system, their stability in the bloodstream, and their accumulation in the tumour microenvironment (TME). They should also be able to penetrate the tumour interstitial fluid, which is under high pressure, and exclusively interact with cancer cells at the active site (Omidi and Barar 2014). Significant research has focused on designing tumour-targeted drug delivery systems, which can be separated into passive targeting. Here, drugs are delivered to the target site by exploiting the enhanced permeability and retention (EPR) effect. Active targeting involves the use of targeting molecules to guide chemotherapeutic agents to their intended targets (Attia, Anton et al. 2019).

In cancer cure, the tumour microenvironment is a key area of study for developing new treatments. It has a unique acidic physiological nature (pH=6.5–6.8). Several studies have leveraged this characteristic by developing drug-loaded, pH-dependent nanoparticles that release their therapeutic payload in acidic environments. Thereby, it enhances drug access and reduces side effects. Tumour blood vessels become more permeable under conditions such as hypoxia and inflammation, characterized by leaky walls due to rapid and abnormal angiogenesis. This leads to enhanced permeability for macromolecules larger than 40 kDa, including nanoparticles, which accumulate within tumour tissue. The absence or dysfunction of lymphatic vessels in tumours further contributes to inefficient drainage, resulting in the prolonged retention of nanoparticles within the tumour. This passive phenomenon is known as the EPR effect. It depends on factors such as the size of the carrier and tumour biology, including vascularity and leakiness (Matsumura and Maeda 1986). The size of nanomedicine, with an optimal range of 20–200 nm, is crucial for effective tumour permeation and retention, as well as excretion through the renal system (for particles less than 6 nm) or the reticuloendothelial system (for particles larger than 500 nm) (Kobayashi, Watanabe et al. 2014).

However, passive targeting faces several challenges, including inter-individual variability, non-specific drug distribution, the non-universal presence of the EPR effect, and variations in blood vessel permeability across different tumours. These factors contribute to low bioavailability and secondary toxicity due to poor selectivity. Active targeting engages in modifying the nanoparticle exterior with specific ligands or antibodies, which facilitates the internalization of nanoparticles and the subsequent release of therapeutic drugs into tumour cells. These ligands bind strongly to receptors on target cells, boosting the specificity and efficacy of drug delivery while minimizing harmful effects on healthy cells. Various proteins or small peptides can be used to modify the nanoparticle surface, improving selectivity and the amount of chemotherapeutic drug released into the target cell. Common targeting moieties include glycoproteins, polypeptides, monoclonal antibodies, folate, and carbohydrates. Additionally, active targeting of nanocarriers with small molecules has shown a potential to bypass *P*-glycoprotein-mediated drug efflux and conquer MDR (Kamaly, Xiao et al. 2012).

Nanoparticles used in therapeutic strategies can also be combined with immunotherapy and gene therapy to synergistically augment the biodistribution and pharmacokinetics of anticancer drugs. Thereby, it helps to improve their effectiveness against MDR cancer cells. Multimodal nanoformulations, which may include materials such as iron, gold, or quantum dots, along with targeting molecules or peptides, have demonstrated improved outcomes in cancer treatment. Chemotherapy regimens often show poor efficacy when administered as a single drug or fixed dose. Various nanoparticles, such as polymeric nanoparticles, liposomes, and metallic nanoparticles, can be employed to formulate drug delivery systems tailored to specific types of cancer cells. Phytochemicals encapsulated into nanocarriers have shown impending health benefits in treating cardiovascular issues, infectious, neurodegenerative diseases, and cancer therapy (Koklesova, Jakubikova et al. 2023). Encapsulation protects therapeutic agents from external degradation by the carrier system. Polymeric nanoparticles are both safe and efficient, often prepared using biocom patible polymers. For instance, Poly(lactic-co-glycolic) acid (PLGA) mediated nanoparticles provide enhanced support for the delivery of phytochemicals.

Liu et al. developed novel multifunctional "nano-actinias" for delivering icariin and CUR to tumour cells and cancer stem cells. In this, new polymer oligomeric hyaluronic acid-hydrazone bond-folic acid-biotin has been used. It self-assembles into nanomicelles. These "nano-actinias" demonstrated greater cytotoxicity against targeted MCF-7 cells and inhibited cancer cell invasion (Liu, Wang et al. 2019). Camptothecin has poor aqueous solubility and toxic effects. It was successfully encapsulated into hydrophobically modified glycol chitosan (HGC) nanoparticles by Min et al. It confirmed that these nanoparticles had prolonged blood circulation and elevated accumulation in MDA-MB231 human breast cancer cells, indicating the high efficacy of the nanoparticles (Min, Park et al. 2008). Similarly, erlotinib and QUE containing solid lipid nanoparticles showed anticancer potential via amplified cellular uptake, enhancing apoptosis induction in erlotinib-resistant A549/ER cells (Ganthala, Alavala et al. 2022). To further improve cellular uptake, nanoparticles are often surface-modified with polyethylene glycol (PEG).

Researchers have developed various polymeric nanoparticles laden with anticancer drugs and *P*-glycoprotein (*P*-gp) inhibitors. Studies suggest that these formulations can modulate adenosine triphosphate (ATP)-binding cassette (ABC) efflux transporters and augment the intracellular accumulation of chemotherapeutic drugs in MDR tumour cells (Kapse-Mistry, Govender et al. 2014; Wei, Song et al. 2021). Li et al. fabricated liponanoparticles loaded with DOX to circumvent the *P*-gp efflux mechanism in DOX-resistant MCF-7/ADR breast cancer cells, significantly improving drug accumulation in resistant cells (Li, Xu et al. 2012). Ersoz and co-authors developed QUE nanoparticles using a PLGA carrier, achieving higher cellular uptake at lower IC50 values, which reduced resistance in brain tumours by decreasing oxidative stress (Ersoz, Erdemir et al. 2020).

Liposomes are specialized microscopic lipid vesicles characterized by a lipid bilayer that separates an internal aqueous core from the external environment. Wang and co-authors developed liposomes containing L-arginine (L-Arg) and DOX, loaded into gold@copper sulfide yolk-shell structures. It showed the targeted release of DOX in drug-resistant cancers (Wang, Jiang et al. 2022). Recently, Zhang et al. prepared a pH-sensitive multifunctional siRNA and DOX-based drug delivery system. It showed the improved anticancer potential in drug-resistant breast cancer (Zhang, Du et al. 2018). The liposomal delivery of CUR has been revealed to inhibit the expression of human epidermal growth factor, enhancing its anticancer effects. Specifically, CUR-loaded liposomes effectively blocked the expression of this growth factor (Karimi, Gheybi et al. 2020). When combined with ultrasound and microbubbles, the cellular uptake of CUR within cancer cells was significantly improved, leading to enhanced cell apoptosis (Radha, Paul et al. 2024). Similarly, nanoliposomal formulations of CUR, such as PG-LipCUR and LipCUR, have demonstrated advanced antitumour potential by actively targeting and eradicating breast adenocarcinoma cell lines (TUBO and 4T1) (Karimi, Gheybi et al. 2020).

Loo and co-workers developed lyotropic liquid crystalline nanoparticles (LCNs) using monoolein, poloxamer 407, PEG-400, and Transcutol® HP to enhance the solubility of berberine (Loo, Madheswaran et al. 2020). Their study revealed that LCNs are promising carriers for improving the solubility and anticancer efficacy of berberine against breast cancer. Liu et al. co-administered QUE (a *P*-glycoprotein (*P*-gp) inhibitor) with PTX in chondroitin sulfate-coated mesoporous silica nanoparticles to reverse *P*-gp-mediated MDR (Liu, Wang et al. 2019). Their results indicated that *P*-gp expression was downregulated. As well, the drug release was influenced by the redox environment within MCF-7/ADR drug-resistant cells. Fan and co-workers developed β-cyclodextrin (β-CD)-based temperature-sensitive supramolecular nanoparticles for the delivery of PTX and DOX, both *in vitro* and *in vivo*. These novel nanoparticles helped overcome cancer resistance by enhancing cellular uptake, improving intracellular drug retention, and inhibiting pump-mediated drug resistance.

Nanodiamonds, a form of carbon nanoparticles, have also been shown to alter tumour efflux pumps and provide binding sites for therapeutic agents. Chow et al. demonstrated that a complex of nanodiamonds and DOX overcame drug efflux, significantly increasing apoptosis and inhibiting tumour growth in a choice of human and murine breast cancer-resistant cells, as well as in a xenograft model (Chow, Zhang et al. 2011). Similarly, Singh and co-workers highlighted the efficiency of folic acid-conjugated nanoparticles encapsulating RES and docetaxel for treating MDR prostate cancer. Their results indicated the downregulation of anti-apoptotic gene expression and inhibition of ABC-transporter markers (Tabassum, Verma et al. 2018). In another study, *Curcuma wenyujin*-loaded gold nanoparticles were prepared to assess their anticancer potential against a breast cancer cell line. Quantum dots (QDs), known for their unique optical and chemical properties, have been explored for drug-delivery systems with therapeutic and diagnostic applications. Investigators found that graphene QDs (GQDs) effectively reversed resistance to DOX in selected MCF-7/ADR cells (Kairdolf, Smith et al. 2013).

Gene therapies based on RNA interference, such as the transfer of small interfering RNAs (siRNAs), small hairpin RNAs (shRNAs), and microRNAs (miRNAs) into tumour cells, hold promise for cancer treatment. However, siRNAs are speedily degraded by RNases and filtered out by the kidneys. To address this challenge, various delivery formulations have been developed, including those based on inorganic nanoparticles, polymers, and/or cationic lipids, to facilitate the systemic delivery of siRNA to solid tumours (Kim, Kim et al. 2016). Remarkably, growing research suggests that by increasing oxidative stress within cells, chemodynamic therapy (CDT) can effectively destroy MDR cancer cells or render them more susceptible to other chemotherapeutic agents. The development of multifunctional stimuli-responsive nanotherapeutics has shown promise in enhancing the penetration of anticancer drugs into tumour cells and reversing MDR. For instance, researchers have demonstrated that DOX-loaded gallic acid-ferrous ions @$CaCO_3$ nanoparticles exhibit potent anticancer effects against DOX-resistant 4T1 breast cancer cells through a combination of chemodynamic and chemotherapeutic approaches (Dong, Hao et al. 2020). Noteworthy progress has been made in overcoming MDR cancers, largely due to advancements in nanocarriers and nano-based formulations that address the limitations of traditional drug delivery systems. However, it is imperative to recognize that most of these studies have been conducted using *in vivo* animal models, which may not accurately reflect the complexity of cancer progression in the human body.

23.6 APPLICATIONS OF PHYTOMOLECULES LOADED NANO-DELIVERY SYSTEM IN RESISTANT CANCER

The integration of phytomolecules into nano-delivery systems has emerged as a transformative strategy in combating MDR cancer cells (Kumar, Yadav et al. 2022). By harnessing the unique properties of these advanced delivery platforms, researchers have significantly enhanced the therapeutic potential of natural compounds, which often face challenges such as poor bioavailability and rapid metabolism (Dewi, Chaerunisaa et al. 2022). Here, this section explores the practical applications of phytomolecule-loaded nano-delivery systems. It emphasizes how these innovative approaches address the limitations associated with resistant cancers. Various case studies and recent advancements demonstrate the effectiveness of these systems in improving drug delivery, precisely targeting cancer cells, and overcoming resistance mechanisms. As such, nano-delivery systems represent a promising frontier in the development of more effective and sustainable cancer therapies. Extensive investigations over the years have led to the screening of a wide array of phytomolecules for their potential efficacy against resistant cancers. This screening process involves evaluating various natural compounds derived from plants for their ability to overcome the challenges posed by MDR cancer cells. These studies aim to identify phytomolecules that not only demonstrate potent anticancer properties but also possess the capability to circumvent the resistance mechanisms that limit the effectiveness of conventional treatments. Through this rigorous screening, researchers have uncovered several promising phytomolecules with the potential to enhance therapeutic outcomes and contribute to more effective cancer treatment strategies.

The literature reported that the CUR is a natural phenolic compound derived from the turmeric plant (*Curcuma longa*). It exhibits potent anticancer activity across various tumour types. It has shown efficacy both as a standalone treatment and as an adjuvant therapy in combination with other antitumour drugs. In this, CUR is a powerful compound extracted from turmeric. It has been widely studied for its anticancer and chemopreventive properties, particularly in breast cancer. It influences cancer development through a complex network of molecular signalling pathways, including those related to cell proliferation, estrogen receptors (ER), and human epidermal growth factor receptor 2 (HER-2). Research has demonstrated that CUR not only affects these critical pathways but also plays a role in regulating apoptosis and the expression of cell cycle-related genes and microRNAs in breast cancer cells. This multifaceted approach highlights CUR's potential as a significant player in both the prevention and treatment of breast cancer. Several studies have been conducted to explore CUR's role in combating resistant breast cancer (Wang, Yu et al. 2016).

One intriguing study investigated a novel nanocarrier, quaternary ammonium palmitoyl glycol chitosan (GCPQ), to address MDR in breast cancer treatment. The GCPQ-based formulation, termed DCG, successfully encapsulated DOX and CUR, offering sustained release profiles. The nanoformulation demonstrated a particle size of 187 ± 1.36 nm and a zeta potential of -28 ± 0.48 mV, confirming its stability. *In vitro* studies revealed enhanced cytotoxicity (88%) in CAL-51 breast cancer cells and induced apoptosis with a 93% apoptotic effect. *In-vivo* biodistribution studies underscored the safety and tumour-targeting ability of DCG nanomicelles, with an IC_{50} of 80.10 µg/mL and stable storage for up to 3 months. These findings suggest that DCG nanomicelles hold significant promise in overcoming MDR in breast cancer, offering the impending for personalized and more effective therapies (Nurjis, Sarwar et al. 2024).

Another research group found that water-soluble PLGA CUR nanoparticles (CUR-NPs) induced apoptosis in cisplatin-resistant human oral cancer cells (CAR cells) while exhibiting low cytotoxicity to normal cells. In this, CUR-NPs triggered DNA condensation, fragmentation, and apoptosis in CAR cells. The nanoformulation also increased the production of reactive oxygen species (ROS) and upregulated the protein expression levels of cleaved caspase-3/caspase-9, AIF, Apaf-1, cytochrome c, and Bax, while downregulating Bcl-2, showing promising results against cisplatin-resistant human oral cancer (Chang, Peng et al. 2013). In another study, CUR-loaded PLGA nanoparticles (CUR-NPs) demonstrated significant anticancer effects on human osteosarcoma U2OS cells. The CUR-NPs induced apoptosis, leading to DNA fragmentation and the formation of apoptotic bodies. Treatment with these nanoparticles superior the activity and expression of caspases-3/-7 and caspase-9 and upregulated cleaved caspase-3/caspase-9, cytochrome c, Apaf-1, and Bad, while downregulating p-Akt (Peng, Lee et al. 2013).

Recent studies have highlighted the potential of lipid nanoparticles (DOX/CUR-NPs) for co-delivering DOX and CUR in the treatment of hepatocellular carcinoma (HCC). This formulation demonstrated a significant reduction in MDR in HCC, with CUR contributing to the down-regulation of MDR1, Bcl-2, HIF-1α, and P-gp at both the mRNA and protein levels. In HCC cells, DOX/CUR-NPs enhanced cytotoxicity, lowered the IC_{50}, and reduced the resistance index compared to DOX alone, indicating a synergistic effect that effectively targets cancer cells and overcomes drug resistance (Zhao, Chen et al. 2015). Another study on cisplatin (CDDP) and CUR co-loaded liposomes

(CDDP/CUR-Lip) showed significant anticancer effects and the ability to overcome drug resistance in HCC. The co-delivery system achieved a synergistic effect, resulting in a notably lower IC_{50} of 0.62 µM against HepG2 cancer cells compared to single-drug treatments. Additionally, CDDP/CUR-Lip effectively increased intracellular ROS levels, contributing to enhanced cytotoxicity in cancer cells.

The formulation also exhibited prolonged retention time and improved antitumour efficacy in both mouse and human HCC models, demonstrating its potential to overcome the limitations of traditional CDDP therapy and address drug resistance in HCC (Cheng, Zhao et al. 2018). The co-encapsulation of DOX and CUR in poly(butyl cyanoacrylate) nanoparticles (PBCA-NPs) demonstrated effective anticancer activity and the reversal of MDR. The dual-agent PBCA-NPs showed comparable cytotoxicity to the combination of two single-agent PBCA-NPs and outperformed both free drug combinations and single-agent-loaded PBCA-NPs. Notably, the simultaneous delivery of DOX and CUR via PBCA-NPs achieved the highest reversal efficacy of MDR in MCF-7/ADR cells, a breast cancer cell line resistant to adriamycin, by significantly downregulating *P*-glycoprotein. This approach effectively enhanced anticancer efficacy and overcame resistance (Duan, Mansour et al. 2012). A study on folic acid-modified nanoparticles ((DOX+CUR)-FA-NPs) demonstrated significant improvements in cancer treatment and overcoming MDR. These nanoparticles exhibited effective tumour targeting, high cell apoptosis, and enhanced cellular uptake in MCF-7/ADR cells, a model for MDR. *In vivo*, (DOX+CUR)-FA-NPs showed significant anticancer efficacy and reduced damage to normal tissues. The targeted delivery of DOX and CUR successfully reversed MDR, making this approach a promising strategy for improving breast cancer therapies (Guo, Yu et al. 2021). Kumbhar et al. developed CUR nanomicelles (CUR-NMs) using DSPE-PEG-2000 to address the challenges of CUR's poor solubility and bioavailability in treating cisplatin-resistant oral cancer. The CUR-NMs, with a size of 32.60±4.2 nm and elevated entrapment efficiency (82.2%), demonstrated significantly enhanced cellular uptake and cytotoxicity in resistant oral cancer cell lines compared to plain CUR. Increased apoptosis was observed with CUR-NMs, suggesting their potential as an effective carrier system to improve CUR's therapeutic impact against resistant oral cancer (Kumbar, Muddapur et al. 2022).

Artemisinin is a natural molecule derived from the sweet wormwood plant (*Artemisia annua*). It has garnered significant interest in cancer therapy due to its remarkable anticancer properties. Initially recognized for its potent antimalarial effects, artemisinin and its derivatives have also shown promise in targeting cancer cells through various mechanisms (Augustin, Staines et al. 2020). To address the limitations and enhance the therapeutic effectiveness of artemisinin against resistant cancer cells, various carrier systems have been explored. Guo et al. highlighted the potential of magnetic nanoparticles (MNP) loaded with dihydroartemisinin (DHA), an artemisinin derivative, for enhanced CDT in cancer treatment. Their study demonstrated that MNPs loaded with DHA significantly increased the generation of ROS through a Fenton-like reaction, leading to improved therapeutic effects. Among the tested formulations, MNP-DHA exhibited the most potent cytotoxicity against breast cancer cells (MCF-7) and was notably effective against drug-resistant breast cancer cells (MCF-7/ADR), showcasing its capability to conquer MDR. Furthermore, MNP-DHA was found to inhibit aggressive breast cancer cell lines, including MDA-MB-231 and MDA-MB-453, underscoring its potential for treating intractable breast cancers. The study also revealed that MNPs can produce ferrous ions in the acidic TME, which catalyse DHA to produce high levels of ROS, contributing to significant cancer cell death (Guo, Yao et al. 2020).

The PIP is an extract from the seeds of plants in the *Piperaceae* family. It has demonstrated anti-metastatic performance and the ability to reverse drug resistance. However, its therapeutic applications are limited due to concerns such as immunotoxicity, reproductive toxicity, and poor aqueous solubility. A study by Pachauri et al. highlights several significant cancer-related findings involving PIP. When encapsulated in PEG-PLGA nanoparticles (P-PEG-PNP), PIP effectively induced apoptosis in MCF-7 breast cancer cells (Figure 23.2). Notably, this research is the foremost to demonstrate the targeted delivery of PIP to MCF-7 cells using P-PEG-PLGA nanoparticles conjugated with an aptamer

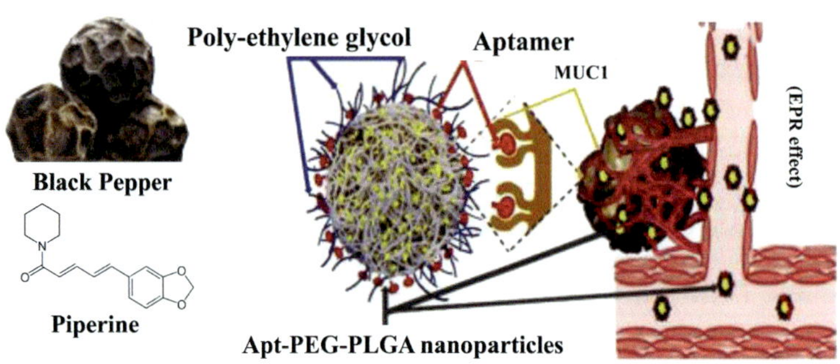

FIGURE 23.2 Presentation of targeted delivery of PIP to MCF-7 cells using Apt-P-PEG-PLGA (Apt-PEG-PLGA) nanoparticles conjugated with an aptamer against mucin1. [Reprinted (adapted) with permission from ref. (Pachauri, Gupta et al. 2015). Copyright © 2015 Elsevier B.V. All rights reserved.]

against mucin1 (Apt-P-PEG-PNP). Additionally, the amalgamation of this aptamer-conjugated P-PEG-PNP with PTX significantly reduced the dosage of PTX required to inhibit the growth of PTX-resistant MCF-7 cells. These findings suggest that P-PEG-PNP could be a promising and safe formulation for enhancing chemotherapy for drug-resistant breast cancers (Pachauri, Gupta et al. 2015).

In another study by Xu et al., the potential of PIP as an inhibitor of *P*-glycoprotein (*P*-gp) and a sensitizer for PTX was investigated, particularly for its application in treating triple-negative breast cancer (TNBC). Here, PIP was found to enhance the intracellular accumulation of PTX and increase cell sensitivity to the drug. Molecular docking simulations, RT-qPCR, and Western blot analysis revealed that PIP competes with PTX for binding to P-gp, affects ATPase activity, and downregulates the MDR1 gene and P-gp expression. The study also evaluated albumin nanoparticles co-encapsulating PTX and PIP (PP@AN), which improved drug accumulation, reduced P-gp expression in tumours, and enhanced antitumour efficacy *in vivo* while minimizing hepatotoxicity and hemotoxicity (Xu, Xiao et al. 2023).

Literature reported that RES is a polyphenolic compound recognized for its potential therapeutic benefits, including anticancer properties. It has been studied for its ability to enhance the efficacy of chemotherapeutic agents, particularly in overcoming drug resistance in tumours. One notable study investigated RES co-encapsulated with PTX in PEGylated liposomes to create a carrier system designed for treating drug-resistant tumours. These liposomes, with an average diameter of 50 nm and over 50% encapsulation efficiency, exhibited potent cytotoxic effects against drug-resistant MCF-7/Adr tumour cells *in vitro*. In vivo, the liposomes improved the bioavailability and tumour retention of both drugs and effectively inhibited tumour growth in mice ($p<0.01$) without significantly increasing toxicity. These findings recommend that the blend of RES and PTX in a nanocarrier system could be a capable approach for enhancing the treatment of drug-resistant tumours (Meng, Guo et al. 2016).

In a study by Liping et al., a novel glutathione (GSH) - responsive polymer (PRES) was synthesized using RES and 3,3′-dithiodipropionic acid through condensation polymerization. This polymer demonstrated dual functions: it inhibited tumour cell growth and self-assembled into nanoparticles (~93 nm) capable of delivering antitumour drugs like PTX. The PRES-based nanoparticles (PTX@PRES NPs) achieved high drug loading (~7%) and effectively overcame MDR. Here, the *in vitro* studies showed that these nanoparticles were stable in systemic circulation but degraded efficiently in the high GSH environment of tumour cells. Moreover, the cytotoxicity tests confirmed that PTX@PRES NPs suppressed the growth of both cancer cells (A549) and drug-resistant cells (A549/PTX). *In vivo* studies demonstrated significant tumour inhibition and enhanced therapeutic efficacy in A549/PTX tumour-bearing mice. In conclusion, it highlights the potential of GSH-responsive polymers for effective drug delivery and overcoming MDR in cancer treatment (Liping, Jian et al. 2022). Xiong et al. developed RES nanoparticles (Pep-1-PEG3.5k-PCL4k@RESs) to improve bioavailability and targeting in anaplastic thyroid cancer (ATC). The nanoparticles significantly delayed tumour growth and induced apoptosis, unlike free RES. Docetaxel/DOX, while effective, caused severe side effects and resistance, which RES nanoparticles could help overcome (Xiong, Lin et al. 2021).

QUE is a naturally occurring bioflavonoid present in a wide range of fruits, vegetables, and grains. It is recognized for its powerful antioxidant, anti-inflammatory, and anticancer effects. Studies have shown that quercetin can slow down the growth of cancer cells, trigger apoptosis, and boost the effectiveness of chemotherapy treatments. Its potential to overcome drug resistance, particularly in cancer treatments, has been explored in several studies. Wang et al. investigated QUE's ability to reverse PTX resistance in non-small cell lung cancer (NSCLC). They found that QUE synergistically enhanced the cytotoxicity of PTX in A549 and PTX-resistant A549/Taxol cells. Here, QUE combined with PTX significantly inhibited Akt and ERK phosphorylation, which were not affected by PTX alone. To address PTX's hydrophobicity and toxicity, biodegradable and targeted cetuximab chitosan nanoparticles (Cet-CTS NPs) were developed for the co-delivery of PTX and QUE. These nanoparticles (290 nm), provided sustained drug release and effectively reduced tumour growth in PTX-resistant A549/Taxol cells and xenografts. The study concluded that QUE-loaded Cet-CTS NPs offer a promising strategy for treating PTX-resistant lung cancer (Wang, Yu et al. 2021).

In another study, PTX and QUE were co-loaded into chondroitin sulphate (ChS)-coated nanoparticles (MSNs-ChS@PQ) to combat MDR in breast cancer (Figure 23.3). These nanoparticles, approximately 227 nm in size, effectively targeted MCF-7/ADR cells, reduced IC_{50} values, increased apoptosis, and enhanced sensitivity to PTX. Additionally, they down-regulated *P*-glycoprotein (*P*-gp) and demonstrated strong tumour-targeting abilities with minimal toxicity, offering a hopeful approach to improving chemotherapy efficacy in drug-resistant breast cancer (Liu, Fu et al. 2020).

In a fascinating approach, researchers evaluated the effects of QUE and its nanoemulsion on MDR and non-MDR cells. The results demonstrated that QUE, particularly when combined with vincristine, effectively induced cytotoxicity, apoptosis, and DNA damage. It also altered ABCB1 gene expression and inhibited efflux activity. The nanoemulsion of QUE maintained its cytotoxic effects and improved bioavailability, highlighting its potential as a capable approach to conquering drug resistance in cancer dealing (Marques, Machado et al. 2021). Emodin (EMO) is a naturally occurring anthraquinone derivative originating in various plants, including *Rheum palmatum* and *Polygonum cuspidatum*. It has been extensively studied for its broad spectrum of pharmacological activities. Importantly, it includes antioxidant,

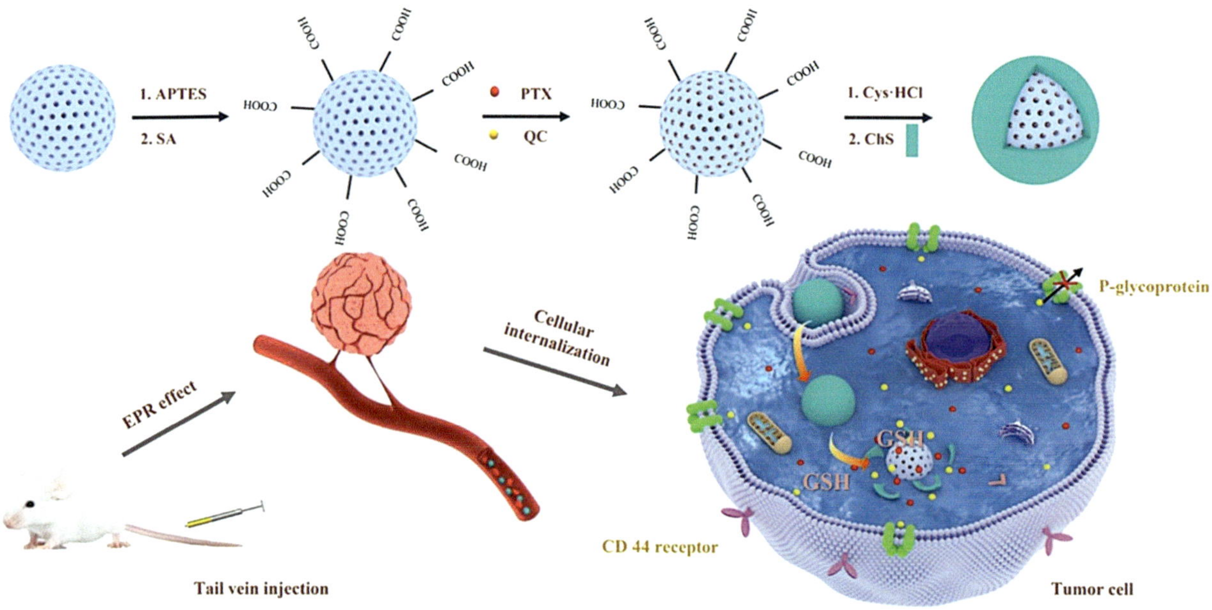

FIGURE 23.3 Presentation of PTX and QUE was co-loaded into chondroitin sulfate-coated nanoparticles (MSNs-ChS@PQ) to combat MDR in breast cancer. [Reprinted (adapted) with permission from ref. (Liu, Fu et al. 2020). Copyright © 2020 Elsevier B.V. All rights reserved]. [(3-aminopropyl)triethoxysilane: APTES; Quercetin: QC; cystamine dihydrochloride: Cys·HCl; succinic anhydride: SA.]

anti-inflammatory, and particularly anticancer effects. Here, MEO anticancer activity has been observed in various cancer types, where it induces apoptosis, inhibits cell proliferation, and disrupts cell cycle progression. Moreover, it has shown potential in reversing MDR in cancer cells, an important challenge in chemotherapy, particularly in breast cancer. However, its limited activity hampers its application as an anticancer agent. In this context, a study explored how EMT contributes to MDR in breast cancer and how EMO-polymer lipid hybrid nanoparticles (E-PLNs) can reverse this resistance. E-PLNs were characterized using techniques including dynamic light scattering, infrared spectroscopy, X-ray diffraction, and differential scanning calorimetry. The results showed that interleukin-6 (IL-6) promotes proliferation, EMT, invasion, and MDR in MCF-7/ADR cells through activation of the JAK2/STAT3 signalling pathway. This pathway was effectively inhibited by the JAK2 inhibitor AG490 or by E-PLNs combined with DOX. The study concludes that E-PLNs may be a promising agent for reversing MDR in breast cancer, potentially improving the efficacy of chemotherapy treatments (Gu, Meng et al. 2024).

The study by Zou et al. focused on overcoming MDR in breast cancer by utilizing polymer-lipid hybrid nanoparticles encapsulating EMO (E-PLNs). The research demonstrated that E-PLNs effectively enhanced the cellular uptake of EMO and restored the sensitivity of drug-resistant MCF-7/ADR breast cancer cells to DOX by inhibiting EMT. The study found that E-PLNs reduced the expression of the drug-resistance protein P-gp as well as apoptosis-related proteins, thereby improving the efficacy of DOX in treating breast cancer. This combination therapy offers a promising strategy to prevent or delay clinical drug resistance (Zou, Lan et al. 2021).

Fisetin (FIS) is a naturally occurring flavonoid with significant anticancer potential, particularly against breast cancer. Its anticancer effects are attributed to its ability to induce apoptosis, inhibit tumour growth, and modulate diverse signalling pathways concerned with cancer development. However, its clinical application has been narrow due to poor aqueous solubility, leading to the exploration of nano-formulation approaches. In a notable study, FIS was complexed with soy phosphatidylcholine and cholesterol to create a novel phytosomal system called "cholephytosomes." The system has been further modified with stearylamine (mPHY) and hyaluronic acid (HPHY) to enhance targeting and cellular uptake (Figure 23.4). Both modified cholephytosomes exhibited high FIS complexation efficiency and improved solubility, with a sustained drug release over 24 hours. Here, the *in vitro* studies using the MDA-MB-231 breast cancer cell line showed that these modified vesicles had significantly lower IC50 values compared to free FIS and conventional drug-phospholipid complexes, indicating superior antitumour activity. The modified cholephytosomes also demonstrated increased cytotoxicity in preclinical models and effectively targeted signalling pathways related to tumorigenesis. This novel phytosomal approach holds a guarantee for improving the clinical utility of FIS in cancer treatment (Talaat, Elnaggar et al. 2024).

Silibinin (SIL) is a chief component of silymarin derived from milk thistle. It is known for its potent antioxidant, anti-inflammatory, and anticancer properties. It has demonstrated effectiveness in inhibiting the growth and

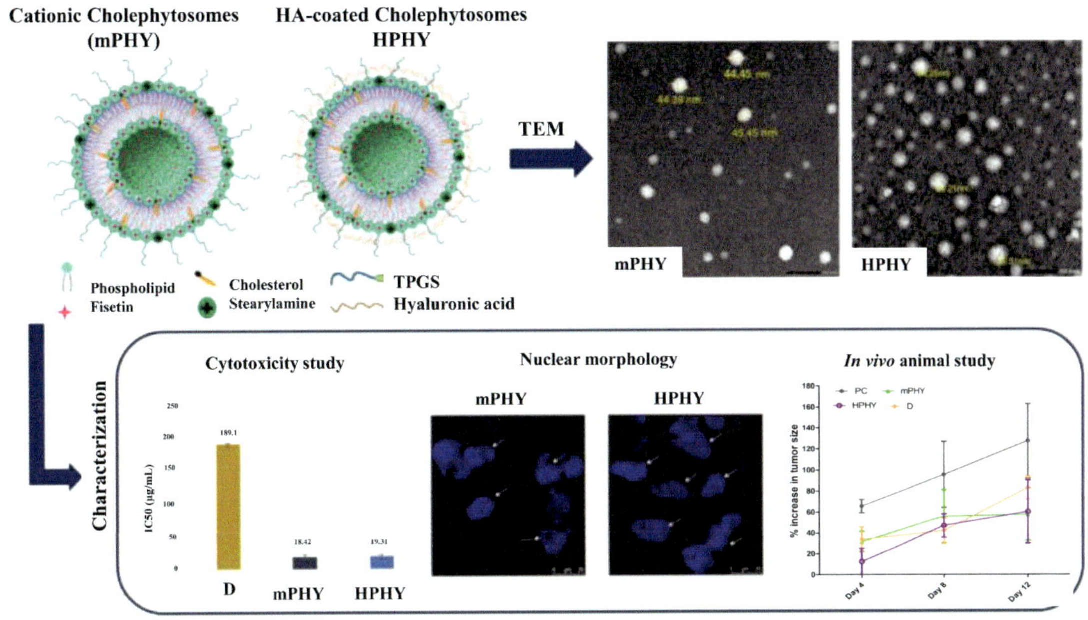

FIGURE 23.4 Presentation of fisetin containing PEGylated cholephytosomes for treatment of MDA-MB-231 breast cancer. [Reprinted (adapted) with permission from ref. (Talaat, Elnaggar et al. 2024). Copyright © 2023, The authors] [D-α-tocopheryl polyethylene glycol succinate: TPGS, Transmission electron microscopy: TEM.]

proliferation of various cancer cells, including colon cancer, through mechanisms such as inducing apoptosis, inhibiting cell cycle progression, and suppressing tumour growth and metastasis. Despite its promising anticancer effects, SIL's clinical use has been restricted due to its poor solubility and bioavailability. To address this, niosome nanocarriers were developed to co-load SIL and methotrexate (MTX) for combination therapy against colon cancer. The niosomes, prepared using the thin film hydration (TFH) method, yielded spherical nanoparticles with an average size of ~100 nm and a zeta potential of −10 mV. The Nio@MS nanoparticles exhibited a significantly lower IC_{50} value (~2.6 μg/mL) compared to free drugs and their combinations, indicating enhanced cytotoxicity. Additionally, Nio@MS decreased the cell adhesion density, promoted apoptosis, and modulated the expression of key apoptosis-related genes, showing improved anti-cancer potential and reduced drug dose requirements by overcoming MDR (Sharifi-Azad, Zenjanab et al. 2024).

Another study investigated the synergistic anticancer effects of metformin (MET) and SIL when co-loaded into PLGA-PEG nanoparticles against T47D breast cancer cells. The PLGA-PEG NPs exhibited round morphology and consistent nanosize distribution. The nanoparticles showed a burst release of the drugs within the first 8 hours. After that, it is followed by sustained release over 7 days. Here, MET/SIL-loaded NPs demonstrated the most effective synergistic cytotoxicity against free drug combinations and single drug-loaded NPs. Additionally, these NPs significantly modulated the expression of apoptotic markers (Bax, Bcl-2, caspase 3 and 9) and hTERT, enhancing the anticancer efficacy of the treatment (Amirsaadat, Jafari-Gharabaghlou et al. 2021).

Ellagic acid (EA) is a polyphenolic molecule. It is found in various fruits, vegetables, and nuts. Known for its antioxidant, anti-inflammatory, and anticancer properties, EA has garnered significant interest in cancer research. It exerts its antitumour effects by modulating multiple cellular pathways, including the inhibition of cell proliferation, induction of apoptosis, and suppression of angiogenesis. Moreover, EA is particularly noted for its ability to interfere with the hedgehog signalling pathway, which is implicated in cancer stem cell maintenance and resistance to conventional therapies. To address drug resistance in cancer therapy, a novel approach was developed using EA and copper ions (Cu^{2+}) to create nanoscale metal-organic frameworks (EA-Cu). These frameworks were loaded with DOX and targeted with chondroitin sulfate (CS) to form the CS/E-C@DOX nanoparticles (CS/NPs). EA inhibited the hedgehog signalling pathway. Thereby, it reduced cancer stemness, while Cu^{2+} disrupted mitochondrial metabolism and induced cuproptosis, enhancing DOX's effectiveness. The CS modification enabled specific tumour targeting, and the EA-Cu framework provided pH/GSH responsive drug release. This strategy significantly increased chemotherapy sensitivity and demonstrated great potential in overcoming cancer chemoresistance (Lu, Tian et al. 2024).

Naringenin is a natural flavonoid found in citrus fruits. It has garnered attention for its potential anticancer properties due to its anti-inflammatory, antioxidant, and anticancer effects. Despite these promising attributes, naringenin's clinical application is hampered by its poor water solubility, low bioavailability, and non-specific distribution. In this context, a study introduced a novel formulation of naringenin, termed NAR-cell penetrating peptide-galactose nanoparticles (NCG), designed to enhance its therapeutic efficacy for HR+ breast cancer. The NCG nanoparticles demonstrated specific liver targeting. Moreover, it improved intestinal barrier permeability in cell and zebrafish xenotransplantation models. Following oral administration in mouse breast cancer xenografts, NCG exhibited targeted liver delivery and effective enterohepatic circulation. This formulation outperformed both free naringenin and the standard treatment, tamoxifen, in terms of cancer inhibition. NCG treatment led to boosted hepatic EST expression and lowered estradiol levels in the blood, liver, and tumour tissue, with minimal side effects obtained. These findings position NCG as a promising candidate for enhancing endocrine therapy, emphasizing the importance of targeting hepatic EST in the management of HR+ breast cancer (Zhao, Tan et al. 2024). In summary, the different types of phytomolecules loaded with advanced nanoformulations demonstrate improved anticancer activity in resistant cancer cells. Phytochemical-based nano-formulations to overcome MDR are summarized in Table 23.1.

TABLE 23.1
Summary of Phytochemicals Integrated in Nanocarriers to Overcome MDR

Drug	Type of Nanoformulation	Type of Cancer	Findings	References
DOX and CUR	GCPQ nanocarrier (DCG)	Breast cancer	Effective release, high cytotoxicity, significant apoptosis	Nurjis, Sarwar et al. (2024)
CUR	PLGA nanoparticles (Cur-NPs)	Cisplatin-resistant human oral cancer (CAR cells)	CUR-NPs induced apoptosis in CAR cells with low cytotoxicity to normal cells. They increased ROS production, upregulated apoptotic proteins, and downregulated anti-apoptotic proteins.	Chang, Peng et al. (2013)
CUR	PLGA CUR Nanoparticles (CUR-NPs)	Osteosarcoma (U2OS Cells)	Induces apoptosis, DNA fragmentation, elevated caspase activity, upregulated cleaved caspases and cytochrome c, downregulated p-Akt	Peng, Lee et al. (2013)
DOX, CUR	Lipid nanoparticles (DOX/CUR-NPs)	Hepatocellular carcinoma (HCC)	Reduced MDR in HCC. Downregulated MDR1, Bcl-2, HIF-1α, and P-gp. Enhanced cytotoxicity, lowered IC50.	Zhao, Chen et al. (2015)
CDDP and CUR	Co-loaded liposomes (CDDP/CUR-Lip)	Hepatocellular carcinoma (HCC)	Synergistic effect, IC50: 0.62 μM. Increased ROS, prolonged retention, improved efficacy.	Cheng, Zhao et al. (2018)
DOX and CUR	Poly(butyl cyanoacrylate) Nanoparticles (PBCA-NPs)	Breast cancer (MCF-7/ADR Cells)	Effective MDR reversal, downregulated P-glycoprotein. Outperformed single-agent PBCA-NPs	Duan, Mansour et al. (2012)
DOX and CUR	(DOX+CUR)-FA-NPs	Breast cancer (MCF-7/ADR Cells)	Enhanced tumour targeting, high cell apoptosis, and improved uptake. Effective MDR reversal reduced normal tissue damage.	Guo, Yu et al. (2021)
CUR	Nanomicelles	Cisplatin-resistant oral cancer	CUR-NMs showed enhanced cellular uptake, cytotoxicity, and apoptosis compared to plain CUR.	Kumbar, Muddapur et al. (2022)
DHA	MNP	Breast cancer (MCF-7, MCF-7/ADR, MDA-MB-231, MDA-MB-453)	MNP-DHA showed strong cytotoxicity and overcame MDR.	Guo, Yao et al. (2020)
PIP	P-PEG-PNP	Breast cancer (MCF-7, PTX-resistant MCF-7)	P-PEG-PNP induced apoptosis in MCF-7 cells and reduced PTX dosage needed for resistant cells.	Pachauri, Gupta et al. (2015)
PIP	Albumin nanoparticles co-encapsulating PTX and PIP (PP@AN)	Triple-negative breast cancer (TNBC)	Enhanced PTX accumulation, increased cell sensitivity, inhibited P-gp, and improved antitumour effects with reduced toxicity.	Xu, Xiao et al. (2023)
RES and PTX	PEGylated liposomes	Drug-resistant tumours	Co-delivery enhanced drug cytotoxicity, bioavailability, and tumour retention, effectively inhibiting tumour growth in mice with reduced toxicity.	Meng, Guo et al. (2016)
PTX and RES	GSH-responsive PRES nanoparticles (~93 nm)	Non-small cell lung cancer (A549)	PTX@PRES NPs effectively overcame MDR, suppressed tumour growth, and enhanced drug efficacy *in vivo*.	Liping, Jian et al. (2022)

(Continued)

TABLE 23.1 (*Continued*)
Summary of Phytochemicals Integrated in Nanocarriers to Overcome MDR

Drug	Type of Nanoformulation	Type of Cancer	Findings	References
RES	Pep-1-PEG3.5k-PCL4k nanoparticles	Anaplastic thyroid carcinoma	RES nanoparticles delayed tumour growth and induced apoptosis, while free resveratrol did not. Docetaxel/DOX caused severe side effects and resistance, overcome by RES.	Xiong, Lin et al. (2021)
QUE	Cetuximab chitosan NPs	Non-small cell lung cancer	QUE enhanced PTX cytotoxicity, inhibited Akt and ERK phosphorylation, and the Cet-CTS NPs reduced tumour growth in PTX-resistant cells and xenografts.	Wang, Yu et al. (2021)
PTX and QUE	MSNs-ChS@PQ NPs	Breast cancer (MCF-7/ADR)	Improved PTX sensitivity, reduced P-gp expression, enhanced tumour targeting, and minimal toxicity.	Liu, Fu et al. (2020)
QUE and Vincristine	Nanoemulsion	MDR and non-MDR cells	Enhanced cytotoxicity, apoptosis, DNA damage, altered ABCB1 expression, and increased bioavailability.	Marques, Machado et al. (2021)
EMO	Polymer lipid hybrid nanoparticles	Breast cancer (MCF-7/ADR)	E-PLNs combined with DOX reversed MDR by inhibiting the JAK2/STAT3 pathway, suppressing EMT, and reducing cell proliferation and invasion.	Gu, Meng et al. (2024)
EMO+DOX	Polymer lipid hybrid nanoparticles (E-PLNs)	Breast cancer (MCF-7/ADR)	E-PLNs enhance EMO uptake and restore DOX sensitivity in drug-resistant breast cancer cells by inhibiting EMT, reducing P-gp expression, and promoting apoptosis.	Zou, Lan et al. (2021)
FIS	Cholephytosomes (mPHY, HPHY)	Breast cancer	Modified cholephytosomes enhanced cytotoxicity, targeting, and inhibition of tumour-related pathways compared to free FIS.	Talaat, Elnaggar et al. (2024)
SIL	Niosomes (Nio@MS)	Colon cancer (CT26)	Nio@MS showed improved cytotoxicity (IC50 ~2.6 µg/mL), reduced cell adhesion, promoted apoptosis, and modulated apoptosis-related gene expression compared to free SIL, MTX, and their combinations.	Sharifi-Azad, Zenjanab et al. (2024)
MET and SIL	PLGA-PEG NPs	Breast cancer	MET/SIL-loaded NPs showed synergistic cytotoxicity, burst release followed by sustained release, and altered expression of apoptosis markers.	Amirsaadat, Jafari-Gharabaghlou et al. (2021)
EA, DOX	EA-Cu/CS nanoplatform	Various cancers	Enhanced chemotherapy sensitivity by reducing stemness, inducing cuproptosis, and providing targeted, pH/GSH responsive release.	Lu, Tian et al. (2024)
Naringenin	Naringenin with cell-penetrating peptide-galactose nanoparticles (NCG)	Hormone receptor-positive (HR+) breast cancer	NCG showed specific liver targeting, improved intestinal barrier permeability, and superior cancer inhibition compared to free NAR and tamoxifen. It increased hepatic EST expression and reduced estradiol levels with minimal side effects.	Zhao, Tan et al. (2024)

23.7 CURRENT CHALLENGES, PROSPECTS, AND CONCLUSION

Overcoming drug resistance to cancer is one of the most pressing challenges faced by the scientific community today. Numerous cancer cells exhibit resistance to conventional synthetic anticancer drugs, posing a significant obstacle to effective cancer therapy. The current research landscape is increasingly focused on leveraging phytomolecules, naturally occurring plant-derived compounds, as potential treatments for drug-resistant cancer cells. Phytomolecules such as CUR, QUE, RES, and PIP have shown promise in enhancing anticancer activity against resistant cancer cells. Advanced nanocarriers have been designed to deliver these phytomolecules directly to resistant cancer cells, and combinations of phytomolecules with synthetic drugs have demonstrated notable improvements in cytotoxicity against these cells. However, several challenges remain. A critical challenge is the incomplete understanding of the underlying mechanisms of drug resistance, which can undermine the efficacy of phytomolecules. Moreover, the toxicity and potential side effects of these phytomolecules must be thoroughly evaluated. To address these issues, it is crucial to develop suitable nanoformulations that minimize side effects and toxicity while improving the bioavailability and therapeutic efficacy of phytomolecules. The poor dissolution, solubility, and low bioavailability of many phytomolecules limit their pharmacological effectiveness, restricting their clinical applications. Additionally, ensuring the specificity and targeting of phytomolecules to resistant cancer cells without harming normal cells is another major challenge. In the future, efforts must focus on overcoming these limitations to enhance the clinical utility of phytomolecules. Advanced nanosystems need to be designed to improve the solubility, stability, and targeted delivery of phytomolecules with minimal toxicity. Combination therapy, where phytomolecules are used alongside conventional anticancer drugs, holds potential for enhancing therapeutic outcomes against resistant cancer types. Furthermore, the use of stimuli-responsive polymers for the delivery of phytomolecules could enable controlled release specifically at the site of resistant cancer cells, thereby enhancing the anticancer potential of these natural compounds.

In conclusion, incorporating phytomolecules into advanced nano-delivery systems offers a promising approach to overcoming drug resistance in cancer. Widely studied phytomolecules such as CUR, QUE, PIP, EA, RES, and others have demonstrated improved anticancer efficacy against resistant cancers in both in vitro and in vivo studies. Nanosystems provide significant advantages, including improved solubility, bioavailability, targeted delivery, and reduced side effects, all of which contribute to enhanced therapeutic outcomes against resistant cancer cells. Moreover, published studies suggest that phytomolecule-based nanodelivery systems can be effectively used to address multidrug resistance when combined with conventional treatments, offering synergistic benefits in treating resistant cancers. Continued development of phytomolecule-incorporated nanodelivery systems could provide viable options for patients with drug-resistant tumours, offering new hope in the ongoing battle against cancer.

ACKNOWLEDGEMENT

The authors are thankful to Dr Namdeo R. Jadhav, Dean, Krishna Institute of Pharmacy, Krishna Vishwa Vidyapeeth (Deemed to be University), Karad, Maharashtra state, India, for his assistance during the writing of this book chapter.

REFERENCES

Aggarwal, B. B., A. B. Kunnumakkara, K. B. Harikumar, S. T. Tharakan, B. Sung and P. Anand (2008). Potential of spice-derived phytochemicals for cancer prevention. *Planta Medica* 74(13): 1560–1569.

Aggarwal, B. B., C. Sundaram, N. Malani and H. Ichikawa (2007). Curcumin: The Indian solid gold. *Advances in Experimental Medicine and Biology* 595: 1–75.

Almatroodi, S. A., M. A. Alsahli, A. Almatroudi, A. K. Verma, A. Aloliqi, K. S. Allemailem, A. A. Khan and A. H. Rahmani (2021). Potential therapeutic targets of quercetin, a plant flavonol, and its role in the therapy of various types of cancer through the modulation of various cell signaling pathways. *Molecules* 26(5): 1315.

Amirsaadat, S., D. Jafari-Gharabaghlou, S. Alijani, H. Mousazadeh, M. Dadashpour and N. Zarghami (2021). Metformin and Silibinin co-loaded PLGA-PEG nanoparticles for effective combination therapy against human breast cancer cells. *Journal of Drug Delivery Science and Technology* 61: 102107.

Anand, U., A. Dey, A. K. S. Chandel, R. Sanyal, A. Mishra, D. K. Pandey, V. De Falco, A. Upadhyay, R. Kandimalla and A. Chaudhary (2023). Cancer chemotherapy and beyond: Current status, drug candidates, associated risks and progress in targeted therapeutics. *Genes & Diseases* 10(4): 1367–1401.

Anwanwan, D., S. K. Singh, S. Singh, V. Saikam and R. Singh (2020). Challenges in liver cancer and possible treatment approaches. *Biochimica et Biophysica Acta (BBA)-Reviews on Cancer* 1873(1): 188314.

Asgharian, P., A. P. Tazekand, K. Hosseini, H. Forouhandeh, T. Ghasemnejad, M. Ranjbar, M. Hasan, M. Kumar, S. M. Beirami and V. Tarhriz (2022). Potential mechanisms of quercetin in cancer prevention: Focus on cellular and molecular targets. *Cancer Cell International* 22(1): 257.

Attia, M. F., N. Anton, J. Wallyn, Z. Omran and T. F. Vandamme (2019). An overview of active and passive targeting strategies to improve the nanocarriers efficiency to tumour sites. *Journal of Pharmacy and Pharmacology* 71(8): 1185–1198.

Augustin, Y., H. M. Staines and S. Krishna (2020). Artemisinins as a novel anti-cancer therapy: Targeting a global cancer pandemic through drug repurposing. *Pharmacology & Therapeutics* 216: 107706.

Bagheri, E., K. Abnous, S. A. Farzad, S. M. Taghdisi, M. Ramezani and M. Alibolandi (2020). Targeted doxorubicin-loaded mesenchymal stem cells-derived exosomes as a versatile platform for fighting against colorectal cancer. *Life Sciences* 261: 118369.

Banik, K., A. M. Ranaware, V. Deshpande, S. P. Nalawade, G. Padmavathi, D. Bordoloi, B. L. Sailo, M. K. Shanmugam, L. Fan and F. Arfuso (2019). Honokiol for cancer therapeutics: A traditional medicine that can modulate multiple oncogenic targets. *Pharmacological Research* 144: 192–209.

Berman, A. Y., R. A. Motechin, M. Y. Wiesenfeld and M. K. Holz (2017). The therapeutic potential of resveratrol: A review of clinical trials. *NPJ Precision Oncology* 1(1): 35.

Bray, F., J. Ferlay, I. Soerjomataram, R. L. Siegel, L. A. Torre and A. Jemal (2018). Global cancer statistics 2018: GLOBOCAN estimates of incidence and mortality worldwide for 36 cancers in 185 countries. *CA: A Cancer Journal for Clinicians* 68(6): 394–424.

Bray, F., M. Laversanne, H. Sung, J. Ferlay, R. L. Siegel, I. Soerjomataram and A. Jemal (2024). Global cancer statistics 2022: GLOBOCAN estimates of incidence and mortality worldwide for 36 cancers in 185 countries. *CA: A Cancer Journal for Clinicians* 74(3): 229–263.

Brown, J. S., S. R. Amend, R. H. Austin, R. A. Gatenby, E. U. Hammarlund and K. J. Pienta (2023). Updating the definition of cancer. *Molecular Cancer Research* 21(11): 1142–1147.

Bruni, D., H. K. Angell and J. Galon (2020). The immune contexture and immunoscore in cancer prognosis and therapeutic efficacy. *Nature Reviews Cancer* 20(11): 662–680.

Cao, B., H. Chen, Y. Gao, C. Niu, Y. Zhang and L. Li (2015). CIP-36, a novel topoisomerase II-targeting agent, induces the apoptosis of multidrug-resistant cancer cells in vitro. *International Journal of Molecular Medicine* 35(3): 771–776.

Carter, L. G., J. A. D'Orazio and K. J. Pearson (2014). Resveratrol and cancer: Focus on in vivo evidence. *Endocrine-Related Cancer* 21(3): R209–R225.

Chang, P.-Y., S.-F. Peng, C.-Y. Lee, C.-C. Lu, S.-C. Tsai, T.-M. Shieh, T.-S. Wu, M.-G. Tu, M. Y. Chen and J.-S. Yang (2013). Curcumin-loaded nanoparticles induce apoptotic cell death through regulation of the function of MDR1 and reactive oxygen species in cisplatin-resistant CAR human oral cancer cells. *International Journal of Oncology* 43(4): 1141–1150.

Chauhan, A., A. U. Islam, H. Prakash and S. Singh (2022). Phytochemicals targeting NF-κB signaling: Potential anti-cancer interventions. *Journal of Pharmaceutical Analysis* 12(3): 394–405.

Chen, Y., X.-Q. Wang, Q. Zhang, J.-Y. Zhu, Y. Li, C.-F. Xie, X.-T. Li, J.-S. Wu, S.-S. Geng and C.-Y. Zhong (2017). (−)-Epigallocatechin-3-gallate inhibits colorectal cancer stem cells by suppressing Wnt/β-catenin pathway. *Nutrients* 9(6): 572.

Cheng, Y., P. Zhao, S. Wu, T. Yang, Y. Chen, X. Zhang, C. He, C. Zheng, K. Li and X. Ma (2018). Cisplatin and curcumin co-loaded nano-liposomes for the treatment of hepatocellular carcinoma. *International Journal of Pharmaceutics* 545(1–2): 261–273.

Cho, C. J., C. P. Yu, C. L. Wu, J. Y. Ho, C. W. Yang and D. S. Yu (2021). Decreased drug resistance of bladder cancer using phytochemicals treatment. *The Kaohsiung Journal of Medical Sciences* 37(2): 128–135.

Chow, E. K., X.-Q. Zhang, M. Chen, R. Lam, E. Robinson, H. Huang, D. Schaffer, E. Osawa, A. Goga and D. Ho (2011). Nanodiamond therapeutic delivery agents mediate enhanced chemoresistant tumour treatment. *Science Translational Medicine* 3(73): 73ra21–73ra21.

Chu, D.-T., T. T. Nguyen, N. L. B. Tien, D.-K. Tran, J.-H. Jeong, P. G. Anh, V. V. Thanh, D. T. Truong and T. C. Dinh (2020). Recent progress of stem cell therapy in cancer treatment: Molecular mechanisms and potential applications. *Cells* 9(3): 563.

Dadwal, A., A. Baldi and R. Kumar Narang (2018). Nanoparticles as carriers for drug delivery in cancer. *Artificial Cells, Nanomedicine, and Biotechnology* 46(sup2): 295–305.

Dehelean, C. A., I. Marcovici, C. Soica, M. Mioc, D. Coricovac, S. Iurciuc, O. M. Cretu and I. Pinzaru (2021). Plant-derived anticancer compounds as new perspectives in drug discovery and alternative therapy. *Molecules* 26(4): 1109.

Deli, T., M. Orosz and A. Jakab (2020). Hormone replacement therapy in cancer survivors: Review of the literature. *Pathology & Oncology Research* 26(1): 63–78.

Dewi, M. K., A. Y. Chaerunisaa, M. Muhaimin and I. M. Joni (2022). Improved activity of herbal medicines through nanotechnology. *Nanomaterials* 12(22): 4073.

Dhyani, P., C. Quispe, E. Sharma, A. Bahukhandi, P. Sati, D. C. Attri, A. Szopa, J. Sharifi-Rad, A. O. Docea and I. Mardare (2022). Anticancer potential of alkaloids: A key emphasis to colchicine, vinblastine, vincristine, vindesine, vinorelbine and vincamine. *Cancer Cell International* 22(1): 206.

Dong, Z., Y. Hao, Q. Li, Z. Yang, Y. Zhu, Z. Liu and L. Feng (2020). Metal-polyphenol-network coated $CaCO_3$ as pH-responsive nanocarriers to enable effective intratumoural penetration and reversal of multidrug resistance for augmented cancer treatments. *Nano Research* 13: 3057–3067.

Duan, J., H. M. Mansour, Y. Zhang, X. Deng, Y. Chen, J. Wang, Y. Pan and J. Zhao (2012). Reversion of multidrug resistance by co-encapsulation of doxorubicin and curcumin in chitosan/poly (butyl cyanoacrylate) nanoparticles. *International Journal of Pharmaceutics* 426(1–2): 193–201.

Durinikova, E., K. Buzo and S. Arena (2021). Preclinical models as patients' avatars for precision medicine in colorectal cancer: Past and future challenges. *Journal of Experimental & Clinical Cancer Research* 40(1): 185.

Efferth, T., O. Kadioglu, M. E. Saeed, E.-J. Seo, A. T. Mbaveng and V. Kuete (2021). Medicinal plants and phytochemicals against multidrug-resistant tumour cells expressing ABCB1, ABCG2, or ABCB5: a synopsis of 2 decades. *Phytochemistry Reviews* 20: 7–53.

Elgendy, S. M., S. K. Alyammahi, D. W. Alhamad, S. M. Abdin and H. A. Omar (2020). Ferroptosis: An emerging approach for targeting cancer stem cells and drug resistance. *Critical Reviews in Oncology/Hematology* 155: 103095.

Elumalai, K., S. Srinivasan and A. Shanmugam (2024). Review of the efficacy of nanoparticle-based drug delivery systems for cancer treatment. *Biomedical Technology* 5: 109–122.

Ersoz, M., A. Erdemir, S. Derman, T. Arasoglu and B. Mansuroglu (2020). Quercetin-loaded nanoparticles enhance cytotoxicity and antioxidant activity on C6 glioma cells. *Pharmaceutical Development and Technology* 25(6): 757–766.

Fowler, H., A. Belot, L. Ellis, C. Maringe, M. A. Luque-Fernandez, E. N. Njagi, N. Navani, D. Sarfati and B. Rachet (2020). Comorbidity prevalence among cancer patients: A population-based cohort study of four cancers. *BMC Cancer* 20: 1–15.

Fulda, S. (2010). Resveratrol and derivatives for the prevention and treatment of cancer. *Drug Discovery Today* 15(17–18): 757–765.

Ganthala, P. D., S. Alavala, N. Chella, S. B. Andugulapati, N. B. Bathini and R. Sistla (2022). Co-encapsulated nanoparticles of Erlotinib and Quercetin for targeting lung cancer through nuclear EGFR and PI3K/AKT inhibition. *Colloids and Surfaces B: Biointerfaces* 211: 112305.

Garg, P., J. Malhotra, P. Kulkarni, D. Horne, R. Salgia and S. S. Singhal (2024). Emerging therapeutic strategies to overcome drug resistance in cancer cells. *Cancers* 16(13): 2478.

Gavas, S., S. Quazi and T. M. Karpiński (2021). Nanoparticles for cancer therapy: Current progress and challenges. *Nanoscale Research Letters* 16(1): 173.

Ghosh, B. and S. Biswas (2021). Polymeric micelles in cancer therapy: State of the art. *Journal of Controlled Release* 332: 127–147.

Gillet, J.-P. and M. M. Gottesman (2010). Mechanisms of multidrug resistance in cancer. *Methods in Molecular Biology* 596: 47–76.

Giovannucci, E. (1999). Tomatoes, tomato-based products, lycopene, and cancer: Review of the epidemiologic literature. *Journal of the National Cancer Institute* 91(4): 317–331.

Gomes, H. I., C. S. Martins and J. A. Prior (2021). Silver nanoparticles as carriers of anticancer drugs for efficient target treatment of cancer cells. *Nanomaterials* 11(4): 964.

Graham, J., M. Quinn, D. Fabricant and N. Farnsworth (2000). Plants used against cancer: An extension of the work of Jonathan Hartwell. *Journal of Ethnopharmacology* 73(3): 347–377.

Gu, H., F. Meng, H. Sun, L. Yang, Q. Li, Z. Chen, T. Cai, Z. Yang and Y. Cai (2024). Polymer lipid hybrid nanoparticles encapsulated with Emodin combined with DOX reverse multidrug resistance of breast cancer via IL-6/JAK2/STAT3 signaling pathway. *Cancer Nanotechnology* 15(1): 10.

Guo, F., N. Yu, Y. Jiao, W. Hong, K. Zhou, X. Ji, H. Yuan, H. Wang, A. Li and G. Wang (2021). Star polyester-based folate acid-targeting nanoparticles for doxorubicin and curcumin co-delivery to combat multidrug-resistant breast cancer. *Drug Delivery* 28(1): 1709–1721.

Guo, S., X. Yao, Q. Jiang, K. Wang, Y. Zhang, H. Peng, J. Tang and W. Yang (2020). Dihydroartemisinin-loaded magnetic nanoparticles for enhanced chemodynamic therapy. *Frontiers in Pharmacology* 11: 226.

Gupta, S. C., R. Kannappan, S. Reuter, J. H. Kim and B. B. Aggarwal (2011). Chemosensitization of tumours by resveratrol. *Annals of the New York Academy of Sciences* 1215(1): 150–160.

Gwinn, M. and V. Vallyathan (2006). Nanoparticles: Health effects-pros and cons. *Environmental Health Perspectives* 114: 1818–1825. Find this article online.

Haibe, Y., M. Kreidieh, H. El Hajj, I. Khalifeh, D. Mukherji, S. Temraz and A. Shamseddine (2020). Resistance mechanisms to anti-angiogenic therapies in cancer. *Frontiers in Oncology* 10: 221.

Haider, T., V. Pandey, N. Banjare, P. N. Gupta and V. Soni (2020). Drug resistance in cancer: Mechanisms and tackling strategies. *Pharmacological Reports* 72(5): 1125–1151.

Hassanpour, S. H. and M. Dehghani (2017). Review of cancer from perspective of molecular. *Journal of Cancer Research and Practice* 4(4): 127–129.

Hirmerová, J., J. Seidlerová, I. Šubrt and Z. Hajšmanová (2022). Prevalence of cancer in patients with superficial vein thrombosis and its clinical importance. *Journal of Vascular Surgery: Venous and Lymphatic Disorders* 10(1): 26–32.

Holohan, C., S. Van Schaeybroeck, D. B. Longley and P. G. Johnston (2013). Cancer drug resistance: An evolving paradigm. *Nature Reviews Cancer* 13(10): 714–726.

Housman, G., S. Byler, S. Heerboth, K. Lapinska, M. Longacre, N. Snyder and S. Sarkar (2014). Drug resistance in cancer: An overview. *Cancers* 6(3): 1769–1792.

Joo, J.-H., S.-S. Hong, Y.-R. Cho and D.-W. Seo (2016). 10-Gingerol inhibits proliferation and invasion of MDA-MB-231 breast cancer cells through suppression of Akt and p38MAPK activity. *Oncology Reports* 35(2): 779–784.

Kairdolf, B. A., A. M. Smith, T. H. Stokes, M. D. Wang, A. N. Young and S. Nie (2013). Semiconductor quantum dots for bioimaging and biodiagnostic applications. *Annual Review of Analytical Chemistry* 6(1): 143–162.

Kamaly, N., Z. Xiao, P. M. Valencia, A. F. Radovic-Moreno and O. C. Farokhzad (2012). Targeted polymeric therapeutic nanoparticles: Design, development and clinical translation. *Chemical Society Reviews* 41(7): 2971–3010.

Kapse-Mistry, S., T. Govender, R. Srivastava and M. Yergeri (2014). Nanodrug delivery in reversing multidrug resistance in cancer cells. *Frontiers in Pharmacology* 5: 159.

Karimi, M., F. Gheybi, P. Zamani, M. Mashreghi, S. Golmohammadzadeh, S. A. Darban, A. Badiee and M. R. Jaafari (2020). Preparation and characterization of stable nanoliposomal formulations of curcumin with high loading efficacy: In vitro and in vivo anti-tumour study. *International Journal of Pharmaceutics* 580: 119211.

Katt, M. E., A. L. Placone, A. D. Wong, Z. S. Xu and P. C. Searson (2016). In vitro tumour models: Advantages, disadvantages, variables, and selecting the right platform. *Frontiers in Bioengineering and Biotechnology* 4: 12.

Khalid, K., X. Tan, H. F. Mohd Zaid, Y. Tao, C. Lye Chew, D.-T. Chu, M. K. Lam, Y.-C. Ho, J. W. Lim and L. Chin Wei (2020). Advanced in developmental organic and inorganic nanomaterial: A review. *Bioengineered* 11(1): 328–355.

Khan, N., F. Afaq, M. Saleem, N. Ahmad and H. Mukhtar (2006). Targeting multiple signaling pathways by green tea polyphenol (−)-epigallocatechin-3-gallate. *Cancer Research* 66(5): 2500–2505.

Khatoon, E., K. Banik, C. Harsha, B. L. Sailo, K. K. Thakur, A. D. Khwairakpam, R. Vikkurthi, T. B. Devi, S. C. Gupta and A. B. Kunnumakkara (2022). Phytochemicals in cancer cell chemosensitization: Current knowledge and future perspectives. *Seminars in Cancer Biology*, 80: 306–339.

Kim, H. J., A. Kim, K. Miyata and K. Kataoka (2016). Recent progress in development of siRNA delivery vehicles for cancer therapy. *Advanced Drug Delivery Reviews* 104: 61–77.

Klochkov, S. G., M. E. Neganova, V. N. Nikolenko, K. Chen, S. G. Somasundaram, C. E. Kirkland and G. Aliev (2021). Implications of nanotechnology for the treatment of cancer: Recent advances. *Seminars in Cancer Biology*, 69: 190–199.

Kobayashi, H., R. Watanabe and P. L. Choyke (2014). Improving conventional enhanced permeability and retention (EPR) effects; what is the appropriate target? *Theranostics* 4(1): 81.

Koklesova, L., A. Liskova, M. Samec, C. Buhrmann, S. M. Samuel, E. Varghese, M. Ashrafizadeh, M. Najafi, M. Shakibaei and D. Büsselberg (2020). Carotenoids in cancer apoptosis: The road from bench to bedside and back. *Cancers* 12(9): 2425.

Koklesova, L., J. Jakubikova, D. Cholujova, M. Samec, A. Mazurakova, M. Šudomová, M. Pec, S. T. Hassan, K. Biringer and D. Büsselberg (2023). Phytochemical-based nanodrugs going beyond the state-of-the-art in cancer management: Targeting cancer stem cells in the framework of predictive, preventive, personalized medicine. *Frontiers in Pharmacology* 14: 1121950.

Kola, I. (2008). The state of innovation in drug development. *Clinical Pharmacology & Therapeutics* 83(2): 227–230.

Kotsakis, A., A. Matikas, F. Koinis, N. Kentepozidis, I. Varthalitis, V. Karavassilis, E. Samantas, P. Katsaounis, E. Dermitzaki and D. Hatzidaki (2016). A multicentre phase II trial of cabazitaxel in patients with advanced non-small-cell lung cancer progressing after docetaxel-based chemotherapy. *British Journal of Cancer* 115(7): 784–788.

Kumar, P., N. Yadav, B. Chaudhary, V. Jain, V. M. Balaramnavar, K. S. Alharbi, S. K. Alenezi, W. H. Al-Malki, M. M. Ghoneim and S. Alshehri (2022). Promises of phytochemical based nano drug delivery systems in the management of cancer. *Chemico-Biological Interactions* 351: 109745.

Kumbar, V. M., U. Muddapur, A. Bin Muhsinah, S. A. Alshehri, M. M. Alshahrani, I. A. Almazni, M. S. Kugaji, K. Bhat, M. R. Peram and M. H. Mahnashi (2022). Curcumin-encapsulated nanomicelles improve cellular uptake and cytotoxicity in cisplatin-resistant human oral cancer cells. *Journal of Functional Biomaterials* 13(4): 158.

Kumbhar, P., O. Patil, P. Karade, R. Bhokare, G. Gupta, S. K. Singh, K. Dua, J. Disouza and V. Patravale (2024). Metal-based nanoparticles in the treatment of infectious diseases. *Nanoscience & Nanotechnology-Asia* 14(2): 69–78.

Labrie, M., J. S. Brugge, G. B. Mills and I. K. Zervantonakis (2022). Therapy resistance: Opportunities created by adaptive responses to targeted therapies in cancer. *Nature Reviews Cancer* 22(6): 323–339.

Lan, J. (2022). Overview of application of nanomaterials in medical domain. *Contrast Media & Molecular Imaging* 2022(1): 3507383.

Li, B., H. Xu, Z. Li, M. Yao, M. Xie, H. Shen, S. Shen, X. Wang and Y. Jin (2012). Bypassing multidrug resistance in human breast cancer cells with lipid/polymer particle assemblies. *International Journal of Nanomedicine* 7: 187–197.

Li, X., J. F. Lovell, J. Yoon and X. Chen (2020). Clinical development and potential of photothermal and photodynamic therapies for cancer. *Nature Reviews Clinical Oncology* 17(11): 657–674.

Liping, Y., H. Jian, T. Zhenchao, Z. Yan, Y. Jing, Z. Yangyang, G. Jing and Q. Liting (2022). GSH-responsive poly-resveratrol based nanoparticles for effective drug delivery and reversing multidrug resistance. *Drug Delivery* 29(1): 229–237.

Liu, M., B. Wang, C. Guo, X. Hou, Z. Cheng and D. Chen (2019). Novel multifunctional triple folic acid, biotin and CD44 targeting pH-sensitive nano-actiniaes for breast cancer combinational therapy. *Drug Delivery* 26(1): 1002–1016.

Liu, M., M. Fu, X. Yang, G. Jia, X. Shi, J. Ji, X. Liu and G. Zhai (2020). Paclitaxel and quercetin co-loaded functional mesoporous silica nanoparticles overcoming multidrug resistance in breast cancer. *Colloids and Surfaces B: Biointerfaces* 196: 111284.

Liu, Y., E.-O. Nwafor, Z. Li, J. Wang, X. Feng, H. Li, B. Jia, H. Ma, J. He and J. Pi (2022). Effects of berberine on liver cancer. *Natural Product Communications* 17(5): 1934578X221102032.

Liu, Z., Q. Liu, B. Xu, J. Wu, C. Guo, F. Zhu, Q. Yang, G. Gao, Y. Gong and C. Shao (2009). Berberine induces p53-dependent cell cycle arrest and apoptosis of human osteosarcoma cells by inflicting DNA damage. *Mutation Research/Fundamental and Molecular Mechanisms of Mutagenesis* 662(1–2): 75–83.

Loo, Y. S., T. Madheswaran, R. Rajendran and R. J. Bose (2020). Encapsulation of berberine into liquid crystalline nanoparticles to enhance its solubility and anticancer activity in MCF7 human breast cancer cells. *Journal of Drug Delivery Science and Technology* 57: 101756.

Lu, S., H. Tian, B. Li, L. Li, H. Jiang, Y. Gao, L. Zheng, C. Huang, Y. Zhou and Z. Du (2024). An ellagic acid coordinated copper-based nanoplatform for efficiently overcoming cancer chemoresistance by cuproptosis and synergistic inhibition of cancer cell stemness. *Small* 20(17): 2309215.

Ma, Y., B. He, M. Jiang, Y. Yang, C. Wang, C. Huang and L. Han (2020). Prevalence and risk factors of cancer-related fatigue: A systematic review and meta-analysis. *International Journal of Nursing Studies* 111: 103707.

Maleki Dana, P., F. Sadoughi, Z. Asemi and B. Yousefi (2022). The role of polyphenols in overcoming cancer drug resistance: A comprehensive review. *Cellular & Molecular Biology Letters* 27: 1–26.

Mansoori, B., A. Mohammadi, S. Davudian, S. Shirjang and B. Baradaran (2017). The different mechanisms of cancer drug resistance: A brief review. *Advanced Pharmaceutical Bulletin* 7(3): 339.

Marques, A. V. L., Ruginsk, B. E., de Oliveira Prado, L., de Lima, D. E., Daniel, I. W., Moure, V. R., & Valdameri, G. (2024). The association of ABC proteins with multidrug resistance in cancer. *Biochimica et Biophysica Acta (BBA)-Molecular Cell Research* 1872: 119878.

Marques, M. B., A. P. Machado, P. A. Santos, M. Carrett-Dias, G. S. Araújo, B. da Silva Alves, B. S. de Oliveira, F. M. da Silva Júnior, C. L. Dora and A. D. Cañedo (2021). Anti-MDR effects of quercetin and its nanoemulsion in multidrug-resistant human leukemia cells. *Anti-Cancer Agents in Medicinal Chemistry (Formerly Current Medicinal Chemistry-Anti-Cancer Agents)* 21(14): 1911–1920.

Matsumura, Y. and H. Maeda (1986). A new concept for macromolecular therapeutics in cancer chemotherapy: Mechanism of tumouritropic accumulation of proteins and the antitumour agent smancs. *Cancer Research* 46(12_Part_1): 6387–6392.

Meng, J., F. Guo, H. Xu, W. Liang, C. Wang and X.-D. Yang (2016). Combination therapy using co-encapsulated resveratrol and paclitaxel in liposomes for drug resistance reversal in breast cancer cells in vivo. *Scientific Reports* 6(1): 22390.

Michaelis, M., M. N. Wass and J. Cinatl Jr (2019). Drug-adapted cancer cell lines as preclinical models of acquired resistance. *Cancer Drug Resistance* 2(3): 447.

Min, K. H., K. Park, Y.-S. Kim, S. M. Bae, S. Lee, H. G. Jo, R.-W. Park, I.-S. Kim, S. Y. Jeong and K. Kim (2008). Hydrophobically modified glycol chitosan nanoparticles-encapsulated camptothecin enhance the drug stability and tumour targeting in cancer therapy. *Journal of Controlled Release* 127(3): 208–218.

Moosavi, M. A., A. Haghi, M. Rahmati, H. Taniguchi, A. Mocan, J. Echeverría, V. K. Gupta, N. T. Tzvetkov and A. G. Atanasov (2018). Phytochemicals as potent modulators of autophagy for cancer therapy. *Cancer Letters* 424: 46–69.

Muley, H., R. Fado, R. Rodriguez-Rodriguez and N. Casals (2020). Drug uptake-based chemoresistance in breast cancer treatment. *Biochemical Pharmacology* 177: 113959.

Nurjis, F., U. Sarwar, J. S. Ali and M. Fayyaz (2024). Doxorubicin and curcumin-loaded nanomicelles targeting multidrug resistance in cancer. BioNanoScience 14: 1–11.

Omidi, Y. and J. Barar (2014). Targeting tumour microenvironment: Crossing tumour interstitial fluid by multifunctional nanomedicines. *BioImpacts: BI* 4(2): 55.

Ostańska, E., D. Aebisher and D. Bartusik-Aebisher (2021). The potential of photodynamic therapy in current breast cancer treatment methodologies. *Biomedicine & Pharmacotherapy* 137: 111302.

Pachauri, M., E. D. Gupta and P. C. Ghosh (2015). Piperine loaded PEG-PLGA nanoparticles: Preparation, characterization and targeted delivery for adjuvant breast cancer chemotherapy. *Journal of Drug Delivery Science and Technology* 29: 269–282.

Palazzolo, S., S. Bayda, M. Hadla, I. Caligiuri, G. Corona, G. Toffoli and F. Rizzolio (2018). The clinical translation of organic nanomaterials for cancer therapy: A focus on polymeric nanoparticles, micelles, liposomes and exosomes. *Current Medicinal Chemistry* 25(34): 4224–4268.

Peng, S.-F., C.-Y. Lee, M.-J. Hour, S.-C. Tsai, D.-H. Kuo, F.-A. Chen, P.-C. Shieh and J.-S. Yang (2013). Curcumin-loaded nanoparticles enhance apoptotic cell death of U2OS human osteosarcoma cells through the Akt-Bad signaling pathway. *International Journal of Oncology* 44(1): 238–246.

Piktel, E., K. Niemirowicz, M. Wątek, T. Wollny, P. Deptuła and R. Bucki (2016). Recent insights in nanotechnology-based drugs and formulations designed for effective anti-cancer therapy. *Journal of Nanobiotechnology* 14: 1–23.

Pistritto, G., D. Trisciuoglio, C. Ceci, A. Garufi and G. D'Orazi (2016). Apoptosis as anticancer mechanism: Function and dysfunction of its modulators and targeted therapeutic strategies. *Aging* 8(4): 603.

Prakash, O., A. Kumar and P. Kumar (2013). Anticancer potential of plants and natural products. *American Journal of Pharmacological Sciences* 1(6): 104–115.

Prasad, S., V. R. Yadav, B. Sung, S. Reuter, R. Kannappan, A. Deorukhkar, P. Diagaradjane, C. Wei, V. Baladandayuthapani and S. Krishnan (2012). Ursolic acid inhibits growth and metastasis of human colorectal cancer in an orthotopic nude mouse model by targeting multiple cell signaling pathways: Chemosensitization with capecitabine. *Clinical Cancer Research* 18(18): 4942–4953.

Radha, R., V. Paul, S. Anjum, A. Bouakaz, W. G. Pitt and G. A. Husseini (2024). Enhancing Curcumin's therapeutic potential in cancer treatment through ultrasound mediated liposomal delivery. *Scientific Reports* 14(1): 10499.

Rezayatmand, H., M. Razmkhah and I. Razeghian-Jahromi (2022). Drug resistance in cancer therapy: The Pandora's Box of cancer stem cells. *Stem Cell Research & Therapy* 13(1): 181.

Safhi, M. M., S. M. Sivakumar, A. Jabeen, F. Zakir, F. Islam, T. Anwer, U. S. Bagul, M. E. Elmobark, G. Khan and R. Siddiqui (2017). Nanoparticle system for anticancer drug delivery: Targeting to overcome multidrug resistance. In: Grumezescu, A. M. (Ed.) *Multifunctional Systems for Combined Delivery, Biosensing and Diagnostics*, Elsevier, pp. 159–169.

Sajjad, H., S. Imtiaz, T. Noor, Y. H. Siddiqui, A. Sajjad and M. Zia (2021). Cancer models in preclinical research: A chronicle review of advancement in effective cancer research. *Animal Models and Experimental Medicine* 4(2): 87–103.

Sharifi-Azad, M., M. K. Zenjanab, M. Shahpouri, M. A. Adili-Aghdam, M. Fathi and R. Jahanban-Esfahlan (2024). Codelivery of methotrexate and silibinin by niosome nanoparticles for enhanced chemotherapy of CT26 colon cancer cells. *Biomedical Materials* 19(5): 055015.

Sharma, R. A., A. J. Gescher and W. P. Steward (2005). Curcumin: The story so far. *European Journal of Cancer* 41(13): 1955–1968.

Sharma, S. V., D. Y. Lee, B. Li, M. P. Quinlan, F. Takahashi, S. Maheswaran, U. McDermott, N. Azizian, L. Zou and M. A. Fischbach (2010). A chromatin-mediated reversible drug-tolerant state in cancer cell subpopulations. *Cell* 141(1): 69–80.

Situmourang, P. C., S. Ilyas, S. E. Nugraha, R. A. Syahputra and N. M. A. Nik Abd Rahman (2024). Prospects of compounds of herbal plants as anticancer agents: A comprehensive review from molecular pathways. *Frontiers in Pharmacology* 15: 1387866.

Story, E. N., R. E. Kopec, S. J. Schwartz and G. K. Harris (2010). An update on the health effects of tomato lycopene. *Annual Review of Food Science and Technology* 1(1): 189–210.

Tabassum, N., V. Verma, M. Kumar, A. Kumar and B. Singh (2018). Nanomedicine in cancer stem cell therapy: From fringe to forefront. *Cell and Tissue Research* 374: 427–438.

Talaat, S. M., Y. S. Elnaggar, M. A. Gowayed, S. O. El-Ganainy, M. Allam and O. Y. Abdallah (2024). Novel PEGylated cholephytosomes for targeting fisetin to breast cancer: In vitro appraisal and in vivo antitumoural studies. *Drug Delivery and Translational Research* 14(2): 433–454.

Talib, W. H., A. R. Alsayed, M. Barakat, M. I. Abu-Taha and A. I. Mahmod (2021). Targeting drug chemo-resistance in cancer using natural products. *Biomedicines* 9(10): 1353.

Tsakiroglou, P., N. E. VandenAkker, C. Del Bo', P. Riso and D. Klimis-Zacas (2019). Role of berry anthocyanins and phenolic acids on cell migration and angiogenesis: An updated overview. *Nutrients* 11(5): 1075.

Tuorkey, M. J. (2015). Cancer therapy with phytochemicals: Present and future perspectives. *Biomedical and Environmental Sciences* 28(11): 808–819.

Ulldemolins, A., J. Seras-Franzoso, F. Andrade, D. Rafael, I. Abasolo, P. Gener and S. Schwartz Jr (2021). Perspectives of nano-carrier drug delivery systems to overcome cancer drug resistance in the clinics. *Cancer Drug Resistance* 4(1): 44.

Urruticoechea, A., R. Alemany, J. Balart, A. Villanueva, F. Vinals and G. Capella (2010). Recent advances in cancer therapy: An overview. *Current Pharmaceutical Design* 16(1): 3–10.

Vasan, N., J. Baselga and D. M. Hyman (2019). A view on drug resistance in cancer. *Nature* 575(7782): 299–309.

Vila-Leahey, A., S. A. Oldford, P. A. Marignani, J. Wang, I. D. Haidl and J. S. Marshall (2016). Ranitidine modifies myeloid cell populations and inhibits breast tumour development and spread in mice. *Oncoimmunology* 5(7): e1151591.

Wang, J., N. Seebacher, H. Shi, Q. Kan and Z. Duan (2017). Novel strategies to prevent the development of multidrug resistance (MDR) in cancer. *Oncotarget* 8(48): 84559.

Wang, K. and J. E. Tepper (2021). Radiation therapy-associated toxicity: Etiology, management, and prevention. *CA: A Cancer Journal for Clinicians* 71(5): 437–454.

Wang, K., L. Jiang and L. Qiu (2022). Near infrared light triggered ternary synergistic cancer therapy via L-arginine-loaded nanovesicles with modification of PEGylated indocyanine green. *Acta Biomaterialia* 140: 506–517.

Wang, X., H. Zhang and X. Chen (2019). Drug resistance and combating drug resistance in cancer. *Cancer Drug Resistance* 2(2): 141.

Wang, Y., H. Yu, S. Wang, C. Gai, X. Cui, Z. Xu, W. Li and W. Zhang (2021). Targeted delivery of quercetin by nanoparticles based on chitosan sensitizing paclitaxel-resistant lung cancer cells to paclitaxel. *Materials Science and Engineering: C* 119: 111442.

Wang, Y., J. Yu, R. Cui, J. Lin and X. Ding (2016). Curcumin in treating breast cancer: A review. *Journal of Laboratory Automation* 21(6): 723–731.

Wattenberg, L. W., Sparnins, V. L., & Barany, G. (1989). Inhibition of N-nitrosodiethylamine carcinogenesis in mice by naturally occurring organosulfur compounds and monoterpenes. *Cancer Research*, 49(10), 2689–2692.

Wei, X., M. Song, W. Li, J. Huang, G. Yang and Y. Wang (2021). Multifunctional nanoplatforms co-delivering combinatorial dual-drug for eliminating cancer multidrug resistance. *Theranostics* 11(13): 6334.

WHO (2020). *Global Health Estimates 2020: Deaths by Cause, Age, Sex, by Country and by Region, 2000–2019*, World Health Organization, Geneva.

Xiong, L., X.-M. Lin, J.-H. Nie, H.-S. Ye and J. Liu (2021). Resveratrol and its nanoparticle suppress doxorubicin/docetaxel-resistant anaplastic thyroid cancer cells in vitro and in vivo. *Nanotheranostics* 5(2): 143.

Xu, W., Y. Xiao, L. Zheng, M. Xu, X. Jiang and L. Wang (2023). Enhancing paclitaxel efficacy with piperine-paclitaxel albumin nanoparticles in multidrug-resistant triple-negative breast cancer by inhibiting P-glycoprotein. *Pharmaceutics* 15(12): 2703.

Yager, J. D. (2012). Catechol-O-methyltransferase: Characteristics, polymorphisms and role in breast cancer. *Drug Discovery Today: Disease Mechanisms* 9(1–2): e41–e46.

Yahya, E. B. and A. M. Alqadhi (2021). Recent trends in cancer therapy: A review on the current state of gene delivery. *Life Sciences* 269: 119087.

Yang, C. S., J. Y. Chung, G.-Y. Yang, S. K. Chhabra and M.-J. Lee (2000). Tea and tea polyphenols in cancer prevention. *The Journal of Nutrition* 130(2): 472S–478S.

Zahavi, D. and L. Weiner (2020). Monoclonal antibodies in cancer therapy. *Antibodies* 9(3): 34.

Zhang, H. and J. Chen (2018). Current status and future directions of cancer immunotherapy. *Journal of Cancer* 9(10): 1773.

Zhang, J., Z. Du, S. Pan, M. Shi, J. Li, C. Yang, H. Hu, M. Qiao, D. Chen and X. Zhao (2018). Overcoming multidrug resistance by codelivery of MDR1-targeting siRNA and doxorubicin using EphA10-mediated pH-sensitive lipoplexes: In vitro and in vivo evaluation. *ACS Applied Materials & Interfaces* 10(25): 21590–21600.

Zhang, L., Q. Lu and C. Chang (2020). Epigenetics in health and disease. In: Chang, C. and Q. Lu (Eds.) *Epigenetics in Allergy and Autoimmunity*, Springer Singapore, pp. 3–55.

Zhao, X., Q. Chen, Y. Li, H. Tang, W. Liu and X. Yang (2015). Doxorubicin and curcumin co-delivery by lipid nanoparticles for enhanced treatment of diethylnitrosamine-induced hepatocellular carcinoma in mice. *European Journal of Pharmaceutics and Biopharmaceutics* 93: 27–36.

Zhao, Y., H. Tan, J. Zhang, D. Zhan, B. Yang, S. Hong, B. Pan, N. Wang, T. Chen and Y. Shi (2024). Developing liver-targeted naringenin nanoparticles for breast cancer endocrine therapy by promoting estrogen metabolism. *Journal of Nanobiotechnology* 22(1): 122.

Zou, T., M. Lan, F. Liu, L. Li, T. Cai, H. Tian and Y. Cai (2021). Emodin-loaded polymer-lipid hybrid nanoparticles enhance the sensitivity of breast cancer to doxorubicin by inhibiting epithelial–mesenchymal transition. *Cancer Nanotechnology* 12: 1–15.

Zubair, M., S. Wang and N. Ali (2021). Advanced approaches to breast cancer classification and diagnosis. *Frontiers in Pharmacology* 11: 632079.

Zugazagoitia, J., C. Guedes, S. Ponce, I. Ferrer, S. Molina-Pinelo and L. Paz-Ares (2016). Current challenges in cancer treatment. *Clinical Therapeutics* 38(7): 1551–1566.

24 Prevention Is Better than Cure
The Role of Phytomolecules in the Prevention of Cancer Development

Sandesh Lodha, Gajanan Kalyankar, Shrikant Joshi, Rutvi Vaidya, and Nikita Panchal
Maliba Pharmacy College, Uka Tarsadia University, Tarsadi, India

24.1 INTRODUCTION

Plants produce phytomolecules, also known as secondary metabolites. These chemicals are important in plant growth, defence systems, and environmental interactions. They include a variety of chemicals, including flavonoids, phenolics, carotenoids, and polyphenols (Bacciottini et al., 2007; Zhang et al., 2015).

Phytomolecules are renowned for their potential health advantages, such as antioxidant and anti-inflammatory (Jang & Lee, 2023) characteristics. They are found in various foods, including vegetables, fruits, nuts, teas, and wines, and are being studied for their potential to prevent diseases such as Cancer (Ballout et al., 2019; Bhosale et al., 2020; Bishayee, 2011), Diabetes (Ansari et al., 2022), Cardiovascular disease (Bachheti et al., 2022), and Neurodegenerative disorders (Ratheesh et al., 2017; Velmurugan et al., 2018).

Secondary metabolites, often called specialized metabolites or natural products, are organic substances produced by various species, including plants, bacteria, fungi, and mammals. Secondary metabolites, as opposed to primary metabolites, are not directly involved in an organism's growth, development, or reproduction. Instead, secondary metabolites frequently serve important roles in ecological interactions, including defence mechanisms against herbivores, diseases, and competition, as well as attracting pollinators and seed dispersers. Secondary metabolites include alkaloids, flavonoids, terpenoids, and phenolic compounds. Because of their various biological functions, these chemicals have piqued the interest of researchers in medicine, agriculture, and industry (Chiocchio et al., 2021; Kumar et al., 2022).

Phytomolecules, as secondary metabolites, have various benefits for plants, human health, and industry. They serve as defence mechanisms against herbivores, attract pollinators, and inhibit competition. For example, alkaloids can deter herbivores due to their bitter taste and toxicity. They also offer health benefits, such as antioxidant properties, which can prevent chronic diseases such as cancer and cardiovascular diseases. Phytomolecules are used in various industries like pharmaceuticals, cosmetics, and food due to their bioactive properties (Feng et al., 2019; Thawabteh et al., 2019; Vishesh Verma et al., 2022).

Phytomolecules are essential in pharmacology due to their diverse biological activities and potential therapeutic benefits. They have antioxidant properties, which help neutralize free radicals and prevent cellular damage, which is crucial in managing diseases like cancer, cardiovascular diseases, and neurodegenerative disorders. They also exhibit anti-inflammatory effects, such as curcumin (Huang et al., 1997) and resveratrol (Du et al., 2012), which can treat chronic inflammatory conditions. Some phytomolecules have antimicrobial properties, making them effective against various pathogens. They can also inhibit cancer cell growth, improve cardiovascular health, and have neuroprotective effects, such as ginkgo biloba extract. They also protect the liver from toxins and oxidative stress (Boghdady, 2013; Silva & Martins, 2022).

Green tea compounds like epigallocatechin gallate (EGCG) and Broccoli sulforaphane are being researched for cancer prevention and therapy (Nandini et al., 2020). The ginkgo biloba extract is known for its cognitive-enhancing effects (Gold et al., 2002; Oken et al., 1998), while garlic allicin has antibacterial, antiviral, and antifungal activities (Ankri & Mirelman, 1999; Nakamoto et al., 2019).

These are different classes of phytomolecules with the source and biological activity.

1. Flavonoids are found in fruits, vegetables, tea, and wine. They have antioxidant, anti-inflammatory, and anticancer properties, lowering the risk of chronic diseases like heart disease and cancer (Farhan et al., 2023; Khan et al., 2021; Kopustinskiene et al., 2020).
2. Alkaloids' sources include plants such as coffee, tea, and certain therapeutic herbs. Biological Properties included analgesic (pain-relieving) and stimulating properties. Examples include coffee and morphine (Ballout et al., 2019; Thawabteh et al., 2019).
3. Terpenoids are found in essential oils from plants such as eucalyptus, mint, and citrus fruits.

 DOI: 10.1201/9781003611516-30

They have antimicrobial, anti-inflammatory, and anticancer properties and are utilized in aromatherapy for therapeutic purposes (Bishayee, 2011).

4. Polyphenols are abundant in foods such as berries, nuts, seeds, and dark chocolate. Having Strong antioxidants that protect cells against oxidative damage and reduce inflammation (Bhosale et al., 2020; Du et al., 2012).
5. Saponins are found in beans, legumes, and some herbs, that show antibacterial and immune-boosting capabilities. They may also help to decrease cholesterol levels (Elekofehinti et al., 2021; Zhu et al., 2023).
6. Sources of carotenoids include carrots, sweet potatoes, and leafy greens, Well-known for their antioxidant capabilities and function in eye health. Beta-carotene, a form of carotenoid, is a precursor of vitamin A (Koklesova et al., 2020; Rowles & Erdman, 2020; Shafiq et al., 2024).
7. Phenolic acids are found in coffee, fruits, and whole grains, and include antioxidant and anti-inflammatory actions. They help to lower the risk of chronic diseases such as cardiovascular disease and cancer (Abotaleb et al., 2020; Wahle et al., 2010).

24.2 MECHANISM OF CANCER GROWTH

Uncontrolled cell growth and proliferation are hallmarks of the complex disease known as cancer. Several mechanisms contribute to the growth and progression of cancer cells. Here are some key mechanisms:

1. *Genetic mutations:* By changing the normal function of genes involved in the regulation of cell growth, genetic mutations significantly contribute to the development of cancer. Examples include:
   - Activation of oncogenes like KRAS in pancreatic cancer.
   - Inactivation of tumour suppressor genes like TP53 in various cancers.
2. *Epigenetic alterations:* Histone alterations and DNA methylation are examples of epigenetic modifications that can either activate oncogenes or mute tumour suppressor genes. Examples include:
   - Hypermethylation of the BRCA1 gene in breast cancer.
   - Histone modifications in prostate cancer affecting gene expression.
3. *Angiogenesis:* Angiogenesis is the process by which cancer cells promote the development of new blood vessels to obtain oxygen and nutrients for their proliferation. Examples include:
   - Vascular endothelial growth factor (VEGF) signalling in promoting angiogenesis in solid tumours.
   - Anti-angiogenic therapies targeting angiogenesis pathways in cancer treatment.
4. *Immune evasion:* Cancer cells can evade immune surveillance by altering checkpoints or expressing immune-suppressive molecules. Examples include:
   - Programmed death-ligand 1 (PD-L1) expression in cancer cells to inhibit T cell activity.
   - Immune checkpoint inhibitors (e.g., anti-PD-1 antibodies) in immunotherapy for cancer.
5. *Metastasis:* Cancer cells develop the capacity to invade adjacent tissues, travel through the blood or lymphatic system, and develop into secondary tumours in other locations. Among the examples are:
   - Epithelial-mesenchymal transition (EMT) promotes cancer cell migration and invasion.
   - Circulating tumour cells (CTCs) spread cancer to distant organs.

These mechanisms collectively contribute to the growth and spread of cancer cells. Understanding these processes is crucial for developing targeted therapies and improving outcomes in cancer treatment.

24.3 ROLE OF PHYTOCHEMICALS IN CANCER PREVENTION

Phytochemicals are natural molecules present in plants, which have been demonstrated to play an essential role in cancer prevention (Mollakhalili Meybodi et al., 2017).

24.3.1 Various Mechanisms for the Prevention of Cancer Are as Follows

24.3.1.1 Antioxidant Properties

Numerous phytochemicals have antioxidant properties that protect cells against reactive oxygen species and free radicals, two things that can lead to cancer. The technique involves directly scavenging free radicals, inducing antioxidant enzymes including superoxide dismutase, catalase, and glutathione, and activating Nrf2, which turns on genes for antioxidant and detoxifying enzymes (Muscolo et al., 2024; Zhang et al., 2015).

24.3.1.2 Inhibiting Carcinogen Formation

Certain phytochemicals, such as the sulphur-containing substance sulforaphane, can block the development of procarcinogens into active, DNA-damaging forms. This keeps carcinogens from initiating cancer (Thomas et al., 2015).

24.3.1.3 Blocking Carcinogen-Induced Mutations

Phytochemicals can inhibit specific metabolic processes that activate carcinogens. They also improve detoxification and trap reactive oxygen species to avoid carcinogen-induced mutations (George et al., 2021).

24.3.1.4 Suppressing Tumour Growth

By inhibiting cell division and causing apoptosis, autophagy, or necrosis in cancer cells, phytochemicals can slow the growth of tumours. They can also improve traditional cancer treatments by blocking angiogenesis, which stops the blood flow to tumours. Phytochemicals can improve the efficacy of chemotherapy and radiotherapy by controlling numerous signalling pathways. Consuming phytochemical-rich foods may assist in preventing subsequent tumour spread (George et al., 2021). The following are examples of the different classes of phytochemicals having chemopreventive effects.

24.3.2 Classes of Phytochemicals Having Chemopreventive Effects

24.3.2.1 Alkaloids

Alkaloids, naturally occurring chemicals found in plants, are promising agents for cancer prevention and treatment. It focuses on the various alkaloids that have shown anti-cancer activity both in vitro and in vivo. Vinblastine, vinorelbine, vincristine, and vindoline are examples of successful anticancer medications. Alkaloids attack cancer through a variety of strategies, including slowing cell proliferation, causing apoptosis (programmed cell death), and targeting specific signalling pathways implicated in cancer development (Ballout et al., 2019; Mondal et al., 2019).

24.3.2.2 Polyphenols

Anthocyanins, Epigallocatechin gallate (EGCG), Resveratrol, Quercetin, and Curcumin are polyphenols, naturally occurring chemicals in plants, fruits, and vegetables. They have been intensively researched for their possible health advantages, particularly in terms of cancer prevention and therapy. Polyphenols' anticancer effects are mediated by several pathways, including antioxidant characteristics, anti-inflammatory properties, and modulation of signalling pathways, and they can induce apoptosis in cancer cells. Polyphenols can also prevent the development of new blood vessels that supply malignancies (Bhosale et al., 2020; Moga et al., 2016).

One of the most prominent examples is curcumin, which is found in turmeric and has anti-inflammatory, anti-proliferative, and antioxidant properties that may help prevent cancer. By altering signalling pathways like NF-kB and PI3K/Akt/mTOR, it reduces the growth of cancer cells and promotes apoptosis. Preclinical studies suggest its potential, notably for colorectal and breast malignancies; however, clinical trials have produced inconsistent results due to curcumin's low bioavailability. Ongoing research intends to increase its effectiveness through improved formulations (Huang et al., 1997).

24.3.2.3 Carotenoids

Carotenoids, which are naturally occurring pigments found in fruits, vegetables, and plants, include beta-carotene, lutein, zeaxanthin, beta-cryptoxanthin, and lycopene. They may be used to treat or prevent cancer. Because of their antioxidant properties, carotenoids counteract harmful free radicals that may be a factor in the development of cancer. They can change the signalling pathways inside cells that regulate cell division, growth, and death. Angiogenesis, the formation of new blood vessels necessary for tumour growth and metastasis, may be inhibited by carotenoids (Koklesova et al., 2020; Rowles & Erdman, 2020; Zare et al., 2021). A lower risk of prostate cancer has been linked to higher blood and dietary levels of lycopene; beta-carotene and lutein may help lower the risk of breast cancer; and beta-cryptoxanthin may help prevent cancer by promoting the expression of the anti-oncogene RB gene (Kotecha et al., 2016; Rudzińska et al., 2023; Shafiq et al., 2024). Because of its strong antioxidant properties, which combat oxidative stress and inhibit the proliferation of cancer cells, lycopene, a carotenoid found in tomatoes and other red fruits, has been associated with the prevention of cancer. Studies have demonstrated that lycopene lowers the risk of colorectal, breast, and prostate cancer by affecting many cancer-related processes, including apoptosis and cell cycle regulation. Although epidemiological studies point to possible connections, more comprehensive clinical trials are needed to prove their effectiveness (Chen et al., 2015; Sadeghian et al., 2021).

24.3.2.4 Flavonoids

Tea is one of several fruits, vegetables, and beverages that contain flavonoids, a broad class of phytonutrients. Their anti-inflammatory, anti-cancer, and antioxidant properties have generated a lot of attention for their potential importance in cancer prevention. Flavonoids can reduce oxidative stress, which is connected to the development of cancer, and neutralize free radicals. Because they shield cells from oxidative stress, they might help lower the risk of cancer. Chronic inflammation is a known risk factor for cancer, and many flavonoids contain anti-inflammatory properties that can help lower it. This may help stop cancer from developing and spreading. Additionally, flavonoids can modify signalling pathways linked to the development of cancer, such as those that regulate metastasis, apoptosis (programmed cell death), and cell proliferation. Detoxification enzymes, which help the body rid itself of potential carcinogens, can be stimulated by some flavonoids. Quercetin, found in tea, apples, and onions, has been shown to cause apoptosis and reduce the growth of cancer cells. Research suggests that it could help prevent breast and prostate cancer. Catechins, which are found in green tea, have demonstrated potent anti-cancer effects in preclinical studies. They are believed to inhibit the growth of cancer cells and stop them from spreading. Products made from soybeans are the main source of genistein. Through regulating the activation of estrogen receptors, it has been studied for its possible role in preventing hormone-related cancers, including breast and prostate cancer

(Farhan et al., 2023; Khan et al., 2021; Kopustinskiene et al., 2020; Wang et al., 2022a).

A flavonoid found in propolis and honey, chrysin may have anticancer properties, including causing apoptosis in several cancer types and preventing the growth of cancer cells. Chrysin lowers oxidative stress and suppresses NF-κB, two processes that are crucial to the development of cancer. Its ability to engage with estrogen receptors suggests a role in cancers linked to hormones. The preclinical studies produced positive findings (Shahbaz et al., 2023; Stompor-Gorący et al., 2021).

24.3.2.5 Isothiocyanates

Sulphur-containing compounds, Isothiocyanates generated from glucosinolates, are hypothesized to be responsible for the chemoprotective effects of consuming more cruciferous vegetables. Evidence suggests that isothiocyanates exert their effects through a variety of separate but interrelated signalling pathways critical for preventing carcinogenesis, such as those involved in detoxification, inflammation, apoptosis, cell cycle, and epigenetic regulation, among others (Mitsiogianni et al., 2019). Many studies have recently shown that a higher intake of cruciferous vegetables is associated with a lower risk of developing various forms of cancers, primarily due to a variety of effects, including (1) metabolic activation and detoxification, (2) inflammation, (3) angiogenesis, (4) metastasis, and (5) regulation of the epigenetic machinery (Navarro et al., 2011).

24.4 MECHANISM OF ACTION OF PHYTOMOLECULES IN THE PREVENTION OF CANCER

Cancer, a multifactorial disease, is associated with various genetic and environmental factors. Among the numerous approaches to mitigate cancer risk, dietary phytomolecules have gained significant attention due to their preventive potential. Phytomolecules, naturally occurring compounds found in plants, possess potent antioxidant properties, which play a pivotal role in preventing the initiation and progression of cancer by neutralizing oxidative stress. Oxidative stress, often driven by an overproduction of reactive oxygen species (ROS), is a crucial factor in cancer development, leading to DNA damage, mutations, and subsequent tumour formation.

24.4.1 Alkaloids: Role in Cancer Prevention

Alkaloids are compounds containing nitrogen, where anticancer activity is noted mainly due to their antioxidant mechanisms. Alkaloids scavenge free radicals, causing cellular damage and ultimately cancer development. Berberine is an isoquinoline alkaloid found in many medicinal plants and has antioxidant effects by modulating mitochondrial functions and reducing ROS levels. It inhibited lipid peroxidation. It is a process wherein lipid oxidation may cause membrane damage leading to cancerous mutations (Mondal et al., 2019b).

For example, *Berberis aristata*, commonly known as the Indian barberry, contains berberine, which possesses protective activity against oxidative stress in hepatic cancer models. The alkaloid neutralizes ROS through overexpression of endogenous antioxidant enzymes such as SOD and catalase so that the cells are not damaged by oxidative stress-induced DNA damage. Likewise, vinblastine and vincristine, alkaloids extracted from Catharanthus roseus, prevent the carcinogenesis that is mediated by ROS by inhibiting oxidative phosphorylation in addition to the inhibition of the proliferation of cancerous cells.

Alkaloids are nitrogenous compounds that exhibit antioxidant activity. They neutralize free radicals that cause oxidative damage to the biomolecules like DNA, proteins, and lipids. The major antioxidant mechanisms of alkaloids include

- *Scavenging free radicals*: Direct scavengers of ROS, berberine and vinblastine alkaloids donate electrons or hydrogen atoms to scavenge free radicals, including superoxide anions and hydroxyl radicals, that cause oxidative stress.
- *Alkaloids inhibition of lipid peroxidation*: These are the processes through which ROS attacks polyunsaturated fatty acids found within the cellular membranes, resulting in damage to the cells. Through this inhibition process, alkaloids prevent lipid peroxidation and therefore protect the integrity of the cell.
- *Upregulation of antioxidant enzymes*: Alkaloids may upregulate the activity of endogenous antioxidant enzymes like superoxide dismutase (SOD), catalase, and glutathione peroxidase. These enzymes are essential for detoxifying ROS and thus maintaining cellular redox balance.

For example, the antioxidant property of berberine from *Berberis aristata* is shown because the increase in SOD and catalase levels has reduced oxidative stress in cancer cells (Atpadkar et al., 2023; Chaudhary et al., 2023; Gan et al., 2017; Olofinsan et al., 2023).

24.4.2 Carotenoids: Role in Cancer Prevention

Carotenoids are those pigments that owe some of their vivid colours to β-carotene, lycopene, and lutein present in many fruits and vegetables. They have some of the most intensive antioxidative activity utilizing quenching free singlet oxygen together with other ROS, thereby contributing protection to cellular components under conditions of oxidative stress that contributes to cancer prevention risk. Lycopene occurs as one of the common carotenoids naturally available in tomatoes. Scientific work on lycopene continues to be significant both in its role in contributing to prostate cancer prevention and otherwise lowering oxidative damage to the DNA through quenching free singlet oxygen, plus it also inhibits the onset of lipid peroxidation.

For instance, another quite effective antioxidant is β-carotene, which itself is a precursor of vitamin A. It scavenges peroxyl radicals and decreases oxidative damage within tissues. Much documented research is present that highlights that β-carotene diminishes the risk for lung cancer, especially amongst the high consumers of those β-carotene-ripest foods, i.e., carrots and sweet potatoes. β-Carotene enhances cellular defences against oxidative damage by modulating the activity of antioxidant enzymes and preventing the initiation of carcinogenesis (Zare et al., 2021).

Carotenoids are natural pigments that have a potent antioxidant effect, primarily through the following mechanisms:

- *Quenching singlet oxygen*: The reactive form of oxygen, known as singlet oxygen, can be quenched by carotenoids like lycopene and β-carotene. Carotenoids quench singlet oxygen by transferring its energy to a carotenoid molecule, leaving it excited but stable.
- *Repression of lipid peroxidation*: Lipid peroxidation can be repressed by the action of carotenoids. Carotenoids consume peroxyl radicals straight from the lipid oxidative degeneration. That decreases oxidation effects on cell membranes.
- It controls the antioxidant enzymes like SOD and glutathione peroxidase and therefore increases its capacity to scavenge ROS.

For example, lycopene found in tomatoes scavenges ROS and, therefore, protects cells by inhibiting lipid peroxidation, prevents oxidative damage, and, as such, lowers the risk of prostate cancer (Fiedor & Burda, 2014; Pérez-Gálvez et al., 2020; Rowles & Erdman, 2020).

24.4.3 Flavonoids: Role in Cancer Prevention

As diversified phytomolecules exhibit vast biological activities, flavonoids are antioxidants, anti-inflammatories, and potential anticancer agents. Indeed, fruits, vegetables, and tea contain flavonoids, including quercetin, kaempferol, and catechins, that act powerfully as free radical scavengers. By donating the hydrogen atoms to neutralize ROS, flavonoids protect DNA from oxidative degradation, which is a point of critical importance to cancer prevention.

Quercetin is an important flavonoid that occurs in apples, onions, and berries, for example, and possesses powerful antioxidant activity. Quercetin directly scavenges superoxide radicals and acts to inhibit the generation of ROS by blocking NADPH oxidase, which plays a key role in generating oxidative stress. It can also upregulate glutathione and catalase, endogenous antioxidants that can increase the defense capacity of the body against DNA damage caused by oxidation.

This characteristic also hides the catalytic activity of readily accessible transition metals like iron and copper, which are proven promoters of the Fenton reaction that generates the free, highly reactive hydroxyl radicals, which eventually contribute to oxidative damage to DNA, etc. Likewise, catechins that consist of epigallocatechin gallate (EGCG) in green tea are rich in this chelating property and, accordingly, confer their anticancer behaviour (Pyo et al., 2024).

Polyphenols are a wide category of compounds that have antioxidant activity and play a role in the prevention of cancer through the following:

- *Scavenging of free radicals by direct interaction with free radicals*: The above-mentioned flavonoids, quercetin, kaempferol, and catechins, scavenge free radicals, including superoxide anions, hydroxyl radicals, and peroxyl radicals by donating a hydrogen atom or electrons to the radicals, thus changing their structures into more stable and less harmful compounds.
- *Metal chelation*: The transition metals, iron and copper, among others, can catalyse highly reactive hydroxyl radicals due to the Fenton reaction. Chelating flavonoids limit their availability and thus prevent oxidative damage.
- *Flavonoids inhibit major enzymes:* responsible for the production of ROS, like NADPH oxidase and xanthine oxidase, thereby reducing the generation of ROS at its source.

The main one is quercetin from apples and onions, as the strongest antioxidant, which will scavenge superoxide radicals and inhibit the activity of NADPH oxidase to prevent ROS (Khan et al., 2021; Ponte et al., 2021).

24.4.4 Polyphenols: Role in Cancer Prevention

Some of the major phytomolecules possessing potent antioxidant activity include polyphenols, which comprise resveratrol, curcumin, and ellagic acid. The compounds protect cells from oxidative stress by scavenging free radicals, inhibiting oxidative enzymes, and upregulating endogenous antioxidant defences. Polyphenols are abundant in fruits, vegetables, tea, and red wine.

Resveratrol, a polyphenol found in grapes and berries, is an antioxidant that scavenges ROS and stimulates the activity of antioxidant enzymes such as glutathione peroxidase. Its mechanisms of cancer prevention include its ability to suppress the formation of ROS and inhibit the activation of pro-oxidant enzymes such as cyclooxygenase (COX) and lipoxygenase. In addition, resveratrol modulates gene expression, particularly those responsible for DNA repair and apoptosis, thereby lowering the incidence of cancer.

Curcumin is another polyphenol obtained from the spice turmeric that exerts potent antioxidant and anticancer activity. Curcumin scavenges free radicals and inhibits ROS-producing enzymes, including xanthine oxidase. Curcumin also prevents oxidative damage to DNA by increasing the expression of antioxidant proteins, including heme oxygenase-1 (HO-1) and nuclear factor erythroid 2-related factor 2 (Nrf2), which play a significant role in cellular defence mechanisms.

Polyphenols are a wide category of compounds that have antioxidant activity and play a role in the prevention of cancer through the following:

- *Free radical scavenging*: Compounds such as resveratrol, curcumin, and ellagic acid can scavenge ROS by donating hydrogen atoms or electrons to stabilize the free radicals, preventing them from damaging DNA and cellular structures.
- *Enhanced level of antioxidant enzymes*: The phenols activate the upregulated endogenous antioxidant enzymes such as glutathione peroxidase and catalase, which help in increasing the cellular activity to neutralize the level of ROS, inhibiting oxidative stress.
- *Inhibition of oxidative enzymes*: These polyphenols inhibit some enzymes, which include the cyclooxygenase and lipoxygenase as part of the metabolic or inflammation processes that cause the production of ROS. With this inhibition, polyphenols reduce oxidative stress.

For instance, resveratrol in grapes protects the cells by increasing the activity of glutathione peroxidase and by inhibiting the activity of the COX enzymes, all of which will decrease its ability to be involved with cancerous changes (De Silva & Alcorn, 2019; Maleki Dana et al., 2022; Zhou et al., 2016).

24.4.5 Isothiocyanates: Role in Cancer Prevention

Isothiocyanates are sulfur compounds produced from glucosinolates in cruciferous vegetables and have shown chemopreventive activity. Sulforaphane and PEITC have free radical scavenging activity and can modulate detoxification enzymes.

Sulforaphane, which is found in broccoli and Brussels sprouts, is a potent activator of the Nrf2 pathway, which controls the expression of detoxification and antioxidant enzymes like glutathione *S*-transferase (GST) and NAD(P)H oxidoreductase 1 (NQO1). Sulforaphane increases cellular antioxidant capacity and protects against oxidative DNA damage and inhibits the initiation of cancer (Mitsiogianni et al., 2019).

Isothiocyanates are sulphur-containing compounds that are derived from glucosinolates. Such compounds can act as antioxidants and inhibit oxidative stress-mediated cellular damage through:

- *Induction of the Nrf2 pathway*: Isothiocyanates, such as sulforaphane, activate the Nrf2 pathway, which upregulates the expression levels of antioxidant and detoxification enzymes, including GST and NAD(P)H

  NQO1, oxidoreductase 1. These enzymes neutralize ROS and enhance cellular defences against oxidative stress.

- *Formation of ROS inhibition*: Isothiocyanates cause the inhibition of the generation of ROS. It does so either by affecting mitochondrial function or by reducing the activities of the enzymes that generate ROS.
- *Induction of phase II detoxification enzymes*: Isothiocyanates induce phase II detoxification enzymes that help clear cells of ROS and other harmful metabolites.

For example, sulforaphane from broccoli increases the transcription of GST and NQO1 through the Nrf2 pathway, providing strong antioxidant protection coupled with a reduction in the incidence of initiation of cancer (Abbaoui et al., 2018; Gupta et al., 2014; Navarro et al., 2011).

All alkaloids, carotenoids, flavonoids, polyphenols, and isothiocyanates are known to inhibit oxidative stress-induced damage responsible for causing cancer through antioxidant mechanisms. These compounds have strong protective effects against cellular oxidative stress, which has been considered to decrease the risk of cancer by scavenging free radicals, inhibiting oxidative enzymes, chelating transition metals, and enhancing endogenous antioxidant defences. Such mechanisms emphasize the role of phytomolecules in preventing cancer and suggest dietary approaches or therapeutic interventions intended to reduce oxidative damage and carcinogenesis.

24.5 MODULATION OF CELL SIGNALING PATHWAYS IN CANCER PREVENTION BY PHYTOMOLECULES

Regulation of cell growth, differentiation, survival, and apoptosis occurs through cellular signalling pathways. Dysregulation is considered a hallmark of cancer due to its association with uncontrollable proliferation, avoidance of apoptosis, and the development of metastatic cancer. Many phytomolecules from plants could, however, regulate key cellular pathways that have been determined to be abnormal in various forms of cancers, and it is with this kind of regulation that cancer could be prevented in its development or progression. Most major signalling pathways that may be involved include the PI3K/Akt pathway, the MAPK, NF-κB pathways, the Wnt/β-catenin pathway, and the p53 pathways in cancer prevention that may be modulated by phyto-molecules (Almatroodi et al., 2021).

24.5.1 PI3K/Akt Pathway

The PI3K/Akt pathway is involved in cell survival, proliferation, and metabolism. This pathway is typically overactivated in many cancers, thus contributing to enhanced cell survival and growth. The modulation of this pathway by phytomolecules leads to apoptosis and prevention of cancer progression (He et al., 2021; Moghbeli, 2024).

- Resveratrol is a polyphenol found in grapes and berries that has been established to be an inhibitor of the PI3K/Akt pathway through the inhibition of the phosphorylation of Akt, thereby causing apoptosis in cancer cells. Other downstream targets include the mammalian target of rapamycin, and resveratrol decreases cell proliferation and induces autophagy (Venkatadri et al., 2016).
- The Curcumin, isolated from the turmeric root, was shown to impinge upon the PI3K/Akt pathway via Akt dephosphorylation with consequent induction of apoptosis and decreased cell survival of cancer cells. Modulation of this pathway was noted with curcumin in all three breast, prostate, and colon cancers; therefore, a wide-spectrum anticancer action (Zoi et al., 2024).

24.5.2 MAPK Pathway

This pathway is involved in cell growth, differentiation, and apoptosis. The dysregulation of this pathway is involved in the development of cancer, especially through the activation of oncogenic kinases such as ERK, JNK, and p38 MAPKs. These phytomolecules may influence the MAPK pathway to suppress cancer progression (Yuan et al., 2020).

- *EGCG:* This is a green tea-flavonoid that was shown to inhibit both ERK and JNK pathways, reducing cell proliferation as well as increasing the rate of apoptosis. The importance of the inhibition of MAPK by EGCG extends to skin, lung, and prostate carcinomas (Kciuk et al., 2023; Khan et al., 2021).
- Lycopene is a carotenoid found in tomatoes which inhibits the phosphorylation of the ERK1/2 MAPK pathway and, in doing so, inhibits inhibiting cell growth cancer. Lycopene also reduces the expression of MMPs by regulating the MAPK pathway, which facilitates metastasis activity (Han et al., 2019).

24.5.3 NF-κB Pathway

The NF-κB pathway is very important in inflammation, immune response, and cell survival. Activation of NF-κB for a long time contributes to tumourigenesis through cell proliferation, induction of apoptosis, and enhancement of metastasis. Most of the phytomolecules have been proven to have inhibitory effects on the activation of NF-κB, thereby reducing cancer risk (Hoesel & Schmid, 2013).

- Curcumin is known for its ability to suppress the NF-κB signalling pathway. This is accomplished by preventing IκB phosphorylation and degradation, which inhibits NF-κB translocation to the nucleus. This inhibition lowers the expression of proinflammatory cytokines and anti-apoptotic genes, both of which are necessary for cancer cell survival and growth (Jobin et al., 1999).
- Quercetin, a flavonoid found in apples and onions, suppresses NF-κB activation by inhibiting IKK activity. This inhibition reduces inflammation, slows the proliferation of cancer cells, and enhances their sensitivity to apoptosis, making it a promising compound for cancer prevention and therapy (Cho et al., 2003).

24.5.4 Wnt/β-Catenin Pathway

Wnt/β-catenin signalling is well known for the regulation of cell proliferation, differentiation, and stem cell renewal. It has been elucidated that aberrant activation of this pathway causes the development of cancer, with β-catenin acting as an oncogene within colon, liver, and breast cancers, in which β-catenin undergoes nuclear accumulation and activates oncogenic target genes; thus, phytomolecules hold promise for use in modulating this pathway in cancer prevention (He & Gan, 2023; Zhang & Wang, 2020).

Sulforaphane is an isothiocyanate found in cruciferous vegetables that acts in the Wnt/β-catenin pathway: it downregulates β-catenin nuclear translocation and suppresses target gene expression of Wnt. It has been determined that it can block the proliferative activity of cancer stem cells, which are typically associated with cancer initiation and recurrence (Li et al., 2017).

Genistein, a soy isoflavone, inhibits the Wnt/β-catenin signalling cascade by lowering the expression of β-catenin and its downstream targets. This modulation reduces cancer cell growth and increases apoptosis, indicating its potential for cancer prevention and treatment (Borrás et al., 2006).

24.5.5 p53 Pathway

The critical mechanisms by which p53, the tumour suppressor protein, prevents cancer include cell cycle arrest, DNA repair, and apoptosis. Mutations in the p53 gene are prevalent in many cancers and cause a loss of tumour-suppressive function. Some phytomolecules restore or enhance p53 activity and thereby prevent cancer development (Marei et al., 2021).

- Resveratrol was found to cause activation of p53 as a mechanism of increased half-life and transactivation potency. Subsequently, genes targeted by p53 can be expressed to carry on apoptosis and cell cycle arrest, leading to the inhibition of tumour growth (Ferraz Da Costa et al., 2017).
- Curcumin also activates the p53 pathway by enhancing the phosphorylation of p53, leading to the stabilization of the protein and enhancement of its tumour-suppressive functions. Activation of p53 triggers apoptosis and senescence in cancer cells, hindering their proliferation (Jin et al., 2016).
- These substances modulate critical signalling pathways involved in the genesis of cancer, such as the PI3K/Akt, MAPK, NF-κB, Wnt/β-catenin, and p53 pathways. Since it acts to inhibit over-activated pathways responsible for oncogenesis as well as restoration of some of the lost tumour suppressor pathways, the whole potential for preventing malignancies is associated with their administration. In all such actions, a little more could be put to the term modulation through the action of chemopreventive abilities. Further research on the modulation of cell signalling by phytomolecules will further elaborate their role in cancer prevention and therapeutic applications (Almatroodi et al., 2021).

24.6 ROLE OF PHYTO MOLECULES IN THE PREVENTION OF CANCER: EPIGENETIC MODIFICATIONS

Cancer is an intricate ailment distinguished by genetic abnormalities and aberrant epigenetic alterations. Epigenetics pertains to inheritable but reversible modifications in gene expression or chromosome organization that do not necessitate modifications in the DNA sequence (Khan et al., 2024).

It is a complex field at the intersection of genetics and environmental factors and provides key insights into the regulation of gene expression and cellular function. Unlike traditional genetics, which focuses on DNA's linear sequence, epigenetics examines reversible modifications to DNA bases or histones that influence gene activity without altering the DNA sequence itself. These epigenetic modifications are essential for understanding the molecular foundations of cell differentiation, development, environmental responsiveness, and the wide-ranging phenotypic variations in human health and disease (Prasanth et al., 2024).

Epigenetic mechanisms encompass DNA methylation, post-translational modifications of histones such as acetylation, methylation, and phosphorylation, as well as the influence of non-coding RNAs (e.g., microRNAs) that modulate gene expression by impeding translation or degrading mRNAs. Moreover, scientific studies indicate that alterations in gene expression patterns, namely those related to uniparental disomy, play a role in the development of abnormalities in growth and maturation, neurological ailments, and cancer (Cheng et al., 2019; Ghosh & Saadat, 2023).

Recent research highlights the significance of factors like maternal behaviour, food, and other internal and external impacts on gene expression, all of which are connected to epigenetic control. Due to the ability of epigenetic alterations to be reversed, current research is primarily focused on the identification and development of medications that can selectively target these processes. Recently, natural bioactive chemicals have attracted interest because of their inherent safety and cost-efficiency. Phytochemicals (active substances present in herbs, spices, vegetables, and fruits) have anti-inflammatory, antioxidant, antibacterial, anticancer, and antidiabetic activities. These chemicals can affect several biological targets, such as cell cycle proteins, transcription factors, cell adhesion molecules, protein kinases, and anti-apoptotic factors (Aggarwal et al., 2015; Montgomery & Srinivasan, 2019).

Furthermore, phytochemicals could control modified metabolic pathways in cancer, including glycolysis, the pentose phosphate pathway, and lipid and amino acid metabolism. Also, phytomolecules can regulate epigenetic pathways, which play a crucial role in the advancement of cancer. It has generated considerable interest in their potential for cancer prevention (Aggarwal et al., 2015; Montgomery & Srinivasan, 2019).

24.6.1 DNA Methylation

DNA methylation involves the addition of a methyl group to cytosine bases in DNA, often leading to gene silencing. Hypermethylation of promoter regions can lead to the silencing of tumour suppressor genes, while hypomethylation can cause oncogene activation (Zhang & Xu, 2017).

In mammals, this methylation predominantly occurs at CpG sites and is facilitated by DNA methyltransferases (DNMTs: DNMT1, DNMT3a, and DNMT3B), which transfer methyl groups from the key donor *S*-adenosylmethionine (SAM) (Campagna & Vignini, 2023; Patnaik et al., 2023). The enzyme nicotinamide *N*-methyltransferase (NNMT) also impacts DNA methylation by regulating SAM levels. NNMT catalyses the methylation of nicotinamide (NAM), which affects the balance of methyl donors, influencing both DNA and histone methylation. By altering SAM concentrations, NNMT plays a critical role in modulating gene expression, as reduced DNA CpG methylation leads to increased gene activity. Because of its effects on gene regulation, NNMT is being explored as a potential therapeutic target in cancer, with several NNMT inhibitors under investigation (Roberti et al., 2021; Wang et al., 2022b).

DNA demethylation occurs actively through proteins known as ten-eleven translocation (TET) dioxygenases or passively as methyl groups are lost over successive cell divisions without DNMT1 activity (Kumar et al., 2018). Histone modification is another major regulatory mechanism, involving proteins that package DNA into chromatin. These modifications, including acetylation, methylation, and phosphorylation, can either facilitate or inhibit gene expression, depending on their nature and location. Histone acetylation, managed by histone acetyltransferases (HATs), generally loosens chromatin structure, allowing gene transcription. Conversely, histone deacetylases (HDACs) remove acetyl groups, compacting chromatin and reducing gene expression. Alongside HATs and HDACs, histone methyltransferases (HMTs) and demethylases (HDMs), respectively, add and remove methyl groups on histones, further influencing gene regulation (Alaskhar Alhamwe et al., 2018; Milazzo et al., 2020).

24.6.1.1 Phytochemicals Influencing DNA Methylation

- *Curcumin:* derived from Curcuma longa, is widely known for its ability to suppress cancer by specifically targeting DNA methylation machinery and other epigenetic machinery. One system that adapts to a wide range of environmental stimuli, including nutrition, is the dynamic DNA methylation process. The precise DNA methylation pattern found in all healthy tissues is altered throughout the development of cancer, creating a design unique to that malignancy. Almost all forms of cancer are characterized by global DNA demethylation, hypomethylation of oncogenes and prometastatic genes, and hypermethylation of tumour suppressor genes (TSGs). Curcumin has the potential to prevent cancer or improve the results of traditional anti-cancer therapy because it has been demonstrated to modify epigenetic events that are dysregulated in cancer cells. Curcumin has demonstrated the ability to reduce the hypermethylation of tumour suppressor genes such as *p16*, *RAR-β*, and *MGMT* in various cancer models. Curcumin inhibits DNA methyltransferases (DNMTs), the enzymes responsible for adding methyl groups to DNA, thereby reactivating the expression of silenced genes (Jiang et al., 2015; Yang, 2011).
- *Epigallocatechin-3-gallate (EGCG):* from green tea (*Camellia sinensis*) has been shown to inhibit DNMT activity, preventing hypermethylation of genes like *p16INK4a* and *RAR-β2* in cancer cells, promoting re-expression of these tumour suppressor genes (Fang et al., 2003).
- *Genistein:* a soy isoflavone, can decrease global DNA methylation levels and inhibit DNMTs, making it effective in the demethylation and re-expression of silenced genes such as *PTEN* and *p53* in breast and prostate cancers (Adjakly et al., 2011; Vardi et al., 2010).

Phytochemicals that influence DNA methylation are listed in Table 24.1.

24.6.2 Histone Modifications

Acetylation, methylation, phosphorylation, and ubiquitination are examples of post-translational modifications that occur in histones, the proteins that encircle DNA. Gene expression and chromatin structure are controlled by these alterations. Oncogenes and tumour suppressor genes may express differently as a result of dysregulated histone changes, which can accelerate the development of cancer (Khan et al., 2024; Prasanth et al., 2024).

24.6.2.1 Phytochemicals Targeting Histone Modifications

- *Sulforaphane:* from cruciferous vegetables like broccoli has been shown to inhibit histone deacetylases (HDACs), enzymes that remove acetyl groups from histones, causing tumour suppressor genes such as p21 to become active again and *Bax* in colorectal cancer cells. This results in cell cycle arrest and apoptosis (Meeran et al., 2010).
- *Resveratrol:* found in grapes and berries, modulates histone acetylation by inhibiting HDACs and activating histone acetyltransferases (HATs), promoting the acetylation of histones associated with p53 and other tumour suppressor genes, increasing their transcription (Karimi Dermani et al., 2017).
- *Quercetin:* a flavonoid found in fruits and vegetables, influences both histone acetylation and methylation, restoring normal chromatin architecture. It has been shown to downregulate HDAC activity and modulate histone methyltransferases (HMTs), which regulate gene expression through histone methylation patterns (Kedhari Sundaram et al., 2019; Tao et al., 2015; Zhao et al., 2019).

Table 24.1 enlists phytochemicals that target histone methylation.

24.6.3 Influence on Non-coding RNAs

Non-coding RNAs (ncRNAs), such as long non-coding RNAs (lncRNAs) and microRNAs (miRNAs), are essential for post-transcriptional regulation of gene expression. Dysregulated expression of miRNAs and lncRNAs has been linked to various cancers. Phytochemicals can influence the expression of these ncRNAs, contributing to cancer prevention.

24.6.3.1 Phytochemicals Modulating Non-coding RNAs

- *Curcumin:* has been found to modulate the expression of several oncogenic and tumour-suppressive miRNAs. For instance, curcumin downregulates oncogenic miR-21 and upregulates tumour-suppressive miR-200 in colorectal cancer, affecting pathways related to apoptosis and metastasis

TABLE 24.1
Effects of Phytomolecules on Epigenetic Modification in Various Cancers

Phytomoleculs	Type of Cancer	Mechanism	Effects	Models/Samples	References
Curcumin	Lung cancer	DNA methylation	WIF-I promoter hypomethylation and NrF2 promoter demethylation	SPC-A-1, H460, and A549 cells	Yang (2011)
			Demethylates the RARβ promoter and decreases DNMT3b.	BALB/c nude mice, A549, LTEP-a-2, and NCI-H460 cells	Jiang et al. (2015)
		Histone modification	Suppresses PRMT5 and MEP50 expression	A549 cells	Chatterjee et al. (2019)
			Increases histone H2A.X and p53 phosphorylation.	NCI-H460 cells	Ting et al. (2015)
		Non-coding RNA expression	Suppresses LIN-28A and raises miR-98 expression, which prevents cell invasion and migration.	A549 cells	Liu et al. (2017b)
			Suppresses lncRNA-UCA1, decreases Thus, Wnt 3a, Wnt 5a, and β-catenin decrease cyclin D1 expression and promote apoptosis.	A549 cells	Wei-Hua Wang et al. (2018)
			Enhances miR-206 expression, suppresses mTOR and AKT phosphorylation, and hence prevents cell invasion and migration.	A549 cells	Wang et al. (2020)
			Promotes apoptosis, inhibits XIAP, and increases miR-192-5p and miR-215.	A549 cells	Ye et al. (2015)
			Causes apoptosis, inhibits cell division, raises PTEN, and elevates lncRNA-MEG3.	A549 and H520 cells	Gao et al. (2021)
	Colorectal cancer	DNA methylation	Upregulates DLEC1 and reduces CpG methylation	HT29, SW480 and HCT116 cells	Guo et al. (2015)
			This causes particular CpG loci to become demethylated.	HCT116, HT29, and RKO cells	Link et al. (2013)
			Restores DNA methylation of Tnf	AOM- and DSS-induced colon tumour model (C57BL/6)	Guo et al. (2018)
		Histone modification	This causes the SIRT1 protein to degrade by ubiquitination.	HCT-116, HCT-15, DLD-1 cells	Lee et al. (2018)
		Non-coding RNA expression	Inhibits cell proliferation, raises Nkd2, and decreases miR-130a.	SW480 cells	Dou et al. (2017)
			Causes cell cycle arrest and death, inhibits c-Myc and CyclinD1, raises miR-34a and miR-27a, and stops proliferation.	HCT116, RKO, SW480, SW620, HT29 and Caco2 CRC cells	Toden et al. (2015)
			Inhibits cell proliferation, raises lncRNA-NBR2, and lowers p-AMPK and p-ACC phosphorylation.	HCT116, HCT8, SW480, and SW620 cells	H. Yu et al. (2019)
			Promotes apoptosis, raises miR-497 and Bcl-2, and decreases lncRNA-KCNQ1OT1.	HCT9 cells	Zheng et al. (2021)
	Gastric cancer	DNA methylation	Causes DNA demethylation and inhibits the expression of DNMT1.	MGC-803 cells	Tong et al. (2020)
		Non-coding RNA expression	MiR-21 inhibition reduces PI3K, AKT, MMP-2, MMP-9, and inhibits PTEN and P21. Causes apoptosis and inhibits migration, invasion, and proliferation.	AGS cells	Liu et al. (2018)
			Suppresses lncRNA-H19, raises p53, stops cell division, and triggers death.	SGC7901 cells, GES-1 cells	Liu et al. (2016)

(Continued)

TABLE 24.1 *(Continued)*
Effects of Phytomolecules on Epigenetic Modification in Various Cancers

Phytomoleculs	Type of Cancer	Mechanism	Effects	Models/Samples	References
	Breast cancer	DNA methylation	Enhances the expression of BRCA1 via decreasing promoter methylation	HCC-38, UACC-3199, and T47D cells	Al-Yousef et al. (2020)
			All DNMTs decrease, which reverses the hypermethylation of GSTP1.	MCF7 cells	Du et al. (2012), U. Kumar et al. (2017), and Liu et al. (2017c)
		Histone modification	Reduces the glutathionylation of the H3 histone, increases histone acetylation, and promotes acetylation	MCF7 and MDA-MB-231 cells	Mukherjee et al. (2012)
		Non-coding RNA expression	Reduces Axl, Slug, CD24, and Rho-A, and raises miR-34a preventing cell migration and invasion.	MCF-10F cells and MDA-MB-231	Gallardo et al. (2019)
			Reduces Snail, lncRNA-H19, and inhibits invasion and migration of cells.	TAMR-MCF-7 cell	Cai et al. (2020)
	Ovarian cancer	DNA methylation	Suppresses the expression of the DNMT3a protein and total DNMT activity.	ATCC HTB-77 cells	Yen et al. (2019)
		Non-coding RNA expression	Reduces miR-320a, raises SMG1, stops cell division, and encourages death		Sun and Fang (2021)
	Prostatic cancer	DNA methylation	Inhibition of total DNMT activity and DNA methylation	PC3 and LNCaP cells	Sharma et al. (2016)
		Non-coding RNA expression	Increases miR-143, decreases PGK1, Reduces cell proliferation and migration	DU145, PC3, and RPWE-1 cells	Cao et al. (2017)
			Increases OCT4, lowers lncRNA-ROR, raises miR-145, and stops the cell cycle.		Liu et al. (2017a)
	Leukaemia	DNA methylation	Down-regulates DNMT1 Expression	K562, MV4–11, HL-60, ML-1, Kasumi-1 and THP-1 cells	Yu et al. (2013)
		Histone modification	Increases acetylation of H3 and H4	HL-60 cells	Chen et al. (2010)
		non-coding RNA expression	Decreases lncRNA-HOTAIR, raises miR-20a-5p, and lowers WT1.	HL-60 cells	Liu et al. (2021)
Epigallocatechin-3-gallate (EGCG)	Esophageal cancer	DNA methylation	Reversal of retinoic acid receptor beta (RARbeta), *O*(6)-methylguanine methyltransferase (MGMT), human mutL homologue 1 (hMLH1), and p16(INK4a) hypermethylation.	KYSE 510 cells	Fang et al. (2003)
			Reversal of Hypermethylation and Reactivation of RARβ,MGMT, p16INK4a, and hMLH1	YSE 510 and KYSE 150 wer	
	Human endothelial cells	Histone modification	Increases histone acetylation (H3K9/14ac, H3ac), and methylation of both active and repressive (H3K4me3) chromatin marks	HMEC-1 and HUVECs Cell lines	Ciesielski et al. (2020)
	Lung cancer	Non-coding RNA expression	A total of 960 lncRNAs and 1,434 mRNAs were significantly altered	A549 and NCI-H460 cells	Hu et al. (2019)
			Altered expression of 35 circRNAs	Mice liver	Yoshitomi et al. (2024)

(Continued)

TABLE 24.1 (*Continued*)
Effects of Phytomolecules on Epigenetic Modification in Various Cancers

Phytomoleculs	Type of Cancer	Mechanism	Effects	Models/Samples	References
Genistein	Breast cancer	DNA methylation	Interact DNMT1 reduced DNA methylation in the promoter region of several tumour suppressor genes (TSGs), including mammary serpin peptidase inhibitor (SERPINB5), phosphatase and tensin homolog (PTEN), adenomatous polyposis coli (APC), and ataxia telangiectasia mutant (ATM).	MCF-7 and MDA-MB-231	Xie et al. (2014)
	Prostatic cancer	DNA methylation	Decrease methylation of BRCA1, GSTP1 and EPHB2 promoters thereby increases expression	DU-145 and PC-3	Adjakly et al. (2011)
			Demethylation of GSTP1 and EPHB2 promoter thereby increases expression	PC-3, DU-145, LNCaP cells	Vardi et al. (2010)
		Histone modification	Increased levels of acetylated histones 3, 4, 2H3K4, 3H3K4, and Pol II	LNCaP, PC3 and RWPE-1 cells	Majid et al. (2010)
		Non-coding RNA expression	Lower expression of miR-29a and miR-1256	LNCaP, VCaP, PC-3, C4–2B, and ARCaPM (Novicure) prostate cancer (PCa) cells	Y. Li et al. (2012)
			miRNA expression	PC-3, DU145, LNCaP	Rabiau et al. (2011)
			Upregulation of miR-574-3p	PCa cells	Chiyomaru et al. (2013)
	Esophageal squamous cell carcinoma	DNA methylation	Reversed DNA hypermethylation and reactivated RARbeta, p16INK4a, and MGMT	KYSE 510	Fang et al. (2005)
Resveratrol	Colon cancer	DNA methylation	Interacted DNMT, Reduce activation of NF kappa B	SW480 cells	Tili et al. (2010)
		Histone modification	Modulation of TNFα, IL-8, RBP/SIRT1,	SW480 cells	Tili et al. (2010)
		Modulating MicroRNA	Increased expression miR-200c	HCT-116	Karimi Dermani et al. (2017)
Indole-3-carbinol (I3C) and Diindolylmethane (DMI)		Histone modification	Modification of COX-2/HDAC thereby modifies nuclear transcription factors including Sp1, estrogen receptor, NF kappa B and aryl hydrocarbon receptor	HT-29, SW620, RKO, LS174T, and HCT-116 cells	Y. Li et al. (2010)
		Non-coding RNA expression	miRNA expression, modify ZEB1, EGFR expression	HT-29, SW620, RKO, LS174T, and HCT-116 cells	Li et al. (2010)
Sulforaphane	Breast cancer	DNA Methylation	Interact with DNMT1, inhibit the growth of tumour cells, and induce apoptosis in cancer cells.	MCF-7 and MDA-MB-231 cells	Meeran et al. (2010)
		Histone modification	H3 and H4 acetylation, H3K9ac, H3K9me3, HBD-2, H3K27me3, RARβ, HBD-2, p21, BAX/HDAC, inhibit the growth of tumour cells and induce apoptosis in cancer cells.	MCF-7 and MDA-MB-231 cells	Meeran et al. (2010)
	Pancreatic cancer		Increase expression of miR-135b-3p modulating RASAL2 gene	BxPC-3, PANC-1, AsPC-1	Yin et al. (2019)
	Colorectal		Decrease expression of miR-21	RKO	Martin et al. (2017)

(*Continued*)

TABLE 24.1 (*Continued*)
Effects of Phytomolecules on Epigenetic Modification in Various Cancers

Phytomoleculs	Type of Cancer	Mechanism	Effects	Models/Samples	References
Berberine	Hepatocellular carcinoma	DNA methylation	DNMT inhibitor	Hep3B	Kim et al. (2020)
Daidzein	Prostate cancer	DNA methylation	DNMT inhibitor, Promoter demethylation	DU-145, PC-3	Adjakly et al. (2011)
Luteolin	Colorectal	Histone modification	Decreased HDAC activity	HCT116	Zuo et al. (2018)
		DNA methylation	DNMT inhibitor, Promoter demethylation targeting P16INK4a, Nrf2 genes	BE colorectal cancer cells, HCT116	Krifa et al. (2014)
	Colon cancer		DNMT inhibitor, Promoter demethylation targeting Nrf2, p53 genes	HT-29, SNU-407	Kang et al. (2019)
Lycopene	Prostate cancer	DNA methylation	DNMT inhibitor, Promoter demethylation targeting GSTP1 genes	PC3	Fu et al. (2014)
		Non-coding RNA expression	Increased miR-let-7f-1 expression		D. Li et al. (2016)
Quercetin	Bladder cancer	DNA methylation	DNMT inhibitor, Promoter demethylation targeting P16INK4a, Er-beta, RASSF1A genes	EJ, J28, T24	Ma (2006)
	Leukaemia	DNA methylation	DNMT inhibitor, Promoter demethylation targeting DAPK1, BCL2L11, BAX, BNIP3, BNIP3L, APAF1 genes	HL60, U937	Alvarez et al. (2018)
		Histone modification	Increased HAT activity targeting DAPK1, BCL2L11, BAX, APAF1, BNIP3, BNIP3L genes	HL-60	Alvarez et al. (2018)
	Cervix cancer	DNA methylation	DNMT inhibitor, Promoter demethylation targeting APC, CDH1, CDH13, DAPK1, FHIT, GSTP1, MGMT, MLH1, PTEN, RARB, RASSF1, SOC51, TIMP3, VHL genes	HeLa	Kedhari Sundaram et al. (2019)
		Histone modification	Decreased HDAC activity, Increased H3 and H4 acetylation, Decreased HMT activity	HeLa	Kedhari Sundaram et al. (2019)
	Oral cancer	MicroRNA modulation	Increased expression of miR-16 targeting HOXA 10	HSC-16, SCC-9	Zhao et al. (2019)
	Breast cancer		Increased expression of miR-146a	MCF-7, MDA-MB-231	Tao et al. (2015)
Rosmarinic acid	Prostate cancer	DNA methylation	Inhibition of HDAC expression targeting p53, Bax, Bcl-2, PARP-1 genes	PC-3, DU145	Y.-G. Jang et al. (2018)
Triptolide	Lung cancer	DNA methylation	Inhibition of HDAC expression targeting p53, Bax, Bcl-2, PARP-1 genes	A549, H460	Nardi et al. (2018)
Thymoquinone	Breast cancer	DNA methylation	Inhibition of HDAC expression targeting p21, Maspin, Bax and Bcl-2 genes	MCF-7	Parbin et al. (2016)
Allicin	Mouse erythroleukemia	Histone modification	Increased H4 acetylation	DS19	Link et al. (2013)
Allyl isothiocyanate					Lea et al. (2001)
Apigenin	Breast cancer	Histone modification	Decreased of HDAC expression, Increased H3 and H4 acetylation targeting p21/waf1 gene	MDA-MB-231	Tseng et al. (2017)
	Prostate cancer		Decrease of HDAC expression	PC-3, 22Rv	Pandey et al. (2012)
	Glioma	MicroRNA modulation	Increased miR-16 targeting MMP-9 gene	U87	Chen et al. (2016b)

(Cao et al., 2017; Liu et al., 2017a; Sun & Fang, 2021; Wang et al., 2020, 2018).

- *EGCG:* influences miRNA expression by downregulating oncogenic miRNAs such as miR-92a and upregulating tumour-suppressive miRNAs like miR-34a. This alters processes related to metastasis, apoptosis, and cell division (Hu et al., 2019).
- *Genistein:* regulates miRNAs such as miR-1260b, which target tumour suppressor genes. It also upregulates miR-34a, a potent tumour suppressor, which is silenced in many cancers, thereby restoring the normal expression of tumour suppressor pathways (Chiyomaru et al., 2013).

24.6.4 Restoration of Epigenetic Balance

Cancer is often characterized by the disruption of epigenetic homeostasis, leading to aberrant gene expression profiles. Phytomolecules can restore this balance by simultaneously modulating multiple epigenetic pathways, promoting the expression of tumour suppressor genes, and inhibiting oncogenes (Abraham et al., 2023).

24.6.4.1 Phytochemicals Restoring Epigenetic Balance

- *Garlic-derived organosulfur compounds:* such as allyl sulfides, have been shown to have broad effects on DNA methylation and histone modification pathways, restoring the expression of tumour suppressor genes like *p21* and *GADD45*, which are often silenced in cancer cells (Choi & Friso, 2010).
- *Apigenin:* found in parsley, celery, and chamomile, restores the expression of *p53* and *p21* by inhibiting both DNMTs and HDACs, thereby reversing hypermethylation and histone deacetylation in cancer cells (Chen et al., 2016b).
- In breast and prostate cancers, indole-3-carbinol (I3C) from cruciferous vegetables corrects aberrant epigenetic alterations and restores normal cellular development and apoptotic mechanisms by modulating both DNA methylation and histone acetylation pathways (Vrânceanu et al., 2022).

Phytomolecules present a promising avenue for cancer prevention by targeting and reversing aberrant epigenetic modifications. These naturally occurring compounds influence DNA methylation, histone modifications, and non-coding RNA expression, restoring the epigenetic balance essential for normal cellular functions. The combination of these epigenetic-modulating properties not only holds therapeutic potential but also emphasizes the importance of diet and phytochemical intake in cancer prevention strategies. Future research will likely expand the understanding of these mechanisms and lead to the development of novel epigenetic-based cancer therapies.

24.7 CLINICAL EVIDENCE AND EPIDEMIOLOGICAL STUDIES

24.7.1 Overview of Clinical Trials

Clinical trials are an authoritative process to identify any molecule as a potent therapeutic agent. In the case of phytomolecules, globally, very few trials have been completed identifying them as preventive or curative options. Moreover, the majority of them have data on already diagnosed or high-risk individuals and not for the sake of absolute prevention. The major shortcomings identified for the current scenario are as follows (Ingram et al., 1997):

Limited trial data: Among the registered and conducted trials, results are reported for around 20% only. As per the status, remaining trials faced pharmacodynamic-kinetic or idiopathic issues.

Bioavailability issues: Limited efficacy due to challenges in absorption and less bioavailability were reported for molecules like curcumin, genistein, resveratrol, lycopene, and green tea in human trials.

Toxicity concerns: Tocopherols and retinoids, which are vitamin derivatives, have shown toxicity, leading to abruptness in the trial process.

Efficacy gaps: Trials with molecules like curcumin and green tea show limited success alone or in combination for modulating cancer biomarkers or symptoms.

24.7.2 Epidemiological Data Supporting Phytochemical Benefits

24.7.2.1 Flavonoids

In observational case control studies for breast cancer carried out in China and New York, phytoestrogens like isoflavones lowered risk, especially in high-intake groups (Dai et al., 2002; Ingram et al., 1997). Whereas in the case of postmenopausal females, high flavonoid intake of isoflavones and lignans is seen as inversely associated with breast cancer risk (Chen et al., 2014; Fink et al., 2006; Hui et al., 2013). Some clinical trials show slower prostate-specific antigen (PSA) progression with soy flavonoid intake in the Asian population (Kurahashi et al., 2007; Lee et al., 2003; Van Die et al., 2014). In studies like the European EPIC trial and Korean case-control research for GIT cancers, it was found that the risk decreases with flavonoid intake, especially in women (Woo et al., 2014; Zamora-Ros et al., 2012). In esophageal adenocarcinoma, anthocyanins have shown protective effects (Petrick et al., 2015). In the Spanish population, a reduced risk of colorectal cancer was seen with flavonoids. While certain flavonoids reduced colorectal cancer risk in Spanish populations (Zamora-Ros et al., 2013). Trials on lung cancer showed reduced risk with flavonoid-rich foods like onions and apples, specifically in smokers (Le Marchand et al., 2000). Through meta-analyses, it is concluded that higher flavonoid consumption (20 mg/day) results in 10% decreased risk of lung cancer (Tang et al., 2009).

24.7.2.2 Phytosterol

Protective effects of phytosterols have been identified in studies carried out in Uruguay and Mexico for breast and gynaecological cancers, especially intake through plant-based and Mediterranean diet (Ronco et al., 1999; Toledo et al., 2015). Similar reduction in risk and incidence was identified in stomach, pancreatic and prostate cancer, where subjects were consuming phytosterols in the form of vegetarian diets and dried fruits (Fraser, 1999). On the contrary to benefits seen in various preclinical studies, the Netherlands Cohort Study showed no significant reduction in risk or incidence but studies in China and in Seventh-day Adventists showed lower cancer risks with higher phytosterol and legume consumption in prostate cancer as well as lung cancer especially adenocarcinoma(Huang, Xu, et al., 2017; Normén et al., 2001).

24.7.2.3 Phenolic Acid

Amongst phenolic acids, caffeic acid and ferulic acid have been most studied clinically (Damasceno et al., 2017). A study in southern Italy identified risk reduction with high intake of caffeic and ferulic acids in prostate cancer (Touaibia et al., 2011). A cohort study in Spain found that hydroxycinnamic acids in coffee, fruits, and vegetables are correlated with lowering the risk of breast cancer incidence in postmenopausal women (Russo et al., 2017). On the basis of ethnopharmacology, Guangzhou University studied traditional Chinese medicine with phenolic acids and reported improved quality of life in progressive lung cancer (Romanos-Nanclares et al., 2020). Even the topical formulations of phenolic acid with antioxidant vitamins decrease DNA mutation in skin cancer, as reported by Duke University (Zhou et al., 2022). The Chinese trials of caffeic acid on advanced oesophageal squamous cell carcinoma (ESCC) are expected to be promising, but no results are reported yet (GASC1 Inhibitor Caffeic Acid for Squamous Esophageal Cell Cancer (ESCC) (GiCAEC), 2020).

24.7.2.4 Stilbenes

Resveratrol, documented for influencing various cellular pathways like wnt, is the most studied stilbene. Till now, the protective effects of resveratrol have been identified against neuroendocrine, breast, colorectal, prostate, and multiple myeloma cancer. Some of the cellular mechanisms have also been discovered for resveratrol. A study conducted in 2009 on patients receiving resveratrol in the form of grape extract showed significant changes in gene expression. In the study performed by the University of Wisconsin, resveratrol-mediated Notch-1 protein activation, inhibiting tumour growth, was studied in neuroendocrine tumour patients (A Biological Study of Resveratrol's Effects on Notch-1 Signaling in Subjects With Low Grade Gastrointestinal Tumours, n.d.). Currently, the impact of resveratrol on IGF-II levels in healthy high-risk African American women is being studied by Loma University (Nagaraj, 2023). In colorectal cancer, resveratrol significantly reduced risk through the Wnt pathway (Aires et al., 2013). However, muscadine grape extract (MPX) showed no significant effect on prostate cancer progression, probably due to low doses (Paller et al., 2018). Higher doses of resveratrol may be necessary to achieve therapeutic benefits. Moreover, A trial for multiple myeloma patients was terminated due to severe adverse effects where resveratrol was used in combination with bortezomib (Popat et al., 2013).

24.7.2.5 Carotenoids

Carotenoids, being potent antioxidants, reduce inflammation and emerge as preventive agents for various types of cancer, including lung, breast, prostate, GIT, and colorectal. Lycopene intake leads to reduced prostate cancer risk (Rowles et al., 2017) while β-carotene and α-carotene show a protective effect in breast cancer (J. He et al., 2018). In case of lung cancer, studies indicate risk lowering in non-smokers with use of carotenoids, whereas in smokers, high-dose β-carotene supplements may increase the risk due to pro-oxidant effects in the tumour microenvironment (Abar et al., 2016). β-Carotene and lycopene lower risks of GIT cancers. In case of colorectal cancer, the protective effect is seen with all carotenoids and specifically risk is reduced with α-carotene, β-cryptoxanthin, and lycopene (Chen et al., 2016a; Huang, Lu, et al., 2017). Lutein shows risk reduction in skin cancer, but β-carotene does show a significant effect for the same (Kimball, 2012). Gynecological cancers, like ovarian and cervical cancers, may also benefit from carotenoid intake, with some studies indicating a protective role against cervical cancer (Woźniak et al., 2021).

24.8 CHALLENGES IN TRANSLATING RESEARCH TO PRACTICE

Considering changeability in many aspects related to phytomolecules, translating research on phytomolecules into clinical practice for cancer prevention and treatment poses significant challenges. Variability in phytomolecule composition: Depending on species, geographical location, growth conditions, collection, and extraction, phytomolecules possess a wide range of variation. This default variability in standardization of doses and consistency assurance makes application difficult in clinical practice. Limited clinical evidence: Despite promising results in pre-clinical and evidence-based approaches in clinical practice, the safety-efficacy confirming robust clinical trials is limited. In the clinical trials, limitations of molecules are revealed, which might have remained disguised at the previous stage, reducing the probability of translating findings into practice.

Poor bioavailability and stability: Being a complex natural molecule, many phytomolecules have low bioavailability, as they are not easily absorbed or utilized by the body and have limited efficacy. Additionally, some are highly reactive or unstable and may get degraded before reaching target sites, making novel drug delivery a must for their clinical applicability.

Dosage determination and toxicity: Phytomolecules can be therapeutic for one indication at certain doses yet parallelly can be toxic to vital organs. For example, while some phytomolecules exhibit anticancer effects at specific concentrations, at the same time can show adverse effects on the kidney or with slight increase in dose can show severe complications, causing difficulty in guideline formation for clinical use.

Complex mechanisms of action: Phytomolecules have the nature of working through multiple mechanisms and carrying out interacting with other innate molecules of biochemical pathways. In some cases, this complexity and multiple effects may enhance their anticancer potential, while in others, it makes side effects or interaction prediction difficult.

Regulatory and commercial barriers: Regulatory requirements for drugs - especially anticancer agents - are stringent, making it difficult for variable phytomolecules. Moreover, unlike synthetic drugs, phytochemicals often face extra scrutiny due to variations in composition and concentration, and obtaining approval can be a lengthy and costly process.

Potential interactions with conventional therapies: In the current scenario, phytomolecules are not yet established as an individual or monotherapy option to prevent or treat cancer. As they are always given with conventional cancer treatments, such as chemotherapy or radiotherapy, chances of interaction increase, potentially diminishing the effectiveness of either agent or increasing toxicity. However, a well-designed clinical trial can solve the issue.

Public perception and misuse: The popularity of natural remedies often leads to misconceptions that phytomolecules are inherently safe or effective, and patients may consume them without professional guidance. This can lead to misuse or reliance on unproven supplements instead of evidence-based treatments, potentially affecting patient outcomes.

To overcome these challenges, a multi-disciplinary approach involving pharmacologists, clinicians, and regulatory agencies is needed to support evidence-based integration of phytomolecules into cancer care (Choudhari et al., 2020; Das & Agarwal, 2024; Fakhri et al., 2024).

24.9 FUTURE PROSPECTS AND CONCLUSION

With a vast range of options, phytomolecules can target multiple pathways involved in cancer development and progression. This diversity increases the likelihood of discovering effective preventive agents. Moreover, if the dose is tapered well, compared to conventional chemotherapeutic agents, phytomolecules precipitate fewer side effects, and hence can be used as a preventive agent for long term. When poly-therapy is concerned, studies have shown synergistic effects too with synthetic molecules. If the limitations regarding bioavailability, dose or toxicity is set within limits, phytomolecules can be identified as potentially good options, reducing treatment burden and improving quality of life. This and further discoveries can be done by emerging advanced technologies of biotech and computation. The efforts put together can open an arena of personalized medicine for the prevention of cancer with reduced side effects, in an individual's escalating quality of life. The findings from clinical trials will help establish guidelines for the clinical practice of phytomolecules as preventive agents of cancer (Huang et al., 2021; Pavithra et al., 2024).

REFERENCES

A Biological Study of Resveratrol's Effects on Notch-1 Signaling in Subjects with Low Grade Gastrointestinal Tumours (No. NCT01476592). (n.d.).

Abar, L., Vieira, A. R., Aune, D., Stevens, C., Vingeliene, S., Navarro Rosenblatt, D. A., Chan, D., Greenwood, D. C., & Norat, T. (2016). Blood concentrations of carotenoids and retinol and lung cancer risk: An update of the WCRF – AICR systematic review of published prospective studies. *Cancer Medicine*, 5(8), 2069–2083.

Abbaoui, B., Lucas, C. R., Riedl, K. M., Clinton, S. K., & Mortazavi, A. (2018). Cruciferous vegetables, isothiocyanates, and bladder cancer prevention. *Molecular Nutrition & Food Research*, 62(18), 1800079.

Abotaleb, M., Liskova, A., Kubatka, P., & Büsselberg, D. (2020). Therapeutic potential of plant phenolic acids in the treatment of cancer. *Biomolecules,* 10(2), 221.

Abraham, M. J., El Sherbini, A., El-Diasty, M., Askari, S., & Szewczuk, M. R. (2023). Restoring Epigenetic reprogramming with diet and exercise to improve health-related metabolic diseases. *Biomolecules*, 13(2), 318.

Adjakly, M., Bosviel, R., Rabiau, N., Boiteux, J.-P., Bignon, Y.-J., Guy, L., & Bernard-Gallon, D. (2011). DNA methylation and soy phytoestrogens: Quantitative study in DU-145 and PC-3 human prostate cancer cell lines. *Epigenomics*, 3(6), 795–803.

Aggarwal, R., Jha, M., Shrivastava, A., & Jha, A. K. (2015). Natural compounds: Role in reversal of epigenetic changes. *Biochemistry (Moscow)*, 80(8), 972–989.

Aires, V., Limagne, E., Cotte, A. K., Latruffe, N., Ghiringhelli, F., & Delmas, D. (2013). Resveratrol metabolites inhibit human metastatic colon cancer cells progression and synergize with chemotherapeutic drugs to induce cell death. *Molecular Nutrition & Food Research*, 57(7), 1170–1181.

Alaskhar Alhamwe, B., Khalaila, R., Wolf, J., Von Bülow, V., Harb, H., Alhamdan, F., Hii, C. S., Prescott, S. L., Ferrante, A., Renz, H., Garn, H., & Potaczek, D. P. (2018). Histone modifications and their role in epigenetics of atopy and allergic diseases. *Allergy, Asthma & Clinical Immunology*, 14(1), 39.

Almatroodi, S. A., Alsahli, M. A., Almatroudi, A., Verma, A. K., Aloliqi, A., Allemailem, K. S., Khan, A. A., & Rahmani, A. H. (2021). Potential therapeutic targets of quercetin, a plant flavonol, and its role in the therapy of various types of cancer through the modulation of various cell signaling pathways. *Molecules*, 26(5), 1315.

Alvarez, M. C., Maso, V., Torello, C. O., Ferro, K. P., & Saad, S. T. O. (2018). The polyphenol quercetin induces cell death in leukemia by targeting epigenetic regulators of pro-apoptotic genes. *Clinical Epigenetics*, 10(1), 139.

Al-Yousef, N., Shinwari, Z., Al-Shahrani, B., Al-Showimi, M., & Al-Moghrabi, N. (2020). Curcumin induces re-expression of BRCA1 and suppression of γ synuclein by modulating DNA promoter methylation in breast cancer cell lines. *Oncology Reports*, 43, 827–838.

Ankri, S., & Mirelman, D. (1999). Antimicrobial properties of allicin from garlic. *Microbes and Infection*, 1(2), 125–129.

Ansari, P., Akther, S., Hannan, J. M. A., Seidel, V., Nujat, N. J., & Abdel-Wahab, Y. H. A. (2022). Pharmacologically active phytomolecules isolated from traditional antidiabetic plants and their therapeutic role for the management of diabetes mellitus. *Molecules*, 27(13), 4278.

Atpadkar, P. P., Gopavaram, S., & Chaudhary, S. (2023). Natural–product–inspired bioactive alkaloids agglomerated with potential antioxidant activity: Recent advancements on structure-activity relationship studies and future perspectives. *Vitamins and Hormones*, 121, 355–393. Elsevier.

Bacciottini, L., Falchetti, A., Pampaloni, B., Bartolini, E., Carossino, A. M., & Brandi, M. L. (2007). Phytoestrogens: Food or drug? *Clinical Cases in Mineral and Bone Metabolism: The Official Journal of the Italian Society of Osteoporosis, Mineral Metabolism, and Skeletal Diseases*, 4(2), 123–130.

Bachheti, R. K., Worku, L. A., Gonfa, Y. H., Zebeaman, M., Deepti, Pandey, D. P., & Bachheti, A. (2022). Prevention and treatment of cardiovascular diseases with plant phytochemicals: A review. *Evidence-Based Complementary and Alternative Medicine*, 2022, 1–21.

Ballout, F., Habli, Z., Monzer, A., Rahal, O. N., Fatfat, M., & Gali-Muhtasib, H. (2019). Anticancer alkaloids: Molecular mechanisms and clinical manifestations. In A. K. Sharma (Ed.), *Bioactive Natural Products for the Management of Cancer: From Bench to Bedside* (pp. 1–35). Springer Singapore.

Bhosale, P. B., Ha, S. E., Vetrivel, P., Kim, H. H., Kim, S. M., & Kim, G. S. (2020). Functions of polyphenols and its anticancer properties in biomedical research: A narrative review. *Translational Cancer Research*, 9(12), 7619–7631.

Bishayee, A. (2011). Triterpenoids as potential agents for the chemoprevention and therapy of breast cancer. *Frontiers in Bioscience*, 16(1), 980.

Boghdady, N. A. E. (2013). Antioxidant and antiapoptotic effects of proanthocyanidin and ginkgo biloba extract against doxorubicin-induced cardiac injury in rats. *Cell Biochemistry and Function*, 31(4), 344–351.

Borrás, C., Gambini, J., Gómez-Cabrera, M. C., Sastre, J., Pallardó, F. V., Mann, G. E., Viña, J., Borrás, C., Gambini, J., Gómez-Cabrera, M. C., Sastre, J., Pallardó, F. V., Mann, G. E., & Viña, J. (2006). Genistein, a soy isoflavone, up-regulates expression of antioxidant genes: Involvement of estrogen receptors, ERK1/2, and NFκB. *The FASEB Journal*, 20(12), 2136–2138.

Cai, J., Sun, H., Zheng, B., Xie, M., Xu, C., Zhang, G., Huang, X., & Zhuang, J. (2020). Curcumin attenuates lncRNA H19-induced epithelial-mesenchymal transition in tamoxifen-resistant breast cancer cells. *Molecular Medicine Reports*, 23(1), 1–1.

Campagna, R., & Vignini, A. (2023). NAD+ Homeostasis and NAD+-consuming enzymes: Implications for vascular health. *Antioxidants*, 12(2), 376.

Cao, H., Yu, H., Feng, Y., Chen, L., & Liang, F. (2017). Curcumin inhibits prostate cancer by targeting PGK1 in the FOXD3/miR-143 axis. *Cancer Chemotherapy and Pharmacology*, 79(5), 985–994.

Chatterjee, B., Ghosh, K., Suresh, L., & Kanade, S. R. (2019). Curcumin ameliorates PRMT5-MEP50 arginine methyltransferase expression by decreasing the Sp1 and NF-YA transcription factors in the A549 and MCF-7 cells. *Molecular and Cellular Biochemistry*, 455(1–2), 73–90.

Chaudhary, P., Janmeda, P., Docea, A. O., Yeskaliyeva, B., Abdull Razis, A. F., Modu, B., Calina, D., & Sharifi-Rad, J. (2023). Oxidative stress, free radicals and antioxidants: Potential crosstalk in the pathophysiology of human diseases. *Frontiers in Chemistry*, 11, 1158198.

Chen, J., Jiang, W., Shao, L., Zhong, D., Wu, Y., & Cai, J. (2016a). Association between intake of antioxidants and pancreatic cancer risk: A meta-analysis. *International Journal of Food Sciences and Nutrition*, 67(7), 744–753.

Chen, J., Wang, G., Wang, L., Kang, J., & Wang, J. (2010). Curcumin p38-dependently enhances the anticancer activity of valproic acid in human leukemia cells. *European Journal of Pharmaceutical Sciences*, 41(2), 210–218.

Chen, M., Rao, Y., Zheng, Y., Wei, S., Li, Y., Guo, T., & Yin, P. (2014). Association between soy isoflavone intake and breast cancer risk for pre- and post-menopausal women: A meta-analysis of epidemiological studies. *PLoS One*, 9(2), e89288.

Chen, P., Zhang, W., Wang, X., Zhao, K., Negi, D. S., Zhuo, L., Qi, M., Wang, X., & Zhang, X. (2015). Lycopene and risk of prostate cancer: A systematic review and meta-analysis. *Medicine*, 94(33), e1260.

Chen, X.-J., Wu, M.-Y., Li, D.-H., & You, J. (2016b). Apigenin inhibits glioma cell growth through promoting microRNA-16 and suppression of BCL-2 and nuclear factor-κB/MMP-9. *Molecular Medicine Reports*, 14(3), 2352–2358.

Cheng, Y., He, C., Wang, M., Ma, X., Mo, F., Yang, S., Han, J., & Wei, X. (2019). Targeting epigenetic regulators for cancer therapy: Mechanisms and advances in clinical trials. *Signal Transduction and Targeted Therapy*, 4(1), 62.

Chiocchio, I., Mandrone, M., Tomasi, P., Marincich, L., & Poli, F. (2021). Plant secondary metabolites: An opportunity for circular economy. *Molecules*, 26(2), 495.

Chiyomaru, T., Yamamura, S., Fukuhara, S., Hidaka, H., Majid, S., Saini, S., Arora, S., Deng, G., Shahryari, V., Chang, I., Tanaka, Y., Tabatabai, Z. L., Enokida, H., Seki, N., Nakagawa, M., & Dahiya, R. (2013). Genistein up-regulates tumour suppressor microRNA-574-3p in prostate cancer. *PLoS One*, 8(3), e58929.

Cho, S., Park, S., Kwon, M., Jeong, T., Bok, S., Choi, W., Jeong, W., Ryu, S., Do, S., Lee, C., Song, J., & Jeong, K. (2003). [No title found]. *Molecular and Cellular Biochemistry*, 243(1/2), 153–160.

Choi, S.-W., & Friso, S. (2010). Epigenetics: A new bridge between nutrition and health. *Advances in Nutrition*, 1(1), 8–16.

Choudhari, A. S., Mandave, P. C., Deshpande, M., Ranjekar, P., & Prakash, O. (2020). Phytochemicals in cancer treatment: From preclinical studies to clinical practice. *Frontiers in Pharmacology*, 10, 1614.

Ciesielski, O., Biesiekierska, M., & Balcerczyk, A. (2020). Epigallocatechin-3-gallate (EGCG) alters histone acetylation and methylation and impacts chromatin architecture profile in human endothelial cells. *Molecules*, 25(10), 2326.

Dai, Q., Franke, A. A., Jin, F., Shu, X.-O., Hebert, J. R., Custer, L. J., Cheng, J., Gao, Y.-T., & Zheng, W. (2002). Urinary excretion of phytoestrogens and risk of breast cancer among Chinese women in Shanghai. *Cancer Epidemiology, Biomarkers & Prevention*, 11(9), 815–821.

Damasceno, S. S., Dantas, B. B., Ribeiro-Filho, J., Araújo, D. A. M., & Galberto M. Da Costa, J. (2017). Chemical properties of caffeic and ferulic acids in biological system: Implications in cancer therapy: A review. *Current Pharmaceutical Design*, 23(20), 3015–3023.

Das, A. P., & Agarwal, S. M. (2024). Recent advances in the area of plant-based anti-cancer drug discovery using computational approaches. *Molecular Diversity*, 28(2), 901–925.

De Silva, S. F., & Alcorn, J. (2019). Flaxseed lignans as important dietary polyphenols for cancer prevention and treatment: Chemistry, pharmacokinetics, and molecular targets. *Pharmaceuticals*, 12(2), 68.

Dou, H., Shen, R., Tao, J., Huang, L., Shi, H., Chen, H., Wang, Y., & Wang, T. (2017). Curcumin suppresses the colon cancer proliferation by inhibiting Wnt/β-catenin pathways via miR-130a. *Frontiers in Pharmacology*, 8, 877.

Du, G.-J., Zhang, Z., Wen, X.-D., Yu, C., Calway, T., Yuan, C.-S., & Wang, C.-Z. (2012). Epigallocatechin gallate (EGCG) is the most effective cancer chemopreventive polyphenol in green tea. *Nutrients*, 4(11), 1679–1691.

Du, L., Xie, Z., Wu, L., Chiu, M., Lin, J., Chan, K. K., Liu, S., & Liu, Z. (2012). Reactivation of RASSF1A in breast cancer cells by curcumin. *Nutrition and Cancer*, 64(8), 1228–1235.

Elekofehinti, O. O., Iwaloye, O., Olawale, F., & Ariyo, E. O. (2021). Saponins in cancer treatment: Current progress and future prospects. *Pathophysiology*, 28(2), 250–272.

Fakhri, S., Moradi, S. Z., Moradi, S. Y., Piri, S., Shiri Varnamkhasti, B., Piri, S., Khirehgesh, M. R., Bishayee, A., Casarcia, N., & Bishayee, A. (2024). Phytochemicals regulate cancer metabolism through modulation of the AMPK/PGC-1α signaling pathway. *BMC Cancer*, 24(1), 1079.

Fang, M. Z., Chen, D., Sun, Y., Jin, Z., Christman, J. K., & Yang, C. S. (2005). Reversal of hypermethylation and reactivation of p16INK4a, RARbeta, and MGMT genes by genistein and other isoflavones from soy. *Clinical Cancer Research: An Official Journal of the American Association for Cancer Research*, 11(19 Pt 1), 7033–7041.

Fang, M. Z., Wang, Y., Ai, N., Hou, Z., Sun, Y., Lu, H., Welsh, W., & Yang, C. S. (2003). Tea polyphenol (-)-epigallocatechin-3-gallate inhibits DNA methyltransferase and reactivates methylation-silenced genes in cancer cell lines. *Cancer Research*, 63(22), 7563–7570.

Farhan, M., Rizvi, A., Aatif, M., & Ahmad, A. (2023). Current understanding of flavonoids in cancer therapy and prevention. *Metabolites*, 13(4), 481.

Feng, G., Chen, M., Ye, H.-C., Zhang, Z.-K., Li, H., Chen, L.-L., Chen, X.-L., Yan, C., & Zhang, J. (2019). Herbicidal activities of compounds isolated from the medicinal plant *Piper sarmentosum*. *Industrial Crops and Products*, 132, 41–47.

Ferraz Da Costa, D., Fialho, E., & Silva, J. (2017). Cancer chemoprevention by resveratrol: The p53 tumour suppressor protein as a promising molecular target. *Molecules*, 22(6), 1014.

Fiedor, J., & Burda, K. (2014). Potential role of carotenoids as antioxidants in human health and disease. *Nutrients*, 6(2), 466–488.

Fink, B. N., Steck, S. E., Wolff, M. S., Britton, J. A., Kabat, G. C., Schroeder, J. C., Teitelbaum, S. L., Neugut, A. I., & Gammon, M. D. (2006). Dietary flavonoid intake and breast cancer risk among women on Long Island. *American Journal of Epidemiology*, 165(5), 514–523.

Fraser, G. E. (1999). Associations between diet and cancer, ischemic heart disease, and all-cause mortality in non-Hispanic white California Seventh-day Adventists. *The American Journal of Clinical Nutrition*, 70(3), 532S–538S.

Fu, L.-J., Ding, Y.-B., Wu, L.-X., Wen, C.-J., Qu, Q., Zhang, X., & Zhou, H.-H. (2014). The effects of lycopene on the methylation of the GSTP1 promoter and global methylation in prostatic cancer cell lines PC3 and LNCaP. *International Journal of Endocrinology*, 2014, 1–9.

Gallardo, M., Kemmerling, U., Aguayo, F., Bleak, T., Muñoz, J., & Calaf, G. (2019). Curcumin rescues breast cells from epithelial-mesenchymal transition and invasion induced by anti-miR-34a. *International Journal of Oncology*, 56, 480–493.

Gan, J., Feng, Y., He, Z., Li, X., & Zhang, H. (2017). Correlations between antioxidant activity and alkaloids and phenols of maca (*Lepidium meyenii*). *Journal of Food Quality*, 2017, 1–10.

Gao, L., Shao, T., Zheng, W., & Ding, J. (2021). Curcumin suppresses tumour growth of gemcitabine-resistant non-small cell lung cancer by regulating lncRNA-MEG3 and PTEN signaling. *Clinical and Translational Oncology*, 23(7), 1386–1393.

GASC1 Inhibitor Caffeic Acid for Squamous Esophageal Cell Cancer (ESCC) (GiCAEC) (Clinical Trial No. NCT04648917). (2020). Hospital of Henan University of Science and Technology.

George, B. P., Chandran, R., & Abrahamse, H. (2021). Role of phytochemicals in cancer chemoprevention: Insights. *Antioxidants*, 10(9), 1455.

Ghosh, P., & Saadat, A. (2023). Neurodegeneration and epigenetics: A review. *Neurología*, 38(6), e62–e68.

Gold, P. E., Cahill, L., & Wenk, G. L. (2002). Ginkgo biloba: A cognitive enhancer? *Psychological Science in the Public Interest*, 3(1), 2–11.

Guo, Y., Shu, L., Zhang, C., Su, Z.-Y., & Kong, A.-N. T. (2015). Curcumin inhibits anchorage-independent growth of HT29 human colon cancer cells by targeting epigenetic restoration of the tumour suppressor gene DLEC1. *Biochemical Pharmacology*, 94(2), 69–78.

Guo, Y., Wu, R., Gaspar, J. M., Sargsyan, D., Su, Z.-Y., Zhang, C., Gao, L., Cheng, D., Li, W., Wang, C., Yin, R., Fang, M., Verzi, M. P., Hart, R. P., & Kong, A.-N. (2018). DNA methylome and transcriptome alterations and cancer prevention by curcumin in colitis-accelerated colon cancer in mice. *Carcinogenesis*, 39(5), 669–680.

Gupta, P., Kim, B., Kim, S., & Srivastava, S. K. (2014). Molecular targets of isothiocyanates in cancer: Recent advances. *Molecular Nutrition & Food Research*, 58(8), 1685–1707.

Han, H., Lim, J. W., & Kim, H. (2019). Lycopene inhibits activation of epidermal growth factor receptor and expression of cyclooxygenase-2 in gastric cancer cells. *Nutrients*, 11(9), 2113.

He, J., Gu, Y., & Zhang, S. (2018). Vitamin A and breast cancer survival: A systematic review and meta-analysis. *Clinical Breast Cancer*, 18(6), e1389–e1400.

He, K., & Gan, W.-J. (2023). Wnt/β-Catenin signaling pathway in the development and progression of colorectal cancer. *Cancer Management and Research*, 15, 435–448.

He, Y., Sun, M. M., Zhang, G. G., Yang, J., Chen, K. S., Xu, W. W., & Li, B. (2021). Targeting PI3K/Akt signal transduction for cancer therapy. *Signal Transduction and Targeted Therapy*, 6(1), 425.

Hoesel, B., & Schmid, J. A. (2013). The complexity of NF-κB signaling in inflammation and cancer. *Molecular Cancer*, 12(1), 86.

Hu, D.-L., Wang, G., Yu, J., Zhang, L.-H., Huang, Y.-F., Wang, D., & Zhou, H.-H. (2019). Epigallocatechin-3-gallate modulates long non-coding RNA and mRNA expression profiles in lung cancer cells. *Molecular Medicine Reports*, 19(3), 1509–1520.

Huang, J., Lu, M., Fang, Y., Xu, M., Huang, W., Pan, Z., Chen, Y., & Zhang, C. (2017). Serum carotenoids and colorectal cancer risk: A case-control study in Guangdong, China. *Molecular Nutrition & Food Research*, 61(10), 1700267.

Huang, J., Xu, M., Fang, Y.-J., Lu, M.-S., Pan, Z.-Z., Huang, W.-Q., Chen, Y.-M., & Zhang, C.-X. (2017). Association between phytosterol intake and colorectal cancer risk: A case–control study. *British Journal of Nutrition*, 117(6), 839–850.

Huang, M. T., Newmark, H. L., & Frenkel, K. (1997). Inhibitory effects of curcumin on tumourigenesis in mice. *Journal of Cellular Biochemistry, Supplement*, 27, 26–34.

Huang, M., Lu, J.-J., & Ding, J. (2021). Natural products in cancer therapy: Past, present and future. *Natural Products and Bioprospecting*, 11(1), 5–13.

Hui, C., Qi, X., Qianyong, Z., Xiaoli, P., Jundong, Z., & Mantian, M. (2013). Flavonoids, flavonoid subclasses and breast cancer risk: A meta-analysis of epidemiologic studies. *PLoS One*, 8(1), e54318.

Ingram, D., Sanders, K., Kolybaba, M., & Lopez, D. (1997). Case-control study of phyto-oestrogens and breast cancer. *The Lancet*, 350(9083), 990–994.

Jang, J.-H., & Lee, T.-J. (2023). Mechanisms of phytochemicals in anti-inflammatory and anti-cancer. *International Journal of Molecular Sciences*, 24(9), 7863.

Jang, Y.-G., Hwang, K.-A., & Choi, K.-C. (2018). Rosmarinic acid, a component of rosemary tea, induced the cell cycle arrest and apoptosis through modulation of HDAC2 expression in prostate cancer cell lines. *Nutrients*, 10(11), 1784.

Jiang, A., Wang, X., Shan, X., Li, Y., Wang, P., Jiang, P., & Feng, Q. (2015). Curcumin reactivates silenced tumour suppressor gene RARβ by reducing DNA methylation. *Phytotherapy Research*, 29(8), 1237–1245.

Jin, H., Lian, N., Zhang, F., Chen, L., Chen, Q., Lu, C., Bian, M., Shao, J., Wu, L., & Zheng, S. (2016). Activation of PPARγ/P53 signaling is required for curcumin to induce hepatic stellate cell senescence. *Cell Death & Disease*, 7(4), e2189–e2189.

Jobin, C., Bradham, C. A., Russo, M. P., Juma, B., Narula, A. S., Brenner, D. A., & Sartor, R. B. (1999). Curcumin blocks cytokine-mediated NF-kappa B activation and proinflammatory gene expression by inhibiting inhibitory factor I-kappa B kinase activity. *Journal of Immunology*, 163(6), 3474–3483.

Kang, K. A., Piao, M. J., Hyun, Y. J., Zhen, A. X., Cho, S. J., Ahn, M. J., Yi, J. M., & Hyun, J. W. (2019). Luteolin promotes apoptotic cell death via upregulation of Nrf2 expression by DNA demethylase and the interaction of Nrf2 with p53 in human colon cancer cells. *Experimental & Molecular Medicine*, 51(4), 1–14.

Karimi Dermani, F., Saidijam, M., Amini, R., Mahdavinezhad, A., Heydari, K., & Najafi, R. (2017). Resveratrol inhibits proliferation, invasion, and epithelial–mesenchymal transition by increasing miR-200c expression in HCT-116 colorectal cancer cells. *Journal of Cellular Biochemistry*, 118(6), 1547–1555.

Kciuk, M., Alam, M., Ali, N., Rashid, S., Głowacka, P., Sundaraj, R., Celik, I., Yahya, E. B., Dubey, A., Zerroug, E., & Kontek, R. (2023). Epigallocatechin-3-gallate therapeutic potential in cancer: Mechanism of action and clinical implications. *Molecules*, 28(13), 5246.

Kedhari Sundaram, M., Hussain, A., Haque, S., Raina, R., & Afroze, N. (2019). Quercetin modifies 5′CpG promoter methylation and reactivates various tumour suppressor genes by modulating epigenetic marks in human cervical cancer cells. *Journal of Cellular Biochemistry*, 120(10), 18357–18369.

Khan, A. U., Dagur, H. S., Khan, M., Malik, N., Alam, M., & Mushtaque, M. (2021). Therapeutic role of flavonoids and flavones in cancer prevention: Current trends and future perspectives. *European Journal of Medicinal Chemistry Reports*, 3, 100010.

Khan, A., Khan, A., Khan, M. A., Malik, Z., Massey, S., Parveen, R., Mustafa, S., Shamsi, A., & Husain, S. A. (2024). Phytocompounds targeting epigenetic modulations: An assessment in cancer. *Frontiers in Pharmacology*, 14, 1273993.

Kim, D.-Y., Kim, S.-H., Cheong, H.-T., Ra, C.-S., Rhee, K.-J., & Jung, B. D. (2020). Berberine induces p53-dependent apoptosis through inhibition of DNA methyltransferase3b in Hep3B cells. *The Korean Journal of Clinical Laboratory Science*, 52(1), 69–77.

Kimball, A. (2012). Correlation between Skin Carotenoid Levels and Previous History of Skin Cancer (No. NCT00836342). Massachusetts General Hospital (Responsible Party).

Koklesova, L., Liskova, A., Samec, M., Buhrmann, C., Samuel, S. M., Varghese, E., Ashrafizadeh, M., Najafi, M., Shakibaei, M., Büsselberg, D., Giordano, F. A., Golubnitschaja, O., & Kubatka, P. (2020). Carotenoids in cancer apoptosis: The road from bench to bedside and back. *Cancers*, 12(9), 2425.

Kopustinskiene, D. M., Jakstas, V., Savickas, A., & Bernatoniene, J. (2020). Flavonoids as anticancer agents. *Nutrients*, 12(2), 457.

Kotecha, R., Takami, A., & Espinoza, J. L. (2016). Dietary phytochemicals and cancer chemoprevention: A review of the clinical evidence. *Oncotarget*, 7(32), 52517–52529.

Krifa, M., Leloup, L., Ghedira, K., Mousli, M., & Chekir-Ghedira, L. (2014). Luteolin induces apoptosis in BE colorectal cancer cells by downregulating calpain, UHRF1, and DNMT1 expressions. *Nutrition and Cancer*, 66(7), 1220–1227.

Kumar, S., Chinnusamy, V., & Mohapatra, T. (2018). Epigenetics of modified DNA bases: 5-Methylcytosine and beyond. *Frontiers in Genetics*, 9, 640.

Kumar, S., Saini, R., Suthar, P., Kumar, V., & Sharma, R. (2022). Plant secondary metabolites: Their food and therapeutic importance. In A. K. Sharma & A. Sharma (Eds.), *Plant Secondary Metabolites* (pp. 371–413). Springer Nature Singapore.

Kumar, U., Sharma, U., & Rathi, G. (2017). Reversal of hypermethylation and reactivation of glutathione S-transferase pi 1 gene by curcumin in breast cancer cell line. *Tumour Biology*, 39(2), 101042831769225.

Kurahashi, N., Iwasaki, M., Sasazuki, S., Otani, T., Inoue, M., Tsugane, S., & Japan Public Health Center–Based Prospective Study Group. (2007). Soy product and isoflavone consumption in relation to prostate cancer in Japanese men. *Cancer Epidemiology, Biomarkers & Prevention*, 16(3), 538–545.

Le Marchand, L., Murphy, S. P., Hankin, J. H., Wilkens, L. R., & Kolonel, L. N. (2000). Intake of flavonoids and lung cancer. *JNCI: Journal of the National Cancer Institute*, 92(2), 154–160.

Lea, M. A., Randolph, V. M., Lee, J. E., & desBordes, C. (2001). Induction of histone acetylation in mouse erythroleukemia cells by some organosulfur compounds including allyl isothiocyanate. *International Journal of Cancer*, 92(6), 784–789.

Lee, M. M., Gomez, S. L., Chang, J. S., Wey, M., Wang, R.-T., & Hsing, A. W. (2003). Soy and isoflavone consumption in relation to prostate cancer risk in China. *Cancer Epidemiology, Biomarkers & Prevention: A Publication of the American Association for Cancer Research, Cosponsored by the American Society of Preventive Oncology*, 12(7), 665–668.

Lee, Y.-H., Song, N.-Y., Suh, J., Kim, D.-H., Kim, W., Ann, J., Lee, J., Baek, J.-H., Na, H.-K., & Surh, Y.-J. (2018). Curcumin suppresses oncogenicity of human colon cancer cells by covalently modifying the cysteine 67 residue of SIRT1. *Cancer Letters*, 431, 219–229.

Li, D., Chen, L., Zhao, W., Hao, J., & An, R. (2016). MicroRNA-let-7f-1 is induced by lycopene and inhibits cell proliferation and triggers apoptosis in prostate cancer. *Molecular Medicine Reports*, 13(3), 2708–2714.

Li, Q.-Q., Xie, Y.-K., Wu, Y., Li, L.-L., Liu, Y., Miao, X.-B., Liu, Q.-Z., Yao, K.-T., & Xiao, G.-H. (2017). Sulforaphane inhibits cancer stem-like cell properties and cisplatin resistance through miR-214-mediated downregulation of c-MYC in non-small cell lung cancer. *Oncotarget*, 8(7), 12067–12080.

Li, Y., Kong, D., Ahmad, A., Bao, B., Dyson, G., & Sarkar, F. H. (2012). Epigenetic deregulation of miR-29a and miR-1256 by isoflavone contributes to the inhibition of prostate cancer cell growth and invasion. *Epigenetics*, 7(8), 940–949.

Li, Y., Li, X., & Guo, B. (2010). Chemopreventive agent 3,3′-diindolylmethane selectively induces proteasomal degradation of Class I histone deacetylases. *Cancer Research*, 70(2), 646–654.

Link, A., Balaguer, F., Shen, Y., Lozano, J. J., Leung, H.-C. E., Boland, C. R., & Goel, A. (2013). Curcumin modulates DNA methylation in colorectal cancer cells. *PLoS One*, 8(2), e57709.

Liu, G., Xiang, T., Wu, Q.-F., & Wang, W.-X. (2016). Curcumin suppresses the proliferation of gastric cancer cells by downregulating H19. *Oncology Letters*, 12(6), 5156–5162.

Liu, J.-M., Li, M., Luo, W., & Sun, H.-B. (2021). Curcumin attenuates Adriamycin-resistance of acute myeloid leukemia by inhibiting the lncRNA HOTAIR/miR-20a-5p/WT1 axis. *Laboratory Investigation*, 101(10), 1308–1317.

Liu, T., Chi, H., Chen, J., Chen, C., Huang, Y., Xi, H., Xue, J., & Si, Y. (2017a). Curcumin suppresses proliferation and in vitro invasion of human prostate cancer stem cells by ceRNA effect of miR-145 and lncRNA-ROR. *Gene*, 631, 29–38.

Liu, W., Huang, M., Zou, Q., & Lin, W. (2018). Curcumin suppresses gastric cancer biological activity by regulation of miRNA-21: An in vitro study. *International Journal of Clinical and Experimental Pathology*, 11(12), 5820–5829.

Liu, W.-L., Chang, J.-M., Chong, I.-W., Hung, Y.-L., Chen, Y.-H., Huang, W.-T., Kuo, H.-F., Hsieh, C.-C., & Liu, P.-L. (2017b). Curcumin inhibits LIN-28A through the activation of miRNA-98 in the lung cancer cell line A549. *Molecules*, 22(6), 929.

Liu, Y., Zhou, J., Hu, Y., Wang, J., & Yuan, C. (2017c). Curcumin inhibits growth of human breast cancer cells through demethylation of DLC1 promoter. *Molecular and Cellular Biochemistry*, 425(1–2), 47–58.

Ma, L. (2006). Growth inhibitory effects of quercetin on bladder cancer cell. *Frontiers in Bioscience*, 11(1), 2275.

Majid, S., Dar, A. A., Shahryari, V., Hirata, H., Ahmad, A., Saini, S., Tanaka, Y., Dahiya, A. V., & Dahiya, R. (2010). Genistein reverses hypermethylation and induces active histone modifications in tumour suppressor gene B-Cell translocation gene 3 in prostate cancer. *Cancer*, 116(1), 66–76.

Maleki Dana, P., Sadoughi, F., Asemi, Z., & Yousefi, B. (2022). The role of polyphenols in overcoming cancer drug resistance: A comprehensive review. *Cellular & Molecular Biology Letters*, 27(1), 1.

Marei, H. E., Althani, A., Afifi, N., Hasan, A., Caceci, T., Pozzoli, G., Morrione, A., Giordano, A., & Cenciarelli, C. (2021). P53 signaling in cancer progression and therapy. *Cancer Cell International*, 21(1), 703.

Martin, S. L., Kala, R., & Tollefsbol, T. O. (2017). Mechanisms for the inhibition of colon cancer cells by sulforaphane through epigenetic modulation of MicroRNA-21 and human telomerase reverse transcriptase (hTERT) down-regulation. *Current Cancer Drug Targets*, 18(1), 97–106.

Meeran, S. M., Patel, S. N., & Tollefsbol, T. O. (2010). Sulforaphane causes epigenetic repression of hTERT expression in human breast cancer cell lines. *PLoS One*, 5(7), e11457.

Milazzo, G., Mercatelli, D., Di Muzio, G., Triboli, L., De Rosa, P., Perini, G., & Giorgi, F. M. (2020). Histone deacetylases (HDACs): Evolution, specificity, role in transcriptional complexes, and pharmacological actionability. *Genes*, 11(5), 556.

Mitsiogianni, M., Koutsidis, G., Mavroudis, N., Trafalis, D. T., Botaitis, S., Franco, R., Zoumpourlis, V., Amery, T., Galanis, A., Pappa, A., & Panayiotidis, M. I. (2019). The role of isothiocyanates as cancer chemo-preventive, chemo-therapeutic and anti-melanoma agents. *Antioxidants*, 8(4), 106.

Moga, M., Dimienescu, O., Arvatescu, C., Mironescu, A., Dracea, L., & Ples, L. (2016). The role of natural polyphenols in the prevention and treatment of cervical cancer: An overview. *Molecules*, 21(8), 1055.

Moghbeli, M. (2024). PI3K/AKT pathway as a pivotal regulator of epithelial-mesenchymal transition in lung tumour cells. *Cancer Cell International*, 24(1), 165.

Mollakhalili Meybodi, N., Mortazavian, A. M., Bahadori Monfared, A., Sohrabvandi, S., & Aghaei Meybodi, F. (2017). Phytochemicals in cancer prevention: A review of the evidence. *Iranian Journal of Cancer Prevention*, *0*(1), e7219.

Mondal, A., Gandhi, A., Fimognari, C., Atanasov, A. G., & Bishayee, A. (2019). Alkaloids for cancer prevention and therapy: Current progress and future perspectives. *European Journal of Pharmacology*, 858, 172472.

Montgomery, M., & Srinivasan, A. (2019). Epigenetic gene regulation by dietary compounds in cancer prevention. *Advances in Nutrition*, 10(6), 1012–1028.

Mukherjee, S., Sarkar, R., Biswas, J., & Roy, M. (2012). Curcumin inhibits histone deacetylase leading to cell cycle arrest and apoptosis via upregulation of p21 in breast cancer cell lines. *International Journal of Green Nanotechnology*, 4(2), 183–197.

Muscolo, A., Mariateresa, O., Giulio, T., & Mariateresa, R. (2024). Oxidative stress: The role of antioxidant phytochemicals in the prevention and treatment of diseases. *International Journal of Molecular Sciences*, 25(6), 3264.

Nagaraj, G. (2023). Effect of Resveratrol on Serum IGF2 among African American Women (No. NCT04266353). Loma Linda University.

Nakamoto, M., Kunimura, K., Suzuki, J., & Kodera, Y. (2019). Antimicrobial properties of hydrophobic compounds in garlic: Allicin, vinyldithiin, ajoene and diallyl polysulfides (Review). *Experimental and Therapeutic Medicine*, 19, 1550–1553.

Nandini, D., Rao, R., Deepak, B., & Reddy, P. (2020). Sulforaphane in broccoli: The green chemoprevention!! Role in cancer prevention and therapy. *Journal of Oral and Maxillofacial Pathology*, 24(2), 405.

Nardi, I., Reno, T., Yun, X., Sztain, T., Wang, J., Dai, H., Zheng, L., Shen, B., Kim, J., & Raz, D. (2018). Triptolide inhibits Wnt signaling in NSCLC through upregulation of multiple Wnt inhibitory factors via epigenetic modifications to Histone H3. *International Journal of Cancer*, 143(10), 2470–2478.

Navarro, S. L., Li, F., & Lampe, J. W. (2011). Mechanisms of action of isothiocyanates in cancer chemoprevention: An update. *Food & Function*, 2(10), 579.

Normén, A. L., Brants, H. A., Voorrips, L. E., Andersson, H. A., Van Den Brandt, P. A., & Goldbohm, R. A. (2001). Plant sterol intakes and colorectal cancer risk in the Netherlands Cohort Study on Diet and Cancer. *The American Journal of Clinical Nutrition*, 74(1), 141–148.

Oken, B. S., Storzbach, D. M., & Kaye, J. A. (1998). The efficacy of Ginkgo biloba on cognitive function in Alzheimer disease. *Archives of Neurology*, 55(11), 1409.

Olofinsan, K., Abrahamse, H., & George, B. P. (2023). Therapeutic role of alkaloids and alkaloid derivatives in cancer management. *Molecules*, 28(14), 5578.

Paller, C. J., Zhou, X. C., Heath, E. I., Taplin, M.-E., Mayer, T., Stein, M. N., Bubley, G. J., Pili, R., Hudson, T., Kakarla, R., Abbas, M. M., Anders, N. M., Dowling, D., King, S., Bruns, A. B., Wagner, W. D., Drake, C. G., Antonarakis, E. S., Eisenberger, M. A., … Carducci, M. A. (2018). Muscadine grape skin extract (MPX) in men with biochemically recurrent prostate cancer: A randomized, multicenter, placebo-controlled clinical trial. *Clinical Cancer Research*, 24(2), 306–315.

Pandey, M., Kaur, P., Shukla, S., Abbas, A., Fu, P., & Gupta, S. (2012). Plant flavone apigenin inhibits HDAC and remodels chromatin to induce growth arrest and apoptosis in human prostate cancer cells: In vitro and in vivo study. *Molecular Carcinogenesis*, 51(12), 952–962.

Parbin, S., Shilpi, A., Kar, S., Pradhan, N., Sengupta, D., Deb, M., Rath, S. K., & Patra, S. K. (2016). Insights into the molecular interactions of thymoquinone with histone deacetylase: Evaluation of the therapeutic intervention potential against breast cancer. *Molecular BioSystems*, 12(1), 48–58.

Patnaik, E., Madu, C., & Lu, Y. (2023). Epigenetic modulators as therapeutic agents in cancer. *International Journal of Molecular Sciences*, 24(19), 14964.

Pavithra, R., Khan, M. R., & Khan, M. S. (2024). Recent advancements in natural compounds for cancer therapy and prevention. *Phytochemistry Reviews*, 23, 1835–1859.

Pérez-Gálvez, A., Viera, I., & Roca, M. (2020). Carotenoids and chlorophylls as antioxidants. *Antioxidants*, 9(6), 505.

Petrick, J. L., Steck, S. E., Bradshaw, P. T., Trivers, K. F., Abrahamson, P. E., Engel, L. S., He, K., Chow, W.-H., Mayne, S. T., Risch, H. A., Vaughan, T. L., & Gammon, M. D. (2015). Dietary intake of flavonoids and oesophageal and gastric cancer: Incidence and survival in the United States of America (USA). *British Journal of Cancer*, 112(7), 1291–1300.

Ponte, L. G. S., Pavan, I. C. B., Mancini, M. C. S., Da Silva, L. G. S., Morelli, A. P., Severino, M. B., Bezerra, R. M. N., & Simabuco, F. M. (2021). The hallmarks of flavonoids in cancer. *Molecules*, 26(7), 2029.

Popat, R., Plesner, T., Davies, F., Cook, G., Cook, M., Elliott, P., Jacobson, E., Gumbleton, T., Oakervee, H., & Cavenagh, J. (2013). A phase 2 study of SRT 501 (resveratrol) with bortezomib for patients with relapsed and or refractory multiple myeloma. *British Journal of Haematology*, 160(5), 714–717.

Prasanth, M. I., Sivamaruthi, B. S., Cheong, C. S. Y., Verma, K., Tencomnao, T., Brimson, J. M., & Prasansuklab, A. (2024). Role of epigenetic modulation in neurodegenerative diseases: Implications of phytochemical interventions. *Antioxidants*, 13(5), 606.

Pyo, Y., Kwon, K. H., & Jung, Y. J. (2024). Anticancer potential of flavonoids: Their role in cancer prevention and health benefits. *Foods*, 13(14), 2253.

Rabiau, N., Trraf, H.-K., Adjakly, M., Bosviel, R., Guy, L., Fontana, L., Bignon, Y.-J., & Bernard-Gallon, D. J. (2011). miRNAs differentially expressed in prostate cancer cell lines after soy treatment. *In Vivo* (Athens, Greece), 25(6), 917–921.

Ratheesh, G., Tian, L., Venugopal, J. R., Ezhilarasu, H., Sadiq, A., Fan, T.-P., & Ramakrishna, S. (2017). Role of medicinal plants in neurodegenerative diseases. *Biomanufacturing Reviews*, 2(1), 2.

Roberti, A., Fernández, A. F., & Fraga, M. F. (2021). Nicotinamide N-methyltransferase: At the crossroads between cellular metabolism and epigenetic regulation. *Molecular Metabolism*, 45, 101165.

Romanos-Nanclares, A., Sánchez-Quesada, C., Gardeazábal, I., Martínez-González, M. Á., Gea, A., & Toledo, E. (2020). Phenolic acid subclasses, individual compounds, and breast cancer risk in a Mediterranean Cohort: The SUN project. *Journal of the Academy of Nutrition and Dietetics*, 120(6), 1002–1015.e5.

Ronco, A., De Stefani, E., Boffetta, P., Deneo-Pellegrini, H., Mendilaharsu, M., & Leborgne, F. (1999). Vegetables, fruits, and related nutrients and risk of breast cancer: A case-control study in Uruguay. *Nutrition and Cancer*, 35(2), 111–119.

Rowles, J. L., & Erdman, J. W. (2020). Carotenoids and their role in cancer prevention. *Biochimica et Biophysica Acta (BBA) - Molecular and Cell Biology of Lipids*, 1865(11), 158613.

Rowles, J. L., Ranard, K. M., Smith, J. W., An, R., & Erdman, J. W. (2017). Increased dietary and circulating lycopene are associated with reduced prostate cancer risk: A systematic review and meta-analysis. *Prostate Cancer and Prostatic Diseases*, 20(4), 361–377.

Rudzińska, A., Juchaniuk, P., Oberda, J., Wiśniewska, J., Wojdan, W., Szklener, K., & Mańdziuk, S. (2023). Phytochemicals in cancer treatment and cancer prevention: Review on epidemiological data and clinical trials. *Nutrients*, 15(8), 1896.

Russo, G., Campisi, D., Di Mauro, M., Regis, F., Reale, G., Marranzano, M., Ragusa, R., Solinas, T., Madonia, M., Cimino, S., & Morgia, G. (2017). Dietary consumption of phenolic acids and prostate cancer: A case-control study in Sicily, Southern Italy. *Molecules*, 22(12), 2159.

Sadeghian, M., Asadi, M., Rahmani, S., Sadeghi, N., Hosseini, S. A., & Zare Javid, A. (2021). Lycopene does not affect prostate-specific antigen in men with non-metastatic prostate cancer: A systematic review and meta-analysis of randomized controlled trials. *Nutrition and Cancer*, 73(11–12), 2796–2807.

Shafiq, M. A., Gul, R., & Nazik, G. (2024). Dietary carotenoids and their multifaceted roles in cancer prevention. *Journal of Health and Rehabilitation Research*, 4(1), 863–868.

Shahbaz, M., Naeem, H., Imran, M., Ul Hassan, H., Alsagaby, S. A., Al Abdulmonem, W., Waqar, A. B., Ghorab, A. H., Abdelgawad, M. A., Ghoneim, M. M., Hussain, M., Al Jbawi, E., & Ihsan, A. (2023). Chrysin a promising anticancer agent: Recent perspectives. *International Journal of Food Properties*, 26(1), 2294–2337.

Sharma, V., Kumar, L., Mohanty, S. K., Maikhuri, J. P., Rajender, S., & Gupta, G. (2016). Sensitization of androgen refractory prostate cancer cells to anti-androgens through re-expression of epigenetically repressed androgen receptor: Synergistic action of quercetin and curcumin. *Molecular and Cellular Endocrinology*, 431, 12–23.

Silva, H., & Martins, F. (2022). Cardiovascular activity of Ginkgo biloba: An insight from healthy subjects. *Biology*, 12(1), 15.

Stompor-Gorący, M., Bajek-Bil, A., & Machaczka, M. (2021). Chrysin: Perspectives on contemporary status and future possibilities as pro-health Agent. *Nutrients*, 13(6), 2038.

Sun, S., & Fang, H. (2021). Curcumin inhibits ovarian cancer progression by regulating circ-PLEKHM3/miR-320a/SMG1 axis. *Journal of Ovarian Research*, 14(1), 158.

Tang, N.-P., Zhou, B., Wang, B., Yu, R.-B., & Ma, J. (2009). Flavonoids intake and risk of lung cancer: A meta-analysis. *Japanese Journal of Clinical Oncology*, 39(6), 352–359.

Tao, S., He, H., & Chen, Q. (2015). Quercetin inhibits proliferation and invasion acts by up-regulating miR-146a in human breast cancer cells. *Molecular and Cellular Biochemistry*, 402(1–2), 93–100.

Thawabteh, A., Juma, S., Bader, M., Karaman, D., Scrano, L., Bufo, S., & Karaman, R. (2019). The biological activity of natural alkaloids against herbivores, cancerous cells and pathogens. *Toxins*, 11(11), 656.

Thomas, R., Butler, E., Macchi, F., & Williams, M. (2015). Phytochemicals in cancer prevention and management? *Nutrients*, 8(2), a815.

Tili, E., Michaille, J.-J., Alder, H., Volinia, S., Delmas, D., Latruffe, N., & Croce, C. M. (2010). Resveratrol modulates the levels of microRNAs targeting genes encoding tumour-suppressors and effectors of TGFβ signaling pathway in SW480 cells. *Biochemical Pharmacology*, 80(12), 2057–2065.

Ting, C.-Y., Wang, H.-E., Yu, C.-C., Liu, H.-C., Liu, Y.-C., & Chiang, I.-T. (2015). Curcumin triggers DNA damage and inhibits expression of DNA repair proteins in human lung cancer cells. *Anticancer Research*, 35(7), 3867–3873.

Toden, S., Okugawa, Y., Buhrmann, C., Nattamai, D., Anguiano, E., Baldwin, N., Shakibaei, M., Boland, C. R., & Goel, A. (2015). Novel evidence for curcumin and boswellic acid–induced chemoprevention through regulation of miR-34a and miR-27a in colorectal cancer. *Cancer Prevention Research*, 8(5), 431–443.

Toledo, E., Salas-Salvadó, J., Donat-Vargas, C., Buil-Cosiales, P., Estruch, R., Ros, E., Corella, D., Fitó, M., Hu, F. B., Arós, F., Gómez-Gracia, E., Romaguera, D., Ortega-Calvo, M., Serra-Majem, L., Pintó, X., Schröder, H., Basora, J., Sorlí, J. V., Bulló, M., … Martínez-González, M. A. (2015). Mediterranean diet and invasive breast cancer risk among women at high cardiovascular risk in the PREDIMED trial: A randomized clinical trial. *JAMA Internal Medicine*, 175(11), 1752.

Tong, R., Wu, X., Liu, Y., Liu, Y., Zhou, J., Jiang, X., Zhang, L., He, X., & Ma, L. (2020). Curcumin-induced DNA demethylation in human gastric cancer cells is mediated by the DNA-damage response pathway. *Oxidative Medicine and Cellular Longevity*, 2020, 1–13.

Touaibia, M., Jean-Francois, J., & Doiron, J. (2011). Caffeic acid, a versatile pharmacophore: An overview. *Mini-Reviews in Medicinal Chemistry*, 11(8), 695–713.

Tseng, T., Chien, M., Lin, W., Wen, Y., Chow, J., Chen, C., Kuo, T., & Lee, W. (2017). Inhibition of MDA-MB-231 breast cancer cell proliferation and tumour growth by apigenin through induction of G2/M arrest and histone H3 acetylation-mediated $p21^{WAF1/CIP1}$ expression. *Environmental Toxicology*, 32(2), 434–444.

Van Die, M. D., Bone, K. M., Williams, S. G., & Pirotta, M. V. (2014). Soy and soy isoflavones in prostate cancer: A systematic review and meta-analysis of randomized controlled trials. *BJU International*, 113(5b), E119–E130.

Vardi, A., Bosviel, R., Rabiau, N., Adjakly, M., Satih, S., Dechelotte, P., Boiteux, J.-P., Fontana, L., Bignon, Y.-J., Guy, L., & Bernard-Gallon, D. J. (2010). Soy phytoestrogens modify DNA methylation of GSTP1, RASSF1A, EPH2 and BRCA1 promoter in prostate cancer cells. *In Vivo (Athens, Greece)*, 24(4), 393–400.

Velmurugan, B. K., Rathinasamy, B., Lohanathan, B. P., Thiyagarajan, V., & Weng, C.-F. (2018). Neuroprotective role of phytochemicals. *Molecules*, 23(10), 2485.

Venkatadri, R., Muni, T., Iyer, A. K. V., Yakisich, J. S., & Azad, N. (2016). Role of apoptosis-related miRNAs in resveratrol-induced breast cancer cell death. *Cell Death & Disease*, 7(2), e2104–e2104.

Verma, V., Sharma, S., Gaur, K., & Kumar, N. (2022). Role of vinca alkaloids and their derivatives in cancer therapy. *World Journal of Advanced Research and Reviews*, 16(3), 794–800.

Vrânceanu, M., Galimberti, D., Banc, R., Dragoş, O., Cozma-Petruţ, A., Hegheş, S.-C., Voştinaru, O., Cuciureanu, M., Stroia, C. M., Miere, D., & Filip, L. (2022). The anticancer potential of plant-derived nutraceuticals via the modulation of gene expression. *Plants*, 11(19), 2524.

Wahle, K. W. J., Brown, I., Rotondo, D., & Heys, S. D. (2010). Plant phenolics in the prevention and treatment of cancer. In M. T. Giardi, G. Rea, & B. Berra (Eds.), *Bio-Farms for Nutraceuticals* (Vol. 698, pp. 36–51). Springer US.

Wang, M., Yu, F., Zhang, Y., Chang, W., & Zhou, M. (2022). The effects and mechanisms of flavonoids on cancer prevention and therapy: Focus on gut microbiota. *International Journal of Biological Sciences*, 18(4), 1451–1475.

Wang, N., Feng, T., Liu, X., & Liu, Q. (2020). Curcumin inhibits migration and invasion of non-small cell lung cancer cells through up-regulation of miR-206 and suppression of PI3K/AKT/mTOR signaling pathway. *Acta Pharmaceutica*, 70(3), 399–409.

Wang, W., Yang, C., Wang, T., & Deng, H. (2022). Complex roles of nicotinamide N-methyltransferase in cancer progression. *Cell Death & Disease*, 13(3), 267.

Wang, W.-H., Chen, J., Zhang, B.-R., Lu, S.-J., Wang, F., Peng, L., Dai, J.-H., & Sun, Y.-Z. (2018). Curcumin inhibits proliferation and enhances apoptosis in A549 cells by downregulating lncRNA UCA1. *Die Pharmazie*, 7, 402–407.

Woo, H., Lee, J., Choi, I., Kim, C., Lee, J., Kwon, O., & Kim, J. (2014). Dietary flavonoids and gastric cancer risk in a Korean population. *Nutrients*, 6(11), 4961–4973.

Woźniak, M., Krajewski, R., Makuch, S., & Agrawal, S. (2021). Phytochemicals in gynecological cancer prevention. *International Journal of Molecular Sciences*, 22(3), 1219.

Xie, Q., Bai, Q., Zou, L.-Y., Zhang, Q.-Y., Zhou, Y., Chang, H., Yi, L., Zhu, J.-D., & Mi, M.-T. (2014). Genistein inhibits DNA methylation and increases expression of tumour suppressor genes in human breast cancer cells. *Genes, Chromosomes & Cancer*, 53(5), 422–431.

Yang, H.-P. (2011). Hypomethylation effects of curcumin, demethoxycurcumin and bisdemethoxycurcumin on WIF-1 promoter in non-small cell lung cancer cell lines. *Molecular Medicine Reports*, 4(4), 675–679.

Ye, M., Zhang, J., Zhang, J., Miao, Q., Yao, L., & Zhang, J. (2015). Curcumin promotes apoptosis by activating the p53-miR-192–5p/215-XIAP pathway in non-small cell lung cancer. *Cancer Letters*, 357(1), 196–205.

Yen, H. Y., Tsao, C.-W., Lin, Y.-W., Kuo, C.-C., Tsao, C.-H., & Liu, C.-Y. (2019). Regulation of carcinogenesis and modulation through Wnt/β-catenin signaling by curcumin in an ovarian cancer cell line. *Scientific Reports*, 9(1), 17267.

Yin, L., Xiao, X., Georgikou, C., Luo, Y., Liu, L., Gladkich, J., Gross, W., & Herr, I. (2019). Sulforaphane induces miR135b-5p and its target gene, RASAL2, thereby inhibiting the progression of pancreatic cancer. *Molecular Therapy - Oncolytics*, 14, 74–81.

Yoshitomi, R., Kumazoe, M., Lee, K.-W., Marugame, Y., Fujimura, Y., & Tachibana, H. (2024). Regulatory effect of epigallocatechin-3-O-gallate on circular RNA expression in mouse liver. *The Journal of Nutritional Biochemistry*, 124, 109506.

Yu, H., Xie, Y., Zhou, Z., Wu, Z., Dai, X., & Xu, B. (2019). Curcumin regulates the progression of colorectal cancer via LncRNA NBR2/AMPK pathway. *Technology in Cancer Research & Treatment*, 18, 1533033819870781.

Yu, J., Peng, Y., Wu, L.-C., Xie, Z., Deng, Y., Hughes, T., He, S., Mo, X., Chiu, M., Wang, Q.-E., He, X., Liu, S., Grever, M. R., Chan, K. K., & Liu, Z. (2013). Curcumin down-regulates DNA methyltransferase 1 and plays an anti-leukemic role in acute myeloid leukemia. *PLoS One*, 8(2), e55934.

Yuan, J., Dong, X., Yap, J., & Hu, J. (2020). The MAPK and AMPK signalings: Interplay and implication in targeted cancer therapy. *Journal of Hematology & Oncology*, 13(1), 113.

Zamora-Ros, R., Agudo, A., Luján-Barroso, L., Romieu, I., Ferrari, P., Knaze, V., Bueno-de-Mesquita, H. B., Leenders, M., Travis, R. C., Navarro, C., Sánchez-Cantalejo, E., Slimani, N., Scalbert, A., Fedirko, V., Hjartåker, A., Engeset, D., Skeie, G., Boeing, H., Förster, J., … González, C. A. (2012). Dietary flavonoid and lignan intake and gastric adenocarcinoma risk in the European Prospective Investigation into Cancer and Nutrition (EPIC) study. *The American Journal of Clinical Nutrition*, 96(6), 1398–1408.

Zamora-Ros, R., Not, C., Guinó, E., Luján-Barroso, L., García, R. M., Biondo, S., Salazar, R., & Moreno, V. (2013). Association between habitual dietary flavonoid and lignan intake and colorectal cancer in a Spanish case–control study (the Bellvitge Colorectal Cancer Study). *Cancer Causes & Control*, 24(3), 549–557.

Zare, M., Norouzi Roshan, Z., Assadpour, E., & Jafari, S. M. (2021). Improving the cancer prevention/treatment role of carotenoids through various nano-delivery systems. *Critical Reviews in Food Science and Nutrition*, 61(3), 522–534.

Zhang, W., & Xu, J. (2017). DNA methyltransferases and their roles in tumourigenesis. *Biomarker Research*, 5(1), 1.

Zhang, Y., & Wang, X. (2020). Targeting the Wnt/β-catenin signaling pathway in cancer. *Journal of Hematology & Oncology*, 13(1), 165.

Zhang, Y.-J., Gan, R.-Y., Li, S., Zhou, Y., Li, A.-N., Xu, D.-P., & Li, H.-B. (2015). Antioxidant phytochemicals for the prevention and treatment of chronic diseases. *Molecules*, 20(12), 21138–21156.

Zhao, J., Fang, Z., Zha, Z., Sun, Q., Wang, H., Sun, M., & Qiao, B. (2019). Quercetin inhibits cell viability, migration and invasion by regulating miR-16/HOXA10 axis in oral cancer. *European Journal of Pharmacology*, 847, 11–18.

Zheng, Z., You, H., Feng, Y., & Zhang, Z. (2021). LncRNA KCNQ1OT1 is a key factor in the reversal effect of curcumin on cisplatin resistance in the colorectal cancer cells. *Molecular and Cellular Biochemistry*, 476(7), 2575–2585.

Zhou, Y., Wu, C., Qian, X., Zhou, J., Li, C., Jiao, Y., Li, Y. Y., & Zhao, L. (2022). Multitarget and multipathway regulation of Zhenqi Fuzheng granule against non-small cell lung cancer based on network pharmacology and molecular docking. *Evidence-Based Complementary and Alternative Medicine*, 2022, 1–16.

Zhou, Y., Zheng, J., Li, Y., Xu, D.-P., Li, S., Chen, Y.-M., & Li, H.-B. (2016). Natural polyphenols for prevention and treatment of cancer. *Nutrients*, 8(8), 515.

Zhu, M., Sun, Y., Bai, H., Wang, Y., Yang, B., Wang, Q., & Kuang, H. (2023). Effects of saponins from Chinese herbal medicines on signal transduction pathways in cancer: A review. *Frontiers in Pharmacology*, 14, 1159985.

Zoi, V., Kyritsis, A. P., Galani, V., Lazari, D., Sioka, C., Voulgaris, S., & Alexiou, G. A. (2024). The role of curcumin in cancer: A focus on the PI3K/Akt pathway. *Cancers*, 16(8), 1554.

Zuo, Q., Wu, R., Xiao, X., Yang, C., Yang, Y., Wang, C., Lin, L., & Kong, A. (2018). The dietary flavone luteolin epigenetically activates the Nrf2 pathway and blocks cell transformation in human colorectal cancer HCT116 cells. *Journal of Cellular Biochemistry*, 119(11), 9573–9582.

Index

β-carotene, 251, 257
Δ^9-tetrahydrocannabinol, 206

A

Active targeting, 29, 68, 79, 89, 106, 122, 134, 154, 171, 182, 200, 208, 224, 236, 252, 266, 283
Albumin nanoparticles, 200
Alkaloids, 324, 326, 327
Alkylating agents, 12, 276, 278
Aloe vera, 43
Alpha-pinene, 242
Angiogenesis, 7, 132, 197, 207, 325
Anthracyclines, 219–227
Antiangiogenic activity, 103, 131
Anti-apoptotic factors, 251
Antibiotics, 276
Antibodies, 69, 284–285
Anti-inflammatory, 132
Antimetabolites, 276
Antimutagenic activity, 233
Antioxidant, 132, 141, 149
Antiproliferative action, 233
Apigenin, 47, 337
Apoptosis, 4, 7, 13, 26, 64, 73, 101, 103, 130, 152, 182, 197, 205, 206, 251, 278, 279
Aptamers, 31, 68, 285
Aromatherapy, 233
Aromatic compounds, 232
Artemisinin, 99, 296, 312
ATP-binding cassette (ABC), 303
Autophagy, 13, 73, 104, 197

B

Berberine, 34, 72, 280, 306
Biomarkers of cancer, 8

C

Camptothecin, 27, 63, 65, 280
Cancer, 42
Cancer Cell Metabolism, 102
Cancer stem cells (CSCs), 303
Cannabidiol, 204–211
Cannabinoid receptor, 205
Cannabinoids, 204–212
Capsaicin, 42, 281
Carbohydrate nanoparticles, 154
Carbon dots, 79
Carbon nanotubes, 90
Carotenes, 253, 257
Carotenoids, 250–258, 308, 325–327, 337
Carvacrol, 234, 235, 239
Caspase pathway, 197
Cell cycle, 3, 74, 103, 137
Cell cycle checkpoints, 4
Cell death pathways, 103
Cell growth, 26
Cell signalling pathways, 13, 141
Cephalotaxus, 280
Chemodynamic therapy, 310
Chemosensitizers, 282
Chemotherapy, 12, 26, 99, 301
Cinnamaldehyde, 241
Cisplatin, 279
Citral, 232, 234, 242
Colchicine, 280
Combination therapy, 80
Conventional drug delivery, 182
Copolymers, 295
Core-shell nanoparticles, 266
Crocetin, 254, 281
Cubosomes, 170, 266
Curcumin, 16, 27, 34, 47, 181–190, 241, 242, 250, 279, 281, 284, 285, 296, 297, 304, 311, 331
Cyclins, 3, 4
Cyclodextrin Nanoparticles, 77
Cytarabine, 303
Cytokines, 75

D

Daunorubicin, 220
De-differentiation, 7
Dendrimers, 52, 78, 91, 255, 297
Dextran/Chitin Conjugates, 169
DNA damage, 102
DNA methylation, 331–332
Docetaxel, 262, 266–268, 286
Doxorubicin, 220, 282, 284, 285, 304, 315
Dual targeting, 31

E

E-cadherin, 72
Ellagic acid, 315
Ellipticine, 280
Emodin, 19
Endocannabinoids, 205
Endocytosis, 162
Enhanced Permeability and Retention (EPR) effect, 32, 68, 252
Epigallocatechin gallate (EGCG), 18, 34, 47, 149, 306, 307, 324
Epigenetic alterations, 142, 325
Epigenetic effects, 75
Essential oils, 231–243
Eugenol, 233, 234, 238, 240, 241, 279
Extrinsic resistance, 303

F

Ferroptosis, 104
Ferulic acid, 17
Fisetin, 314
Flavanone glycoside, 129, 135
Flavones/Flavonoids, 281
Flavonoids, 324, 326, 328, 337
Folic acid, 134
Fucoxanthin, 251
Functionalised Nanoparticles, 241

G

Gallic acid, 26
Gelatin nanoparticles, 155
Gene therapies, 310
Genetic mutations, 325
Genistein, 17
Geraniol, 232, 242
Gingerol, 19, 281, 308
Gold nanoparticles, 134, 153, 221, 224, 237
G-protein coupled receptors, 205
Graphene-Based Nanocarriers, 80

H

Hedgehog Pathway, 14
Hesperidin, 129
Histone modification, 332
Hormone therapy, 301
Hormone-based drugs, 277
Hybrid nanoparticles, 255, 304
Hydroxyapatite-based ceramic nanoparticles, 122

I

Immune evasion, 325
Immunomodulatory, 142
Immunosuppression, 8, 102
Immunotherapy, 12, 75, 301, 304
Inclusion complex, 239
Inflammatory Pathways, 102
Inorganic Nanoparticles, 66
Intrinsic pathway, 264
Intrinsic resistance, 303
Isothiocyanates, 327, 329

L

Lactones, 44, 232
Limonene, 242, 279
Linalool, 232, 234, 237
Lipid-based nanoparticles, 254, 265
Liposomes, 48, 50, 65, 66, 78, 80, 89, 118, 121, 122, 155, 163–164, 200, 210, 220, 237, 240–241, 253, 285, 296
Lutein, 255, 257
Lycopene, 18, 34, 250, 251, 255, 257, 281

M

Magic Bullet Theory, 29
Magnetic nanohybrid, 185
Magnetic nanoparticles, 198, 312
Malignant tumours, 6, 9
MAPK Pathway, 330
Marine organisms, 12
Matrix metalloproteinase, 72
Matrix metalloproteinase-9, 250–251
Mesoporous silica nanoparticles, 77, 80, 172
Metal Nanoparticles, 78
Metastasis, 7, 73, 102, 142, 207, 325
Micelles, 51, 69, 78, 169, 183, 210, 265, 267, 295
Microbubbles, 52
Microemulsions, 294
microRNA, 75
Microtubule targeting agents, 119
Microtubules, 264

Mitosis, 3
Multi-drug resistance, 118, 219, 275, 303, 306, 310
Multi-drug therapy, 279

N

Nano-actinias, 309, 310
Nanocapsules, 122, 170, 172, 198, 210, 264
Nanoclusters, 171
Nanodiamonds, 90, 171
Nanoemulsion, 89, 133, 168, 237, 239, 254, 264
Nanogels, 171
Nanolipid carrier, 78
Nanoparticles, 29, 47, 51, 68, 91, 165–168, 208, 237, 293, 308, 309
Nanosponge, 171
Nanostructured Lipid Carriers, 154, 210, 237, 242, 265
Nanotechnology, 65
Naringenin, 315
Natural product, 11
N-cadherin, 72
NF-κB pathway, 330
Niosomes, 122, 170, 296
Non-coding RNAs, 332
Notch Signalling Pathway, 14

O

Oncosis, 104
Overexpressed receptors, 6
Oxidative stress, 102

P

p53 Pathway, 330
Paclitaxel, 47, 48, 87, 119, 262, 284, 307, 313
Passive targeting, 32, 67, 77, 88, 106, 121, 133, 143, 152, 162, 183, 198, 207, 223, 236, 252, 263, 264, 282, 285, 309
Patient-derived xenografts, 304, 305
Peptides, 31, 69, 285
P-glycoprotein, 46
Phenolic acids, 337
Phlorotannin, 12
Photodynamic therapy, 12, 301
pH-sensitive mesoporous nanoparticles, 80
pH-sensitive nanoformulations, 122, 185–187, 224, 295, 297
Phytocannabinoids, 205
Phytochemicals, 42
Phytoestrogens, 197
Phytosomes, 50, 170
Phytosterols, 337
PI3K/Akt Pathway, 330
Piperine, 18, 304, 312
Platinum-based compounds, 276, 278
Podophyllotoxin, 43
Polymeric nanoparticles, 79, 133, 183, 184, 194, 198, 199, 221, 223, 224, 254, 286, 304
Polyphenols, 325, 326, 328
Pro-apoptotic factors, 251
Pro-nanolipospheres, 211
Proto-oncogenes, 7

Q

Quantum dots, 92, 185, 310
Quercetin, 17, 35, 48, 161–173, 297, 306, 313, 332

R

Radiation therapy, 301
Reactive Oxygen Species (ROS), 152, 251
Receptor-mediated endocytosis, 198
Redox-responsive, 295
Resveratrol, 17, 35, 194–201, 279, 281, 306, 313, 331, 332
Reticuloendothelial system, 252
Retinoblastoma, 3–5

S

Saponins, 325
Secondary metabolites, 324
Self-emulsifying drug-delivery systems, 211
Silibinin, 314
Silymarin, 19, 44
Smart Drug Delivery, 32
Solid Lipid Nanoparticles, 77, 153, 170, 183, 184, 186, 187, 199, 210, 221, 242
Sparstolonin B, 308
Stem cell transplantation, 301
Stilbenes, 337
Sulforaphane, 19, 35, 332
Synthetic cannabinoids, 205

T

Taxanes, 262, 280, 305
Terpenes & terpenoids, 231, 232, 324
Thermo-responsive, 295
Thermosensitive hydrogel, 200
Thymol, 234
Thymoquinone, 44, 48, 140
Topoisomerase I, 64
Topoisomerase II, 219–222, 276
Tumour Microenvironment, 75

U

Ursolic acid, 232

V

Vinca alkaloids, 43, 117, 280
Vincristine, 47, 307
Vinflunine, 117, 122
Vinorelbine, 117, 122

W

Warburg effect, 197
Wnt Signalling Pathway, 14
Wnt/β-catenin pathway, 330

X

Xanthophylls, 253

Z

Zeaxanthin, 250, 257
Zingiberene, 48, 232